EXCITATION IN HEAVY
PARTICLE COLLISIONS

WILEY-INTERSCIENCE SERIES IN ATOMIC AND MOLECULAR COLLISIONAL PROCESSES

Advisory Editor

C. F. BARNETT

ION-MOLECULE REACTIONS *by E. W. McDaniel, V. Čermák, A. Dalgarno, E. E. Ferguson, L. Friedman*

THEORY OF CHARGE EXCHANGE *by Robert A. Mapleton*

DISSOCIATION BY HEAVY PARTICLE COLLISIONS *by G. W. McClure, J. M. Peek*

EXCITATION BY HEAVY PARTICLE COLLISIONS *by E. W. Thomas*

In preparation:

Ionization by Heavy Particle Collisions by J. W. Hooper, F. W. Bingham
Charge Exchange in Gases and Solids by C. F. Barnett, I. A. Sellin

EXCITATION IN HEAVY PARTICLE COLLISIONS

E. W. Thomas

Georgia Institute of Technology

WILEY-INTERSCIENCE

a Division of John Wiley & Sons, Inc.
New York · London · Sydney · Toronto

Library of Congress Catalog Card Number: 79-37939

ISBN 0-471-85890-0

Printed in the United States of America.

10 9 8 7 6 5 4 3 2 1

SERIES PREFACE

The proliferation of scientific literature in the past two decades has exceeded the point where an individual scientist can either absorb or keep up with all of the publications even in narrow fields. In recognition of this situation, the Federal Council for Science and Technology, in 1963, established the National Standard Reference Data System with responsibility for its implementation assumed by the National Bureau of Standards of the Department of Commerce. In 1968, the existing authorities of the Department of Commerce to operate the NSRDS were supplemented by special legislation passed by the Congress of the United States. Within NBS, the standard reference data program is administered by the Office of Standard Reference Data which provides coordination and planning for data compilation efforts in the field of the physical sciences. As part of this effort, the Office of Standard Reference Data and the Division of Research of the United States Atomic Energy Commission have sponsored jointly the establishment of the Atomic and Molecular Processes Information Center at the Oak Ridge National Laboratory. The objectives of this Center are the collection, evaluation, and critical review of data in specialized topics of heavy particle atomic and molecular collisions. This monograph, *Excitation in Heavy Particle Collisions*, is the fourth of a series of critical reviews. Grateful acknowledgement is made to Professor E. W. Thomas in planning and writing this book.

C. F. Barnett

PREFACE

Excitation induced by collisions between atoms and ions has been a subject of detailed scientific research for some four decades. Impetus for such studies is provided by the visual observation of light emission in aurora, plasmas, and discharges. Most collisional excitation experiments involve spectrophotometric recording of light emission, but a few recent experiments are based on direct determinations of excited state formation. There is little unanimity in the data published by the separate research groups, even when they employ nominally the same experimental techniques. A scientist or engineer attempting to understand a physical situation is hampered by the confusion in the published data. The purpose of this monograph is to compile the available data and to attempt some reconciliation of the diverse results.

My approach to this task has involved three distinct steps. First, I considered the philosophy of the experimental techniques and derived a series of criteria that must be fulfilled in order for an experiment to be deemed both a valid and accurate measurement of the stated physical quantity. In the second step, I cataloged and collected the data. My third and final step was to examine the experimental techniques against the criteria and thereby identify the deficiencies that exist in a particular study. These three distinct steps are reflected in the structure of the monograph. The first three chapters are devoted to formulating the criteria; Chapter Four is a catalog of the data; the remaining chapters contain the detailed compilation of data and examination of the experimental procedures.

Few of the measurements reviewed here can be regarded completely without criticism. It is therefore impossible for me to provide the reader with sets of data that are reliable within definite limits of accuracy. In a few cases, I have given guidance as to which data are highly reliable and as to which are grossly misleading; in most situations the discussion is restricted to an analysis of the sources of inadequacy without an objective assessment of the magnitude of the errors. It is my hope that this monograph will stimulate a greater attention to detail, so that the next generation of experiments will be more amenable to critical review.

In constructing this monograph I have had in mind the needs of scientists and engineers who are not specialists in the field of collisional excitation. To provide them with a compilation of data and some guidance as to the validity of the results has been my objective. Also, I hope that my colleagues in the field of collisional excitation research will find value in the data tabulations and be stimulated by my criticisms; I do not anticipate that they will necessarily concur with my arguments.

The coverage of the monograph is complete up to January of 1970. There is also an appendix containing a partial listing of more recent publications. It is my intention to update this work by periodic reviews in the journal "Atomic Data."

Many of the tables in the monograph have been developed from the original numerical data supplied to me by the researchers whose work is reviewed. The provision of this tabular data eliminates errors that occur in the plotting and reading of

published graphs; therefore the accuracy of the data compilation is greatly improved. The authors that have assisted in this manner are too numerous to mention by name, however, special acknowledgment is due to Drs. Dufay, de Heer, Geballe, and Hughes who have gone to considerable trouble to provide me with tabulations of the extensive data generated in their own research programs.

I wish to extend my sincere thanks to C. F. Barnett of the Oak Ridge National Laboratory for his continued encouragement during the writing of this monograph. A particular debt of gratitude is owed to Mrs. E. Carter for painstaking editorial work on the final manuscript. I am also most grateful to Professor H. B. Gilbody of the Queen's University of Belfast for his review and criticisms at various stages in the writing of this monograph.

E. W. THOMAS

Atlanta, Georgia
November 1971

CONTENTS

EXCITATION IN HEAVY PARTICLE COLLISIONS

CHAPTER 1

INTRODUCTION

A beam of heavy particles traversing a gaseous target will be attenuated by collisional scattering. Some of the collision events will cause a change in the internal energy of either, or both, the colliding systems; such events are termed inelastic collisions. This monograph is concerned with inelastic collision events that involve the formation of excited states; these may be accompanied by rearrangement of atomic electrons or changes in molecular association. The objective of the work is to tabulate the available experimental information and to provide a critical assessment of the data's validity.

There is no general method for the detection of specific excited states, consequently few direct measurements of excitation probabilities exist. Most of the information comes from the spectrometric analysis of collisionally induced photon emission. The probability of forming an excited state can be deduced from the quantitative measurement of emission but the procedure may be complicated, requiring experimental or theoretical assessment of the probabilities for all processes that populate and depopulate the excited state of interest.

The quantitative information on collisionally induced emission is of practical importance. Data from measurements under controlled laboratory conditions can be used to diagnose the processes that give rise to light emission from such macroscopic situations as auroral phenomena and plasmas. Diagnosis by measurement of emission is a particularly attractive tool; the detector of radiation may be remote from the region of interest and will not disturb the phenomenon under study.

The theoretical predictions of probabilities for the occurrence of inelastic events are generally unsatisfactory. For complex systems the calculations are intractable. Much theoretical effort is devoted to the formulation of approximations that facilitate calculations; the effectiveness of such approximations must be checked by comparison with experimental data. Collision cross-section information, required for the understanding of physical phenomena, must generally be obtained by direct experimental measurement.

Most of the early quantitative studies of excitation by heavy particle collisions were carried out during the decade 1930 to 1940. They developed from an interest in the mechanisms that give rise to the emission of light from discharges. Many of the early experiments were carried out with ion beams of uncertain composition; the low sensitivity of the light detectors then available often required target gas pressures which were too high to ensure that the measurements were not severely distorted by secondary excitation mechanisms. The early experiments relied on photographic recording of collisionally induced emission. Photographic emulsions are not suitable

for the quantitative recording of low-level photon fluxes, since they exhibit poor efficiency and nonlinearity of response; moreover, their characteristics are greatly dependent on the processing condition. The data of that era are of historical interest only and cannot be regarded as providing information on the relative probabilities of specific microscopic processes. These data are not considered in this monograph and the reader is referred to two reviews for details: one made by Maurer[1] in 1939, and a second by Massey and Burhop[2] in 1952.

From the time of the Second World War until the late 1950s, there was little work done on this subject. Then interest in collisional excitation phenomena was renewed, primarily through the study of the emission resulting from proton bombardment of atmospheric gases, a mechanism which contributes to auroral phenomena. Experimental studies were greatly facilitated by the development of sensitive stable photomultipliers that allowed the rapid and convenient quantitative recording of weak, collisionally induced emissions. Recently techniques have been developed for direct measurement of the excitation probabilities without recourse to the intermediate step of recording spontaneous emission; this provides valuable checks on the more usual optical methods. There are now some 1000 individual quantitative measurements of collisionally induced emission and excitation that provide information on the microscopic behavior of collision processes. Recent reviews of limited areas of this field have been provided by Ghosh and Srivastava in 1961[3] and 1962,[4] and by de Heer in 1966.[5]

1-1 SCOPE

The objective of this monograph is the tabulation and critical assessment of all measurements, published between 1945 and January 1, 1970, on the formation of specific excited states of atomic and molecular systems as a result of collisions between atoms, ions, and molecules. Areas included in this compendium are (1) processes of direct excitation; (2) simultaneous removal of one or more electrons and excitation of an electron that remains bound; (3) transfer of an electron from one of the collision systems into the excited state of the other; and (4) dissociation of molecules giving rise to excited atomic and molecular fragments. Data are presented either as measurements of collisionally induced emission or as formation probabilities for specific excited states. Both types of data give information on the fundamental process of excitation, although in the case of emission measurements the relationship may involve quantities that are not directly obtainable from the experiment.

Studies of the angular distribution of scattered particles associated with excitation events are not considered here. Reference is made to the available publications, but such data are neither discussed in detail nor tabulated.

In principle, an experimental measurement provides a value for a quantity that is accurate within certain limitations assessed by the author of the work. The discrepancies among independent experimental measurements are far greater than expected. An attempt has been made to examine the details of all experimental procedures, to list their apparent inadequacies, and to conciliate the diverse results.

In general, it is not possible to make a logical choice as to the most accurate experimental determination of a particular quantity. In a few cases there are obvious

errors in experimental techniques that necessarily invalidate the results; such measurements are discussed in this monograph but are generally excluded from the data tabulations on the grounds that they might be misleading. Minor errors in technique are identified in many experiments, but their influence on the accuracy of the data can rarely be assessed in retrospect. One may attempt a subjective choice, from a group of measurements, of the data most likely to be accurate by considering the extent of these minor errors. Rarely can this be done in a completely satisfactory manner. As a result it is impossible to arrive at "best" values for experimental measurements.

This monograph is designed to cover only processes whose nature is unambiguously defined by the experiment. It, therefore, does not include data on the formation of excited states by multiple collision processes. All of the data presented here may be related quite directly to single collision interactions. Chapter 2 is devoted to an operational definition of the collision cross section and its relation to microscopic quantities.

Studies of excitation processes have been made for a wide variety of impact energies and collision combinations. A lower energy limit of 10-eV was established for data to be included in this compilation. This energy is of the same order as the threshold for most inelastic processes that give rise to the formation of excited states. The few experiments at lower energies are devoted to multiple collision effects and do not fulfill the requirement that the data should bear a known unambiguous relationship to single collision cross sections. In practice, no experiments have been done at impact energies above 1.5-MeV, and most are below 0.5-MeV. There are no reliable data for molecules including more than six atoms, and most work is limited to mono- and diatomic molecules.

No discussion of the theoretical predictions nor interpretations of collisional-excitation probabilities are included.

In general, only information that has been published in the open literature—refereed journals and published proceedings of conferences—is included. All other material is generally excluded. Nevertheless, when "unpublished" information or private communications shed additional light on the validity of experimental results they have been also considered here.

A detailed search of literature published between 1945 and January 1, 1970, was made by surveying the major published journals and various abstracting and bibliographic tools. Then a list of publications to be included was circulated to the principal authors with a request for information on any of their publications that were missing. A catalog of the data obtained in this manner is included in Chapter 4.

The data contained in this monograph are presented in tabular form; this method has the dual virtues of precision and economy of space utilization. Some 80% of all the data presented are the numerical data supplied by authors of the original publications. The remaining data have been obtained by reading published graphs.

All published data are discussed in the text. A few measurements, however, are excluded from the tabular compendium on the grounds that the authors have superseded the data with revised measurements, making the original publications redundant, or on the grounds that the experiment is contrary to the procedures expected in an accurate measurement, making the data both wrong and misleading.

This monograph will fulfill the dual roles of data compendium and critical review. Chapter 2 is devoted to the relationship between the microscopic concept of collision

cross section and the macroscopic observables that are measured empirically. For convenience, the results of these discussions are summarized in the form of a set of criteria that must be fulfilled in order that an experiment should be a valid measurement of the microscopic behavior of the collision process. In Chapter 3 problems involved in the empirical measurement of the macroscopic parameters are considered; again, the conclusions are summarized in a convenient set of criteria that must be fulfilled in order that a measurement should be accurate. Chapter 4 is a tabular index of available data with notations as to which of the specified criteria appear to be violated in the experiments; there are cross references to the detailed discussion of the individual experiments and to the tabulation of the empirical data in the succeeding eight chapters.

1-2 CLASSIFICATION OF COLLISION MECHANISMS

The passage of a homogeneous beam of ions, atoms, or molecules through a target composed of a gas or of another beam of particles leads to both elastic and inelastic collisions among the incident and target particles. Here an *elastic collision* is defined as one in which there is no change in the energy of internal motion associated with the target or projectile and in which the system's kinetic energy is conserved. An *inelastic collision*, on the other hand, results in a transformation of kinetic energy into internal energy, or vice versa, for either the struck particle or the projectile, or both. This transfer of energy results in the excitation of internal motion in the particle receiving the energy, or in an increase in its kinetic energy. Some of the more important collisional excitation processes which are considered in this monograph are shown in Table 1-1.

In order to unambiguously specify a collision mechanism, it is necessary to determine the post- and prior-collision states of ionization and excitation for both target and projectile systems. Most experiments presume that the prior-collision states of target and projectile are known; errors that sometimes occur in practice are discussed in Sections 3-1 and 3-2. The complete specification of the postcollision conditions cannot be achieved without detecting simultaneously the states of ionization and excitation for both postcollision products. For example, if an experiment detects only photon emission for the target system, it will be impossible to distinguish between excitation in a charge transfer process 10/01* and excitation occurring simultaneously with an ionization event, 10/11*e. With a single exception,† all published excitation experiments have involved the determination of the postcollision condition of excitation and ionization for only one of the colliding systems. The empirical data therefore represent the probability for forming the detected excited state summed over all possible states of excitation and ionization for the other system. Returning to our previous example of measurement of emission from singly charged ions produced in a target, the data will give the sum of probabilities for all such processes as 10/01*, 10/11*e, 10/0*1*, 10/1*1*e, 10/21*2e, etc. The absence of information about the

---

† This one experiment by Young et al.[7] is for the simultaneous formation of excited projectile and target in the process

$$H^+ + N_2 \rightarrow H^* + N_2^{+*}.$$

A detailed discussion is in Section 8-3-C(4).

TABLE 1-1   IMPORTANT REACTIONS

(i)	Direct target excitation	$A + B \rightarrow A + B^*$	$[00/00^*]^a$
		$A^+ + B \rightarrow A^+ + B^*$	$[10/10^*]$
(ii)	Direct projectile excitation	$A + B \rightarrow A^* + B$	$[00/0^*0]$
		$A^+ + B \rightarrow A^{+*} + B$	$[10/1^*0]$
(iii)	Simultaneous ionization and excitation	$A + B \rightarrow A + B^{+*} + e$	$[00/01^*e]$
		$A + B \rightarrow A^+ + B^* + e$	$[00/10^*e]$
		$A^+ + B \rightarrow A^+ + B^{+*} + e$	$[10/11^*e]$
(iv)	Charge transfer into an excited state	$A^+ + B \rightarrow A^* + B^+$	$[10/0^*1]$
(v)	Charge transfer with simultaneous excitation of the target	$A^+ + B \rightarrow A + B^{+*}$	$[10/01^*]$
(vi)	Dissociation of molecular targets	$A + BC \rightarrow A + B^* + C$	$[00/0(0^*0)]$
		$A + BC \rightarrow A + B^{+*} + C + e$	$[00/0(1^*e)]$

^{a} This convention, due to Hasted,[6] is introduced as a shorthand notation for simple collision processes. A reaction $A^{a+} + B^{b+} \rightarrow A^{m+} + B^{n+} + (m + n)e$ involving collision between A and B with a, b and m, n as prior- and post-collision charge states, is written as $ab/mn(m + n)e$. An asterisk indicates that the system is excited.

state of one of the postcollision products is generally not a serious disadvantage when using emission data as a diagnostic tool. Problems do occur in the comparison of theory with experiment, since it is necessary to carry out the calculations for all the possible postcollision situations that contribute to the signal and to compare their sum with the empirical data.

It should be noted that, in a few situations, the reaction may be sufficiently simple, so that no ambiguity occurs. For example, in the important class of problems in which protons excite neutral targets with no simultaneous removal of electrons $(10/10^*)$, there is no ambiguity because the proton has no atomic structure, cannot be ionized nor excited, and therefore maintains the same condition before and after the collision.

CHAPTER 2

RELATIONSHIP BETWEEN CROSS SECTION
AND EXPERIMENTAL PARAMETERS

The concept of collision cross section is a microscopic property of the interacting atomic systems related to the probability that a particular event will occur. For practical experiments, an operational definition of a collision is developed in terms of macroscopic observables, such as particle beam flux and target density. Measurements are valid only under experimental conditions in which the operational definition of cross section is consistent with the microscopic concept. Furthermore, the accuracy of the cross section determination is directly related to the accuracy with which the experimental parameters are determined.

Although the fundamental concept of a collision cross section is a property of the atomic systems, the operational definition in terms of observed parameters is related to the geometrical arrangement of the apparatus utilized for the measurement. Two basic geometrical arrangements have been used. The most frequently used method involves a projectile-beam incident on a static gas target. In the second arrangement, employed for cases where neither of the atomic systems is stable, it is usual to form both the projectile beam and the static target into beams, direct them to some common point, and measure the process in the region of intersection. The operational definitions of cross section in these two circumstances are discussed in Sections 2-1-B and 2-1-C, respectively. In practice, some 98% of the measurements discussed in this monograph are of total cross sections determined in the beam-gas experimental configuration.

Most experiments do not detect directly the formation of an excited state but measure the photons emitted as the result of the spontaneous radiative decay of the excited collision products. In principle, the formation of all excited states, except those which are metastable or exhibit autoionization, may be studied in this manner. The results are expressed in terms of an emission cross section, a concept defined in Section 2-1-B. The cross section for photon emission is related to all processes that populate the excited state, not to collisional excitation alone. Thus the emission cross section is not, strictly speaking, a microscopic quantum mechanical property of the colliding systems, although in principle it is possible to derive from it a value of the excitation cross section. In Section 2-2 we consider in general terms the relationship that exists between emission and excitation cross sections and the circumstances under which this relationship is unambiguous. Some 98% of all the measurements documented in this monograph involve the experimental determination of an emission cross section rather than an excitation cross section.

It will be clear from the discussions of emission cross sections that, in general, it is only possible to derive level-excitation cross sections if there are accurate subsidiary data on the branching ratios and spontaneous transition probabilities that are applicable to the situation. Section 2-3 is devoted to a brief discussion of the sources of such information and the accuracies that may be anticipated.

This chapter is concluded in section 2-4 with a summary of the conditions under which the operational definitions of cross sections are valid and the relationships between emission and excitation cross sections are unambiguous.

2-1 OPERATIONAL DEFINITION OF CROSS SECTION

A. General Formulation

Consider a specific process ab/cd in which the initial and final states are completely defined. To characterize the situation, suppose that an excited state, i, is formed in one or other of the colliding systems and that a cross section is to be defined for the formation of this final postcollision system. For convenience, the projectiles will be taken as moving at a velocity of v with respect to a Cartesian coordinate system, and the target will be assumed to be at rest with respect to this system. In Section 2-1-C, the concept is generalized to include the case where both particles are moving with respect to the observer's frame of reference.

When a single projectile moves through an element of distance dx in a gas of density N_a (particles/cc), the probability of making a collision that results in the formation of the excited state i is proportional to N_a and dx. Introducing a constant of proportionality σ_i^*, the probability of the collision occurring is given by

$$P_i = N_a \sigma_i^* \, dx. \tag{2-1}$$

Now suppose that a projectile beam of density N_b (particles/cc) and velocity v is incident on the target. The number of excited atoms formed per second in an elementary volume $dA \, dx$, is $dN_i = N_a N_b v \sigma_i^* \, dx \, dA$. It follows that the number density of excited atoms formed per unit time is

$$N_i = N_a N_b v \sigma_i^*. \tag{2-2}$$

Here, N_i may be the density of excited atoms in the target N_{ai} or the density in the beam N_{bi}. These more specific designations will be used later when specific situations are discussed. The constant or proportionality σ_i^* is called the *total cross section* for the process. Equation 2-2 forms the basis for the operational definition of a total cross section in terms of observable parameters.

In general, the projectile experiences only a small change in energy and direction as a result of the collision, and the target recoils with only a small velocity at an angle of approximately $90°$ to the projectile beam direction. Thus to a first approximation, the trajectory, the speed of the projectile, and also the position of the target system are essentially unchanged by the collision. The validity of these generalizations should be examined carefully in the context of each experiment. They will not be accurate, for example, when collisions are at impact energies close to threshold.

B. Beam-Static Target Gas Configuration

Most experimental measurements have been carried out using a beam of particles that traverses a target gas, which is assumed to be at rest. Observations are made of the number of excited atoms formed in the projectile beam or in the static target. This technique is generally used for targets that are stable in a gaseous form at room temperatures. Measurements have also been carried out, however, with highly dissociated hydrogen and metallic vapours by the use of target cells at high temperatures.

   1. Total Excitation Cross Sections. An operational definition may be achieved by developing Eq. 2-2. The number of excited particles produced per unit time as a beam of cross sectional area A traverses a finite distance l in the target will be $\mathcal{N}_i$, such that

$$\mathcal{N}_i = \int_l \int_A dN_i = \int_l \int_A N_a N_b v \sigma_i^* \, dx \, d\mathrm{A}. \qquad (2\text{-}3)$$

If no projectile undergoes more than one collision in the finite path l and the target density is uniform, then

$$\mathcal{N}_i = \sigma_i^* l N_a \int_A N_b v \, d\mathrm{A}.$$

The remaining integral is equal to the flux of the projectile beam I_b (particles/sec). Therefore, an operational definition of *total cross section* is

$$\sigma_i^* = \frac{\mathcal{N}_i}{N_a I_b \, l}. \qquad (2\text{-}4)$$

Strictly speaking, this development is not accurate since the densities of target atoms and projectiles which have experienced no impact are depleted by the collision process. The term N_a will decrease with increasing I_b, and I_b decreases with increasing N_a. By use of sufficiently small values of I_b and N_a, the respective depletion of target density and projectile-beam current in the observation region can be reduced to negligible proportions, and Eq. 2-4 will be applicable. The circumstances under which $\mathcal{N}_i$ is linearly dependent on N_a and I_b are called single-collision conditions. It is imperative that every determination of a total excitation cross section by use of Eq. 2-4 must demonstrate that the experiment is operating under single-collision conditions. Under these circumstances, I_b and N_a are constant throughout the apparatus and may be monitored at any convenient point.

   2. Total Emission Cross Section. Most experiments do not determine the number of excited atoms $\mathcal{N}_i$ directly; rather they measure the number of photons emitted as a result of the spontaneous decay of the excited state. It is convenient to define a cross section for photon emission in a manner analogous to the definition of an excitation cross section. Take J_{ij} as the number of photons emitted per second in the transition $i \rightarrow j$ from a length l of the beam path through the target region. The operational definition of total emission cross section may be stated as

$$\sigma_{ij}^* = \frac{J_{ij}}{N_a I_b \, l}. \qquad (2\text{-}5)$$

Again, the assumption is made that N_a and I_b are both sufficiently tenuous to cause one to neglect the depletion of these quantities as a result of collisions.

   The relationship between emission and excitation is complicated and depends greatly on the particular event under study. The value of σ_{ij}^* is related to all processes

that populate and depopulate the excited state i, not just related to the direct collisional excitation mechanism. Also the lifetime of the excited state may be sufficiently
long that the atom will drift a considerable distance from the point of excitation
before emission takes place; this is particularly important when projectiles are excited.
Under these circumstances the value of σ_{ij}^*, as defined above, may also depend on
the point at which observations are made. In Section 2-2, discussion is devoted to
considering the relationship between the direct excitation cross section and the so-
called emission cross section. Strictly speaking, the emission cross section should
not be regarded as a microscopic property of the colliding systems. Nevertheless, it
does constitute the most frequently measured result in an excitation experiment. The
position is adopted that emission cross section measurements constitute valid fundamental data only when the experiment includes sufficient information so that the
emission and excitation cross sections may, in principle, be related to each other.
When this cannot be done, the experimental data are peculiar to the apparatus and
are of no fundamental value. A specific example of this is when an emission cross
section is a function of experimental geometry not properly recorded in the published
reports.

The definition of emission cross section by Eq. 2-5 makes the assumption that the
depletion of I_b and N_a may be neglected; however, this does not inherently require
J_{ij} to be linearly dependent on N_a, since some of the secondary mechanisms, which
contribute to the population of the state i, are dependent on target density. In Section
2-2-A this dependence is considered in detail, and it is concluded that the only
measurement of σ_{ij}^* that has fundamental value is at low target densities, where J_{ij}
is linearly dependent on N_a. Under these circumstances the contributions to σ_{ij}^* from
factors related to apparatus geometry are negligible. Thus measurements of σ_{ij}^*
should be demonstrated to be independent of N_a and, therefore, independent of
apparatus geometry and collisional depletion of the projectile-beam current.

It is emphasized that the emission cross section is defined in terms of J_{ij} the rate
of emission of photons in all directions from the observation region. Most experiments
involve the detection of photons emitted into some small solid angle $d\omega$ about some
mean angle θ to the beam direction. In the event that the emission is anisotropic
such a measurement, using a limited collection angle, will not represent the fraction
$d\omega/4\pi$ of the total emission. In general, emitted spectral lines will exhibit polarization
that will be related to anisotropy in the emission. The problem of how to measure
the total flux of photons J_{ij} in the presence of anisotropy related to polarization is
considered in Section 3-4-D.

These considerations are very important in the discussion of collisional excitation
data, since most of the information has been obtained in experiments that measure
the emission cross section. The excitation cross section is inferred from the emission
cross section by establishing a model for the processes that populate and depopulate
the excited state. The methods by which these models are established and the tests
which must be made for their validity in order to derive excitation cross sections are
discussed at length in Section 2-2.

C.   Crossed-Beam Configurations

In the crossed-beam configuration, the apparatus determines the number of excited
atoms formed as two particle beams intersect each other. A major advantage of this
approach is that one may study atomic species which are not stable under normal

conditions and, therefore, unsuitable for the beam-static gas configuration. The major disadvantage of the technique, compared with the beam-static gas arrangement, is that the densities of both colliding species are low and the signal to noise ratios in typical systems are very poor. The use of the cross-beam configuration has been confined to the study of the interaction of H^+, He^+, Ne^+, and Ar^+ with atomic hydrogen, the data for which are reviewed in Chapter 5.

Neither of the colliding species is at rest in the laboratory frame of reference, and the choice of which beam is the projectile and which is the target is arbitrary. For convenience, one generally considers the target to be the slower of the two beams. The densities of the projectile and target beams will be taken as N_b and N_a, respectively, with the projectile velocity as v_b in the $+y$ direction, and the target velocity as v_a in the $+x$ direction. The angle between the target and projectile beams will be taken as $90°$ since this is used in the existing experiments.

By analogy to Eq. 2-2, the number of excited states formed per unit volume of the interaction region will be $N_i(xyz) = N_a N_b v' \sigma_i^*$. Here $v' = (v_a^2 + v_b^2)^{1/2}$ is the relative velocity of the colliding systems. In a target beam that is neither convergent nor divergent, the density N_a may be a function of the y and z coordinates but not of x. Similarly, in a parallel projectile beam, N_b is dependent on x and z but not on y. Multiplying $N_i(xyz)$ by an element of volume $dx\, dy\, dz$ and integrating over the region of x and y coordinates, where both N_a and N_b are not zero, we arrive at an expression $\mathcal{N}_i'\, dz$, which represents the number of intersections occurring in a thin slab of the interaction region between z and $z + dz$.

$$\mathcal{N}_i'\, dz = \int_{\Delta x}\int_{\Delta y} N_i\, dx\, dy\, dz = v'\sigma_i^*\, dz \int_{\Delta y} N_a(y,z)\, dy \int_{\Delta x} N_a(x,z)\, dx$$

$$= v'\,\sigma_i^*\, dz\, \frac{i_a(z)}{v_a}\, \frac{i_b(z)}{v_b}. \tag{2-6}$$

Here $i_a(z)/v_a = \int_{\Delta y} N_a(y,z)\, dy$ and $i_b(z)/v_b = \int_{\Delta x} N_b(x,z)\, dx$ have the physical meaning of target and projectile fluxes contained between the coordinates z and $z + dz$. Finally, integrating $\mathcal{N}_i'\, dz$ over the coordinates z to get $\mathcal{N}_i$, the total number of excitation collisions occurring in the interaction region per second, one arrives at the following operational definition of *cross section* σ_i^*:

$$\sigma_i^* = \frac{\mathcal{N}_i}{I_a I_b}\, \frac{v_a v_b}{v'}\, F. \tag{2-7}$$

Here $I_a = \int_{\Delta z} i_a\, dz$ and $I_b = \int_{\Delta z} i_b\, dz$ are the total fluxes of projectile beams. The so-called *form factor* F is equal to $I_a I_b (\int_{\Delta z} i_a i_b\, dz)^{-1}$ and represents of the overlap of the density distributions existing in the two beams. If the density of one or both beams is uniform, then the form factor is simply equal to Δz, the height of the region of interaction between the two beams.

The development of Eq. 2-7 requires that the collisional depletion of the projectile and target beam-currents be negligible. This may be guaranteed by ensuring the proportionality of $\mathcal{N}_i$ to both I_a and I_b. It is also necessary that the form factor be evaluated directly and that both beams be demonstrated to be neither divergent nor convergent. In practice, it is necessary that these requirements be tested for each energy at which the apparatus is operated since beam divergence and the form factor may change appreciably as the projectile's energy is varied.

Note that both targets and projectiles are particle beams, and the measurement of pressure is not required in the performance of a crossed-beam experiment. In view of the numerous errors that may arise in pressure determinations (see Section 3-2), the absence of pressure measurement constitutes an advantage of the crossed-beam approach.

The crossed-beam configuration has only been applied to the study of the excitation of atomic hydrogen by H^+, He^+, Ne^+, and $Ar^{+8,9}$ impact. The experiment detects spontaneous emission from the region where the two beams intersect. By analogy with Eq. 2-5, one may define an emission cross section for a crossed-beam experiment by the following equation:

$$\sigma_{ij}^* = \frac{J_{ij}}{I_a I_b} \frac{v_a v_b}{v'} F.$$

Here J_{ij} is now the total photon emission into all directions from all atoms excited in the region of intersection.

D.   Coincidence Experiments

In order that a collision process ab/cd should be completely defined, it is necessary that the postcollision states of both c and d be determined. In general, experiments are designed to determine the excitation of *either* the projectile *or* the target. Therefore, the measured cross sections will be for the formation of the state detected, integrated over all possible states of excitation and ionization of the other particle in the collision. With the important exception of certain simple collision combinations (see Section 1-2), a reaction is only completely specified by the simultaneous analysis of both collision products in coincidence.

At the time of writing, only one coincidence experiment had been carried out in excitation studies.[7] This experiment detects simultaneously the emission of a Balmer-alpha and an N_2^+ first positive photon, both resulting from the reaction $H^+ + N_2 \rightarrow H^* + N_2^{+*}$. The experiment involved the conventional beam-static gas target configuration and is discussed in detail in Section 8-3-C(4). The operational definition of emission cross section can be approached in precisely the same manner as in Section 2-1-B, and an equation analogous to Eq. 2-5 is derived

$$\sigma^*\begin{pmatrix} i & j \\ k & l \end{pmatrix} = \frac{J\begin{pmatrix} i & j \\ k & l \end{pmatrix}}{N_a I_b \, l}. \tag{2-8}$$

Here

$$\sigma^*\begin{pmatrix} i & j \\ k & l \end{pmatrix}$$

represents the cross section for the simultaneous emission of a photon from the target $(i \rightarrow j)$ and from the projectile $(k \rightarrow l)$. The term

$$J\begin{pmatrix} i & j \\ k & l \end{pmatrix}$$

is the rate of emission of such photons in coincidence from a length l of the ion beam path through the chamber. The general considerations developed in Section 2-1-B

apply equally well to the present case. Some care is necessary to ensure that J does truly represent the desired count rate; in particular, the lifetime of the excited states may be quite long, requiring large resolving times in the electronics and increasing the possibility of accidental coincidence counts.

2-2   DETERMINATION OF CROSS SECTION FROM MEASURED EMISSION

The detection of the collisionally excited ions or atoms by their spontaneous radiative decay is a very powerful technique. The wavelength of the emission characterizes uniquely the excited state under observation. The intensity of the emission is directly proportional to the rate at which the atoms are excited, and the polarization of the emission may give information on the relative populations of the degenerate magnetic sublevels of the excited state. Thus, the experimental problem reduces to the analysis and quantitative detection of radiation and its polarization. Techniques for accomplishing this are sensitive and, for the most part, well understood.

The optical technique cannot readily be applied to the detection of metastable and other long-lived excited states that will travel to the confining walls of any practical experimental system long before they decay with the emission of radiation. In general, such states must be handled by special techniques, some of which will be discussed in Section 3-3. The formation of the metastable state of hydrogen or hydrogen-like ion may, however, be detected by applying an external field, to cause mixing of the $2s$ state with the $2p$ state, which then decays radiatively. Since the detection problem for this particular case reduces to the measurement of photon emission, the method is included in the present discussion.

The total cross sections for excitation and emission have been defined in Eqs. 2-4 and 2-5. The excitation cross section is the microscopic quantity of fundamental interest, whereas the emission cross section is the quantity that is usually derived from experiment. Emission is related to all the processes that populate and depopulate a given state, not just the direct excitation process. The relative importance of the various mechanisms and the method by which their influence is assessed depend critically on the nature of the experimental geometry.

An additional parameter of interest is the polarization of the emission. In a beam-static gas configuration, the direction of the incident beam provides an axis of quantization for the system of colliding particles. The polarization of the emission is related to the relative population of the energy degenerate magnetic sublevels of an excited state. Measurement of polarization will give some information on the relative values of these cross sections and, in certain cases, can be combined with the total cross section for the level to give a measurement of the cross sections for the formation of individual magnetic quantum number states. Polarization constitutes valid cross section information in the same manner as an emission cross section does. Both quantities are directly related to microscopic collision cross sections, but it is not always possible to derive microscopic cross sections from the macroscopic data.

In the development of Eqs. 2-4 and 2-5, it has been assumed that the target density and projectile flux are both tenuous, and consequently neither is appreciably depleted by the collisions that occur. This guarantees that the measured excitation cross

section σ_i^* is independent of these two quantities; however, this does not necessarily ensure the independence of the emission cross section σ_{ij}^* from the target and projectile densities. The relationships between emission and excitation cross sections will be derived with the assumption that depletion of N_a and I_b are negligible but that the density of excited atoms may still be influenced by these factors.

Three basic experimental configurations, which require separate treatments of the relationship between emission measurements and excitation cross section, can be recognized. (1) In a beam-static gas experimental configuration most of the excited target particles will recoil with energies that do not exceed an electron volt and will, therefore, emit at a point that is close to the position where collisional excitation took place; the emission cross section will be invariant with position along the beam path through the target. (2) When the fast projectile is excited, it will generally move an appreciable distance before emission occurs; the emission will, therefore, vary with position along the projectile path through the target. (3) In the case of the metastable particle detected by the application of an electric field, the emission is localized in the region of the quenching field.

The methods for deriving collisional excitation cross sections from the measurement of emission for these three separate situations will be discussed in Sections 2-2-A, 2-2-B, and 2-2-C. Although, in principle, the secondary processes leading to the population and depopulation of the excited state are similar in all three cases, their relative importance and the methods by which they are estimated will be different. It is necessary to formulate this problem in some detail since certain experimental determinations have involved inadequate or inaccurate consideration of the relevant secondary mechanisms. In Section 2-2-D, the relationship of polarization to excitation cross section will be discussed.

A.   Spontaneous Decay of the Target

Under conditions in which no projectile undergoes more than one collision as it traverses the target ("single collision conditions"), five mechanisms contribute to the population and depopulation of the excited state. Inevitably, the state is populated by direct collisional excitation and cascade from higher levels, while depopulation occurs through spontaneous radiative decay. Under some circumstances, the population of an excited state is also dependent on absorption of resonant photons and collisional transfer of excitation energy. The relative importance of these latter mechanisms varies from one excited state to another and gives rise to an apparent dependence of emission cross section on target density.

The rate of increase of the density N_{ai} of target atoms in the excited state i, by direct collisional excitation, is given by $N_a N_b v \sigma_i^*$. Here N_a is the particle density of the target; N_b is the particle density of the incident projectile beam of velocity v, and σ_i^* is the relevant cross section. The excited atoms will decay radiatively to lower states j, giving rise to a rate of depopulation $N_{ai}\sum_{j<i}A_{ij}$, where $\sum_{j<i}A_{ij}$ is the radiative transition probability for the decay $i \rightarrow j$, summed over all lower states. Similarly, higher states k decay into the state i, giving rise to a rate of population by cascade equal to $\sum_{k>i}N_{ak}A_{ki}$.

Population by the absorption of photons is peculiar to excited states that couple optically to the ground state. A fraction f_i of the resonant photons is subsequently absorbed within the target region causing a repopulation of the level i at a rate of

$f_i A_{i0} N_{ai}$; here A_{i0} is the radiative transition probability for the decay of the state i to the ground state. Obviously, f_i will depend on the density of the target gas and the geometry of the experimental system.

An excited target particle may undergo a second collision with a target atom before radiative decay takes place; in this collision the state of excitation may be changed. Such processes cause the transfer of excitation energy from one atom to another and may lead to both the population and depopulation of the excited state i. The rate of population of the state i by collisional transfer from all states x may be written as

$$\sum_x \sigma_{xi} v_x N_a N_{ax} . \tag{2-9}$$

Here σ_{xi} is the cross section for transfer of excitation energy from state x, excited by a primary collision, to state i. The velocity of the target particle after the collision v_x is often assumed to be thermal, but there is evidence that the struck target particles receive an appreciable amount of kinetic energy from the impact and, indeed, that the effective value of v_x may vary appreciably with projectile energy.[10] For convenience, let $c_{xi} = \sigma_{xi} v_x N_a$, this representing the collision frequency.

Considering all the mechanisms described above, one can formalize an equation for the rate of increase of excited state density

$$\frac{dN_{ai}}{dt} = v N_a N_b \sigma_i^* + \sum_{k>i} N_{ak} A_{ki} + \sum_x c_{xi} N_{ax} + f_i A_{i0} N_{ai} , \tag{2-10}$$

and an equation for the rate of depopulation

$$-\frac{dN_{ai}}{dt} = N_{ai} \sum_{j<i} A_{ij} + \sum_x N_{ai} c_{ix} . \tag{2-11}$$

After equilibrium conditions have been established, the rates of population and depopulation of the state i will be equal, the right hand side of Eqs. 2-10 and 2-11 may be equated, and an expression for the equlibrium excited state density N_{ai} evaluated. The number of photons emitted per unit time into all directions from a length l of the beam path, where the cross sectional area of the beam is A, will be given by $J_{ij} = N_{ai} A l A_{ij}$. From the original definition of the emission cross section (Eq. 2-5) and the steady-state value of N_{ai} found from Eqs. 2-10 and 2-11, the following expression is obtained:

$$\sigma_{ij}^* = \frac{A_{ij} l A}{l N_a I_b} \left[\frac{v N_a N_b \sigma_i^* + \sum_{k>i} N_{ak} A_{ki} + \sum_x c_{xi} N_{ax}}{\sum_{j<i} A_{ij} - f_i A_{i0} + \sum_x c_{ix}} \right]. \tag{2-12}$$

For convenience, the population of the state x can be expressed in terms of a new factor $r_x = A N_{ax} (N_a I_b)^{-1}$. The cascade term $N_{ak} A_{ki} l A (N_a I_b l)^{-1}$ may be replaced by σ_{ki}^*, the emission cross section for the cascade transition and Eq. 2-12 rearranged to give

$$\sigma_{ij}^* = A_{ij} \frac{[\sigma_i^* + \sum_{k>i} \sigma_{ki}^* + \sum_x r_x c_{xi}]}{[\sum_{j<i} A_{ij} - f_i A_{i0} + \sum_x c_{ix}]} . \tag{2-13}$$

In this expression the emission cross section is related to terms describing photo-absorption and collisional transfer, each being dependent on the construction of the apparatus and operating conditions of the experiment. In the fundamental definition

of emission cross section [Section 2-1-B(2)], it was suggested that a measurement of this quantity, which is dependent upon the parameters of the apparatus, is of little fundamental value. In the form described by Eq. 2-13, the parameter σ_{ij}^* can best be regarded as an "apparent" emission cross section. The photoabsorption and transfer terms are both dependent on target density; therefore, at sufficiently low pressures they become negligible. The emission cross section is then independent of the apparatus and has a fundamental value;

$$\sigma_{ij}^* = \frac{A_{ij}}{\sum_{j<i} A_{ij}} \left[\sigma_i^* + \sum_{k>i} \sigma_{ki}^* \right]. \tag{2-14}$$

Two additional forms of this equation are often of value. It is noted that ratios of the cross sections for emission of two lines with a common upper level must be in the ratio of the relevant transition probabilities. As a result one may replace σ_{ki}^*, the cascade term of Eq. 2-14, by $\sigma_{kl}^* A_{ki}/A_{kl}$, where l is any level lower than k. Hence,

$$\sigma_i^* = \sigma_{ij}^* \frac{\sum_{j<l} A_{ij}}{A_{ij}} - \sum_{k>i} \sigma_{kl}^* \frac{A_{ki}}{A_{kl}}. \tag{2-15}$$

Similarly, one may replace σ_{ij}^* of Eq. 2-14 by $A_{ij} \sum_{j<i} \sigma_{ij}^*/\sum_{j<i} A_{ij}$; this, with some rearrangement, gives

$$\sigma_i^* = \sum_{j<i} \sigma_{ij}^* - \sum_{k>i} \sigma_{ki}^*. \tag{2-16}$$

Equation 2-16 shows that a level-excitation cross section may be determined by measurement of the emission cross sections for all transitions into and out of the state i. For most cases, the limited spectral range of a practical detection system precludes measurement of all the required transitions. Equation 2-15 indicates that, when the ratios of transition probabilities are known, σ_i^* may be determined through the measurement of emission cross sections for one transition out of the state i, and one transition out of each of the levels k that contributes to the cascade population. Note that the transition from the level k need not be the cascade line itself; any convenient transition may be used. Accurate values of transition probability are obtainable from theory for only a very few atomic and molecular systems (see Section 2-3).

Only in a relatively small number of cases is it possible to evaluate σ_i^* explicitly, either through measurement of all relevant transitions (Eq. 2-16) or by the use of accurate branching ratios determined from theoretical considerations (Eq. 2-15). Therefore, a substantial part of all the available collisional excitation data is left in the form of emission cross section measurements.

The above treatment involves the important assumption that the excited target atom emits at the same place where it was excited. At present, there is no evidence to indicate that this assumption is violated for projectile impact on atomic targets. This may not be true, however, for excited fragments produced by the dissociation of molecules. For example, an H atom formed in the $3s$ state as a result of the dissociation of an H_2 molecule might recoil with some 5-eV of energy and exhibit a decay length of approximately 5 mm. The limited field of view of an optical system might result in part of the emission not being detected and the consequential intro-duction of an instrumental error. In reported experiments, no attempt has been made to explore the spatial distribution of the emitters. The only effective procedure for

eliminating this problem is to detect the photons from the whole volume from which the emission is significant. This point will be raised again in later sections when assessing data on the emission from long-lived states of light dissociation fragments. It is possible that the problem is only relevant in 'such cases.

Emission cross sections determined under pressure-independent conditions are of definite intrinsic value. They are of direct importance in the application of emission measurements to diagnostic problems. Furthermore, in many cases the cascade terms may be shown to be quite small; therefore, by Eqs. 2-15 and 2-16, the emission cross sections are approximately proportional to the level-excitation cross section. Measurements of emission cross sections under pressure-independent conditions will be included in the compilation of data. For such cases, the relationship between emission and excitation cross sections will be given by Eqs. 2-14, 2-15, and 2-16. Emission cross sections determined under conditions in which pressure independence has not been demonstrated are of no fundamental value because they are functions of the geometry of the experimental system. Such data are excluded from the compilations.

B. Spontaneous Decay of the Projectile

Excited states of the projectile may be produced by direct excitation; also they may be created in collisions involving a simultaneous process of charge transfer, dissociation, or stripping. The excited projectile has an appreciable velocity and will move a considerable distance before decaying to a lower level with the emission of radiation. The intensity of photon emission will, therefore, vary with penetration through the target, in contrast to the emission from excited target systems. The mechanisms leading to the population and depopulation of the excited state are essentially the same as detailed for target excitation in Section 2-2-A. In a projectile beam, however, the number density of the emitting species is small. Absorption of resonance photons and collisional transfer of excitation energy to other projectiles have not been identified in any of the experiments under consideration here and are perhaps important only for higher projectile beam densities than are presently employed.

The following mathematical developments will be predicated on the assumption that target and projectile densities are sufficiently tenuous to prevent any projectiles from undergoing more than one collision in the apparatus and that neither the density of unexcited projectiles nor the density of target atoms are appreciably depleted by the collisions. This is the same preliminary condition that has been assumed in the development of all the other operational definitions of cross section. This may be far more difficult to achieve in the present case. The *decay length of the excited particle* may be defined as the product of the projectile's velocity and the lifetime of the excited state. Under some circumstances this decay length may be tens of centimeters requiring apparatus dimensions of this order of size. With such long targets, the density must be quite low to ensure negligible probability that the projectile will suffer two collisions. It should be noted that the collision cross sections for excited atoms may be much higher than for the ground state projectiles.[11] The target density must be sufficiently low so that these excited projectiles are not destroyed by a second collision.

It is important to note that the wavelength of the emission from a projectile may be appreciably changed by the Doppler effect. In addition to a shift in wavelength

the apparent line width of the emission will be broadened because of the finite acceptance angle of any practical detection system. The formulations established in this section are in terms of the rate of an emissive transition between the levels $i \rightarrow j$ rather than in terms of the emission rate at a specific wavelength. The influence of the Doppler effect may be regarded as a problem in the definition of detection sensitivity and will, therefore, be treated as such in Section 3-4-F.

Two allied experimental configurations have been used for the measurement of projectile excitation. In one configuration the emission is observed from the path of the projectile beam through the target region; the intensity increases asymptotically towards some constant value at large target penetration. In the second situation a beam is passed through a gas cell into an evacuated flight tube where the decay of emission is observed as a function of distance beyond the exit from the cell. The advantage of this second approach is that there is no target emission in the observation region. The disadvantage lies in the use of a gas cell that inevitably has poorly defined boundaries, possibly causing errors because projectiles scattered through large angles may be intercepted by the exit orifice. Consideration of these configurations is divided into two sections. The first deals at length with the general problems of relating projectile emission to an excitation cross section, with particular reference to emission from the beam traversing the gaseous target. The second part considers more briefly the few matters in which an experiment, involving a determination of decay in an evacuated flight tube, differs from one involving measurement of emission in the target region.

1. Decay within the Target Region. In a manner similar to our previous consideration of target excitation, the rate of population of the excited state i in the projectile beam is given by

$$\frac{dN_{bi}}{dt} = vN_bN_a\sigma_i^* + \sum_{k>i} N_{bk}A_{ki} - N_{bi}\sum_{j<i} A_{ij}. \tag{2-17}$$

As before, N_b is the particle density in the projectile beam; N_a is the target density; v is the projectile velocity; and σ_i^* is the excitation cross section. The first term describes the rate of formation by excitation from the ground state of the projectile; the second term introduces population by cascade from higher states k; and the third term describes the depopulation by spontaneous decay to some lower state j. The term A_{ij} is the radiative transition probability for the decay $i \rightarrow j$. Absorption of resonance radiation and collisional transfer has been neglected.

As a first approximation to the solution of Eq. 2-17, the cascade term is neglected. The time variable is converted to distance by the relation $x = vt$, where x is the distance penetrated through the target by a projectile in a time t. The equation is integrated with the boundary condition that $N_{bi} = 0$ at $x = 0$, giving

$$N_{bi} = \frac{vN_aN_b\sigma_i^*}{\sum_{j<i} A_{ij}} \left\{ 1 - \exp - \frac{x}{v\tau_i} \right\}. \tag{2-18}$$

Here, τ_i in the exponential is equal to $[\sum_{j<i} A_{ij}]^{-1}$ and is the lifetime of the state i. The number of photons J_{ij} emitted from a beam of cross sectional area A in a finite path length L to $L + l$ in the gas is given by

$$J_{ij} = \int_L^{L+l} AN_{bi}A_{ij}\,dx. \tag{2-19}$$

The integration of Eq. 2-19 gives the following result:

$$\frac{J_{ij}}{N_a I_b\, l} = \frac{A_{ij}\sigma_i^*}{\sum_{j<i} A_{ij}}\left[1 - \frac{v\tau_i}{l}\left(1 - \exp-\frac{l}{v\tau_i}\right)\exp-\frac{L}{v\tau_i}\right]. \tag{2-20}$$

In practice, for many experimental situations $l \ll v\tau_i$, and Eq. 2-20 is simplified to

$$\frac{J_{ij}}{N_a I_b l} = \frac{A_{ij}\sigma_i^*}{\sum_{j<i} A_{ij}}\left[1 - \exp-\frac{L}{v\tau_i}\right]. \tag{2-21}$$

It is important to note that the emission varies with position in the target gas. When $L \gg v\tau_i$ then J_{ij} becomes independent of position; under such circumstances one may legitimately define an emission cross section as

$$\sigma_{ij}^* = \frac{J_{ij}}{N_a I_b\, l} = \frac{A_{ij}\sigma_i^*}{\sum_{j<i} A_{ij}}. \tag{2-22}$$

The emission cross section has been defined under conditions in which the competing processes of collisional formation of the excited state and the spontaneous decay by emission are equal. It would be confusing to define an emission cross section that was dependent on L. The emission cross section may be found either by operating at large L, where Eq. 2-22 is applicable, or by fitting an equation of the form of 2-21 to the intensity buildup curve in order to derive a value of the quantity σ_{ij}^* defined by Eq. 2-22. Except for the absence of a cascade term, Eq. 2-22 is the same as the corresponding relationship for target excitation that is expressed in Eq. 2-14. In certain experimental determinations the effect of finite lifetime has been overlooked, and the emission cross section has been related to the excitation cross section using the above equation. This is incorrect since Eq. 2-22 only applies when observations are made at a sufficiently large distance from the entrance to the chamber.

The formulae obtained are first-order approximations because cascade has been neglected. The introduction of cascade terms into Eqs. 2-21 and 2-22 is a complex procedure that has only been fully carried out in one experiment.[12] Basically, the procedure for a first-order cascade correction is to take an expression, such as 2-18, as the population of the state k with the cascade correction neglected and substitute this into Eq. 2-17; the integrations are then repeated. The cascade correction introduces terms involving the lifetimes of the upper states into Eq. 2-20. These lifetimes will generally be different and often longer than those of the lower state i. The importance of cascade may be assessed by measuring the variation of emission cross section with penetration and attempting to fit it by Eq. 2-20 or 2-21. If the experimental results cannot be explained by a single function of this form, then contributions to the emission are coming from more than one level, probably by cascade. The assessment of cascade contribution in general terms is complicated in that it is dependent not only on the population of the higher state but also on the penetration through the chamber at which the measurement is being made. It is convenient to consider the nature and extent of the problem separately for each experimental situation. At sufficiently large penetrations, the cascade also becomes independent of penetration, and Eqs. 2-14, -15, and -16, which were developed for excitation of the target, will again become applicable. In an experimental measurement, it is required either that cascade should be shown to be negligible or, alternatively, that the penetration L utilized is sufficiently large so that the emission cross section, including cascade contribution, is independent of L.

2. Decay in an Evacuated Flight Tube. Suppose that the gas cell is of length X and that the rate of emission at a point L centimeters from the exit orifice of the gas cell is to be determined. Equation 2-18, evaluated at $x = X$, gives the excited-state density of the beam as it exits from the gas cell. Beyond the exit from the cell, the excited state population decreases exponentially with distance. Provided again that $l \ll v\tau_i$ (i.e., the length of path from which photons are observed is much less than the decay length of the excited state), then the rate of photon emission J_{ij} from an observation region of length l is given by

$$\frac{J_{ij}}{N_a l I_b} = \frac{A_{ij}}{\sum_{j < i} A_{ij}} \sigma_i^* \left[1 - \exp - \frac{X}{v\tau_i} \right] \exp - \frac{L}{v\tau_i}. \qquad (2-23)$$

This expression is analogous to Eq. 2-21 that is derived for emission from the projectile as it traverses the target region. By fitting Eq. 2-23 to the observed emission, it is possible to derive a value for $\sigma_i^* A_{ij} [\sum_{j < i} A_{ij}]^{-1}$, which one may refer to as the apparent emission cross section by analogy with Eq. 2-22. Cascade has been neglected from the present formulation but may be included by the same successive approximation method described in the discussion of projectile emission in the collision region [see Section 2-2-B(1)]. In general it is to be expected either that the cascade contribution to emission be shown to be negligible, or, alternatively, that it must be assessed directly.

Any experiment that employs a gas cell as the collision region may suffer errors in two ways: (1) due to the interception of part of the excited projectile flux by the exit orifice from the cell and (2) due to pressure gradients at both the exit and entrance of the gas cell. Tests must be made to ensure that the measured cross sections are independent of exit aperture size and of gas cell length.

C.   Induced Decay of Metastable Projectiles

Certain metastable excited states may be detected by applying an electric field in order to shift the energy levels by the Stark effect and thus cause mixing with an adjacent state that will decay radiatively. The technique is best known for the detection of the metastable $2s$ state of hydrogen, which, by the application of a suitable field, is mixed with the $2p$ state and subsequently decays to the ground state with the emission of Lyman-alpha radiation. The time required for the $2s \to 2p \to 1s$-decay sequence varies from about 20 years in the zero field case to twice the $2p$-state lifetime $(3 \times 10^{-9}$ sec) for large fields. The lifetime of the $2s$ state as a function of applied field has been investigated both theoretically[12,13,14,15] and experimentally.[16,17] The only experiments using this type of detection, which are relevant to this monograph, have been concerned with the formation of metastable hydrogen atoms in a fast beam of ions traversing either a static target gas or target beam.[18] Therefore, comments will be confined to the specific problems of this case.

Recent work by Fite et al.[19] and by Sellin and co-workers[20] indicates that field-induced emission is polarized and that the degree of polarization varies with the field strength. A practical photon detection system will have only a limited angular acceptance, and the anisotropy must be taken into account when relating measured signal to the rate of photon emission. Previous publications purporting to show that emission is isotropic[21] are incorrect.

Experiments are divided into two classes that correspond closely to the arrangements used for measuring the population of short-lived excited states. The beam may be allowed to traverse a gas cell and then be detected by quenching in an evacuated flight tube. Alternatively, the field may be applied in the collision region itself. The latter technique suffers from the disadvantage that the detector will also record Lyman-alpha emission from normal radiative decay of the $2p$ state and possibly some emission from excitation of the target gas. Spurious background signals may be identified by taking the difference between the signals observed with the quenching field off and on. The effect of the applied field is the reduction of the lifetime of the excited state, and emission does not necessarily occur at the point where the particle first enters the field. Therefore, many of the considerations previously developed for states that normally decay radiatively must also be introduced here. Furthermore, considerable care is necessary to exclude stray electric and magnetic fields that will cause quenching outside the field of view.

When using the target-gas-cell approach, the quenching field will typically be applied some distance beyond the exit from the cell, at a point where the majority of the nonmetastable states have decayed. A field of about 100 to 500 V/cm is applied perpendicular to the projectile beam, and the resulting photons are detected in a suitable manner. Tests must be made to ensure that no significant attenuation of the beam of excited atoms takes place before detection. The cross sections for collisional loss of the excited atom by such processes as stripping (0*0/10e), de-excitation (0*0/00), and excitation (0*0/0**0) may be very high.[17, 22] Losses by collision with background gas must be avoided. If the target chamber is X cm long, the target particle density N_a, the beam current I_b (particles/sec), and the cross section for forming the excited state is σ_i^*, then—neglecting cascade population—the current of excited states in the beam emerging from the cell is $I_{bi} = N_a I_b X \sigma_i^*$. If a field is applied to this beam to reduce the lifetime of the excited state to τ_i where the beam velocity is v, then the rate of photon emission from a small observation length l at a distance L from the point at which the beam first enters the field is

$$J_{ij} = N_a I_b X \sigma_i^* \left(1 - \exp - \frac{l}{v \tau_i} \right) \exp - \frac{L}{v \tau_i}. \tag{2-24}$$

Because of the influence of fringing fields, the value of L is difficult to determine, and it is desirable to plot the variation of intensity with position and to determine an integrated intensity that will be proportional to the number of excited states in the beam emergent from the cell. For low velocity beams and high quenching fields, the product $v \tau_i$ may be sufficiently small to cause essentially all the decay to take place within the field of view of the detector.

The application of a quenching field within the collision region itself will cause a deflection of charged projectiles and result in complications in the collision geometry. Jaecks et al.[16] solved this problem by the very simple procedure of having the field alternate in space. It should again be recognized that the effect of the field was to reduce lifetime to the order of magnitude of the $2p$ state. In general, the effective lifetime of the $2s$ state in the field is short, and the intensity rapidly builds up to an equilibrium value at a small penetration through the target. The measurement of the equilibrium emission J_{ij} (photons/sec) observed from l cm of beam path for an incident beam current I_b (particles/sec) and target density N_a (particles/cc) allows a direct determination of excitation cross section σ_i^*

$$J_{ij} = N_a I_b \sigma_i^* l. \tag{2-25}$$

In no case covered by the present monograph has cascade into the $2s$ level been determined accurately. The cascade transitions $np \to 2s$ may be measured directly, but completely different detection systems are required since the emission lies in the visible spectrum. A simple approach for assessing the importance of cascade[16] utilizes the fact that higher states have appreciably longer lifetimes than the $2p$ level, and the spatial dependence of Lyman-alpha emission, which arises from cascade, will therefore be different from that due to the quenching process. An examination of the spatial dependence will indicate whether or not cascade is significant. It is to be noted that the application of an electric field not only causes mixing of the $n = 2$ levels but also causes mixing of the $n = 3$ and higher states, leading to an effective lifetime of the mixed states close to that of the np level. Clearly, cascade into the $2s$ state may be appreciably enhanced by the applied field at the expense of cascade into the $2p$ and should be determined under applied field conditions.

D.   Polarization of Emission

Transitions from different magnetic quantum sublevels of excited states give rise to emissions of definite polarization. In collision experiments the particles are excited to states that have a definite component of angular momentum in the direction of the projectile beam. The observed emission in a transition $i \to j$ will exhibit polarization related to the relative population of the various substates of different angular momentum components along the projectile's direction of motion.

Consider a projectile beam moving in the x direction. The radiation from an excited atom may be regarded as due to an electric dipole in the x direction and two equal dipoles in the y and z directions. Let $I_{ij}(\theta)$ be the rate of photon emission from a length l of the ion beam path per unit solid angle in a direction θ with respect to the projectile beam direction. Also let $I_{ij}^{\parallel}$ and $I_{ij}^{\perp}$ be the rates of emission of photons from the length l per unit solid angle perpendicular to the beam direction with planes of polarization respectively parallel and perpendicular to the beam axis. The polarization fraction is defined as

$$P = \frac{I_{ij}^{\parallel} - I_{ij}^{\perp}}{I_{ij}^{\parallel} + I_{ij}^{\perp}}. \tag{2-26}$$

The index ij is used to indicate that attention is directed to a transition between nondegenerate states $i \to j$. It may be shown[23] that the angular distribution of emission is given by

$$I_{ij}(\theta) = J_{ij} \frac{3}{4\pi} \frac{(1 - P \cos^2\theta)}{(3 - P)}. \tag{2-27}$$

If one multiplies $I_{ij}(\theta)$ by an element of solid angle $d\Omega = \sin\theta\, d\theta\, d\phi$ and integrates over all θ and ϕ, then the total emission in all directions from the length l of the projectile beam path is J_{ij}. This is the quantity used in the operational definition of emission cross section in Sections 2-2-A, 2-2-B, and 2-2-C. Equation 2-27 has been experimentally confirmed by McFarland[24] in the study of excitation of helium by electrons.

Polarization is important in collisional excitation experiments for three reasons. First, it is directly related to collisional excitation cross sections and therefore provides valuable information on the microscopic processes. Second, it gives rise to anisotropy

in the emission that must be taken into account when measuring an emission cross section. Third, the detection sensitivity of certain types of equipment depends on the polarization of the emission that is involved. The last two considerations influence the determination of optical detection efficiency and are therefore considered in depth in Chapter 3. The present discussions are confined to the interpretation of the polarization fraction in terms of excitation cross sections.

It should be noted that certain emissions are inherently unpolarized. For example, with an atom where LS coupling is applicable, emission from states of zero angular momentum are always unpolarized (i.e., $P = 0$). An important example of this is emission from n^1S and n^3S states of helium.

Any measurement of polarization fraction must include an assessment of the influence of stray magnetic fields in the apparatus. Larmor precession about residual magnetic fields may result in rotation of the polarization axis. If there is an appreciable component of a stray field perpendicular to the direction of motion, then the polarization measurement will be in error. Thus, a criterion for the validity of a polarization measurement is that it must be carried out in a field-free region of space.

The polarization fraction defined by Eq. 2-26 is expressed in terms of the observed intensities and therefore constitutes an operational definition of this parameter. The intensities are dependent on all populating and depopulating processes that affect the excited state, just as emission cross sections are dependent on a multitude of processes. Therefore, one must logically accept a polarization measurement as being valid only under the same conditions as an emission cross section. In particular, one requires that P be measured under conditions in which it is independent of both target density and projectile beam density. Even with this requirement, the polarization fraction is still affected by cascade, and emissions resulting from population by this process will be unpolarized. There has been no general method suggested in the literature by which the component of P arising from cascade may be accurately determined. In some cases, the total cascade population of a level may be shown to be negligible, but this is not generally true. Therefore, theoretical expressions for P written only in terms of cross sections for the direct excitation of specific sublevels are not necessarily equal to the operational definition of polarization fraction that is given in Eq. 2-26.

A complete treatment of the relationship between polarization fraction and level-excitation cross sections for atomic systems is given by Percival and Seaton.[25] They consider the excitation from the ground state to an upper level i and the subsequent emission of a photon to reach the level j. Cascade population of the state i is ignored. It is assumed that the projectile beam and the target have completely random distribution of spin directions. Moreover, the discussion is simplified by considering only initial states that have zero orbital angular momentum. Therefore, anisotropy is introduced into the problem entirely through the motion of the incident particle. Published in this paper[25] are tables of results for an Oppenheimer—Penney[26,27] treatment, which is applicable to excitation of atoms for which fine and hyperfine structure energies are large. In particular, this allows the calculation of polarization of collisionally induced emissions from helium in terms of the cross section for exciting the magneic sublevels. The reader is referred to the original paper[25] for the full list of relationships for different transitions.

As an example of a specific case in which polarization can be used to determine cross sections, consider the excitation of a helium n^1P state that has three sublevels

($m_l = +1, 0, -1$). Taking $\sigma_{\pm 1}$ as the cross section for the $m_l = +1$ and -1 states and σ_0 as the cross section for the $m_l = 0$ state, then the polarization fraction of the $n^1P \rightarrow n^1S$ emission is

$$P_{ij} = \frac{\sigma_0 - \sigma_{\pm 1}}{\sigma_0 + \sigma_{\pm 1}}. \tag{2-28}$$

Also, the total cross section for exciting the level is $\sigma_i^* = \sigma_0 + 2\sigma_{\pm 1}$. From measurements of σ_i^* and P, the individual cross sections σ_0 and $\sigma_{\pm 1}$ may be determined; this is a relatively simple example. Emission in the $n^1D \rightarrow n^1P$ transition will involve three cross sections, and it is impossible to separate out the individual components. It is again emphasized that errors may occur in this approach due to the impossibility of removing the cascade contribution to the measured value of P.

There has been little study of the polarization of emission resulting from the excitation of molecular systems. It seems unlikely that emission from dissociation fragments will exhibit polarization. The fragments will generally be moving with an appreciable velocity corresponding to potential energy released in the dissociation process; also there will be a distribution of the directions of motion. Recently Gaily et al.[28] have shown experimentally that the polarization of Lyman-alpha radiation induced by the collisional dissociation of H_2^+ ions is in fact zero; however there is every reason to suppose that the molecular line emissions will exhibit polarization. Jette and Cahill[29] have considered the relationships between polarization fraction and level-excitation cross sections for the case of a diatomic molecule using a direct extension of the atomic theory by Percival and Seaton. There are no systematic experimental investigations of the polarization of molecular emission.

Polarization fractions are directly related to emission cross sections and constitute valuable information. Accordingly, these measurements are included in the present monograph. In a few cases, it may prove possible to use measured emission cross sections and polarization fractions to arrive at estimates for the cross sections of specific m_l substates. The validity of such a procedure must be assessed for each case in which it is used. The measured values of polarization are given in the literature either as a fraction or as a percentage; in this text, we utilize the latter form of presentation.

2-3 BRANCHING-RATIO INFORMATION

It is clear from the preceding discussions that it is possible to relate emission and excitation cross sections only if subsidiary information on branching ratios is available.

Some useful compendia of information are available in the case of atomic transitions. The National Standard Reference Data System (NSRDS), under the auspices of the National Bureau of Standards (NBS), has published a listing of sources of data;[30] critical data compilations are in preparation.[31] Only in the case of hydrogen-like systems and in a few lower levels of helium are the transition probabilities known with high accuracy. A frequent practice in the interpretation of excitation experiments is to neglect the limited accuracy of the theoretical transition probabilities when considering the overall error estimates. The NSRDS compilations include an estimate of the absolute accuracy of the values listed but do not give any indication of the

accuracy of branching ratios determined from these values. Since a branching ratio involves two or more transitions emanating from a common upper level, it is possible that the ratios of transition probabilities may be more accurate than the transition probabilities themselves.

In the case of molecular systems, the theoretical prediction of Franck-Condon factors is fairly accurate for most common systems. Bates[32] has provided a set of tables from which relative transition probabilities may be calculated, provided that the basic molecular constants are known from spectroscopic data.[33] No convenient listing of data sources exists for molecular transition probabilities; therefore reference must be made to the original publications. Confusion has sometimes arisen in the interpretation of collisional excitation of molecules through the incorrect assumption that the ratio of radiative transition probabilities is in the ratio of the relevant Franck-Condon factors.† The ratio of the wavelengths of the transitions and the ratios of the electronic transition moments must also be included. Cases where such confusion has occurred will be noted when appropriate in the discussions of experimental data.

2-4   SUMMARY

In the preceding sections operational definitions have been established for excitation and emission cross sections in various experimental configurations. An experimental measurement must include a demonstration of the required functional relationship between the various parameters in the formulae that provide the operational definition of cross section. Furthermore, the experiment must demonstrate independence of factors omitted from the definition. Fulfilling such tests ensures the validity of the measurement. The accuracy of a determination is related to the measurement of the various experimental parameters and is discussed in Chapter 3. A cross section that was determined under conditions at variance with the operational definition is generally useless and may be misleading. A cross section that was determined under conditions of poor accuracy can still be of value if the limitations are appreciated.

Considerable discussion has been devoted to the concept of the cross section for emission of photons. Unlike the collisional excitation cross section, this is not a fundamental microscopic property of the collision combination; however these two quantities are of course closely related. Most of the experiments reviewed in this monograph have involved the determination of an emission cross section. Only in a relatively small number of cases is transition-probability information available in order to permit the deduction of level-excitation cross sections. It has been shown that, even under conditions where projectiles undergo no more than one collision while traversing the target, the emission may still be dependent on target density and even geometry. Conditions were established for which the measured emission cross sections are independent of the geometry of the apparatus and the pressure of the target. These conditions constitute criteria that must be fulfilled if the emission data are to be considered as fundamental data.

The question of polarization has also been considered, and it is concluded that polarization fractions constitute valuable cross section information. Since the

---

† In Ref. 33 (pages 20 and 199) and Ref. 34 (page 152) there are detailed explanations of the relationship between intensity of emission and the Franck-Condon factors for a transition.

polarization fraction defined by Eq. 2-26 involves only the ratios of intensities, most of the criteria on the validity of emission cross sections are unimportant. Criteria for the validity of a measurement of polarization and for the subsequent derivation of cross sections are summarized in the following list.

The preceding discussions are summarized in Tables 2-1 and 2-2.

TABLE 2-1   CRITERIA FOR THE VALIDITY OF THE OPERATIONAL DEFINITIONS OF EXCITATION AND EMISSION CROSS SECTIONS

1. The measured cross sections should be demonstrated to be independent of target density.

2. The measured cross sections should be demonstrated to be independent of projectile flux.

3. The measured cross sections should be demonstrated to be independent of the extent of the observation region (total cross sections only).

4. Full geometrical definition of observation region should be specified.

5. Projectile beams should be shown to be neither divergent nor convergent (crossed beams only).

6. The form factor should be evaluated (crossed beams only).

7. Cascade corrections should be accurately applied (only if the data are presented as a level-excitation cross section derived from an emission measurement).

8. Accurate transition probabilities should be used (only if the data are presented as a level-excitation cross section derived from an emission measurement).

9. Emission recorded from an excited projectile must be independent of the distance traveled through the target.

10. The cascade contribution to emission from a projectile must be assessed by one of the following methods: (a) by showing it to be negligible, (b) by measuring emission at a point where the cascade is independent of projectile penetration through the target, or (c) by direct measurement.

11. The measured cross section should be shown to be independent of the quenching field (quenching techniques only).

12. Measurements made on a projectile beam that has traversed a gas cell must be shown to be independent of the cell length and the size of the exit aperture.

13. Loss of excited atoms by processes of collisional destruction before detection must be negligible.

TABLE 2-2   CRITERIA FOR THE VALIDITY OF THE OPERATIONAL DEFINITION OF POLARIZATION FRACTION AND THE DEDUCTION OF LEVEL-EXCITATION CROSS SECTION FOR M_L SUBSTATES

1. The polarization fraction should be demonstrated to be independent of target density.

2. The polarization fraction should be demonstrated to be independent of projectile flux.

3. The polarization fraction should be demonstrated to be independent of stray magnetic fields in the observation region.

4. In order to derive level-excitation cross sections, the influence of cascade must be assessed.

5. The relationship used to derive level-excitation cross section from polarization must be correct.

6. The level-excitation cross section used with the measured polarization to derive substate excitation cross sections must be valid in accordance with the specified criteria for such cross sections.

CHAPTER 3

DETERMINATION OF EXPERIMENTAL PARAMETERS

The operational definitions of cross section are in terms of macroscopic observables: beam flux, target thickness, excited-state density, and rate of photon emission. The accuracy of a cross section is governed by the accuracy with which these macroscopic parameters are recorded. Although the required experimental techniques are well established, there remain many discrepancies among experimental data; these must be ascribed to erroneous measurement.

The determination of each relevant parameter requires first its definition. The target and projectile must be of a known composition, preferably a single chemical species in a single state of excitation and ionization; the excited state of interest must be completely resolved from neighboring levels. The densities of these defined systems must then be quantitatively measured.

The excited-state density is the most difficult parameter to determine. There is no general technique that provides good resolution while at the same time giving reliable quantitative measurements. The most popular approach is the determination of emission cross sections. This technique uses optical spectroscopy to provide sufficient resolution to adequately define the source of the signal; however, in practice the optical photometry required to quantitatively measure the signal is difficult to carry out accurately and appears to give unreliable data. The direct measurements of excited-state density are less frequently used: the detection reduces to measurement of current, which can be done accurately, but the resolution is poor, and the signal may involve contributions from a number of excited states.

The present chapter reviews in depth the various techniques that are available for detection of beam flux (Section 3-1) and measurement of target density (Section 3-2). Separate consideration is given to the direct detection of excited-state formation (Section 3-3) and to the monitoring of photon emission to arrive at an emission cross section (Section 3-4). The conclusions of each section are summarized in the form of criteria that should be fulfilled if an experimental measurement is to be considered accurate.

3-1 PROJECTILE-BEAM DEFINITION AND DETECTION

A. Role of Beam Contaminants

Each separate constituent of a projectile beam will exhibit a different cross section for the production of the signal that the experiment is designed to detect. The measurements represent a mean of the cross sections for the various contributory processes, each weighted according to the proportion of the various beam constituents. This quantity is of little value unless the relative proportions of the different projectile states are established by separate measurements.

The impurity content of a projectile beam will be dependent on the source of the projectiles and the nature of the beam handling system. Most ion beams are produced in electron impact or discharge sources. In order to produce a maximum output ion current it is necessary to operate these sources at electron energies of some hundreds of electron volts. This energy exceeds the threshold for production of excited states and multiply charged ions. Other types of ion sources have been used in a few experiments; each has its own peculiarities regarding impurity content of the resulting beam. In a typical experiment, the ion source is followed by an acceleration region and a mass spectrometer that will sort the mass and charge states of the projectiles. Neutral projectiles are generally formed by charge transfer neutralization of fast ions emerging from the mass spectrometer section of the apparatus. The ions are passed through a gas cell, and the residual flux of ions in the emergent beam are removed by an electric field. The projectile's velocity is changed by only a small amount as a result of the charge transfer process, and it is generally assumed that this change is negligible.

In principle, the uncertainty of the chemical composition, charge state, and energy may be removed by using a momentum selector between the ion source and the experiment. The influence of excited states generally presents a very difficult problem that has only been subject to complete analysis in a few cases.

1. Chemical Contamination. Outright chemical contamination of projectile beams should be most unlikely in modern experiments. Almost every ion accelerator includes a mass analyzer that defines the charge-to-mass ratio and velocity of the incident projectile. It is also satisfactory to momentum-analyze the projectiles after they have passed through the interaction region to confirm that the beam has a unitary composition.

2. Charge-State Contamination. The output of an ion source may contain various states of ionization, depending on the energy of the bombarding electrons in the source. Provision must be made for selecting one state of ionization only. In general this is accomplished satisfactorily by using a mass spectrometer before the projectiles enter the interaction region. Neutral projectile beams produced by charge transfer neutralization must have the residual ion flux removed by electrostatic deflection before entering the interaction region.

There is always a possibility that the projectile beam will undergo changes in composition as it traverses the distance between the mass spectrometer and the point where it is detected. Changes that occur while the beam negotiates the flight path between the mass spectrometer and the collision chamber will be dependent on the residual gas pressure in that region and can be reduced to negligible proportions by suitable use of high vacuum techniques. It is to be expected that a report of an experiment will quote the operating pressure of this part of the apparatus to prove that the

changes in beam composition may be neglected. There is also a danger that the composition of the projectile beam will change appreciably as it traverses the target region itself. Signals produced from collisions involving projectiles that have undergone a charge rearrangement prior to the collisional excitation will vary as the square of the target pressure. It has previously been established [see Section 2-1-B(1)] that the operational definition of cross section is only valid when there is negligible change in the beam current or composition. This condition can be obtained by operating the experiment at a sufficiently low target density that the signal varies linearly with pressure. This condition will occur at target pressures in the region of 10^{-3} torr or less. It is possible to arrive at a linear dependence of emission on target density over a limited range of conditions at much higher pressures. This situation is generally associated with the beam acquiring an equilibrium distribution of charge state components; such a distribution will be unaffected by small changes of pressure. Clearly, the equilibrium conditions must be avoided.

3. Energy Spread. For any practical ion source the extracted beam will have a finite energy spread, and the measured cross section will be a weighted mean value over the range of energies involved. If appreciable cross section changes occur in an energy interval less than the energy spread of the beam, then the detail of such variations will be distorted or obliterated entirely. The energy spread of ions extracted from a discharge source may be as much as 500-eV. Suitably designed electron impact sources will produce distributions of less than one electron volt, but the intensity of the ion beam is often too low to allow this type of source to be used in excitation experiments. A mass spectrometer between the source and the experiment will reduce the effective velocity distribution of ions in the projectile beam.

All existing evidence suggests that, for projectile energies above a few kiloelectron volts, there will be no occurrence of rapid cross section variations that might be obscured by energy resolutions of less than one percent of the mean projectile energy. In general, such resolutions will be readily achieved with a magnetic mass spectrometer. Recent measurements at lower energies[35] do indicate quite extreme cross section variations, which can only be fully resolved with energy resolutions of the order of one electron volt. It is to be expected that an experimental determination will include an estimate of the energy resolution of the projectile beam; at low impact energies a statement of resolution is essential.

Note that the accuracy with which the energy of the projectiles is determined has an important bearing on the accuracy of the experimental data as a whole. The accuracy of the energy measurement should be quoted as well as the estimated accuracy of the cross section determination itself.

4. Control of Excited State Population. There are severe difficulties in assaying the excited-state contamination of the projectile beam, since there is no general method whereby excited and ground state projectiles may be separated.

Excited ions are produced either by electron impact excitation in the source or by the impact of the ion on the source gas as the ion is accelerated by the extraction field. Excited states can also be produced by ion molecule reactions in the source. Neutral projectile beams will include excited constituents formed by electron pick up into an excited state during passage through the neutralizer cell.

Generally, excited states decay exponentially in time with a rate characteristic of the wave function of the excited atom and that of the lower state to which the decay is possible. Many states have decay rates of the order of 10^8 sec^{-1}, and times of flight

of ions from source to collision chamber are often of the order of 10^{-8} sec so that the excited ions decay in flight to lower states. In order to calculate the population of states in a beam at the entrance to the collision chamber, it is necessary to know the population emergent from the ion source, the lifetime of each state initially excited, and the lifetime of each state above the ground state that is populated by transitions in flight from higher states. Fairly weak electric and magnetic fields, often interposed along the beam path by accident or by design, can sometimes drastically alter the normal decay rate of a state. For example, the metastable 2S state of hydrogen is perturbed by a field of the order of 500 V/cm to such an extent that it will decay at a rate nearly equal to that of the 2P state. Also, in the H atom, the higher angular momentum states that have relatively long lives can be mixed with lower angular momentum states in very weak fields to produce a change in their decay rates. While the lower excited states of a system are generally preferentially populated, the higher states should not be ignored since they have longer lifetimes and may possess disproportionately large cross sections for certain processes, such as stripping or inducing excitation or ionization in a target molecule. Many of the above factors have been considered in more detail by McClure[36] in an experimental study of ionization and electron transfer in collisions of two H atoms, one of which was electronically excited. In this work, special steps were taken to remove excited states because of indirect evidence that the ionization cross section might vary as the square of the principal quantum number and strongly influence the results.

Molecule-ions such as H_2^+, CO^+, etc., are usually produced in a number of different vibrationally and rotationally excited states. Since the vibrational levels of molecules are fairly closely spaced (0.27-eV for H_2^+ and less in heavier molecules), relatively distant collisions can cause transitions and modify the population of the vibrational and rotational levels. The excited-state populations at the exit from the ion source might therefore be very much dependent on source pressure. Excited vibrational states of the lowest electronic states of homopolar molecules have no dipole moment and hence have very long lifetimes—so long in fact that the proportion may be unmodified during flight from the ion source to the collision chamber in an apparatus of typical dimensions.

The importance of the excited-state composition of the beam will vary greatly with the nature of the process under study. There is considerable evidence that the state of projectile excitation influences stripping, charge transfer, and dissociation.[37] There is no evidence as yet that either electronic, vibrational, or rotational excitation of atomic or molecular projectiles will appreciably affect the cross section for collisional excitation of a target.

Ideally, one would deal always with particles in a single well-defined state or a precisely known population of states. These goals have rarely been achieved except in the following cases: (1) the projectile has no internal structure (for example, H^+, D^+, He^{++} ions, or other stripped nuclei); (2) the projectile has only one state; (3) the projectile is produced in a collision process in which the available energy is sufficient to produce only one state, and this one state is not depopulated by secondary collisions in the ion source or elsewhere along the route of the ion from the ion source to the collision region.

An alternative approach is to show that the cross section under investigation is independent of the conditions under which the projectiles are formed. Presumably, excited-state content is dependent on source operating conditions. Independence of

signals from source conditions would suggest either that there are no excited states surviving the passage to the experimental region or that the process under investigation is not greatly affected by the state of excitation of the projectile. Under no circumstances does this approach give adequate proof that excited state content may be neglected.

The control of excited-state content in projectile beams is in its infancy. Little attention has been paid to the problem in the cases where it is expected to occur.

B. Detection

The major requirements on the beam-detection system are that it should measure beam flux accurately and that it should prevent secondary emitted particles and reflected projectiles from returning to the observation region. A wide range of detectors has been developed, but attention will be confined to those that have been used in collisional excitation studies. In general, beam currents must be quite large in order to permit detection of photon emission, and it is relatively simple to detect the projectiles as a current. Therefore, it is possible to omit consideration of particle "counters" that are designed to detect individual projectiles and are not suitable for high fluxes.

Every type of detector involves projectile beam impact on a surface. Light emission will occur at the point of impact due to excitation of the solid and of atoms ejected from the surface.[38] In an emission experiment it is imperative that the detector does not view this region.

1. Ion Beams. The customary device for measuring beam fluxes of ions is the Faraday cup. This name is applied to many different device configurations. Basically, it consists of an insulated aperture followed by a deep receptacle or cup. The cup is connected to a current meter such as an electrometer.

Precautions must be taken to prevent appreciable fluxes of particles being ejected from the base of the cup. Any flux of charged particles leaving the Faraday cup will give rise to an erroneous measurement of beam current. Any particles of appreciable kinetic energy, whether neutral or charged, that return to the observation region may cause excitation of the target and provide additional false contributions to the signal. Carter and Colligon[38] have provided a detailed discussion of all processes whereby particles are reflected or ejected from metallic surfaces. Their review shows that for some mechanisms the flux of such particles may represent an appreciable fraction of the incident beam flux. Ejected electrons are the major source of worry since the secondary emission coefficient is high, typically of the order unity for incident projectile energies of a few kiloelectron volts.[39] The coefficient for ejection of secondary ions is quite small, typically 10^{-2},[40] so this process does not constitute such a serious problem.

Two basic procedures are used to prevent errors due to particles ejected and reflected from the base of the Faraday cup. First, the cup should be provided with an aperture that subtends only a small angle at the base, and geometrically restricts the escape of particles; (10^{-2} steradians is suggested as an upper limit to the angle). Secondly, it is important to provide complete electrical suppression of secondary electrons. With proper electrical suppression of secondary electrons and partial suppression of heavy particles by suitable geometry, the errors inherent in measuring a beam current with a Faraday cup should be substantially less than one percent.

The only customary test of the Faraday cup operation is to measure the apparent beam current as a function of the electron suppressor voltage. A range of operation must be established for which the apparent beam current is independent of this voltage.

The Faraday cup is generally regarded as an absolute device, and most experiments assume that any accuracy limitations are set by the current meter rather than the cup. The possibility of small errors due to reflection and secondary emission are generally negligible, although often no tests are made.

2. Neutral Beams. Only a few of the experiments reported in this monograph have involved the detection of neutral beam fluxes. All have employed secondary emission techniques in which the beam is incident on a metal surface and the secondary electrons are drawn off by a potential to a second plate where they are measured as a current. There is no problem due to secondary emitted electrons returning to the experimental region. It is desirable to inhibit heavy particles from returning to the collision region by inserting an aperture that subtends an angle of less than 10^{-2} steradians at the secondary emitting surface.

The secondary emission coefficients of a metal vary with the condition of the surface and with the velocity, mass, and charge state of the incident projectiles. The detection efficiency of the secondary emission device will vary likewise and must be determined by frequent calibration. Frequently the efficiency of detection for a given neutral projectile is determined by using a beam of singly charged ions of the same mass and velocity as the neutrals; the ion current is measured directly using the secondary emitting surface as a Faraday cup; the ratio of the secondary emitting current to the incident ion beam current is used to give a measure of efficiency. Provided that the secondary electron emission coefficients for the charged and neutral projectile are equal, then the measured efficiency is applicable to the detection of the neutral projectile. Such equality, however, does not occur; the secondary emission coefficients for heavy ions and neutral projectiles may differ by as much as 10%[39] while the coefficients for H impact exceed those for H^+ by up to 62%.[41, 42] It must be concluded that the secondary electron detector is not absolute, and its efficiency for neutral species cannot be accurately calibrated by using a beam of ions.

In some experiments the calibration of secondary emission detectors for neutral atomic hydrogen projectiles has been carried out by using a proton beam and by assuming that the difference between secondary emission coefficients for ions and atoms was the same as that measured by Barnett and Reynolds.[42] This procedure is not necessarily correct, since the secondary emission coefficients will be dependent on the nature, condition, and orientation of the surface. There is no reason to believe that the data of Barnett and Reynolds will have universal applicability.

The detection of neutral beams by thermal effects does not suffer from the defects of the secondary emission detector. A small piece of metal foil may be inserted into the beam path and its temperature measured using a thermocouple or thermistor. The foil and electrical leads are chosen to minimize both heat capacity and radiation loss in order to ensure small thermal time constants and high sensitivity. The temperature rise of the foil is dependent on the power of the projectile beam. Although the device is not absolute, the efficiency may be determined using an ion beam since the response is independent of the charge state of the projectile. The major disadvantages of the thermal detector are poor sensitivity and the long response time.

The secondary emission detector continues to be favored for neutral beam detection

while the thermal device has never yet been employed in an excitation experiment, although it has found many applications in charge rearrangement studies. Since excitation experiments normally employ quite high beam currents, there would appear to be no reason why the poor sensitivities of thermal detectors should preclude their use.

C. Summary

The preceding discussions are summarized by Table 3-1 which gives a list of criteria that should be fulfilled in an accurate experimental measurement.

TABLE 3-1   CRITERIA FOR THE ACCURATE DEFINITION OF BEAM COMPOSITION AND DETECTION EFFICIENCY

1. The charge state of the projectile must be defined.
2. The chemical species of the projectile must be defined.
3. The projectile must be in a single known state of excitation, or the distribution of excited states must be known.
4. The energy of the projectiles should be known accurately.
5. The energy resolution of the projectile beam must be adequate for the problem being studied and should be quoted.
6. Secondary particles from the beam monitor must be suppressed.
7. Reflected projectiles from the beam monitor must be suppressed by the use of limiting apertures.
8. The efficiency with which projectiles are detected must be known.
9. Light produced at the detector surface must be prevented from entering the photon detector.

3-2 TARGET DEFINITION AND DENSITY MEASUREMENT

The same basic considerations which are applicable to the production and detection of projectiles are also valid for the production of a gaseous target and the measurement of its density. Clearly, the target must be free of contamination from chemical impurities and be in a known state of excitation. Furthermore, the density of the target must be uniform in the region over which observations are made. The measurement of target density is generally achieved through the measurement of pressure and the use of the ideal gas equation. The accurate measurement of low pressures is a difficult procedure, and many large errors have occurred in collisional excitation studies due to inadequate attention to this point.

A. Role of Contaminants

Contaminants of the target produce ambiguity as to what the measured cross section represents, just as definitely as do contaminants in the projectile beam. Every target species will exhibit a different cross section, and the measured signal will be a

mean value for various processes weighted according to the relative proportions of the species present. In practice, uncertainty in target composition is a less severe problem than uncertainties in projectile beam composition.

1. Chemical Contamination. Chemical contamination of the target is unlikely to be a source of trouble. It is to be hoped that a report of an experiment will document the purity of the gas and any precautions taken to remove contaminants. In general, target gases of high purities are available from commercial sources. Experiments on collisionally-induced emission have a built-in diagnostic system that helps in the elimination of errors due to chemical contamination. The detection system may be set to receive only wavelengths that correspond to emissions from the target and thereby automatically exclude emission from impurity atoms. The impurities may still cause errors in pressure measurement, but these are likely to be very small. A further consideration is that the detection system may also be used to carry out a spectroscopic analysis of the emission with the specific intention of detecting the presence of impurity atoms. For example, the cross sections for emission of certain lines in the spectra of N_2^+ and O_2^+ induced by proton impact on the parent gases are very large (see Chapter 8), and this emission may be readily utilized to detect partial pressures of N_2 and O_2 as low as 10^{-7} torr.

A special case of contamination occurs for experiments that utilize an atomic H target produced by catalytic dissociation of H_2 at a temperature of about $2400°K$. Some 6% or more of the molecular hydrogen remains undissociated; this constitutes a chemical impurity. Its influence is assessed by assaying the fraction of undissociated H_2, using a technique such as the double charge transfer method of Lockwood et al.[43] and measuring directly the relevant cross section for a room temperature H_2 target. The contribution from H_2 can thereby be calculated and subtracted from the observed signals.

2. Excited-State Contamination. Most experiments have utilized target gases that are normally at room temperature. There are a very small number of cases in which elevated temperatures have been used in order to produce targets of metal vapors or of atomic hydrogen.

Consider first room temperature experiments. Provided that the temperature of the collision chamber is uniform and that the target gas is in thermodynamic equilibrium with the walls of the chamber, there will be a Boltzmann distribution of population among the various excited states. Thus, the relative proportion of various excited states of the target may be estimated quite unambiguously. For normal room temperatures the population of excited electronic states of the permanent atomic and molecular gases will be quite negligible. The population of excited vibrational states of molecules will also be very small, but the population of the rotational states will be quite significant. Thus, measured cross sections for excitation of molecular gases are a sum over all possible prior-collision rotational states weighted according to the Boltzmann distribution. It is to be expected that experiments with molecular targets will include a measurement of the target temperature. There is some evidence that, for high projectile energies, the cross section for collisionally induced excitation of electronic and vibrationally excited levels of molecules is independent of the rotational state of the molecule. At low energies however, where the time for a rotational transition is comparable with the time taken for the collision, this is no longer true. The available experimental data on rotational excitation are considered in Section 8-3-F.

For metallic targets produced by thermal vaporization of the solid material, the same considerations hold as for experiments at room temperatures. There are some additional practical difficulties in ensuring that the apparatus is at uniform temperature. For all situations used in practice, the thermal population of excited states can be ignored. Note, however, that certain materials such as boron, carbon, and aluminum have two or more values of total angular momentum associated with the lowest lying electronic state each of which has a slightly different energy. The population of these states might be quite significant. In practice, materials with this characteristic have not yet been used in excitation experiments.

There remains the special case of an atomic hydrogen target produced by catalytic dissociation of H_2 in a furnace at about $2400°K$. The thermal population of excited states of H will be negligible under these circumstances.

B.   Spatial Pressure Gradients

It is important that there be no gradients of pressure in the observation region nor between this region and the pressure measuring device. It is clearly necessary that the collision region be at uniform temperature to eliminate this problem. The influence of cold traps that are sometimes placed between the pressure measuring device and the observation region is considered in Section 3-2-D.

It is necessary to avoid pressure gradients that may occur close to the aperture through which the projectile beam enters the collision chamber. The apparatus used in emission measurements will generally observe only a very small region of the collision chamber. If this region of observation is located immediately behind the aperture, then there will be uncertainty as to the pressure in this region.

C.   Thermal Motion of the Target

It is generally assumed that in a single-beam experiment the individual atoms of the gaseous target are at rest—an assumption which obviously introduces considerable simplification. It may be shown that, for a given velocity of impact, a total cross section is the same whether measured in the laboratory or in the center-of-mass frames of reference. If the total cross section remains essentially constant over the range of impact velocities defined by the vector sum of the incident particle velocity and the distribution of target velocities, then the experimentally measured total cross section should not be influenced by thermal motion. Only at threshold or other sharp structures associated with inelastic processes, should consideration of the thermal motion be important. For the range of energies covered in the present monograph (greater than 10-eV), the influence of thermal motion on a total cross section determination may be neglected.

D.   Measurement of Target Pressure

The number density of a gaseous target may, in principle, be determined by the measurement of the gas pressure and the assumption of the ideal gas law, which is expected to be accurate at low pressures. Unfortunately it would seem that serious errors have occurred in the measurement of gas pressures in the range 10^{-6} to 10^{-2} torr, which cast doubt on the results of many otherwise accurate experiments.

Of the various techniques available for the measurement of pressure in the region 10^{-6} to 10^{-2} torr, the McLeod gauge has long been regarded as the standard instrument. Although the use of this instrument is inconvenient, it is often employed for the direct measurement of pressure in experimental systems. Most other pressure gauges are relative instruments, depending on some thermal or ionization property of the gas under study, and their calibration for each gas of interest may be traced back to a comparison with a McLeod gauge. Such a calibration is essentially dependent on the McLeod gauge and will exhibit any systematic errors that the standard device may have. The McLeod gauge works on the principle of compressing a sample of the test gas through a known ratio to a final pressure, which may be conveniently read off as the difference in height of two mercury columns. The ideal gas law is then used to calculate the original gas pressure. The gauge has many practical problems associated with its use, such as the depression of mercury in the capillary tubes, sticking of mercury, and cleanliness; nevertheless with considerable care and labor these may be reduced to negligible proportions.[44] The gauge remains inherently unsuitable for condensible vapors or gases whose coefficient of adsorption on glass is high; moreover it is not a continuously reading instrument and cannot be conveniently automated. Despite these limitations, the McLeod gauge has been frequently employed in single-beam atomic collision experiments, both for the direct measurement of pressure, and also for the *in situ* calibration of other gauge types. Recently, doubt has arisen about the accuracy of such measurements. It has been shown that the necessary precaution of inserting a cold trap between the gauge and the experimental system, to prevent the contamination of the latter with mercury, results in the introduction of the secondary mechanisms of thermal transpiration and "pumping," both of which may significantly affect the measurement of pressure.

The thermal transpiration effect was originally discussed in a paper by Rusch[45] in 1932, although little attention was paid to it until comparatively recently.[44, 46] Dushman[47] shows that, if two vessels at different temperatures T_a and T_b are connected together by tubing with a diameter smaller than the mean free path of the molecules (molecular flow conditions), the equilibrium pressures in the two chambers will not be equal but will be given by $p_a/p_b = (T_a/T_b)^{1/2}$. When the mean free path of the molecules is considerably less than the diameter of the connecting tube (viscous flow conditions), the pressures will be equal. The pressure regions in which these relations apply and the nature of the effect in the transition region are sensitive functions of the connecting tube diameter. A typical experimental system will consist of three parts: collision chamber and McLeod gauge, both at room temperature, and the cold trap at considerably lower temperature. The connecting tubes between these three parts will, in general, be of different radii. If the experimental conditions fall entirely into the viscous or molecular flow regions, the pressure in the experimental system will be the same as in the gauge, although the pressure in the cold trap may be different. For conditions in the intermediate region, the pressures in the experimental system and gauge may very well be different. Errors as high as 10% have been reported[45] at pressures between 10^{-3} and 10^{-1} torr. This pressure range probably represents the transition region for a typical experimental system; for pressures much above or below this range, the thermal transpiration effect may become negligible. The most obvious solution to the problem is to use an ordinary U-tube trap or similar configuration where the two connecting tubes may be made of equal diameters. Rusch[45] has also shown empirically that, for certain ratios of tube diameters, the effect may be made

negligible. It is to be noted that the thermal transpiration effect may make the measured cross sections either too high or too low, depending on the precise geometrical configuration employed.

The so-called "pumping" effect refers to the continuous streaming of mercury from the gauge reservoir onto the cold trap, which acts as a mercury diffusion pump, reducing gas pressure above the mercury reservoir and so causing erroneous pressure measurements. Although the effect was first mentioned by Gaede[48] in 1915, the consequences were ignored until it was rediscovered by Ishii and Nakayama.[49] Its explanation was placed on a firm theoretical foundation by Takaishi.[50] It may be shown that the effect will be at a constant maximum in the molecular flow pressure region below about 10^{-3} torr, and decreases to zero in the viscous flow region above 10^{-1} torr. The magnitude of the error depends sensitively on the temperature of the mercury reservoir, the radii of connecting tubing, and the nature of the gas whose pressure is to be measured. For light gases such as hydrogen and helium the collision cross sections between the streaming mercury and the gas atoms are quite small, and the predicted pressure measurement error is of the order 2 to 4% in most experimental configurations. The effect can become very serious for the heavier atoms, and values of 10 to 30% have been quoted in some situations. Although it is impossible to give correction factors of general applicability, the following are quoted as typical values calculated using Ishii's theoretical expressions. For the specific case of a McLeod gauge operated at a room temperature of 27°C, with a 1-cm diameter tubulation to the cold trap that is held at liquid nitrogen temperature, the measured cross sections would have to be reduced in accordance with: H_2, 2%; He, 3%; Ne, 7%; CO, 10%; O_2, 10%; N_2, 10%; and Ar, 12%. It is impossible in most cases to return to published experimental data and apply corrections because the all important parameters of reservoir temperature and connecting tube radii are not recorded. There is also some continuing uncertainty as to the magnitude of the correction. Although various techniques have been suggested for eliminating the pumping error,[44, 49] none is entirely successful and all have the disadvantage of making the use of the McLeod gauge more cumbersome.

Most experiments are carried out at target pressures below 10^{-3} mm where molecular flow conditions are expected. Only the pumping effect will be important, leading to cross sections that are too high. The required correction will be almost invariant with pressure, and in the case of helium and hydrogen the correction may be negligible in certain systems. A few experiments have involved target pressures in the transition region between molecular and viscous flow conditions where both thermal transpiration and pumping errors are expected. The pressure dependence of each of the effects leads to a total correction, which for particular circumstances might be either positive or negative. In the high-pressure molecular flow region, which is rarely used in atomic collision experiments, both mechanisms will give rise to a negligible error.

In the field of low-pressure measurement a recent development, which may eventually replace the McLeod gauge, is the sensitive differential capacitance manometer. A very thin metallic diaphragm forms the common plate of two parallel plate capacitors in adjacent arms of an ac bridge circuit. A pressure difference across the diaphragm will cause a deflection and a corresponding change in capacitance, which is detected by the bridge circuit. The commercially available devices are frequently calibrated at high pressures in the 0.1 to 1 torr range, using a dead weight technique; the calibration

is extrapolated to pressures down to 10^{-4} torr, under the assumption that the device behaves linearly. The accuracy of such extrapolation may be questioned; however, Utterback[51] has shown that it is quite good in the case of one commercially available device. Two virtually untried null techniques are available for removing the necessity for extrapolating a calibration made at high pressures. A null condition may be achieved by the application of an equal pressure of hydrogen on the opposite side of the diaphragm to the test gas in order to restore it to its undeflected conditions. The pressure of the hydrogen may be measured by a McLeod gauge with very little pumping effect error. A second technique, suggested by Romy in unpublished reports, is that the diaphragm could be returned to its equilibrium condition by the application of a dc field across one set of capacitor plates. The pressure of the test gas would then be measured in terms of the applied voltage and the physical dimensions of the capacitors, independent of the use of any other pressure measuring instrument. At the present time, the capacitance manometer must be considered as being in a developmental stage; certain problems of stability and calibration still require solution. Nevertheless, it does offer the possibility of eliminating the use of the McLeod gauge and instead referring the calibration either to the weight technique or to the measurement of electrical quantities. Further advantages of the capacitance manometer are that it may be operated to give continuous readings; that it introduces no contamination into the system; and that its calibration is independent of the nature of the gas. Of considerable importance to atomic collision experiments is that the capacitance manometer allows the use of condensible vapors as static targets, a field which has hitherto been much neglected because of the difficulty of measuring the pressure of such vapors.

In conclusion, it is clear that whenever a collision experiment has been carried out at low target densities using a McLeod gauge to measure pressure, the pumping error has influenced the results. The intermediate and high-pressure regions are not often used in such experiments; therefore the thermal transpiration effect is of lesser importance. Without an intimate knowledge of the geometry and temperature of an apparatus, it is impossible to return to the older published data to make corrections for these effects. In much recent work the problem has been eliminated by the use of one or more of the available techniques, although many data are still being produced without the necessary corrections. There is some evidence that the use of theoretically predicted corrections is inaccurate.

E. Summary

The preceding discussions have documented the types of errors that have occurred in the target-density definition and in the method by which these errors may be overcome. Although the precautions that are needed may seem complicated, it appears that there is no insurmountable problem in accurately defining target conditions. It is certainly true that many recently published data have involved some errors principally due to problems associated with McLeod gauges. It is also clear that if published records of experimental measurements had included complete specifications for the gas handling and pressure measurement systems, then it would generally have been possible to make good retrospective estimates of the magnitudes of any errors that had occurred. The problems of target definition are summarized by Table 3-2.

1. The chemical composition of the target must be known.

2. The temperature of the target must be known and should be specified in a published report.

3. The temperature of the target must be uniform.

4. The observation region should not be located close to an orifice.

5. The effect of target thermal motion should be assessed.

6. The measurement of gas pressure must be free of pumping and thermal transpiration errors due to cold traps.

7. The gas must obey the ideal gas law if a compression technique is to be used in the pressure measurement.

8. All necessary precautions must be taken to eliminate errors due to sticking, capillary depression, etc. if a McLeod gauge is used.

3-3   EXCITED-STATE DENSITY: DIRECT DETERMINATION

There have been a number of techniques developed for the direct determination of excited-state densities without the intermediate step of measuring spontaneous emission. Each method has a specialized application to the detection of a particular excited state or group of states. None is of general applicability to the measurement of excitation cross sections. These techniques have only been applied to about 2% of the papers reviewed in this monograph.

The first technique is the Lorentz ionization method that has been applied to the detection of H,[52] H$_2$,[53] and He$^+$ [54] in a state of high principal quantum number formed in projectile beams traversing a target. This technique involves the removal of the excited electron with a high field and the detection of the resulting ion. The second method is the monitoring of the flux of projectiles that lose an energy appropriate to the formation of a particular excited state of the target or projectile system. The third method is the detection of fast metastable atoms by means of the different charge rearrangement cross sections for ground and metastable states.

There are a number of common features in these methods. All involve a projectile beam that traverses a target cell and emerges into a flight tube where the analysis is carried out. It is therefore necessary to ensure that no appreciable fraction of the projectiles is intercepted by the exit aperture from the target cell. In all three methods the detection of the excited atoms is reduced to the measurement of a current. The techniques for measuring current are inherently simpler and potentially more accurate than the quantitative measurement of photons required in emission measurements. The operational definition of cross section will be in the form of the ratio of the current of detected excited atoms to the current of beam particles. Therefore, there is also a tendency for any systematic errors in current measurement procedure to cancel out. The energy-loss method is applicable to measurements of excitation in both projectile and target systems. The other two methods can only handle formation of excited projectiles.

A.   Field Ionization

An electron bound in an atom that moves with a velocity v through external electric field E_0 and a magnetic field of induction B experiences a force given by the expression $F = eE_0 + e(\mathbf{v} \times \mathbf{B})$. The quantity $(\mathbf{v} \times \mathbf{B})$ is sometimes known as the *equivalent electric field*. The applied field E_0 or the "equivalent" electric field will produce a lowering of the potential barrier experienced by the electron in the excited atom. This can lead to the ionization by transitions both over and through the potential barrier. The overall transition probability is dependent on the time that the excited atom resides in the field region and also on the magnitude of this field. The applied field also produces a Stark splitting of the levels, and therefore for each value of n there is a range of fields that will remove electrons from the sub-levels.

This technique has been used for the detection of H and He^+ in high n states[52, 54] and also for molecular H_2 in high Rydberg states of electronic excitation.[53] Two basic experimental configurations are found in the literature. One technique utilizes an electric field parallel to the direction of motion to cause field ionization,[56, 57] and the other uses a transverse magnetic field.[55] Both systems utilize a weak field to remove charged particles from the beam before it enters the detection region. Monitors are arranged to detect the ions produced in the field ionization region and also to detect the transmitted neutral beam, which includes atoms in the lower excited and ground states.

In practice, the range of states that can be detected in this manner is limited. The field required to ionize excited states of high principal quantum number is small, and the excited atoms may be removed accidentally by ionization in stray fields existing in the apparatus. In many experiments, the limiting upper value of n was determined by the necessity for using a separate magnetic or electrostatic deflection system to remove ions from the projectile beam before it entered the detector. The electric and magnetic fields available are limited to the order of 10^5 V/cm, which governs the lower limit to the principal quantum number of states that can be detected.

This technique is inherently unable to completely resolve states of different principal quantum number for any but the lowest values of n. Inevitably, any attempt to obtain a cross section for the formation of a particular state must involve some estimate of the overlap with signals from neighboring levels. Corrections must be applied for the loss of excited atoms by normal radiative decay during the flight between the detection system and the cell where the projectiles were excited. In most experiments this correction is made by using theoretical lifetimes and assuming a statistical population of the various substates of a given principal quantum number. In optical studies of the relative population of substates (Chapter 9), the data clearly indicate that a statistical population does not occur. The best method of correcting for this loss is to measure it directly by determining the decrease in the excited state population as a function of the distance between the gas cell and detector.

The Lorentz ionization technique is unable to resolve substates of different angular momentum quantum numbers and, at best, will give an approximate sum of all cross sections for forming a state of a particular principal quantum number; the technique is inherently inferior to optical methods. It has therefore been decided to exclude detailed consideration of this type of experiment from this monograph. Sources of data are listed in the comprehensive table of Chapter 4 and are mentioned in the text where appropriate. The data, however, are not reproduced here. It should be

noted that, in most experiments of this type, the data are not presented in the form of a cross section for the formation of a specific excited state.

B.   Energy-Loss Techniques

The kinetic energy lost by a projectile during collision will be uniquely determined by the nature of the inelastic process. The measurement of the flux of projectiles I_i (particles/sec) that have lost an energy appropriate to the excitation of a state i, as a beam of total flux I_b (particles/sec) traverses a target of length l and density N_a (particles/cc) will allow the determination of an excitation cross section $\sigma_i = I_i/N_a I_b l$.

This experimental approach will require the provision of an electrostatic energy analyzer at the exit from the gas cell. This device must transmit 100% of the projectiles that experience the necessary energy loss and completely reject all particles of different energies.

In principle, the energy-loss technique offers definite advantages over photon emission and should be subject to fewer systematic errors. The cross section involves the ratio of scattered and incident projectile currents, thus eliminating any systematic errors in the measurement of these quantities. Population of the excited state through secondary processes such as cascade, photoabsorption, and collisional transfer, which all contribute to photoemission cross sections, does not contribute to this measurement.

In order for this technique to be used, it is necessary that the energy spread of the projectile beam and the resolution of the scattered particle analyzer both be considerably narrower than the energy separation between adjacent excited states. This is very difficult to achieve with ion beams, even at rather low energies. It is likely that this requirement will limit the use of the technique to the investigation of the lowest excited states of atomic and molecular systems, where the energy separations are comparatively large.

There have been two experiments of this type. The first by Lorents et al.[58, 59] is used to measure the excitation of a helium target to the 2^3S state by impact of He^+ ions. The electrostatic detection system was arranged to rotate about the scattering cell and thus permitted the measurement of differential scattering cross sections. The differential measurement was integrated to provide a total cross section for which an overall accuracy of $\pm 50\%$ was claimed. This accuracy is comparable to that obtained in many determinations based on optical emission techniques. The second experiment of this type is by Park et al.[60] who measured a total cross section for the excitation of helium to the 2^1S and 2^1P states by proton impact. The energy-loss spectrum is determined by retardation and energy analysis of the projectiles after exit from the target; resolution of 2.5-eV has been achieved with projectile energies of 125-keV.

C.   Detection of Metastables by Charge Rearrangement

Every excited state will exhibit different charge rearrangement cross sections. For example, the cross section for stripping an electron in a metastable excited state will be different from the cross section for stripping the ground state. This may provide a basis for distinguishing the relative populations of the two states. To clarify this type of technique, consider the method of Gilbody et al.[61, 62] for determination of the cross section for forming metastable helium as a He^+ beam traverses a target. The ions in the emergent beam are removed by an electric field, and the neutral component

is passed into a gas cell in which a transverse electric field is applied. Electrons are stripped from the neutral helium atoms, and the resulting ions are removed by the field. The attenuation of the neutral flux is monitored as a function of gas pressure in the detector. Each constituent of the projectile beam will decrease exponentially with density. The measured signal can be deconvoluted into the sum of two exponential decays and hence the relative proportions of each constituent determined.

This type of method can be applied only to assaying the excited state content of beams that include a small number of constituents. In practice, the distance between the formation and detection regions is kept long so that all short-lived excited states will decay; the beam is composed only of a ground state and metastable state component. The metastable content will include all relevant cascade contributions. The technique will work only if there are large differences in the cross sections for charge rearrangement of the constituents and if the metastable atoms are an appreciable fraction of the total flux.

This type of approach has been used by Gilbody et al.,[61, 62] Miers et al.,[63] Haugsjaa et al.,[64] and Utterback.[65] It is not possible to carry out a complete discussion of this type of method because each application is essentially a different procedure. Detailed consideration will be given to these techniques when considering the data in later chapters.

D. Summary

In the preceding section, no attempt has been made to provide a completely comprehensive discussion of the methods for direct determination of excited state density. None of the techniques has found wide application to this type of experiment. Field ionization does not really have sufficient resolution to provide unambiguous detection of a particular state. Energy loss methods show more promise but their use will be restricted to low-lying excited states that are well separated from adjacent levels. The charge-rearrangement methods have application only to certain specialized problems and would seem to be of value only for the detection of the metastable content of a beam. The criteria against which experiments of this type may be assessed are given in Table 3-3.

TABLE 3-3   CRITERIA FOR THE ACCURATE DIRECT MEASUREMENT OF EXCITED STATE DENSITY

---

1. Measurements must be shown to be independent of target cell geometry.
2. Complete resolution of individual excited states must be demonstrated.
3. Efficiency of detection must be shown to be 100%.
4. Corrections must be evaluated for loss of excited atoms by normal radiative decay.
5. Lifetime against ionization must be small compared with the particle's residence time in the field (for field-ionization techniques).

3-4 PHOTON EMISSION: QUANTITATIVE MEASUREMENT

The great value of the optical techniques is that they provide the high resolution necessary to measure separate cross sections for excited levels of only slightly differing energies; however, they suffer from two important disadvantages. In the first place,

the data obtained by the experiment are emission cross sections. The relationship of these quantities to the level-excitation cross sections has been discussed at length in Section 2-2. It was shown that only in a few cases is it possible to derive level-excitation cross sections from emission measurements. The second limitation is more practical. The calibration of the quantum detection efficiency of the optical system is quite a complicated procedure, and the accuracy achieved is poor. Consequently, large systematic differences occur between the emission cross sections measured by various workers.

This section will begin with a brief discussion of the techniques available for analysis of spectra and measurement of photon flux (Section 3-4-A), followed by a discussion of how the detection efficiency of the optical systems may be determined absolutely (Section 3-4-B). Polarization of emission is of intrinsic interest since it is related to the population of the various energy degenerate sublevels of the emitting level. The methods for determining polarization fraction are also discussed in detail (Section 3-4-C). A number of problems that influence the calibration of detection sensitivity under some circumstances are identified. Polarization of emission causes anisotropy of emission and also influences calibration when the detection system exhibits instrumental polarization (Section 3-4-D). A further complication, arising when observing emission from a fast projectile, is that the emission may be appreciably Doppler shifted, which requires an energy-dependent correction to the calibration of detection sensitivity (Section 3-4-E). Most situations involve detection of a single wavelength of emission. In a few cases it is desirable to determine the sum of two or more unresolved emissions simultaneously, for example, the unresolved rotational structure of a molecular emission. This situation requires additional attention in order to determine an effective value of the detection sensitivity (Section 3-4-F). The quantitative determination of photon emission is a complicated procedure with many opportunities for systematic error. In the conclusion of this topic there are included in Section 3-4-G a list of criteria that must be obeyed if a measurement is to be accurate, and a brief survey of the manner in which the data will be affected by neglect of the required precautions.

A.   Optical Detection Techniques

The optical system is required to perform the functions of isolating the transitions of interest from adjacent emissions and allowing quantitative measurement of intensity. Collisional excitation experiments are hampered by low signals, and fine spectral resolution is often sacrificed in order to obtain useful signal-to-noise ratios.

The basic detection system consists of three items: (1) a large aperture optical system to collect as much of the light emitted from the collision region as is possible; (2) an analyzing device to separate the spectral line of interest from all other emissions; and (3) a photon detector to make a quantitative measurement. Considerable care is required to ensure high efficiency of light collection and detection.

Much of the older work in this field was carried out using photographic spectrometers. This technique is now rarely used, being of poor sensitivity and relatively inaccurate for quantitative measurement.

Wavelength tables are required for identifying the transitions that give rise to the emission of a particular spectral line from the collision chamber. The extensive

tabulations by Moore[66, 67] provide sufficient data on atomic systems. For molecular emissions the situation is very poor, as no complete listing of wavelengths and corresponding transitions exists. The work by Pearse and Gaydon[68] provides a source of identification for prominent band systems only. Alternatively, one must turn to the original publications of results as listed by Herzberg.[33] A particularly glaring deficiency exists in the case of molecular hydrogen because existing publications of wavelengths[69] and term schemes[70] are not particularly convenient for the identification of the transitions in an observed spectrum.

1. Visible and Near-Ultraviolet Regions.   The near-ultraviolet and visible spectral regions (2000–7000 Å) are convenient for experiments, since relatively simple transmission optics may be used. For most atomic systems emissions at these wavelengths represent transitions between fairly high principal quantum number states. The transitions between first excited and ground states will generally lie in the far ultraviolet.

The photomultiplier is the standard device for photon detection in this region. It is sensitive, detecting up to 25% of the light falling on the photocathode, and provides high gain of up to 10^7. Frequently, thermal emission from the photocathode provides the greatest noise source in the apparatus. It is often useful to improve the signal-to-noise ratio by the well-known techniques of cooling the photocathode or pulsing the light flux and utilizing phase sensitive detection.[71] For small light intensities, such as those obtained from the collision region, the output of a photomultiplier is a linear function of the input signal; however, if the spectral sensitivity of the detection system is calibrated against a high intensity source such as a blackbody radiator, the linearity of response must be confirmed.

The most versatile optical system is a large-aperture scanning monochromator used in conjunction with a photomultiplier. Auxiliary optical devices may be used to focus an image of the collision region onto the entrance slit of the monochromator, thus ensuring that the full optical aperture of the system is utilized. The monochromator possesses the distinct advantage that the wavelength of transmission and the resolution may be readily changed, allowing the whole emitted spectrum to be investigated. This feature is particularly important during the initial stages of an experiment when there is uncertainty as to precisely what emissions are occurring. A monochromator may exhibit spurious transmission of light due to scattering from internal surfaces. This is generally unimportant when analyzing emission from the collision region since the integrated intensity of the whole spectrum is small; however, the intensity of a standard continuum source used for calibrating detection sensitivity will vary by many orders of magnitude from one end of the spectrum to the other. Considerable errors due to scattered light may occur at the wavelengths where the standard source is weak [see Section 3-4-B(1)].

The greatest photon detection efficiency may be obtained by utilizing a photomultiplier and interference filter. Auxiliary optical systems must be used to ensure that the light traverses the filter parallel to its axis. It must be shown that the filter has no spurious transmission bands and that no appreciable transmission occurs at unexpected spectral systems, such as excited background gas or even light emitted as a beam strikes a surface.

The choice between the use of a filter or a monochromator will be dictated primarily by the sensitivity required. In both cases, attention must be paid to the possibility of scattered light or spurious transmissions, particularly when carrying out calibrations

utilizing standard sources of high intensity. Jacquinot[72] has discussed the relative luminosities of various types of optical systems in some detail.

2. Vacuum Ultraviolet Spectral Regions. The emissions of greatest interest and intensity in atomic collision experiments are those between the lowest excited and ground states. Unfortunately, for many atoms and ions these states lie in the vacuum ultraviolet, and their study has often been neglected because of experimental difficulties. Comprehensive reviews of the techniques of vacuum ultraviolet spectroscopy and the optical characteristics of materials in this spectral region may be found in the works of Samson[73] and Schreider.[74]

Photomultipliers operating in this region have been constructed with windows of sapphire or of various fluorites which transmit down to about 1100 Å. Below this wavelength, the energy of the photon is sufficiently great to eject an electron directly from a metal surface; nude multipliers may be used with no photocathode surface nor window, and the photons are allowed to eject electrons directly from the first dynode. This has the particular advantage that the device will not be sensitive to visible light. Certain fluorescent materials such as sodium salycilate have been employed to convert the ultraviolet photons into visible light that may be detected directly by a conventional photomultiplier. Photoionization detectors are also used in this region.[75] When filled with a rare gas, such a detector may be constructed to have a 100% detection efficiency[74]—a considerable advantage when quantitative measurements are to be made.

It is very important to keep the light losses on reflection and transmission as low as possible. At wavelengths between 1100 and 2000 Å various materials such as certain fluorites and sapphire will transmit rather well, and the reflectivity of most surfaces is quite high. Thus, windows and lenses may be utilized to separate various parts of the optical system, and various types of grating spectrometers may be employed. The light path between the collision chamber and the detector must be evacuated. At wavelengths below 1100 Å, the most suitable dispersive device is the grazing-incidence grating spectrometer. Again, the light path must be evacuated, but here no suitable window materials are available, and differential pumping systems must be provided between the target gas chamber and the spectrometer. Such systems will have only a very limited optical aperture. Thin metallic films might be used as "windows,"[73] although there are formidable difficulties in preparing unsupported films of large area. Only in one series of experiments has extensive use been made of vacuum ultraviolet spectrometers.[76, 77, 78]

Although interference filters are not available for low wavelengths, certain other selective detection systems have been produced for investigating specific spectral lines. A device that has found considerable application in atomic collision experiments is the Lyman-alpha photon detector described by Brackmann et al.[79] This consists of a filter of molecular oxygen contained between lithium fluoride plates, followed by a photoionization detector operated as a self-quenching Geiger counter. The lithium fluoride window does not transmit below 1050 Å, and the iodine vapor filling of the counter will not detect photons of wavelengths above 1330 Å. Within this range, the molecular oxygen has seven sharp transmission bands, one of which happens to coincide with the Lyman-alpha wavelength. Thus, the filter—detector combination responds to a relatively small number of narrow bands of wavelengths. The detection efficiency of the counter at the Lyman-alpha wavelength is only a few percent. As with any filter detector combination, care must be taken to ensure that only the radiation

of interest is being detected. The narrowness of the transmission band of the Lyman-alpha counter has two important consequences when this detector is applied to collision experiments. Lyman-alpha emission from deuterium is displaced by 0.3 Å from that emitted by hydrogen because of the different reduced masses of the two systems. Therefore, the absorption of the filter for deuterium emission is greater than that for the hydrogen line, and the detection sensitivities are different. Secondly, the emission from a fast particle will be Doppler-shifted, and the detector certainly should not be used under circumstances in which such a shift is appreciable. This latter problem is discussed in Section 3-4-E.

The vacuum ultraviolet spectral region remains relatively unexplored in collisional excitation experiments. The majority of the effort has been devoted to considering the emission of Lyman-alpha radiation using the oxygen-filtered counter. All work using the oxygen-filtered counter for the detection of Lyman-alpha radiation traces its calibration back to a normalization of an experimental measurement of collisional excitation to a theoretical estimate based upon the Born approximation. Definite discrepancies exist between measurements using the counter and determinations using a spectrometer whose sensitivity was calibrated directly[78] [see further discussion in Section 3-4-B(2)].

3. Infrared Spectral Region. There have been a few studies of collisionally induced emission at wavelengths of about 1μ, which technically fall into the classification of the infrared region of the spectrum. The techniques used for this work were the same as for the visible region with the exception that special photomultipliers with good sensitivities in the infrared were employed as detectors. There has been no study at wavelengths beyond 1μ.

Transitions giving rise to emission in this wavelength region will generally be between states of high excitation. The cross sections for forming such states will be small, and the branching ratios will favor decay with emission of visible and ultraviolet lines rather than the infrared. Thus, it is expected that the emission cross sections will be small.

The major technical problem in studying infrared emission will be the absence of detectors that have sensitivities comparable to the photomultipliers used in the visible and ultraviolet regions. No detector that permits counting of single photons is available. The combined influence of low cross sections and poor detection sensitivity would make this a very difficult region to explore in studies of collisionally induced emission. Since no work has been done in this region, the topic will not be discussed further.

B. Calibration of Detection Sensitivity

In order to determine an absolute value of an emission cross section it is necessary to calibrate the efficiency with which the optical system collects and detects the emission from the collision region. Extensive attention is given to the consideration of methods for determining detection sensitivity because this undoubtedly represents the least accurate step in a cross section determination by optical methods. Apart from the monographs by Bauer[80] and Samson,[73] which consider the visible and ultraviolet regions respectively, the published literature provides few comprehensive discussions of the methods used for the quantitative measurement of light intensity.

Smit[23] shows that when the collisionally induced emission is polarized, the angular

distribution of photons is anisotropic; therefore, a determination of an absolute cross section requires an integration of emission over all angles. For clarity, it will be assumed here that the radiation is unpolarized. In Section 3-4-D the influence of polarization on the determination of absolute cross sections is considered in detail.

A practical detection system views a section of the beam path between x and $x + l$ cm from the entrance to the collision chamber and collects light emitted into a solid angle $\omega(x)$ defined by the entrance aperture of the optical system. We may define a *detection sensitivity* $K(\lambda)$ at the wavelength λ of the $i \to j$ transition in terms of the output signal of the detection system S_{ij} produced by the emission of J_{ij} photons per second in all directions from the length l of the beam path through the collision chamber

$$S_{ij} = \frac{J_{ij} K(\lambda)}{l \, 4\pi} \int_x^{x+l} \omega(x) \, dx. \tag{3-1}$$

For small field of view the solid angle ω is invariant with position, and the integral may be replaced by the product ωl. As defined here the factor $K(\lambda)$ includes such general features of the optical system as the transmission of the optical components, quantum efficiency of the detector, and the gain of the electronic circuits. The determination of ω and l is a geometric problem. It will be shown that by adopting certain procedures these factors do not have to be explicitly measured.

An obvious technique for determining the absolute detection sensitivity is to measure the response of the system to a standard source whose output is known from basic theoretical considerations. Two such sources are the blackbody radiator and electron synchrotron radiation. The emission from the blackbody peaks sharply in the infrared regions, but the intensity is too low to be useful at wavelengths below about 3500 Å. The emission from the circular orbit in an electron synchrotron can be calculated in terms of the electron current, and, although it peaks in the far ultraviolet, the intensity remains useful up to visible wavelengths. Synchrotron radiation exhibits extreme directionality and high polarization. Because of practical difficulties in utilizing the device, this source has not yet been employed in the calibration of a collisional excitation experiment. A further approach to the determination of $K(\lambda)$ is to employ a detector, such as a rare gas photoionization counter,[74] that has an absolute efficiency known accurately from theoretical considerations. Additional techniques have been devised for limited spectral regions, but the procedures may be traced to the use of a standard source or to an absolute detector.

The problems of calibration in the visible region and in the ultraviolet region are considered separately. It will be assumed throughout that the emission is monochromatic. For emission from an atomic target, this is quite valid. Emission from a fast moving projectile may exhibit Doppler broadening, and emission from molecular states may have unresolved rotational structure that extends over wide ranges of wavelengths. The influence of these two problems on calibration is considered in Sections 3-4-E and 3-4-F, respectively. In some experiments the absolute magnitudes of data have been established by normalization to previously determined cross sections for collisionally induced emission. Strictly speaking, this is not an absolute calibration since the data that provide a basis for normalization do not constitute a standard source whose output is known from basic theoretical considerations; however, the technique has been frequently employed, and a discussion will be appended (Section 3-4-G) to outline the precautions that must be observed.

1. Calibration in Visible Regions. (a) STANDARD LAMP. The blackbody radiator provides the basic standard in this spectral region. The emission of a tungsten strip filament lamp is invariably used as a transfer standard. A suitable lamp has a filament of known dimensions contained in a glass envelope with a flat window for viewing the emission and a rear surface shaped to prevent spurious reflections. Accurate control and stabilization of heating current is mandatory, since emission increases as approximately the seventh power of current.[81] Because of the gradual evaporation of tungsten from the filament and changes in the emissivity of the tungsten surface, the characteristics of the lamp are not stable with time. Operating at low temperatures reduces evaporation, and the use of ac heating power is thought to inhibit changes in the crystalline structure of the filament.[85] It is necessary to confirm the stability of operation by periodic calibration.

It is convenient to have the lamp calibrated directly against the emission from a blackbody by a standards laboratory. Lamp calibration may be expressed in the form of spectral distributions of emission at a series of accurately specified heating currents. An alternative technique is to measure the temperature of the lamp directly using an optical pyrometer and to determine the emissive power as a function of wavelength from the emissivity measurements of Larrabee[82] or de Vos.[83] The temperature determined by using the pyrometer is a brightness temperature and must be converted to a true temperature, making a correction for window transmission.[84] Requirements for this second technique are that a choice be made of a set of emissivity tables from a number of available published values, and that measurements be made of window transmission and filament temperature. The direct comparison of the lamp with the blackbody is preferable because it obviates the need for determining these parameters and also eliminates certain other possible sources of systematic error.[85]

(b) EXPERIMENTAL CONFIGURATION. In Eq. 3-1, the detection efficiency $K(\lambda)$ is defined in terms of the output signal produced by the light emitted from a section of the beam path through the chamber. The standard lamp filament has a similar geometrical configuration, being also a "strip" of emission, and it may be conveniently oriented parallel to the experimental source and on the axis of the optical system. Care must be taken to restrict the observed part of the filament to a region where temperature is uniform.[85] Suppose that the effective solid angle from which radiation is detected is ω' and the length of filament observed l'. The emissive power $E_{\lambda'}$, expressed in photons of wavelength λ' emitted per angstrom bandwidth per unit time from unit area of the filament into one steradian, is known from the calibration of the lamp. The observed signal in this case is S', such that

$$S' = b \int_{x'}^{x'+l'} \omega' \, dx' \int_{\lambda_1}^{\lambda_2} E_{\lambda'} \, K'(\lambda') \, d\lambda'. \tag{3-2}$$

The second integral expresses the fact that the sensitivity of the system $K'(\lambda')$ to continuum radiation will vary over the transmission band $\lambda_1 \to \lambda_2$ of the optical system. The width of the lamp filament is b and is assumed to be considerably smaller than the field of view of the system.

There are definite advantages to placing the lamp at the position in the optical system normally occupied by the experimental source of emission. This may be achieved by physically removing the collision experiment and substituting the lamp with no change in the positions of the optical components. Alternatively, an image of the lamp filament may be formed in the position occupied by the ion beam and

correction applied for losses due to the auxiliary lens and windows. Lamp and collision region are now viewed under precisely the same optical conditions. Therefore

$$\int_{l}^{x+l} \omega \, dx = \int_{x'}^{x'+l'} \omega' \, dx',$$

and by taking Eqs. 3-1 and 3-2

$$\frac{J_{ij}}{l} = \frac{4\pi b S_{ij}}{K(\lambda)S'} \int_{\lambda_1}^{\lambda_2} E_{\lambda'} K'(\lambda') \, d\lambda'. \tag{3-3}$$

The determination of the rate of emission of photons from 1 cm of the projectile beam path (J_{ij}/l) does not require the measurement of the solid angle $\omega(x)$ and the length l. It is emphasized that, if the substitution technique is not employed, then it is necessary to separately determine the geometrical parameters $\omega(x)$ and l both for the standard lamp configuration and also for the arrangement used when viewing the collision region.

The useful area of the standard lamp filament is very small, and the field of view of the optical system must be restricted. This may be achieved by using a lens or mirror system to form an image of the collision region on the entrance to the detection system. Diaphragms may be introduced at this point to define the field of view. The same optical system must of course be used for viewing the lamp and also for viewing the collision region. The introduction of lenses does not necessarily reduce signal strength because the reduction in field of view will be compensated by an increase in the effective solid angle from which light is collected.

The evaluation of the integral in Eq. 3-3 is now considered for the separate cases of a monochromator and for an interference filter detection system.

(c) CALIBRATION OF A MONOCHROMATOR. If a lens is used to form an image of the collision region and of the standard lamp filament on the entrance slit of the monochromator, then the field of view of the optical system will be defined by the width and height of this slit. If no auxiliary optics are used, the field of view and the acceptance angle of the system will be defined by the optical characteristics of the monochromator, probably by the grating or prism area.

Generally, entrance and exit slits of the monochromator are kept at the same width d, as a best compromise between requirements of luminosity and resolution. With a dispersion of D (mm/Å), the spectral slit width of the device is $\beta = d/D$. For a single-dispersion instrument the spectral radiation distribution transmitted from a continuum source will have the form of a triangle with a width at half maximum of β, as shown in Fig. 3-1. The center of the transmitted band would be set at the wavelength λ of the line radiator whose intensity is to be determined.

For most experimental systems the characteristics of the various lenses, mirrors, and windows will not vary greatly over the small wavelength range 2β; therefore, the sensitivity of the optical system to the continuum radiation $K'(\lambda')$ will have the same triangular form as is shown in Fig. 3-1. At the center of the transmission band $\lambda' = \lambda$, and $K'(\lambda')$ is equal to the efficiency of the system at the wavelength of the line emitter $K(\lambda)$. Generally, the emissive power E_λ will vary only slightly over the wavelength range λ_1 to $\lambda_1 + 2\beta$ and, therefore, may be put equal to the value E_λ at the wavelength of the line emitter. Thus

$$\int_{\lambda_1}^{\lambda_2} K'(\lambda') E_\lambda \, d\lambda' = E_\lambda \int_{\lambda_1}^{\lambda_1 + 2\beta} K'(\lambda') \, d\lambda'. \tag{3-4}$$

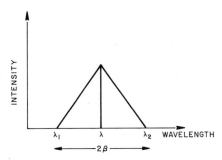

Fig. 3-1   Intensity distribution of light transmitted by a monochromator with a spectral slit width of β.

The remaining integral is the area under the curve shown in Fig. 3-1 with $K(\lambda)$ the sensitivity at the peak. Therefore

$$E_\lambda \int_{\lambda_1}^{\lambda_1 + 2\beta} K'(\lambda') \, d\lambda' = E_\lambda \frac{K(\lambda)}{2} 2\beta = E_\lambda K(\lambda)\beta. \tag{3-5}$$

Equation 3-3 may now be rewritten

$$\frac{J_{ij}}{l} = \frac{S_{ij}4\pi}{S'} bE_\lambda \frac{d}{D}. \tag{3-6}$$

The emission per unit length J_{ij}/l is expressed in terms of a ratio of two signals and certain well defined parameters of the monochromator.

(d)   Calibration of a Filter-Detector Combination.   A suitable optical configuration is to use two lenses to image the emission onto the surface of the detector, the filter being placed between the lenses where the light rays are arranged to be parallel to the optical axis. A diaphragm must be used to reduce the effective area of the detector sufficiently to ensure a small field of view. Provided the transmission band of the filter is small, $E_{\lambda'}$ may again be considered constant. Hence, Eq. 3-3 becomes

$$\frac{J_{ij}}{l} = \frac{S_{ij}b4\pi E_\lambda}{S'K(\lambda)} \int_{\lambda_1}^{\lambda_2} K'(\lambda') \, d\lambda'. \tag{3-7}$$

In general, the variation of $K(\lambda)$ with wavelength over the small range λ_1 to λ_2 is dependent only on the transmission τ_λ of the filter. We may write $K(\lambda) = \tau_\lambda C$, where C is invariant over the range λ_1 to λ_2. Then Eq. 3-7 becomes

$$\frac{J_{ij}}{l} = \frac{S_{ij}4\pi}{S'} bE_\lambda \frac{\int_{\lambda_1}^{\lambda_2} \tau_{\lambda'} \, d\lambda'}{\tau_\lambda}. \tag{3-8}$$

Measurement of the transmission as a function of wavelength λ may be made quite readily using a continuum source and a spectrometer. Transmission at the wavelength of the discrete spectral line is τ_λ, and the integral is the area under the transmission curve of the filter.

(e)   General Considerations.   The arguments pertaining to the calibration of a spectrometer system and of a filter-detector system [see (c) and (d) above] both involve a number of important approximations. Among these are the assumptions

that detector sensitivity, optical transmission, and emissive power of the standard source are all invariant with wavelength over the transmission band of the optical system. A full general treatment, which does not make these assumptions but is too extensive to be repeated here, has been given by Merril and Layton.[86] It must suffice to state that these assumptions are quite well obeyed under most circumstances employed in collisional excitation experiments.

A general treatment has been developed for a rather restricted situation in which the standard (or its image) may be substituted directly for the collisionally induced emission. This technique has great advantages through eliminating the necessity for measuring geometrical parameters and ensuring that calibration is carried out under precisely the same conditions as the experimental measurement. Although this approach has often been utilized, some workers choose to employ different conditions for viewing the lamp and the collision chamber. In such cases the effective values of ω, ω', l, and l' must be determined geometrically. Furthermore, it is most important that tests are made to ensure that there is no variation of sensitivity over the optical system field of view. Where the fields of view are not the same, one certainly cannot assume that the sensitivity of the apparatus during the calibrations is the same as during the measurement of the line emission.

In the design of certain detection systems based on optical spectrometers, the dispersive element, grating or prism, provides either the field or aperture stop for the whole optical system. In most scanning spectrometers the change in transmitted wavelength is achieved by rotating the dispersive element. In such a case, if the solid angle ω' or the observation region l' is defined by the grating or prism, then it will vary with wavelength. The geometrical parameters must then be measured at each wavelength. Again, the substitution technique obviates this problem.

Certain precautions must be observed because of the high intensity of the emission from the standard lamp. Since the light flux will be many orders of magnitude greater than that from the experimental system, it is necessary to confirm the linearity of the detection system. In the event that the linear dynamic range of the detector is insufficient, calibrated neutral density filters may be utilized to reduce the intensity of the standard. A further source of possible error is scattered light and spurious transmissions at wavelengths other than that which the equipment is designed to transmit. In particular, the light emission from the standard at red wavelengths is of very high intensity, and a small spurious transmission of the red region may be comparable to the actual intensity at blue and near-ultraviolet wavelengths, where the lamp emission is some orders of magnitude lower. Scattered light is particularly troublesome in spectrometer systems. Transmission by interference at high orders will be possible in diffraction grating systems and when using interference filters. In general, the problem of spurious transmissions can be neglected when viewing the collision region because the aggregate intensity over the whole spectrum is small. The problems of scattered light in grating spectrometer systems have been discussed by Penchina[87] and also by Watanabe and Tabisz.[88]

2. Calibration in the Ultraviolet Spectral Regions. Various techniques have been used at lower wavelengths, but none exhibits the inherent simplicity of a comparison with a blackbody radiator. Here attention will be confined to those techniques that have actually been utilized in excitation experiments. For a full discussion of available methods, the reader is referred to the comprehensive reviews of Samson[73] and Shreider.[74]

The direct determination of the overall detection sensitivity of the optical system may be conveniently carried out using the so-called "branching-ratio" method, suggested by Griffin and McWhirter.[89] Simultaneous observations are made of the intensities of two spectral lines from a single upper state of the same atom, one lying in the visible region and the other in the ultraviolet. If the relative transition probabilities for the two transitions are known and the intensity of the visible transition is determined absolutely by using a detection system calibrated against a blackbody, then the response of the ultraviolet detection system may be calculated. Only for helium and hydrogen are the transition probabilities known with high accuracy. In hydrogen, levels with different angular momentum quantum numbers are degenerate, and the technique cannot be used unless there is knowledge of the separate populations of these levels. Van Eck and de Heer[90] have shown that the source for this calibration may be conveniently provided by the collisional excitation experiment. Recently these branching-ratio techniques have been extended by the use of molecular emissions and in the configuration of a collision experiment.[91, 92]

The detection efficiency of the optical system $K(\lambda)$ is the product of the transmission of the optical system $T(\lambda)$ and the quantum efficiency $D(\lambda)$ of the detector. The direct determination of $K(\lambda)$ is preferable if the system is complex, such as a spectrometer—detector combination. A suitable standard of emission may not be available, however, and alternative techniques may be devised by considering the separate determination of transmission and quantum efficiency. Transmission may be determined by the ratio of the intensity of monochromatic light incident on the optical system to the intensity of the light transmitted, and does not require the knowledge of a detection sensitivity.

The quantum efficiency of certain types of detectors is relatively insensitive to wavelength, and a determination of efficiency made at visible wavelengths may be accurate at ultraviolet regions. Johnston and Madden[93] have shown that with certain precautions the response of a thermopile will be invariant with wavelength; however, this instrument is not sufficiently sensitive for use as a detector of emission from a collision experiment. The sensitivity of a photomultiplier coated in sodium salicylate, although not strongly dependent on wavelength, is certainly not constant,[94] and will vary with the age of the coating.

The photoionization chamber provides a very convenient device for detecting photons in the ultraviolet region. It has been shown that the photoelectric yield of rare gases is unity[94, 95] and that photoionization chambers of known efficiency may be constructed.[94, 96] Rare gas fillings may be used only at wavelengths below 1022 Å, the ionization potential of zenon. Since there are no solid materials that transmit efficiently in this spectral region, the light must enter the gas-filled chamber through an open aperture, and it is necessary to provide a differentially pumped region between the detector and the collision experiment. Andreev et al.[75] have described the use of an NO-filled counter with lithium fluoride windows to detect Lyman-alpha radiation from a collision experiment. Unlike the rare gases, the photoelectric yield of NO is not unity, and it is necessary to use the experimental measurements of this quantity by Watanabe.[97] The work of Samson[94] indicates that the thermopile, used as a calibrated detector by Watanabe, will have almost the same response at the Lyman-alpha wavelength as for the visible wavelengths where it was calibrated. Therefore, the determination of the photoelectric yield of NO is related to a measurement of detector sensitivity at visible wavelengths.

Much of the work carried out in the ultraviolet spectral regions is concerned solely with the emission of Lyman-alpha radiation at 1215 Å, using the oxygen-filtered Geiger counter as a detector.[79] The measurements are frequently normalized to a cross section for the emission of "countable uv" from the electron impact excitation of molecular hydrogen determined by Fite and Brackmann.[98] The absolute magnitude of the "countable uv" emission was established by comparison with the cross section for electron impact excitation of the $2p$ level of atomic hydrogen measured in the same apparatus; the magnitude of the $2p$ cross section was, in turn, established by normalization to theoretical predictions. The "countable uv" contained a small,[99] but unknown, proportion of molecular emissions in addition to Lyman-alpha radiation. Both the intensity of the emissions from the molecule and the relative variation of detector sensitivity with wavelength are unknown. Therefore, it is legitimate to normalize to this measurement only when using an oxygen-filtered counter of the same design as that employed by Fite and Brackmann, including a filter of the same thickness. The procedure is not very accurate since the measurements by Fite and Brackmann are reliable only to within $\pm 30\%$.

C. Determination of Polarization Fraction

The emission from a collisional excitation experiment will, in general, be polarized. In Section 2-2-D, a discussion was given about the relationship of polarization to the cross sections when exciting various energy degenerate sublevels of an atom. Polarization is of importance for its own sake since it may, in principle, be related to excitation cross section. Polarization is also of importance through its influence on the calibration of detection efficiency (see Section 3-4-D).

The polarization fraction is defined in terms of the rate of emission of photons polarized in a plane parallel to the beam axis $I^{\parallel}$ and the emission polarized perpendicular to the axis $I^{\perp}$.

$$P = \frac{I^{\parallel} - I^{\perp}}{I^{\parallel} + I^{\perp}} \tag{3-9}$$

These intensities are defined in terms of the rate of photon emission at 90° to the beam direction from unit length of the beam path. Since the expression for P may be reduced to the ratios of intensities, the precise definition of solid angle and extent of observation region is unimportant.

For visible and near-ultraviolet wavelengths polarization analyzers, such as the Glan—Thomson and Nicol prisms, may be used to measure P. The analyzer must be located between the experiment and the detection systems at a point where the light beam is parallel. The beam axis, $I^{\parallel}$ and $I^{\perp}$ may then be separately measured by placing the principal section of the analyzer successively perpendicular and parallel to the ion beam path. Care must be taken to ensure that the transmission of light polarized perpendicularly to the principal section is small. Attenuation will generally be 100% when using an analyzer that depends on the birefringent properties of a material, provided that the light beam is parallel to the axis of the prism. Analyzers such as Polaroid, which depend on dichroism (selective absorption), do not exhibit 100% transmission of light polarized in the preferred plane nor 100% suppression of the perpendicular component. The use of dichroic materials in large aperture systems may sometimes be unavoidable, but care must be taken to investigate the transmission properties.

In most optical systems, particularly those employing grating spectrometers, the efficiency of the device for transmission and detection will depend on the plane of polarization of the incident radiation. Thus, the relative values of $I^{\parallel}$ and $I^{\perp}$ must be corrected for the different detection efficiencies. An excellent technique for eliminating this problem entirely is to place a quarter-wave plate behind the polarization analyzer. With a suitable orientation of this device with respect to the plane of the polarization analyzer, the light transmitted to the spectrometer will be circularly polarized. Clearly, the efficiency of the subsequent optical components is then independent of the orientation of the polarization analyzer. An alternative approach is to measure the different sensitivities. The detection system is arranged to view an unpolarized source, such as an incandescent lamp, with a polarizer placed between the source and spectrometer. Any change of signal—as the polarizer is rotated—must be due to variation in the detection system sensitivity with the plane of polarization. If $S^{\parallel}$ and $S^{\perp}$ are the measured signals from the collision system when the plane of transmission of the analyzer is, respectively, parallel and perpendicular to the ion beam path, and $K^{\parallel}$ and $K^{\perp}$ are the corresponding detection sensitivities, then

$$P = \frac{S^{\parallel}/K^{\parallel} - S^{\perp}/K^{\perp}}{S^{\parallel}/K^{\parallel} + S^{\perp}/K^{\perp}}. \tag{3-10}$$

Rearrangement of Eq. 3-10 will show that only the ratio of the sensitivities is required to determine the polarization fraction P.

In the vacuum ultraviolet regions simple transmission polarization analyzers are not available. It is quite convenient to investigate the angular dependence of the emission and to apply Eq. 2-27 to determine P. Although there is no need to determine the absolute detection efficiency of the apparatus, it certainly is necessary to make a correction for the change of sensitivity with angle θ. The length of the beam path observed at an angle θ is greater than that observed at 90°. A correction may be applied by multiplying the signal observed at the angle θ by sine θ to compensate for this change. The effective value of the solid angle into which radiation is accepted might change with θ, particularly when using a large field of view. Also, problems might occur due to spurious internal reflections in the collision chamber and divergence of the ion beam. If this technique is used at visible wavelengths, then an experimental correction for all these effects may be achieved by multiplying the measured signal at an angle θ by the ratio $S'(90)/S'(\theta)$ for an $n^1S \to 2^1P$ transition of helium excited by proton impact, for which the polarization must be zero.

When measuring polarization for its intrinsic interest, considerable care must be taken to minimize the influence of mechanisms that tend to depolarize the emission. Stray magnetic fields are particularly important because a field component perpendicular to the axis of the ion beam will cause a Larmor precession of the excited electrons around the direction of the magnetic field and hence a change in the polarization of the emission. The influence of the external field increases with the lifetime of the excited states as well as with the strength of the field. The effect may, in principle, be reduced by adequate magnetic shielding. Alternatively, an axial magnetic field may be deliberately applied so that the resultant field almost coincides with the beam axis, and the depolarizing component of field perpendicular to the beam is proportionally insignificant. It is necessary to test that measured polarization fractions are independent of target and projectile beam densities, thus insuring a negligible influence of

collisional transfer and photoabsorption. Cascade cannot be eliminated, but in most of the cases that have been studied, its effect was small.

It is interesting to note that certain emissions, for example, the $n^1S \to n^1P$ transitions excited in helium, must have a polarization fraction of zero. A good check on the operation of equipment designed to determine polarization fraction is to attempt to measure the polarization of such a transition and show that $P = 0$.

D. Influence of Polarization on the Measurement of Emission Cross Sections

The measurement of an emission cross section requires the determination of the total flux of photons J_{ij} emitted per second in all directions for a length l of the beam path. The discussion of methods by which the sensitivity of the apparatus may be determined is based on the assumption that the emission is isotropic and that the detection sensitivity of the system is independent of the planes of polarization of the emitted photons. These assumptions are not necessarily correct. If the emission is polarized, then the distribution of emitted photons will be anisotropic (Eq. 2-27). Furthermore, the detection sensitivity [defined in Eq. 3-1] of the apparatus to light polarized parallel to the beam direction $K^{\parallel}$ may be different from that for light polarized perpendicular to the beam direction $K^{\perp}$, giving rise to an "instrumental polarization" $\alpha = K^{\parallel}/K^{\perp}$.

The maximum error involved in neglect of anisotropy can be readily estimated. Suppose that a measurement of emission cross section is made at an angle of 90° to the projectile beam on the assumption that polarization is zero. The true cross section will be lower than the measurement by 50%, if $P = -1$ and higher by 50%, if $P = +1$. Thus the maximum error is $\pm 50\%$, since P cannot exceed a magnitude of unity. The error will generally be less than this estimate since a value of P of ± 1 requires that only one m_l state is populated and that the others have a negligible cross section; apart from the threshold of an excitation cross section, this will be a most unlikely occurrence.

It must be emphasized that not all emissions are polarized. For example, emissions for $L = 0$ states of atoms that exhibit LS coupling are inherently unpolarized. Furthermore, in some collision combinations the polarization measured, using the techniques described in Section 3-4-C, may turn out to be zero at all the impact energies used in the experiment. Under circumstances where the polarization fraction is zero, the considerations of the present section may be neglected entirely.

A measurement made at a single angle θ to the beam direction with no correction for anisotropy in emission does represent valid information. One might regard the resulting information as a "cross section for the emission of photons at an angle θ to the beam direction." This is not the same as the total cross section for emission in all directions as defined by Eq. 2-5; neither does it relate to level-excitation cross sections through equations derived in Sections 2-2-A, 2-2-B, and 2-2-C. The data are related both to the emission cross section σ_{ij} and to the polarization fraction P_{ij}. Provided the measurements are made under conditions in which the operational definitions of both of these quantities are valid (see Section 2-4), then the measurement is independent of the apparatus and may be regarded as valuable information related to the microscopic cross section. If the instrumental polarization is neglected, however, then the measured output is peculiar to the apparatus and does not constitute fundamental information. Neglect of anisotropy makes the relationship between

emission and microscopic cross section more complicated, but the data are still valid. Neglect of instrumental polarization is an error.

The polarization will be dependent on the energy of the projectile. Therefore, neglect of correction for polarization will produce an energy-dependent distortion of the data.

In the following discussions equations will be developed that allow general corrections to be made to the calibrations of detection sensitivity described in Section 3-4-B of this chapter. Methods of operation will be described that obviate the necessity for these corrections.

1. Formulation of a General Correction Factor. Suppose that the apparatus is set to observe emission from the transition $i \to j$ at an angle θ to the beam direction. Let $K^{\parallel}(\lambda)$ be the instrumental sensitivity to radiation of wavelength λ polarized parallel to the beam direction and $K^{\perp}(\lambda)$ be the sensitivity to radiation polarized perpendicular to the beam direction. The ratio $K^{\parallel}(\lambda) \div K^{\perp}(\lambda) = \alpha$ is called the *instrumental polarization*. The sensitivity to unpolarized radiation $K(\lambda)$ is given by

$$K(\lambda) = \frac{K^{\parallel}(\lambda) + K^{\perp}(\lambda)}{2} = \frac{(\alpha + 1)K^{\perp}(\lambda)}{2}. \qquad (3\text{-}11)$$

The photon emission per second per unit solid angle from a length l of beam path at angle θ will be $I_{ij}(\theta)$. The components of this emission polarized parallel and perpendicular to the beam direction will be taken as $I_{ij}^{\parallel}(\theta)$ and $I_{ij}^{\perp}(\theta)$, respectively. Then

$$I_{ij}(\theta) = I_{ij}^{\parallel}(\theta) + I_{ij}^{\perp}(\theta) = I_{ij}^{\perp}(\theta)\,(C(\theta) + 1), \qquad (3\text{-}12)$$

where $C(\theta) = I_{ij}^{\parallel}(\theta) \div I_{ij}^{\perp}(\theta)$. The signal received by an apparatus with an entrance aperture ω set at an angle θ to the beam path will be given by the following relationship, which is analogous to Eq. 3-1:

$$S_{ij}(\theta) = \frac{I_{ij}^{\parallel}(\theta)K^{\parallel} + I_{ij}^{\perp}(\theta)K^{\perp}}{l} \int_{x}^{x+l} \omega(x)\,dx$$

$$= \frac{I(\theta)2K(\lambda)(C\alpha + 1)}{l(C + 1)(\alpha + 1)} \int_{x}^{x+l} \omega(x)\,dx. \qquad (3\text{-}13)$$

Using Eq. 2-27 for the anisotropy in emission, one arrives at

$$S_{ij}(\theta) = \frac{J_{ij}}{l}\frac{1}{4\pi}\frac{3(1 - P\cos^2\theta)}{(3 - P)}\frac{2K(\lambda)\,(C\alpha + L)}{(C + 1)(\alpha + 1)} \int_{x}^{x+l} \omega(x)\,dx. \qquad (3\text{-}14)$$

This reduces to Eq. 3-1 if the polarization fraction is zero and the instrumental polarization is unity (i.e., the sensitivity is independent of polarization). It follows that the observed signal $S_{ij}(\theta)$ at an angle θ must be multiplied by a factor

$$\gamma = \frac{(3 - P)}{3(1 - P\cos^2\theta)}\frac{(C + 1)(\alpha + 1)}{2(C\alpha + 1)}, \qquad (3\text{-}15)$$

to correct for the effects of polarization and to allow the determination of the quantity J_{ij}, which is used in the definition of emission cross section (Eq. 2-5). This correction will be required in all the derivations of Sections 3-4-B and should be inserted into Eqs. 3-3 and 3-8. Clearly, for such conditions as $\theta = 90°$ and $P = 0$, or $C(\theta) = 1$ and $P = 0$, this correction is unity. This was the situation assumed in the derivations of Section 3-4-B.

This correction factor may be serious in some cases. Taking a value of 0.6 for the instrumental polarization, which is in accord with estimates for some spectrometers,[100, 101] and assuming a situation in which polarization fraction P is 0.30, then the value of the correction factor γ when $\theta = 90°$ will be 0.86. Thus, neglect of the correction introduces an error of 14 % in this case.

2. Methods of Eliminating the Correction. The best method of handling the problem of polarization is to remove its influence entirely. Woolsey and McConkey[101] have suggested that the problem may be tackled by operating at $\theta = 90°$ and arranging α to be $\frac{1}{2}$. Here P will be equal to $(C - 1)/(C + 1)$, and γ will reduce to unity. It is possible to arrange for α to be $\frac{1}{2}$ for any particular wavelength by placing a suitable oriented polarizer in front of the entrance slit of the monochromator or other detection system; the orientation of the polarizer may need to be changed as a function of wavelength. The sensitivity of the system to unpolarized radiation, $K(\lambda)$, which will be determined by the methods described in Section 3-4-B, must be evaluated with the polarizer in place. Clout and Heddle[102] have discussed similar techniques that involve no loss of signal strength.

Frequently it has been the practice to set the apparatus at a "magic angle" of $\theta = 54°46'$ where $\cos^2 \theta = \frac{1}{3}$. This simplifies the correction factor γ but does *not* reduce it to unity. When using an optical system that exhibits no instrumental polarization, such as a filter and photomultiplier combination, then, $\alpha = 1$ and the correction factor γ does reduce to unity.

3. Methods for Determining the Correction. All optical systems based on spectrometers will exhibit some instrumental polarization. Grating spectrometers are particularly bad. The reflectivity of the grating will be different for the transverse electric and transverse magnetic components of the radiation falling on it; moreover, it will vary as the grating is rotated to scan a spectrum. Additional polarization may arise at mirrors and lenses. When such arrangements are used for determination of emission cross sections, a correction must be made for the effect of instrumental polarization. It should be noted that instrumental polarization will be a function of wavelength.

Certain optical systems will exhibit no instrumental polarization, and the coefficient α will be unity. Particular examples of these are photomultiplier filter combinations and photoionization detectors. It is unnecessary to correct for instrumental effects. The anisotropy of emission must be taken into account if the measurements are to be accurately related to an emission cross section.

The instrumental polarization of an optical system may be determined by using an incandescent lamp or other unpolarized source of emission. A polarization analyzer is used to successively transmit light polarized parallel to the plane that the projectile beam normally occupies, and light polarized perpendicular to this plane. The ratio of the two signals $S^{\parallel}/S^{\perp}$ will equal the ratio of the instrumental sensitivities $K^{\parallel}/K^{\perp} = \alpha$. Such a determination is required for each separate wavelength to be detected in the collision experiment. Care must be taken that the polarizing element exhibits zero transmission of light polarized perpendicular to the transmission plane; of necessity this requires that the light be parallel as it traverses this element.

4. Summary. The influence of polarization on absolute cross section measurements has received inadequate attention in most experiments. Moiseiwitsch and Smith,[100] concluded that failure to make adequate allowance for instrumental polarization is the most serious general fault in electron impact excitation measurements.

A similar conclusion is applicable to heavy-particle excitation studies. Errors of this nature will cause inaccuracy in the absolute magnitudes of the measured cross sections and may distort the energy dependence. Neglect of anisotropy in the emission is less serious. The measurements may be regarded as "apparent" cross sections for the emission of light into a particular direction; however, such data cannot be related to a level-excitation cross section without a measurement of polarization.

E.   Measurements on Many-Lined Sources

The formulation of expressions for detection sensitivity have been devised for monochromatic emission from the experiment. Sometimes it is impossible to provide sufficient resolution to allow separate measurements on two close spectral lines. In Fig. 3-2, is shown the detection efficiency of a single dispersion monochromator as a

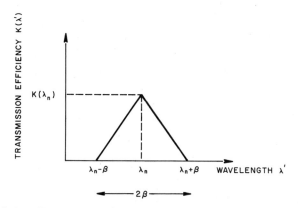

Fig. 3-2   Transmission efficiency of a single dispersion monochromator with equal exit and entrance slits set at a central wavelength λ_n with a resolution of β (Å).

function of wavelength for equal entrance and exit slit widths. It is quite obvious that the effective sensitivities for two different lines transmitted simultaneously are not the same. Two major areas of difficulty are anticipated. The first situation is where two or more individual spectral lines are detected simultaneously at a single setting of the detector system. Such a situation may, in principle, be eliminated by using sufficient spectrographic resolution and detecting each line separately. The customary reason for not resolving the lines is that at the required high resolution the signal strengths will be unacceptably low. The second situation is in molecular emission where a complex rotational band structure may extend over a wide wavelength range. In this case, it is desired to measure all the components simultaneously in order to arrive at a total cross section for emission of the whole structure. These two situations will be considered separately.

1.   Line Sources.   For simplicity, consider two atomic spectral lines λ_1 and λ_2. If they cannot be resolved, the objective is to make an accurate measurement of the sum of the intensities of the two lines that can then be directly related to the sum of the two emission cross sections. The monochromator will be taken to have an entrance slit width a cm, exit slit width b, and a dispersion $D(mm/A)$. In the plane of the exit slit, two images of the entrance slit will be formed, each of width a cm and centered, respectively, on wavelengths λ_1 and λ_2. Only if the exit slit width is equal to or greater

than $a + (\lambda_2 - \lambda_1) D$ will both images be completely transmitted through the exit slit. In many experiments, entrance and exit slits are kept equal, and the above conditions cannot be attained. The percentage error decreases as slit widths are enlarged. Clearly, no definite assessment of error can be made without knowing the intensities of the two lines. In general, the only satisfactory procedure is to have separate control of the slit widths and to satisfy the condition $b \geq a + (\lambda_2 - \lambda_1) D$.

One must reconsider the determination of the absolute sensitivity of the apparatus. For simplicity, suppose that the sensitivity is to be determined by comparison with a standard lamp as in Section 3-4-B(1). In Eq. 3-1, which relates the signal to the total photon emission, we must replace $J_{ij}K(\lambda)$ by $J_{ij}K(\lambda_1) + J_{i'j'}K(\lambda_2)$ to indicate the sum of photon emission rates from the two lines λ_1 and λ_2, corresponding to the transitions $i \rightarrow j$ and $i' \rightarrow j'$. The net signal, by analogy with Eq. 3-1, is the sum of contributions from the two lines

$$S_{ij} + S_{i'j'} = \frac{(J_{ij}K(\lambda_1) + J_{i'j'}K(\lambda_2))}{l \, 4\pi} \int_x^{x+l} \omega(x) \, dx. \tag{3-16}$$

The detection sensitivities of the system at the two wavelengths may be determined by closing the exit slit so that $b = a$ and by measuring the signal from a standard lamp when the wavelengths λ_1 and λ_2 are set to be central in the transmission band of the system. Using Eq. 3-2 and invoking the approximations made in Eqs. 3-4 and 3-5, one may arrive at expressions for $K(\lambda_1)$ and $K(\lambda_2)$. These sensitivities, determined at the center of the triangularly shaped transmission bands when $a = b$, will be the same as the effective sensitivities at the same wavelengths λ_1 and λ_2 when the exit slit is opened so that $b = a + (\lambda_2 - \lambda_1) D$. For small ranges of wavelengths, the sensitivities $K(\lambda_1)$ and $K(\lambda_2)$ will be approximately equal, and the transmission band of the spectrometer when $b = a + (\lambda_2 - \lambda_1) D$ is a trapezoid instead of the triangular shape that occurs when $b = a$ (see Fig. 3-2). A mean value of the two sensitivities may be substituted in Eq. 3-16 to allow determination of $J_{ij} + J_{i'j'}$ in terms of measured signals and certain other parameters. This can be related to the emission cross section for the two lines added together, in the normal manner. In the event that the sensitivities at λ_1 and λ_2 are not approximately equal, then it is impossible to determine $J_{ij} + J_{i'j'}$ without knowledge of the relative intensities of the two components.

The methods outlined above have been made specific to the case of a monochromator system. Similar formulations may be written for a photomultiplier interference filter arrangement. For this type of detection system it is most unlikely that the detection efficiencies at the two relevant wavelengths are the same; therefore, the measurements will not accurately represent the sum of the line-emission cross sections.

2. **Molecular Emissions.** The rotational structure of molecular emission may sometimes extend over some hundreds of angstroms. In order to measure cross sections for the formation of individual rotational states, it will be necessary to provide high resolution and to monitor individual peaks in the rotational structure. Most measurements, however, are concerned with determining the emission for a transition between two definite vibrational states. This requires integration of the light intensity over the whole rotational structure.

In some cases, the rotational structure is sufficiently compact to allow a monochromator with wide exit slit to be used to measure the whole structure directly. This is accomplished in the manner described above for atomic line sources by setting the

exit slit width b to be greater than $a + (\lambda_2 - \lambda_1) D$, where $\lambda_2 - \lambda_1$ is the spectral range covered by the rotational structure. If the detector sensitivity is invariant over the interval $\lambda_2 - \lambda_1$, this procedure will give a reliable measurement. The approach is restricted to relatively small rotational structures extending over, say 10—20 Å; broader bands must be handled by an integration technique.

The integration procedure involves measuring the signal with the monochromator's pass band centered on a series of wavelengths $\lambda_1, \lambda_2, \ldots \lambda_n, \ldots \lambda_N$, which encompass the complete rotational structure of interest. The intervals between wavelength settings are made equal to the instrumental resolution β (i.e., $\lambda_{n+1} = \lambda_n + \beta$). The sum of the signals $\sum_{n=1}^{n=N} S_n$ is taken as the total signal associated with the transition S_{ij} and is used in Eq. 3-1 to arrive at a measurement of emission cross section in the same manner as described for line sources in Section 3-4. This procedure involves some inherent assumptions that can best be illustrated by considering the procedure in more complete mathematical detail.

Suppose that $f(\lambda') d\lambda'$ represents the number of photons in the spectral range $\lambda' \to \lambda' + d\lambda'$ emitted per second in all directions from a length l of the beam path through a collision chamber. The objective of the integration procedure is to determine the total number of photons emitted in the transition $i \to j$ integrated over the whole rotational structure. That is to say, it is necessary to determine

$$\int_{\lambda_1}^{\lambda_N} f(\lambda') \, d\lambda'$$

and to set it to equal J_{ij}, as used in Eq. 3-1. For a monochromator with equal entrance and exit slit widths the overall efficiency of the detection system $K(\lambda')$ as a function of wavelength has the triangular form shown in Fig. 3-2. This involves the quite reasonable assumption that the efficiency of the detector itself is constant over the small range of wavelength encompassed by the instrumental resolution. By analogy with Eq. 3-1, the observed signal when the monochromator is set at the wavelength λ_n is given by

$$S(\lambda_n) = \frac{\int_x^{x+l} \omega(x) \, dx}{l4\pi} \int_{\lambda_n - \beta}^{\lambda_n + \beta} f(\lambda')K(\lambda') \, d\lambda'. \tag{3-17}$$

Consider the second integral in the expression. If it is assumed that $f(\lambda')$, the emission is constant over the spectral range $\lambda_n - \beta \to \lambda_n + \beta$ and is therefore set equal to $f(\lambda_n)$; the remaining integral reduces to the area under the efficiency curve shown in Fig. 3-2. The integral is therefore equal to $f(\lambda_n)K(\lambda_n)\beta$. Hence,

$$S(\lambda_n) = \frac{\int_x^{x+l} \omega(x) \, dx}{l4\pi} f(\lambda_n)K(\lambda_n)\beta. \tag{3-18}$$

The next step in the procedure is to sum the signals obtained at each wavelength setting λ_n over the whole wavelength range of the rotational structure. If $K(\lambda_n)$, the detection efficiency at the centre of the transmission band, is the same for all settings between λ_1 and λ_n, then the summation of Eq. 3-18 over the whole wavelength range is

$$S = \sum_{n=1}^{n=N} S(\lambda_n) = \int_x^{x+l} \frac{\omega(x) \, dx}{l4\pi} \cdot K(\lambda) \sum_{n=1}^{N} f(\lambda_n)\beta. \tag{3-19}$$

In the limit as $\beta \to 0$, $\sum_n f(\lambda_n)\beta = \int f(\lambda') \, d\lambda'$, which is the total photon output in the transition, J_{ij}.

$$\text{limit } \beta \to 0 \sum_{n=1}^{n=N} f(\lambda_n)\beta = J_{ij}. \tag{3-20}$$

Hence, Eq. 3-17 reduces to the Eq. 3-1 defined for a line emitter except that the signal is here taken as the sum $\sum S(\lambda_n)$ of the signals at each wavelength setting in the integration procedure.

Thus the procedure of measuring the signal in segments and adding them together is valid, but only if the following requirements are obeyed.

(1) The intensity of emission $f(\lambda')$ exhibits no appreciable change over the pass band of the monochromator.

(2) The detection sensitivity at the central frequency of the transition band is independent of wavelength setting over the whole range of the rotational structure.

(3) The detection system produces a triangular response function over the pass band at all settings.

(4) The measurements are made in the limit as $\beta \to 0$.

If these requirements are not fulfilled the integration procedure is not valid. The second of these conditions may be checked by reference to the calibration of detection efficiency of the system. The third condition may be assured by proper attention to slit widths. A suitable procedure for ensuring that the other two conditions are obeyed in practice is to show that the signal S_{ij} is independent of the resolution of the monochromator.

3. Summary. It is clearly desirable that an experiment should utilize a sufficient resolution to separate spectral lines completely from their neighbors. This requires that the transmission band width of the monochromator, or filter, be less than the separation of the adjacent spectral lines. In some cases this may prove to be impossible and two or more lines may be measured together. The problem is particularly severe when dealing with unresolved rotational structure in molecular emission. The techniques outlined above allow the determination of a sum of two or more line-emission cross sections or an integrated intensity over a rotational band. As formulated above, the technique requires that the detector sensitivity be invariant with wavelength over the spectral range of interest. An important test of the procedure is that the final result should be independent of the resolution of the monochromator.

F. Influence of the Doppler Shift

Because of the Doppler effect, the radiation emitted by a particle moving with a speed v at an angle θ with the line of sight of the detector is given by

$$\lambda = \lambda_0 \frac{(1 + \beta \cos \theta)}{(1 - \beta^2)^{1/2}}, \tag{3-21}$$

where λ_0 is the wavelength of the emission from the particle when stationary with respect to the detector, and $\beta = v/c$. When viewing at an angle of 90° to the particle's motion, the detector will record a shift in wavelength given approximately by $\lambda - \lambda_0 \approx \dfrac{\beta^2}{2} \lambda_0$. With a practical detection system having an acceptance angle δ,

the observed line will be Doppler broadened to $\Delta\lambda \approx \beta\lambda_0\delta$. Doppler shifts and broadening are small and completely unimportant for the excitation of target systems and also for the majority of projectile excitation studies. Nevertheless, they can become very significant when observing emission from low-mass projectiles at high energies. For example, the Balmer-alpha line emitted by a fast H atom will be shifted by approximately 0.21 Å at 30-keV and 6.9 Å at 1-MeV. More serious perhaps is the Doppler broadened width of this line. When viewed perpendicular to the beam path, using a detection system with an acceptance angle of 0.1 radian, the width will range from 5.3 Å at 30-keV to 30.0 Å at 1-MeV.

One may regard the Doppler effect as influencing the calibration of the system. The calibration problem has been formulated in terms of a line emitter of fixed wavelength and negligible intrinsic line width. When Doppler effects are significant, this is no longer the case. A monochromator set at a fixed wavelength exhibits an appreciable variation of sensitivity as a function of wavelength in its pass band (see Fig. 3-2). The transmission of an interference filter will also vary sharply with wavelength. Neglect of the Doppler broadening will cause the measured cross section to be erroneously low by an amount that increases systematically with increasing velocity.

In some situations it may be possible to reduce the effect of Doppler shift and broadening to negligible proportions. The acceptance angle δ of the optical system may be restricted to reduce the Doppler broadened width $\Delta\lambda$ as much as possible. If a monochromator is used for dispersion, the exit slit may be opened sufficiently wide to transmit the whole of the Doppler-broadened line. The calibration of detection sensitivity must then be along the lines suggested in Section 3-4-D. The central wavelength of the transmission band should be adjusted to allow for the shift of the central wavelength with impact velocity.

Such procedures will not be possible at high impact velocities because the required restriction in acceptance angle δ will seriously reduce the available signal level.

For situations where the Doppler width cannot be reduced to a negligible size some other procedure must be adopted. Light emitted at an angle θ will be detected by the optics with an efficiency $K(\lambda, \theta)$. This factor is related to the geometry of the system through the angle θ and also through the wavelength λ of the emission at that angle. This sensitivity must be assessed as a function of both variables and integrated over the whole effective aperture of the system in order to derive an effective sensitivity. Since the wavelength of emission at an angle θ is dependent on velocity (Eq. 3-21), the procedure must be repeated at each impact energy. Clearly, the required formulations will be very much dependent on the geometry of the system.

This problem is relevant to only a small number of the published measurements, and unfortunately it has been completely neglected in most of these cases. Since the effect is essentially instrumental, no broad assessment of error limits will be made, but the problem will be considered as it arises in the review of available experimental data.

A criterion for accurate experimental measurement is that the detection sensitivity must be established over the whole Doppler width of the spectral line. Clearly, this is required only for the measurement of emission from projectiles.

G. Normalization to Other Cross-Section Data

There have been a number of experiments in which the absolute magnitudes of cross sections have been assigned by normalization to theoretical predictions or to previous experimental measurements. Strictly speaking, such experiments have

performed only a relative measurement, and the normalization of the data is a method for displaying the data and not part of the experiment itself. In view of the frequency with which this is used, it is desirable to formulate a few general rules that should be followed.

It is clearly necessary that the normalization should be made to accurate information. If theory is used for normalization purposes, then this must be proven to be accurate by independent test. Experimental data should be used for normalization only if the conditions under which they were obtained are consistent with the criteria for valid and accurate measurement that have been established in Chapters 2 and 3.

It is vitally important when normalizations are made to previous experimental data that the conditions of observation are the same as those for the original measurement. For example, if the standard data were obtained without correction for anisotropy associated with polarization of emission, then the experiment must utilize the same direction of observation. Also, it is incorrect to normalize emission data that are uncorrected for polarization effects or cascade population to a theoretical prediction of a level-excitation cross section. Normalization procedures of this type are invalid. In general, the requirement of similar operating conditions is superfluous if both experiments are completely free of errors; unfortunately this is not often the case.

It is not appropriate to consider this problem in further detail here since each normalization procedure involves different pitfalls. In discussions of data in the later chapters, particular attention will be directed at the validity of normalization procedures in the light of the brief comments made above.

H. Summary

The measurement of emission cross sections using optical techniques provides detailed information about the processes that lead to the population of excited states. At first sight the method appears quite simple; however, the detailed understanding of the operation of the optical equipment is complicated. It must be admitted that in the majority of the experimental measurements of emission reviewed in this monograph insufficient attention has been paid to the points raised in the preceding sections. In the review of experimental data in later chapters, it will be noted that many discrepancies exist among data from various sources. Undoubtedly, many of the differences can be traced to inadequacies in the optical techniques.

In view of the complexity of the preceding discussions, it is valuable to consider briefly how the inadequate attention to detail will affect the various experimental measurements.

For the case of excitation of the target system, the detection sensitivity is not directly related to the energy of the incident projectile. The only errors that are likely to vary with energy are those associated with inadequate attention to polarization. Certain emissions are inherently unpolarized, and certain experimental arrangements are inherently insensitive to polarization. In such cases it is to be expected that most of the experimental errors will be associated with the measurement of detection efficiencies. Errors in this will cause the measurements to be wrong by a constant factor; moreover, the calibration errors are likely to be somewhat independent of wavelength. The measurement of geometrical parameters is certainly independent of λ, and

the shape of the spectral distribution from standard lamps does not change greatly with small errors in the measurement of temperature. Therefore, it is to be expected that an error will be propagated through all the data measured with a particular arrangement. It will be found when considering experimental data that measurements by certain groups of workers show a very large but consistent systematic difference that is independent of wavelength and collision combination. These differences, some as large as 300 %, are undoubtedly due to systematic errors in calibration of sensitivity. Such errors do not affect the energy dependence of the data. The problems are more complicated when the emission is polarized because neglect of anisotropy and instrumental polarization may cause energy-dependent errors in the measurements.

It is concluded that the energy dependence of measured cross sections should be reasonably reliable since they are not greatly influenced by errors in determination of sensitivity. This fact will be illustrated in later chapters by taking groups of measurements that differ by 200 to 300 % and normalizing them together. Such a procedure often indicates agreement in the energy dependence to within $\pm 10\%$ despite the great differences in magnitudes. This point is very important when using emission data to check the validity of cross section calculations. Comparisons of energy dependence provide a very good test of theory; whereas comparisons of magnitudes may be very inconclusive due to errors in the experimental procedures.

Measurements of emission from projectiles are more difficult than those from targets. Inadequate attention to the variation of emission with penetration (Section 2-2) and Doppler shifts may cause very large energy-dependent errors. Fortunately, however, it is relatively simple to detect such inadequacies by a retrospective assessment of the published procedures.

In summary of the preceding sections, Tables 3-4 and 3-5 list the criteria that should be obeyed if an experiment is to be considered accurate.

TABLE 3-4   Criteria for Accurate Measurement of Photon Emission

A. General Requirements:

1. The detection system must have sufficient resolution to isolate the spectral features of interest.

2. Scattered light and other spurious transmissions must be shown to be negligible.

3. The detection system should be shown to have a linear response with signal intensity throughout the dynamic range employed.

4. If the output data are expressed as a cross section for emission or as a level-excitation cross section, then corrections must be applied for anisotropy of emission related to polarization. Alternatively, the experiment may be operated at the "magic angle" of 54°46′, at which the data will be independent of anisotropy from this cause.

5. If the emission is polarized, then corrections are required for instrumental polarization.

6. When measurements are made on unresolved spectral features occupying a band width comparable to the transmission band of the system, then procedures must be adopted to ensure that the measured cross section is independent of spectral resolution.

7. The emission must be integrated over the whole Doppler-width of the line and attention paid to sensitivity variations over the whole wavelength interval.

B. Additional Criteria when Using a Standard Source for Calibration:

8. The emission of the standard must be known accurately.

9. The standard should be checked regularly for stability.

TABLE 3–4 *(continued)*

---

10. The calibrations should be carried out with the same field of view as the experiment itself. Alternatively, the sensitivity must be shown to be uniform over the whole field of view employed in calibration and in viewing the collision region.

11. The relevant geometrical parameters must be measured accurately.

C. Additional Criteria when Using a Detector of Known Sensitivity (e.g., Photoionization Chamber with a Rare Gas Filling):

12. The detector response should be known absolutely.

13. The detector response should be stable.

14. The transmission of any additional optical components should be accurately determined.

D. Additional Criteria when the Data Are Normalized to a Previous Cross Section Determination:

15. The previous determination should be in accordance with the criteria established for accuracy of the cross section measurements.

16. The conditions of experimental measurement must be essentially the same as those for which the standard cross section was measured experimentally or calculated theoretically.

TABLE 3-5   CRITERIA FOR ACCURATE MEASUREMENT OF POLARIZATION FRACTION

---

1. The polarization analyzer must exhibit 100% rejection of light polarized perpendicular to its principal section. Alternatively, the effect of any transmission must be assessed.

2. Light passing through the polarization analyzer must be parallel to the axis of the device.

3. Corrections must be applied for instrumental polarization.

4. Measurements must be independent of secondary population processes other than cascade, which cannot be eliminated (i.e., independent of target pressure and beam density).

CHAPTER 4

SUMMARY AND CRITIQUE OF THE
EXPERIMENTAL DATA

4-1 INTRODUCTION

A catalog of the available measurements of collisionally induced excitation and emission is presented. This includes a summary of the scope of the experimental measurements, an assessment of how well the experimental procedures fulfill the criteria that have been established in Chapters 2 and 3, and an index to the location of a detailed discussion of a particular measurement in the following Chapters 5 through 10. This catalog, based on references published between 1945 and January 1, 1970, includes all known data on absolute or relative magnitudes of collisional excitation and emission cross sections. For convenience, the catalog is given in tabular form (Table 4-1).

The tabular catalog may be considered as being in two parts. The left-hand side of the table gives the purely factual information on an entry: the reactants, scope of the measurement, publication reference, and reference to the location of the detailed discussion of the data in Chapters 5 through 12. The right-hand side contains a listing of those criteria for valid and accurate measurement, established in Chapters 2 and 3, which are violated in the experiment described in the table entry. The violated criteria are indicated by reference to the summary lists shown as Tables 2-1, 2-2 and Tables 3-1 through 3-5.

It is impossible to carry out a detailed analysis of a set of data by listing criteria that appear to be violated; in practice, every experimental measurement can be criticized through lack of attention to some detail. It is necessary to carry out a detailed discussion of each case to determine the magnitude and nature of the error that is introduced in the experimental data. In many cases neglect of a particular point may provide errors that are negligible in comparison with the random errors inherent in the experimental measurement. The detailed discussion of data, comparison of measurements by various authors, and attempts to reconcile differences are carried out in the following six chapters of this book. The critique carried out in the tables of this chapter represents a summary assessment of the experimental procedure. It does not in any way seek to establish the limitations of accuracy of the various empirical measurements. Criteria are listed as being violated irrespective of whether the inadequacy of the procedure is likely to produce significant errors to the data.

It proves to be impossible to assess all experiments against the criteria in a

completely objective manner. Published reports often contain insufficient information to allow a retrospective judgment of whether a particular procedure was carried out accurately. For the purposes of assessing the validity of experimental procedures, some use has been made of unpublished reports, theses, and private communications with the original authors. In those cases where no definite information could be elucidated, a subjective judgment was made, generally the criterion being listed as violated.

4-2   CATALOG OF AVAILABLE DATA

The table is arranged in alphabetical order of the projectiles. For each projectile, the targets are also listed alphabetically. For cases where data are available for various states of ionization of the projectile, the most highly ionized state is listed first. A key to the layout of the various columns is provided.

The reader's attention is also drawn to Appendix I where there is a listing of work published after the cutoff date for inclusion in the body of the monograph.

TABLE 4-1   CATALOG OF AVAILABLE DATA

Key to Column Headings

Proj.	Projectiles (alphabetical order).
Tar.	Target (alphabetical order under each projectile).
Exc.	Excited particle.
State	Multiplet designations of excited states.
Energy	Energy range of measurements, in Kiloelectronvolts.
Data	Type of data published. σ_i and σ_{ij}, level-excitation and line-emission cross sections; (θ), measurements made as a function of angle; (R), measurements are of relative cross sections only. Data are classified as "relative" if they are (a) measured in arbitrary units; (b) normalized to theoretical predictions; (c) normalized to measurements by a different author. T_i, threshold energy determined—no cross section information; P_{ij}, emission polarization.
Reference	Publication reference. Includes only major reports from which data are reproduced; minor reports are cited in the text.
Orig.	Origin of data. T, tabular data published; A, data from author by private communication or from unpublished report; G, read from published graphs.
Disc.	Section where data are discussed in detail (Chapters 5 through 10).
Tab. Data	Section where data are tabulated (Chapters 5 through 10).
Op. Def.	Criteria for the operational definition of cross section and polarization that are violated. σ_i and σ_{ij} are for level-excitation and line-emission cross sections (criteria in Table 2-1, page 25); P_{ij}, for polarization data (criteria in Table 2-2, page 25).
N_B	Measurement and definition of projectile beam (criteria in Table 3-1, page 32).
N_A	Measurement and definition of target density (criteria in Table 3-2, page 38).
N_i	Direct measurement of excited state density (criteria in Table 3-3, page 41).
J_{ij}	Measurement of photon emission rate (criteria in Table 3-4, page 63).
P_{ij}	Measurement of polarization fraction (criteria in Table 3-5, page 64).

TABLE 4–1 *(continued)*

Notes

[a] Includes work with D projectile.

[b] Includes work with D^+ projectile.

[c] Includes work with D_2^+ projectile.

[d] Includes work with D_2 target.

[e] Excited state formed in the projectile.

[f] Excited state formed in the target.

[g] Cross sections are the sum of emissions from target and projectile.

[h] Measurement of photons emitted from projectile and target in coincidence.

[i] Data tabulations include unpublished revised values provided by original authors.

[j] Data not reproduced on grounds that either accuracy or resolution is poor and the data may therefore be misleading (see text).

[k] Measurements on high principal quantum number states by the technique of field ionization. Not reproduced (see Section 3-3-A).

[l] Data represent relative intensities of emission lines. Not reproduced (see text).

[m] Published information is so inadequate that an assessment of this factor is impossible.

[n] Published data are of relative cross section measurements and accuracy in the determination of this parameter does not influence the validity of the data.

[o] Published information includes no quantitative cross section information.

[p] Data measured as a function of projectile scattering angle; not discussed or reproduced in this monograph.

TABLE 4-1

										Op. Def.						
Proj.	Tar.	Exc.	State	Energy	Data	Reference	Orig.	Disc.	Tab. Data	$\sigma_i-\sigma_{ij}$	P_{ij}	N_B	N_A	N_i	J_{ij}	P_{ij}
Ar^{2+}	H_2	H	2p	0.1–0.5	$\sigma_{ij}(R)$	Dunn et al. (99)	j	8-2-G	j	3		3,6,7			1,4,15	
Ar^+	Ar	Ar^{+g}	$^2P,{}^2F,{}^4P$	5–25	σ_{ij}	Sluyters et al. (103)	j	7-3-F	j	1,2,10		1,3	6		1,2,4,5,6,7,10	
Ar^+	Ar	Ar^{+g}	$^2F,{}^4P$	0.4–3	$\sigma_{ij}(R)$	Neff (104)	A	7-3-F	7-3-H	10		3,7	6		1,2,4,5,6,15,16	
Ar^+	Ar	Ar^{+g}	$^2P,{}^2F,{}^4P$	30–400	σ_{ij}	Thomas et al. (71)	A	7-3-F	7-3-H	10		3	6		1,2,4,5,6,7	
Ar^+	Ar	Ar^{+f}	^4P	120–420	σ_{ij}	Robinson et al. (105)	A	7-3-F	7-3-H	10		3			1,2,5,6,7	
Ar^+		Ar^{+e}	^4P	130–420	σ_{ij}		A	10-3-A	10-3-B	10		3			1,2,5,6,7	
Ar^+	Ar	Ar^{+e}	^2P	0.022–0.048	$\sigma_i(R)$	Haugsjaa et al. (64)	j	10-3-A	j	o		o	o	o		
Ar^+	Ar	Ar^{+f}	Unspec.	0.5–3.0	$\sigma_i(\theta)$	Barat et al. (106)	j	12-3	j	o		o	o	o		
Ar^+	CO	CO^+	Unspec.	2.5	σ_{ij}	Bayes et al. (107)	ℓ	8-5	ℓ	m		m	m		m	
Ar^+	CS_2	CS_2^+	Unspec.	2.5	σ_{ij}	Bayes et al. (107)	ℓ	8-5	ℓ	m		m	m		m	
Ar^+	H	H	2p	2–3.2	$P_{ij}(R)$	Young et al. (9)	A	5-3-C	5-3-D		3	m	m		m	
Ar^+	H_2	H	n=3,4	0.2–30	$\sigma_{ij}(R)$	Gusev et al. (109)	A	8-2-G	8-2-H	3		3,5	n		2,4,5,15	
Ar^+	D_2	D_2	$X^1\Sigma$	0.011–0.017	$\sigma_i(R)$	Moran et al. (108)	ℓ	8-2-G	ℓ	o		o	o	o		
Ar^+	DBr	DBr^+	Unspec.	2.5	σ_{ij}	Bayes et al. (107)	ℓ	8-5	ℓ	m		m	m		m	
Ar^+	HBr	HBr^+	Unspec.	2.5	σ_{ij}	Bayes et al. (107)	ℓ	8-5	ℓ	m		m	m		m	
Ar^+	HCl	HCl^+	Unspec.	2.5	σ_{ij}	Bayes et al. (107)	ℓ	8-5	ℓ	m		m	m		m	
Ar^+	He	Ar^+	$^2P,{}^2F,{}^4P$	5–22.5	σ_{ij}	Sluyters et al. (103)	j	10-3-A	j	1,10		1,3	6		1,2,4,5,6,7,10	
Ar^+	He	He	^3D	150–380	σ_{ij}	Thomas et al. (71)	G	6-6	6-6			3	6		2,4,5	
Ar^+		Ar^+	^2P	30–350	σ_{ij}		A	10-3-A	10-3-B	10		3	6		1,2,4,5,6,7	
Ar^+		Ar^+	$^2F,{}^4P$	110–340	σ_{ij}		A	10-3-A	10-3-B	10		3	6		1,2,4,5,6,7	
Ar^+	Kr	Ar^+	^4P	5–24	σ_{ij}	Sluyters et al. (103)	j	10-3-A	j	1,10		1,3	6		1,2,4,5,6,7,10	
Ar^+	Kr	Ar^+	$^2P,{}^2F,{}^4P$	110–400	σ_{ij}	Thomas et al. (71)	A	10-3-A	10-3-B	10		3	6		1,2,5,6,7	
Ar^+	N_2	N_2^+	$B^2\Sigma$	0.4–2	$\sigma_{ij}(R)$	Neff (104)	A	8-3-E	8-3-G	10		3,7	6		2,4,5,6,15,16	
Ar^+	N_2	N_2^+	$B^2\Sigma$	1–37	$\sigma_{ij}(R)$	Sluyters et al. (103)	ℓ	8-3-E	ℓ			1,3,5	6		2,4,5	
Ar^+	N_2	N_2^+	$B^2\Sigma$	1–30	$\sigma_{ij}(R)$	Polyakova et al. (110, 111)	j	8-3-E	j			1,2,3,4	6		4,5,6,15	
Ar^+	N_2O	N_2O^+	Unspec.	2.5	σ_{ij}	Polyakova et al. (112)	ℓ	8-5	ℓ	m		m	m		m	
Ar^+	Ne	Ar^+	$^2P,{}^2F,{}^4P$	5–24	σ_{ij}	Bayes et al. (107)	j	10-3-A	j	1,10		1,3	6		1,2,4,5,6,7,10	
Ar^+	Ne	Ar^+	^2P	30–300	σ_{ij}	Sluyters et al. (103)	A	10-3-A	10-3-B	10		3	6		1,2,4,5,6,7	
Ar^+	Ne	Ar^+	$^2F,{}^4P$	11–300	σ_{ij}	Thomas et al. (71)	A	10-3-A	10-3-B	10		3	6		1,2,4,5,6,7	
Ar^+	Xe		Unspec.	7.5–24	σ	Sluyters et al. (102)	j	7-5-F	j	1		1,3	6		2,4,5,10	

TABLE 4-1 (Continued)

Proj.	Tar.	Exc.	State	Energy	Data	Reference	Orig.	Disc.	Tab. Data	Op. Def. σ_i-σ_{ij}	Op. Def. P_{ij}	N_B	N_A	N_i	J_{ij}	P_{ij}
Ar	H_2	H	n=3,4	0.2–30	$\sigma_{ij}(R)$	Gusev et al. (109)	A	8-2-G	8-2-H	3		3,5	n		2,4,5,15	
Ar	N_2	N_2^+	$B^2\Sigma$	1–30	$\sigma_{ij}(R)$	Polyakova et al. (112)	j	8-3-E	j			1,2,3,4	6		4,5,6,15	
Ar	N_2	N_2^+	$B^2\Sigma$	3–37	$\sigma_{ij}(R)$	Gusev et al. (111)	ℓ	8-3-E	ℓ			1,3,5			2,4,5	
Ba^+	N_2	N_2^+	$B^2\Sigma$	328–2130	$\sigma_{ij}(R)$	Kurzweg et al. (113)	A	8-3-E	8-3-G			3			1,2,4,6	
Ba^{2+}	N_2	N_2^+	$B^2\Sigma$	804–4160	$\sigma_{ij}(R)$	Kurzweg et al. (113)	A	8-3-E	8-3-G			3			1,2,4,6	
Ca^+	N_2	Ca	$^1P, ^1F, ^3D, ^3F$	0.4–2	$\sigma_{ij}(R)$	Neff (104)	A	10-3-A	10-3-B	10		3,7	6		1,2,4,5,6,15,16	
		Ca^+	$^2S, ^2P$	0.4–2	$\sigma_{ij}(R)$		A	10-3-A	10-3-B	10		3,7	6		1,2,4,5,6,15,16	
Ca	N_2	Ca	^1P	2.0	σ_{ij}	Neff (104)	T	10-3-A	10-3-B	1,10		1,3,7	6		2,4,5,15,16	
Cs^+	Ar	Ar	Unspec.	4–75	$\sigma_{ij}(R)$	Andersen et al. (114)	j	7-3-G	j			3	6		1,2,4,5,6	
		Ar^+	Unspec.	4–75	$\sigma_{ij}(R)$		j	7-3-G	j			3	6		1,2,4,5,6	
		Cs^+	6p	6–85	$\sigma_{ij}(R)$		A	10-3-A	10-3-B	9,10		m	n		1,2,4,5,6	
Cs^+	He	Cs^+	6s,5d,5d'	4–5.5	T_i	Matveyev et al. (115)	T	10-3-A	10-3-B	0		0	0		0	
Cs^+	He	Cs^+	6s	4–12	$\sigma_{ij}(R)$	Bobashev et al. (116)	G	10-3-A	10-3-B	9,10		2	n		4,5	
Cs^+	He	Cs^+	6p	4–75	$\sigma_{ij}(R)$	Andersen et al. (114)	A	10-3-A	10-3-B	9,10		m	n		1,2,4,5,6	
Cs^+	Ne	Cs^+	6s,5d,5d'	3–11	$\sigma_{ij}(R), T_i$	Matveyev et al. (115)	G	10-3-A	10-3-B	9,10		2	n		4,5	
H^+	Ar	Ar^+	^2S	10–35	σ_{ij}	Van Eck et al. (76)	G	7-3-B	7-3-H	7,10					2,4,5,15	
H^+[b]	Ar	H	2s,2p	0.5–24	$\sigma_{ij}(R)$	Pretzer et al. (117) Jaecks et al. (16)	A	9-3-C	9-3-K	7,10			m		4,7,15,16	
H^+	Ar	H	$9 \leq n \leq 16$	10–180	$\sigma_i(R)$	Il'in et al. (118, 119)	k	9-3-C	k	7,12			n	1,2,4		
H^+	Ar	H	3s,4s	4–115	σ_{ij}	Hughes et al. (120, 121)	A	9-3-C	9-3-K	7,10		6,7	6		7,10	
H^+	Ar	H	2s,2p	8–42	σ_{ij}	Andreev et al. (75)	G	9-3-C	9-3-K	7,10			6		4,5	
H^+	Ar	H	3p	10–40	σ_{ij}, σ_i	Andreev et al. (122)	G	9-3-C	9-3-K	7,10			6		4,5	
			3p	10–30	P_{ij}		j	9-3-C	j		3					4
H^+	Ar	H	3s,3p,3d	10–30	σ_{ij}	Andreev et al. (123)	G	9-3-C	9-3-K	7,10			6		4,5	
H^+	Ar	Ar^+	$^2P, ^2F, ^4P$	2–10	σ_{ij}	Jaecks et al. (124)	A	7-3-B	7-3-H	7,12			6		2,4,5,10	
H^+	Ar	Ar	4p ^2P	40–600	σ_{ij}	Dufay et al. (125)	A	7-3-B	7-3-H	1,2			6		2,4,5,10	
		Ar^+	2p ^2P	25–500	σ_{ij}		A	7-3-B	7-3-H	1,2			6		2,4,5,10	
H^+	Ar	H	2p	0.6–12	$P_{ij}, \sigma_i(R)$	Gaily et al. (28)	A	9-3-C	9-3-K		4,6					
H^+	Ar	Ar^+	$2p^5$	120–300	σ_i	Volz et al. (126)	A	7-3-B	7-3-H					2,4		
H^+	Ar	Ar^{2+}	$2p^5 3p^5$	120–300	σ_i		A	7-3-B	7-3-H					2,4		

TABLE 4-1 (Continued)

Proj.	Tar.	Exc.	State	Energy	Data	Reference	Orig.	Disc.	Tab. Data	Op. Def. $\sigma_i-\sigma_{ij}$	Op. Def. P_{ij}	N_B	N_A	N_i	J_{ij}	P_{ij}
H^+	Ar	Ar	5p	150–1000	σ_{ij}	Thomas (127)	A	7-3-B	7-3-H						1,4,5,6	
H^+[b]		Ar^+	4P	150–1000	σ_{ij}		A	7-3-B	7-3-H						1,4,5,6	
H^+	Ar	H	2p	0.4–26	P_{ij}	Teubner et al. (128)	G	9-3-C	9-3-K		3					
H^+	Ar	H	3s,3d	168	σ_i	Thomas et al. (129)	A	9-3-C	9-3-K	7,10					4	
H^+	Ar	H	2s	2–70	σ_{ij}	Bayfield (130)	G	9-3-C	9-3-K	7					4,12	
H^+	Ar	H	2s	3–20	$\sigma_{ij}(\theta,R)$	Jaecks et al. (131)	p	p	p	p			p			
H^+	Ba	H[e]	n=11,14	6–30	$\sigma_i(R)$	McFarland et al. (132)	k	9-3-J	k	5,6,7,12			n	1,2		
H^+	C_8F_{16}	H[e]	n=11,14	4–25	$\sigma_i(R)$	McFarland et al. (132)	k	9-3-I	k	5,6,7,12			n	1,2		
H^+	CH_4	CH	$B\,^2\Sigma, A\,^2\Delta$	30–500	σ_{ij}	Carre et al. (133)	A	8-5-C	8-5-D				6		1,2,4,5,6,10	
H^+		H[f]	n=3,4,5	30–120	σ_{ij}		A	8-5-C	8-5-D	3			6		1,2,4,5,10	
H^+	C_2H_2	CH	$B\,^2\Sigma, A\,^2\Delta$	30–600	σ_{ij}	Carre et al. (133)	A	8-5-C	8-5-D				6		1,2,4,5,6,10	
H^+		H[f]	n=3,4,5	30–120	σ_{ij}		A	8-5-C	8-5-D	3			6		1,2,4,5,10	
H^+	C_2H_2	H[e]	$n \geq 9$	20–150	$\sigma_i(R)$	Le Doucen et al. (134)	k	9-3-I	k	7,12			n	1,2,4		
H^+	C_6H_6	H[e]	$n \geq 9$	20–150	$\sigma_i(R)$	Le Doucen et al. (134)	k	9-3-I	k	7,12			n	1,2,4		
H^+	CO	CO^+	$A\,^3\Pi, B\,^2\Sigma$	30–600	σ_{ij}, σ_i	Poulizac et al. (135)	A	8-5-A	8-5-D	7,8			6		2,4,5,6,10	
		C	$^3P, ^1D$	30–140	σ_{ij}		A	8-5-A	8-5-D				6		2,4,5,6,10	
		C^+	$^2P, ^2F$	20–140	σ_{ij}		A	8-5-A	8-5-D				6		1,2,4,5,6,10	
		O	$^3P, ^5P, ^5D$	30–120	σ_{ij}		A	8-5-A	8-5-D				6		1,2,4,5,6,10	
		H	n=3	20–100	σ_{ij}		j	9-3-I	j	9,10			6		2,4,5,7,10	
H^+	CO_2	H	n=10	60–180	$\sigma_i(R)$	Il'in et al. (57)	k	9-3-I	k	7,12			n	1,2,4		
H^+	Cd	H	$9 \leq n \leq 16$	10–180	σ_i	Oparin et al. (136)	k	9-3-J	k	7,12			n	1,2,4		
H^+	Cl_2CF_2	H[e]	n=11,14	6–30	$\sigma_i(R)$	McFarland et al. (132)	k	9-3-I	k	5,6,7,12			n	1,2		
H^+	Cs	H[e]	2s	0.16–3.2	$\sigma_{ij}(R)$	Donnally et al. (137)	G	9-3-J	9-3-K	5,6,7		7			4,16	
H^+	Cs	H	$9 \leq n \leq 16$	10–180	$\sigma_i(R)$	Il'in et al. (118)	k	9-3-J	k	7,12			n	1,2,4		
H^+	Cs	H	2s	0.5–5	σ_{ij}	Cesati et al. (138)	T	9-3-J	9-3-K	7,10,12,13					4,16	
H^+	Cs	H	2s	2–30	$\sigma_{ij}(R)$	Sellin et al. (139)	G	9-3-J	9-3-K	7,10,12		1,2,4,7			4,16	
H^+	H	H[f]	2p	0.6–30	$\sigma_{ij}(R)$	Stebbings et al. (8) Young (140)	A	5-3-A	5-3-D	5,6,7			m		15	
H^+	H	H[f]	2p	2–21	P_{ij}	Kauppaila et al. (141)	G	5-3-C	5-3-D		3					2

Table 4-1 (*Continued*)

Proj.	Tar.	Exc.	State	Energy	Data	Reference	Orig.	Disc.	Tab. Data	Op. Def. $\sigma_i-\sigma_{ij}$	Op. Def. P_{ij}	N_B	N_A	N_i	J_{ij}	P_{ij}
H+	H	H[e]	2p	0.6–30	$\sigma_{ij}(R)$	Stebbings et al. (8) Young (140)	A	5-3-B	5-3-D	5,6,7,10					15	
H+	H	H[e]	2s	44–120	$\sigma_{ij}(R)$	Ryding et al. (142)	G	5-3-B	5-3-D	3,7					16	
H+	H	H[e]	2s	3–70	σ_{ij}	Bayfield (143)	A	5-3-B	5-3-D	7					4,15	
H+	H_2	H[e]	2s	10	σ_{ij}	Madansky et al. (144)	j	9-3-F	j	o			o		o	
H+	H_2	H[g]	2p	0.5–3	$\sigma_{ij}(R)$	Dunn et al. (99)	j	8-2-C	j	3,9,10		6,7			1,4,7,15	
H+	H_2	H[e]	2s	9–40	$\sigma_{ij}(R)$	Colli et al. (145, 146)	A[i]	9-3-F	9-3-K	7,10,12,13		1,2,4,7	m		4,16	
H+	H_2	H[e]	$3\le n\le5$	5–200	σ_{ij}	Hughes et al. (147)	j	9-3-F	j	7,9,10		6,7	6		2,4,5,7,10	
H+	H_2	H[f]	$3\le n\le5$	5–130	σ_{ij}		A	8-2-C	8-2-H	3		6,7	6		2,4,5,10	
H+	H_2	H[e]	2s	15	$\sigma_{ij}(R)$	Sellin (17)	T	9-3-F	9-3-K	7,10,12			6		4,16	
H+	H_2	H[e]	n=14	25–100	$\sigma_i(R)$	Riviere et al. (56)	k	9-3-F	k	7,12			n	1,2		
H+	H_2	H[e]	$9\le n\le16$	10–180	$\sigma_i(R)$	Il'in et al. (118, 119, 57)	k	9-3-F	k	7,12			n	1,2,4		
H+	H_2	H[e]	2s	30–120	$\sigma_{ij}(R)$	Ryding et al. (142)	G	9-3-F	9-3-K	7,10			6		4,7,16	
H+	H_2	H[e]	3s,4s	5–120	σ_{ij}	Hughes et al. (120, 121)	A	9-3-F	9-3-K	7,10		6,7	6		2,7,10	
H+	H_2	H[e]	2s,2p	10–34	σ_{ij}	Andreev et al. (148)	G	9-3-F	9-3-K	7,10			6		4,5	
H+	H_2	H[e]	3p	10–34	σ_i		G	9-3-F	9-3-K	7,10			6		4,5	
H+	H_2	H[f]	2s,2p,3p	10–35	σ_{ij},σ_i	Andreev et al. (148)	G	8-2-C	8-2-H	3,7			6		4,5	
H+[b]	H_2[d]	H[g]	2p	0.4–25	$\sigma_{ij}(R)$	Van Zyl et al. (149)	A	8-2-C	8-2-H	3,9,10			6		1,4,7,15,16	
H+	H_2	H[f]	n=3,4,5	150–1000	σ_{ij}	Edwards et al. (150)	A	8-2-C	8-2-H	3					4,5	
		H_2	$3d\,^1\Pi$	150–1000	σ_{ij}		A	8-2-C	8-2-H						1,4,5,6	
		H_2	$3d\,^1\Sigma$	150–1000	σ_{ij}		A	8-2-C	8-2-H						1,4,5,6	
H+	H_2	H[g]	2p	13–130	$\sigma_{ij}(R)$	Dahlberg et al. (151)	A	8-2-C	8-2-H	3,9,10			6		4,5,7,15,16	
		H_2	$B\,^1\Sigma$	20–130	$\sigma_{ij}(R)$		A	8-2-C	8-2-H						4,5,6,15,16	
H+	H_2	H[e]	2s	2–70	σ_{ij}	Bayfield (130)	G	9-3-F	9-3-K	7					4,12	
H+	H_2	H[e]	$11\le n\le14$	5–30	$\sigma_i(R)$	McFarland et al. (132)	k	9-3-F	k	5,6,7,12			n	1,2		
H+	H_2	H[e]	$9\le n$	20–150	$\sigma_i(R)$	Le Doucen et al. (134)	k	9-3-F	k	7,12			n	1,2,4		
H+	H_2	H_2	$X\,^1\Sigma$	0.1–0.6	$\sigma_i(\theta)$	Moore et al. (152)	p	p	p	p		p	p	p		
H+	H_2	H_2	$X\,^1\Sigma$	0.1–0.6	σ_i		ℓ	8-2-C	ℓ	o		o	o	o		
D+	H_2	D	2s	4.5–34	$\sigma_i(\theta,R)$	Dose et al. (153)	p	p	p	p		p	p			
H+	H_2O	H[e]	n=11,14	5	$\sigma_i(R)$	McFarland et al. (132)	k	9-3-I	k	5,6,7,12			n	1,2		

TABLE 4-1 (Continued)

Proj.	Tar.	Exc.	State	Energy	Data	Reference	Orig.	Disc.	Tab. Data	Op. Def. $\sigma_i-\sigma_{ij}$	Op. Def. P_{ij}	N_B	N_A	N_i	J_{ij}	P_{ij}
H^+	He	He	$^1S, ^1P$	200	σ_{ij}	Hughes et al. (154)	j	6-1-B, 6-2-A	j	1,7		6,7	6		2,4,5,10	
H^+	He	H	$n=4$	200	σ_{ij}		j	9-3-A	j	7,9,10		6,7	6		2,4,5,7,10	
H^+	He	He	$^1S, ^1D$	10–35	σ_{ij}	Van Eck et al. (155)	j	6-1-B	j				6		2,5,10	
H^+	He	H	$n=4,5$	15–35	σ_i, P_{ij}		G	6-2-E	6-2-G		1		6		2,5,10	2,3
H^+	He	H	2s	7–39	σ_{ij}	Colli et al. (145, 146)	A^i	9-3-A	9-3-K	7,10,12,13		1,2,4,7	m		4,16	
H^+	He	He^+	$^1S, ^1P, ^1D$	5–100	σ_i	Van Eck et al. (10)	j	6-1-B	j				6		2,10	
H^+	He	He^+	$^1P, ^1D$	5–35	P_{ij}		T	6-2-F	6-2-G		1		6			2
H^+	He	He^+	$^1S, ^1P, ^1D$	20–130	σ_i, σ_i	Dodd and Hughes (156)	A	6-2-A,B,C	6-2-G	1		6,7	6		2,4,5,10	
H^+	He	He^+	$n=4$	20–130	σ_{ij}		A	6-2-E	6-2-G			6,7	6		2,4,5,10	
H^+	He	H	2p,3p	10–35	σ_{ij}	De Heer et al. (78)	G	9-3-A	9-3-K	7,10			6		2,4,5,10	
H^+	He	H	$3 \le n \le 7$	10–30	σ_{ij}, σ_i	Bobashev et al. (157, 158)	j	9-3-A	j	7,9,10			6		2,4,5,7	
H^+	He	H	2s	3–23	$\sigma_{ij}(R)$	Pretzer, Jaecks, Gaily et al. (117, 16, 28)	A^i	9-3-A	9-3-K	7,10			m		4,7,15,16	
H^+	He	H	2p	0.25–28	$\sigma_{ij}(R)$		A^i	9-3-A	9-3-K	7,10			m		4,7,15,16	
H^+	He	H	$9 \le n \le 16$	10–180	$\sigma_i(R)$	Il'in et al. (118, 119, 57)	k	9-3-A	k	7,12			n	1,2,4		
H^+	He	H	3s,4s	5–120	σ_{ij}, σ_i	Hughes et al. (120, 121)	A	9-3-A	9-3-K	7,10		6,7	6		7,10	
H^+	He	H	2s,2p	3–71	$\sigma_{ij}(R)$	Dose (159)	G	9-3-A	9-3-K	7,10			6		4,5	
H^+	He	H	2s	45–200	$\sigma_{ij}(R)$	Ryding et al. (142)	G	9-3-A	9-3-K	7,10			6		4,7,16	
H^+	He	H	2s,2p	10–40	$\sigma_{ij}(R)$	Andreev et al. (75)	G	9-3-A	9-3-K	7,10			6		4,5	
H^+[b]	He	H	2s	2–60	$\sigma_{ij}(\theta,R)$	Dose et al. (153)	p	p	p	p		8	p		p	
H^+	He	H	3p	10–40	σ_i, σ_i	Andreev et al. (122)	G	9-3-A	9-3-K	7,10			6		4,5	
H^+	He	H	3s,3p,3d	14–30	σ_{ij}	Andreev et al. (123)	G	9-3-A	9-3-K	7,10			6		4,5	
H^+	He	He^+	$^1S, ^1P, ^1D$	20–500	σ_i	Denis et al. (160)	A	6-2-A,B,C	6-2-G	1			6		2,4,5,10	
H^+	He	He^+	$^1S, ^1P, ^1D$	75–1000	σ_i	Thomas et al. (161)	A	6-2-A,B,C	6-2-G				6		4,5	
H^+	He	He^+	$n=4$	150–1000	σ_{ij}		A	6-2-E	6-2-G				6		4,5	
H^+	He	He^+	$^1S, ^1D$	60–350	σ_i	Robinson et al. (162)	T	6-2-A,C	6-2-G	7			6		2,5	
H^+	He	He^+	$n=4$	30–150	σ_i	Moustafa Moussa et al. (163)	T	6-2-E	6-2-G				6		2,4,5,10	
H^+	He	He	$^1S, ^1D$	1.5–20	$\sigma_{ij}(R), P_{ij}$	Krause et al. (164)	G	6-2-A,C,F	6-2-G		3		n		4,5	1,2,3
H^+	He	He	$^3S, ^3D$	2–20	$\sigma_{ij}(R), P_{ij}$		G	6-2-D,F	6-2-G	1	1,3		n		4,5	1,2,3
H^+	He	He	$^1S, ^1P, ^1D$	1–150	σ_i	Van den Bos et al. (165)	A	6-2-A,B,C	6-2-G	1			6		2,10	
H^+	He	He	$^1P, ^1D$	1–150	P_{ij}		A	6-2-F	6-2-G		1		6			2

TABLE 4-1 (*Continued*)

Proj.	Tar.	Exc.	State	Energy	Data	Reference	Orig.	Disc.	Tab. Data	Op. Def.		N_B	N_A	N_i	J_{ij}	P_{ij}
										σ_i–σ_{ij}	P_{ij}					
H^{+b}	He	H	2p	0.6–12	$P_{ij},\sigma_i(R)$	Gaily et al. (28)	A	9-3-A	9-3-K		4,6					
H^+	He	He	$^1P,^1D$	120–1000	P_{ij}	Scharmann et al. (166)	A	6-2-F	6-2-G		1		n			
H^+	He	He	$^1S,^1P,^1D$	120–1000	$\sigma_{ij}(R)$	Scharmann et al. (167)	G	6-2-A,B,C	6-2-G				n			
H^+	He	H	3s,3d	150	σ_i	Thomas et al. (129)	A^i	9-3-A	9-3-K	7,10					4	
H^+	He	H	2s	5–20	$\sigma_{ij}(\theta,R)$	Jaecks et al. (131)	p	p	p	p			p			
H^+	He	Hg	$^1S,^1P,^3S$	5–36	σ_{ij}	Bobashev et al. (168)	G	7-6-B	7-6-C	1					2,4,5	
		Hg^{2+}	$5d^8 6s^2$	5–36	σ_{ij}		G	7-6-B	7-6-C	1					2,4,5	
		H	n=4	5–35	σ_{ij}		i	9-3-J	j	1,7,9,10					2,4,5,7	
H^+	K	H	$9\leq n\leq 16$	10–180	$\sigma_i(R)$	Il'in et al. (118, 119)	k	9-3-J	k	7,12			n	1,2,4		
H^+	K	H	2s	10–30	$\sigma_{ij}(R)$	Sellin et al. (139)	G	9-3-J	9-3-K	7,10,12					4,16	
H^+	K	H	n=11,14	6–30	$\sigma_i(R)$	McFarland et al. (132)	k	9-3-J	k	5,6,7,12			n	1,2		
H^+	Kr	Kr^+	^2S	10–35	σ_{ij}	Van Eck et al. (76)	G	7-4-B	7-4-E						2,4,5,15	
H^{+b}	Kr	H	2p	0.10–22	$\sigma_{ij}(R)$	Pretzer et al. (117, 169)	j	9-3-D	j	7,10		m			1,4,7,15,16	
H^+	Kr	H	2s,2p	10–40	σ_{ij}	Andreev et al. (75)	G	9-3-D	9-3-K	7,10		6			4,5	
H^+	Kr	H	3p	10–40	σ_{ij}	Andreev et al. (122)	G	9-3-D	9-3-K	7,10		6			4,5	
H^+	Li	H	$9\leq n\leq 16$	10–180	$\sigma_i(R)$	Il'in et al. (118, 119)	k	9-3-J	k	7,12			n	1,2,4		
H^+	Mg	H	$8\leq n$	15–180	$\sigma_i(R)$	Il'in et al. (118, 54)	k	9-3-J	k	7,12			n		1,2,4	
H^+	Mg	H	$9\leq n\leq 16$	10–180	$\sigma_i(R)$	Oparin et al. (136)	k	9-3-J	k	7,12			n	1,2,4		
H^+	Mg	H	n=11,14	6–30	$\sigma_i(R)$	McFarland et al. (132)	k	9-3-J	k	5,6,7,12			n	1,2,4		
H^+	Mg	H	n=6	5–70	$\sigma_{ij}(R),\sigma_i(R)$	Berkner et al. (170)	j	9-3-J	j	7,10,12			n	1,2	7,15,16	
D^+	Mg	D	$10\leq n$	10–40	$\sigma_i(\theta)$	Kingdon et al. (171)	p	p	p	p	p	p	p	p		
H^+	N_2	N_2^+	$B^2\Sigma,A^2\Pi$	1.5–33	σ_{ij}	Carleton et al. (172, 173)	G	8-3-C	8-3-G			6	6		4,10	
H^+	N_2	N	^4P	1.5–33	σ_{ij}		G	8-3-C	8-3-G			6	6		4,10	
H^+	N_2	H	n=4	1.5–4	σ_{ij}		j	9-3-G	j	7,9,10		6	6		4	
H^+	N_2	H	n=4	3–30	σ_{ij}	Sheridan et al. (173)	j	9-3-G	j	7,9,10			6		4	
H^+	N_2	H	n=3,4	5–200	σ_{ij}	Philpot et al. (174, 175)	j	9-3-G	j	7,9,10		6,7	6		2,4,5,7,10	
H^+	N_2	H	2p	0.25–25	$\sigma_{ij}(R)$	Dunn et al. (99)	j	9-3-G	j	7,10			m		1,4,6,7	
						Van Zyle et al. (149)	j	9-3-G	j	7,10			m			

TABLE 4-1 (*Continued*)

Proj.	Tar.	Exc.	State	Energy	Data	Reference	Orig.	Disc.	Tab. Data	Op. Def. $\sigma_i - \sigma_{ij}$	Op. Def. P_{ij}	N_B	N_A	N_i	J_{ij}	P_{ij}
H^+	N_2	N_2^+	$B\,^2\Sigma$	5–130	σ_{ij}, σ_i	Philpot et al. (175)	A	8-3-C	8-3-G			6,7	6		2,4,5,6,10	
		N_2^+	$^3F, ^3D$	10–130	σ_{ij}		A	8-3-C	8-3-G			6,7	6		1,2,4,5,6,10	
H^+	N_2	N_2^+	$B\,^2\Sigma$	10–65	$\sigma_{ij}(R)$	Sheridan et al. (176)	j	8-3-C	j				6		2,3,4,5	
	H	$n=3$	10–65	σ_{ij}		j	9-3-G	j	7,9,10			6		2,3,4,5,7,10		
H^+	N_2	H	$n=10$	60–180	$\sigma_i(R)$	Il'in et al. (57)	k	9-3-G	k	7,12			n	1,2,4		
H^+	N_2	H	$n=3$	2–25	σ_i, P_{ij}	Murray et al. (177)	j	9-3-G	j	1,7,10			6		2,4,7,10	
H^+	N_2	N_2^+	$B\,^2\Sigma, A\,^2\Pi$	25–1000	σ_{ij}, σ_i	Dufay et al. (178, 179)	A^i	8-3-C	8-3-G			1	6		2,4,5,6,10	
	N	$^2D, ^4D, ^4F$	40–120	σ_{ij}		A	8-3-C	8-3-G				6		1,2,4,5,6,10		
	N^+	$^1P, ^1D, ^3D, ^3F$	25–1000	σ_{ij}		A^i	8-3-C	8-3-G				6		1,2,4,5,6,10		
	H	$n=3,4$	20–120	σ_{ij}		j	9-3-G	j	9,10			6		2,4,5,7,10		
H^+	N_2	H	$3s,4s$	5–120	σ_{ij}, P_{ij}	Hughes et al. (120, 121, 180)	A	9-3-G	9-3-K	7,10		6,7	6		2,7,10	
		$3p,3d$	10–35			G	9-3-G	9-3-K	7,10	3	6,7	6		2,4,7,10		
H^+	N_2	N_2^+	$B\,^2\Sigma$	60–400	σ_{ij}, σ_i	Robinson et al. (162)	T	8-3-C	8-3-G						2,5,6	
	N^+	^3F	60–300	σ_{ij}		T	8-3-C	8-3-G						1,2,5,6		
H^+	N_2	N_2^+	$B\,^2\Sigma$	20–100	σ_{ij}	Baker et al. (181)	G	8-3-C	8-3-G			6,7	6		2,4,5,6,10	
	N^+	^3D	20–100	σ_{ij}		G	8-3-C	8-3-G			6,7	6		1,2,4,5,6,10		
H^+	N_2	N_2^+	$B\,^2\Sigma$	10–130	$\sigma_{ij}(R), \sigma_i(R)$	Dahlberg et al. (182)	A	8-3-C	8-3-G	7,10			6		4,5,6,15	
H^{+b}		N_2	$C\,^3\Pi$	14–100	$\sigma_{ij}(R)$	Thomas et al. (183)	A	8-3-C	8-3-G	1					4,5,6,15	
	N	^2P	15–130	$\sigma_{ij}(R)$		A	8-3-C	8-3-G						4,5,15,16		
	N^+	^3F	15–130	$\sigma_{ij}(R)$		A	8-3-C	8-3-G						1,4,5,6,15		
	H	$2p$	20–130	$\sigma_{ij}(R)$		A	9-3-G	9-3-K	7,10					2,4,5,7,15,16		
H^{+b}	N_2	N_2^+	$B\,^2\Sigma$	150–1000	σ_{ij}, σ_i		A	8-3-C	8-3-G	1					4,5,6	
	N_2	$C\,^3\Pi$	150–900	σ_{ij}		A	8-3-C	8-3-G						1,4,5,6		
	N^+	$^3D, ^3F$	150–900	σ_{ij}		A	8-3-C	8-3-G						1,4,5,6		
H^+	N_2	N_2^{+h}	$B\,^2\Sigma$	1.5–30	σ_{ij}	Young et al. (7)	G	8-3-C	8-3-G	3,7,10		5,6,7	6		m	
	H^h	$3p,3d$	1.5–30	σ_{ij}		G	9-3-G	9-3-K	3,7,10			6		m		
H^+	N_2^+	N_2^+	$B\,^2\Sigma$	0.1–13.5	$\sigma_{ij}(R)$	Moore et al. (184)	ℓ	8-3-C	ℓ	7,10					4,5	
H^+	N_2	H	$3s,3d$	168	σ_i	Thomas et al. (129)	A^i	9-3-G	9-3-K	7,10			n		4	
H^+	N_2	H	$n=11,14$	5–30	$\sigma_i(R)$	McFarland et al. (132)	k	9-3-G	k	5,6,7,12		1,3,5	n	1,2		
H^+	N_2	N_2^+	$B\,^2\Sigma$	3–37	$\sigma_{ij}(R)$	Gusev et al. (111)	ℓ	8-3-C	ℓ				o		2,4,5	
H^+	N_2	N_2	$X\,^2\Sigma$	0.1–0.6	σ_i	Moore et al. (152)	ℓ	8-3-C	ℓ	o		o	o			
H^+	N_2					M.F... (133)	k	9-3-I	k	5,6,7,12			n	1,2		

TABLE 4-1 (Continued)

Proj.	Tar.	Exc.	State	Energy	Data	Reference	Orig.	Disc.	Tab. Data	σ_i-σ_{ij}	P_{ij}	N_B	N_A	N_i	J_{ij}	P_{ij}
H^+	NO	N^+	$^3F, ^5P$	60–300	σ_{ij}	Robinson et al. (162)	T	8-5-B	8-5-D						1,2,5,6	
		O^+	^4D	80–300	σ_{ij}		T	8-5-B	8-5-D						1,2,5,6	
H^+	Na	H	$8 \leqslant n \leqslant 16$	10–180	$\sigma_i(R)$	Il'in et al. (118, 119, 54)	k	9-3-J	k	7,12			n	1,2,4		
H^+	Ne	Ne	3s,3s'	13–35	σ_{ij}	Van Eck et al. (76)	G	7-2-B	7-2-E						2,4,5,6,15	
		Ne^+	^2S	10–35	σ_{ij}		G	7-2-B	7-2-E						2,4,5,6,15	
H^+	Ne	Ne	3p',4d	5–35	σ_{ij}, P_{ij}	De Heer et al. (77)	G	7-2-B	7-2-E		3				2,4,5,10	2
		Ne^+	$^4P, ^2D$	7.5–35	σ_{ij}		G	7-2-B	7-2-E						2,4,5,10	
H^+	Ne	H	2p,3p	7.5–35	σ_i, σ_i	De Heer et al. (78)	G	9-3-B	9-3-K	7,10			6		2,4,5,10	
H^{+b}	Ne	H	$3 \leqslant n \leqslant 7$	10–30	σ_i, σ_i	Bobashev et al. (157, 158)	j	9-3-B	j	7,9,10			6		2,4,5,7	
H^{+b}	Ne	H	2s,2p	0.25–23	$\sigma_{ij}(R)$	Pretzer et al. (117) Jaecks et al. (16)	A^i	9-3-B	9-3-K	7,10			m		4,7,15,16	
H^+	Ne	H	$9 \leqslant n \leqslant 16$	10–180	$\sigma_i(R)$	Il'in et al. (118, 119)	k	9-3-B	k	7,12			n	1,2,4		
H^+	Ne	H	3s,4s	2–120	σ_{ij}, σ_i	Hughes et al. (120, 121)	A	9-3-B	9-3-K	7,10		6,7	6		7,10	
H^+	Ne	H	2s,2p	10–40	σ_{ij}	Andreev et al. (75)	G	9-3-B	9-3-K	7,10			6		4,5	
H^+	Ne	Ne	3p',3p	25–600	σ_{ij}	Dufay et al. (125)	A	7-2-B	7-2-E	1,2			6		2,4,5,10	
		Ne^+	$^2P, ^4D$	30–500	σ_{ij}		A	7-2-B	7-2-E	1,2			6		1,2,4,5,6,10	
H^+	Ne	H	3p	10–40	σ_{ij}, σ_i	Andreev et al. (122)	G	9-3-B	9-3-K	7,10			6		4,5	
H^+	Ne	H	3s,3p,3d	12–30	σ_{ij}	Andreev et al. (123)	G	9-3-B	9-3-K	7,10			6		4,5	
H^{+b}	Ne	H	2p	0.6–15	$P_{ij}, \sigma_i(R)$	Gaily et al. (28)	A	9-3-B	9-3-K		4,6		6			
H^+	Ne	H	n=6	15–60	$\sigma_i(R), \sigma_i(R)$	Berkner et al. (170)	j	9-3-B	j	7,10,12			6		7,15,16	
H^+	O_2	O_2^+	$b\,^4\Sigma$	5–130	σ_{ij}, σ_i	Hughes et al. (185)	A	8-4-B	8-4-E	8		6,7	6		2,4,5,6,10	
		O	^2D	10–130	σ_{ij}		A	8-4-B	8-4-E			6,7	6		1,2,4,5,6,10	
H^+	O_2	O_2^+	$b\,^4\Sigma, a\,^2\Pi$	20–1000	σ_{ij}, σ_i	Dufay et al. (178)	A	8-4-B	8-4-E				6		1,2,4,5,6,10	
		O	$^3P, ^5P, ^5D$	30–1000	σ_{ij}		A	8-4-B	8-4-E				6		1,2,4,5,6,10	
		O^+	$^2D, ^2F, ^2G$	40–500	σ_{ij}		A	8-4-B	8-4-E				6		1,2,4,5,6,10	
		O^+	^4D	30–1000	σ_{ij}		A	8-4-B	8-4-E				6		1,2,4,5,6,10	
		H	n=3,4	20–120	σ_{ij}		j	9-3-H	j	9,10			6		2,4,5,7,10	
H^+	O_2	H	3s,4s	5–120	σ_i	Hughes et al. (120, 121)	A	9-3-H	9-3-K	7,9,10		6,7	6		2,7,10	
H^+	O_2	O_2^+	$b\,^4\Sigma$	20–100	σ_{ij}	Baker et al. (181)	G	8-4-B	8-4-E			6,7	6		2,4,5,6	
		O^+	^2D	20–100	σ_{ij}		G	8-4-B	8-4-E			6,7	6		1,2,4,5,6	

75

TABLE 4-1 (*Continued*)

Proj.	Tar.	Exc.	State	Energy	Data	Reference	Orig.	Disc.	Tab. Data	σ_i–σ_{ij} (Op. Def.)	P_{ij} (Op. Def.)	N_B	N_A	N_i	J_{ij}	P_{ij}
H⁺	O_2	O_2^+	$b\,^4\Sigma$	150–900	σ_{ij},σ_i	Thomas et al. (230)	A	8-4-B	8-4-E	8					4,5,6	
		O^+	^2D	150–1000	σ_{ij}		A	8-4-B	8-4-E						1,4,5,6	
H⁺	Rb	H	2s	5–32	$\sigma_{ij}(R)$	Sellin et al. (139)	G	9-3-J	9-3-K	7,10,12					4,16	
H⁺	Tℓ	H	n=11,14	6–30	σ_i	McFarland et al. (132)	k	9-3-G	k	5,6,7,12			n	1,2		
H⁺[b]	Xe	H	2s,2p	0.3–22	$\sigma_{ij}(R)$	Pretzer et al. (117,16) Jaecks et al.	A	9-3-E	9-3-K	7,10			m		4,7,15,16	
H⁺	Xe	H	2s,2p	10–40	σ_{ij}	Andreev et al. (75)	G	9-3-E	9-3-K	7,10			6		4,5	
H⁺	Xe	Xe⁺	6p	24–610	σ_{ij}	Dufay et al. (125)	G	7-5-B	7-5-F	1,2			6		2,4,5,10	
		Xe⁺	^4D	30–510	σ_{ij}		G	7-5-B	7-5-F	1,2			6		2,4,5,10	
H⁺	Xe	Xe⁺	$^4D,^4F$	1–10	σ_{ij}	Jaecks et al. (124)	A	7-5-B	7-5-F				6		2,4,5,10	
H⁺	Xe	H	3p	10–40	σ_{ij},σ_i	Andreev et al. (122)	G	9-3-E	9-3-K	7,10			6		4,5	
H⁺	Xe	Xe	Unspec.	2.0	$\sigma_i(\theta,R)$	Baudon et al. (186)	p	p	p	p		p	p			
H⁺	Zn	H	$9 \leq n \leq 16$	10–180	$\sigma_i(R)$	Oparin et al. (136)	k	9-3-J	k	7,12			n	1,2,4		
H	Ar	H	2s,2p,3p	5–40	$\sigma_{ij}(R)$	Orbeli et al. (187)	G	9-4-A	9-4-C	7,10			6		4,5,7,15	
H[a]	Ar	H	2p	2–55	$\sigma_{ij}(R)$	Dose et al. (188)	G	9-4-A	9-4-C	3,7,10		3	n		4,8,15	
H[a]	Ar	H	2p	2–55	P_{ij}	Dose et al. (189)	G	9-4-B	9-4-C		3	8	n			
H	H_2	H[e]	2s	15	$\sigma_{ij}(R)$	Sellin (17)	T	9-4-A	9-4-C	7,10,12		1,3			4,15,16	
H	H_2	H[g]	2p	13–120	$\sigma_{ij}(R)$	Dahlberg et al. (151)	A	8-2-D	8-2-H	3,9,10		3,8			4,5,7,15,16	
		H_2	$B\,^1\Sigma$	50	$\sigma_{ij}(R)$		A	8-2-D	8-2-D			3,8			4,5,7,15,16	
H	He	H_2	$^1S,^1P,^1D$	5–35	σ_{ij}	Van Eck et al. (10,190)	A[i]	6-3-A	6-3-C			3,8			2,10	
		H_2	$^3S,^3P,^3D$	5–35	σ_{ij}		A[i]	6-3-A	6-3-C			3,8			2,10	
			$^1D,^3P,^3D$	5–35	P_{ij}		A	6-3-B	6-3-C							2
H	He	He	^1S	60–150	σ_i	Robinson et al. (162)	T	6-3-A	6-3-C	7		3			2	
H	He	H	2s,2p	5–40	σ_{ij}	Ankudinov et al. (191)	G	9-4-A	9-4-C	7,10		3	6		4,5,7,15	
H	He	H	$^3S,^3D$	4–20	$\sigma_{ij}(R)$	Krause et al. (164)	G	6-3-A	6-3-C	1		1,3	n	1,3	4,5	1,2,3,4
			^3D	5–20	P_{ij}		G	6-3-B	6-3-C		1,3	3	n	3		
H[a]	He	H	2p	2.5–55	$\sigma_{ij}(R)$	Dose et al. (188)	G	9-4-A	9-4-C	3,7,10		8	n		4,8,15	
H[a]	He	H	2p	2–55	P_{ij}	Dose et al. (189)	G	9-4-B	9-4-C		3	8	n			
H	He	H	2s,2p,3p	5–40	σ_{ij}	Orbeli et al. (187)	G	9-4-A	9-4-C	7,10		3	6		4,5,7,15	
H	Kr	H	2s,2p	5–40	σ_{ij}	Ankudinov et al. (191)	G	9-4-A	9-4-C	7,10		3	6		4,5,7,15	
H						Orbeli et al. (187)	G	9-4-A	9-4-C	7,10		3	6		4,5,7,15	

TABLE 4-1 (Continued)

Proj.	Tar.	Exc.	State	Energy	Data	Reference	Orig.	Disc.	Tab. Data	Op. Def. σ$_i$–σ$_{ij}$	Op. Def. P$_{ij}$	N$_B$	N$_A$	N$_i$	J$_{ij}$	P$_{ij}$
H	N$_2$	N$_2^+$	$C\,^3\Pi$	1–5	σ_{ij}	Carleton et al. (172)	j	8-3-D	j	1					4,10	
H	N$_2$	N$_2^+$	$B\,^2\Sigma$	10–130	$\sigma_{ij}(R)$, $\sigma_i(R)$	Dahlberg et al. (182)	A	8-3-D	8-3-G			3,8	6		4,5,6,15	
		N$_2$	$C\,^3\Pi$	14–100	$\sigma_{ij}(R)$		A	8-3-D	8-3-G			3,8			4,5,6,15	
		N	^2P	15–90	$\sigma_{ij}(R)$		A	8-3-D	8-3-G			3,8			4,5,15,16	
		N$^+$	^3F	15–120	$\sigma_{ij}(R)$		A	8-3-D	8-3-G			3,8			1,4,5,6,15	
		H	2p	20–130	$\sigma_{ij}(R)$		A	9-4-A	9-4-C	7,10		3,8			4,5,7,15,16	
H	N$_2$	N$_2^+$	$B\,^2\Sigma$	60–120	σ_{ij},σ_i	Robinson et al. (162)	T	8-3-D	8-3-G			3			2,5,6	
	N$_2$	N	^3F	60–120	σ_{ij}		T	8-3-D	8-3-G			3			1,2,5,6	
H	N$_2$	N$_2^+$	$B\,^2\Sigma$	3–37	$\sigma_{ij}(R)$	Gusev et al. (111)	ℓ	8-3-D	ℓ			1,3,5			2,4,5	
H	Ne	H	2s,2p	5–40	σ_{ij}	Ankudinov et al. (191)	G	9-4-A	9-4-C	7,10		3	6		4,5,7,15	
H[a]	Ne	H	2p	1.8–55	$\sigma_{ij}(R)$	Dose et al. (188)	G	9-4-A	9-4-C	3,7,10		8	n		4,8,15	
H[a]	Ne	H	2p	2–55	P_{ij}	Dose et al. (189)	G	9-4-B	9-4-C		3		n			
H	Ne	H	2s,2p,3p	5–40	σ_{ij}	Orbeli et al. (187)	G	9-4-A	9-4-C	7,10		3	6		4,5,7,15	
H$^-$	Ar	H	2s,2p	5–40	σ_{ij}	Andreev et al. (192)	G	9-5-A	9-5-B	7,10			6		4,5,7,15	
D$^-$	H$_2$	D	$6 \leq n \leq 9$	20,000	$\sigma_i(R)$	Berkner et al. (193)	k	9-5-A	k	1,7,10,12			n	1,2		
H$^-$	He	H	2s,2p	5–40	σ_{ij}	Andreev et al. (192)	G	9-5-A	9-5-B	7,10			6		4,5,7,15	
H$^-$	Kr	H	2s,2p	5–40	σ_{ij}	Andreev et al. (192)	G	9-5-A	9-5-B	7,10			6		4,5,7,15	
H$^-$	Ne	H	2s,2p	5–40	σ_{ij}	Andreev et al. (192)	G	9-5-A	9-5-B	7,10			6		4,5,7,15	
H$^-$	Xe	H	2s,2p	5–40	σ_{ij}	Andreev et al. (192)	G	9-5-A	9-5-B	7,10			6		4,5,7,15	
H$_2^{+}$[c]	Ar	H	2s,2p	0.25–25	$\sigma_{ij}(R)$	Van Zyl et al. (194) Jaecks et al. (195)	A	9-6-A	9-6-E	7,10		3	m		4,7,15,16	
H$_2^+$	Ar	Ar$^+$	$^2P,\,^2F,\,^4P$	2–10	σ_{ij}	Jaecks et al. (124)	A	7-3-C	7-3-H	7,10		3			2,4,5,10	
H$_2^+$	Ar	H	3s	20–120	σ_{ij}	Hughes et al. (196)	G	9-6-A	9-6-E	7,10		3,6,7	6		7,15,16	
H$_2^+$	H$_2$	j	$3 \leq n \leq 5$	200	σ_{ij}	Hughes (197)	j	9-6-B	j	9,10		3,6,7	6		2,4,5,7,10	
H$_2^+$	H$_2$	H[e]	$3 \leq n \leq 5$	5–130	σ_{ij}	Hatfield et al. (198)	j	9-6-B	j	9,10		3,6,7	6		2,4,5,7,10	
		H[f]	$3 \leq n \leq 5$	10–130	σ_{ij}		A	8-2-E	8-2-H	3		3,6,7	6		2,4,5,10	
H$_2^+$	H$_2$	H[e]	$6 \leq n \leq 9$	10,000	$\sigma_i(R)$	Berkner et al. (193)	k	9-6-B	k	1,7,10,12			n	1,2		
H$_2^+$	H$_2$	H[f]	2s,2p,3p	10–30	σ_{ij},σ_i	Andreev et al. (148)	G	8-2-E	8-2-H	3,7		3	6		4,5,15	
H$_2^+$	H$_2$	H[e]	2s,2p,3p	12–28	σ_{ij},σ_i		G	9-6-B	9-6-E	7,10		3	6		4,5,7,15	
H$_2^{+}$[c]	H$_2$[d]	H[g]	2p	1–25	$\sigma_{ij}(R)$	Van Zyl et al. (149)	A	8-2-E	8-2-H	3,9,10		3			1,4,7,15,16	

TABLE 4-1 (Continued)

Proj.	Tar.	Exc.	State	Energy	Data	Reference	Orig.	Disc.	Tab. Data	Op. Def. σ_i–σ_{ij}	Op. Def. P_{ij}	N_B	N_A	N_i	J_{ij}	P_{ij}
H_2^+	H_2	H^e	3s	20–120	σ_{ij}	Hughes et al. (196)	G	9-6-B	9-6-E	7,10		3,6,7	6		7,15,16	
H_2^+	He	H	n=3	2–18	$\sigma_{ij}(R)$	Hanle et al. (199)	j	9-6-A	j	1,9,10		1,3,6	6		2,4,5,7	
H_2^+	He	H	n=3	130 & 200	σ_{ij}	Hughes (197)	j	9-6-A	j	9,10		3,6,7	6		2,4,5,7,10	
H_2^{+c}	He	H	2s,2p	0.25–25	$\sigma_{ij}(R)$	Jaecks et al. (195) Van Zyl et al. (194)	A	9-6-A	9-6-E	7,10		3	m		4,7,15,16	
H_2^+	He	H	3≤n≤7	10–30	σ_{ij}	Bobashev et al. (157, 158)	j	9-6-A	j	9,10		3	6		2,4,5,7	
H_2^+	He	He	$^1S,^1P,^1D$	1–150	σ_i	Van den Bos et al. (200)	T	6-4-A	6-4-C			3			2	
			$^3S,^3P,^3D$	1–150	σ_i		T	6-4-A	6-4-C			3			2	
			$^1P,^1D,^3P,^3D$	1–150	P_{ij}		T	6-4-B	6-4-C		3	3				2
H_2^+	He	He	^1S	200–800	σ_i	Thomas et al. (201)	A	6-4-A	6-4-C	7		3	6			
H_2^+	He	H	2p	3 & 6	P_{ij}	Gaily et al. (28)	T	9-6-D	9-6-D			3	n			
H_2^+	He	H	3s	20–120	σ_{ij}	Hughes et al. (196)	G	9-6-A	9-6-E	7,10		3,6,7	6		7,15,16	
H_2^+	He	He	^1D	120–1100	P_{ij}	Scharmann et al. (166)	G	6-4-B	6-4-C			3	n			
H_2^+	Hg	Hg	$^1S,^1P,^3S$	11–36	σ_{ij}	Bobashev et al. (168)	A	7-6-B	7-6-C	1		3			2,4,5	
	Hg^{2+}		$5d^8 6s^2$	11–36	σ_{ij}		A	7-6-B	7-6-C	1		3			2,4,5	
		H	n=4	5–35	σ_{ij}		j	9-6-C	j	1,9,10		3			2,4,5,7	
H_2^{+c}	Kr	H	2p	0.25–26	$\sigma_{ij}(R)$	Van Zyl et al. (194)	A	9-6-A	9-6-E	7,10		3			1,4,7,15,16	
H_2^+	N_2	H	2p	1–25	$\sigma_{ij}(R)$	Dunn et al. (99) Van Zyl et al. (149)	j	9-6-B	j	7,10		3	m		1,4,6,7,15	
H_2^+	N_2	N_2^+	$B^2\Sigma$	80–500	σ_{ij}	Dufay et al. (202)	j	8-3-E	j			3	6		2,4,5,6,10	
		N^+	^3F	10–600	σ_{ij}		A	8-3-E	8-3-G			3	6		1,2,4,5,6,10	
		H	n=4	70 & 100	σ_{ij}		j	9-6-B	j	9,10		3	6		2,4,5,7,10	
H_2^+	N_2	N_2^+	$B^2\Sigma$	400	$\sigma_{ij}(R)$	Thomas et al. (201)	T	8-3-E	8-3-E			3			4,5,6	
H_2^+	N_2	H	3s	100	σ_{ij}	Hughes et al. (196)	T	9-6-B	9-6-E	7,10		3,6,7	6		7,15,16	
H_2^+	N_2	N_2^+	$B^2\Sigma$	0.1–13.5	$\sigma_{ij}(R)$	Moore et al. (184)	ℓ	8-3-E	ℓ			3,5,6,7			4,5	
H_2^{+c}	Ne	H	2s,2p	0.3–25	$\sigma_{ij}(R)$	Van Zyl et al. (194) Jaecks et al. (195)	A	9-6-A	9-6-E	7,10		3	m		4,7,15,16	
H_2^+	Ne	H	3≤n≤7	10–30	σ_{ij}	Bobashev et al. (157, 158)	j	9-6-A	j	9,10		3	6		2,4,5,7	
H_2^+	Ne	H	3s	20–120	σ_{ij}	Hughes et al. (196)	G	9-6-A	9-6-E	7,10		3,6,7	6		7,15,16	
H_2^+	O_2	O^+	^2D	50–70	σ_{ij}	Dufay et al. (202)	T	8-4-C	8-4-C			3	6		1,2,4,5,6,10	
H_2^+		H	n=4	40–80	σ_{ij}		j	9-6-B	j	9,10		3	6		2,4,5,7,10	

TABLE 4-1 *(Continued)*

Proj.	Tar.	Exc.	State	Energy	Data	Reference	Orig.	Disc.	Tab. Data	Op. Def. $\sigma_i-\sigma_{ij}$	Op. Def. P_{ij}	N_B	N_A	N_i	J_{ij}	P_{ij}
H_2^+[c]	Xe	H	2s,2p	0.2–25	$\sigma_{ij}(R)$	Van Zyl et al. (194)	A	9-6-A	9-6-E	10		3			4,7,15,16	
						Jaecks et al. (195)										
H_2^+	Xe	Xe^+	$^4D,^4F$	1–10	σ_{ij}	Jaecks et al. (124)	A	7-5-C	7-5-F			3			2,4,5,10	
H_3^+	Ar	H	3s	20–120	σ_{ij}	Hughes et al. (196)	G	9-7-A	9-7-C	7,10		3,6,7	6		7,15,16	
H_3^+	H_2	H[g]	2p	0.6–5.5	$\sigma_{ij}(R)$	Dunn et al. (99)	A	8-2-E	8-2-H	3,9,10		3			1,4,7,15,16	
H_3^+	H_2	H[e]	$3\le n\le5$	200	σ_i	Hughes (197)	j	9-7-B	j	9,10		3,6,7	6		2,4,5,7,10	
H_3^+	H_2	H[f]	$3\le n\le5$	20–130	σ_{ij}	Hatfield et al. (198)	A	8-2-E	8-2-H	3		3,6,7	6		2,4,5,10	
H_3^+	H_2	H[e]	$3\le n\le5$	20–130	σ_{ij}		j	9-7-B	j	9,10		3,6,7	6		2,4,5,7,10	
H_3^+	He	He	3s	20–120	$\sigma_{ij}(R)$	Hughes et al. (196)	G	9-7-B	9-7-C	7,10		3,6,7	6		7,15,16	
H_3^+	He	H	n=3	2–18	$\sigma_{ij}(R)$	Hanle et al. (199)	j	9-7-A	j	1,9,10		1,3,6	6		2,4,5,7	
H_3^+	He	H	n=3	200	σ_{ij}	Hughes (197)	j	9-7-A	j	9,10		3,6,7	6		2,4,5,7,10	
H_3^+	He	H	2p	0.7–4	$\sigma_{ij}(R)$	Dunn et al. (99)	A	9-7-A	9-7-C	7,10		3	m		7,15	
H_3^+	He	H	$3\le n\le7$	10–30	σ_{ij}	Bobashev et al. (157, 158)	j	9-7-A	j	9,10		3	6		2,4,5,7	
H_3^+	He	He	$^1S,^1P,^1D$	1–150	σ_i	Van den Bos et al. (200)	T	6-4-A	6-4-C			3			2	
H_3^+			$^3S,^3P,^3D$	1–150	σ_i		T	6-4-A	6-4-C			3			2	
			$^1P,^1D,^3P,^3D$	1–150	P_{ij}		T	6-4-B	6-4-C		3					2
H_3^+	He	He	^1S	300 & 600	σ_i	Thomas et al. (201)	A	6-4-A	6-4-C	7			6		7,15,16	
H_3^+	He	H	3s	20–120	σ_{ij}	Hughes et al. (196)	G	9-7-A	9-7-C	7,10		3,6,7	6		1,4,6,7,15	
H_3^+	N_2	H	2p	0.5–4.5	$\sigma_{ij}(R)$	Dunn et al. (99)	j	9-7-B	j	7,10		3			4,5,6	
H_3^+	N_2	N_2^+	$B\,^2\Sigma$	600	$\sigma_{ij}(R)$	Thomas et al. (201)	T	8-3-E	8-3-E			3	m		2,4,5,7	
H_3^+	Ne	H	$3\le n\le7$	10–30	σ_{ij}	Bobashev et al. (157, 158)	j	9-7-A	j	9,10		3	6		7,15,16	
H_3^+	Ne	H	3s	20–120	σ_{ij}	Hughes et al. (196)	G	9-7-A	9-7-C	7,10		3,6,7	6		2,4,5,15	
He^+	Ar	Ar^+	^2S	5–35	σ_{ij}	Van Eck et al. (76)	A	7-3-D	7-3-A			3			2,4,5,15	
He^+	Ar	Ar	4s,4s'	0.3–10	σ_{ij}	De Heer et al. (203)	A	7-3-D	7-3-H			3			2,4,5,15,16	
He^+	Ar	Ar^+	$2s,^2P,^2D$	0.3–10	σ_{ij}		A	7-3-D	7-3-H			3			1,2,4,5,6,15,16	
He^+	Ar	Ar^+	$^2P^2,^2D^2,^2F$	0.3–35	σ_{ij}	Jaecks et al. (124)	A	7-3-D	7-3-H			3			2,4,5,10	
He^+	Ar	Ar^+	$^4P,^4D$	0.3–35	σ_{ij}		A	7-3-D	7-3-H			3			2,4,5,10	
He^+	Ar	Ar^+	^2F	150–325	σ_{ij}	Thomas et al. (71)	G	7-3-D	7-3-H			3	6		1,2,4,5,6	
He^+	Ar	Ar^+	^2P	0.015–0.5	$\sigma_{ij}(R)$	Lipeles et al. (204)	G	7-3-D	7-3-H				n		1,4,5,6,12,14	
He^+	Ar	Ar^+	Unspec.	0.05–0.3	$\sigma_{ij}(R)$	Schlumbohm (205)	j	7-3-D	j				m		1,4,5,6,15	

TABLE 4-1 (Continued)

Proj.	Tar.	Exc.	State	Energy	Data	Reference	Orig.	Disc.	Tab. Data	σ_i-σ_{ij}	P_{ij}	N_B	N_A	N_i	J_{ij}	P_{ij}
He$^+$	Ar	He	^3S	10–30	σ_i	Miers et al. (63)	j	10-2-A	j	7,10,12		1,3	6	1,2		
He$^+$	Ar	Ar$^+$	Unspec.	0.094	$\sigma_i(\theta)$	Champion et al. (206)	p	p	p	p		p	p	p		
He$^+$	Ar	Ar	Unspec.	0.5–3.0	$\sigma_i(\theta)$	Baudon et al. (207)	p	p	p	p		m	m	p		
He$^+$	CO	CO$^+$	^2F	0.1–1.0	$\sigma_{ij}(R)$	Lipeles (208)	j	8-5-A	j			m			1,4,5,6	
He$^+$	Cs	He	^3S	1.5–25	σ_i	Schlachter et al. (209)	j	10-2-A	j	7,10,12		1,3		1,2		
He$^+$	H	H	^2p	0.5–31	$\sigma_{ij}(R)$	Young et al. (9)	A	5-3-A	5-3-D	5,6,7,10					1,15,16	
He$^+$	H	H	^2p	1.9–25	P_{ij}		A	5-3-C	5-3-D		3					
He$^+$	H$_2$	H	n=3	2–18	$\sigma_{ij}(R)$	Hanle et al. (199)	j	8-2-F	j	1		1,3,6	6		2,4,5	
He$^+$	H$_2$	He	^3D	2–15	$\sigma_{ij}(R)$		j	10-2-B	j	1,10		1,3,6	6		1,2,4,5	
He$^+$	H$_2$	He	^1P	5–35	σ_i	De Heer et al. (78)	G	10-2-A	10-2-C	7,10		3	6		4,5	
He$^+$	H$_2$	H	$3 \leq n \leq 7$	5–35	σ_{ij}	Andreev et al. (210)	j	8-2-F	8-2-H	3		3	6			
He$^+$	H$_2$[d]	H	^2p	0.03–25	$\sigma_{ij}(R)$	Van Zyl et al. (149)	A	8-2-F	8-2-H	3		3,5	6		1,4,15,16	
He$^+$	H$_2$	H	2s,2p,3p	8–30	σ_{ij},σ_i	Andreev et al. (148)	G	8-2-F	8-2-H	3,7		3	6		4,5	
He$^+$	H$_2$	H	3s,3p,3d	10–30	σ_i	Ankudinov et al. (211)	G	8-2-F	8-2-H	3,7		3	6		2	
He$^+$	H$_2$	H	^2p	0.4–30	$\sigma_{ij}(R)$	Young et al. (9)	A	8-2-F	8-2-H	5,6,7		3			1,15,16	
He$^+$	H$_2$	He	$^1S,^3S$	80–250	σ_i	Gilbody et al. (61)	T	10-2-A	10-2-C	7		1,3		1,2		
He$^+$	H$_2$	He	^3S	10–30	σ_i	Miers et al. (63)	j	10-2-A	j	7,10,12		3,5			2,4,5,15	
He$^+$	H$_2$	H	n=3,4	0.2–30	$\sigma_{ij}(R)$	Gusev et al. (109)	A	8-2-F	8-2-H	3			6			
He$^+$	H$_2$		$^1S,^3S$	0.044–0.066	$\sigma_i(R)$	Utterback (65)	j	10-2-B	j	o		o	o		o	
He$^+$	He	He[e]	$^1S,^1P,^1D$	5–100	σ_{ij}	De Heer et al. (212)	T	6-5-A	6-5-D			3			1,2,10	
He$^+$	He	He[f]	$^3S,^3P,^3D$	5–100	σ_{ij}		T	6-5-A	6-5-D			3			1,2,10	
He$^+$	He		$^1P,^1D,^3P,^3D$	7.5–100	P_{ij}		T	6-5-B	6-5-D		3	3			o	2
He$^+$	He	He[e]	$^1S,^1P,^1D$ / $^3S,^3P,^3D$	5–90	σ_i,P_{ij}	De Heer et al. (12, 213)	A[i]	10-2-A	10-2-C	7,10		3	6		2,4,5,10	
He$^+$	He	He[e]	$^1S,^1P,^1D$ / $^3S,^3P,^3D$	20–120	σ_i	Head et al. (214)	A	10-2-A	10-2-C	7,10		3,6,7	6		4,5,7,10	
He$^+$	He	He[f]	^3S	0.6	$\sigma_i(\theta)$	Lorents et al. (58)	p	p	p	p		p	p			
He$^+$	He	He[g]	$^1S,^1P,^1D$	0.04–5	$\sigma_{ij}(R)$	Dworetsky et al. (35)	G	6-5-C	6-5-D	1,9,10		5,7	n		4,5,7	
He$^+$	He		$^3S,^3P,^3D$	0.04–5	$\sigma_{ij}(R)$		G	6-5-C	6-5-D	1,9,10		5,7	n		4,5,7	
He$^+$	He	He[f]	^3S	0.075–0.6	$\sigma_i(\theta)$	Coffey et al. (59)	p	p	p	p		p	p			

TABLE 4-1 (*Continued*)

Proj.	Tar.	Exc.	State	Energy	Data	Reference	Orig.	Disc.	Tab. Data	Op. Def. σ_i–σ_{ij}	Op. Def. P_{ij}	N_B	N_A	N_i	J_{ij}	P_{ij}
He$^+$	He	He^g	^{1}P,^{3}P	0.05–4	$\sigma_{ij}(R)$	Dworetsky et al. (215)	G	6-5-C	6-5-D	1,9,10		5,7	n		4,5,7	
He$^+$	He	He^e	^{3}S	10–30	σ_i	Miers et al. (63)	j	10-2-A	j	7,10,12		1,3	6	1,2		
He$^+$	He–	He^f	Unspec.	0.5–3.0	$\sigma_i(\theta)$	Baudon et al. (216)	p	p	p	p		p	p	p		
He$^+$	He	He^{+e}	Unspec.	0.5–3.0	$\sigma_i(\theta)$		p	p	p	p		p	p	p		
He$^+$	Kr	Kr$^+$	^{2}S	5–35	σ_{ij}	Van Eck et al. (76)	G	7-4-C	7-4-E			3			2,4,5,15	
He$^+$	Kr	He	^{1}P	10–35	σ_i	De Heer et al. (78)	G	10-2-A	10-2-C	7,10		3	6		2,4,5,10	
He$^+$	Kr	Kr$^+$	^{2}P,^{2}D,^{2}F	0.3–10	σ_{ij}	Jaecks et al. (124)	A	7-4-C	7-4-E	7,10		3			2,4,5,10	
He$^+$	Kr	Kr$^+$	^{4}P,^{4}D	0.3–10	σ_{ij}		A	7-4-C	7-4-E			3			2,4,5,10	
He$^+$	Kr	Kr$^+$	5s,5s'	0.3–10	σ_{ij}	De Heer et al. (203)	j	7-4-C	j	1		3			2,4,5,15,16	
He$^+$	Kr	Kr$^+$	^{2}S,^{2}P,^{2}D	0.3–10	σ_{ij}		A	7-4-C	7-4-E			3			2,4,5,6,15,16	
He$^+$	Kr	Kr$^+$	^{2}F,^{4}P	0.3–10	σ_{ij}		A	7-4-C	7-4-E			3			2,4,5,15,16	
He$^+$	Kr	Unspec.	Unspec.	0.05–0.5	σ_{ij}	Lipeles et al. (204)	j	7-4-C	j	m					1,4,5,6,12,14	
He$^+$	Kr	Kr$^+$	Unspec.	0.05–0.3	$\sigma_{ij}(R)$	Schlumbohm (205)	j	7-4-C	j				m		1,4,5,6,15	
He$^+$	Kr	Kr	Unspec.	0.5–3.0	$\sigma_i(\theta)$	Baudon et al. (207)	p	p	p	p		p	p	p		
He$^+$	Mg	He	n$\geq$8	30–180	σ_i	Il'in et al. (54)	k	10-1	k	7,12		3		1,2,4		
He$^+$	Mg	He	n$\geq$10	30	$\sigma_i(R)$	Solov'en et al. (217)	k	10-1	k	7,12		3		1,2,4		
He$^+$	N$_2$	N$_2^+$	B$^2\Sigma$	10–65	σ_i	Sheridan et al. (176)	j	8-3-E	j			3			2,3,4,5	
He$^+$	N$_2$	He	^{1}D,^{3}D	10–120	σ_i	Head et al. (214)	A	10-2-A	10-2-C	7,10		3,6,7	6		4,5,7,10	
He$^+$	N$_2$	N$_2^+$	B$^2\Sigma$	0.1–13.5	$\sigma_{ij}(R)$	Moore et al. (184)	ℓ	8-3-E	ℓ			3,5,6,7	6		4,5	
He$^+$	N$_2$	N$_2^+$	B$^2\Sigma$	0.5–37	$\sigma_{ij}(R)$	Polyakova et al. (110, 111)	ℓ	8-3-E	ℓ			1,3,5			2,4,5	
He$^+$	N$_2$	N$_2^+$	B$^2\Sigma$	1–30	$\sigma_{ij}(R)$	Polyakova et al. (112)	j	8-3-E	j			1,2,3,4	6		4,5,6,15	
He$^+$	Na	He	n$\geq$8	30–180	σ_i	Il'in et al. (54)	k	10-1	k	7,12		3		1,2,4		
He$^+$	Ne	Ne$^+$	3s	5–35	σ_{ij}	Van Eck et al. (76)	A	7-2-D	7-2-E			3			2,4,5,6,15	
He$^+$	Ne	Ne$^+$	2S	10–35	σ_{ij}		A	7-2-D	7-2-E			3			2,4,5,6,15	
He$^+$	Ne	He	^{1}S,^{1}P,^{1}D	5–35	σ_i,P$_{ij}$	De Heer et al. (78, 213)	G^i	10-2-A	10-2-C	7,10		3	6		2,4,5,10	2
He$^+$	Ne	Ne	3s	5–35	σ_{ij}	De Heer et al. (77)	j	7-2-D	j		3	3			2,4,5,6,15	
He$^+$	Ne	Ne$^+$	3p,3p',4d	5–35	σ_{ij},P$_{ij}$		G	7-2-D	7-2-E			3			2,4,5,10	
He$^+$	Ne	Ne$^+$	^{2}S,^{2}D	5–35	σ_{ij}		j	7-2-D	j			3			2,4,5,6,15	
He$^+$	Ne	Ne$^+$	^{4}P,^{2}P,^{2}D	5–35	σ_{ij}		G	7-2-D	7-2-E			3			2,4,5,10	
He$^+$	Ne	Unspec.	Unspec.	0.05–0.3	σ_{ij}	Lipeles et al. (204)	j	7-2-D	j	m			n		1,4,5,6,12,14	

TABLE 4-1 (Continued)

Proj.	Tar.	Exc.	State	Energy	Data	Reference	Orig.	Disc.	Tab. Data	Op. Def. σ_i-σ_{ij}	Op. Def. P_{ij}	N_B	N_A	N_i	J_{ij}	P_{ij}
He^+	Ne	Ne	$3p'$	0.3–30	σ_{ij}	Jaecks et al. (124)	A	7-2-D	7-2-E			3			2,4,5,10	
		Ne	$^2D, {}^2F$	0.3–30	σ_{ij}		A	7-2-D	7-2-E			3			2,4,5,6,10	
		Ne	$^4P, {}^4D, {}^4F$	0.3–30	σ_{ij}		A	7-2-D	7-2-E			3			2,4,5,10	
He^+	Ne	Ne	$3s$	0.3–6	σ_{ij}	De Heer et al. (203)	A	7-2-D	7-2-E			3			2,4,5,15,16	
		Ne	$3p'$	0.3–6	σ_{ij}		A	7-2-D	7-2-E			3			2,4,5,10	
He^+	Ne	He	^3P	0.024–6	$\sigma_{ij}(R)$	Tolk et al. (218)	G	10-2-A	10-2-C	9,10			n		2,4,5,16	
He^+	Ne	He	$n \geq 8$	30–180	σ_i	Il'in et al. (54)	k	10-1	k	7,12		3		1,2,4		
He^+	Ne	He	Unspec.	0.5–3.0	$\sigma_i(\theta)$	Baudon et al. (207)	p	p	p	p		p	p	p		
		Ne	Unspec.	0.5–3.0	$\sigma_i(\theta)$		p	p	p	p		p	p	p		
He^+	Ne	Ne	$3s$	0.026–0.6	$\sigma_i(\theta)$	Coffey et al. (219)	p	p	p	p		p	p	p		
He^+	O_2	He	^3D	20–120	σ_i	Head and Hughes (214)	A	10-2-A	10-2-C	7,10		3,6,7	6		4,5,7,10	
He^+	Xe	Unspec.	Unspec.	0.05–0.5	$\sigma_{ij}(R)$	Lipeles et al. (204)	j	7-5-D	j	m			n		1,4,5,6,12,14	
He^+	Xe	Xe^+	$^2P, {}^2D, {}^2F$	0.2–10	σ_{ij}	Jaecks et al. (124)	A	7-5-D	7-5-F			3			1,2,4,5,6,10	
		Xe^+	$^4P, {}^4D, {}^4F$	0.2–10	σ_{ij}		A	7-5-D	7-5-F			3			1,2,4,5,6,10	
He^+	Xe	Xe	$6s, 6s'$	0.2–10	σ_{ij}	De Heer et al. (203)	j	7-5-D	j	1		3			2,4,5,15,16	
		Xe^+	$^2S, {}^2P, {}^2D, {}^4P$	0.2–10	σ_{ij}		A	7-5-D	7-5-F			3			1,2,4,5,6,16	
He^+	Xe	He	^3S	10–30	σ_i	Miers et al. (63)	j	10-2-A	j	7,10,12		1,3	6	1,2	2,4,5,15	
He^+	Xe	Xe	Unspec.	0.5–3.0	$\sigma_i(\theta)$	Baudon et al. (207)	p	p	p	p		p	p	p		
He	H_2	H	$n=3,4$	0.2–30	$\sigma_{ij}(R)$	Gusev et al. (109)	A	8-2-F	8-2-H	3		3,5			2,4,5,15	
He	N_2	N_2^+	$B\,^2\Sigma$	1–30	$\sigma_{ij}(R)$	Polyakova et al. (112)	j	8-3-E	j			1,2,3,4	6		4,5,6,15	
He	N_2	N_2^+	$B\,^2\Sigma$	3–37	$\sigma_{ij}(R)$	Gusev et al. (111)	ℓ	8-3-E	ℓ			1,3,5			2,4,5	
He	Ne	Ne	$3p'$	10–35	σ_{ij}	De Heer et al. (203)	G	7-2-C	7-2-E			3,8			2,4,5,15,16	
K^+	Ar	K^+	$4s, 3d$	0.28	T_i	Matveyev et al. (115)	o	10-3-A	o	o		o	o	o		
K^+	He		^1P	0.228	T_i	Matveyev et al. (115)	T	6-6	6-6			o	o	o		
K^+	He	K^+	$4s, 3d, 4s'$	0.8	T_i		o	10-3-A	o	o		o	o	o		
K^+	He	K^+	$4s, 3d, 4s'$	0.8–9	$\sigma_{ij}(R)$	Matveyev et al. (220)	G	10-3-A	10-3-B	9,10		2	n		4,5	
K^+	Kr	K^+	$4s'$	0.29	T_i	Matveyev et al. (115)	o	10-3-A	o	o		o	o	o		
K^+	N_2	N_2^+	$B\,^2\Sigma$	0.4–2	$\sigma_{ij}(R)$	Neff (104)	A	8-3-E	8-3-G	10		3,7	6		1,2,4,5,6,15,16	
		K	^2P	0.4–2	$\sigma_{ij}(R)$		A	10-3-A	10-3-B			3,7	6		1,2,4,5,15,16	
K^+	N_2	N_2^+	$B\,^2\Sigma$	1–20	$\sigma_{ij}(R)$	Polyakova et al. (110)	ℓ	8-3-E	ℓ			5			2,4,5	

TABLE 4-1 (*Continued*)

Proj.	Tar.	Exc.	State	Energy	Data	Reference	Orig.	Disc.	Tab. Data	Op. Def. $\sigma_i-\sigma_{ij}$	Op. Def. P_{ij}	N_B	N_A	N_i	J_{ij}	P_{ij}
K$^+$	Ne	K$^+$	4s'	1.5–6	$\sigma_{ij}(R)$	Matveyev et al. (115)	G	10-3-A	10-3-B	9,10		2	n		4,5	
K$^+$	Xe	K$^+$	3d	0.56	T_i	Matveyev et al. (115)	o	10-3-A	o	o			o		o	
K	Ar	K	$4\,^2P$	0.025–0.6	$\sigma_{ij}(R)$	Anderson et al. (221)	j	10-3-A	j	1,2,10		3,8			1,4,6	
K	Cl$_2$	K	$4\,^2P$	0.025–0.6	$\sigma_{ij}(R)$	Anderson et al. (221)	j	10-3-A	j	1,2,10		3,8			1,4,6	
K	H$_2$	K	$4\,^2P$	0.025–0.6	$\sigma_{ij}(R)$	Anderson et al. (221)	j	10-3-A	j	1,2,10		3,8			1,4,6	
K	He	K	$4\,^2P$	0.025–0.6	$\sigma_{ij}(R)$	Anderson et al. (221)	j	10-3-A	j	1,2,10		3,8			1,4,6	
K	Kr	K	$4\,^2P$	0.025–0.6	$\sigma_{ij}(R)$	Anderson et al. (221)	j	10-3-A	j	1,2,10		3,8			1,4,6	
K	N$_2$	K	$4\,^2P$	0.025–0.6	$\sigma_{ij}(R)$	Anderson et al. (221)	j	10-3-A	j	1,2,10		3,8			1,4,6	
K	Ne	K	$4\,^2P$	0.025–0.6	$\sigma_{ij}(R)$	Anderson et al. (221)	j	10-3-A	j	1,2,10		3,8			1,4,6	
K	O$_2$	K	$4\,^2P$	0.025–0.6	$\sigma_{ij}(R)$	Anderson et al. (221)	j	10-3-A	j	1,2,10		3,8			1,4,6	
Kr$^+$	Ar	Kr$^+$	$^2F,\,^4P,\,^4D$	110–210	σ_{ij}	Thomas et al. (71)	A	10-3-A	10-3-B	9,10		3	6		1,2,4,5,6,7	
Kr$^+$	He	Kr$^+$	$^2F,\,^4P,\,^4D$	110–210	σ_{ij}	Thomas et al. (71)	A	10-3-A	10-3-B	9,10		3	6		1,2,4,5,6,7	
Kr$^+$	Kr	Kr^f	Unspec.	0.5–3.0	$\sigma_i(\theta)$	Barat et al. (106)	p	p	p	p		p	p	p		
Kr$^+$	Ne	Kr$^+$	$^2F,\,^4P,\,^4D$	110–210	σ_{ij}	Thomas et al. (71)	A	10-3-A	10-3-B	9,10		3	6		1,2,4,5,6,7	
Li$^+$	H$_2$	H	n=4	5–20	σ_{ij}	Van Eck et al. (222)	j	8-2-G	j	1,3		1,3	6		2,4,5,10	
Li$^+$	H$_2$	H$_2$	Unspec.	0.01–0.05	$\sigma_i(R)$	Schöttler et al. (223)	ℓ	8-2-G	ℓ	o		o	o	o		
Li$^+$	He	He	^1P	10–25	σ_{ij}	Van Eck et al. (222)	j	6-6	j	1		1,3	6		2,4,5,10	
Li$^+$	Li	Li^f	^2S	0.05–0.2	$\sigma_{ij}(\theta)$	Aberth et al. (224)	p	p	p	p		p	p	p		
Li$^+$	N$_2$	N$_2^+$	$B\,^2\Sigma$	0.7–5	$\sigma_{ij}(R)$	Polyakova et al. (110)	ℓ	8-3-E	ℓ	p		5	ℓ		2,4,5	
Li$^+$	N$_2$	N$_2^+$	Unspec.	0.360	$\sigma_i(\theta,R)$	Aberth et al. (225)	p	p	p	p		p	p	p		
Li$^+$	N$_2$	N$_2^+$	Unspec.	0.360	$\sigma_i(\theta,R)$	Aberth et al. (225)	p	p	p	p		p	p	p		
Li$^+$	O$_2$	O$_2^+$	Unspec.	0.328	$\sigma_i(\theta,R)$		p	p	p	p		p	p	p		
Li$^+$	O$_2$	O$_2^+$	Unspec.	0.328	$\sigma_i(\theta,R)$		p	p	p	p		p	p	p		
Mg$^+$	N$_2$	Mg	$^3S,\,^3D$	0.4–2.0	$\sigma_{ij}(R)$	Neff (104)	A	10-3-A	10-3-B	10		3,7	6	p	1,2,4,5,6,15,16	
Mg$^+$	N$_2$	Mg$^+$	^2F	0.4–2.0	$\sigma_{ij}(R)$		A	10-3-A	10-3-B	10		3,7	6	p	1,2,4,5,6,15,16	
N$^+$	Mg	N	n $\geq$ 10	105	$\sigma_i(R)$	Solov'en et al. (217)	k	10-1	k	7,12		3		1,2,4		
N$^+$	N$_2$	N$_2^+$	$B\,^2\Sigma$	10–65	$\sigma_{ij}(R)$	Sheridan et al. (176)	j	8-3-E	j			3	6		2,3,4,5	
N$^+$	N$_2$	N$_2^+$	$B\,^2\Sigma$	0.1–13.5	$\sigma_{ij}(R)$	Moore et al. (184)	ℓ	8-3-E	ℓ			3,5,6,7		p	4,5	
N$^+$	N$_2$	N$_2^+$	$B\,^2\Sigma$	32–2000	$\sigma_{ij}(R)$	Kurzweg et al. (113)	A	8-3-G	8-3-G			3		p	1,2,4,6	

TABLE 4-1 (Continued)

Proj.	Tar.	Exc.	State	Energy	Data	Reference	Orig.	Disc.	Tab. Data	Op. Def. $\sigma_i-\sigma_{ij}$	Op. Def. P_{ij}	N_B	N_A	N_i	J_{ij}	P_{ij}
N_2^+	Ar	N_2^+	$B\,^2\Sigma$	0.4–2	$\sigma_{ij}(R)$	Neff (104)	A	10-3-A	10-3-B	10		3,7	6		1,2,4,5,6,15,16	
N_2^+	CO	CO^+	$A\,^2\Pi$	0.011–0.3	$\sigma_{ij}(R)$	Utterback et al. (226)	A	8-5-A	8-5-D	1		6,7	6		1,4,5,6	
N_2^+	Mg	N	$n \geqslant 10$	≈150	$\sigma_i(R)$	Solov'en et al. (217)	k	11-1	k	7,12		3		1,2,4		
N_2^+	N_2	$N_2^{+}g$	$B\,^2\Sigma$	0.4–3	$\sigma_{ij}(R)$	Neff (104)	A	8-3-E	8-3-G			3,7	6		2,4,5,6,15,16	
N_2^+	N_2	$N_2^{+}g$	$B\,^2\Sigma$	1–10	$\sigma_{ij}(R)$	Doering (227)	A	8-3-E	8-3-G			3			15	
N_2^+	N_2	N_2^{g}	^3F	3–11	$\sigma_{ij}(R)$	Doering (227)	A	8-3-E	8-3-G			3			15	
N_2^+	N_2	$N_2^{+}g$	$B\,^2\Sigma$	28–2000	$\sigma_{ij}(R)$	Kurzweg et al. (113)	A	8-3-E	8-3-G	9		3			1,2,4,6	
N_2^+	N_2	$N_2^{+}g$	$B\,^2\Sigma$	1	$\sigma_{ij}(R)$	Doering (227)	T	8-3-E	8-3-E			3			15	
Na^+	N_2	N_2^+	$B\,^2\Sigma$	1–2	$\sigma_{ij}(R)$	Neff (104)	A	8-3-E	8-3-G			3,7	6		2,4,5,6,15,16	
Na^+	N_2	N^+	$^1D,^3F$	1–2	$\sigma_{ij}(R)$		A	8-3-E	8-3-G			3,7	6		1,2,4,5,6,15,16	
Na^+	N_2	Na	$^2P,^2D$	0.4–2	$\sigma_{ij}(R)$		A	10-3-A	10-3-B	10		3,7	6		1,2,4,5,15,16	
Na^+	N_2	N_2^+	$B\,^2\Sigma$	1–30	$\sigma_{ij}(R)$	Polyakova et al. (110)	ℓ	8-3-E	ℓ			5			2,4,5	
Ne^+	Ar	Ar^+	$^2F,^4P,^4D$	100–380	σ_{ij}	Thomas et al. (71)	G	7-3-E	7-3-H		3	3	6		1,2,4,5,6	
Ne^+	Ar	Ar^+	Unspec.	0.05–0.3	$\sigma_{ij}(R)$	Schlumbohm (205, 228)	j	7-3-E	j				m		1,4,5,6,15	
Ne^+	CO_2	CO_2^+	$A\,^2\Pi$	0.003–0.24	$\sigma_{ij}(R)$	Schlumbohm (229)	j	8-5-A	j			6,7	6,7		1,2,4,6,15	
Ne^+	CO_2	CO^+	$A\,^2\Pi$	0.004–0.28	$\sigma_{ij}(R)$		j	8-5-A	j			6,7	6,7		1,2,4,6,15	
Ne^+	H	H	2p	4.2–16.7	P_{ij}	Young et al. (9)	A	5-3-C	5-3-D		3				4,5	
Ne^+	H_2	H	$3 \leqslant n \leqslant 7$	5–35	σ_{ij}	Andreev et al. (210)	G	8-2-G	8-2-H	3		3	6			
Ne^+	H_2	H	n=3,4	0.2–30	$\sigma_{ij}(R)$	Gusev et al. (109)	A	8-2-G	8-2-H	3		3,5			2,4,5,15	
Ne^+	He	^3p	^3p	0.10–4.5	$\sigma_{ij}(R)$	Tolk et al. (218)	G	6-6	6-6	1		5,7	n		4,5,16	
Ne^+	He	He	^3D	100–350	$\sigma_{ij}(R)$	Thomas et al. (71)	A	6-6	6-6			3	6		2,4,5	
Ne^+	He	He	n=4	150–390	$\sigma_{ij}(R)$		A	6-6	6-6			3	6		2,4,5	
Ne^+	He	He^+	Unspec.	0.5–3.0	$\sigma_i(\theta)$	Baudon et al. (207)	p	p	p	p		p	p	p		
Ne^+	He	He^+	Unspec.	0.5–3.0	$\sigma_i(\theta)$		p	p	p	p		p	p	p		
Ne^+	Kr	Unspec.	Unspec.	0.05–0.3	$\sigma_i(R)$	Schlumbohm (110)	j	7-4-E	j				m		1,4,5,6,15	
Ne^+	Mg	Ne	$n \geqslant 10$	150	$\sigma_i(R)$	Solov'en et al. (217)	k	10-1	k	7,12		3	6	1,2,4		
Ne^+	N_2	N_2^+	$B\,^2\Sigma$	0.4–2	$\sigma_{ij}(R)$	Neff (104)	A	8-3-E	8-3-G			3	6		2,4,5,6,15,16	
Ne^+	N_2	N_2^+	$^1D,^3P,^3D,^3F$	0.4–2	$\sigma_{ij}(R)$		A	8-3-E	8-3-G			3	6		1,2,4,5,6,15,16	
Ne^+	Ne	Ne	3p	0.4–2	$\sigma_{ij}(R)$		A	10-3-A	10-3-B	10		3,7	6		1,2,4,5,15,16	
Ne^+	N_2	N_2^+	$B\,^2\Sigma$	10–65	$\sigma_{ij}(R)$	Sheridan et al. (176)	j	8-3-E	j			3	6		2,3,4,5	
Ne^+															1,2,4,6,15,16	

TABLE 4-1 (*Continued*)

Proj.	Tar.	Exc.	State	Energy	Data	Reference	Orig.	Disc.	Tab. Data	Op. Def. $\sigma_i-\sigma_{ij}$	Op. Def. P_{ij}	N_B	N_A	N_i	J_{ij}	P_{ij}
Ne$^+$	N$_2$	N$_2^+$	B$^2\Sigma$	1–37	$\sigma_{ij}(R)$	Polyakova et al. (110, 111)	ℓ	8-3-E	ℓ			1,3,5			2,4,5	
Ne$^+$	N$_2$	N$_2^+$	B$^2\Sigma$	0.1–13.5	$\sigma_{ij}(R)$	Moore et al. (184)	ℓ	8-3-E	ℓ			3,6,5,7			4,5	
Ne$^+$	N$_2$	N$_2^+$	B$^2\Sigma$	1–30	$\sigma_{ij}(R)$	Polyakova et al. (112)	j	8-3-E	j			1,2,3,4	6		4,5,6,15	
Ne$^+$	Ne	Ne^f	Unspec.	0.5–3.0	$\sigma_i(\theta)$	Barat et al. (106)	p	p	p			p	p	p		
Ne$^+$	O$_2$	O$_2^+$	b$^4\Sigma$	0.005–0.237	$\sigma_{ij}(R)$	Schlumbohm (229)	j	8-4-D	j			p	6	p	1,2,4,6,15,16	
Ne	H$_2$	H	n=3,4	0.2–30	$\sigma_{ij}(R)$	Gusev et al. (109)	A	8-2-G	8-2-H	3		3,5	6		2,4,5,15	
Ne	N$_2$	N$_2^+$	B$^2\Sigma$	3–37	$\sigma_{ij}(R)$	Gusev et al. (111)	ℓ	8-3-E	ℓ			1,3,5			2,4,5	
Ne	N$_2$	N$_2^+$	B$^2\Sigma$	1–30	$\sigma_{ij}(R)$	Polyakova et al. (112)	j	8-3-E	j			1,2,3,4	6		4,5,6,15	
O$^+$	Mg	O	n $\geqslant$ 10	120	$\sigma_i(R)$	Solov'en et al. (217)	k	10-1	k	7,12		3		1,2,4		
O$^+$	N$_2$	N$_2^+$	B$^2\Sigma$	30–2000	$\sigma_{ij}(R)$	Kurzweg et al. (113)	A	8-3-E	8-3-G			3			1,2,4,6	
O$_2^+$	Mg	O	n $\geqslant$ 10	$\approx$150	$\sigma_i(R)$	Solov'en et al. (217)	k	10-1	k	7,12		3		1,2,4		
Rb$^+$	He	Rb$^+$	5s,4d,5s'	1.8$_{\div}$2	T$_i$	Matveyev et al. (115)	o	10-3-A	o	o		o	o		0	
Xe$^+$	N$_2$	N$_2^+$	B$^2\Sigma$	274–1810	$\sigma_{ij}(R)$	Kurzweg et al. (113)	A	8-3-E	8-3-G			3			1,2,4,6	

CHAPTER 5

EXPERIMENTS INVOLVING ATOMIC HYDROGEN TARGETS: TARGET EXCITATION AND CHARGE TRANSFER EXCITATION OF PROJECTILES

5-1 INTRODUCTION

Two excitation mechanisms are considered in this chapter: the process of direct excitation of an atomic hydrogen target by a projectile, and the formation of excited atomic hydrogen by charge transfer as protons traverse the atomic hydrogen target. In the first mechanism,

$$X^+ + H \rightarrow [X^+] + H^*, \qquad (5\text{-}1)$$

the square brackets indicate that observation of the target excitation gives no information on the postcollision state of excitation nor ionization of the projectile. Experiments of this type are due to Young, Stebbings, and co-workers[8, 9] and also to Kauppila et al.[141] Most of the work involved protons as the projectiles, so there was no ambiguity as to the postcollision state of excitation nor ionization. There is also a limited series of measurements on excitation by beams of He^+, Ne^+, and Ar^+. In the second type of process,

$$H^+ + H \rightarrow H^* + H^+, \qquad (5\text{-}2)$$

there is no ambiguity as to the postcollision states of excitation nor to the ionization of the colliding systems. Work on this latter process has been carried out by Stebbings et al.,[8] Ryding et al.,[142] and Bayfield.[231,232,143] Although the processes of direct excitation and of charge transfer are inherently different, there are definite similarities in experimental procedures; in one case, an experiment was designed to study both processes simultaneously.

There are also a number of experiments on the formation of excited atomic hydrogen by impact on fast H atoms on various neutral target gases, a process which may be represented by the equation,

$$H + X \rightarrow H^* + X. \qquad (5\text{-}3)$$

Although the mechanism is very similar to the direct excitation process described by Eq. 5-1, the details of the available experiments are quite different since the published work involves only stable targets. Processes of this type are considered in Chapter 9.

Measurements on the formation of excited atomic hydrogen as a result of the impact of H^+ on H have a fundamental importance. The wave functions of the post and priorcollision states are known exactly, and a comparison of theory with experiment provides a direct test of the validity of the approximations made in the theoretical predictions. Studies of excitation induced by collisions between more complex atomic structures are of less value for this purpose since a discrepancy between theory and experiment cannot be readily attributed to inadequacies of the theoretical formulation; such discrepancies may be attributable to errors in the wave functions.

There are numerous experimental difficulties in the pursuit of problems involving atomic hydrogen targets because hydrogen is not stable in the atomic form under normal conditions of temperature. The conventional experimental technique for producing atomic hydrogen is to use the catalytic dissociation of the molecular species on a hot tungsten surface; the hydrogen gas is heated in a tungsten furnace to a temperature of around 2700°K; the projectiles are either directed across the furnace, giving rise to a "beam-gas" type of experimental geometry, or the atomic hydrogen may be allowed to effuse from the source in the form of a beam, which is traversed by the projectiles in a crossed-beam configuration. All the experiments involving the "beam-gas" configuration have been designed to sample the metastable atom content of the fast beam emerging from the exit of the cell. The mestastable flux is determined by a conventional technique: the field-induced mixing of the $2s-$ and $2p$-levels with the detection of the resulting emitted Lyman-alpha photon. Any attempt to monitor spontaneous emission from the atomic target itself would probably be frustrated by large background signals from the incandescent walls of the furnace, excitation of the target gas, and emission caused by fast projectiles and secondary electrons striking surfaces. The "beam-gas" configuration has, therefore, been applied only to the charge transfer processes described by Eq. 5-2. Whereas, with the crossed-beam configuration it is possible to detect light emitted spontaneously from the interaction region or to sample the metastable excited-state content of either beam after it has left the region of interaction. This technique has been applied to the determination of the cross sections for production of $2p$-excited states both in the target atomic hydrogen beam by direct excitation and also in the projectile beam by charge transfer.

5-2   EXPERIMENTAL TECHNIQUES

A.   General Considerations

Typically, in a tungsten furnace some 90% of the contents undergoes catalytic dissociation at a temperature of 2700°K. It is necessary that the remaining fraction of molecular hydrogen be measured, and the contribution to the signal from the

interaction of projectiles with the H_2 must be assessed. Each experiment described here includes a technique for this purpose.

For any collisional excitation experiment, the major sources of inaccuracy in the experimental procedures are the determination of target density and the efficiency of excited-state detection. These problems will affect the absolute magnitudes of the data but will not influence the relative variation of cross section with impact energy. These same generalities also hold in the case of experiments carried out with atomic hydrogen targets. It is very difficult to determine accurately and absolutely the density of the atomic hydrogen in either the target furnace or in the target beam. The measurement of target pressure in the furnace itself is out of the question because of the high temperature of operation; the measurement of density in a target beam is hampered because there is no method for absolute determination of a neutral atomic hydrogen flux. In any experiment the determination of absolute photon detection efficiency is a very difficult procedure. Because of these difficulties the various experiments that utilize atomic hydrogen targets have in fact concentrated on the determination of relative cross sections. Absolute magnitudes have been assigned either by normalization to theoretical predictions or by normalization to some previous cross section measurements. None of the available experiments is inherently absolute.

Some problems, which are peculiar to this type of experiment, arise from using, within the apparatus, a furnace heated by a current through its walls. The magnetic field produced by this very large heating current influences the measurements, particularly through the quenching of the metastable H atoms. The high temperature of the furnace will cause substantial emission of electrons and photons that may contribute to the background problems. Proper attention to the elimination of spurious signals is to be expected in any accurate experimental measurement.

B. Experimental Arrangements

1. The Work of Stebbings et al.   Stebbings, Young, and co-workers have studied the emission of Lyman-alpha radiation $(2p \rightarrow 1s)$ induced by the impact of H^+ on H[8,9] and He^+ on H.[9] This work included measurements of charge transfer into the excited state (Eq. 5-2) and excitation of the target (Eq. 5-1). Studies were also made of the polarization of the Lyman-alpha emission induced by the impact of He^+, Ne^+, and Ar^+ on atomic hydrogen.[9] The apparatus, based on the crossed-beam configuration, was described first by Stebbings et al.[8] and remained essentially unchanged for the later work by Young et al.[9]

The handling system for the atomic hydrogen beam was contained in three large vacuum tanks. In the first tank was located the furnace, operating at a temperature of $2700°K$. The beam of hydrogen issued through an aperture in the wall of the furnace, traversed the second vacuum chamber, which acted as a vacuum buffer, and entered the third chamber, which constituted the observation region. In the third chamber, the flux of the neutral beam and the fractional dissociation were monitored; the beam was crossed with electrons, and the H^+ and H_2^+ currents were separately measured with a mass spectrometer. To determine the dissociation fraction it was necessary to utilize the ratios of cross sections for the electron-impact-ionization of H and H_2 as determined by Fite and Brackmann[98] in a previous crossed-beam experiment.

The projectile ions were extracted from a duoplasmatron source, mass analyzed, collimated, and directed to cross the neutral beam at an angle of 90°. Beam flux was determined by monitoring the current on a conventional Faraday cup. When using beams of He^+, Ne^+, and Ar^+, no attempt was made to assess the possible influence on the measurements of long-lived and metastable states that might exist in the projectile beams.

The experiment was designed to measure spontaneously emitted Lyman-alpha radiation. The detector was an iodine-filled Geiger counter, sensitive to radiation within the range 1080 to 1260-Å, and arranged to observe light emitted from the region where the two beams intersected. In front of the detector was a cell containing dry molecular oxygen that acted as a wavelength selective filter. Oxygen has seven narrow transmission windows one of which happens to coincide with the wavelengths of the Lyman-alpha line. The other windows do not coincide with the wavelength of photons that can be produced in collisions of H^+, Ne^+, or Ar^+ with H. The 1215-Å line of He^+ $(4 \rightarrow 2)$ may be excited in the $He^+ + H$ experiments and will coincide exactly with the transmission band of the filter. Young et al. did not take any account of spurious signals that might be produced in this manner. It is likely that the error will be small; the cross section for exciting the $n = 4$ state will not be large; moreover, there will be a tendency for the projectile to drift downstream from the observation region before the emission of the photon takes place.

The signals due to the collision of the neutral beam with residual gases in the vacuum system constituted significant background. This was removed by chopping the neutral beam with a rotating-toothed wheel and identifying the true signal in the detector output by its specific frequency and phase. There is a danger that a modulation technique will introduce background signals that are coherent with the chopper. There is a flow of molecular hydrogen gas from the furnace chamber to the observation chamber, due to the pressure differentials between these regions; this flow will also be modulated by the chopper. There is a danger that the background pressure of the apparatus, which depends on the flow rate of the neutral beam into the observation region, will also be modulated. These two problems were taken into account by the techniques developed earlier by Fite and Brackmann in a similar crossed-beam experiment[98] and constitute only a small uncertainty in the measurement.

For all determinations of emission cross sections induced by the impact of He^+ on a target of atomic hydrogen[9] the photon detector was placed either at an angle of 54.5 or 125.5° to the direction of the projectile beam. At these angles the intensity of the emission is independent of anisotropy in the angular distribution of emission that is associated with polarization [see Section 3-4-D(2)].

Polarization measurements were made for the emission induced by the impact of He^+, Ne^+, and Ar^+ on H.[9] The intensity of emission was measured at the angles of 50° and 90° to the direction of the fast projectiles; the polarization fraction was determined through Eq. 2-27—the prediction of the variation of intensity with angle. Measurements of polarization involve only the ratios of light intensities; therefore the absolute efficiency of the detector is unimportant.

For the study of collisionally induced emission in the $H^+ + H$ problem, there is a complication in that the Lyman-alpha photons are emitted both by the target, as a result of direct excitation, and also by the projectile, as a result of charge transfer

into the excited state. These two contributions must be separated by the Doppler shift of emission from the fast projectile. In the Stebbings et al.[8,9] experiments observations were made at an angle of 90° to the projectile beam to determine the sum of emissions from projectile and target. At an angle of 54.5° the emission from the fast projectile is shifted in wavelength so that it is absorbed in the oxygen filter; consequently the detector response is primarily due to direct excitation of the slow target beam. The difference between the signals at 90° and 54.5° gives the measurement of capture into the excited state. Clearly, the measurements at 90° will include a distortion due to anisotropy of emission associated with polarization, whereas those at 54.5° will not. Allowance was made by Stebbings et al. for anisotropy by use of a theoretical prediction of polarization. This is unsatisfactory since there is no evidence that the predictions are correct; however, since the correction for the effect of polarization never amounted to more than 10%, the influence of any error on the measured cross section will be quite small.

An error arose in the use of Doppler shift to separate target and projectile emission in the $H^+ + H$ experiment. At low energies the Doppler shift is insufficient to carry the wavelength of projectile emission outside the transmission band of the molecular oxygen filter. Stebbings and his co-workers evaluated a correction, for the remaining component of projectile emission, based on the absorption coefficients measured by Watanabe.[233] The pressure of oxygen in the filter was higher than that at which Watanabe had provided data. Stebbings et al. predicted the absorption of their filter on the assumption that absorption coefficients for all wavelengths close to the Lyman-alpha line increased with pressure at a rate that could be extrapolated from Watanabe's low pressure work at 1215.67-Å. Recently Gaily[234] has shown that the variation of absorption coefficients with pressure is very dependent on the precise value of wavelength. Thus, the assumptions made by Stebbings et al. to arrive at the absorption profile of their filter are incorrect. In their most recent paper, Young, Stebbings, and McGowan[9] have taken this problem into account and published revised values for proton impact excitation of the $2p$ state in the atomic target. The correction above 10-keV impact energy is within the random error of the determinations; at 1.0-keV it amounts to 25%. The corrected data from the recent publication[9] are used here. The revised values for the charge transfer into the $2p$ state in the same collision combination are not yet published; in this monograph corrected values obtained by private communication are displayed.[140]

The excited $2p$ states formed in the fast projectile beam will have a sufficiently long lifetime to cause an appreciable proportion to be lost from the field of view before emission takes place. A correction up to 26% at high impact energies was evaluated using the theoretical value of $2p$ state lifetime. The cross section data were all determined relative to the emission cross section at 10-keV impact energy. This in turn was normalized to the cross section for excitation of H to the $2p$ state by impact of electrons of 300-eV energy. An electron gun was used to cross the neutral beam at right angles at a point downstream from the intersection of the ion and neutral beams. The photon detector was moved to view this second intersection region. A correction was made for the change in density of the neutral beam between the two points on the assumption that the flow from the furnace was effusive, and consequently the product of the length of the projectile's path through the beam and the beam density dropped off inversely with distance from the furnace. No evidence

was presented to confirm this behavior of the beam. The cross section for the spontaneous emission of Lyman-alpha induced by the impact of electrons on atomic hydrogen was taken to be 0.37 πa_0^2 at the electron energy of 300-eV. This value for the electron impact cross section, obtained from the work of Fite and Brackmann,[98] is essentially the theoretical value of the excitation cross section predicted by the first Born approximation.[235] It should be noted that the experimental values include cascade contributions, whereas the theoretical value, to which the data are normalized, does not contain cascade.

There is a danger that stray fields in the apparatus will cause field quenching of the metastable $2s$ state to the $2p$ state and thus enhance the intensity of the Lyman-alpha emission. It was shown that such effects produced a negligible contribution to the signal.

Throughout these experiments no attempt was made to determine the form factor that describes the density distributions of the two interacting beams (see Section 2-1-C); no arguments were presented that would justify neglecting it. The gas effused from an aperture in the furnace, and a small part of the angular distribution was selected by the collimation system to form the target beam. Provided that the flow is isotropic for the small range of angles selected, the beam density will be uniform, although decreasing with distance from the furnace. Under these circumstances knowledge of the form factor was probably not important.

It was estimated by Stebbings et al. that the errors in the relative variation of cross section with energy did not amount to more than $\pm 15\%$ with respect to the reference energy of 10-keV. The error in normalization at the energy of 10-keV was estimated to be less than $\pm 20\%$.[9] The data included no correction for cascade population of the excited states.

2. The Work of Ryding et al.   Ryding, Wittkower, and Gilbody[142] have studied the formation of the $2s$ state by charge transfer as protons are incident on an atomic hydrogen target. The experimental arrangement was of the "beam-gas" configuration.

The mass-analyzed proton beam of energy 40—200-keV was directed along the axis of the cylindrical tungsten furnace. Target temperature was 2600°K. The projectile flux was monitored with a secondary emission detector whose response had been shown to be the same for protons and neutral hydrogen atoms of the same velocity.[236]

Metastable atoms were monitored by the conventional technique of field quenching to the $2p$ state and measuring the Lyman-alpha emission that resulted. The detector was an electron multiplier with a LiF window that responded to all wavelengths from 1050 to 1300-Å. The detection region was located downstream from the furnace at a point where most excited states had decayed by spontaneous emission and where the beam was composed of metastable and ground state hydrogen atoms. It was therefore unnecessary to provide the detector with a filter to isolate the Lyman-alpha line from other emissions. The signal represents the rate of formation of the $2s$ state by direct excitation plus all cascade populations of that state. A correction was made for that small fraction of excited atoms that traversed the observation region without decaying. It was assumed that the detection efficiency was independent of projectile velocity; since the spatial distribution of emission will be dependent on velocity, this assumption requires that the detector response is uniform over the field of view. It was not demonstrated that this requirement was fulfilled, and it is not possible

in retrospect to assess whether this might influence the accuracy of the data.

Some quenching of the metastables was caused by the magnetic field associated with the current passing through the furnace. A correction was evaluated by using argon at constant density in the furnace; the decrease of Lyman-alpha count with increase in furnace temperature directly indicates the loss of signal due to field quenching.

Measurements obtained from this experiment were in the form of the ratio of the cross sections for the formation of fast metastables by H^+ impact on H to that for the formation of the same state due to H^+ impact on H_2. Experimentally, this ratio was expressed in terms of the signal from the metastable atom detector when using a furnace at high temperatures, where it contains both H and H_2, and a signal at low temperatures, where the target is entirely H_2. The measurement required corrections for the incomplete dissociation of H_2 in the hot target and also corrections for the difference between the density of the atomic H in the hot target and the density of the molecular H_2 in the cold target. Their correction for residual H_2 was determined by measuring the ratio of the fluxes of neutral Ar formed by two electron-charge-transfers as a beam of Ar^{2+} traversed the target under the hot and cold conditions. Under single-collision conditions, neutral Ar can be produced only by charge transfer on a target particle that has two electrons, namely H_2; the ratio of the fluxes is directly proportional to the ratio of H_2 densities for these two operating conditions. In the work being considered the ratio of the densities of atomic H with the hot furnace and molecular H_2 with the cold furnace was determined in an indirect manner. A previous experiment[236] in the same apparatus had studied the ratio of the total cross sections for charge transfer into all states of the fast atomic particle for impact of H^+ on H and H^+ impact on H_2. By taking the ratio of these two cross sections to be 0.48 at 120-keV impact energy, a value consistent with two independent determinations, it was possible for Ryding et al. to determine the ratio of the two densities.

As stated above, the output data from this experiment were in the form of the ratio of cross sections for formation of H(2s) by impact of H^+ on targets of H and of H_2. The experiment was also used to measure directly the relative variation with impact energy of the cross section for the formation of metastables by charge transfer of H^+ on H_2; this is simply the relative variation of signal with impact energy for a constant H_2 target density at room temperature. From these two sets of data it has been possible to determine the relative variation of the cross section for the formation of metastable H by impact of H^+ on H with energy.

The cross sections were placed on an absolute basis by normalization to Mapleton's[237] predictions of the cross section for formation of H(2s) by charge transfer as a result of the impact of 100-keV H^+ on H. Here the data are presented in the form of the cross section determined by the above procedures. Data on the formation of metastable H by impact of H^+ on H_2 have been separately assessed in Chapter 9.

The accuracy of data from this experiment is very dependent on the validity of the various normalization procedures utilized to determine the magnitude of the cross section. Measurements of the relative variation with energy of the ratio of the cross sections for charge transfer formation of H(2s) by impact of H^+ on H and H_2 is valid, independent of any of the assumptions concerning normalization. It should be appreciated that this ratio represents the total cross sections for the formation

of the 2s state by all processes including cascade population. The relative variation with energy of the cross section for forming H(2s) by impact of H^+ on H_2 is determined directly. It follows that the derived values of the relative variation of cross section in the $H^+ + H$ collision with impact energy will also be valid. Any error in the previous measurements of total cross sections that are used to establish relative target densities for the hot and cold furnace will influence the absolute magnitude of the cross section ratio as well as influence the absolute magnitude of the final cross section. Similarly, the validity of normalization to Mapleton's theoretical predictions will influence only the accuracy of the absolute numbers. In the latter connection, it should be noted that Mapleton's predictions are cross sections for the formation of the 2s state by direct charge transfer, whereas the Ryding experiment measured the sum of all populating processes including cascade.

In conclusion, the relative variation of their cross section data with impact energy is a valid measurement independent of the accuracy of the various normalization procedures, whereas the absolute magnitude of the cross section involves the use of theoretical and experimental data from other sources that might prove to be of doubtful accuracy. The relative values of the cross sections determined by this method were estimated to be accurate to within $\pm 13\%$. The limits of accuracy for the absolute values will of course be directly dependent on the validity of the various normalization procedures and cannot be assessed in an objective manner.

3. The Work of Bayfield. Bayfield has determined the cross section for the formation of the metastable state of hydrogen by charge transfer of protons on an atomic hydrogen target at energies from 3—70-keV. Preliminary reports of this work[231,232] were superseded and much extended by a more complete publication.[143] Many of the experimental details of detection systems were described earlier in a paper dealing with charge transfer into the 2s state induced by $H^+ + H_2$ and $H^+ + Ar$ collisions.[130] Details of the target region have been discussed separately.[238]

The projectile beam was produced in a radio frequency source, accelerated, mass analyzed, and finely collimated to have an angular divergence of less than 0.09°. The projectiles traversed the tungsten furnace target along its axis. Two and one-half meters downstream from the interaction region the metastable flux was monitored by detection of Lyman-alpha photons induced by field quenching. The flight distance between target and observation region was sufficient to allow most excited states to decay either into the ground or into metastable states by the time the projectile entered the metastable detector. Therefore, the signal represented the formation of the metastable state by direct excitation and all relevant cascade processes.

The cylindrical furnace was operated at a temperature of 2400°K and achieved a 96% dissociation fraction.[238] Appreciable quenching of the metastable atoms by the magnetic field of the furnace was observed. The influence of this problem on the data was removed by gating both the furnace and projectile beam on and off. For approximately 40% of the operating time the furnace was gated off; during this period the projectile beam was gated on and measurements were made; the repetition rate of the cycle was ten per second. Successful removal of the quenching effect was demonstrated by showing that the observed metastable signal with argon in the target was independent of target temperature.

Care was taken to ensure a negligible loss of metastable atoms between the target and detection regions. Spurious fields that might quench the metastables were

eliminated; particular attention was directed at removing the influence of the photon detector's focusing magnets. Measurements of angular distribution of the metastables showed that for low projectile energies an appreciable fraction of the excited projectiles were intercepted by the exit aperture of the target cell; a small correction was evaluated for this loss.

The experimental procedure required measurement of the rate of gas flow into the furnace and the dissociation fraction within the hot target; there was no need for a direct determination of target density. The undissociated H_2 target gas was supplied from an isolated reservoir, the volume of which was decreased by a piston arrangement at a rate that kept supply pressure constant. The rate of gas flow into the target was proportional to the rate of change of reservoir volume.[130] The determination of dissociation fraction was based on a method by Lockwood et al.[43] in which the flux of fast H^- formed as H^+ traverses the target is used as a measure of the density of residual H_2. Under single-collision conditions H^- may be formed only by two electron charge transfer on an H_2 molecule, a process for which the cross section is known. The dissociation fraction is evaluated from the measurement of H^- formation and knowledge of the rate at which H_2 is supplied to the target cell.[238]

Measurements were made, at a particular projectile energy, of the metastable signals that were produced with two targets, hydrogen at elevated temperatures and argon. The ratio of the two signals, each divided by the rate of gas flow into the apparatus, was related to the cross sections for charge transfer on H, on H_2, and on Ar as well as to the dissociation fraction existing in the hydrogen target. Cross sections for charge transfer into the metastable state on H_2 and Ar had been determined by Bayfield in a previous experiment,[130] which will be reviewed in Section 9-9-A(12). From these data the charge transfer cross section on H was evaluated.

It is valuable to consider briefly the important features of Bayfield's experiment that provided the H_2 and Ar cross sections to which the present work with a H target is normalized. Except for the fact that the target was at room temperatures, the apparatus was exactly the same as that used for atomic hydrogen. The target thickness was determined from a measurement of the fractional neutralization of an H^+ beam, and knowledge of total charge transfer cross sections from other experimental work; efficiency of the metastable detector was estimated directly. Errors that might influence the dependence of cross section on energy were generally less than $\pm 15\%$; systematic errors in the absolute values were estimated to be less than $\pm 55\%$. In the overall procedure, the major error was the neglect of anisotropy related to polarization that occurs in emission induced by field quenching.[239,240,20] Bayfield's work is in excellent agreement with data by two other groups who utilized essentially different methods [see Sections 8-3-C(1) and 9-3-F(1)]. It is concluded that the basis for the normalization of the data by Bayfield for the H target, as provided by his own independent experiments, is both valid and well substantiated by the work of other groups.

In the case of the atomic hydrogen target, measurements of relative variation of cross section with impact energy are believed to be accurate to within $\pm 10\%$ above 10-keV and $\pm 15\%$ below 10-keV. Absolute magnitudes are no more accurate than the data to which they are normalized, $\pm 55\%$. Neglect of the anisotropy of the field-induced emission causes these data to be too small by approximately 10%.

4. The Work of Kauppila et al. Kauppila et al. have presented preliminary measurements on the polarization of the spontaneously emitted Lyman-alpha radiation induced through excitation of atomic hydrogen by proton impact.[141] Proper assessment of the validity of these data derived from this work is not possible because of an inadequate report of it. The experimental arrangement, based on the crossed-beam configuration, utilized atomic hydrogen formed by effusion from a tungsten furnace. Light from the interaction region was reflected by a LiF plate into the photon detector—an iodine vapor-filled Geiger counter preceded by a molecular oxygen filter. The reflector was arranged so that light was incident on it at the Brewster's angle; and by rotation of the counter and mirror about the observation axis it was possible to separately monitor light polarized parallel and perpendicular to the direction of the primary projectile.

It was necessary to eliminate the emissions from the excited projectiles produced by charge transfer. By observing at an angle of 54.5° to the direction of the projectile beam and carefully adjusting the pressure of the oxygen filter, it was possible to insure that the narrow band of the filter transmitted only some 2% of the Doppler-shifted projectile emission. This figure was checked by observing Lyman-alpha radiation induced in a $H^+ + Ar$ collision in which all the relevant emission must be from the projectile and therefore exhibits Doppler shift.

The intensity of radiation polarized parallel to the projectile beam path, I_p, and the component polarized perpendicular to the beam path, I_s, were monitored at an angle of 54.5° to the beam direction. The polarization fraction, P, is defined in terms of $I^{\parallel}$ and $I^{\perp}$—the intensities of emission polarized parallel and perpendicular to the projectiles beam's direction when observed at an angle of 90°

$$P = \frac{I^{\parallel} - I^{\perp}}{I^{\parallel} + I^{\perp}}. \qquad (5\text{-}4)$$

It may be shown that the intensities of emission measured at the angle of 54.5° are related to the polarization fraction as follows:

$$P = \frac{I_p - I_s}{I_p + I_s/3}. \qquad (5\text{-}5)$$

This expression may be written in terms of ratios of I_p and I_s so that the determination of polarization fraction requires no knowledge of the absolute efficiency of the detector, nor the densities of either the target or the projectile beams. Opportunities for systematic errors are few; however, the statistical accuracy of the polarization fraction is likely to be poor since Eq. 5-5 involves the difference between two signals, both of similar magnitude and of limited statistical accuracy.

In the report of these preliminary measurements,[141] no mention was made of tests to ensure that the results are not influenced by the existence of stray fields in the apparatus. Weak magnetic fields can seriously distort a measurement of polarization by causing a Larmor precession of the excited electron around the direction of the magnetic field.

No estimate was made of whether these data will be appreciably affected by cascade population of the $2p$ state. Neither were estimates of possible systematic error presented even though error of this type would be small. The random error amounts to an uncertainty in the polarization of about $\pm 10\%$ (i.e., $P = X\% \pm 10\%$).

5-3 CROSS-SECTION MEASUREMENTS

A. Excitation of the Atomic Hydrogen Target

In Data Table 5-3-1 is shown the cross section for emission of the Lyman-alpha $(2p \rightarrow 1s)$ line of H induced by impact of H^+ on an atomic hydrogen target. These data were presented by Young et al.[9] and are the revisions of the original measurements described by Stebbings et al.[8] [see Section 5-2-B(1)]. There is no experimental evidence from which cascade contributions to the emission might be assessed; however, one may use the theoretical predictions by Bates and Griffing[241] to estimate that cascade population from higher ns states might amount to approximately 7% at an impact energy of 10-keV. This figure, while not an accurate prediction, does serve to show that cascade population may not be negligible. Therefore, these data are line-emission cross sections and are not equal to level-excitation cross sections.

Data Table 5-3-2 contains the emission cross section of the Lyman-alpha $(2p \rightarrow 1s)$ line induced by the impact of He^+ on an atomic hydrogen target as measured by Young et al.[9] On the basis that cascade population of the $2p$ state as a result of the $H^+ + H$ collision may be quite appreciable, as argued above, it seems reasonable to suppose that the cascade populations of the $2p$ state may be significant in the case of $He^+ + H$. It follows therefore that these data should not be equated to level-excitation cross sections.

In the experimental determination of the relative variation of cross section with impact energy, Young et al.[9] estimated random errors to be less than $\pm 15\%$; the absolute magnitudes, established by normalization to theoretical predictions [see Section 5-2-B(1)], were said to be accurate to $\pm 20\%$; these estimates of limitations of accuracy would appear to be valid. In the course of these experiments, a study of the emission of Lyman-alpha induced by the impact of He^+ on a H_2 target was shown to be in good agreement with two other completely independent determinations (see Section 8-2-F), suggesting that the measurements for the $He^+ + H$ and $H^+ + H$ collisions are also of good accuracy.

B. Formation of Excited Projectiles by Charge Transfer

There are measurements on the formation of the metastable state of H by charge transfer as a result of the impact of H^+ on an H target. These data by Bayfield[143] and by Ryding et al.[142] shown in data Table 5-3-3, represent the cross section for the formation of the $2s$ state by direct excitation and by all relevant cascade processes. It should be recalled [see Section 5-2-B(3)] that the method used by Bayfield to establish a cross section involved normalization to previous cross section measurements that appear to come from reliable sources. The work of Ryding et al. is normalized to a theoretical prediction, of which the accuracy has not been demonstrated. Under the circumstances the large discrepancy between the data by Bayfield and that by Ryding et al. is not surprising. Within the estimated limits of accuracy the dependence of cross section on energy, in these two measurements, is in quite good agreement. The differences in absolute magnitudes may indicate only the inaccuracy of the theoretical predictions to which Ryding et al. normalized their data.

There is good reason to assume (1) that the data of Bayfield represent the cross section within the limits of accuracy assigned to this experiment (2) and that the work of Ryding et al. confirms the energy dependence of the data and serves to extend this to higher energies.

There are no experimental data from which the cascade contribution to the 2s state population might be estimated. It is possible, however, to make a rough assessment of cascade from the theoretical predictions of the cross sections for the charge transfer process by Bates and Dalgarno.[242] On this basis, cascade into the 2s state from higher np levels will amount to 6% or more at an energy of 25-keV; this figure is comparable with the estimated limits of accuracy for the measured relative variation of cross section with impact energy.

In Data Table 5-3-4 are shown the cross sections for the spontaneous emission of Lyman-alpha $(2p \rightarrow 1s)$ from excited projectiles formed by charge transfer in the $H^+ + H$ collision. These are revised values, obtained by private communication,[140] of the data originally published by Stebbings et al.[8] [see Section 5-2-B(1)]. The methods utilized by Stebbings et al. normalize the data to the theoretical cross section prediction for forming the 2p state by impact of electrons on atomic hydrogen. These theoretical predictions are expected to be quite valid, and the overall accuracy of the normalization procedure is estimated to be better than $\pm 20\%$. The data are a line-emission cross section and include no assessment of cascade. The theoretical cross section predictions by Bates and Dalgarno[242] suggest that cascade contributions at 25-keV might amount to 20% of the signal. It is probable that cascade contributions will be considerably attenuated by the construction of the apparatus; cascade comes from long-lived ns states that will tend to drift out of the region of observation before decaying to the 2p state. Nevertheless, cascade might prove to be an important factor, and it would be incorrect to regard these measurements as representing the cross sections for the formation of the 2p state.

C. Polarization of Emission

Young et al.[9] and Kauppila et al.[141] have carried out measurements of the polarization of the Lyman-alpha $(2p \rightarrow 1s)$ emission resulting from excitation of the atomic hydrogen target by the impact of H^+, He^+, Ne^+, and Ar^+. All of these measurements are of very poor statistical accuracy. Systematic errors were not quoted in any of the studies but are expected to be small.

Polarization of emission induced in the $H^+ + H$ collision as measured by Kauppila et al.[141] is shown in Data Table 5-3-5; the random error bars amount to $\pm 10\%$ (i.e., $P = X\% \pm 10\%$). Polarization fractions of emission induced by the impact of He^+, Ne^+, and Ar^+, respectively, as measured by Young et al.[9] are shown in Data Tables 5-3-6 through 5-3-8. These latter data are fragmentary and do not give a good picture of how polarization fraction is behaving. The statistical errors represent the following polarization uncertainties in three impact cases: 10% for He^+ (i.e., $P = X\% \pm 10\%$); 10 to 20% for Ne^+; and about 16% for Ar^+.

There are no other data with which these polarization measurements may be compared. Within the very wide error limits assigned to these data, the measurements may be regarded as reliable.

D. Data Tables

5-3-1 Cross Sections for the Emission of the 1216-Å-Lyman-Alpha Line of HI ($2p \rightarrow 1s$) Induced by the Impact of H^+ on an Atomic Hydrogen Target. [These data represent excitation of the atomic hydrogen target only.] They are revisions of the work previously published by Stebbings et al.[8]

5-3-2 Cross Sections for the Emission of the 1216-Å Lyman-Alpha Line of HI ($2p \rightarrow 1s$) Induced by the Impact of He^+ on an Atomic Hydrogen Target.

5-3-3 Cross Sections for the Formation of the $2s$ State of H in the Projectile as a Result of Charge Transfer as Protons Traverse a Target of Atomic Hydrogen. These data represent the sum of cross sections for formation of the $2s$ state by direct excitation and by cascade.

5-3-4 Cross Sections for the Spontaneous Emission of the 1216-Å Lyman-Alpha Line of HI ($2p \rightarrow 1s$) Induced by the Impact of H^+ on an Atomic Hydrogen Target. These data represent formation of an excited projectile by charge transfer. They are revised values of the measurement published previously by Stebbings et al.[8]

5-3-5 Polarization Fraction of 1216-Å Lyman-Alpha Line of HI ($2p \rightarrow 1s$) Induced by the Impact of H^+ on an Atomic Hydrogen Target. These data represent emission from the atomic hydrogen target only.

5-3-6 Polarization Fraction of the 1216-Å Lyman-Alpha Line of HI ($2p \rightarrow 1s$) Induced by the Impact of He^+ on an Atomic Hydrogen Target.

5-3-7 Polarization Fraction of the 1216-Å Lyman-Alpha Line of HI ($2p \rightarrow 1s$) Induced by the Impact of Ne^+ on an Atomic Hydrogen Target.

5-3-8 Polarization Fraction of the 1216-Å Lyman-Alpha Line of HI ($2p \rightarrow 1s$) Induced by the Impact of Ar^+ on an Atomic Hydrogen Target.

5-3-1
YOUNG ET AL. (9)

ENERGY (KEV)	CROSS SECTION (SQ. CM)
5.75E-01	3.70E-18
1.25E 00	3.30E-17
2.5CE 00	2.50E-17
5.00E 00	3.20E-17
1.00E 01	3.65E-17
2.00E 01	4.30F-17
3.00E 01	5.60E-17

5-3-2
YOUNG ET AL. (9)

ENERGY (KEV)	CROSS SECTION (SQ. CM)
5.00E-01	9.00E-18
1.00E 00	2.70E-17
1.0CE 00	2.85E-17
2.00E 00	2.90E-17
2.0CE 00	2.75E-17
3.00E 00	3.80E-17
4.0CE 00	3.65E-17
5.00E 00	3.50E-17
7.00E 00	4.20E-17
8.0CE 00	3.55E-17
1.C0E 01	3.95E-17
1.00E 01	3.85E-17
1.50E 01	4.35E-17
2.00E 01	4.90E-17
2.0CE 01	4.70E-17
2.50E 01	4.80E-17
3.00E 01	3.85E-17
3.10E 01	3.80E-17

5-3-3A
BAYFIELD ET AL. (143)

ENERGY (KEV)	CROSS SECTION (SQ. CM)
3.CCE 00	4.60E-19
3.50E 00	1.30E-18
4.00E 00	1.30E-18
5.00E 00	3.40E-18
5.50E 00	4.90E-18
6.CCE 00	5.00E-18
7.00E 00	1.00E-17
8.0CE 00	1.55E-17
9.0CE 00	1.70E-17
1.00E 01	2.35E-17
1.05E 01	2.30E-17
1.20E 01	2.42E-17
1.3CE 01	2.63E-17
1.40E 01	2.93E-17
1.53E 01	3.15E-17
1.6CE 01	3.20E-17
1.75E 01	3.40E-17
1.85E 01	3.50E-17
2.00E 01	3.58E-17
2.1CE 01	3.40E-17
2.25E 01	3.12E-17
2.40E 01	3.55E-17
2.60E 01	3.35E-17
2.80E 01	3.14E-17
3.0CE 01	3.18E-17
3.20E 01	2.82E-17
3.70E 01	2.28E-17
4.00E 01	2.02E-17
4.50E 01	1.72E-17
5.0CE 01	1.30E-17
5.50E 01	9.90E-18
6.0CE 01	7.30E-18
6.50E 01	5.90E-18
7.00E 01	4.20E-18

5-3-3B
RYDING ET AL. (142)

ENERGY (KEV)	CROSS SECTION (SQ. CM)
4.40E 01	8.10E-18
5.00E 01	5.83E-18
6.00E 01	4.46E-18
7.00E 01	2.82E-18
8.00E 01	2.25E-18
9.0CE 01	1.64E-18
1.00E 02	1.06E-18
1.20E 02	5.15E-19

5-3-4
YOUNG ET AL. (140)

ENERGY (KEV)	CROSS SECTION (SQ. CM)
6.00E-01	6.17E-17
1.25E 00	2.65E-17
2.5CE 00	3.48E-17
5.00E 00	4.10E-17
1.00E 01	3.90E-17
2.00E 01	3.00E-17
3.00E 01	1.30E-17

5-3-5
KAUPFILA ET AL. (141)

ENERGY (KEV)	POLARIZATION (PERCENT)
2.00E 00	-2.50E 00
3.00E 00	5.90E 00
3.50E 00	1.02E 01
4.0CE 00	-4.00E-01
5.00E 00	-4.00E-01
6.0CE 00	-7.90E 00
7.00E 00	-1.11E 01
8.0CE 00	-6.20E 00
1.00E 01	1.04E 01
1.50E 01	1.22E 01
2.09E 01	1.69E 01

5-3-6
YOUNG ET AL. (9)

ENERGY (KEV)	POLARIZATION (PERCENT)
1.86E 00	1.50E 01
5.19E 00	0.0
2.51E 01	1.10E 01

5-3-7
YOUNG ET AL. (9)

ENERGY (KEV)	POLARIZATION (PERCENT)
4.18E 00	4.40E 01
6.53E 00	6.00E 00
1.67E 01	1.00E 00

5-3-8
YOUNG ET AL. (9)

ENERGY (KEV)	POLARIZATION (PERCENT)
2.07E 00	2.00E 01
3.23E 00	1.50F 01

CHAPTER 6

EXCITATION OF A HELIUM TARGET

6-1 INTRODUCTION

The excitation of a helium target has received considerable attention. A wide variety of ionic and neutral projectiles has been employed, and in some cases the existing measurements cover an energy range from 1 to 1000-keV. There is particular interest in comparing theoretical predictions of these cross sections with the experimental measurements.

The general features common to all studies of the excitation of a helium target are discussed first (Section 6-1), including a brief review of the experimental arrangements employed by the various investigators. This is followed by a compilation and review of the available data for excitation by impact of various projectiles.

A. General Considerations

The formation of an excited neutral atom induced by proton impact is of considerable importance:

$$H^+ + He \rightarrow H^+ + He^*. \tag{6-1}$$

There is considerable theoretical interest in processes of this type because they involve comparatively simple atomic systems where the wave functions are accurately known. Experimentally, the post– and priorcollision conditions of the charge state and excitation are completely defined through a measurement of emission from the excited target. Transition probabilities are known accurately from theory,[243, 31] and the measured emission cross sections may be readily converted to level-excitation cross sections.

The formation of an excited helium ion by the simultaneous removal of one electron and excitation of the other is a two-electron process and, hence, exhibits theoretical complexity:

$$H^+ + He \rightarrow [H^+ + e] + He^{+*}. \tag{6-2}$$

The experimental measurements of He II emission does not readily provide data on a completely defined process. There is no information as to whether the ejected electron is captured by the projectile or liberated into a continuum state; the square brackets of Eq. 6-2 indicate that the measurement of emission gives no information on the state of ionization nor the excitation of the structures contained within them.

All groups of transitions such as $n(S,P,D) \rightarrow 3(S,P,D)$ exhibit essentially the same wavelength for any principal quantum number n, and for this reason cannot be separated by conventional spectroscopic methods. Hence, it is not possible to derive level-excitation cross sections in an unambiguous manner, and the data can only be presented as an emission function.

Incident projectiles of greater internal complexity than H^+ induce processes of the same type as those described by Eqs. 6-1 and 6-2, except that changes may also occur in the electron arrangement around the projectile. The measurement of helium emission may be related to a cross section for excitation of the helium target summed over all possible final states of the projectile. An experimental ambiguity might arise if a significant proportion of the projectile beam is in an excited state before the collision. Most experiments are at impact energies considerably greater than the internal energy of excitation of the projectile, and it is not expected that the measured cross section for transitions in the target will be appreciably affected by a small proportion of excited projectiles in the beam. In certain experiments at very low impact energies, however, the excited states might contribute an appreciable error. Such situations will be noted specifically in the discussions of the relevant data.

In all the presently available work on the formation of excited states of neutral helium, the measured emission cross sections have been converted into level-excitation cross sections, using Eq. 2-15 and employing the theoretical transition probabilities tabulated by Gabriel and Heddle.[243] If the critically evaluated transition probabilities by Wiese et al.[31] were to be used instead, the cross section for the excitation of the 4^1P level would be increased by 5.4% and those for the 5^1P, 6^1P, and 4^3S states reduced by, respectively, 2.5, 2.8, and 4.1%. All other discrepancies are less than 2% and may be considered negligible. It should be noted that none of the authors whose work is reviewed here has included any consideration of the possible errors in the transition probabilities when arriving at an estimate of the confidence limits of the data. In the present work, the data are shown as originally published, based on the transition probabilities of Gabriel and Heddle.[243]

Any incident projectile may cause the excitation of a singlet state: the n^1P levels by "dipole allowed" transitions, and the n^1S and n^1D levels by "quadrupole allowed" transitions. The excitation of the triplet states by a proton involves a change of spin that is formally forbidden by the Wigner spin conservation rule. Nevertheless, weak triplet emission has been observed in two experiments.[155, 164] Incident projectiles which have a bound electron may cause the excitation of triplet levels through the mechanism of electron exchange.

The secondary mechanism of collisional transfer (see Section 2-2-A) contributes to the population of the ^1D state and to transfer between various triplet levels;[243] the absorption of resonance radiation (see Section 2-2-A) influences the population of the ^1P state. Both these secondary populating mechanisms must be eliminated by maintaining the target pressure sufficiently low. It is necessary that the independence of the measured cross section from target pressure be demonstrated over the full range of energies for each experiment. In Section 6-2-B an assessment is made of certain other techniques with an analytical basis that have also been applied to the assessment of resonance absorption effects. It is concluded there that such techniques are not adequate.

Cascade population of the n^1S and n^1D states is generally quite small,[10, 165] but may be as high as 50% of the level population for triplet and n^1P states[165, 200]

at low energies. It should be noted that cascade into the n^1S states is fed from the n^1P levels and is, therefore, affected by resonance absorption effects.

Since the orbital angular momentum of the n^1S and n^3S states is zero, the emission from these levels is unpolarized. Therefore, problems associated with instrumental polarization and anisotropy of emission do not occur in these cases. For all the other states, appreciable polarization has been observed, and it is to be expected that experimental measurements will include a correction for instrumental polarization and anisotropy of emission. The former has generally been neglected, although many experiments have paid some attention to anisotropy.

When attempting to assess the agreement between theoretical predictions and experiments, considerable weight should be attached to the comparison of the functional dependence of the cross sections on energy and the comparison of the values of polarization fractions. Both of these quantities will be free of the major sources of systematic error that influence experimental measurements.

The measurements involving the largest uncertainties in any experiment that is designed for quantitative measurement of collisionally induced emission are the determinations of target density and optical detection efficiency. The functional dependence of emission cross section on projectile energy is independent of these quantities and, therefore, should be a reliable measurement. In Section 6-2 the functional dependence of cross sections for excitation of helium by protons, as measured by a number of different groups, is examined in some detail. It is shown that, although absolute measurements of cross sections differ by as much as a factor of two, there is very good agreement among the independent determinations of the functional dependence of cross section on impact energy. For all of the data on excitation of helium by impact of various projectiles the energy dependence of cross section will be, generally speaking, a reliable quantity, whereas the absolute magnitudes of cross sections are, unfortunately, subject to considerable errors.

These considerations also have some bearing on the validity of measurements of polarization fractions. Since the polarization of emission may be related directly to the ratio of two light intensities, the systematic errors in the determination of target density and photon detection efficiency cancel out. Thus, polarization can be determined with high systematic accuracy. Unfortunately, the random errors in the polarization fraction, defined by Eq. 2-32, become quite large when the intensities of light that are polarized parallel and perpendicular to the path of the projectile beam reach comparable intensity.

B. Experimental Arrangements

1. The Work of de Heer et al. The most comprehensive collection of work on the excitation of helium by various projectiles has been by de Heer, van Eck, Van den Bos, and others. The work is characterized by extreme attention to detail. The measurements on the excitation of helium by various projectiles now range over energies from 1 to 150-keV.

Over a period of time, this work has seen many improvements, primarily in the quality of the absolute calibration. In particular, much of the early work by van Eck[10, 155, 244] has been subjected to numerous revisions,[165, 213] some of which are not published in the open literature. The most recent and accurate data are used in this review, and the sources of such data are summarized as follows.

(a) $H^+ + He$. The first publication on emission of neutral helium lines by van Eck et al.[155] was superseded by a more extensive investigation[10] and a conference paper.[244] These measurements were revised by Van den Bos et al.,[165] and the later data are used here. A few measurements on polarization of emission published by van Eck et al.[10] were not repeated elsewhere. Since the measurement of polarization is independent of calibration of detection sensitivity, the data by van Eck et al.[10] are expected to be accurate and are retained here. Measurements on the emission of He II lines from the helium ion by van Eck et al.[155] were not repeated by Van den Bos et al. We retain the original measurements here and include also additional measurements by Moustafa Moussa and de Heer[163] at somewhat higher energies. The absolute magnitudes of the He II emissions by van Eck et al.[155] are probably of very poor accuracy.

(b) $H + He$. The first publication by van Eck et al.[155] was superseded by a more extensive investigation[10] and by a conference paper.[244] These measurements have all been renormalized by de Heer and Van den Bos,[190] and the corrected values are used here.

(c) H_2^+ AND $H_3^+ + He$. The values from the most recent published paper[200] are used here.

(d) $He^+ + He$. The data of de Heer and Van den Bos[212] are considered accurate and are used here.

In addition to the work referred to above, certain preliminary measurements were presented in a review article,[245] but are to be completely disregarded. The first work by van Eck et al.[155] utilized an apparatus designed previously by Sluyters et al.[103] for heavy particle impact experiments [reviewed in Section 7-3-A(2)]. Of the measurements retained for the present review, only the data by van Eck et al.[155] on He II emission were taken with this apparatus. All later experiments by Van den Bos, Moustafa Moussa, and de Heer et al. use the same basic experimental configuration which is described below.

The apparatus was based on a Leiss monochromator with a Czerny—Turner grating mounting using photomultiplier detection. No lens was used, and light was detected from a length of beam path defined by the size of the diffraction grating. In most of the work, the monochromator remained perpendicular to the ion beam path. Corrections for anisotropy in the emission were made by direct measurement of the polarization fraction and application of Eq. 2-27. A need for correction due to the instrumental polarization was demonstrated and applied, as described in Section 3-4-D(1).[10, 165] Measurements were also made at an angle of 56° in some of the earlier work and also in the $He^+ + He$ studies, where the Doppler effect was used to separate target and projectile emissions.

All measurements were carried out at sufficiently low pressures so that the emission could be demonstrated to vary linearly with target pressure. Target pressure was measured with a McLeod gauge designed to reduce gauge pumping errors (see Section 3-2-D) to a minimum.

No lens nor other imaging system was used between the monochromator and the collision region. The dimensions of the entrance slit and grating defined the region from which light was collected. For calibration purposes, the standard lamp was placed a considerable distance behind the collision chamber, and diaphragms were used to ensure that only the central part of the filament was viewed. Therefore, when observing the standard lamp, the solid angle from which light was collected and the

field of view were both much less than when measurements were made of emissions from the collision region. It is not satisfactory to rely on simple geometrical measurements to allow for these differences. In the earlier papers by this group, no mention is made of tests to ensure that the response of the system was independent of the solid angle and field of view; however, recent work by Van den Bos et al.[246], Moustafa Moussa and de Heer[163] include some investigations which show, over the range of conditions used in these various experiments, that the response is almost independent of solid angle and field of view.

Inadequate attention was paid to the problem of scattered light. A correction was determined directly for wavelengths below 3300-Å, and the magnitude of the scattered light flux was assumed to be constant at higher wavelengths. This assumption is not necessarily correct because most of the scattered light comes from the zero-order diffraction by the grating, and the path of the zero-order diffracted light ray will vary with grating position. No corrections were made at wavelengths above 3300-Å. It was stated that errors "up to" 20% may have occurred for the absolute calibration of the He ($3^1S \rightarrow 2^1P$) line at 7281-Å, due to scattered light.

The measurement of polarization was achieved using a Glan—Thomson prism placed between the collision region and the monochromator, in a region where the light beam was slightly divergent. Use of a polarization analyzer with a divergent light beam is not good practice, but tests were made[213] to ensure that there was a 100% rejection of light polarized perpendicular to the principal plane; the results are unpublished.

A detailed discussion of errors is given by Van den Bos et al.[15, 165, 246] It is estimated that, in general, systematic errors range from 6.5 to 9%, and random errors are about 5%. In certain cases, the error limits are higher due to special consideration, such as poor detection efficiency. The measurement of the polarization fraction is stated to be accurate within $\pm 10\%$. The authors choose not to add the error contributions from various causes in a linear sum but rather to determine a "standard deviation" of the errors; this can hardly be considered correct under circumstances where only a few error sources are identified. This suggests that the error limits ascribed to the programs of van Eck et al. and Van den Bos et al., both calculated on the basis of a standard deviation of the contributing parts, are perhaps conservative. Corrections that have been made to the published data over a period of years have sometimes been outside the estimated range of errors of the previous work. One might therefore conclude that the assessment of uncertainties is not entirely adequate, particularly in the early publications.

2. The Work of Hughes et al. Measurements by Hughes, Dodd, and others on excitation by proton impact are found in two papers.[154, 156] The earliest of these[154] includes preliminary measurements made only at a single impact energy of 200-keV; because of poor accuracy the data are not considered further here. Attention will be directed entirely to the more comprehensive series of measurements by Dodd and Hughes.[156]

Their experimental arrangement[154, 147] utilized a ½-meter Ebert—Fastie monochromator fitted with a photomultiplier as a detector. The axis was at 30° to the ion beam path. The monochromator was situated *under* the ion beam path with the slit parallel to the beam. A lens was used to focus a section of the ion beam path on the monochromator entrance slit, and the entrance window was defined by the slit height and width. Such an arrangement has many disadvantages; it is necessary that the slit

be maintained sufficiently wide so that the entrance window is larger than the image of the ion beam. With a 1/16 in. diameter ion beam, this optical arrangement necessitated the use of spectrometer slit widths that precluded resolutions better than 25-Å. A 2-cm length of the beam was in view, and due to the inclination of the optical axis, much of this was out of focus. As a further complicating feature, the Ebert—Fastie spectrometer exhibits curvature of the spectral line in the exit plane[247] which results in some loss of light unless curved slits are used. This might be particularly serious in the present case because the full length of the entrance slit was employed.

Calibration of the detection system involved the use of a standard of emission for which the stability may have been in some doubt since it was not subjected to regular tests. Calibration procedures involved the use of completely different solid angles from those employed when viewing the collision region. Some tests, for which the results are unpublished,[248] showed that the response of the system was constant over the field of view for both experimental and calibration situations.

No attention was paid to the possible influence of polarization on the measurement of the total excitation cross section. From the work by Van den Bos,[165] it seems likely that the highest polarization fraction encountered in this work would be approximately 46% for the $4^1D \to 2^1P$ transition at 15-keV. This would cause an error of about 23%. The magnitude of the error will vary with polarization and therefore with projectile energy. It must be concluded that the energy dependence of the data may exhibit appreciable errors due to neglect of polarization.

The collision chamber was insulated from ground, and the chamber itself served as the Faraday cup. Some tests were made to ensure that electrons and ions ejected from the end of the chamber and returning to the observation region did not produce an appreciable fraction of the observed excitation; the results are unpublished.[248]

Throughout this work, quite high target pressures were maintained. Independence of cross section from target density was adequately demonstrated for the ^1S states. The 4^1D level was observed to be strongly populated by collisional transfer. Although an emission measurement at a specific pressure was published, Dodd and Hughes point out that this measurement could not be accurately related to the excitation cross section. Therefore, it is not presented here. In dealing with the ^1P levels, account was taken of the absorption of resonance radiation in the manner of Gabriel and Heddle[243] who applied the theoretical treatment of Phelps.[249] This procedure is not entirely satisfactory and will be discussed at greater length in Section 6-2-B.

The absolute accuracy of the work by Dodd and Hughes was quoted as being within ±40%.

3. The Work of Thomas and Bent. Excitation by proton impact,[161,250] with a related brief study of H_2^+ and H_3^+ impact excitation,[201] is the subject of Thomas' and Bent's work. It has demonstrated that protons and deuterons of equal velocities exhibit the same cross sections for target excitation. The equality was assumed to hold at velocities somewhat lower than those where it had been tested, and a limited series of measurements for deuteron impact were plotted as though they were for protons of equal velocity. This assumption is continued here.

The experimental arrangement utilized a $\frac{1}{2}$-meter Ebert—Fastie monochromator fitted with photomultiplier detection, placed at an angle of 90° to the ion beam path. For calibration, a standard tungsten strip filament lamp was placed at the point normally occupied by the ion beam in order to ensure that the field of view was the same during calibration as for the experiment. The temperature of the lamp filament

was measured directly with an optical pyrometer, and the relevant values of emissive power were obtained from tables.[83] This procedure is less accurate than a direct comparison between a standard lamp and a blackbody.

Insufficient attention was paid to errors introduced by polarization through anisotropy of the emission and instrumental polarization effects. A brief investigation was made of the polarization fraction of emission from ^1D and ^1P states; from this it was concluded that the polarization fraction did not exceed 8% in the worst case, and therefore the assumption of isotropy did not introduce an error exceeding 4%. For most excited states the estimated error was somewhat less than this; however, this estimate of polarization fraction is at variance with the more accurate measurements by Van den Bos et al.[165, 246] and Scharmann and Schartner[166] on the $4^1D \rightarrow 2^1P$ emissions. If these more accurate investigations are accepted, then the data by Thomas and Bent on the 4^1D state might be in error by up to 9% at the extremities of the energy range used. The effect of instrumental polarization was neglected completely.

In general, target gas densities were maintained sufficiently low to ensure a linear relationship between emission and target pressure; however, in the case of the ^1P levels, the range over which the pressure independence was established was rather small for certain impact energies. This problem is considered further in our detailed discussion of data on the ^1P transitions (Section 6-2-B).

During calibration of the detection sensitivity of the system, particular care was taken to eliminate the effects of stray light within the monochromator. When calibrating at the blue end of the spectrum and in the near ultraviolet, a calibrated band pass filter was placed in front of the standard lamp to block intense emissions in the red end of the spectrum. This precaution was found to be necessary at wavelengths below 4300-Å.

The authors estimate that for the He n^1S, n^1D, and He$^+$ measurements systematic errors did not exceed $\pm 25\%$ and random errors $\pm 10\%$. For the n^1P case, where absorption of resonance photons is a problem, the systematic error was increased to $\pm 35\%$.

4. The Work of Robinson and Gilbody. Robinson and Gilbody[251, 162] have included excitation by protons as well as fast hydrogen atoms in their work. A large prism spectrometer fitted with photomultiplier detection was used with its axis at 54.5° to the direction of the primary beam. At this angle, the measurements should be independent of anisotropy due to polarization of the emission. Tests, for which the results are unpublished,[252] indicated that the errors introduced by the instrumental polarization of the detection system were small compared with other uncertainties in the measurement. No corrections were made to eliminate this problem.

A lens was used to form an image of the collision region on the entrance slit of the spectrometer. Therefore, emissions are detected only from a short section of the ion beam path. The region of observation was located 10 cm from the beam entrance aperture to the collision chamber, and the measurements may be presumed to be unaffected by any pressure gradients in the region of that aperture.

Calibration was carried out using a standard lamp placed in the position normally occupied by the ion beam. This procedure ensures that calibration and experiment employed precisely the same optical arrangement. No mention is made of checks for the influence of scattered light. The linearity of the response of the detection system

was stated to be very good.[252] Robinson and Gilbody quote possible systematic error of $\pm 20\%$ with an additional random uncertainty of $\pm 10\%$ due to reading errors.

5. The Work of Dufay et al. The measurements by Denis, Dufay, and Gaillard on the excitation of helium by protons are reported in one rather short note.[160] This gives inadequate details for a review, so reference is also made to more complete reports that describe work involving other targets[178] and the same apparatus. Many references have also been made to Denis's unpublished thesis[253] in which the study of the collisional excitation of helium is fully documented.

Denis's experimental system utilized a Czerny-Turner grating monochromator with photomultiplier detection as the major optical element. A quartz lens formed an image of the collision region on the spectrometer entrance slit. The optical axis of the system was at $90°$ to the beam path.

No account was taken of polarization, either as regards anisotropy of emission or instrumental polarization (Section 3-4-D). Using the published polarization measurements,[165] it is estimated that neglect of anisotropy will cause an error up to 17% in the measurement of the 4^1D level and somewhat less for the other ^1D and ^1P levels. It is not possible to estimate the error caused by the neglect of the instrumental polarization.

The calibration of the detection sensitivity was achieved by using a standard lamp placed some distance from the collision region. Inevitably the field of view and solid angle used during calibration were different from that used during experimental measurement. No tests were made to ensure that the response was the same under these two circumstances. No attention was paid to the possible influence of scattered light in the monochromator.

Target pressures used in this experiment were far too high, and, except for a few measurements on the n^1S cross sections at high energies, a correction for pressure effects was always necessary. At lower energies, the projectile beam composition changed during its passage through the chamber, and corrections were applied for this effect. Population of n^1P and n^1D states by photoabsorption and collisional transfer was observed and assessed through extrapolation procedures. The removal of all these pressure-dependent corrections was not very satisfactory and may have contributed numerous errors to the determination.

The authors' quoted limits of accuracy in this work are $\pm 50\%$ possible systematic error and $\pm 10\%$ random error.

6. The Work of Krause and Soltysik. These experiments concern the excitation of helium by protons and neutral hydrogen atoms in the impact energy range 1.5 to 20-keV. A brief conference report[254] was superseded by a more extensive publication[164] with some revisions of data. The work is restricted to relative measurements of emission from the 3^1S, 3^1D, 3^3S, and 3^3D states with determinations of the polarization of the emission.

A projectile beam of H^+ ions was produced in an rf source, mass analyzed, collimated, and directed through a short target gas cell. The beam traversed the cell, emerged into an evacuated region, and the current was monitored on a Faraday cup. A large aperture plano-cylindrical lens was used to collect light from the collision region and to direct it onto a detector. A spectrometer was used for an initial survey of the emission, but all quantitative measurements were carried out using a cooled

photomultiplier and suitable interference filters. No attempt was made at absolute cross section measurements, and all emission data are presented as relative emission cross sections.

The polarization analysis employed a Glan—Thomson prism placed between the lens and the interference filter. A cylindrical lens was used in the apparatus and produced focusing only in one plane. Therefore, the light was not parallel to the axis of the prism. No test was made to ensure that this arrangement provided total rejection of light polarized perpendicularly to the optic axis. It was reported that the polarization fractions of the 3^1S and 3^3S emissions were about 7% and almost invariant with energy in contradiction to the definite theoretical predictions of zero polarization. It is possible that the incorrect use of the Glan—Thomson prism produced an instrumental polarization that rendered the polarization fractions erroneous. Although polarization fractions for the $3^1S \rightarrow 2^1P$ and $3^3S \rightarrow 2^3P$ lines are presented in the published reports, they are omitted from the present compendia of data since they are probably incorrect.

All of the emission measurements were carried out at an angle of 90° to the beam path, and they do not take account of anisotropy in emission due to polarization. Therefore measurements on the emission from the 3^1D and 3^3D states do not accurately represent a total emission cross section (see Section 3-4-D).

Measurements of H^+ impact excitation of the 3^1S and 3^1D states were carried out in a straightforward manner. Graphs of emission as a function of target pressure were presented; these indicated linearity throughout the pressure range employed.* Appreciable attenuation of the H^+ beam occurred through charge transfer neutralization. A correction was made for this in order to calculate the actual proton beam current in the center of the collision region. The linearity of emission with pressure indicated that excitation of the singlet states by the small amount of neutral hydrogen in the beam must be insignificant compared with direct proton excitation.

Measurements on excitation of the 3^3S and 3^3D states by protons and neutral hydrogen were carried out in an indirect manner. The emission in the $3^3S \rightarrow 2^3P$ and $3^3D \rightarrow 2^3P$ transitions induced by protons on helium was observed to have a nonlinear dependence on target density. A plot of emission divided by pressure versus pressure was linear with a finite intercept, suggesting that the emission was composed of two parts, one of which depended linearly on target pressure and the other, on the square of target pressure. It was concluded that the component dependent on the square of pressure was caused by neutral hydrogen atoms formed by a charge transfer event undergoing a second collision with the excitation of a target helium atom. It was claimed that the linear component was direct excitation of the triplet states of helium by protons, a process which is formally forbidden by the laws of conservation of spin. Analysis of the pressure dependence of the emission permitted a determination of the cross sections for these two processes. An indirect measurement of this type must be interpreted with extreme caution. It is not consistent with the

---

* In the published reports extensive use is made of graphs showing intensity as a function of pressure. These graphs must be interpreted with caution. The pressure measurements quoted are actually ion gauge readings in the collimation region. This pressure was about 67 times lower than that in the target cell. Moreover, the readings were not corrected for the calibration constant of the ion gauge. Actual target pressures ranged from "a few" microns to some tenths of a micron. These pressures are not quoted in the report.[164]

operational definitions of cross section that are established in Chapter 2. In the event that the model, which describes the processes occurring in the collision region, is incorrect or incomplete, the derived cross sections may be erroneous.

Complete discussion of the data derived from this experiment is given later in the relevant sections. It is suggested that all measurements of triplet emission must be considered with caution due to possible inadequacies in the model assumed for the collision processes.

In the original publications the estimated limits of accuracy of the data were shown as error bars. In general the emission cross section measurements were accurate to $\pm 2\%$, except for the triplet states induced by proton impact where they rose as high as $\pm 25\%$ due to the low intensity of emission. The polarization fraction had large uncertainties amounting to as much as ± 10 in the percentage polarization of the proton induced triplet emissions.

7. The Work of Thomas and Gilbody. Thomas and Gilbody[71] were primarily concerned with emissions induced by collisions between heavy rare gas ions and atoms, but they have included some work with a helium target. They have made measurements of He I and He II emission induced by the impact of Ne^+ and Ar^+ at energies from 100 to 400-keV. Data are presented as emission cross sections.

The experimental system utilized a large aperture prism spectrometer fitted with a photomultiplier as the detection system. A lens was used to focus emission onto the spectrometer entrance slit. Due to low projectile currents, the intensities of emission were weak, and the effects of photomultiplier dark current were reduced by mechanically chopping the light flux at the entrance slit and identifying the required signal in the photomultiplier output by its specific frequency and phase.

Determination of the detection efficiency of the system was made by using a standard tungsten strip filament lamp placed at the point normally occupied by the ion beam. In this manner, the optical system was the same during the experiment as for the calibration, and it was unnecessary to determine precisely the positions and apertures of the various parts of the optical system. The linearity of the detection system's response to the intense standard lamp was checked. To measure the lamp temperature an optical pyrometer was used, and the emissive power was obtained from the measurements by de Vos.[83] This is less satisfactory than using a lamp whose emissive power has been obtained by direct comparison with a blackbody (Section 3-4-B). The spectrometer was placed at an angle of $90°$ to the ion beam path, and no attention was paid to the possible influence of polarization. Therefore, errors may be present due to anisotropy of emission and instrumental polarization effects.

The ion beam was accurately monitored by using a rather deep parallel plate collector with suitable suppression voltage. Target pressure was monitored with a Pirani gauge that was frequently calibrated against a liquid nitrogen-trapped McLeod gauge. Linearity of intensity of emission with target pressure was ensured for all experimental measurements.

Random errors were estimated as $\pm 15\%$ and systematic errors as $\pm 12\%$.

8. The Work of Dworetsky et al. Dworetsky, Novick, Smith, and Tolk[35,255,215] have been concerned with the measurement of He I emission induced by the impact of low energy He^+ ions on a helium target. The measurements give the sum of emission from the target by direct excitation, and from the projectile by charge transfer. The primary objective of the investigation was to make a phenomenological study of the problem. The data are in the form of relative variations of emission cross section

with impact energy. No attempt was made to determine absolute emission cross sections.

The first work by this group was a series of measurements of the emission of visible radiation induced by the impact of He^+ on He. Two reports are available[35, 255] containing identical data. More recent experiments[215] involve studies of the formation of the 2^1P and 2^3P states; the 2^1P level was detected through the emission of the 584-Å $(2^1P \to 1^1S)$ line in the vacuum ultraviolet, and the 2^3P level through the detection of the 10,830-Å $(2^3P \to 2^3S)$ line in the infrared.

The He^+ ions are produced in a simple Nier-type electron impact source and accelerated directly into the collision region. By keeping the energy of the source electrons below 55 eV it was possible to guarantee that no metastable He^+ ions nor multiply charged helium ions could be formed in the source. Ion beam currents were monitored on a Faraday cup. There was no systematic investigation of the variation of intensity of emission with target pressure.

The measurements on spectral lines in the visible region were carried out using an interference filter for resolution and a photomultiplier for detection. It is possible that an appreciable fraction of the emission was due to cascade population. An attempt was made to establish an absolute magnitude for some of the emission cross sections by normalizing to previous measurements by de Heer et al.[12] of the cross sections for the formation of the parent levels of the emissive transitions. This is an inherently incorrect procedure because the data are in the form of emission cross sections, including a cascade contribution, while the previous work by de Heer et al. is corrected for cascade. No reliance can be placed on these absolute values by Dworetsky et al. It should also be noted that the normalizations in the two reports are different: the first[255] involved a mathematical error and is wrong; the second[35] was carried out correctly but is, nevertheless, conceptually incorrect due to the neglect of cascade.

The $2^3P \to 2^3S$ line in the infrared was resolved with a filter and detected with a liquid nitrogen-cooled photomultiplier. Some relative calibrations of detection sensitivity at various wavelengths were carried out using a standard tungsten filament lamp, and measurements were made of transitions that will cause a cascade population of the 2^3P state. It was concluded that the cascade contribution to the 10,830-Å line was less than 10% of its intensity.

The $2^1P \to 1^1S$ transition at 584-Å was measured using an evacuated normal incidence grating monochromator and an open ended multiplier. Care was taken to ensure that this line was not appreciably attenuated by resonance trapping. It was argued that, due to the momentum transfer during the collision process, most of the 584-Å radiation will be Doppler-shifted sufficiently to prevent its reabsorption. Measurements of signal intensity versus pressure yielded a straight line, which is to be expected only if trapping is negligible.

No details were given on the energy spread of the projectile beam. The influence of polarization on the measured emission cross sections was neglected. No estimates were given of the accuracy of the data.

9. The Work of Tolk and White. Tolk and White have studied the emission of the 3888-Å $(3^3P^0 \to 2^3S)$ line of He I induced by the impact of Ne^+ on helium.[218] The one report of this work is very abbreviated and does not allow a proper assessment of the procedures employed.

The ion source was a Nier type of the same design as that used by Dworetsky

et al.[35, 255, 215] for the work described in the preceding section. No mass analysis was used between source and observation region, but by controlling the energy of the source electrons it was possible to eliminate metastable helium ions and doubly charged helium ions from the projectile flux.

The emission was analyzed with a monochromator and detected with a photomultiplier, but the influence of polarization on the measured emission cross section was ignored. No attempt was made to study the variation of emission with target density in order to confirm the existence of single-collision conditions. The data are in the form of a relative variation of emission cross section with projectile energy.

An absolute magnitude was assigned to the data by a normalization procedure. The data were normalized to the cross section for the emission of the same line in the He^+ + He collision combination as determined by Dworetsky et al.[35] The reference data had, in turn, been assigned their absolute magnitude by normalization to the work of de Heer et al.[12] Such a complicated normalization procedure will inevitably be of poor accuracy. Furthermore, it is basically invalid because the measurements by de Heer et al. are of level-excitation cross sections, whereas those of Dworetsky et al.[35] and by Tolk and White[218] are all of line-emission cross sections. De Heer[12] estimates the cascade contribution to the He^+ + He emission to be 11% at a 10-keV impact energy. Normalization procedures, therefore, will inherently involve errors of at least this magnitude. It is concluded that the absolute magnitudes of the data by Tolk and White cannot be considered as reliable.

No details are given of the energy resolution of the projectile beam. No estimates are made of the accuracy of the data.

10. The Work of Scharmann and Schartner. Scharmann and Schartner have performed a series of measurements on the excitation of helium by impact of protons. Their studies have concentrated on establishing the high energy behavior of the excitation process, in the region 0.1 to 1.1-MeV, where it was expected that the theoretical predictions of the First Born Approximation should be accurate. The discussions of experimental technique in the first three chapters of this monograph have indicated that it is quite difficult to carry out an accurate measurement of a cross section. The two major problems that occur are in the determination of the target density and in the calibration of optical detection sensitivity. The measurement of the relative variation of an emission cross section with impact energy and of polarization fraction does not require the determination of these quantities and, therefore, should be quite an accurate quantity. The work of Scharmann and Schartner capitalizes on this fact. Their data include only measurements of the relative variation of cross section with energy and of polarization fractions of emission.

This group has produced five separate publications describing their work. The first two publications concerning the polarization of emission[256, 251] were superseded and greatly extended by a full report which contains the only description of the experimental system.[166] A conference report[258] on the relative variation of emission cross sections with impact energy was also superseded by a more complete report.[167] All the data used here, as well as the experimental details, are drawn from the two most complete discussions.[166, 167]

The projectiles were produced in a radio frequency ion source and accelerated to energies between 100 and 1100-keV in a Van de Graaff accelerator. Mass analysis of the projectile beam allowed selection of the required ion. Projectile beam current was monitored on an elaborate Faraday cup that provided complete suppression of

electrons and inhibited reflection of heavy particles. Target pressures were monitored on a calibrated ionization gauge. Since no attempt was made to measure absolute cross sections, the accuracy of the calibration was unimportant.

Light from the observation region was focused into a parallel beam and then directed onto the sensitive surface of a photomultiplier. Interference filters and Polaroid films were inserted into the parallel region of the light beam and provided definition of specific wavelengths and planes of polarization. The relative variation of emission function with energy for a given spectral line was simply equal to the relative variation of the output signal from the multiplier with a suitable interference filter in place. A Polaroid film was inserted to successively transmit light polarized parallel and perpendicular to the ion beam path; the ratio of the two signals represented the ratio of the intensities of light with these two planes of polarization and was inserted in Eq. 2-26 to determine the polarization fraction P.

The relative values of the emission cross section were determined under pressure-independent conditions. Maximum target pressures were 10^{-4} torr for the 3^1P, 4^1P, 4^1D, 5^1D states, 4.5×10^{-4} torr for the 3^1S and 3^1D states, and 2×10^{-4} torr for the 4^1S and 5^1S states. The emission data were corrected for anisotropy of emission associated with polarization. No attempt was made to assess cascade contributions to the measurements. The statistical accuracy of the data was said to be $\pm 10\%$.

Particular care was taken in the polarization measurements to eliminate the depolarizing effect of stray magnetic fields. An attempt was made to determine the polarization of the $n^1S \to 2^1P$ lines. These gave essentially zero polarization, as expected for a spherically symmetric parent level, and confirmed that the inherent polarization of the detection system was zero. It proved impossible to definitely establish a region where the polarization of the $n^1P \to 2^1S$ lines were independent of pressure even at 1 or 2×10^{-4} torr. For the $n^1D \to 2^1P$ transitions polarization appeared to be independent of pressure below 5×10^{-4} torr, and all data were acquired in this region. The limits of accuracy of the data were assessed as $\pm 30\%$ of the polarization fraction. This figure included a contribution for the depolarization due to cascade population of the excited state. From the general consistency of these data and the attention paid to detail in the experimental procedures, it would appear that this estimate of possible error is certainly an upper limit, and the data may, in fact, be more reliable than the authors state.

11. Other Work. The work of Dieterich,[259] Hanle and Voss,[199] and Šternberg and Tomaš[260] is not considered in detail. The experimental procedures used in these early studies are at variance with many of the criteria that have been established as necessary. Furthermore, all the determinations have since been repeated by other workers under more satisfactory conditions.

6-2 EXCITATION OF HELIUM BY PROTONS AND DEUTERONS

Because of its theoretical importance, the excitation of helium by proton impact has received more attention than any other collisional excitation mechanism. The formation of ^1S, ^1P, and ^1D states is considered in Sections 6-2-A through 6-2-C. In Section 6-2-D, the limited amount of data on the formally forbidden process of

triplet excitation is considered. Formation of the excited helium ion is considered in Section 6-2-E, and all measurements on polarization are included in 6-2-F. All data tabulations are listed in Section 6-2-G.

It is generally expected that deuterons and protons of the same velocities will exhibit the same cross sections. Deuteron beams are sometimes used to extend the lower velocity limit of an investigation. Deuteron data are shown in the tables and on graphs as though they were protons of the same velocity.

A. Excitation of the n^1S Levels of Helium

There are no known secondary processes that significantly populate or depopulate the n^1S excited states at low pressures. Cascade population of the 3^1S state may amount to 7% at 150-keV impact energy. Cascade population of higher states is less than one or two percent for all energies covered by experiments reported here. Polarization of the $n^1S \to 2^1P$ transitions must be equal to zero. It is to be expected that the major sources of systematic errors in an experimental measurement will be due to difficulties in the determination of target density and in the determination of detection efficiency of the optical system. Errors due to these causes will not influence the validity of a measurement of the functional dependence of cross section on impact energy. The data for this series of cases are shown in Data Tables 6-2-1 through 6-2-5.

Most of the data shown here are in the form of level-excitation cross sections. In addition, however, there are two groups of determinations of the relative variation cross sections for emission of spectral lines with projectile impact energy. These measurements include data on the $3^1S \to 2^1P$ transition by Krause and Soltysik[164] and data on the 3^1S, 4^1S and $5^1S \to 2^1P$ transitions by Scharmann and Schartner.[167] Since the cascade contributions to these emissions are less than 7% for the 3^1S state and less than 2% for the other transitions, the measured line emission cross sections will be proportional to the level-excitation cross sections to within an accuracy of 7 and 2%, respectively. In order to facilitate presentation of these data in this monograph, the relative line-emission cross sections have been normalized to level-excitation cross sections determined by other research groups. The data by Krause and Soltysik[164] are normalized to the work of Van den Bos et al.[165] at 10-keV impact energy; the work of Scharmann and Schartner is normalized to that of Thomas and Bent[161] at 140-keV impact energy.

The wavelength of the $3^1S \to 2^1P$ line (7281-Å) lies in a spectral region where photomultipliers are generally not very sensitive. It is likely that the measurement of the 3^1S excitation cross section is the least accurate in this series.

It is pertinent to examine the consistency of the energy dependence of the cross sections as measured by the various authors using different experimental systems. The cross section for the 4^1S state is chosen for this study since it has been measured by many authors, and being large may be expected to be reasonably free of random errors. In Fig. 6-1, the six cross sections that are measured have been normalized to unity at 130-keV impact energy. For clarity, individual data points have been omitted. It is observed that there is excellent agreement among the determinations by Dodd and Hughes,[156] Van den Bos et al.,[165] Thomas and Bent,[161] Robinson and Gilbody,[162] and by Scharmann and Schartner.[167] The weight of evidence suggests that the energy dependence of the cross section is that which has been determined by these authors. In contrast, work by Denis et al.[160] indicates a systematic difference

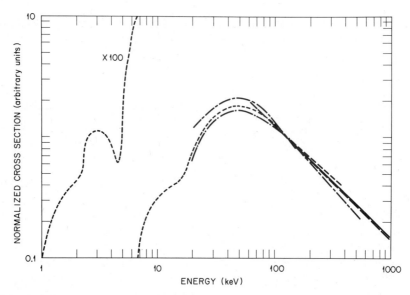

Fig. 6-1   Relative cross sections for the excitation of the 4^1S state of helium by the impact of H^+ and D^+ on a helium target. All measurements have been normalized to unity at 130 keV: _ _ _ Van den Bos et al.[165] (from 1 to 150 keV); ___ . ___ Dodd and Hughes[156] (from 20 to 130 keV); _____ Thomas and Bent[161] [(from 75 to 900 keV) including H^+ and D^+ data]; _ _ _ _ Denis et al.[168] (from 20 to 500 keV); _ _____ _ Robinson and Gilbody[160] (from 60 to 350 keV); ___ ___ Scharmann and Schartner[167] [(from 140 to 910 keV) normalized to the work of Thomas and Bent at 140 keV; strictly speaking, this is a line-emission cross section (see text)].

that increases with decreasing energy. It is recalled [Section 6-1-B(5)] that their measurements at low energies involved a mixed beam of protons and neutral H atoms from which a proton impact excitation cross section was extracted by examining the pressure dependence of the emission. It seems possible that this correction by Denis et al.[160] was not carried out adequately.

This examination of consistency has only been carried out for the 4^1S state. Reference to the higher n^1S states indicates that essentially the same behavior occurs; there is general agreement among the various authors, except Denis et al.[160] whose data lie consistently higher at low energies.

At impact energies above 50-keV, the energy dependence of the excitation cross section is approximately the same for the 3^1S, 4^1S, 5^1S and 6^1S states. This led Van den Bos et al.[165] to suggest making a consistency check by comparing the cross section ratios $\sigma(n)/\sigma(n+1)$ obtained by various researchers. It is to be expected that errors in absolute calibration will generally be invariant with wavelength (see Section 3-4-H). Thus, despite the obvious differences in the measured cross sections, these ratios should be similar. Table 6-1 is taken from the work of Van den Bos et al.[165] to compare the average values of these ratios determined for all energies above 50-keV.

Agreement in the case of the $\sigma(3^1S)/\sigma(4^1S)$ ratio is poor. The $3^1S \rightarrow 2^1P$ line lies in the far red of the visible spectrum where the sensitivity of the detector is poor. Van den Bos et al.[165] experienced problems due to internal reflection within the monochromator. The absolute calibration by Dodd and Hughes[156] required an extrapolation of the calibration curve of their standard lamp. Under these circumstances the

TABLE 6-1   RATIOS OF CROSS SECTIONS AT ENERGIES ABOVE 50-keV

	$\dfrac{\sigma(3^1S)}{\sigma(4^1S)}$	$\dfrac{\sigma(4^1S)}{\sigma(5^1S)}$	$\dfrac{\sigma(5^1S)}{\sigma(6^1S)}$
Van den Bos et al.[165]	3.50	1.78	1.80
Thomas and Bent[161]		1.79	1.89
Robinson and Gilbody[162]		1.92	
Dodd and Hughes[156]	1.65	2.12	
Denis et al.[160]	3.37	2.37	2.04

disagreement between ratios is not surprising. Agreement in the case of the $\sigma(4^1S)/\sigma(5^1S)$ ratio is much better, particularly if one disregards the work of Denis et al.[160] which was not carried out under single-collision conditions. Agreement in the $\sigma(5^1S)/\sigma(6^1S)$ ratio is also quite good.

It is not possible to make an objective judgement as to which of these measurements represents the most accurate absolute value. There are five independent determinations. In no case did an experimental procedure include attention to all of the criteria that have been established for valid, accurate data measurement. The discrepancies among data are larger than the estimated limits of accuracy assigned to the data by the various authors. It seems reasonable to suppose that the five available determinations should bracket the true values of the cross sections.

Scharmann and Schartner[167] have carried out a detailed study of the various published measurements of excitation cross sections. They point out that at energies above 100-keV there is, in fact, a small systematic variation of the functional dependence of cross section on energy as one moves through the n^1S group of states. This dependence on the principal quantum number n is shown quite clearly in the measurements of Scharmann and Schartner[167] which have high statistical accuracy; it is also apparent in the work of Thomas and Bent[161] and in that of Robinson and Gilbody,[162] although in the last two cases the effect is partially obscured by the poorer statistical accuracy of the data. The work of Denis et al.[160] does not show the dependence on n. We have, however, suggested that the work of Denis et al. is less accurate than the work of the other authors; therefore it may be discounted in this comparison. The existence of a dependence of the functional form of the cross section on the principal quantum number for impact energies above 100-keV would appear to be definitely established by the available data.

In conclusion, it is suggested that the energy dependence of the n^1S cross section established by Dodd and Hughes,[156] Van den Bos et al.[165, 246] Thomas and Bent,[161] Robinson and Gilbody,[162] and by Scharmann and Schartner[167] is quite accurate. Except for the work of Denis et al.[160] and all the data for the 3^1S level, the ratios of cross sections given in Table 6-1 are also fairly consistent. These aspects of the data may be compared with theory with some confidence. There remains, however, some considerable ambiguity about the absolute magnitude of the cross sections. It is not possible to make an objective decision as to which measurement represents the most accurate cross section determination, although it does seem reasonable to suggest that the range of measured values does bracket the true value of the cross sections.

The data exhibit reasonable agreement in the functional dependence of cross section on energy and agreement also in the ratios of cross sections for the formation

of different states. The major discrepancy lies in the absolute magnitudes of the data. In Section 3-4-H it was concluded that this observation will indicate that the discrepancies are due primarily to the determination of the efficiency with which the collisionally induced emission is detected.

B.   Excitation of the n^1P Levels of Helium

The excitation of the n^1P levels by the dipole transitions from the 1^1S ground state is of great theoretical interest since the calculation is not very sensitive to ground state wave functions, and the predictions are, therefore, expected to be quite accurate. Unfortunately, the experimental measurement of the cross sections for the excitation of these levels is greatly hampered by the repopulation of ^1P levels by the absorption of resonant photons. In principle, this problem is best handled by working at sufficiently low pressures so that the repopulation is unimportant. Suitable cascade corrections were applied for all the level-excitation measurements presented here.

Emissions in the $n^1P \rightarrow 2^1S$ sequence are polarized, and corrections are expected for the resulting anisotropy of emission and for the influence of instrumental polarization on the determination of detection sensitivity (see Section 3-4-D). These corrections were ignored in all of the experiments discussed here except those by Van den Bos et al.[165] and those by Scharmann and Schartner.[167] It is noted that the experimentally measured values of polarization fraction[165, 166] for the $n^1P \rightarrow 2^1S$ lines are less than 15% at all energies from 25 to 1000-keV. Error due to neglect of anisotropy of emission when polarization is 15% will amount to no more than about 5% in a typical experimental system viewing at 90° to the beam path. It is impossible, in retrospect, to assess the error that might be associated with neglect of instrumental polarization.

The cross section measurements for the 3^1P, 4^1P, 5^1P, and 6^1P states are shown in Data Tables 6-2-6 through 6-2-9 in Section 6-2-G. There are some additional measurements of the cross sections for formation of the 7^1P and 8^1P states by van Eck et al.;[10] these are not displayed here on the grounds of poor accuracy. The original series of measurements on the n^1P states by Van Eck et al.[10] have been superseded and much revised by the more recent work of Van den Bos et al.,[165] but no remeasurement of the 7^1P and 8^1P states was undertaken. In view of the changes in the data for the lower n^1P states, it seems likely that the data of van Eck et al. for the 7^1P and 8^1P levels is very inaccurate and its inclusion would be misleading.

Most authors present their data in the form of level-excitation cross sections, calculated from the measurement of line emission by Eq. 2-15. Cascade population of the n^1P states from higher ^1S and ^1D amounts to as much as 22% at low energies around 20-keV. Cascade decreases to about 5% at 150-keV and to 1% at 900-keV, reflecting the fact that the n^1S and n^1D cross sections decrease more rapidly with increasing energy than do the n^1P cross sections. Scharmann and Schartner[167] present data on relative variation of line-emission cross sections rather than level-excitation cross sections. Since cascade contributions to these lines will be between 1 and 5% in the energy range of their work, the line-emission cross sections will be proportional to the level-excitation cross sections within an accuracy of 1 to 5%, depending on energy. To facilitate the presentation of these data in this monograph

the relative line-emission cross sections have been normalized to level-excitation cross sections as determined by another group. The work of Thomas and Bent[161] is chosen for the normalization since it covers essentially the same energy range as the work of Scharmann and Schartner.[167]

The $5^1P - 2^1S$ (3964-Å) line of He is quite close to the H_ε (3970-Å) line of hydrogen. A small amount of hydrogen emission resulting from charge transfer might be expected. The work of Van den Bos et al.[165] includes specific precautions to exclude this. Since the H_ε emission is small and is expected to decrease rapidly with increasing energy, it is unlikely to produce a significant error in the higher energy work of Thomas and Bent,[161] Denis et al.,[160] and of Scharmann and Schartner.[167]

There are substantial variations among the results of n^1P excitation cross section measurement. These discrepancies are completely different from those exhibited in the n^1S data that, it has been concluded, are due primarily to calibration errors. The discrepancies in the n^1P measurements may be traced to inadequate assessment of reabsorption of resonance photons. The methods used by the various investigators to correct for this problem must be examined in some detail.

Most of the work by Van den Bos et al.[165] was done at pressures where the emission cross section was independent of target density, usually below a few times 10^{-4} torr. At a few projectile energies signal levels were very low, and emission cross section measurements were made at pressures above 10^{-4} torr. For those few cases, a correction was determined from measurements of light signal as a function of pressure at other energies. Since this correction appeared to be independent of projectile energy, it was applied with some confidence. The same procedure was adopted by Thomas and Bent with somewhat poorer accuracy caused by low signal-to-noise ratios. Scharmann and Schartner,[167] who made only relative measurements of emission cross sections, used target pressures of 10^{-4} torr; this provided emission cross sections independent of target density for all projectile energies of the experiment.

Dodd and Hughes,[156] and also Denis et al.[160] attempted to use a theoretical treatment of absorption to determine the level-excitation cross sections. This theoretical method, due to Phelps,[249] was based on Holstein's[261] treatment of photoabsorption in gases. Exact expressions are given for the transport of photons to walls of either an infinite cylinder or an infinite parallel-sided slab. For application to a more practical target chamber, the formulation for a cylinder is used, with a series of test values of cylinder radius and excitation cross section until agreement is obtained between the experimentally measured variation of emission and pressure. The "effective radius," so obtained, is substituted in the theory pertaining to the infinite cylinder and is utilized for correcting data taken at relatively high pressures. Any success of these procedures must clearly depend on there being a unique "effective radius." This approach was successfully used by Gabriel and Heddle[243] for work on the emission from an electron impact experiment. Dodd and Hughes[156] applied it to proton impact studies to obtain the final results presented here, but they stated that the determination of "effective radius" was not reliable. Denis, Dufay, and their co-workers[253] could not determine a unique value of "effective radius" at all; the value that was used to determine the cross section was a radius that fitted the low-pressure data only and did not agree with high-pressure measurements. Thomas and Bent[161] also mention that they attempted this approach and could not derive a unique value of radius to fit all measurements. Clearly, if it proves impossible to determine a unique value of "effective radius," then there is no justification for using the procedure.

In the work of Denis, Dufay, and others the use of a value of "effective radius" peculiar to a very limited range of pressures is obviously an incorrect procedure.

According to the experience of Thomas and Bent,[262] the use of Phelp's theory[249] with an "effective radius" determined from high-pressure measurements would give a value of excitation cross section that is somewhat lower than that determined by direct measurement under pressure-independent conditions. Reference to Data Tables 6-2-6 through 6-2-8 will indicate that the data of Dodd and Hughes,[156] and of Denis et al.[160] lie considerably below the data of Van den Bos et al.,[165] and of Thomas and Bent;[161] the former groups used the absorption theory, while the latter groups used pressure-independent conditions. This is in complete contrast to the situation for the ^1S levels where the systematic differences can be definitely ascribed to errors in calibration. Therefore, the measurements of level-excitation cross section by Van den Bos et al.[165] and those by Thomas and Bent[161] are reliable since independence of cross section from target density is demonstrated; in contrast, the work of Dodd and Hughes[156] and of Denis et al.[160] is invalid because independence of cross section from target density is not demonstrated. Measurements of relative variation of emission cross section with impact energy by Scharmann and Schartner[167] were carried out under pressure-independent conditions and are therefore also valid.

Theoretical calculations of intrinsically high accuracy are available for the 3^1P excitation cross section. Thomas[250] shows that the experimental results of Thomas and Bent[161] are in very close agreement with the distortion approximation predictions of Bell[263] over a very wide range of energies. At the higher end of the range of energies covered by Thomas and Bent,[161] these calculations are likely to be more accurate than any of the existing experimental calculations. Agreement between theory and experiment at high energies gives us confidence in the experimental data of Thomas and Bent; the measurements of Van den Bos at lower energies are in reasonable conformity with those of Thomas and Bent.

Useful comparisons may be made of the functional dependence of cross section on projectile energy determined from the measurements of various groups. Figure 6-2 shows the measured cross sections for the formation of the 3^1P state, normalized to unity at 130-keV. Scharmann and Schartner's[167] results that represent only the emission cross section are approximately proportional to level-excitation cross section; their results normalized to the work of Thomas and Bent[161] at 140-keV are included on this composite diagram. The agreement among the various determinations is rather poor. It is argued that one should neglect the work of Dodd and Hughes[156] and that of Denis et al.[160] on the grounds that the measurements involved the theoretical assessment of the resonance absorption problem, and that the validity of the procedure for the experiments in question was not properly established. It is possible that the error due to this cause will be energy-dependent. The cross section for absorption of photons will be altered if the wavelength of emission is Doppler-shifted due to the excited atom recoiling from the collision with an appreciable velocity; the mean recoil velocity will undoubtedly vary with projectile impact energy. A further justification for neglecting the work of Denis et al.[160] is that the projectile beam traversing the collision region contained an appreciable fraction of neutralized atoms. The only remaining determinations are those by Thomas and Bent,[161] by Van den Bos et al.,[165] and by Scharmann and Schartner.[167] The work of Thomas and Bent agrees with that of Scharmann and Schartner within the combined limits of accuracy of the two determinations. Strictly speaking, both determinations involve

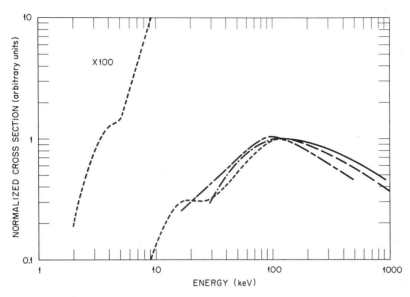

Fig. 6-2   Relative cross sections for the excitation of the 3^1P state of helium by the impact of H^+ and D^+ on a helium target. All measurements have been normalized to unity at 130 keV: _ _ _ Van den Bos et al.[165] (from 2 to 150 keV); ___ . ___ Dodd and Hughes[156] (from 30 to 130 keV); _____ Thomas and Bent[161] [(from 75 to 900 keV) including H^+ and D^+ data]; ___ _ ___ Denis et al. (from 20 to 500 keV); ___ ___ Scharmann and Schartner[167] [(from 140 to 1050 keV) normalized to the work of Thomas and Bent[161] at 140 keV; strictly speaking, this is a line-emission cross section (see text)].

errors; the work of Scharmann and Schartner is not corrected for cascade population, and the work of Thomas and Bent is not corrected for polarization effects. Application of a correction for cascade would, in fact, bring the sets of data into rather good agreement. The effect of polarization cannot be included in retrospect. It is concluded that the functional dependence of cross section on impact energy is established most reliably by the work of Thomas and Bent,[161] of Scharmann and Schartner,[167] and of Van den Bos et al.[165] Comparisons of data for higher n^1P states lead to similar conclusions.

Above 50-keV impact energy, the cross section for the various n^1P states exhibits a similar variation with energy, independent of the principal quantum number. It is useful to compare the average values of the cross section ratios $\sigma(n)/\sigma(n + 1)$ calculated from the data produced by various research groups. This comparison is shown in Table 6-2. The data by Denis et al.[160] were obtained under poor experimental conditions, and the ratios of cross sections vary systematically with energy. The work of Van den Bos et al.[165] agrees with that by Thomas and Bent[161] for the

TABLE 6-2   RATIOS OF CROSS SECTIONS AT ENERGIES ABOVE 50-KEV

	$\dfrac{\sigma(3^1P)}{\sigma(4^1P)}$	$\dfrac{\sigma(4^1P)}{\sigma(5^1P)}$	$\dfrac{\sigma(5^1P)}{\sigma(6^1P)}$
Van den Bos et al.[165]	3.80	2.10	
Thomas and Bent[161]	4.28	2.17	1.59
Denis et al.[160]	4.49	3.86	

$\sigma(4^1P)/\sigma(5^1P)$ ratio but exhibits substantial disagreement in the case of the $\sigma(3^1P)/\sigma(4^1P)$ ratio.

In conclusion, it is suggested that only the absolute magnitudes of the measurements by Van den Bos et al.[165] and those by Thomas and Bent[161] should be regarded as valid; however, the rather good agreement between these two determinations is probably fortuitous. The most reliable estimate of the functional energy dependence of the cross section may also be established by these two series of measurements, plus the relative measurements by Scharmann and Schartner.[167]

C.   Excitation of the n^1D Levels of Helium

The n^1D levels are known to be readily populated by collisional transfer from adjacent levels. With the exception of measurements by Denis et al.,[160] all the work that pertains to the formation of the n^1D levels has been carried out under conditions where secondary populating mechanisms made a negligible contribution to the measurements. Denis et al.[160] carried out measurements as a function of pressure and attempted to extrapolate to zero pressure; this is a procedure likely to be inaccurate. Also, the target pressures utilized by Denis et al.[160] were sufficiently high to make an appreciable neutralization of the projectile beam take place as it traversed the observation region. Consequently, there will be a contribution to the signal from excitation by impact of H atoms in secondary collisions.

Population of n^1D states by cascade from higher ^1P levels is estimated to be less than one percent of the population by direct collisional excitation for all energies covered by available measurements. There have been no experimental measurements of cascade from higher ^1F levels into the n^1D states. According to the theoretical predictions of Van den Bos,[246] the cross sections for collisional excitation of n^1F levels are some two orders of magnitude less than those for the corresponding n^1D levels resulting in cascade populations of less than one percent.

The polarization fractions for the $n^1D \rightarrow 2^1P$ transitions are large, ranging from a maximum in the energy region 10 to 20-keV of 40—50%,[164, 165] and decreasing with energy to -20% in the region of 1000-keV.[166] The experiments by Van den Bos et al.[165] and of Scharmann and Schartner[167] included full corrections for the effect of polarization both in regard to the anisotropy of emission and to the instrumental polarization effects. Robinson and Gilbody[162] allow for anisotropy but do not correct for instrumental polarization; unpublished tests show that the neglect of instrumental polarization produces errors that are comparable to the statistical scatter of the data.[251] In the work of Denis et al.,[160] Thomas and Bent,[161] and Krause and Soltysik[164] the errors involved in neglect of polarization were ignored. Inadequate attention to the influence of polarization may cause errors in both the absolute magnitudes of the data and in the functional dependence of cross section on projectile energy. It is not possible to assess in retrospect the magnitude of the errors due to this cause.

Most of the data on the formation of n^1D states shown in Data Tables 6-2-10 through 6-2-13 are in the form of level-excitation cross sections. In addition, however, there are determinations of relative variation of line-emission cross section with impact energy carried out by Krause and Soltysik,[164] and by Scharmann and Schartner.[167] That cascade population of the n^1D levels is less than one percent of

the population by direct collisional excitation has been established. Hence, line-emission cross sections are proportional to level-excitation cross sections within this accuracy. In order to facilitate comparison of data from various sources, the relative line-emission cross sections have been normalized to level-excitation cross sections as determined by other groups. The data by Krause and Soltysik[164] are normalized to the work of Van den Bos et al.[165] at 10-keV energy. The work of Scharmann and Schartner[167] is normalized to that of Van den Bos et al.[165] at 140-keV impact energy for the 3^1D state, to the work of Thomas and Bent[161] at 145-keV energy for the 4^1D state, and to the work of Robinson and Gilbody[160] at 145-keV for the 5^1D state.

Comparison of the agreement among various experimental determinations of the functional dependence of cross section on projectile impact energy is valuable. Figure 6-3 shows the data for the formation of the 4^1D state with all determinations normalized to unity at a projectile energy of 130-keV. There is considerable agreement among the determinations. The work of Denis et al.[160] has been criticized on the grounds that the experiments were not carried out under single-collision conditions. This work should be deleted from the comparison. The work of Robinson and Gilbody lies above that of two other determinations at high projectile energies, but the discrepancy is comparable to the scatter of data points about the line of best fit. The agreement between the data of Thomas and Bent,[161] and that of Scharmann and Schartner[167] is surprisingly good, particularly because the work of the former group neglected to correct for the effects of polarization of the emission. It is concluded that the functional dependence of cross section on projectile impact energy is most reliably established by the work of Van den Bos et al.,[165] of Scharmann and Schartner,[167] and of Thomas and Bent.[161] The work of Robinson and Gilbody[162] is

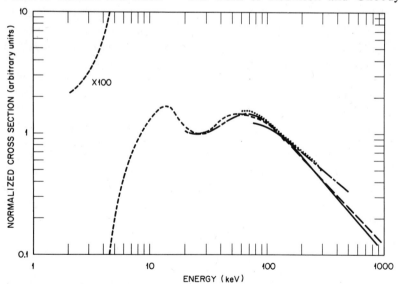

Fig. 6-3 Relative cross sections for the excitation of the 4^1D state of helium by the impact of H^+ and D^+ on a helium target. All measurements have been normalized to unity at 130 keV: _ _ _ Van den Bos et al.[165] (from 2 to 150 keV); _____ Thomas and Bent[161] [(from 75 to 900 keV) including H^+ and D^+ data]; _ _ _ _ _ Denis et al.[160] (from 20 to 500 keV); . . . . Robinson and Gilbody[162] (from 60 to 300 keV); _ _ _ _ _ Scharmann and Schartner[167] [(from 145 to 950 keV) normalized to the work of Thomas and Bent at 140 keV; strictly speaking, this is a line-emission cross section (see text)].

consistent with this functional dependence when allowance is made for the scatter of data points about the line of best fit. Similar conclusions apply to the measurements for other states in the n^1D series.

At impact energies above 50-keV., the energy dependence of the excitation cross sections are approximately the same for the 3^1D, 4^1D, 5^1D, and 6^1D states. It is useful to make a comparison of the cross section ratios $\sigma(n)/\sigma(n + 1)$ obtained by various researchers. Table 6-3, showing the average values of these ratios for impact energies above 50-keV, is drawn from the work of Van den Bos et al.[165] As far as comparisons are possible, the agreement is rather poor.

TABLE 6-3   RATIOS OF CROSS SECTIONS AT ENERGIES ABOVE 50-KEV

	$\dfrac{\sigma(3^1D)}{\sigma(4^1D)}$	$\dfrac{\sigma(4^1D)}{\sigma(5^1D)}$	$\dfrac{\sigma(5^1D)}{\sigma(6^1D)}$
Van den Bos et al.[165]	2.70	1.95	1.75
Robinson and Gilbody[162]		2.5	
Denis et al.[160]		2.12	1.93

It is not possible to make an objective judgement as to which of these measurements represents the most accurate absolute values. All four determinations of absolute magnitudes of cross sections are independent. In no case did an experimental procedure include attention to all of the criteria that have been established for valid, accurate data measurement. The discrepancies among data are larger than the estimated limits of accuracy assigned to the data by the various authors. It seems reasonable to suppose that the four available determinations should bracket the true values of the cross section.

Scharmann and Schartner have studied the various published measurements in detail and conclude that for impact energies above 100-keV there is, in fact, a small systematic variation of the functional dependence of cross section as one moves through the n^1D states. This dependence on the principal quantum number n is shown most clearly in Scharmann and Schartner's work,[167] which has high statistical accuracy; the dependence is also apparent in Thomas and Bent's work.[161] The fact that the data by Denis et al.[160] do not follow this pattern is not understood; perhaps their results should be discounted because inadequate attention was paid to ensuring that the measurements were carried out under single-collision conditions. Dependence of the functional form of the cross section on principal quantum number for impact energies above 100-keV seems to be definitely established by the available data.

In conclusion, it is suggested that the energy dependence of the n^1D cross sections as established by the work of Van den Bos et al.,[164] of Scharmann and Schartner,[167] and of Thomas and Bent[161] is quite accurate. This aspect of the data may be compared to theory with some confidence. It is not possible to make an objective decision as to which measurement represents the most accurate determination of cross section. Two factors—the general agreement about the form of the functional dependence of cross section on projectile energy, and the serious disagreement about magnitudes of cross sections—indicate that the major discrepancy among the various experiments lies in the determination of the detection efficiency of the optical system.

D. Excitation of Triplet States

It is generally believed that an LS coupling scheme correctly describes the low-lying states of helium. If interaction between a helium atom and a proton is given by Coulomb's law, then the incident proton should produce no change in the total spin of the helium atom. Therefore, it is expected that the triplet states will not be excited by proton impact.

Robinson and Gilbody[162] state that they observed no excitation of triplet states. Krause and Soltysik,[164] and van Eck et al.[155] note weak emission in some triplet systems at energies below 30-keV. Because of emission weakness, care should be taken to ensure that the excitation is formed by proton impact and is not due to some spurious effect.

The work of van Eck et al.[155] involved attention to the elimination of neutrals and electrons that might be formed in the projectile beam by H^+ impact on slit edges. Van Eck et al. present a single measurement of the $3^3P \to 2^3S$ emission cross section of 5.8×10^{-21} cm^2 at 30-keV impact energy. This probably represents a fairly reliable upper estimate to the cross section. It is to be compared with the cross section for excitation by impact of neutrals of 4.16×10^{-19} cm^2 at 30-keV (see Section 6-3).

The work of Krause and Soltysik[164] involved an indirect determination of the emission from the 3^3S and 3^3D states in the manner described in Section 6-2-B(6). The data were obtained from an analysis of a nonlinear dependence of intensity on target pressure. No attention was paid to eliminating neutrals and electrons that may have been formed by impact of projectiles on slit edges. The published diagram of the apparatus[164] shows the entrance to the target cell to be of smaller diameter than the so-called "collimating apertures," a feature which will tend to enhance the likelihood of secondary electrons entering the observation region. Such slit-edge effects would produce an emission that is linearly dependent on target pressure. It is concluded that the data of Krause and Soltysik may be unreliable and that further investigation is required with greater attention to the possible influence of fast neutrals and electrons. The measurements on the relative variation of the $3^3S \to 2^3P$ and $3^3D \to 2^3P$ line-emission cross sections are shown in Data Table 6-2-14. The cross sections amount to 1 to 2% of the cross sections for emission of these same lines induced by impact of hydrogen atoms.

It should be noted that the proton-induced emission data presented here include contributions from higher states by cascade. In the emission of triplet lines induced by impact of H, H_2^+, and H_3^+, the cascade contributions amount to 10 or 20% in many cases.[155, 200] It is possible that the emission of triplet lines induced by proton impact represents cascading from higher states and not direct excitation of the $n = 3$ levels.

E. The Formation of Excited He$^+$

The excited He$^+$ ion may be formed either by charge transfer or by ionization, and it is not possible to determine from a measurement of emission which process predominates. The various angular momentum states of He$^+$ have approximately the same energies, and the emitted spectral lines are a composite of all transitions between groups of states with the same principal quantum number. It is not possible to separate the transitions from individual states by spectroscopic methods; therefore

the data presented here are in the form of emission cross sections. The available data on the emission of the 4686-Å ($n = 4 \rightarrow n = 3$) and 3203-Å ($n = 5 \rightarrow n = 3$) lines are displayed in Data Tables 6-2-15 and 6-2-16.

The extensive series of measurements on helium excitation by van Eck et al.[155] includes emission data for the 4686-Å and 3203-Å He II lines. The majority of this first series of measurements was later repeated with improved techniques and with significant changes in the magnitudes of the data.[10, 165] No revisions were made to the He II work; therefore the absolute magnitudes of the original data by van Eck[155] that we present here may be in error by as much as a factor of two. There is a more recent measurement of the $n = 4 \rightarrow n = 3$ (4686-Å) He II emission by Moustafa Moussa and de Heer[163] utilizing the apparatus described by van Eck et al.[10] and by Van den Bos et al.[165] [see Section 6-1-A(1)]. This measurement ranges in energy from 30 to 150-keV and appears to indicate a considerably higher cross section than the earlier measurements by van Eck et al.;[155] these data are also displayed here. It would probably be justifiable to normalize the older data of van Eck et al. to that of Moustafa Moussa and de Heer, although this has not been done here. The original He II data by van Eck et al. are displayed in Fig. 5 of Ref. 155; it should be noted that there is an error in the legend for that figure; these data are "Line-Emission Cross Sections" not "Level-Excitation Cross Sections," as stated. Additional data on the emission of the He II $2p \rightarrow 1s$ and $3p \rightarrow 1s$ ultraviolet transitions are to be found in an unpublished thesis by van Eck.[264]

The other measurements of He II emission cross sections displayed in Data Table 6-2-15 are by Dodd and Hughs,[156] and Thomas and Bent.[161]

The functional energy dependence of the emission cross sections are essentially the same for all measurements. Systematic differences among absolute values are of similar magnitudes to those found in the data on the He I states. It is not possible to make an objective assessment as to which data provides the most accurate absolute value.

F. Polarization of the Emission from Excited Helium

The polarization fraction is determined by the ratio of two light intensities of the same wavelength. There are few opportunities for systematic errors in such a measurement. When the two light intensities are of approximately the same magnitude, the statistical accuracy of the polarization fraction may be very poor. It should be noted that it is inherently impossible for the $n^1S \rightarrow 2^1P$ emissions of He I to be polarized. This was verified experimentally by van Eck et al.[10] Data for a number of emissions from n^1P, n^1D, and 3^3D states of helium are shown in Data Tables 6-2-17 through 6-2-24 in Section 6-2-G.

Van Eck et al.[10] published measurements in the energy range 10—35-keV for a number of different transitions. The $3^1P \rightarrow 2^1S$ and $4^1D \rightarrow 2^1P$ measurements were repeated by Van den Bos et al.[165] using the same experimental arrangement but extended to a greater range of energies. The two series of measurements are in excellent agreement. For the present data compilation, the more recent data by Van den Bos et al.[165] are used for the $3^1P \rightarrow 2^1S$ and $4^1D \rightarrow 2^1P$ transitions, and the earlier data by van Eck et al.[10] are shown for the other emissions.

The work of Krause and Soltysik[164] included measurements of polarization of emission from the 3^1S, 3^1D, 3^3S, and 3^3D states. It was suggested in Section 6-1-B(6) that the apparatus contained an inherent instrumental polarization. The observation

of nonzero polarization in the 3^1S and 3^3S emissions is probably false, so this data is omitted. The 3^1D and 3^3D data are retained but may be in error due to this problem.

The work of Scharmann and Schartner includes measurements on both $n^1D \to 2^1P$ as well as $n^1P \to 2^1S$ lines. The measurements published in two early papers[256, 257] have been revised and extended in the last publication[166] from which the data utilized here are drawn.

In all experiments, proper precautions were taken to ensure that data on $n^1D \to 2^1P$ emissions were independent of pressure. For the $n^1P \to 2^1S$ emissions, which are greatly influenced by absorption of resonance photons, pressure-independent conditions could not be established even as low as 10^{-4} torr. The data on $n^1P \to 2^1S$ emissions presented in the work of van Eck, Van den Bos and co-workers,[10, 165] as well as the work of Scharmann and Schartner,[166] are essentially a "low-pressure" measurement of apparent polarization at about 10^{-4} torr and are not a true polarization fraction. It is not possible to estimate the likely change in polarization fraction if target pressures could be further reduced to attain a region where pressure independence would be exhibited. Scharmann and Schartner provide some interesting measurements of apparent polarization fraction as a function of both target density and projectile energy. The variation of polarization fraction with density exhibits a systematic dependence on the impact energy of the projectile.

The measurements of polarization in the $3^3D \to 2^3S$ transition given by Krause and Soltysik[164] were not obtained under pressure-independent conditions. Excitation of the n^3D states by proton impact is formally forbidden by conservation of spin. It has been suggested [Section 6-1-B(6)] that insufficient attention was directed to removing secondary populating processes, such as collisional excitation by secondary electrons liberated from slit edges. A substantial fraction of the triplet emission was due to two collision processes giving rise to a square law dependence on target density. The results shown here were obtained by extrapolating the apparent polarization fractions to zero pressure.

Van Eck, Van den Bos et al.,[10, 165] and also Scharmann and Schartner[166] document their precautions to ensure that depolarization by stray fields was eliminated. Such precautions were not mentioned in the reports by Krause and Soltysik.[164] In the former two groups of experiments, adequate attention was paid to the removal of instrumental polarization, whereas in the case of Krause and Soltysik[164] this problem was probably present.

It should be noted that in one paper by Scharmann and Schartner[256] unpublished data of Van den Bos were presented; these were subsequently shown to be erroneous. The data presented in this monograph are correct.

There is complete agreement between the work of Van den Bos et al.,[165] and of Scharmann and Schartner[166] over the small range of energies that are common to both determinations. The results for $n^1D \to 2^1P$ transitions should be quite reliable. The data for $n^1P \to 2^1S$ transitions should be treated with caution because pressure-independent conditions were not established in either determination.

G. Data Tables

For some experiments, the range of projectile velocities has been extended by the use of D^+. Such cases are specifically noted, and the data are shown on the tables as though they were protons with the same impact velocity; that is, it is shown as the deuteron energy divided by two.

1. Cross Sections for the Excitation of He States, Induced Impact of H⁺ on a He Target

6-2-1 3^1S State.   [Tables 6-2-1D and 6-2-1E represent relative variation with energy of the 7281-Å line of He I ($3^1S \to 2^1P$). These are shown respectively normalized to 35×10^{-20} cm² at 10-keV and 64×10^{-20} cm² at 140-keV.]

6-2-2 4^1S State.   [Table 6-2-2C is for D⁺ impact. Table 6-2-2G is the relative variation with energy of the 5048-Å line of He I ($4^1S \to 2^1P$) shown normalized to 1.57×10^{-19} cm² at 140-keV.]

6-2-3 5^1S State.   [Table 6-2-3F is the relative variation with energy of the 4438-Å line of He I ($5^1S \to 2^1P$) shown normalized to 9×10^{-20} cm² at 175-keV.]

6-2-4 6^1S State.

6-2-5 7^1S State.

6-2-6 3^1P State.   [Table 6-2-6C is for D⁺ impact. Table 6-2-6F is the relative variation with energy of the 5016-Å line of He I ($3^1P \to 2^1S$) shown normalized to 3.26×10^{-18} cm² at 140-keV.]

6-2-7 4^1P State. [Table 6-2-7D is the relative variation with energy of the 3965-Å line of He I ($4^1P \to 2^1S$) shown normalized to 7.6×10^{-19} cm² at 175-keV.]

6-2-8 5^1P State.

6-2-9 6^1P State.

6-2-10 3^1D State.   [Tables 6-2-10B and 6-2-10C represent relative variation with energy of the 6678-Å line of He I ($3^1D \to 2^1P$). These are shown respectively normalized to 37.4×10^{-20} cm² at 10-keV and 21.4×10^{-20} cm² at 140-keV.]

6-2-11 4^1D State.   [Table 6-2-11B is for D⁺ Impact. Table 6-2-11F is the relative variation with energy of the 4922-Å line of He I ($4^1D \to 2^1P$) shown normalized to 5.3×10^{-20} cm² at 145-keV.]

6-2-12 5^1D State.   [Table 6-2-12D is the relative variation with energy of the 4388-Å line of He I ($5^1D \to 2^1P$) shown normalized to 2.85×10^{-20} cm² at 145-keV impact energy.]

6-2-13 6^1D State.

2. Cross Sections for the Emission of He I Triplet Lines Induced by Impact of H⁺ on a He Target

6-2-14   (A) 7065-Å ($3^3S \to 2^3P$); (B) 5876-Å ($3^3D \to 2^3P$).

3. Cross Sections for the Emission of He II Lines Induced by the Impact of H⁺ on a He Target

6-2-15   4686-Å ($n = 4 \to n = 3$).

6-2-16   3203-Å ($n = 5 \to n = 3$).

4. Polarization Fractions of He I Lines Induced by the Impact of H⁺ on a He Target

6-2-17   5016-Å ($3^1P \to 2^1S$).†

6-2-18   3965-Å ($4^1P \to 2^1S$).†

6-2-19   3614-Å ($5^1P \to 2^1S$).†

6-2-20   6678-Å ($3^1D \to 2^1P$).

6-2-21   4922-Å ($4^1D \to 2^1P$).

6-2-22   4388-Å ($5^1D \to 2^1P$).

6-2-23   4144-Å ($6^1D \to 2^1P$).

6-2-24   5876-Å ($3^3D \to 2^3P$).

† These measurements were not carried out under pressure independent conditions.

6-2-1A
VAN DEN BOS ET AL. (165)

ENERGY (KEV)	CROSS SECTION (SQ. CM)
5.00E 00	7.20E-20
6.00E 00	1.13E-19
7.00E 00	1.65E-19
8.00E 00	2.30E-19
9.00E 00	2.75E-19
1.00E 01	3.50E-19
1.25E 01	4.60E-19
1.50E 01	5.50E-19
2.00E 01	1.08E-18
2.50E 01	1.43E-18
3.00E 01	1.49E-18
3.50E 01	1.69E-18
4.00E 01	1.63E-18
5.00E 01	1.67E-18
6.00E 01	1.43E-18
7.00E 01	1.21E-18
8.00E 01	1.31E-18
9.00E 01	1.14E-18
1.00E 02	1.01E-18
1.10E 02	1.06E-18
1.20E 02	9.93E-19
1.30E 02	8.63E-19
1.40E 02	8.92E-19
1.50E 02	8.13E-19

6-2-1B
DODD ET AL. (156)

ENERGY (KEV)	CROSS SECTICN (SQ. CM)
1.00E 01	4.15E-19
2.00E 01	7.30E-19
3.00E 01	1.06E-18
4.00E 01	1.19E-18
5.00E 01	1.17E-18
6.00E 01	1.04E-18
7.00E 01	1.02E-18
9.00E 01	8.80E-19
1.10E 02	8.00E-19
1.20E 02	7.50E-19

6-2-1C
DENIS ET AL. (160)

ENERGY (KEV)	CROSS SECTION (SQ. CM)
7.00E 01	1.20E-18
8.00E 01	1.10E-18
1.00E 02	9.00E-19
1.50E 02	6.00E-19
2.00E 02	4.30E-19

6-2-1D
KRAUSE ET AL. (164)

ENERGY (KEV)	CROSS SECTION (SQ. CM)
4.00E 00	4.13E-20
6.00E 00	1.27E-19
8.00E 00	2.39E-19
1.00E 01	3.50E-19
1.20E 01	4.44E-19
1.40E 01	5.67E-19
1.60E 01	7.40E-19
1.80E 01	9.52E-19
2.00E 01	1.17E-18

6-2-1E
SCHARMANN ET AL. (167)

ENERGY (KEV)	CROSS SECTION (SQ. CM)
1.25E 02	7.02E-19
1.40E 02	6.40E-19
1.60E 02	5.84E-19
1.85E 02	5.10E-19
2.20E 02	4.21E-19
2.75E 02	3.46E-19
3.15E 02	2.94E-19
3.40E 02	2.72E-19
4.30E 02	2.17E-19
5.30E 02	1.80E-19
6.70E 02	1.42E-19
8.00E 02	1.28E-19
9.10E 02	1.04E-19
1.05E 03	9.16E-20

6-2-2A
VAN DEN BOS ET AL. (165)

ENERGY (KEV)	CROSS SECTION (SQ. CM)
1.00E 00	2.80E-22
1.50E 00	8.50E-22
2.00E 00	1.20E-21
2.50E 00	2.80E-21
3.00E 00	3.00E-21
3.50E 00	2.90E-21
4.00E 00	2.20E-21
4.50E 00	1.70E-21
5.00E 00	4.10E-21
5.50E 00	1.08E-20
6.00E 00	1.85E-20
6.50E 00	2.65E-20
7.00E 00	3.55E-20
7.50E 00	4.70E-20
8.00E 00	5.78E-20
8.50E 00	6.66E-20
9.00E 00	7.68E-20
9.50E 00	8.45E-20
1.00E 01	8.82E-20
1.25E 01	1.04E-19
1.50E 01	1.15E-19
2.00E 01	2.06E-19
2.50E 01	3.38E-19
3.00E 01	4.08E-19
3.50E 01	4.48E-19
4.00E 01	4.89E-19
5.00E 01	4.71E-19
6.00E 01	4.48E-19
7.00E 01	4.51E-19
8.00E 01	4.17E-19
9.00E 01	3.89E-19
1.00E 02	3.51E-19
1.10E 02	3.21E-19
1.20E 02	2.96E-19
1.30E 02	2.73E-19
1.40E 02	2.47E-19
1.50E 02	2.21E-19

6-2-2B
DODD ET AL. (156)

ENERGY (KEV)	CROSS SECTION (SQ. CM)
2.00E 01	2.63E-19
3.00E 01	5.60E-19
4.00E 01	6.80E-19
5.00E 01	6.90E-19
6.00E 01	6.70E-19
7.00E 01	6.10E-19
8.00E 01	5.60E-19
9.00E 01	5.20E-19
1.00E 02	4.90E-19
1.10E 02	4.60E-19
1.30E 02	4.05E-19

6-2-2C
THOMAS ET AL. (161)

ENERGY (KEV)	CROSS SECTION (SQ. CM)
7.50E 01	2.65E-19
1.00E 02	2.13E-19
1.25E 02	1.69E-19
1.50E 02	1.36E-19
1.75E 02	1.16E-19
2.00E 02	1.07E-19

6-2-2D
THOMAS ET AL. (161)

ENERGY (KEV)	CROSS SECTICN (SQ. CM)
1.50E 02	1.27E-19
2.00E 02	1.07E-19
2.50E 02	8.90E-20
3.00E 02	7.39E-20
3.50E 02	6.28E-20
4.00E 02	5.64E-20
4.50E 02	4.89E-20
5.00E 02	4.51E-20
6.00E 02	3.63E-20
7.00E 02	3.07E-20
8.00E 02	2.91E-20
9.00E 02	2.50E-20

6-2-2E
ROBINSON ET AL. (162)

ENERGY (KEV)	CROSS SECTICN (SQ. CM)
6.00E 01	3.40E-19
7.00E 01	3.00E-19
8.00E 01	2.70E-19
9.00E 01	2.70E-19
1.00E 02	2.15E-19
1.30E 02	1.78E-19
1.50E 02	1.65E-19

ENERGY (KEV)	CROSS SECTION (SQ. CM)
1.80E 02	1.30E-19
2.00E 02	1.31E-19
2.30E 02	1.19E-19
2.50E 02	1.06E-19
3.00E 02	8.25E-20
3.50E 02	7.40E-20

6-2-2F
DENIS ET AL. (160)

ENERGY (KEV)	CROSS SECTION (SQ. CM)
2.00E 01	2.40E-19
3.00E 01	3.60E-19
4.00E 01	4.20E-19
5.00E 01	4.20E-19
6.00E 01	4.00E-19
7.00E 01	3.80E-19
8.00E 01	3.40E-19
1.00E 02	2.60E-19
1.50E 02	1.70E-19
2.00E 02	1.25E-19
3.00E 02	8.00E-20
4.00E 02	6.00E-20
5.00E 02	4.60E-20

6-2-2G
SCHARMANN ET AL. (167)

ENERGY (KEV)	CROSS SECTION (SQ. CM)
1.40E 02	1.57E-19
1.75E 02	1.26E-19
2.70E 02	8.46E-20
2.75E 02	8.14E-20
3.15E 02	6.98E-20
3.80E 02	5.62E-20
5.00E 02	4.61E-20
5.80E 02	3.83E-20
7.10E 02	3.25E-20
8.40E 02	2.52E-20
9.10E 02	2.33E-20

6-2-3A
VAN DEN BOS ET AL. (165)

ENERGY (KEV)	CROSS SECTION (SQ. CM)
5.00E 00	1.60E-21
5.50E 00	3.80E-21
6.00E 00	5.90E-21
6.50E 00	7.30E-21
7.00E 00	9.20E-21
7.50E 00	1.50E-20
8.00E 00	2.09E-20
8.50E 00	2.67E-20
9.00E 00	3.23E-20
9.50E 00	3.76E-20
1.00E 01	4.29E-20
1.25E 01	4.63E-20
1.50E 01	5.25E-20
2.00E 01	8.01E-20
2.50E 01	1.65E-19
3.00E 01	2.22E-19
3.50E 01	2.56E-19
4.00E 01	2.77E-19
5.00E 01	2.64E-19
6.00E 01	2.75E-19
7.00E 01	2.55E-19
8.00E 01	2.34E-19
9.00E 01	2.24E-19
1.00E 02	2.05E-19
1.10E 02	1.93E-19
1.20E 02	1.72E-19
1.30E 02	1.68E-19
1.40E 02	1.70E-19
1.50E 02	1.55E-19

6-2-3B
DODD ET AL. (156)

ENERGY (KEV)	CROSS SECTION (SQ. CM)
2.00E 01	1.16E-19
3.00E 01	2.63E-19
4.00E 01	3.20E-19
5.00E 01	3.30E-19
6.00E 01	3.00E-19
7.00E 01	3.00E-19
8.00E 01	2.64E-19
9.00E 01	2.52E-19
1.00E 02	2.30E-19
1.10E 02	2.17E-19
1.30E 02	1.91E-19

6-2-3C
THOMAS ET AL. (161)

ENERGY (KEV)	CROSS SECTION (SQ. CM)
1.50E 02	9.09E-20
2.00E 02	7.22E-20
2.50E 02	5.60E-20
3.00E 02	4.36E-20
3.50E 02	3.78E-20
4.00E 02	3.06E-20
4.50E 02	2.68E-20
5.00E 02	2.50E-20
6.00E 02	2.04E-20
7.00E 02	1.69E-20
8.00E 02	1.55E-20
9.00E 02	1.11E-20

6-2-3D
ROBINSON ET AL. (162)

ENERGY (KEV)	CROSS SECTION (SQ. CM)
6.00E 01	1.90E-19
7.00E 01	1.80E-19
8.00E 01	1.60E-19
9.00E 01	1.45E-19
1.00E 02	1.30E-19
1.30E 02	9.20E-20
1.50E 02	8.60E-20
2.00E 02	6.20E-20
2.50E 02	4.70E-20
3.00E 02	3.60E-20

6-2-3E
DENIS ET AL. (160)

ENERGY (KEV)	CROSS SECTION (SQ. CM)
2.00E 01	9.50E-20
3.00E 01	1.50E-19
4.00E 01	1.80E-19
5.00E 01	1.90E-19
6.00E 01	1.80E-19
7.00E 01	1.60E-19
8.00E 01	1.40E-19
1.00E 02	1.10E-19
1.50E 02	7.00E-20
2.00E 02	5.20E-20
3.00E 02	3.40E-20
4.00E 02	2.50E-20
5.00E 02	2.00E-20

6-2-3F
SCHARMANN ET AL. (167)

ENERGY (KEV)	CROSS SECTION (SQ. CM)
1.40E 02	1.17E-19
1.75E 02	9.00E-20
2.70E 02	6.00E-20
2.75E 02	5.83E-20
3.15E 02	4.91E-20
3.80E 02	4.02E-20
5.00E 02	3.05E-20
5.80E 02	2.59E-20
7.10E 02	2.22E-20
8.40E 02	1.75E-20
9.10E 02	1.62E-20

6-2-4A
VAN DEN BOS ET AL. (165

ENERGY (KEV)	CROSS SECTION (SQ. CM)
1.00E 01	2.20E-20
1.25E 01	2.40E-20
1.50E 01	2.70E-20
1.75E 01	3.00E-20
2.00E 01	3.55E-20
2.50E 01	9.10E-20
3.00E 01	1.19E-19
3.50E 01	1.25E-19
4.00E 01	1.51E-19
5.00E 01	1.51E-19
6.00E 01	1.51E-19
7.00E 01	1.48E-19
8.00E 01	1.36E-19
9.00E 01	1.20E-19
1.00E 02	1.15E-19
1.10E 02	1.08E-19
1.20E 02	1.04E-19
1.30E 02	8.60E-20
1.40E 02	8.00E-20
1.50E 02	7.80E-20

6-2-4B
THOMAS ET AL. (161)

ENERGY (KEV)	CROSS SECTION (SQ. CM)
1.50E 02	6.20E-20
2.00E 02	4.86E-20
2.50E 02	3.45E-20
3.00E 02	2.55E-20
3.50E 02	2.34E-20
4.00E 02	1.92E-20
4.50E 02	1.59E-20
5.00E 02	1.43E-20
6.00E 02	1.21E-20
7.00E 02	1.01E-20
8.00E 02	9.23E-21
9.00E 02	8.30E-21

6-2-4C
DENIS ET AL. (160)

ENERGY (KEV)	CROSS SECTION (SQ. CM)
2.00E 01	4.00E-20
3.00E 01	6.50E-20
4.00E 01	8.00E-20
5.00E 01	9.00E-20
6.00E 01	9.00E-20
7.00E 01	8.00E-20
8.00E 01	7.00E-20
1.00E 02	5.50E-20
1.50E 02	3.60E-20
2.00E 02	2.70E-20
3.00E 02	1.80E-20
4.00E 02	1.30E-20
5.00E 02	1.10E-20

6-2-5
THOMAS ET AL. (161)

ENERGY (KEV)	CROSS SECTION (SQ. CM)
1.50E 02	3.52E-20
2.00E 02	2.75E-20
2.50E 02	1.85E-20
3.00E 02	1.47E-20
3.50E 02	1.30E-20
4.00E 02	1.04E-20
4.50E 02	9.68E-21
5.00E 02	8.65E-21
6.00E 02	6.51E-21
7.00E 02	6.51E-21
8.00E 02	5.11E-21
9.00E 02	4.85E-21

6-2-6A
VAN DEN BOS ET AL. (165)

ENERGY (KEV)	CROSS SECTION (SQ. CM)
2.00E 00	7.60E-21
3.00E 00	2.88E-20
4.00E 00	5.12E-20
5.00E 00	5.52E-20
6.00E 00	1.02E-19
7.00E 00	1.82E-19
8.00E 00	2.51E-19
9.00E 00	3.67E-19
1.00E 01	5.55E-19
1.25E 01	7.76E-19
1.50E 01	1.09E-18
2.00E 01	1.23E-18
2.50E 01	1.22E-18
3.00E 01	1.27E-18
3.50E 01	1.52E-18
4.00E 01	1.66E-18
5.00E 01	2.37E-18
6.00E 01	2.85E-18
7.00E 01	2.86E-18
8.00E 01	3.44E-18
9.00E 01	3.53E-18
1.00E 02	3.80E-18
1.10E 02	4.00E-18
1.20E 02	3.90E-18
1.30E 02	3.98E-18
1.40E 02	4.03E-18
1.50E 02	3.94E-18
4.00E 02	2.43E-18
4.50E 02	2.28E-18
5.00E 02	2.10E-18
6.00E 02	1.99E-18
7.00E 02	1.75E-18
8.00E 02	1.67E-18
9.00E 02	1.47E-18

6-2-6B
CODD ET AL. (156)

ENERGY (KEV)	CROSS SECTION (SQ. CM)
3.00E 01	7.50E-19
4.00E 01	1.16E-18
5.00E 01	1.48E-18
6.00E 01	1.75E-18
7.00E 01	1.95E-18
8.00E 01	2.10E-18
9.00E 01	2.20E-18
1.00E 02	2.30E-18
1.10E 02	2.37E-18
1.20E 02	2.30E-18
1.30E 02	2.32E-18

6-2-6C
THOMAS ET AL. (161)

ENERGY (KEV)	CROSS SECTION (SQ. CM)
7.50E 01	2.88E-18
1.00E 02	3.30E-18
1.25E 02	3.24E-18
1.50E 02	3.26E-18
1.75E 02	3.03E-18
2.00E 02	3.20E-18

6-2-6D
THOMAS ET AL. (161)

ENERGY (KEV)	CROSS SECTION (SQ. CM)
1.50E 02	3.28E-18
2.00E 02	3.17E-18
2.50E 02	2.81E-18
3.00E 02	2.61E-18
3.50E 02	2.78E-18

6-2-6E
DENIS ET AL. (160)

ENERGY (KEV)	CROSS SECTION (SQ. CM)
2.00E 01	3.00E-19
3.00E 01	4.20E-19
4.00E 01	5.50E-19
5.00E 01	7.00E-19
6.00E 01	8.00E-19
7.00E 01	9.00E-19
8.00E 01	1.00E-18
1.00E 02	1.05E-18
1.50E 02	9.20E-19
2.00E 02	8.00E-19
3.00E 02	6.20E-19
4.00E 02	5.30E-19
5.00E 02	4.60E-19

6-2-6F
SCHARMANN ET AL. (167)

ENERGY (KEV)	CROSS SECTION (SQ. CM)
1.40E 02	3.26E-18
1.75E 02	3.02E-18
2.00E 02	2.89E-18
2.35E 02	2.77E-18
2.60E 02	2.86E-18
2.90E 02	2.44E-18
3.60E 02	2.24E-18
4.20E 02	2.03E-18
6.00E 02	1.65E-18
7.20E 02	1.52E-18
8.90E 02	1.30E-18
1.05E 03	1.17E-18

6-2-7A
VAN DEN BOS ET AL. (165)

ENERGY (KEV)	CROSS SECTION (SQ. CM)
1.00E 01	1.20E-19
1.25E 01	2.00E-19
1.50E 01	2.60E-19
2.00E 01	3.20E-19
2.50E 01	3.20E-19
3.00E 01	3.65E-19
3.50E 01	4.20E-19
4.00E 01	4.50E-19
5.00E 01	6.00E-19
6.00E 01	7.00E-19
7.00E 01	7.90E-19
8.00E 01	9.40E-19
9.00E 01	9.10E-19

ENERGY (KEV)	CROSS SECTION (SQ. CM)
1.00E 02	9.50E-19
1.10E 02	1.15E-18
1.20E 02	1.05E-18
1.30E 02	1.08E-18
1.40E 02	1.03E-18
1.50E 02	1.04E-18

6-2-7B
THOMAS ET AL. (161)

ENERGY (KEV)	CROSS SECTION (SQ. CM)
1.50E-01	7.53E-19
2.00E-01	7.48E-19
2.50E-01	7.15E-19
3.00E-01	6.65E-19
3.50E-01	5.96E-19
4.00E-01	5.88E-19
4.50E-01	5.48E-19
5.00E-01	5.11E-19
6.00E-01	4.37E-19
7.00E-01	4.06E-19
8.00E-01	3.40E-19
9.00E-01	3.64E-19

6-2-7C
DENIS ET AL. (160)

ENERGY (KEV)	CROSS SECTION (SQ. CM)
2.00E 01	6.00E-20
3.00E 01	8.50E-20
4.00E 01	1.10E-19
5.00E 01	1.30E-19
6.00E 01	1.60E-19
7.00E 01	1.90E-19
8.00E 01	2.10E-19
1.00E 02	2.50E-19
1.50E 02	2.50E-19
2.00E 02	2.10E-19
3.00E 02	1.60E-19
4.00E 02	1.32E-19
5.00E 02	1.18E-19

6-2-7D
SCHARMANN ET AL. (167)

ENERGY (KEV)	CROSS SECTION (SQ. CM)
1.40E 02	9.05E-19
1.75E 02	7.60E-19
2.00E 02	8.03E-19
2.35E 02	6.83E-19
2.60E 02	7.48E-19
3.60E 02	5.65E-19
4.20E 02	5.28E-19
5.30E 02	4.48E-19
6.00E 02	4.42E-19
7.20E 02	3.86E-19
8.20E 02	3.46E-19
8.90E 02	3.43E-19
1.05E 03	2.93E-19

6-2-8A
VAN DEN BOS ET AL. (165)

ENERGY (KEV)	CROSS SECTION (SQ. CM)
1.00E 01	6.00E-20
1.25E 01	1.40E-19
1.50E 01	1.70E-19
2.00E 01	2.40E-19
2.50E 01	2.20E-19
3.00E 01	2.20E-19
3.50E 01	2.20E-19
4.00E 01	2.20E-19
5.00E 01	2.90E-19
6.00E 01	3.60E-19
7.00E 01	3.90E-19
8.00E 01	4.00E-19
9.00E 01	4.30E-19
1.00E 02	4.40E-19
1.10E 02	4.80E-19
1.20E 02	4.80E-19
1.30E 02	4.80E-19
1.40E 02	5.00E-19
1.50E 02	5.20E-19

6-2-8B
THOMAS ET AL. (161)

ENERGY (KEV)	CROSS SECTION (SQ. CM)
1.50E-01	4.01E-19
2.00E-01	3.70E-19
2.50E-01	3.69E-19
3.00E-01	3.08E-19
3.50E-01	2.86E-19
4.00E-01	2.61E-19
4.50E-01	2.30E-19
5.00E-01	2.17E-19
6.00E-01	2.03E-19
7.00E-01	1.83E-19
8.00E-01	1.68E-19
9.00E-01	1.44E-19

6-2-8C
DENIS ET AL. (160)

ENERGY (KEV)	CROSS SECTION (SQ. CM)
3.00E 01	2.30E-20
4.00E 01	3.00E-20
5.00E 01	3.60E-20
6.00E 01	4.20E-20
7.00E 01	5.00E-20
8.00E 01	5.50E-20
1.00E 02	6.40E-20
1.50E 02	6.00E-20
2.00E 02	5.00E-20
3.00E 02	4.00E-20
4.00E 02	3.40E-20
5.00E 02	3.00E-20

6-2-9
THOMAS ET AL. (161)

ENERGY (KEV)	CROSS SECTION (SQ. CM)
1.50E 02	2.76E-19
2.00E 02	2.26E-19
2.50E 02	2.22E-19
3.00E 02	1.93E-19
3.50E 02	1.58E-19
4.00E 02	1.64E-19
4.50E 02	1.44E-19
5.00E 02	1.52E-19
6.00E 02	1.34E-19
7.00E 02	1.18E-19
8.00E 02	9.90E-20
9.00E 02	9.10E-20

6-2-10A
VAN DEN BOS ET AL. (165)

ENERGY (KEV)	CROSS SECTION (SQ. CM)
3.50E 00	4.50E-20
4.00E 00	6.20E-20
5.00E 00	1.40E-19
6.00E 00	2.44E-19
7.00E 00	3.14E-19
8.00E 00	3.38E-19
9.00E 00	3.50E-19
1.00E 01	3.74E-19
1.25E 01	4.64E-19
1.50E 01	4.84E-19
2.00E 01	3.74E-19
2.50E 01	2.73E-19
3.00E 01	2.93E-19
3.50E 01	3.34E-19
4.00E 01	3.50E-19
5.00E 01	3.67E-19
6.00E 01	3.64E-19
7.00E 01	3.49E-19
8.00E 01	3.22E-19
9.00E 01	2.91E-19
1.00E 02	2.70E-19
1.10E 02	2.50E-19
1.20E 02	2.36E-19
1.30E 02	2.26E-19
1.40E 02	2.14E-19
1.50E 02	2.02E-19

6-2-10B
KRAUSE ET AL. (164)

ENERGY (KEV)	CROSS SECTION (SQ. CM)
1.45E 00	2.10E-21
1.94E 00	6.70E-21
2.46E 00	1.29E-20
2.82E 00	1.57E-20
3.10E 00	2.00E-20
3.30E 00	2.72E-20
3.45E 00	3.39E-20
3.63E 00	4.48E-20

4.00E 00	7.49E-20
4.50E 00	1.06E-19
6.00E 00	2.05E-19
8.00E 00	2.94E-19
1.00E 01	3.74E-19
1.20E 01	3.95E-19
1.40E 01	3.93E-19
1.60E 01	3.60E-19
1.80E 01	3.28E-19
2.00E 01	2.89E-19

6-2-10C
SCHARMANN ET AL. (167)

ENERGY (KEV)	CROSS SECTION (SQ. CM)
1.25E 02	2.35E-19
1.40E 02	2.14E-19
1.60E 02	2.02E-19
1.80E 02	1.65E-19
2.20E 02	1.31E-19
2.75E 02	1.10E-19
3.15E 02	9.32E-20
3.40E 02	9.53E-20
4.30E 02	6.84E-20
5.30E 02	5.81E-20
6.80E 02	4.36E-20
8.00E 02	3.78E-20
9.10E 02	3.27E-20
1.05E 03	2.87E-20

6-2-11A
VAN DEN BOS ET AL. (165)

ENERGY (KEV)	CROSS SECTION (SQ. CM)
2.00E 00	2.30E-21
3.00E 00	3.40E-21
4.00E 00	7.20E-21
5.00E 00	1.90E-20
6.00E 00	4.81E-20
7.00E 00	7.56E-20
8.00E 00	1.02E-19
9.00E 00	1.22E-19
1.00E 01	1.40E-19
1.25E 01	1.85E-19
1.50E 01	1.83E-19
1.75E 01	1.50E-19
2.00E 01	1.29E-19
2.50E 01	1.13E-19
3.00E 01	1.16E-19
3.50E 01	1.32E-19
4.00E 01	1.45E-19
5.00E 01	1.52E-19
6.00E 01	1.58E-19
7.00E 01	1.51E-19
8.00E 01	1.48E-19
9.00E 01	1.42E-19
1.00E 02	1.27E-19
1.10E 02	1.24E-19
1.20E 02	1.14E-19
1.30E 02	1.09E-19
1.40E 02	9.90E-20
1.50E 02	9.10E-20

6-2-11B
THOMAS ET AL. (161)

ENERGY (KEV)	CROSS SECTION (SQ. CM)
7.50E 01	7.45E-20
1.00E 02	7.13E-20
1.25E 02	6.06E-20
1.50E 02	5.06E-20
1.75E 02	4.06E-20
2.00E 02	3.73E-20

6-2-11C
THOMAS ET AL. (161)

ENERGY (KEV)	CROSS SECTION (SQ. CM)
1.50E 02	4.92E-20
2.00E 02	3.82E-20
2.50E 02	3.12E-20
3.00E 02	2.50E-20
3.50E 02	2.11E-20
4.00E 02	1.76E-20
4.50E 02	1.54E-20
5.00E 02	1.34E-20
6.00E 02	1.17E-20
7.00E 02	9.97E-21
8.00E 02	8.75E-21
9.00E 02	7.16E-21

6-2-11D
ROBINSON ET AL. (162)

ENERGY (KEV)	CROSS SECTION (SQ. CM)
6.00E 01	1.15E-19
7.00E 01	1.17E-19
8.00E 01	1.05E-19
9.00E 01	1.03E-19
1.00E 02	8.60E-20
1.30E 02	7.35E-20
1.50E 02	6.50E-20
1.80E 02	6.10E-20
2.00E 02	5.18E-20
2.30E 02	4.90E-20
2.50E 02	4.33E-20
3.00E 02	3.40E-20

6-2-11E
DENIS ET AL. (160)

ENERGY (KEV)	CROSS SECTION (SQ. CM)
2.00E 01	1.00E-19
3.00E 01	9.50E-20
4.00E 01	1.15E-19
5.00E 01	1.30E-19
6.00E 01	1.42E-19
7.00E 01	1.45E-19
8.00E 01	1.31E-19
1.00E 02	1.20E-19
1.50E 02	8.50E-20
2.00E 02	6.70E-20
3.00E 02	4.80E-20
4.00E 02	3.80E-20
5.00E 02	3.20E-20

6-2-11F
SCHARMANN ET AL. (167)

ENERGY (KEV)	CROSS SECTION (SQ. CM)
1.45E 02	5.30E-20
1.85E 02	3.96E-20
2.20E 02	3.40E-20
2.70E 02	2.76E-20
3.10E 02	2.31E-20
4.05E 02	1.86E-20
5.10E 02	1.51E-20
6.20E 02	1.16E-20
8.20E 02	8.62E-21
9.50E 02	7.75E-21

6-2-12A
VAN DEN BOS ET AL. (165)

ENERGY (KEV)	CROSS SECTION (SQ. CM)
2.00E 00	9.60E-22
3.00E 00	2.87E-21
4.00E 00	3.80E-21
5.00E 00	7.80E-21
6.00E 00	1.52E-20
7.00E 00	2.68E-20
8.00E 00	4.36E-20
9.00E 00	5.66E-20
1.00E 01	6.80E-20
1.25E 01	1.06E-19
1.50E 01	1.06E-19
2.00E 01	7.46E-20
2.50E 01	5.98E-20
3.00E 01	5.72E-20
3.50E 01	5.95E-20
4.00E 01	6.75E-20
5.00E 01	8.69E-20
6.00E 01	7.66E-20
7.00E 01	7.56E-20
8.00E 01	7.72E-20
9.00E 01	7.37E-20
1.00E 02	6.86E-20
1.10E 02	6.17E-20
1.20E 02	5.88E-20
1.30E 02	5.18E-20
1.40E 02	4.76E-20
1.50E 02	4.55E-20

6-2-12B
ROBINSON ET AL. (162)

ENERGY (KEV)	CROSS SECTION (SQ. CM)
8.00E 01	4.35E-20
9.00E 01	4.20E-20
1.00E 02	3.85E-20
1.30E 02	3.10E-20

ENERGY (KEV)	CROSS SECTION (SQ. CM)
1.50E 02	2.78E-20
1.80E 02	2.15E-20
2.00E 02	1.95E-20
2.30E 02	1.80E-20

6-2-12C
DENIS ET AL. (160)

ENERGY (KEV)	CROSS SECTION (SQ. CM)
2.00E 01	4.60E-20
3.00E 01	4.20E-20
4.00E 01	5.00E-20
5.00E 01	5.70E-20
6.00E 01	6.50E-20
7.00E 01	7.00E-20
8.00E 01	7.00E-20
1.00E 02	6.00E-20
1.50E 02	4.20E-20
2.00E 02	3.20E-20
3.00E 02	2.30E-20
4.00E 02	1.80E-20
5.00E 02	1.50E-20

6-2-12D
SCHARMANN ET AL. (167)

ENERGY (KEV)	CROSS SECTION (SQ. CM)
1.45E 02	2.85E-20
1.85E 02	2.10E-20
2.20E 02	1.83E-20
2.70E 02	1.45E-20
3.10E 02	1.23E-20
4.05E 02	8.88E-21
5.10E 02	7.33E-21
6.20E 02	5.97E-21
8.20E 02	4.46E-21
9.50E 02	4.06E-21

6-2-13A
VAN DEN BOS ET AL. (165)

ENERGY (KEV)	CROSS SECTION (SQ. CM)
1.00E 01	3.50E-20
1.25E 01	3.61E-20
1.50E 01	4.05E-20
1.75E 01	3.93E-20
2.00E 01	3.32E-20
2.50E 01	2.88E-20
3.00E 01	3.03E-20
3.50E 01	3.34E-20
4.00E 01	3.43E-20
5.00E 01	4.07E-20
6.00E 01	4.39E-20
7.00E 01	4.35E-20
8.00E 01	4.24E-20
9.00E 01	3.95E-20
1.00E 02	3.72E-20
1.10E 02	3.45E-20
1.20E 02	3.14E-20
1.30E 02	3.00E-20
1.40E 02	2.82E-20
1.50E 02	2.64E-20

6-2-13B
DENIS ET AL. (160)

ENERGY (KEV)	CROSS SECTION (SQ. CM)
2.00E 01	2.40E-20
3.00E 01	2.00E-20
4.00E 01	2.30E-20
5.00E 01	2.80E-20
6.00E 01	3.20E-20
7.00E 01	3.40E-20
8.00E 01	3.40E-20
1.00E 02	3.20E-20
1.50E 02	2.30E-20
2.00E 02	1.80E-20
3.00E 02	1.30E-20
4.00E 02	1.05E-20
5.00E 02	9.00E-21

6-2-14A
KRAUSE ET AL. (164)

ENERGY (KEV)	CROSS SECTION (ARB. UNITS)
4.00E 00	2.50E-01
6.00E 00	4.60E-01
8.00E 00	6.60E-01
1.00E 01	1.03E 00
1.20E 01	1.20E 00
1.40E 01	1.38E 00
1.60E 01	1.52E 00
1.80E 01	1.70E 00
2.00E 01	1.88E 00

6-2-14B
KRAUSE ET AL. (164)

ENERGY (KEV)	CROSS SECTION (ARB. UNITS)
2.00E 00	4.70E-01
3.00E 00	9.20E-01
4.00E 00	2.60E 00
6.00E 00	1.07E 01
8.00E 00	1.58E 01
1.00E 01	1.86E 01
1.20E 01	1.98E 01
1.40E 01	1.93E 01
1.60E 01	1.75E 01
1.80E 01	1.49E 01
2.00E 01	1.36E 01

6-2-15A
VAN ECK ET AL. (155)

ENERGY (KEV)	CROSS SECTION (SQ. CM)
1.50E 01	1.26E-20
1.75E 01	2.36E-20
2.00E 01	3.55E-20
2.50E 01	5.17E-20
3.00E 01	6.43E-20
3.25E 01	6.75E-20
3.50E 01	6.93E-20

6-2-15B
CODD ET AL. (156)

ENERGY (KEV)	CROSS SECTION (SQ. CM)
2.00E 01	5.50E-20
3.00E 01	1.03E-19
4.00E 01	1.14E-19
5.00E 01	1.02E-19
6.00E 01	8.50E-20
7.00E 01	7.00E-20
8.00E 01	6.35E-20
9.00E 01	5.35E-20
1.00E 02	4.90E-20
1.10E 02	4.20E-20
1.30E 02	3.75E-20

6-2-15C
THOMAS ET AL. (161)

ENERGY (KEV)	CROSS SECTION (SQ. CM)
1.50E 02	1.09E-20
2.00E 02	8.80E-21
2.50E 02	6.35E-21
3.00E 02	4.81E-21
3.50E 02	4.16E-21
4.00E 02	3.57E-21
4.50E 02	2.92E-21
5.00E 02	2.71E-21
6.00E 02	2.23E-21
7.00E 02	1.95E-21
8.00E 02	1.65E-21
9.00E 02	1.40E-21

6-2-15D
MOUSTAFA ET AL. (163)

ENERGY (KEV)	CROSS SECTION (SQ. CM)
3.00E 01	7.93E-20
3.50E 01	9.22E-20
4.00E 01	9.02E-20
4.50E 01	8.45E-20
5.00E 01	8.52E-20
6.00E 01	7.47E-20
7.00E 01	6.07E-20
8.00E 01	5.08E-20
9.00E 01	4.37E-20
1.00E 02	3.96E-20
1.10E 02	3.37E-20
1.20E 02	3.32E-20
1.30E 02	3.00E-20
1.40E 02	2.72E-20
1.50E 02	2.50E-20

6-2-16
VAN ECK ET AL. (155)

ENERGY (KEV)	CROSS SECTION (SQ. CM)
1.50E 01	3.40E-21
2.00E 01	1.18E-20

2.50E 01	1.78E-20
3.00E 01	2.21E-20
3.50E 01	2.46E-20

6-2-17A
VAN DEN BOS ET AL. (165)

ENERGY (KEV)	POLARIZATION (PERCENT)
2.00E 00	1.00E 00
3.00E 00	-2.00E 00
4.00E 00	-3.00E 00
5.00E 00	-3.00E 00
6.00E 00	-4.00E 00
7.00E 00	-6.00E 00
8.00E 00	-3.00E 00
1.00E 01	1.90E 01
1.50E 01	3.30E 01
2.00E 01	2.00E 01
2.50E 01	6.00E 00
3.00E 01	4.00E 00
3.50E 01	-8.00E 00
4.00E 01	-7.00E 00
5.00E 01	-6.00E 00
6.00E 01	-5.00E 00
7.00E 01	-3.00E 00
8.00E 01	-3.00E 00
9.00E 01	0.0
1.00E 02	-1.00E 00
1.10E 02	2.00E 00
1.20E 02	2.00E 00
1.30E 02	1.00E 00
1.40E 02	3.00E 00
1.50E 02	4.00E 00

6-2-17B
SCHARMANN ET AL. (166)

ENERGY (KEV)	POLARIZATION (PERCENT)
1.00E 02	5.70E 00
2.80E 02	-3.00E 00
4.00E 02	-5.70E 00
4.50E 02	-8.00E 00
6.00E 02	-1.10E 01
8.35E 02	-1.40E 01

6-2-18
VAN ECK ET AL. (10)

ENERGY (KEV)	POLARIZATION (PERCENT)
7.50E 00	1.00E 01
1.00E 01	4.30E 00
1.50E 01	2.70E 01
2.00E 01	3.24E 01
2.50E 01	1.93E 01
3.00E 01	4.30E 00
3.50E 01	-2.80E 00

6-2-19
VAN ECK ET AL. (10)

ENERGY (KEV)	POLARIZATION (PERCENT)
1.00E 01	0.0
1.50E 01	1.25E 01
2.00E 01	1.74E 01
2.50E 01	9.00E 00
3.00E 01	-1.00E 00
3.50E 01	-1.40E 01

6-2-20A
KRAUSE ET AL. (164)

ENERGY (KEV)	POLARIZATION (PERCENT)
1.40E 00	2.69E 01
2.00E 00	3.08E 01
2.50E 00	3.64E 01
2.75E 00	3.74E 01
2.95E 00	3.52E 01
3.00E 00	3.30E 01
3.20E 00	3.47E 01
3.40E 00	3.23E 01
3.70E 00	3.18E 01
4.00E 00	3.44E 01
4.60E 00	3.40E 01
6.00E 00	4.09E 01
8.00E 00	4.59E 01
1.00E 01	4.60E 01
1.20E 01	4.59E 01
1.30E 01	4.58E 01
1.40E 01	4.60E 01
1.60E 01	4.44E 01
1.80E 01	4.15E 01
2.01E 01	3.67E 01

6-2-20B
SCHARMANN ET AL. (166)

ENERGY (KEV)	POLARIZATION (PERCENT)
1.20E 02	2.25E 01
1.50E 02	1.90E 01
2.00E 02	1.25E 01
3.00E 02	3.50E 00
4.00E 02	-3.50E 00
5.80E 02	-1.00E 01
7.50E 02	-1.50E 01
8.10E 02	-1.60E 01
9.00E 02	-1.75E 01
1.00E 03	-1.90E 01

6-2-21A
VAN DEN BOS ET AL. (165)

ENERGY (KEV)	POLARIZATION (PERCENT)
3.00E 00	2.70E 01
4.00E 00	3.30E 01
6.00E 00	3.50E 01
8.00E 00	3.90E 01
1.00E 01	4.00E 01
1.50E 01	4.60E 01
2.00E 01	4.40E 01
2.50E 01	4.20E 01
3.00E 01	3.70E 01
3.50E 01	3.20E 01
4.00E 01	3.30E 01
5.00E 01	3.40E 01
6.00E 01	3.40E 01
8.00E 01	3.10E 01
1.00E 02	2.80E 01
1.20E 02	2.40E 01
1.40E 02	2.10E 01

6-2-21B
SCHARMANN ET AL. (166)

ENERGY (KEV)	POLARIZATION (PERCENT)
1.20E 02	2.25E 01
1.50E 02	2.05E 01
2.20E 02	1.25E 01
3.00E 02	7.00E 00
4.00E 02	-2.00E 00
6.00E 02	-1.00E 01
1.00E 03	-1.45E 01
1.10E 03	-2.00E 01

6-2-22A
VAN ECK ET AL. (10)

ENERGY (KEV)	POLARIZATION (PERCENT)
1.00E 01	3.16E 01
1.50E 01	3.96E 01
1.75E 01	3.81E 01
2.00E 01	3.70E 01
2.50E 01	3.01E 01
3.00E 01	2.85E 01
3.50E 01	2.78E 01

6-2-22B
SCHARMANN ET AL. (166)

ENERGY (KEV)	POLARIZATION (PERCENT)
1.25E 02	2.05E 01
2.00E 02	1.20E 01
3.00E 02	4.50E 00
5.00E 02	-7.50E 00
7.00E 02	-1.20E 01
9.00E 02	-1.75E 01

6-2-23
VAN ECK ET AL. (10)

ENERGY (KEV)	POLARIZATION (PERCENT)
1.00E 01	3.19E 01
1.50E 01	4.27E 01

2.00E 01	3.90E 01
2.50E 01	3.33E 01
3.00E 01	3.16E 01
3.50E 01	3.31E 01

6-2-24

KRAUSE ET AL. (164)

ENERGY (KEV)	POLARIZATION (PERCENT)
2.00E 00	7.60E 00
3.00E 00	1.03E 01
4.00E 00	9.60E 00
5.00E 00	1.55E 01
6.00E 00	2.02E 01
7.00E 00	1.95E 01
8.00E 00	2.34E 01
9.00E 00	2.02E 01
1.00E 01	1.89E 01
1.10E 01	1.70E 01
1.20E 01	1.69E 01
1.30E 01	1.46E 01
1.40E 01	1.48E 01
1.60E 01	1.23E 01
1.80E 01	1.34E 01
2.00E 01	1.10E 01

6-3 EXCITATION OF HELIUM BY H IMPACT

The study of the collisional excitation of helium by impact of neutral hydrogen atoms involves basically the same problems as H^+ impact experiments. In Section 6-1-A, the precautions involved in helium excitation experiments were discussed in general terms.

The projectile beam is produced by charge transfer of protons in a gas cell. The residual ion flux is removed with a magnetic or electric field. A small proportion of the projectile beam may be in an excited state. In a practical experiment, many of the short-lived states of the atomic hydrogen will decay before reaching the collision chamber, and the metastable states will be quenched by the fields used to remove the ions. It is not considered that the projectiles remaining in the excited state are likely to seriously affect the measurement of a cross section for target excitation.

Available data on this collision combination are sparse and primarily due to a single research group. There is no confirmatory evidence of the measurements although there is no reason to doubt their validity. In the discussion of the work done by various groups on excitation by proton impact (see Section 6-2) it was noted that there is considerable agreement among the various determinations about the functional dependence of cross section on energy. Discrepancies tended to arise primarily in the absolute magnitudes of the data. It is expected that the same situation will hold in the present case. There is confidence in the accuracy of the functional energy dependence of the cross sections, but there is less certainty about the absolute magnitudes.

The data on excitation cross sections are discussed in Section 6-3-A and that on polarization of emission in Section 6-3-B. The data are shown in tabular form in Section 6-3-C.

A. Cross Section Measurements

Robinson and Gilbody[162] considered only the formation of the 4^1S state at energies from 60 to 150-keV, utilizing the apparatus described previously in Section 6-1-B(4). It is stated that no measurable triplet emission was detected, indicating that in the energy range covered by their work, the excitation of the triplet levels is less likely than the excitation of the corresponding singlet levels. The data by Robinson and Gilbody[162] exhibit a random scatter up to $\pm 20\%$. The systematic error limit is stated to be $\pm 20\%$.

The extensive work by van Eck et al. on the excitation by H^+ impact has been subjected to revisions by Van den Bos; however, the companion data for H impact,[10, 244] taken at the same time, have not been revised in the published literature. F. J. de Heer and J. Van den Bos[190] have undertaken a re-analysis and renormalization of the data of van Eck et al.[10] These unpublished revised data are displayed here in the data compilations since they supersede the existing published measurements. For the singlet states the revisions amount to only a few percent, while for the triplet states some measurements have changed by a factor of two. The revised data by de Heer and Van den Bos[190] omit all measurements below 10-keV on the grounds that they are of poor statistical accuracy. In Section 6-1-B(1), the apparatus used for this work is discussed in some detail. It is estimated that the random error in these measurements does not exceed 8% and that the systematic error is less than 15%.

Krause and Soltysik[164] present relative measurements of the $3^3S \rightarrow 2^3P$ and $3^3D \rightarrow 2^3P$ emissions determined by the rather indirect procedure described in Section 6-1-B(6). Part of the emission from the triplet states produced as protons traversed the target was found to depend on the square of target pressure. This was ascribed to neutral atoms formed by charge transfer undergoing a second collision that caused the excitation of a target atom. It is also possible that this emission was due to collisional transfer of excitation energy when atoms in an excited singlet state collided with ground state atoms and formed a triplet excited state. Such a process would also show a square law dependence on pressure. Arguments are presented suggesting that this is unlikely, but there is no direct proof that collisional transfer does not occur. Even if the interpretation of the observed behavior is correct, the "projectiles" must be a rather impure beam because they undoubtedly include a significant fraction of excited atoms and possibly a small amount of H^- produced by a two-electron transfer. Consequently, the data presented by Krause and Soltysik must be treated with caution and are certainly of poorer quality than the direct measurements by van Eck et al.

When considering an experiment involving neutral projectiles, some careful attention must be paid to the method by which the flux of neutrals was detected. Both in the work of Robinson and Gilbody[162] and in that of van Eck et al.,[10] secondary emission techniques were used, with the efficiency of detection established by using a proton beam. It is shown[265] that the detector used by Robinson and Gilbody[162] exhibited the same secondary emission coefficient for impact of ions as for neutrals. Therefore, calibration using protons is justifiable. Van Eck et al.[10] make the assumption that the secondary emission coefficients for H and H^+ impact are equal. There is no justification for this assumption. In a review of the mechanisms of secondary electron emission, Carter and Colligon[266] present ample evidence of differences between secondary emission coefficients for impact of fast ions and their parent atoms. According to the data by Stier et al.[41] on secondary emission coefficients, the error introduced by the assumption of equal coefficients would cause the cross section measurements by van Eck et al.[10] to be 10% too small at an energy of 35-keV and to be in error by lesser amounts at lower energies. In the work of Krause and Soltysik the neutral beam component was taken as the attenuation of the proton beam that was observed when gas was admitted into the chamber to the desired pressure. Since the attenuation was in close agreement with that predicted using charge transfer data by Allison,[267] this procedure is probably quite satisfactory.

In the measurements by van Eck et al.[10, 244, 190] adequate attention has been paid to assessing cascade and eliminating possible errors associated with the polarization

of the emission. Only in the case of the n^1P states was there some inadequacy in the treatment of pressure-dependent effects, namely the absorption of resonant photons. The cross section data were taken at low target pressures and extrapolated to zero pressure using a pressure-dependent correction established in a proton impact experiment. It is expected that this procedure is accurate since the absorption process is independent of the mechanism by which the emission is induced.

The cross sections for formation of specific excited states measured by Robinson and Gilbody,[162] and by van Eck et al.[10, 190] are shown in Data Tables 6-3-1 to 6-3-6. The data by Krause and Soltysik[164] are expressed as relative measurements of line-emission cross sections. According to van Eck,[155] cascade population of the 3^3D state is large (about 30%); therefore the emission measurement is not necessarily directly proportional to the level-excitation cross section. The data by Krause and Soltysik are displayed in the same relative units as the original publication in Data Table 6-3-7.

The excitation of the 4^1S state (Data Table 6-3-1) is the only case in which comparison of data from two independent sources is possible. If an extrapolation of the corrected[190] data of van Eck et al.[10] is made toward higher energies, the cross section appears to be less than that measured by Robinson and Gilbody.[162] The data for proton-impact excitation of this same level show a converse behavior, with the measurements of Robinson and Gilbody[162] being appreciably below those of van Eck et al.[10] and Van den Bos et al.[165] (see Section 6-2A). The discrepancies among the independent measurements of cross sections have been ascribed primarily to errors in absolute calibrations. It is not clear why the discrepancies in the H impact work are not the same as for H^+ impact measurements by the same authors.

Relative measurements of the emission function by Krause and Soltysik[164] show the same energy dependence as the level-excitation cross section measured by van Eck et al.[10, 199] for the 3^3D state.

B. Polarization of Emission

Most of the information on polarization is derived from the work of van Eck et al.[10] The emission cross sections determined in that work were later revised,[190] but there were no further measurements of polarization. There is no reason to doubt that the measurements of polarization are accurate. The polarization fraction may be written in terms of a ratio of intensities, and any systematic errors in calibration or the determination of projectile flux will cancel out. Satisfactory precautions were taken to establish that the measurements were independent of pressure, and to eliminate depolarization by stray magnetic fields. The polarization measurements are stated to have a possible random error of $\pm 5\%$ and negligible systematic error.

Krause and Soltysik[164] present data on the polarization of the $3^3S \rightarrow 2^3P$ and $3^3D \rightarrow 2^3P$ emissions. It has been suggested [Section 6-1-B(6)] that there is an instrumental polarization error in these data. It is to be expected that the $3^3S \rightarrow 2^3P$ transition should be unpolarized; therefore the apparent polarization of the emission observed in this experiment is probably a mistake. Consequently, the $3^3S \rightarrow 2^3P$ data are not presented here. The $3^3D \rightarrow 2^3P$ data is retained, but it may be in error due to the instrumental polarization. No precautions are noted to eliminate nor to assess the depolarization of emission by stray magnetic fields.

C. Data Tables

1. Cross Sections for Excitation of He States Induced by Impact of H on a Helium Target

6-3-1 (A) 4^1S State; (B) 4^1S State;† (C) 5^1S State.†

6-3-2 (A) 3^1P State;† (B) 4^1P State.†

6-3-3 (A) 4^1D State;† (B) 5^1D State.†

6-3-4 (A) 4^3S State;† (B) 5^3S State.†

6-3-5 (A) 3^3P State;† (B) 4^3P State.†

6-3-6 (A) 3^3D State;† (B) 4^3D State.†

2. Cross Sections for Emission of He I Triplet Lines I Induced by Impact of H on a Helium Target

6-3-7 (A) 7065-Å ($3^3S \rightarrow 2^3P$); (B) 5876-Å ($3^3D \rightarrow 2^3P$)

3. Polarization Fractions of He I Lines Induced by the Impact of H on a Helium Target

6-3-8 (A) 4922-Å ($4^1D \rightarrow 2^1P$); (B) 3889-Å ($3^3P \rightarrow 2^3S$)

6-3-9 (A) 4472-Å ($4^3D \rightarrow 2^3P$); (B) 5876-Å ($3^3D \rightarrow 2^3P$)

† These data are revisions by Van den Bos[190] of the published measurements by van Eck et al.[10]

6-3-1A
ROBINSON ET AL. (162)

ENERGY (KEV)	CROSS SECTION (SQ. CM)
6.00E 01	1.56E-19
6.50E 01	1.63E-19
7.00E 01	1.73E-19
7.50E 01	1.30E-19
8.00E 01	1.05E-19
9.50E 01	1.07E-19
1.00E 02	7.00E-20
1.10E 02	7.40E-20
1.20E 02	6.00E-20
1.30E 02	6.80E-20
1.50E 02	4.80E-20

6-3-1B
VAN ECK ET AL. (10)

ENERGY (KEV)	CROSS SECTION (SQ. CM)
1.00E 01	1.30E-19
1.50E 01	1.63E-19
2.00E 01	1.85E-19
2.50E 01	1.96E-19
3.00E 01	1.82E-19
3.50E 01	1.58E-19

6-3-1C
VAN ECK ET AL. (10)

ENERGY (KEV)	CROSS SECTION (SQ. CM)
1.00E 01	5.79E-20
1.50E 01	7.15E-20
2.00E 01	8.58E-20
2.50E 01	8.68E-20
3.00E 01	7.95E-20
3.50E 01	7.08E-20

6-3-2A
VAN ECK ET AL. (10)

ENERGY (KEV)	CROSS SECTION (SQ. CM)
1.00E 01	5.60E-20
1.50E 01	1.78E-19
2.00E 01	2.50E-19
2.50E 01	2.93E-19
3.00E 01	3.71E-19
3.50E 01	4.23E-19

6-3-2B
VAN ECK ET AL. (10)

ENERGY (KEV)	CROSS SECTION (SQ. CM)
1.00E 01	2.60E-20
1.50E 01	7.90E-20
2.00E 01	1.25E-19
2.50E 01	1.47E-19
3.00E 01	1.86E-19
3.50E 01	2.12E-19

6-3-3A
VAN ECK ET AL. (10)

ENERGY (KEV)	CROSS SECTION (SQ. CM)
1.00E 01	2.36E-20
1.50E 01	3.76E-20
2.00E 01	4.04E-20
2.50E 01	3.82E-20
3.00E 01	3.93E-20
3.50E 01	4.03E-20

6-3-3B
VAN ECK ET AL. (10)

ENERGY (KEV)	CROSS SECTION (SQ. CM)
1.00E 01	1.14E-20
1.50E 01	1.75E-20
2.00E 01	2.13E-20
2.50E 01	2.07E-20
3.00E 01	1.88E-20
3.50E 01	1.74E-20

6-3-4A
VAN ECK ET AL. (10)

ENERGY (KEV)	CROSS SECTION (SQ. CM)
1.00E 01	1.50E-19
1.50E 01	3.32E-19
2.00E 01	4.07E-19
2.50E 01	3.88E-19
3.00E 01	3.43E-19
3.50E 01	3.13E-19

6-3-4B
VAN ECK ET AL. (10)

ENERGY (KEV)	CROSS SECTION (SQ. CM)
1.00E 01	9.30E-20
1.50E 01	1.87E-19

2.00E 01	2.22E-19
2.50E 01	2.18E-19
3.00E 01	1.57E-19
3.50E 01	1.35E-19

6-3-5A
VAN ECK ET AL. (10)

ENERGY (KEV)	CROSS SECTION (SQ. CM)
1.00E 01	9.61E-19
1.50E 01	1.17E-18
2.00E 01	8.57E-19
2.50E 01	5.99E-19
3.00E 01	4.16E-19
3.50E 01	2.72E-19

6-3-5B
VAN ECK ET AL. (10)

ENERGY (KEV)	CROSS SECTION (SQ. CM)
1.00E 01	3.64E-19
1.50E 01	4.71E-19
2.00E 01	3.68E-19
2.50E 01	2.50E-19
3.00E 01	1.71E-19
3.50E 01	1.25E-19

6-3-6A
VAN ECK ET AL. (10)

ENERGY (KEV)	CROSS SECTION (SQ. CM)
1.00E 01	1.61E-19
1.50E 01	2.33E-19
2.00E 01	1.62E-19
2.50E 01	1.31E-19
3.00E 01	8.89E-20
3.50E 01	6.88E-20

6-3-6B
VAN ECK ET AL. (10)

ENERGY (KEV)	CROSS SECTION (SQ. CM)
1.00E 01	5.75E-20
1.50E 01	7.54E-20
2.00E 01	6.09E-20
2.50E 01	4.93E-20
3.00E 01	3.97E-20
3.50E 01	3.28E-20

6-3-7A KRAUSE ET AL. (164)			6-3-8A VAN ECK ET AL. (10)		6-3-9A VAN ECK ET AL. (10)	
ENERGY (KEV)	CROSS SECTION (ARB. UNITS)		ENERGY (KEV)	POLARIZATION (PERCENT)	ENERGY (KEV)	POLARIZATION (PERCENT)
4.00E 00	3.12E 00		1.00E 01	2.79E 01	1.00E 01	5.80E 00
8.00E 00	4.44E 01		1.50E 01	2.79E 01	1.50E 01	1.01E 01
1.00E 01	6.76E 01		2.00E 01	2.64E 01	2.00E 01	1.04E 01
1.20E 01	8.26E 01		2.50E 01	2.59E 01	2.50E 01	1.00E 01
1.40E 01	8.97E 01		3.00E 01	2.48E 01	3.00E 01	9.10E 00
1.60E 01	9.72E 01		3.50E 01	2.31E 01	3.50E 01	8.50E 00
1.80E 01	1.01E 02					
2.00E 01	1.01E 02					

6-3-7B KRAUSE ET AL. (164)			6-3-8B VAN ECK ET AL. (10)		6-3-9B KRAUSE ET AL. (164)	
ENERGY (KEV)	CROSS SECTION (ARB. UNITS)		ENERGY (KEV)	POLARIZATION (PERCENT)	ENERGY (KEV)	POLARIZATION (PERCENT)
			5.00E 00	-2.20E 00	5.00E 00	1.19E 01
4.00E 00	3.20E 01		7.50E 00	7.00E-01	6.00E 00	1.60E 01
6.00E 00	4.90E 01		1.00E 01	1.90E 00	8.00E 00	1.79E 01
8.00E 00	7.66E 01		1.50E 01	3.80E 00	1.00E 01	1.58E 01
1.00E 01	1.11E 02		2.00E 01	4.90E 00	1.20E 01	1.50E 01
1.20E 01	1.50E 02		2.50E 01	4.70E 00	1.30E 01	1.49E 01
1.40E 01	1.71E 02		3.00E 01	4.10E 00	1.40E 01	1.59E 01
1.60E 01	1.54E 02		3.50E 01	2.90E 00	1.60E 01	1.69E 01
1.80E 01	1.30E 02				1.80E 01	1.48E 01
2.00E 01	1.15E 02				2.00E 01	1.41E 01

6-4 EXCITATION OF HELIUM BY H_2^+ AND H_3^+ IMPACT

The study of the collisional excitation of helium by the impact of molecular hydrogen ions involves basically the same problems as H^+ and H^0 impact experiments. In Section 6-1-B the precautions involved in helium excitation experiments were discussed in general terms.

It is not expected that a small proportion of excited states in the projectile beam will cause appreciable errors in the measurement of cross sections for excitation of the target. Thomas and Bent[201] attempted to assess the possible influence of excited states by varying the conditions in the ion source. No influence of ion source conditions on the measured cross section was found.

Available cross-section data on these cases are primarily due to Van den Bos et al.[200] with a limited series of measurements by Thomas and Bent.[201] Both groups have contributed extensively to the measurement of excitation by proton impact; the discrepancies between their data are primarily in the magnitudes of the cross sections, a problem associated with errors in calibration of detection sensitivity. In general, the functional dependence of a cross section on energy is quite well established. In the present case, there are no opportunities for comparing data from various sources to indicate whether there is a consensus on the measurements of cross section or polarization. But based on the conclusions drawn from the study of excitation by proton impact, it is likely that the energy dependence of the cross section measurements

will be quite accurate. It is not possible to assess the accuracy of the absolute magnitudes.

Attention is first directed to the data on level-excitation cross sections, (Section 6-4-A) and then to data on polarization of the emission (Section 6-4-B); only a limited amount of data on the latter aspect is available.

A. Cross-Section Measurements

The work of Thomas and Bent[201] utilized the apparatus that was evaluated in Section 6-1-B(3). Measurements were restricted to the formation of the 4^1S state at energies from 200 to 800-keV. The extensive collection of data by Van den Bos et al. was obtained using the apparatus described in Section 6-1-B(1); preliminary data presented in a review article[245] have been shown to be incorrect, and the present compilation is drawn from a revised program of measurements.[200]

It would appear that all data were obtained with due attention to all relevant precautions. Cascade was properly assessed, this being quite a large correction for some of the triplet states. Data were obtained at sufficiently low target pressures so that secondary processes were effectively eliminated. The work of Van den Bos et al.[200] included all precautions for dealing with polarization of the emission, while the series of measurements by Thomas and Bent[201] was limited to the $4^1S \rightarrow 2^1P$ emission which is unpolarized. The data by Thomas and Bent[201] are stated to have a random error up to $\pm 10\%$ and a systematic error not exceeding $\pm 25\%$. The data of Van den Bos et al.[200] are stated to be reproducible to within $\pm 5\%$ with a possible systematic error that generally did not exceed $\pm 8\%$, although for certain specific cases the error was higher due to some cause such as low cross sections, poor sensitivity, etc.

The available data for excitation by H_2^+ are shown in Data Tables 6-4-1 through 6-4-6, and those for H_3^+ impact, in Data Tables 6-4-7 through 6-4-12 (Section 6-4-C).

A comparison of the data by Van den Bos et al.[200] and those by Thomas and Bent[201] can only be made by interpolating over the energy gap between the two sets of data. It would appear, however, that the work of Thomas and Bent is lower than the other data by perhaps 25%. In the study of the proton impact, the data by the two groups differed by some 60% for the 4^1S state.

B. Polarization of Emission

Van den Bos et al.[200, 246] have carried out a series of polarization measurements for singlet and triplet emission induced by the impact of both projectiles. Scharmann and Schartner have measured polarization of the $4^1D \rightarrow 2^1P$ line induced by H_2^+ impact; the original publication of this data[256] was superseded and extended by a more recent publication.[166] In both determinations, attention was given to all relevant precautions expected in such work.

The results are shown in Data Tables 6-4-13 through 6-4-16. In the case of the 4^1D line, it is possible to make a comparison between the two sets of data; good agreement is exhibited. We have previously shown in the discussion of proton-impact excitation (Section 6-2-E) that measurements of polarization by these two groups are in rather good agreement.

C. Data Tables

1. Cross Sections for Excitation of He States Induced by Impact of H_2^+ on a Helium Target

6-4-1   (A) 3^1S State; (B) 4^1S State; (C) 4^1S State; (D) 5^1S State; (E) 6^1S State.

6-4-2   (A) 3^1P State; (B) 4^1P State; (C) 5^1P State.

6-4-3   (A) 3^1D State; (B) 4^1D State; (C) 5^1D State; (D) 6^1D State.

6-4-4   (A) 3^3S State; (B) 4^3S State; (C) 5^3S State; (D) 6^3S State.

6-4-5   (A) 3^3P State; (B) 4^3P State; (C) 5^3P State; (D) 6^3P State.

6-4-6   (A) 3^3D State; (B) 4^3D State; (C) 5^3D State.

2. Cross Sections for Excitation of He States Induced by Impact of H_3^+ on a Helium Target

6-4-7   (A) 3^1S State; (B) 4^1S State; (C) 4^1S State; (D) 5^1S State; (E) 6^1S State.

6-4-8   (A) 3^1P State; (B) 4^1P State; (C) 5^1P State.

6-4-9   (A) 3^1D State; (B) 4^1D State; (C) 5^1D State; (D) 6^1D State.

6-4-10   (A) 3^3S State; (B) 4^3S State; (C) 5^3S State; (D) 6^3S State.

6-4-11   (A) 3^3P State; (B) 4^3P State; (C) 5^3P State; (D) 6^3P State.

6-4-12   (A) 3^3D State; (B) 4^3D State; (C) 5^3D State.

3. Polarization Fractions of He I Lines Induced by Impact of H_2^+ on a Helium Target

6-4-13   (A) 5016-Å ($3^1P \rightarrow 2^1S$); (B) 4922-Å ($4^1D \rightarrow 2^1P$); (C) 4922-Å ($4^1D \rightarrow 2^1P$).

6-4-14   (A) 3188-Å ($4^3P \rightarrow 2^3S$); (B) 5876-Å ($3^3D \rightarrow 2^3P$).

4. Polarization Fractions of He I Lines Induced by Impact of H_3^+ on a Helium Target

6-4-15   (A) 5016-Å ($3^1P \rightarrow 2^1S$); (B) 4922-Å ($4^1D \rightarrow 2^1P$).

6-4-16   (A) 3188-Å ($4^3P \rightarrow 2^3S$); (B) 5876-Å ($3^3D \rightarrow 2^3P$).

6-4-1A
VAN DEN BOS ET AL. (200)

ENERGY (KEV)	CROSS SECTION (SQ. CM)
2.00E 00	1.70E-20
3.00E 00	4.30E-20
3.50E 00	5.70E-20
4.00E 00	7.00E-20
4.50E 00	7.80E-20
5.00E 00	8.20E-20
5.50E 00	9.30E-20
6.00E 00	1.09E-19
6.50E 00	1.16E-19
7.00E 00	1.28E-19
8.00E 00	1.39E-19
9.00E 00	1.65E-19
1.00E 01	1.94E-19
1.50E 01	4.27E-19
2.00E 01	4.54E-19
2.50E 01	5.11E-19
3.00E 01	5.92E-19
3.50E 01	6.45E-19
4.00E 01	8.22E-19
5.00E 01	1.33E-18
6.00E 01	1.54E-18
7.00E 01	1.65E-18
8.00E 01	1.68E-18
9.00E 01	1.78E-18
1.00E 02	1.72E-18
1.10E 02	1.58E-18
1.20E 02	1.71E-18
1.30E 02	1.60E-18
1.40E 02	1.54E-18
1.50E 02	1.48E-18

6-4-1B
VAN DEN BOS ET AL. (20C

ENERGY (KEV)	CROSS SECTION (SQ. CM)
1.00E 00	1.60E-21
1.50E 00	1.90E-21
2.00E 00	4.60E-21
2.50E 00	7.30E-21
3.00E 00	6.80E-21
3.50E 00	4.80E-21
4.00E 00	6.10E-21
4.50E 00	7.30E-21
5.00E 00	9.90E-21
6.00E 00	1.20E-20
7.00E 00	1.35E-20
8.00E 00	1.53E-20
9.00E 00	1.80E-20
1.00E 01	2.32E-20
1.25E 01	4.50E-20
1.50E 01	8.80E-20
2.00E 01	1.40E-19
2.50E 01	1.53E-19
3.00E 01	1.60E-19
3.50E 01	1.82E-19
4.00E 01	2.07E-19
5.00E 01	2.78E-19
6.00E 01	3.35E-19
7.00E 01	4.22E-19
8.00E 01	4.85E-19
9.00E 01	5.10E-19
1.00E 02	5.31E-19
1.10E 02	5.54E-19
1.20E 02	5.68E-19
1.30E 02	5.47E-19
1.40E 02	5.01E-19
1.50E 02	4.84E-19

6-4-1C
THOMAS ET AL. (201)

ENERGY (KEV)	CROSS SECTION (SQ. CM)
2.00E 02	3.00E-19
3.00E 02	2.03E-19
4.00E 02	1.41E-19
5.00E 02	1.21E-19
6.00E 02	1.02E-19
8.00E 02	8.10E-20

6-4-1D
VAN DEN BOS ET AL. (200)

ENERGY (KEV)	CROSS SECTION (SQ. CM)
2.50E 00	5.00E-22
3.00E 00	9.00E-22
3.50E 00	1.80E-21
4.00E 00	1.50E-21
4.50E 00	1.40E-21
5.00E 00	1.40E-21
5.50E 00	2.20E-21
6.00E 00	3.10E-21
6.50E 00	3.90E-21
7.00E 00	4.70E-21
8.00E 00	5.20E-21
9.00E 00	5.80E-21
1.00E 01	6.10E-21
1.50E 01	2.28E-20
2.00E 01	5.07E-20
2.50E 01	6.60E-20
3.00E 01	6.90E-20
3.50E 01	1.01E-19
4.00E 01	1.06E-19
5.00E 01	1.52E-19
6.00E 01	1.77E-19
7.00E 01	2.06E-19
8.00E 01	2.50E-19
9.00E 01	3.00E-19
1.00E 02	3.09E-19
1.10E 02	3.14E-19
1.20E 02	2.94E-19
1.30E 02	3.10E-19
1.40E 02	3.03E-19
1.50E 02	3.05E-19

6-4-1E
VAN DEN BOS ET AL. (200)

ENERGY (KEV)	CROSS SECTION (SQ. CM)
3.00E 00	9.00E-22
4.00E 00	1.00E-21
5.00E 00	1.40E-21
6.00E 00	2.20E-21
6.50E 00	2.40E-21
7.00E 00	3.20E-21
8.00E 00	3.70E-21
9.00E 00	4.70E-21
1.00E 01	4.80E-21
1.50E 01	7.50E-21
2.00E 01	2.22E-20
2.50E 01	3.15E-20
3.00E 01	3.61E-20
3.50E 01	5.39E-20
4.00E 01	5.62E-20
5.00E 01	7.88E-20
6.00E 01	9.65E-20
7.00E 01	1.16E-19
8.00E 01	1.31E-19
9.00E 01	1.39E-19
1.00E 02	1.52E-19
1.10E 02	1.36E-19
1.20E 02	1.54E-19
1.30E 02	1.59E-19
1.40E 02	1.59E-19
1.50E 02	1.52E-19

6-4-2A
VAN DEN BOS ET AL. (200)

ENERGY (KEV)	CROSS SECTION (SQ. CM)
1.00E 00	7.30E-21
1.50E 00	2.71E-20
2.00E 00	7.67E-20
2.50E 00	8.85E-20
3.00E 00	7.51E-20
3.50E 00	7.95E-20
4.00E 00	1.23E-19
5.00E 00	2.17E-19
6.00E 00	2.64E-19
7.00E 00	2.68E-19
8.00E 00	2.69E-19
9.00E 00	3.30E-19
1.00E 01	3.40E-19
1.50E 01	5.80E-19
2.00E 01	9.00E-19
2.50E 01	1.32E-18
3.00E 01	1.65E-18
3.50E 01	1.76E-18
4.00E 01	2.01E-18
5.00E 01	2.32E-18
6.00E 01	2.77E-18
7.00E 01	3.02E-18
8.00E 01	3.27E-18
9.00E 01	3.56E-18
1.00E 02	3.91E-18
1.10E 02	4.06E-18
1.20E 02	4.35E-18
1.30E 02	4.47E-18
1.40E 02	4.53E-18
1.50E 02	4.63E-18

6-4-2B
VAN DEN BOS ET AL. (200)

ENERGY (KEV)	CROSS SECTION (SQ. CM)
3.50E 00	1.10E-20
4.00E 00	1.70E-20
5.00E 00	2.60E-20
6.00E 00	3.70E-20
7.00E 00	6.00E-20
8.00E 00	7.70E-20
9.00E 00	9.30E-20
1.00E 01	1.10E-19
1.50E 01	1.59E-19
2.00E 01	2.29E-19
2.50E 01	3.36E-19
3.00E 01	3.66E-19
3.50E 01	4.14E-19
4.00E 01	4.61E-19
5.00E 01	4.92E-19
6.00E 01	5.12E-19
7.00E 01	5.70E-19
8.00E 01	6.35E-19
9.00E 01	6.38E-19
1.00E 02	6.20E-19
1.10E 02	7.50E-19
1.20E 02	7.60E-19
1.30E 02	7.13E-19
1.40E 02	7.55E-19
1.50E 02	8.02E-19

6-4-2C
VAN DEN BOS ET AL. (200)

ENERGY (KEV)	CROSS SECTION (SQ. CM)
3.00E 00	1.18E-20
4.00E 00	1.49E-20
5.00E 00	1.69E-20
6.00E 00	1.91E-20
7.00E 00	3.21E-20
8.00E 00	4.51E-20
9.00E 00	6.28E-20
1.00E 01	6.90E-20
1.50E 01	1.00E-19
2.00E 01	1.50E-19
2.50E 01	2.30E-19
3.00E 01	2.50E-19
3.50E 01	2.70E-19
4.00E 01	3.00E-19
5.00E 01	3.50E-19
6.00E 01	3.60E-19
7.00E 01	3.60E-19
8.00E 01	4.00E-19
9.00E 01	4.00E-19
1.00E 02	4.70E-19
1.10E 02	5.20E-19
1.20E 02	5.60E-19
1.30E 02	5.60E-19
1.40E 02	5.90E-19
1.50E 02	6.40E-19

6-4-3A
VAN DEN BOS ET AL. (200)

ENERGY (KEV)	CROSS SECTION (SQ. CM)
2.00E 00	3.40E-20
3.00E 00	9.90E-20
4.00E 00	1.41E-19
5.00E 00	1.67E-19
6.00E 00	1.96E-19
7.00E 00	2.41E-19
8.00E 00	2.97E-19
9.00E 00	3.58E-19
1.00E 01	4.15E-19
1.50E 01	9.80E-19
2.00E 01	9.60E-19
2.50E 01	8.40E-19
3.00E 01	7.40E-19
3.50E 01	6.30E-19
4.00E 01	6.00E-19
5.00E 01	5.00E-19
6.00E 01	4.80E-19
7.00E 01	4.40E-19
8.00E 01	4.30E-19
9.00E 01	4.80E-19
1.00E 02	4.90E-19
1.10E 02	4.60E-19
1.20E 02	4.60E-19
1.30E 02	4.30E-19
1.40E 02	4.20E-19
1.50E 02	4.10E-19

6-4-3B
VAN DEN BOS ET AL. (200)

ENERGY (KEV)	CROSS SECTION (SQ. CM)
1.00E 00	1.30E-21
2.00E 00	5.10E-21
2.50E 00	8.60E-21
3.00E 00	1.52E-20
3.50E 00	1.88E-20
4.00E 00	1.79E-20
4.50E 00	1.67E-20
5.00E 00	1.75E-20
5.50E 00	1.82E-20
6.00E 00	2.00E-20
7.00E 00	2.47E-20
8.00E 00	3.44E-20
9.00E 00	4.55E-20
1.00E 01	6.03E-20
1.50E 01	1.55E-19
2.00E 01	2.40E-19
2.50E 01	2.39E-19
3.00E 01	2.35E-19
3.50E 01	2.43E-19
4.00E 01	2.06E-19
5.00E 01	1.78E-19
6.00E 01	1.80E-19
7.00E 01	1.87E-19
8.00E 01	1.98E-19
9.00E 01	2.11E-19
1.00E 02	2.19E-19
1.10E 02	2.08E-19
1.20E 02	2.14E-19
1.30E 02	2.03E-19
1.40E 02	2.05E-19
1.50E 02	2.01E-19

6-4-3C
VAN DEN BOS ET AL. (200)

ENERGY (KEV)	CROSS SECTION (SQ. CM)
1.00E 00	7.00E-22
2.00E 00	9.00E-22
2.50E 00	1.50E-21
3.00E 00	2.20E-21
3.50E 00	2.90E-21
4.00E 00	5.70E-21
4.50E 00	7.60E-21
5.00E 00	8.60E-21
5.50E 00	8.60E-21
6.00E 00	8.60E-21
7.00E 00	9.10E-21
8.00E 00	1.01E-20
9.00E 00	1.26E-20
1.00E 01	1.89E-20
1.25E 01	3.99E-20
1.50E 01	6.03E-20
2.00E 01	9.96E-20
2.50E 01	1.11E-19
3.00E 01	1.18E-19
3.50E 01	1.06E-19
4.00E 01	1.01E-19
5.00E 01	9.19E-20
6.00E 01	8.61E-20
7.00E 01	8.65E-20
8.00E 01	9.44E-20
9.00E 01	9.36E-20
1.00E 02	1.00E-19
1.10E 02	1.08E-19
1.20E 02	1.09E-19
1.30E 02	1.01E-19
1.40E 02	1.12E-19
1.50E 02	1.05E-19

6-4-3D
VAN DEN BOS ET AL. (200)

ENERGY (KEV)	CROSS SECTION (SQ. CM)
2.00E 00	3.00E-22
2.50E 00	7.00E-22
3.00E 00	9.00E-22
3.50E 00	9.00E-22
4.00E 00	1.40E-21
4.50E 00	2.20E-21
5.00E 00	3.30E-21
5.50E 00	4.40E-21
6.00E 00	4.60E-21
7.00E 00	4.90E-21
8.00E 00	5.50E-21
9.00E 00	6.30E-21
1.00E 01	7.50E-21
1.50E 01	2.09E-20
2.00E 01	3.46E-20
2.50E 01	4.70E-20
3.00E 01	4.95E-20
3.50E 01	5.64E-20
4.00E 01	5.56E-20
5.00E 01	4.88E-20
6.00E 01	5.20E-20
7.00E 01	5.51E-20
8.00E 01	5.96E-20
9.00E 01	5.60E-20

ENERGY (KEV)	CROSS SECTION (SQ. CM)
1.00E 02	6.00E-20
1.10E 02	5.78E-20
1.20E 02	5.73E-20
1.30E 02	5.82E-20
1.40E 02	5.91E-20
1.50E 02	5.47E-20

6-4-4A
VAN DEN BOS ET AL. (200)

ENERGY (KEV)	CROSS SECTION (SQ. CM)
2.00E 00	2.06E-20
3.00E 00	1.47E-19
4.00E 00	2.61E-19
4.50E 00	3.31E-19
5.00E 00	3.81E-19
6.00E 00	5.35E-19
7.00E 00	7.47E-19
8.00E 00	9.77E-19
9.00E 00	1.07E-18
1.00E 01	1.10E-18
1.50E 01	1.25E-18
2.00E 01	1.44E-18
2.50E 01	2.00E-18
3.00E 01	2.66E-18
3.50E 01	3.19E-18
4.00E 01	3.56E-18
5.00E 01	3.62E-18
6.00E 01	3.32E-18
7.00E 01	2.81E-18
8.00E 01	2.42E-18
9.00E 01	1.94E-18
1.00E 02	1.50E-18
1.10E 02	1.21E-18
1.20E 02	1.09E-18
1.30E 02	9.20E-19
1.40E 02	7.40E-19
1.50E 02	6.10E-19

6-4-4B
VAN DEN BOS ET AL. (200)

ENERGY (KEV)	CROSS SECTION (SQ. CM)
1.00E 00	5.00E-22
2.00E 00	5.90E-21
2.50E 00	1.52E-20
3.00E 00	1.21E-20
3.50E 00	7.30E-21
4.00E 00	9.60E-21
4.50E 00	1.55E-20
5.00E 00	2.33E-20
5.50E 00	2.87E-20
6.00E 00	2.98E-20
6.50E 00	3.11E-20
7.00E 00	3.11E-20
8.00E 00	3.19E-20
9.00E 00	4.29E-20
1.00E 01	5.38E-20
1.50E 01	1.71E-19
2.00E 01	2.35E-19
2.50E 01	3.07E-19
3.00E 01	4.31E-19
3.50E 01	5.76E-19

ENERGY (KEV)	CROSS SECTION (SQ. CM)
4.00E 01	6.32E-19
5.00E 01	7.10E-19
6.00E 01	6.66E-19
7.00E 01	5.95E-19
8.00E 01	5.04E-19
9.00E 01	4.66E-19
1.00E 02	4.19E-19
1.10E 02	3.83E-19
1.20E 02	2.93E-19
1.30E 02	2.38E-19
1.40E 02	2.03E-19
1.50E 02	1.72E-19

6-4-4C
VAN DEN BOS ET AL. (200)

ENERGY (KEV)	CROSS SECTION (SQ. CM)
2.00E 00	6.00E-22
2.50E 00	2.10E-21
3.00E 00	5.90E-21
3.50E 00	1.09E-20
4.00E 00	1.13E-20
4.50E 00	8.70E-21
5.00E 00	7.20E-21
5.50E 00	6.10E-21
6.00E 00	8.30E-21
7.00E 00	1.48E-20
7.50E 00	1.61E-20
8.00E 00	1.64E-20
8.50E 00	1.58E-20
9.00E 00	1.41E-20
1.00E 01	1.16E-20
1.50E 01	3.72E-20
2.00E 01	8.42E-20
2.50E 01	1.51E-19
3.00E 01	1.71E-19
3.50E 01	2.24E-19
4.00E 01	2.51E-19
5.00E 01	2.87E-19
6.00E 01	2.75E-19
7.00E 01	2.58E-19
8.00E 01	2.26E-19
9.00E 01	1.88E-19
1.00E 02	1.59E-19
1.10E 02	1.30E-19
1.20E 02	1.08E-19
1.30E 02	9.39E-20
1.40E 02	7.57E-20
1.50E 02	6.27E-20

6-4-4D
VAN DEN BOS ET AL. (200)

ENERGY (KEV)	CROSS SECTION (SQ. CM)
2.00E 00	4.00E-22
2.50E 00	1.20E-21
3.00E 00	2.40E-21
3.50E 00	2.40E-21
4.00E 00	3.30E-21
4.50E 00	4.60E-21
5.00E 00	5.10E-21
5.50E 00	5.10E-21
6.00E 00	4.80E-21

ENERGY (KEV)	CROSS SECTION (SQ. CM)
6.50E 00	5.40E-21
7.00E 00	6.40E-21
8.00E 00	9.50E-21
8.50E 00	1.09E-20
9.00E 00	1.17E-20
9.50E 00	1.02E-20
1.00E 01	9.20E-21
1.50E 01	1.96E-20
2.00E 01	4.40E-20
2.50E 01	7.85E-20
3.00E 01	1.02E-19
3.50E 01	1.34E-19
4.00E 01	1.50E-19
5.00E 01	1.71E-19
6.00E 01	1.65E-19
7.00E 01	1.55E-19
8.00E 01	1.35E-19
9.00E 01	1.16E-19
1.00E 02	9.48E-20
1.10E 02	7.48E-20
1.20E 02	6.87E-20
1.30E 02	5.98E-20
1.40E 02	4.74E-20
1.50E 02	3.99E-20

6-4-5A
VAN DEN BOS ET AL. (200)

ENERGY (KEV)	CROSS SECTION (SQ. CM)
1.00E 00	1.05E-20
2.00E 00	5.30E-20
3.00E 00	8.10E-20
4.00E 00	1.18E-19
5.00E 00	1.95E-19
6.00E 00	2.52E-19
7.00E 00	2.81E-19
8.00E 00	2.81E-19
9.00E 00	2.61E-19
1.00E 01	2.40E-19
1.50E 01	3.34E-19
2.00E 01	6.90E-19
2.50E 01	1.01E-18
3.00E 01	1.17E-18
3.50E 01	1.33E-18
4.00E 01	1.56E-18
5.00E 01	1.54E-18
6.00E 01	1.34E-18
7.00E 01	1.12E-18
8.00E 01	9.10E-19
9.00E 01	7.30E-19
1.00E 02	5.80E-19
1.10E 02	4.60E-19
1.20E 02	3.76E-19
1.30E 02	2.80E-19
1.40E 02	2.60E-19
1.50E 02	2.30E-19

6-4-5B
VAN DEN BOS ET AL. (200)

ENERGY (KEV)	CROSS SECTION (SQ. CM)
1.50E 00	2.50E-21
2.00E 00	6.60E-21

ENERGY (KEV)	CROSS SECTION (SQ. CM)
2.50E 00	6.20E-21
3.00E 00	1.07E-20
3.50E 00	3.06E-20
4.00E 00	5.03E-20
4.50E 00	5.38E-20
5.00E 00	4.96E-20
5.50E 00	4.84E-20
6.00E 00	5.05E-20
6.50E 00	5.87E-20
7.00E 00	6.47E-20
8.00E 00	9.48E-20
9.00E 00	1.17E-19
1.00E 01	1.44E-19
1.50E 01	1.41E-19
2.00E 01	1.34E-19
2.50E 01	2.55E-19
3.00E 01	4.22E-19
3.50E 01	4.69E-19
4.00E 01	5.81E-19
5.00E 01	6.19E-19
6.00E 01	5.86E-19
7.00E 01	5.34E-19
8.00E 01	4.31E-19
9.00E 01	3.49E-19
1.00E 02	3.06E-19
1.10E 02	2.50E-19
1.20E 02	1.78E-19
1.30E 02	1.53E-19
1.40E 02	1.33E-19
1.50E 02	1.09E-19

6-4-5C
VAN DEN BOS ET AL. (200)

ENERGY (KEV)	CROSS SECTION (SQ. CM)
1.00E 00	5.00E-22
1.50E 00	5.00E-22
2.00E 00	1.10E-21
2.50E 00	2.90E-21
3.00E 00	5.60E-21
3.50E 00	7.80E-21
4.00E 00	3.80E-21
4.50E 00	3.20E-21
5.00E 00	6.30E-21
5.50E 00	1.16E-20
6.00E 00	1.63E-20
7.00E 00	2.15E-20
8.00E 00	2.48E-20
9.00E 00	3.40E-20
1.00E 01	4.37E-20
1.50E 01	5.94E-20
2.00E 01	6.38E-20
2.50E 01	1.03E-19
3.00E 01	1.58E-19
3.50E 01	1.94E-19
4.00E 01	2.25E-19
5.00E 01	2.39E-19
6.00E 01	2.26E-19
7.00E 01	1.95E-19
8.00E 01	1.61E-19
9.00E 01	1.55E-19
1.00E 02	1.30E-19
1.10E 02	1.08E-19
1.20E 02	8.39E-20
1.30E 02	7.15E-20
1.40E 02	6.06E-21
1.50E 02	5.22E-20

6-4-5D
VAN DEN BOS ET AL. (200)

ENERGY (KEV)	CROSS SECTION (SQ. CM)
1.50E 00	3.00E-22
2.00E 00	4.00E-22
2.50E 00	7.00E-22
3.50E 00	9.00E-22
4.00E 00	2.90E-21
4.50E 00	4.80E-21
5.00E 00	3.40E-21
5.50E 00	1.80E-21
6.50E 00	3.30E-21
7.50E 00	6.90E-21
8.50E 00	8.90E-21
9.50E 00	9.60E-21
1.25E 01	1.20E-20
1.75E 01	2.20E-20
2.00E 01	2.70E-20
2.50E 01	3.70E-20
3.00E 01	6.10E-20
3.50E 01	6.30E-20
4.00E 01	1.07E-19
5.00E 01	1.17E-19
6.00E 01	1.07E-19
7.00E 01	9.80E-20
8.00E 01	7.80E-20
9.00E 01	6.60E-20
1.00E 02	5.50E-20
1.10E 02	4.50E-20
1.20E 02	3.50E-20
1.30E 02	3.00E-20
1.40E 02	2.40E-20
1.50E 02	2.10E-20

6-4-6A
VAN DEN BOS ET AL. (200)

ENERGY (KEV)	CROSS SECTION (SQ. CM)
1.00E 00	1.50E-20
2.00E 00	7.20E-20
3.00E 00	2.01E-19
3.50E 00	2.91E-19
4.00E 00	3.29E-19
5.00E 00	3.96E-19
5.50E 00	4.17E-19
6.00E 00	4.27E-19
6.50E 00	4.75E-19
7.00E 00	5.04E-19
8.00E 00	5.54E-19
9.00E 00	5.91E-19
1.00E 01	6.40E-19
1.50E 01	8.60E-19
2.00E 01	9.54E-19
2.50E 01	1.02E-18
3.00E 01	9.60E-19
3.50E 01	8.10E-19
4.00E 01	7.60E-19
5.00E 01	6.10E-19
6.00E 01	4.40E-19
7.00E 01	3.42E-19
8.00E 01	2.80E-19
9.00E 01	2.30E-19
1.00E 02	1.95E-19
1.10E 02	1.60E-19
1.20E 02	1.40E-19
1.30E 02	1.26E-19
1.40E 02	1.16E-19
1.50E 02	1.20E-19

6-4-6B
VAN DEN BOS ET AL. (200)

ENERGY (KEV)	CROSS SECTION (SQ. CM)
1.00E 00	2.70E-21
1.50E 00	4.00E-21
2.00E 00	1.27E-20
3.00E 00	3.64E-20
3.50E 00	4.52E-20
4.00E 00	4.59E-20
4.50E 00	4.27E-20
5.00E 00	4.04E-20
5.50E 00	4.42E-20
6.00E 00	5.34E-20
7.00E 00	7.17E-20
8.00E 00	1.01E-19
9.00E 00	1.25E-19
1.00E 01	1.45E-19
1.50E 01	2.48E-19
2.00E 01	3.44E-19
2.50E 01	3.73E-19
3.00E 01	3.81E-19
3.50E 01	4.12E-19
4.00E 01	3.58E-19
5.00E 01	2.77E-19
6.00E 01	2.08E-19
7.00E 01	1.62E-19
8.00E 01	1.30E-19
9.00E 01	1.00E-19
1.00E 02	9.20E-20
1.10E 02	7.90E-20
1.20E 02	6.60E-20
1.30E 02	6.30E-20
1.40E 02	5.50E-20
1.50E 02	5.00E-20

6-4-6C
VAN DEN BOS ET AL. (200)

ENERGY (KEV)	CROSS SECTION (SQ. CM)
1.00E 00	1.30E-21
2.00E 00	2.90E-21
3.00E 00	8.30E-21
4.00E 00	1.84E-20
4.50E 00	2.53E-20
5.00E 00	3.01E-20
5.50E 00	2.94E-20
6.00E 00	2.67E-20
6.50E 00	2.54E-20
7.00E 00	2.78E-20
8.00E 00	3.34E-20
9.00E 00	4.12E-20
1.00E 01	5.33E-20
1.50E 01	1.30E-19
2.00E 01	1.72E-19
2.50E 01	2.10E-19
3.00E 01	2.10E-19
3.50E 01	2.13E-19

ENERGY (KEV)	CROSS SECTION (SQ. CM)
4.00E 01	2.00E-19
5.00E 01	1.57E-19
6.00E 01	1.18E-19
7.00E 01	8.70E-20
8.00E 01	7.20E-20
9.00E 01	5.30E-20
1.00E 02	5.00E-20
1.10E 02	4.60E-20
1.20E 02	3.70E-20
1.30E 02	3.40E-20
1.40E 02	3.20E-20
1.50E 02	2.80E-20

6-4-7A
VAN DEN BOS ET AL. (200)

ENERGY (KEV)	CROSS SECTION (SQ. CM)
1.00E 00	2.20E-20
2.00E 00	1.16E-19
3.00E 00	1.68E-19
3.50E 00	1.72E-19
4.00E 00	1.84E-19
4.50E 00	1.97E-19
5.00E 00	2.00E-19
5.50E 00	2.18E-19
6.00E 00	2.26E-19
6.50E 00	2.38E-19
7.00E 00	2.45E-19
7.50E 00	2.70E-19
8.00E 00	3.05E-19
9.00E 00	3.32E-19
1.00E 01	3.34E-19
1.50E 01	4.35E-19
2.00E 01	4.94E-19
2.50E 01	5.75E-19
3.00E 01	6.63E-19
3.50E 01	6.91E-19
4.00E 01	8.27E-19
5.00E 01	9.62E-19
6.00E 01	1.07E-18
7.00E 01	1.20E-18
8.00E 01	1.21E-18
9.00E 01	1.27E-18
1.00E 02	1.38E-18
1.10E 02	1.47E-18
1.20E 02	1.53E-18
1.30E 02	1.65E-18
1.40E 02	1.58E-21
1.50E 02	1.85E-21

6-4-7B
VAN DEN BOS ET AL. (200)

ENERGY (KEV)	CROSS SECTION (SQ. CM)
1.00E 00	6.00E-21
2.00E 00	1.85E-20
2.50E 00	2.61E-20
3.00E 00	3.05E-20
3.50E 00	3.22E-20
4.00E 00	3.34E-20
4.50E 00	3.16E-20
5.00E 00	2.89E-20
5.50E 00	2.99E-20
6.00E 00	3.11E-20
7.00E 00	3.34E-20
8.00E 00	3.37E-20
9.00E 00	3.60E-20
1.00E 01	3.69E-20
1.50E 01	8.60E-20
2.00E 01	1.36E-19
2.50E 01	1.83E-19
3.00E 01	2.13E-19
3.50E 01	2.43E-19
4.00E 01	2.81E-19
5.00E 01	3.21E-19
6.00E 01	3.39E-19
7.00E 01	3.52E-19
8.00E 01	3.28E-19
9.00E 01	3.72E-19
1.00E 02	4.18E-19
1.10E 02	4.60E-19
1.20E 02	4.82E-19
1.30E 02	5.10E-19
1.40E 02	5.23E-19
1.50E 02	5.48E-19

6-4-7C
THOMAS ET AL. (201)

ENERGY (KEV)	CROSS SECTION (SQ. CM)
3.00E 02	3.33E-19
4.50E 02	2.14E-19
6.00E 02	1.58E-19

6-4-7D
VAN DEN BOS ET AL. (200)

ENERGY (KEV)	CROSS SECTION (SQ. CM)
1.50E 00	3.00E-21
2.00E 00	2.90E-21
2.50E 00	3.00E-21
3.00E 00	4.60E-21
3.50E 00	7.60E-21
4.00E 00	1.00E-20
4.50E 00	1.34E-20
5.00E 00	1.59E-20
5.50E 00	1.70E-20
6.00E 00	1.78E-20
6.50E 00	1.82E-20
7.00E 00	1.73E-20
7.50E 00	1.57E-20
8.00E 00	1.59E-20
8.50E 00	1.62E-20
9.00E 00	1.65E-20
9.50E 00	1.68E-20
1.00E 01	1.71E-20
1.50E 01	2.11E-20
2.00E 01	3.11E-20
2.50E 01	5.85E-20
3.00E 01	7.70E-20
3.50E 01	1.06E-19
4.00E 01	1.25E-19
5.00E 01	1.44E-19
6.00E 01	1.57E-19
7.00E 01	1.62E-19
8.00E 01	1.70E-19
9.00E 01	1.87E-19
1.00E 02	2.08E-19
1.10E 02	2.21E-19
1.20E 02	2.40E-19
1.30E 02	2.62E-19
1.40E 02	2.82E-19
1.50E 02	2.88E-19

6-4-7E
VAN DEN BOS ET AL. (200)

ENERGY (KEV)	CROSS SECTION (SQ. CM)
2.00E 00	2.80E-21
2.50E 00	3.00E-21
3.00E 00	3.80E-21
3.50E 00	4.40E-21
4.00E 00	6.00E-21
5.00E 00	6.80E-21
6.00E 00	7.70E-21
6.50E 00	9.30E-21
7.00E 00	9.90E-21
8.00E 00	1.02E-20
9.00E 00	1.04E-20
9.50E 00	1.00E-20
1.00E 01	1.02E-20
1.50E 01	1.31E-20
2.00E 01	1.65E-20
2.50E 01	2.39E-20
3.00E 01	3.30E-20
3.50E 01	5.11E-20
4.00E 01	6.49E-20
5.00E 01	7.70E-20
6.00E 01	7.60E-20
7.00E 01	7.56E-20
8.00E 01	7.40E-20
9.00E 01	8.93E-20
1.00E 02	1.02E-19
1.10E 02	1.09E-19
1.20E 02	1.22E-19
1.30E 02	1.28E-19
1.40E 02	1.44E-19
1.50E 02	1.42E-19

6-4-8A
VAN DEN BOS ET AL. (200)

ENERGY (KEV)	CROSS SECTION (SQ. CM)
1.00E 00	1.77E-19
2.00E 00	4.38E-19
3.00E 00	8.06E-19
4.00E 00	8.39E-19
5.00E 00	9.07E-19
6.00E 00	8.90E-19
7.00E 00	1.07E-19
8.00E 00	9.51E-19
9.00E 00	9.04E-19
1.00E 01	8.50E-19
1.50E 01	7.80E-19
2.00E 01	7.80E-19
2.50E 01	8.80E-19

ENERGY (KEV)	CROSS SECTION (SQ. CM)
3.00E 01	1.12E-18
3.50E 01	1.41E-18
4.00E 01	1.66E-18
5.00E 01	2.30E-18
6.00E 01	2.73E-18
7.00E 01	2.93E-18
8.00E 01	3.10E-18
9.00E 01	3.11E-18
1.00E 02	3.31E-18
1.10E 02	3.25E-18
1.20E 02	3.50E-18
1.30E 02	3.61E-18
1.40E 02	3.71E-18
1.50E 02	3.99E-18

6-4-8B
VAN DEN BOS ET AL. (200)

ENERGY (KEV)	CROSS SECTION (SQ. CM)
1.50E 00	8.20E-20
2.00E 00	1.06E-19
2.50E 00	1.23E-19
3.00E 00	1.14E-19
3.50E 00	1.06E-19
4.00E 00	9.00E-20
4.50E 00	8.50E-20
5.00E 00	8.80E-20
5.50E 00	9.10E-20
6.00E 00	9.70E-20
6.50E 00	9.80E-20
7.00E 00	1.08E-19
8.00E 00	1.33E-19
9.00E 00	1.49E-19
1.00E 01	1.61E-19
1.50E 01	1.85E-19
2.00E 01	1.88E-19
2.50E 01	2.01E-19
3.00E 01	2.62E-19
3.50E 01	3.15E-19
4.00E 01	3.82E-19
5.00E 01	4.51E-19
6.00E 01	5.40E-19
7.00E 01	5.82E-19
8.00E 01	6.41E-19
9.00E 01	6.36E-19
1.00E 02	7.33E-19
1.10E 02	8.70E-19
1.20E 02	8.78E-19
1.30E 02	8.07E-19
1.40E 02	8.42E-19
1.50E 02	9.01E-19

6-4-8C
VAN DEN BOS ET AL. (200)

ENERGY (KEV·)	CROSS SECTION (SQ. CM)
1.00E 00	3.00E-20
2.00E 00	3.80E-20
3.00E 00	6.50E-20
4.00E 00	9.00E-20
5.00E 00	1.01E-19
6.00E 00	9.00E-20
7.00E 00	8.20E-20
8.00E 00	8.00E-20
9.00E 00	9.90E-20
1.00E 01	1.06E-19
1.50E 01	1.24E-19
2.00E 01	1.39E-19
2.50E 01	1.66E-19
3.00E 01	1.84E-19
3.50E 01	1.92E-19
4.00E 01	2.30E-19
5.00E 01	2.50E-19
6.00E 01	3.03E-19
7.00E 01	3.32E-19
8.00E 01	3.44E-19
9.00E 01	3.66E-19
1.00E 02	4.10E-19
1.10E 02	4.20E-19
1.20E 02	4.70E-19
1.30E 02	4.70E-19
1.40E 02	4.90E-19
1.50E 02	5.20E-19

6-4-9A
VAN DEN BOS ET AL. (200)

ENERGY (KEV)	CROSS SECTION (SQ. CM)
1.50E 00	7.60E-19
2.00E 00	8.20E-19
2.50E 00	7.80E-19
3.00E 00	8.10E-19
3.50E 00	8.20E-19
4.00E 00	8.70E-19
5.00E 00	9.70E-19
6.00E 00	9.80E-19
7.00E 00	1.00E-18
8.00E 00	1.04E-18
9.00E 00	1.05E-18
1.00E 01	1.10E-18
1.50E 01	1.54E-18
2.00E 01	1.43E-18
2.50E 01	1.33E-18
3.00E 01	1.25E-18
3.50E 01	1.07E-18
4.00E 01	1.04E-18
5.00E 01	9.00E-19
6.00E 01	7.80E-19
7.00E 01	6.60E-19
8.00E 01	5.90E-19
9.00E 01	5.90E-19
1.00E 02	5.50E-19
1.10E 02	5.50E-19
1.20E 02	5.00E-19
1.30E 02	5.10E-19
1.40E 02	5.10E-19
1.50E 02	5.30E-19

6-4-9B
VAN DEN BOS ET AL. (200)

ENERGY (KEV)	CROSS SECTION (SQ. CM)
1.00E 00	7.80E-20
2.00E 00	1.51E-19
3.00E 00	1.79E-19
3.50E 00	2.09E-19
4.00E 00	2.21E-19
5.00E 00	2.12E-19
6.00E 00	2.02E-19
7.00E 00	1.71E-19
8.00E 00	1.64E-19
9.00E 00	1.60E-19
1.00E 01	1.56E-19
1.50E 01	2.00E-19
2.00E 01	2.16E-19
2.50E 01	2.27E-19
3.00E 01	2.44E-19
3.50E 01	2.71E-19
4.00E 01	3.41E-19
5.00E 01	3.61E-19
6.00E 01	3.25E-19
7.00E 01	2.99E-19
8.00E 01	2.78E-19
9.00E 01	2.63E-19
1.00E 02	2.68E-19
1.10E 02	2.69E-19
1.20E 02	2.71E-19
1.30E 02	2.63E-19
1.40E 02	2.63E-19
1.50E 02	2.66E-19

6-4-9C
VAN DEN BOS ET AL. (200)

ENERGY (KEV)	CROSS SECTION (SQ. CM)
1.00E 00	2.14E-20
2.00E 00	4.99E-20
3.00E 00	5.61E-20
4.00E 00	5.78E-20
4.50E 00	5.61E-20
5.00E 00	5.82E-20
5.50E 00	6.36E-20
6.00E 00	6.85E-20
6.50E 00	6.78E-20
7.00E 00	7.02E-20
7.50E 00	7.36E-20
8.00E 00	7.49E-20
9.00E 00	6.96E-20
1.00E 01	6.44E-20
1.50E 01	7.66E-20
2.00E 01	9.83E-20
2.50E 01	1.15E-19
3.00E 01	1.29E-19
3.50E 01	1.40E-19
4.00E 01	1.74E-19
5.00E 01	1.86E-19
6.00E 01	1.52E-19
7.00E 01	1.42E-19
8.00E 01	1.32E-19
9.00E 01	1.34E-19
1.00E 02	1.27E-19
1.10E 02	1.27E-19
1.20E 02	1.26E-19
1.30E 02	1.25E-19
1.40E 02	1.30E-19
1.50E 02	1.27E-19

6-4-9D
VAN DEN BOS ET AL. (200)

ENERGY (KEV)	CROSS SECTION (SQ. CM)
1.00E 00	7.20E-21
1.50E 00	1.65E-20
2.0CE 00	1.62E-20
2.50E 00	2.26E-20
3.0CE 00	2.38E-20
3.50E 00	2.72E-20
4.00E 00	2.72E-20
4.50E C0	2.68E-20
5.00E 00	2.68E-20
6.0CE 00	2.49E-20
7.00E 00	2.75E-20
7.5CE 00	3.14E-20
8.00E 00	3.35E-20
9.00E 00	3.44E-20
1.00E 01	3.53E-20
1.50E 01	3.87E-20
2.00E 01	4.19E-20
2.50E 01	4.51E-20
3.00E 01	6.10E-20
3.50E 01	6.90E-20
4.00E 01	7.40E-20
5.00E 01	8.80E-20
6.00E 01	8.30E-20
7.00E 01	8.30E-20
8.00E 01	7.80E-20
9.00E 01	7.60E-20
1.00E 02	7.50E-20
1.10E 02	7.00E-20
1.20E 02	7.10E-20
1.30E 02	7.20E-20
1.40E 02	7.50E-20
1.50E 02	7.00E-20

6-4-10A
VAN DEN BOS ET AL. (200)

ENERGY (KEV)	CROSS SECTION (SQ. CM)
1.00E 00	1.00E-20
2.00E 00	1.31E-19
3.00E 00	1.88E-19
4.00E 00	2.98E-19
5.00E 00	4.95E-19
6.00E 00	5.36E-19
7.00E 00	7.63E-19
8.00E 00	9.10E-19
9.00E 00	1.06E-18
1.00E 01	1.08E-18
1.50E 01	1.52E-18
2.0CE 01	1.75E-18
2.50E 01	1.67E-18
3.00E 01	1.68E-18
3.50E 01	1.81E-18
4.00E 01	2.11E-18
5.00E 01	2.48E-18
6.00E 01	3.02E-18
7.00E 01	3.41E-18
8.00E 01	3.58E-18
9.00E 01	3.66E-18
1.00E 02	3.65E-18
1.10E 02	3.65E-18
1.20E 02	3.39E-18
1.30E 02	3.18E-18
1.40E 02	2.80E-18
1.50E 02	1.79E-18

6-4-10B
VAN DEN BOS ET AL. (200)

ENERGY (KEV)	CROSS SECTION (SQ. CM)
1.00E 00	2.00E-21
2.0CE 00	5.80E-21
3.00E 00	1.71E-20
4.0CE 00	3.52E-20
4.50E 00	3.45E-20
5.00E 00	2.82E-20
5.5CE 00	2.74E-20
6.00E 00	2.78E-20
6.5CE 00	2.92E-20
7.00E 00	3.71E-20
8.00E 00	4.63E-20
9.0CE 00	4.83E-20
1.00E 01	4.88E-20
1.50E 01	1.10E-19
2.00E 01	2.25E-19
2.50E 01	3.01E-19
3.00E 01	3.65E-19
3.50E 01	3.89E-19
4.00E 01	4.31E-19
5.00E 01	4.88E-19
6.00E 01	5.84E-19
7.00E 01	6.27E-19
8.00E 01	7.69E-19
9.00E 01	7.30E-19
1.00E 02	7.11E-19
1.10E 02	6.41E-19
1.20E 02	6.50E-19
1.30E 02	6.43E-19
1.40E 02	5.65E-19
1.50E 02	5.42E-19

6-4-10C
VAN DEN BOS ET AL. (200)

ENERGY (KEV)	CROSS SECTION (SQ. CM)
2.00E 00	9.00E-22
3.00E 00	3.70E-21
4.00E 00	7.10E-21
4.50E 00	1.27E-20
5.00E 00	1.64E-20
5.50E 00	1.91E-20
6.00E 00	2.31E-20
6.50E 00	2.02E-20
7.00E 00	1.86E-20
7.50E 00	1.56E-20
8.00E 00	1.31E-20
9.00E 00	1.40E-20
1.00E 01	1.59E-20
1.50E 01	2.98E-20
2.00E 01	5.11E-20
2.50E 01	8.77E-20
3.00E 01	1.25E-19
3.50E 01	1.62E-19
4.00E 01	1.82E-19
5.00E 01	2.26E-19
6.00E 01	2.62E-19
7.00E 01	3.06E-19
8.00E 01	3.23E-19
9.00E 01	3.24E-19
1.00E 02	3.31E-19
1.10E 02	3.35E-19
1.20E 02	3.13E-19
1.30E 02	2.89E-19
1.40E 02	2.78E-19
1.50E 02	2.41E-19

6-4-10D
VAN DEN BOS ET AL. (200)

ENERGY (KEV)	CROSS SECTION (SQ. CM)
3.00E 00	3.90E-21
3.5CE 00	5.30E-21
4.00E 00	6.80E-21
5.0CE 00	8.40E-21
6.00E 00	9.90E-21
7.00E 00	1.23E-20
8.00E 00	1.30E-20
8.50E 00	1.32E-20
9.0CE 00	1.26E-20
9.50E 00	1.26E-20
1.00E 01	1.30E-20
1.50E 01	1.59E-20
2.00E 01	1.81E-20
2.50E 01	3.13E-20
3.00E 01	5.46E-20
3.50E 01	1.01E-19
4.00E 01	1.04E-19
5.00E 01	1.31E-19
6.00E 01	1.50E-19
7.00E 01	1.79E-19
8.00E 01	1.86E-19
9.00E 01	1.89E-19
1.00E 02	1.86E-19
1.10E 02	1.72E-19
1.20E 02	1.76E-19
1.30E 02	1.72E-19
1.40E 02	1.55E-19
1.50E 02	1.43E-19

6-4-11A
VAN DEN BOS ET AL. (200)

ENERGY (KEV)	CROSS SECTION (SQ. CM)
1.00E 00	3.60E-20
1.50E 00	6.70E-20
2.00E 00	7.70E-20
3.00E 00	1.35E-19
3.50E 00	1.63E-19
4.00E 00	1.60E-19
4.50E 00	1.52E-19
5.0CE 00	1.40E-19
6.00E 00	1.63E-19
7.00E 00	1.95E-19
8.00E 00	2.28E-19
9.00E 00	2.37E-19
9.50E 00	2.19E-19
1.00E 01	2.11E-19
1.50E 01	2.38E-19

ENERGY (KEV)	CROSS SECTION (SQ. CM)
2.00E 01	2.70E-19
2.50E 01	3.81E-19
3.00E 01	5.31E-19
3.50E 01	7.36E-19
4.00E 01	9.93E-19
5.00E 01	1.22E-18
6.00E 01	1.48E-18
7.00E 01	1.64E-18
8.00E 01	1.83E-18
9.00E 01	1.95E-18
1.00E 02	1.85E-18
1.10E 02	1.69E-18
1.20E 02	1.49E-18
1.30E 02	1.41E-18
1.40E 02	1.22E-18
1.50E 02	1.11E-18

6-4-11B
VAN DEN BOS ET AL. (200)

ENERGY (KEV)	CROSS SECTION (SQ. CM)
1.00E 00	8.60E-21
1.50E 00	1.61E-20
2.00E 00	2.37E-20
2.50E 00	3.17E-20
3.00E 00	3.96E-20
3.50E 00	3.82E-20
4.00E 00	3.77E-20
5.00E 00	6.17E-20
5.50E 00	8.00E-20
6.00E 00	9.95E-20
7.00E 00	9.60E-20
8.00E 00	9.26E-20
9.00E 00	9.23E-20
9.50E 00	9.61E-20
1.00E 01	9.65E-20
1.50E 01	1.23E-19
2.00E 01	1.45E-19
2.50E 01	1.48E-19
3.00E 01	1.58E-19
3.50E 01	1.99E-19
4.00E 01	3.73E-19
5.00E 01	4.81E-19
6.00E 01	5.67E-19
7.00E 01	6.78E-19
8.00E 01	7.08E-19
9.00E 01	7.86E-19
1.00E 02	7.76E-19
1.10E 02	6.84E-19
1.20E 02	6.23E-19
1.30E 02	5.82E-19
1.40E 02	5.45E-19
1.50E 02	4.56E-19

6-4-11C
VAN DEN BOS ET AL. (200)

ENERGY (KEV)	CROSS SECTION (SQ. CM)
1.00E 00	2.30E-21
1.50E 00	4.90E-21
2.00E 00	6.60E-21
2.50E 00	8.30E-21
3.00E 00	9.20E-21

3.50E 00	1.22E-20
4.00E 00	1.49E-20
4.50E 00	1.92E-20
5.00E 00	2.19E-20
5.50E 00	1.90E-20
6.00E 00	1.62E-20
6.50E 00	1.46E-20
7.00E 00	1.51E-20
8.00E 00	2.14E-20
9.00E 00	2.72E-20
9.50E 00	2.87E-20
1.00E 01	2.94E-20
1.50E 01	6.10E-20
2.00E 01	8.00E-20
2.50E 01	8.70E-20
3.00E 01	8.60E-20
3.50E 01	9.30E-20
4.00E 01	1.14E-19
5.00E 01	1.75E-19
6.00E 01	2.23E-19
7.00E 01	2.74E-19
8.00E 01	3.08E-19
9.00E 01	3.30E-19
1.00E 02	3.24E-19
1.10E 02	3.05E-19
1.20E 02	2.79E-19
1.30E 02	2.64E-19
1.40E 02	2.51E-19
1.50E 02	2.10E-19

6-4-11D
VAN DEN BOS ET AL. (200)

ENERGY (KEV)	CROSS SECTION (SQ. CM)
1.00E 00	1.00E-21
2.00E 00	2.40E-21
3.00E 00	3.70E-21
4.00E 00	4.30E-21
5.00E 00	5.50E-21
5.50E 00	8.90E-21
6.00E 00	1.04E-20
7.00E 00	8.10E-21
8.00E 00	6.30E-21
9.00E 00	6.40E-21
9.50E 00	7.10E-21
1.00E 01	7.50E-21
1.50E 01	1.65E-20
2.00E 01	3.30E-20
2.50E 01	4.00E-20
3.00E 01	4.10E-20
3.50E 01	3.70E-20
4.00E 01	4.50E-20
5.00E 01	6.30E-20
6.00E 01	9.10E-20
7.00E 01	1.09E-19
8.00E 01	1.34E-19
9.00E 01	1.41E-19
1.00E 02	1.40E-19
1.10E 02	1.30E-19
1.20E 02	1.20E-19
1.30E 02	1.10E-19
1.40E 02	9.70E-20
1.50E 02	9.60E-20

6-4-12A
VAN DEN BOS ET AL. (200)

ENERGY (KEV)	CROSS SECTION (SQ. CM)
1.00E 00	2.23E-19
1.50E 00	3.70E-19
2.00E 00	4.86E-19
3.00E 00	5.60E-19
4.00E 00	6.60E-19
5.00E 00	7.80E-19
6.00E 00	8.90E-19
6.50E 00	9.50E-19
7.00E 00	9.00E-19
7.50E 00	8.80E-19
8.00E 00	9.20E-19
8.50E 00	8.70E-19
9.00E 00	8.70E-19
9.50E 00	8.70E-19
1.00E 01	8.70E-19
1.50E 01	9.30E-19
2.00E 01	1.13E-18
2.50E 01	1.36E-18
3.00E 01	1.49E-18
3.50E 01	1.48E-18
4.00E 01	1.38E-18
5.00E 01	1.19E-18
6.00E 01	1.10E-18
7.00E 01	9.70E-19
8.00E 01	8.70E-19
9.00E 01	7.40E-19
1.00E 02	6.60E-19
1.10E 02	5.60E-19
1.20E 02	4.90E-19
1.30E 02	4.30E-19
1.40E 02	3.80E-19
1.50E 02	3.50E-19

6-4-12B
VAN DEN BOS ET AL. (200)

ENERGY (KEV)	CROSS SECTION (SQ. CM)
1.00E 00	2.90E-20
2.00E 00	7.10E-20
3.00E 00	1.01E-19
4.00E 00	1.22E-19
5.00E 00	1.38E-19
5.50E 00	1.39E-19
6.00E 00	1.38E-19
7.00E 00	1.32E-19
8.00E 00	1.24E-19
9.00E 00	1.32E-19
9.50E 00	1.41E-19
1.00E 01	1.48E-19
1.50E 01	2.46E-19
2.00E 01	4.09E-19
2.50E 01	4.65E-19
3.00E 01	5.00E-19
3.50E 01	5.30E-19
4.00E 01	5.50E-19
5.00E 01	5.30E-19
6.00E 01	4.90E-19
7.00E 01	4.10E-19
8.00E 01	4.00E-19
9.00E 01	3.60E-19
1.00E 02	3.10E-19

1.10E 02	2.70E-19
1.20E 02	2.40E-19
1.30E 02	2.20E-19
1.40E 02	1.86E-19
1.50E 02	1.64E-19

6-4-12C
VAN DEN BOS ET AL. (200)

ENERGY (KEV)	CROSS SECTION (SQ. CM)
1.00E 00	1.20E-20
2.00E 00	2.30E-20
2.50E 00	2.90E-20
3.00E 00	3.30E-20
4.00E 00	3.90E-20
5.00E 00	4.40E-20
6.00E 00	5.20E-20
6.50E 00	6.00E-20
7.00E 00	6.30E-20
7.50E 00	6.90E-20
8.00E 00	6.00E-20
9.00E 00	5.80E-20
1.00E 01	5.90E-20
1.50E 01	1.46E-19
2.00E 01	1.75E-19
2.50E 01	2.16E-19
3.00E 01	2.48E-19
3.50E 01	2.82E-19
4.00E 01	3.00E-19
5.00E 01	3.20E-19
6.00E 01	3.06E-19
7.00E 01	2.85E-19
8.00E 01	2.34E-19
9.00E 01	2.02E-19
1.00E 02	1.90E-19
1.10E 02	1.81E-19
1.20E 02	1.68E-19
1.30E 02	1.59E-19
1.40E 02	1.41E-19
1.50E 02	1.45E-19

6-4-13A
VAN DEN BOS ET AL. (200)

ENERGY (KEV)	POLARIZATION (PERCENT)
2.00E 00	5.00E 00
3.00E 00	4.00E 00
4.00E 00	-6.00E 00
5.00E 00	-3.00E 00
6.00E 00	-1.00E 00
7.00E 00	-1.00E 00
8.00E 00	0.0
9.00E 00	2.00E 00
1.00E 01	3.00E 00
1.50E 01	6.00E 00
2.00E 01	9.00E 00
2.50E 01	9.00E 00
3.00E 01	1.00E 01
3.50E 01	8.00E 00
4.00E 01	9.00E 00
5.00E 01	6.00E 00
6.00E 01	8.00E 00

7.00E 01	5.00E 00
8.00E 01	3.00E-02
9.00E 01	3.00E-06
1.00E 02	2.00E 00
1.10E 02	2.00E 00
1.20E 02	2.00E 00
1.30E 02	1.00E 00
1.40E 02	1.00E 00
1.50E 02	2.00E 00

6-4-13B
VAN DEN BOS ET AL. (200)

ENERGY (KEV)	POLARIZATION (PERCENT)
2.00E 00	1.80E 01
3.00E 00	2.30E 01
4.00E 00	2.20E 01
5.00E 00	1.80E 01
6.00E 00	1.60E 01
8.00E 00	1.90E 01
1.00E 01	2.80E 01
1.50E 01	3.70E 01
2.00E 01	4.10E 01
2.50E 01	4.30E 01
3.00E 01	3.80E 01
4.00E 01	3.60E 01
5.00E 01	3.20E 01
6.00E 01	3.00E 01
7.00E 01	2.90E 01
8.00E 01	2.80E 01
9.00E 01	2.80E 01
1.00E 02	2.50E 01
1.10E 02	3.20E 01
1.20E 02	2.90E 01
1.30E 02	3.00E 01
1.40E 02	2.70E 01
1.50E 02	2.70E 01

6-4-13C
SCHARMANN ET AL. (166)

ENERGY (KEV)	POLARIZATION (PERCENT)
1.20E 02	3.05E 01
1.50E 02	3.10E 01
2.20E 02	2.50E 01
3.00E 02	2.10E 01
4.00E 02	1.45E 01
6.00E 02	6.50E 00
8.00E 02	1.00E 00
1.00E 03	-5.00E 00
1.10E 03	-6.50E 00

6-4-14A
VAN DEN BOS ET AL. (200)

ENERGY (KEV)	POLARIZATION (PERCENT)
2.00E 00	1.80E 01

3.00E 00	1.20E 01
4.00E 00	1.00E 01
5.00E 00	1.00E 01
7.00E 00	1.10E 01
8.00E 00	1.00E 01
9.00E 00	9.00E 00
1.00E 01	1.00E 01
1.50E 01	4.00E 00
2.00E 01	1.00E 00
2.50E 01	-3.00E 00
3.00E 01	-3.00E 00
3.50E 01	2.00E 00
4.00E 01	-2.00E 00
5.00E 01	-1.00E 00
6.00E 01	2.00E 00
7.00E 01	3.00E 00
8.00E 01	-2.00E 00
9.00E 01	0.0
1.00E 02	0.0
1.10E 02	-2.00E 00
1.20E 02	0.0
1.30E 02	-3.00E 00
1.40E 02	-4.00E 00
1.50E 02	-4.00E 00

6-4-14B
VAN DEN BOS ET AL. (200)

ENERGY (KEV)	POLARIZATION (PERCENT)
1.00E 00	-5.00E 00
2.00E 00	1.00E 01
3.00E 00	1.60E 01
4.00E 00	1.90E 01
5.00E 00	1.20E 01
6.00E 00	1.20E 01
7.00E 00	1.30E 01
8.00E 00	1.10E 01
9.00E 00	1.50E 01
1.00E 01	1.60E 01
1.50E 01	1.60E 01
2.00E 01	1.40E 01
2.50E 01	1.20E 01
3.00E 01	1.20E 01
3.50E 01	1.30E 01
4.00E 01	1.00E 01
5.00E 01	8.00E 00
6.00E 01	9.00E 00
7.00E 01	6.00E 00
8.00E 01	3.00E 00
9.00E 01	2.00E 00
1.00E 02	1.00E 00
1.10E 02	3.00E 00
1.20E 02	4.00E 00
1.30E 02	5.00E 00
1.40E 02	4.00E 00
1.50E 02	5.00E 00

6-4-15A
VAN DEN BOS ET AL. (200)

ENERGY (KEV)	POLARIZATION (PERCENT)
1.00E 00	-8.00E 00

2.00E 00	-1.00E 01
3.00E 00	-9.00E 00
4.00E 00	-7.00E 00
5.00E 00	-8.00E 00
6.00E 00	-8.00E 00
7.00E 00	-5.00E 00
8.00E 00	-3.00E 00
9.00E 00	-1.00E 00
1.00E 01	7.00E 00
1.50E 01	7.00E 00
2.00E 01	4.00E 00
2.50E 01	6.00E 00
3.00E 01	1.00E 00
3.50E 01	0.0
4.00E 01	3.00E 00
5.00E 01	0.0
6.00E 01	1.00E 01
7.00E 01	1.30E 01
8.00E 01	1.30E 01
9.00E 01	1.60E 01
1.00E 02	1.80E 01
1.10E 02	1.30E 01
1.20E 02	1.20E 01
1.30E 02	1.00E 01
1.40E 02	1.00E 01
1.50E 02	9.00E 00

6-4-15B
VAN DEN BOS ET AL. (200)

ENERGY (KEV)	POLARIZATION (PERCENT)
1.00E 00	-1.10E 01
2.00E 00	-9.00E 00
3.00E 00	-7.00E 00
4.00E 00	1.00E 00
5.00E 00	1.00E 01
6.00E 00	1.00E 01
7.00E 00	1.10E 01
8.00E 00	1.10E 01
9.00E 00	1.30E 01

1.00E 01	1.50E 01
1.50E 01	2.90E 01
2.00E 01	3.30E 01
3.00E 01	3.50E 01
4.00E 01	3.30E 01
5.00E 01	3.30E 01
6.00E 01	3.50E 01
7.00E 01	3.30E 01
8.00E 01	3.10E 01
9.00E 01	3.00E 01
1.00E 02	3.00E 01
1.10E 02	2.80E 01
1.20E 02	3.00E 01
1.30E 02	2.90E 01
1.40E 02	3.00E 01
1.50E 02	2.60E 01

6-4-16A
VAN DEN BOS ET AL. (200)

ENERGY (KEV)	POLARIZATION (PERCENT)
2.00E 00	8.00E 00
3.00E 00	1.20E 01
4.00E 00	1.20E 01
5.00E 00	1.40E 01
6.00E 00	1.70E 01
7.00E 00	1.40E 01
8.00E 00	1.30E 01
9.00E 00	1.00E 01
1.00E 01	1.20E 01
1.50E 01	0.0
2.00E 01	0.0
2.50E 01	-1.00E 00
3.00E 01	-1.00E 00
3.50E 01	-2.00E 00
4.00E 01	-2.00E 00
5.00E 01	-2.00E 00
6.00E 01	-2.00E 00
7.00E 01	-2.00E 00
8.00E 01	-5.00E 00
9.00E 01	-3.00E 00

1.00E 02	-5.00E 00
1.10E 02	-6.00E 00
1.20E 02	-6.00E 00
1.30E 02	-5.00E 00
1.40E 02	-6.00E 00
1.50E 02	-8.00E 00

6-4-16B
VAN DEN BOS ET AL. (200)

ENERGY (KEV)	POLARIZATION (PERCENT)
1.00E 00	-5.00E 00
2.00E 00	0.0
3.00E 00	3.00E 00
4.00E 00	7.00E 00
5.00E 00	9.00E 00
6.00E 00	1.00E 01
7.00E 00	1.30E 01
8.00E 00	1.40E 01
9.00E 00	1.30E 01
1.00E 01	1.50E 01
1.50E 01	1.50E 01
2.00E 01	1.60E 01
2.50E 01	1.60E 01
3.00E 01	1.40E 01
3.50E 01	1.70E 01
4.00E 01	1.50E 01
5.00E 01	1.60E 01
6.00E 01	1.40E 01
7.00E 01	1.20E 01
8.00E 01	1.10E 01
9.00E 01	1.00E 01
1.00E 02	9.00E 00
1.10E 02	7.00E 00
1.20E 02	6.00E 00
1.30E 02	6.00E 00
1.40E 02	8.00E 00
1.50E 02	5.00E 00

6-5 EXCITATION OF HELIUM BY He$^+$ IMPACT

The reader is referred to Section 6-1 for a general discussion of the problems involved in the study of the collisional excitation of helium. This particular collision combination will give rise to He I emissions from the projectile as the result of charge transfer, as well as by direct excitation of the target. The latter is of principal interest here. In principle, target and projectile emission may be separated by viewing the collision chamber at an angle to the beam and utilizing the Doppler shift of the projectile emission to allow spectroscopic resolution of the two components. In some of the low energy impact work presented here, the Doppler shift of the emission was so small that resolution was impossible, and the sum of the emission from the target and from the projectile was measured. These data are very important and are included in the present compilation although they do represent a sum of two processes.

For simplicity, first, the excitation of the target alone is considered (Section 6-5-A); second, the polarization of emission from the target alone, (Section 6-5-B); finally,

the very limited amount of information on the sum cross sections for the formation of the excited states of helium both in the target and in the projectile beam, (Section 6-5-C). These data are compiled in Section 6-5-D.

A.　Emission from the Target

De Heer and Van den Bos[212, 268] utilized the equipment reviewed in Section 6-1-B(1). The second[212] of the two reports on this work completely supersedes the first[268] and provides full information on experimental techniques. An interesting procedure was utilized at very low impact velocities where the Doppler shift was insufficient to allow complete resolution of the projectile and target emission. In Section 2-2-B, it has been shown that emission from the projectile will increase exponentially towards some constant value as a function of penetration through the target gas. The observations were made just beyond the entrance to the collision chamber, at a point where projectile emission was much smaller than the target emission. In this manner, although complete resolution was not achieved, the error due to the overlap of the lines was small. At energies below 10-keV, it appears that there was a residual error of the order of 5 to 10% due to inadequate resolution of the lines. At higher energies, complete resolution was possible.

In the review of the techniques used by de Heer and Van den Bos [see Section 6-1-B(1)], it was concluded that the work was done with careful attention to detail. All the measurements presented here were made under pressure-independent conditions with all required corrections for polarization effects and cascade population. In the discussion of excitation by proton impact it has been concluded that the functional dependence of cross section on projectile impact energy is a particularly reliable quantity. This same conclusion should hold also for excitation by He^+ impact. The results are shown in Data Tables 6-5-1 through 6-5-6.

A single determination of the cross section for the excitation of the 2^3S state of the helium target by 600-eV He^+ ions was made by Lorents et al.,[58] using an energy-loss technique. The experiment was designed primarily to measure the inelastic differential scattering cross section. Integration over all angles gave a value of $6.9 \pm 3.5 \times 10^{-18}$ cm^2. This determination is only a little less accurate than the optical emission techniques, and it should not be subject to the same systematic errors that are observed using optical techniques.

B.　Polarization of Emission from the Target

De Heer and Van den Bos[212] have determined polarization fractions for a number of lines induced in this collision combination. For all data at energies below 35-keV, tests were made to ensure that stray magnetic fields did not cause errors in the measurement. At energies above 35-keV, for which a separate accelerator was used, no such tests were made. These data are shown in Data Tables 6-5-7 through 6-5-17. The rapid variations as the energy of impact is varied may not be physically significant.

C.　Sum of Emission from Projectile and Target Measured Together

De Heer et al.[78] made a few measurements of this case as part of a program to determine the formation of fast neutral atoms in the 2^1P and 3^1P states by charge transfer. The target pressures of 8×10^{-4} torr and above are too high for lines that

are affected by absorption of resonant photons. Van Eck et al.[10] showed, using the same apparatus, that pressures less than 2×10^{-4} torr are needed for measurements on the 3^1P state. The accuracy of this work is considered poor; therefore it is not presented here. The cross section for the formation of the 3^1P state in the target was later determined with high accuracy by de Heer and Van den Bos[212] and has been discussed in Section 6-5A.

Dworetsky et al.[35, 255, 215] have been concerned with the structure that occurs in the emission cross section close to threshold for the excitation process. All the data presented here are in the form of relative variation of emission cross sections with projectile energy. The assignment of an absolute magnitude to the cross sections in these papers by normalization to the work of de Heer et al.[12] is invalid because the measurements are of line emission functions, and the work of de Heer et al.[12] is a measurement of level-excitation cross sections. In the discussion of this experiment [Section 6-1-B(8)] it was noted that their data were not corrected for the effects of polarization of emission and insufficient care was taken to ensure that the measurements are independent of the target density. For the case of the 10,830-Å-line $(2^3P \rightarrow 2^3S)$, it was shown that the cascade contribution to the emission is less than 10% of the signal. No such tests were carried out for the other lines, and it is possible that cascade may provide appreciable contributions to some of these data. The measurements are shown in Data Tables 6-5-18 through 6-5-26.

D. Data Tables

1. Cross Sections for Excitation of He States Induced by Impact of H⁺ on a Helium Target (Target Excitation only)

6-5-1 (A) 4^1S State; (B) 5^1S State; (C) 6^1S State.

6-5-2 (A) 3^1P State; (B) 4^1P State; (C) 5^1P State.

6-5-3 (A) 4^1D State; (B) 5^1D State; (C) 6^1D State.

6-5-4 (A) 4^3S State; (B) 5^3S State; (C) 6^3S State.

6-5-5 (A) 3^3P State; (B) 4^3P State; (C) 5^3P State.

6-5-6 (A) 3^3D State; (B) 4^3D State; (C) 5^3D State.

2. Polarization Fractions of He I Lines Induced by Impact of He⁺ on a Helium Target (Target Excitation only)

6-5-7 5016-Å $(3^1P \rightarrow 2^1S)$.

6-5-8 3965-Å $(4^1P \rightarrow 2^1S)$.

6-5-9 4922-Å $(4^1D \rightarrow 2^1P)$.

6-5-10 4388-Å $(5^1D \rightarrow 2^1P)$.

6-5-11 4144-Å $(6^1D \rightarrow 2^1P)$.

6-5-12 3889-Å $(3^3P \rightarrow 2^3S)$.

6-5-13 3188-Å $(4^3P \rightarrow 2^3S)$.

6-5-14 2995-Å $(5^3P \rightarrow 2^3S)$.

6-5-15 5876-Å $(3^3D \rightarrow 2^3P)$.

6-5-16 4472-Å $(4^3D \rightarrow 2^3P)$.

6-5-17 4026-Å $(5^3D \rightarrow 2^3P)$.

3. **Cross Sections for the Emission of He I Lines Induced by Impact of He$^+$ on a Helium Target (All data in arbitrary units—sum of contributions from the target and projectile)**

6-5-18   7281-Å ($3^1S \rightarrow 2^1P$).

6-5-19   584-Å ($2^1P \rightarrow 1^1S$).

6-5-20   5015-Å ($3^1P \rightarrow 2^1S$).

6-5-21   6678-Å ($3^1D \rightarrow 2^1P$).

6-5-22   7065-Å ($3^3S \rightarrow 2^3P$).

6-5-23   4713-Å ($4^3S \rightarrow 2^3P$).

6-5-24   10,830-Å ($2^3P \rightarrow 2^3S$).

6-5-25   3888-Å ($3^3P \rightarrow 2^3S$).

6-5-26   5876-Å ($3^3D \rightarrow 2^3P$).

6-5-1A
DE HEER ET AL. (212)

ENERGY (KEV)	CROSS SECTION (SQ. CM)
5.00E 00	2.20E-20
7.50E 00	3.26E-20
1.00E 01	4.87E-20
1.25E 01	3.56E-20
1.50E 01	3.55E-20
1.75E 01	3.46E-20
2.00E 01	3.96E-20
2.25E 01	4.23E-20
2.50E 01	4.42E-20
2.75E 01	4.67E-20
3.00E 01	4.87E-20
3.25E 01	5.29E-20
3.50E 01	6.38E-20
4.00E 01	7.83E-20
4.50E 01	9.06E-20
5.00E 01	9.79E-20
5.50E 01	9.11E-20
6.00E 01	8.81E-20
6.50E 01	8.01E-20
7.00E 01	7.14E-20
7.50E 01	7.03E-20
8.00E 01	6.93E-20
8.50E 01	6.80E-20
9.00E 01	6.74E-20
9.50E 01	6.71E-20
1.00E 02	7.14E-20

6-5-1B
DE HEER ET AL. (212)

ENERGY (KEV)	CROSS SECTION (SQ. CM)
1.00E 01	7.10E-21
1.25E 01	1.05E-20
1.50E 01	1.20E-20
1.75E 01	1.38E-20
2.00E 01	1.52E-20
2.25E 01	1.86E-20
2.50E 01	1.89E-20
2.75E 01	1.87E-20
3.00E 01	1.91E-20
3.25E 01	2.10E-20
3.50E 01	2.08E-20
4.00E 01	2.53E-20
4.50E 01	2.91E-20
5.00E 01	3.23E-20
5.50E 01	3.15E-20
6.00E 01	2.99E-20
6.50E 01	2.77E-20
7.00E 01	2.62E-20
8.00E 01	2.38E-20
8.50E 01	2.29E-20
9.00E 01	2.35E-20

6-5-1C
DE HEER ET AL. (212)

ENERGY (KEV)	CROSS SECTION (SQ. CM)
1.00E 01	6.10E-21
1.25E 01	7.80E-21
1.50E 01	1.01E-20
1.75E 01	1.00E-20
2.00E 01	9.10E-21
2.25E 01	7.70E-21
2.50E 01	1.04E-20
2.75E 01	9.60E-21
3.00E 01	1.11E-20
3.50E 01	1.23E-20
4.00E 01	1.45E-20
4.50E 01	1.60E-20
5.00E 01	1.67E-20
5.50E 01	1.64E-20
6.00E 01	1.54E-20
7.00E 01	1.44E-20
8.00E 01	1.35E-20

ENERGY (KEV)	CROSS SECTION (SQ. CM)
4.00E 01	3.20E-19
4.50E 01	3.24E-19
5.00E 01	2.91E-19
6.00E 01	2.60E-19
7.00E 01	2.37E-19
8.00E 01	2.15E-19

6-5-2A
DE HEER ET AL. (212)

ENERGY (KEV)	CROSS SECTION (SQ. CM)
7.50E 00	4.29E-19
1.00E 01	5.79E-19
1.25E 01	6.65E-19
1.50E 01	6.49E-19
1.75E 01	5.38E-19
2.00E 01	5.30E-19
2.25E 01	5.46E-19
2.50E 01	6.34E-19
3.00E 01	8.48E-19
3.50E 01	9.07E-19
4.00E 01	9.71E-19
4.50E 01	8.71E-19
5.00E 01	7.86E-19
5.50E 01	7.00E-19
6.00E 01	6.52E-19
6.50E 01	5.20E-19
7.00E 01	5.09E-19
7.50E 01	4.76E-19
8.00E 01	4.10E-19

6-5-2B
DE HEER ET AL. (212)

ENERGY (KEV)	CROSS SECTION (SQ. CM)
1.25E 01	2.12E-19
1.50E 01	2.93E-19
1.75E 01	3.16E-19
2.00E 01	2.94E-19
2.25E 01	2.72E-19
2.50E 01	2.17E-19
2.75E 01	2.42E-19
3.00E 01	2.63E-19
3.25E 01	2.98E-19
3.50E 01	3.26E-19

6-5-2C
DE HEER ET AL. (212)

ENERGY (KEV)	CROSS SECTION (SQ. CM)
1.00E 01	1.59E-19
1.25E 01	1.64E-19
1.50E 01	1.92E-19
1.75E 01	2.00E-19
2.00E 01	2.16E-19
2.25E 01	1.86E-19
2.50E 01	1.72E-19
2.75E 01	1.64E-19
3.00E 01	1.74E-19
3.25E 01	1.66E-19
3.50E 01	1.68E-19
4.00E 01	1.81E-19
4.50E 01	1.85E-19
5.00E 01	1.65E-19
5.50E 01	1.74E-19
6.00E 01	1.61E-19
7.00E 01	1.54E-19
8.00E 01	1.42E-19

6-5-3A
DE HEER ET AL. (212)

ENERGY (KEV)	CROSS SECTION (SQ. CM)
5.00E 00	5.17E-20
7.50E 00	6.24E-20
1.00E 01	7.69E-20
1.25E 01	8.93E-20
1.50E 01	7.00E-20
1.75E 01	7.11E-20
2.00E 01	1.01E-19
2.25E 01	1.53E-19
2.50E 01	2.00E-19
2.75E 01	2.54E-19
3.00E 01	2.84E-19
3.25E 01	2.91E-19
3.50E 01	2.95E-19
4.00E 01	2.90E-19
4.50E 01	2.49E-19
5.00E 01	2.00E-19
5.50E 01	1.57E-19
6.00E 01	1.30E-19
6.50E 01	1.07E-19
7.00E 01	7.83E-20
7.50E 01	7.91E-20
8.00E 01	7.69E-20
9.00E 01	7.43E-20
1.00E 02	6.86E-20

6-5-3B
DE HEER ET AL. (212)

ENERGY (KEV)	CROSS SECTION (SQ. CM)
7.50E 00	2.41E-20
1.00E 01	2.72E-20
1.25E 01	4.43E-20
1.50E 01	5.53E-20
1.75E 01	4.11E-20
2.00E 01	3.46E-20
2.25E 01	3.95E-20
2.50E 01	5.53E-20
3.00E 01	8.84E-20
3.50E 01	1.12E-19
4.00E 01	1.06E-19
4.50E 01	1.01E-19
5.00E 01	8.18E-20
5.50E 01	7.21E-20
6.00E 01	5.18E-20
7.00E 01	3.90E-20
8.00E 01	3.60E-20
9.00E 01	3.07E-20
1.00E 02	3.16E-20

6-5-3C
CE HEER ET AL. (212)

ENERGY (KEV)	CROSS SECTION (SQ. CM)
1.00E 01	9.90E-21
1.25E 01	1.30E-20
1.50E 01	2.31E-20
1.75E 01	2.68E-20
2.00E 01	1.94E-20
2.25E 01	1.52E-20
2.50E 01	1.56E-20
2.75E 01	2.87E-20
3.00E 01	3.37E-20
3.50E 01	4.52E-20
4.00E 01	4.62E-20
4.50E 01	4.49E-20
5.00E 01	3.56E-20
6.00E 01	2.74E-20
7.00E 01	2.04E-20
8.00E 01	1.72E-20

6-5-4A
DE HEER ET AL. (212)

ENERGY (KEV)	CROSS SECTION (SQ. CM)
5.00E 00	2.14E-20
7.50E 00	4.02E-20
1.00E 01	3.40E-20
1.25E 01	5.54E-20
1.50E 01	7.05E-20
1.75E 01	5.16E-20
2.00E 01	3.19E-20
2.25E 01	3.26E-20
2.50E 01	3.42E-20
2.75E 01	4.80E-20
3.00E 01	5.73E-20

3.25E 01	7.32E-20
3.50E 01	7.83E-20
4.00E 01	8.22E-20
5.00E 01	9.25E-20
5.50E 01	1.13E-19
6.00E 01	1.15E-19
6.50E 01	1.23E-19
7.00E 01	1.43E-19
7.50E 01	1.34E-19
8.00E 01	1.21E-19
8.50E 01	1.21E-19
9.50E 01	1.12E-19
1.00E 02	1.11E-19

6-5-4B
DE HEER ET AL. (212)

ENERGY (KEV)	CROSS SECTION (SQ. CM)
5.00E 00	5.90E-21
7.5CE 00	1.70E-20
1.00E 01	1.70E-20
1.25E 01	1.53E-20
1.50E 01	2.77E-20
1.75E 01	2.95E-20
2.00E 01	2.81E-20
2.25E 01	2.32E-20
2.5CE 01	1.76E-20
2.75E 01	1.74E-20
3.00E 01	2.18E-20
3.50E C1	2.41E-20
4.00E 01	3.33E-20
5.0CE 01	3.99E-20
6.00E 01	5.57E-20
7.00E 01	6.71E-20
8.0CE 01	6.69E-20
8.50E 01	6.74E-20
9.00E 01	7.01E-20
9.50E 01	6.81E-20
1.C0E 02	6.34E-20

6-5-4C
CE HEER ET AL. (212)

ENERGY (KEV)	CROSS SECTION (SQ. CM)
1.25E 01	8.40E-21
1.50E 01	1.07E-20
1.75E 01	1.35E-20
2.00E 01	1.81E-20
2.25E 01	1.88E-20
2.50E 01	1.50E-20
3.00E 01	1.45E-20
3.5CE 01	1.43E-20
4.00E 01	1.93E-20
5.00E 01	2.55E-20
6.00E 01	3.03E-20
7.00E 01	3.52E-20
7.50E 01	3.63E-20
8.00E 01	3.94E-20
8.50E 01	4.12E-20
9.00E 01	4.19E-20
1.00E 02	3.70E-20

6-5-5A
DE HEER ET AL. (212)

ENERGY (KEV)	CROSS SECTION (SQ. CM)
5.00E 00	1.50E-19
7.50E 00	2.94E-19
1.00E 01	4.06E-19
1.25E 01	5.63E-19
1.50E 01	5.80E-19
1.75E 01	5.10E-19
2.00E 01	4.41E-19
2.25E 01	4.79E-19
2.50E 01	4.88E-19
3.00E 01	5.57E-19
3.50E 01	6.23E-19
4.00E 01	6.89E-19
4.50E 01	7.33E-19
5.00E 01	7.78E-19
5.50E 01	7.69E-19
6.00E 01	7.52E-19
7.00E 01	7.19E-19
7.50E 01	7.11E-19
8.00E 01	6.58E-19
8.50E 01	6.32E-19
9.00E 01	5.51E-19
9.5CE 01	5.27E-19
1.00E 02	5.16E-19

6-5-5B
DE HEER ET AL. (212)

ENERGY (KEV)	CROSS SECTION (SQ. CM)
7.50E 00	1.33E-20
1.0CE 01	5.79E-20
1.25E 01	9.28E-20
1.5CE 01	1.23E-19
1.75E 01	1.34E-19
2.00E 01	1.35E-19
2.25E 01	1.26E-19
2.50E 01	1.20E-19
2.75E C1	1.29E-19
3.00E 01	1.42E-19
3.50E 01	1.65E-19
4.00E 01	2.03E-19
4.50E 01	2.28E-19
5.00E 01	2.52E-19
5.50E 01	2.64E-19
6.00E 01	2.57E-19
6.50E 01	2.46E-19
7.00E 01	2.47E-19
8.00E 01	2.28E-19
9.00E 01	2.01E-19

6-5-5C
CE HEER ET AL. (212)

ENERGY (KEV)	CROSS SECTION (SQ. CM)
1.00E 01	2.40E-20
1.25E 01	2.63E-20
1.50E 01	3.90E-20

ENERGY (KEV)	CROSS SECTION (SQ. CM)
2.00E 01	4.47E-20
2.50E 01	3.62E-20
2.75E 01	3.44E-20
3.00E 01	3.54E-20
3.25E 01	3.82E-20
3.50E 01	4.35E-20
4.00E 01	6.01E-20
4.50E 01	7.73E-20
5.00E 01	8.38E-20
5.50E 01	8.90E-20
6.00E 01	9.44E-20
6.50E 01	8.96E-20
7.00E 01	8.36E-20
7.50E 01	8.13E-20
8.00E 01	8.65E-20
9.00E 01	6.97E-20

6-5-6A
DE HEER ET AL. (212)

ENERGY (KEV)	CROSS SECTION (SQ. CM)
5.00E 00	7.86E-19
7.50E 00	7.75E-19
1.00E 01	7.98E-19
1.25E 01	6.61E-19
1.50E 01	6.94E-19
1.75E 01	1.01E-18
2.00E 01	1.50E-18
2.25E 01	1.45E-18
2.50E 01	1.38E-18
3.00E 01	1.17E-18
3.50E 01	9.59E-19
4.00E 01	7.69E-19
4.50E 01	7.46E-19
5.00E 01	6.68E-19
5.50E 01	5.84E-19
6.00E 01	5.03E-19
6.50E 01	4.36E-19
7.00E 01	3.70E-19
7.50E 01	3.52E-19
8.00E 01	3.17E-19
8.50E 01	2.95E-19
9.00E 01	2.50E-19
9.50E 01	2.22E-19
1.00E 02	2.02E-19

6-5-6B
DE HEER ET AL. (212)

ENERGY (KEV)	CROSS SECTION (SQ. CM)
7.50E 00	1.30E-19
1.00E 01	1.10E-19
1.25E 01	1.37E-19
1.50E 01	1.30E-19
1.75E 01	1.18E-19
2.00E 01	1.60E-19
2.50E 01	2.72E-19
3.00E 01	3.16E-19
3.50E 01	2.95E-19
4.00E 01	2.64E-19
5.00E 01	2.15E-19
6.00E 01	1.79E-19
7.00E 01	1.48E-19

ENERGY (KEV)	CROSS SECTION (SQ. CM)
8.00E 01	1.16E-19
9.00E 01	8.80E-20
1.00E 02	7.54E-20

6-5-6C
DE HEER ET AL. (212)

ENERGY (KEV)	CROSS SECTION (SQ. CM)
5.00E 00	6.75E-20
7.50E 00	4.71E-20
1.00E 01	3.67E-20
1.25E 01	5.13E-20
1.50E 01	7.27E-20
1.75E 01	5.33E-20
2.00E 01	5.22E-20
2.25E 01	5.05E-20
2.50E 01	8.22E-20
3.00E 01	1.16E-19
3.50E 01	1.33E-19
4.00E 01	1.22E-19
4.50E 01	1.17E-19
5.00E 01	1.05E-19
5.50E 01	1.02E-19
6.00E 01	9.89E-20
7.00E 01	8.60E-20
8.00E 01	6.88E-20
9.00E 01	6.01E-20
1.00E 02	5.04E-20

6-5-7
DE HEER ET AL. (212)

ENERGY (KEV)	POLARIZATION (PERCENT)
1.00E 01	-5.20E 00
1.50E 01	-1.11E 01
2.00E 01	7.80E 00
2.50E 01	7.40E 00
3.00E 01	0.0
3.50E 01	-6.30E 00
4.00E 01	-9.20E 00
4.50E 01	0.0
5.00E 01	-8.70E 00
5.50E 01	-8.10E 00
6.00E 01	-5.30E 00
6.50E 01	-3.60E 00
7.00E 01	-2.60E 00
7.50E 01	5.00E-01
8.00E 01	-7.00E 00
8.50E 01	-8.10E 00
9.00E 01	-5.00E-01

6-5-8
DE HEER ET AL. (212)

ENERGY (KEV)	POLARIZATION (PERCENT)
2.00E 01	3.80E 00
2.50E 01	8.20E 00
3.00E 01	7.40E 00
3.50E 01	5.20E 00

6-5-9
DE HEER ET AL. (212)

ENERGY (KEV)	POLARIZATION (PERCENT)
7.50E 00	2.45E 01
1.00E 01	2.67E 01
1.25E 01	2.53E 01
1.50E 01	2.56E 01
1.75E 01	2.67E 01
2.00E 01	3.15E 01
2.25E 01	2.93E 01
2.50E 01	2.88E 01
2.75E 01	3.00E 01
3.00E 01	2.85E 01
3.25E 01	3.05E 01
3.50E 01	3.00E 01
4.00E 01	3.19E 01
5.00E 01	3.12E 01
5.50E 01	3.24E 01
6.00E 01	2.85E 01
6.50E 01	2.85E 01
7.00E 01	2.85E 01
7.50E 01	2.85E 01
8.00E 01	2.75E 01
8.50E 01	2.95E 01

6-5-10
DE HEER ET AL. (212)

ENERGY (KEV)	POLARIZATION (PERCENT)
1.00E 01	2.24E 01
1.25E 01	2.48E 01
1.50E 01	2.56E 01
1.75E 01	2.42E 01
2.00E 01	2.34E 01
2.25E 01	1.97E 01
2.50E 01	2.62E 01
2.75E 01	2.48E 01
3.00E 01	2.51E 01
3.50E 01	2.51E 01
4.00E 01	2.48E 01
4.50E 01	2.48E 01
5.00E 01	2.56E 01
5.50E 01	2.31E 01
6.00E 01	2.28E 01
7.00E 01	2.37E 01
8.00E 01	2.34E 01

6-5-11
DE HEER ET AL. (212)

ENERGY (KEV)	POLARIZATION (PERCENT)
1.50E 01	2.45E 01
1.75E 01	2.68E 01
2.00E 01	2.03E 01
2.25E 01	2.46E 01
2.50E 01	2.06E 01
3.00E 01	2.40E 01
3.50E 01	2.18E 01
4.00E 01	1.97E 01

ENERGY (KEV)	POLARIZATION (PERCENT)
5.00E 01	1.93E 01
6.00E 01	2.06E 01
7.00E 01	1.97E 01
8.00E 01	2.22E 01

6-5-12
DE HEER ET AL. (212)

ENERGY (KEV)	POLARIZATION (PERCENT)
5.00E 00	1.27E 01
7.50E 00	6.10E 00
1.00E 01	-5.30E 00
1.25E 01	-2.00E 00
1.50E 01	5.70E 00
1.75E 01	5.20E 00
2.00E 01	6.50E 00
2.25E 01	3.80E 00
2.50E 01	5.00E-01
3.00E 01	0.0
3.50E 01	2.60E 00
4.00E 01	-5.00E-01
5.00E 01	-2.00E 00
6.00E 01	5.20E 00
7.00E 01	5.00E-01
8.00E 01	3.40E 00
9.00E 01	5.00E-01

6-5-13
DE HEER ET AL. (212)

ENERGY (KEV)	POLARIZATION (PERCENT)
1.00E 01	4.30E 00
1.50E 01	-5.00E-01
2.00E 01	1.50E 00
2.50E 01	9.90E 00
3.00E 01	9.10E 00
3.50E 01	5.30E 00

6-5-14
DE HEER ET AL. (212)

ENERGY (KEV)	POLARIZATION (PERCENT)
1.00E 01	-2.00E 00
1.25E 01	-6.40E 00
1.50E 01	-5.00E-01
2.00E 01	-1.11E 01
2.50E 01	-8.10E 00
3.00E 01	1.00E 00
3.50E 01	2.00E 00

6-5-15
DE HEER ET AL. (212)

ENERGY (KEV)	POLARIZATION (PERCENT)
7.50E 00	1.19E 01
1.00E 01	1.90E 01
1.25E 01	1.15E 01
1.50E 01	1.60E 01
1.75E 01	1.49E 01
2.00E 01	1.60E 01
2.25E 01	9.50E 00
2.50E 01	1.45E 01
3.00E 01	1.38E 01
3.50E 01	1.53E 01
4.00E 01	1.38E 01
4.5CE 01	1.56E 01
5.00E 01	1.45E 01
5.50E 01	1.56E 01
6.00E 01	1.94E 01
6.50E 01	1.24E 01
7.00E 01	1.42E 01
7.50E 01	1.27E 01
8.00E 01	1.07E 01
8.50E 01	9.50E 00
9.00E 01	6.10E 00

6-5-16
DE HEER ET AL. (212)

ENERGY (KEV)	POLARIZATION (PERCENT)
1.25E 01	1.52E 01
1.50E 01	1.42E 01
1.75E 01	1.42E 01
2.00E 01	1.34E 01
2.50E 01	1.52E 01
3.00E 01	1.45E 01
3.50E 01	1.49E 01
4.00E 01	1.56E 01
5.00E 01	1.56E 01
6.00E 01	1.45E 01
7.00E 01	1.19E 01

6-5-17
DE HEER ET AL. (212)

ENERGY (KEV)	POLARIZATION (PERCENT)
1.00E 01	1.23E 01
1.50E 01	1.56E 01
1.75E 01	1.19E 01
2.00E 01	1.23E 01
2.50E 01	1.19E 01
3.00E 01	1.19E 01
3.50E 01	9.50E 00
4.00E 01	1.42E 01
5.00E 01	1.11E 01
5.50E 01	1.23E 01
6.00E 01	1.38E 01

6-5-18
DWORETSKY ET AL. (35)

ENERGY (KEV)	CROSS SECTION (ARB. UNITS)
4.55E-02	1.00E-02
5.05E-02	1.00E-02
5.60E-02	2.00E-02
6.05E-02	7.00E-02
6.35E-02	4.20E-01
6.68E-02	8.20E-01
6.88E-02	1.02E 00
7.30E-02	1.00E 00
7.80E-02	1.08E 00
8.60E-02	1.34E 00
9.20E-02	1.76E 00
9.40E-02	2.04E 00
9.70E-02	2.16E 00
1.03E-01	2.25E 00
1.06E-01	2.55E 00
1.10E-01	2.90E 00
1.17E-01	2.99E 00
1.21E-01	2.91E 00
1.28E-01	2.39E 00
1.33E-01	2.03E 00
1.36E-01	1.77E 00
1.41E-01	1.64E 00
1.45E-01	1.43E 00
1.51E-01	1.34E 00
1.62E-01	1.43E 00
1.71E-01	1.65E 00
1.83E-01	1.79E 00
1.99E-01	1.67E 00
2.10E-01	1.46E 00
2.19E-01	1.33E 00
2.43E-01	1.23E 00
2.54E-01	1.31E 00
2.58E-01	1.40E 00
2.74E-01	1.52E 00
2.81E-01	1.64E 00
2.93E-01	1.69E 00
3.12E-01	1.60E 00
3.22E-01	1.49E 00
3.26E-01	1.36E 00
3.30E-01	1.35E 00
3.44E-01	1.24E 00
3.60E-01	1.05E 00
3.85E-01	9.70E-01
3.97E-01	1.00E 00
4.50E-01	1.09E 00
4.90E-01	1.20E 00
5.00E-01	1.32E 00
5.50E-01	1.43E 00
6.50E-01	1.49E 00
7.10E-01	1.52E 00
7.70E-01	1.58E 00
8.30E-01	1.55E 00
9.00E-01	1.65E 00
9.70E-01	1.57E 00
1.06E 00	1.46E 00
1.25E 00	1.24E 00
1.40E 00	1.17E 00
1.53E 00	1.19E 00
1.62E 00	1.16E 00
1.73E 00	1.21E 00
2.03E 00	1.25E 00
2.26E 00	1.24E 00
2.50E 00	1.12E 00
2.80E 00	1.05E 00
3.06E 00	1.00E 00
3.57E 00	1.04E 00

4.40E 00   1.23E 00
4.58E 00   1.27E 00
5.10E 00   1.31E 00

6-5-19
DWORETSKY ET AL. (215)

ENERGY (KEV)	CROSS SECTION (ARB. UNITS)
4.97E-02	4.20E-02
5.47E-02	2.03E-01
5.94E-02	2.66E-01
6.97E-02	1.54E-01
6.92E-02	4.97E-01
9.44E-02	3.43E-01
8.57E-02	4.76E-01
8.05E-02	6.16E-01
7.46E-02	6.51E-01
6.50E-02	7.07E-01
1.15E-01	5.32E-01
1.00E-01	6.51E-01
9.46E-02	7.00E-01
8.45E-02	7.42E-01
1.06E-01	7.98E-01
1.30E-01	8.26E-01
9.44E-02	1.00E 00
1.26E-01	1.25E 00
1.40E-01	1.26E 00
1.45E-01	1.48E 00
1.35E-01	1.55E 00
1.40E-01	1.64E 00
1.52E-01	1.74E 00
1.51E-01	1.87E 00
1.58E-01	2.00E 00
1.61E-01	2.23E 00
1.81E-01	2.29E 00
1.99E-01	2.40E 00
1.93E-01	2.67E 00
1.81E-01	2.82E 00
1.71E-01	2.83E 00
2.32E-01	2.91E 00
2.02E-01	3.08E 00
2.32E-01	3.27E 00
2.27E-01	3.45E 00
2.12E-01	3.50E 00
2.74E-01	3.58E 00
2.67E-01	3.68E 00
3.04E-01	3.64E 00
2.44E-01	3.83E 00
3.37E-01	3.75E 00
2.86E-01	3.99E 00
3.77E-01	3.84E 00
3.30E-01	3.98E 00
4.54E-01	3.90E 00
4.04E-01	3.93E 00
3.59E-01	3.99E 00
5.07E-01	3.99E 00
5.86E-01	4.01E 00
5.51E-01	4.19E 00
6.78E-01	4.31E 00
6.50E-01	4.58E 00
7.06E-01	4.71E 00
7.89E-01	4.77E 00
7.62E-01	5.05E 00
8.00E-01	5.14E 00
8.51E-01	5.45E 00
1.12E 00	5.41E 00
9.84E-01	5.51E 00
8.93E-01	5.53E 00
8.99E-01	5.85E 00
1.01E 00	6.04E 00
1.24E 00	6.05E 00
1.42E 00	6.19E 00
1.62E 00	6.66E 00

6-5-20
DWORETSKY ET AL. (35)

ENERGY (KEV)	CROSS SECTION (ARB. UNITS)
4.90E-02	2.00E-02
5.35E-02	4.00E-02
6.00E-02	2.00E-02
6.20E-02	8.00E-02
6.30E-02	2.30E-01
6.55E-02	5.10E-01
6.80E-02	8.50E-01
6.90E-02	1.15E 00
7.15E-02	1.31E 00
7.55E-02	1.16E 00
8.20E-02	1.34E 00
8.60E-02	1.53E 00
8.85E-02	1.71E 00
9.25E-02	1.79E 00
9.60E-02	1.70E 00
9.70E-02	1.54E 00
1.10E-01	1.45E 00
1.14E-01	1.56E 00
1.21E-01	1.49E 00
1.30E-01	1.52E 00
1.32E-01	1.70E 00
1.46E-01	1.74E 00
1.55E-01	1.73E 00
1.58E-01	1.68E 00
1.88E-01	1.62E 00
2.00E-01	1.65E 00
2.16E-01	1.64E 00
2.36E-01	1.61E 00
2.57E-01	1.68E 00
2.77E-01	1.73E 00
2.97E-01	1.68E 00
3.25E-01	1.70E 00
3.50E-01	1.85E 00
3.78E-01	1.92E 00
4.08E-01	1.84E 00
4.40E-01	1.84E 00
4.95E-01	1.87E 00
5.35E-01	1.93E 00
5.73E-01	1.90E 00
6.00E-01	1.81E 00
6.75E-01	1.91E 00
6.80E-01	1.93E 00
7.15E-01	1.88E 00
7.45E-01	1.97E 00
7.75E-01	2.04E 00
8.10E-01	2.03E 00
8.85E-01	2.20E 00
9.20E-01	2.28E 00
9.70E-01	2.24E 00
1.07E 00	2.10E 00
1.11E 00	2.14E 00
1.21E 00	2.03E 00
1.24E 00	2.17E 00
1.40E 00	2.31E 00
1.60E 00	2.39E 00
1.74E 00	2.44E 00
1.83E 00	2.36E 00
1.92E 00	2.24E 00
2.03E 00	2.26E 00
2.20E 00	2.14E 00
2.41E 00	2.08E 00
2.63E 00	2.03E 00
2.86E 00	1.98E 00
3.10E 00	2.12E 00
3.30E 00	2.27E 00
3.47E 00	2.52E 00
4.05E 00	2.48E 00
4.35E 00	2.44E 00
5.00E 00	2.56E 00

6-5-21
DWORETSKY ET AL. (35)

ENERGY (KEV)	CROSS SECTION (ARB. UNITS)
4.00E-02	3.00E-02
4.55E-02	3.00E-02
5.15E-02	3.00E-02
5.75E-02	4.00E-02
6.10E-02	3.00E-02
6.30E-02	1.10E-01
6.45E-02	2.80E-01
6.70E-02	5.80E-01
6.90E-02	1.07E 00
7.15E-02	1.34E 00
7.35E-02	1.50E 00
7.65E-02	1.71E 00
7.70E-02	1.97E 00
7.80E-02	2.11E 00
8.40E-02	2.16E 00
8.90E-02	2.22E 00
9.25E-02	2.51E 00
9.60E-02	2.61E 00
1.02E-01	2.66E 00
1.08E-01	2.54E 00
1.15E-01	2.53E 00
1.24E-01	2.40E 00
1.29E-01	2.53E 00
1.33E-01	2.60E 00
1.36E-01	2.72E 00
1.41E-01	2.90E 00
1.49E-01	3.05E 00
1.56E-01	3.09E 00
1.65E-01	2.97E 00
1.75E-01	2.89E 00
1.85E-01	2.93E 00
1.91E-01	3.09E 00
2.06E-01	3.03E 00
2.25E-01	2.87E 00
2.33E-01	2.78E 00
2.48E-01	2.65E 00
2.68E-01	2.40E 00
2.97E-01	2.23E 00
3.30E-01	2.02E 00
3.75E-01	1.88E 00
4.15E-01	1.97E 00
4.65E-01	2.02E 00
5.20E-01	1.94E 00
5.70E-01	1.92E 00
5.95E-01	1.78E 00
6.40E-01	1.75E 00
7.00E-01	1.72E 00
7.50E-01	1.68E 00
8.25E-01	1.60E 00
9.00E-01	1.61E 00
9.90E-01	1.55E 00
1.09E 00	1.61E 00

ENERGY (KEV)	CROSS SECTION (ARB. UNITS)
1.20E 00	1.63E 00
1.38E 00	1.52E 00
1.59E 00	1.46E 00
1.79E 00	1.42E 00
2.00E 00	1.34E 00
2.22E 00	1.34E 00
2.25E 00	1.35E 00
2.38E 00	1.46E 00
2.79E 00	1.54E 00
2.98E 00	1.57E 00
3.17E 00	1.64E 00
3.65E 00	1.64E 00
3.95E 00	1.60E 00
4.15E 00	1.72E 00
4.50E 00	1.69E 00
4.70E 00	1.78E 00
4.70E 00	1.87E 00
4.85E 00	1.94E 00

6-5-22
DWORETSKY ET AL. (35)

ENERGY (KEV)	CROSS SECTION (ARB. UNITS)
4.00E-02	1.00E-02
4.90E-02	1.00E-02
5.50E-02	1.00E-02
5.70E-02	5.00E-02
5.90E-02	1.50E-01
6.10E-02	5.60E-01
6.60E-02	1.47E 00
6.90E-02	1.55E 00
7.40E-02	1.75E 00
7.90E-02	1.95E 00
8.90E-02	1.94E 00
9.40E-02	2.01E 00
1.00E-01	2.14E 00
1.03E-01	2.25E 00
1.07E-01	2.33E 00
1.13E-01	2.51E 00
1.18E-01	2.75E 00
1.24E-01	2.85E 00
1.32E-01	2.40E 00
1.37E-01	2.23E 00
1.45E-01	2.13E 00
1.58E-01	1.74E 00
1.66E-01	1.20E 00
1.77E-01	1.01E 00
1.98E-01	1.01E 00
2.16E-01	1.41E 00
2.39E-01	1.40E 00
2.93E-01	1.01E 00
3.18E-01	9.20E-01
3.45E-01	1.03E 00
3.65E-01	1.21E 00
3.85E-01	1.35E 00
4.08E-01	1.57E 00
4.40E-01	1.69E 00
4.90E-01	1.65E 00
5.30E-01	1.49E 00
5.85E-01	1.31E 00
6.85E-01	1.04E 00
7.75E-01	9.20E-01
8.75E-01	8.60E-01
9.60E-01	9.10E-01
1.06E 00	8.60E-01
1.17E 00	8.30E-01
1.27E 00	7.70E-01
1.45E 00	6.90E-01
1.77E 00	7.50E-01
1.98E 00	7.90E-01
2.18E 00	8.60E-01
2.67E 00	7.90E-01
3.20E 00	6.80E-01
3.45E 00	6.70E-01
3.95E 00	7.70E-01
4.95E 00	1.03E 00

6-5-23
DWORETSKY ET AL. (35)

ENERGY (KEV)	CROSS SECTION (ARB. UNITS)
4.00E-02	9.00E-02
4.50E-02	1.10E-01
5.05E-02	4.00E-02
5.45E-02	1.00E-02
6.00E-02	3.00E-02
6.35E-02	1.10E-01
6.60E-02	5.90E-01
6.80E-02	1.29E 00
7.10E-02	1.51E 00
7.70E-02	3.59E 00
7.90E-02	4.38E 00
8.00E-02	4.24E 00
8.10E-02	4.13E 00
8.55E-02	2.00E 00
9.00E-02	1.11E 00
9.60E-02	1.15E 00
9.70E-02	1.27E 00
1.03E-01	2.53E 00
1.05E-01	2.69E 00
1.12E-01	2.61E 00
1.14E-01	2.74E 00
1.15E-01	2.95E 00
1.18E-01	3.11E 00
1.24E-01	3.55E 00
1.27E-01	2.90E 00
1.42E-01	1.57E 00
1.49E-01	1.42E 00
1.66E-01	2.83E 00
1.78E-01	3.28E 00
2.04E-01	2.80E 00
2.08E-01	2.68E 00
2.13E-01	2.27E 00
2.22E-01	1.95E 00
2.33E-01	2.03E 00
2.42E-01	2.25E 00
2.52E-01	2.34E 00
2.77E-01	2.09E 00
2.93E-01	1.70E 00
3.07E-01	1.54E 00
3.22E-01	1.35E 00
3.45E-01	1.18E 00
3.60E-01	1.26E 00
3.85E-01	1.48E 00
4.15E-01	1.63E 00
4.45E-01	1.60E 00
4.60E-01	1.49E 00
4.80E-01	1.53E 00
5.05E-01	1.48E 00
5.40E-01	1.57E 00
5.80E-01	1.54E 00
6.15E-01	1.61E 00
6.50E-01	1.71E 00
7.00E-01	1.62E 00
7.50E-01	1.55E 00
8.10E-01	1.80E 00

ENERGY (KEV)	CROSS SECTION (ARB. UNITS)
8.75E-01	2.17E 00
9.00E-01	2.68E 00
9.55E-01	3.04E 00
1.00E 00	3.14E 00
1.09E 00	2.80E 00
1.16E 00	2.44E 00
1.18E 00	2.31E 00
1.25E 00	2.26E 00
1.31E 00	2.29E 00
1.43E 00	2.61E 00
1.53E 00	3.17E 00
1.62E 00	3.68E 00
1.70E 00	3.80E 00
1.83E 00	3.57E 00
1.88E 00	3.26E 00
2.00E 00	2.77E 00
2.12E 00	2.39E 00
2.20E 00	2.18E 00
2.33E 00	2.07E 00
2.44E 00	2.09E 00
2.52E 00	2.29E 00
2.78E 00	2.86E 00
3.00E 00	3.25E 00
3.25E 00	3.33E 00
3.75E 00	3.09E 00
4.00E 00	3.11E 00
4.30E 00	3.06E 00
4.50E 00	3.18E 00
4.75E 00	3.37E 00

6-5-24
DWORETSKY ET AL. (215)

ENERGY (KEV)	CROSS SECTION (ARB. UNITS)
5.86E-02	3.50E-01
6.46E-02	1.11E 00
6.41E-02	1.50E 00
7.46E-02	1.62E 00
6.97E-02	1.86E 00
7.89E-02	1.83E 00
7.94E-02	2.25E 00
8.28E-02	2.84E 00
9.84E-02	2.75E 00
8.28E-02	2.84E 00
8.28E-02	3.04E 00
8.81E-02	3.04E 00
8.75E-02	3.18E 00
9.77E-02	3.12E 00
1.21E-01	3.38E 00
1.46E-01	3.30E 00
1.47E-01	3.49E 00
1.76E-01	4.26E 00
1.98E-01	4.32E 00
2.23E-01	4.24E 00
1.74E-01	4.45E 00
1.98E-01	4.44E 00
2.21E-01	4.44E 00
2.47E-01	5.07E 00
2.45E-01	5.27E 00
2.72E-01	5.23E 00
2.94E-01	5.64E 00
2.96E-01	5.78E 00
3.49E-01	5.89E 00
3.47E-01	6.03E 00
3.28E-01	6.14E 00
3.26E-01	6.42E 00
3.98E-01	6.39E 00
4.24E-01	6.42E 00

3.69E-01	6.59E 00
4.15E-01	6.86E 00
4.48E-01	7.27E 00
5.03E-01	7.40E 00
5.03E-01	7.58E 00
5.47E-01	7.41E 00
6.11E-01	7.13E 00
7.06E-01	7.01E 00
7.11E-01	7.28E 00
8.11E-01	7.28E 00
1.02E 00	6.68E 00
1.01E 00	6.51E 00
1.26E 00	6.21E 00
1.27E 00	5.86E 00
1.52E 00	5.45E 00
1.51E 00	5.18E 00
2.04E 00	5.37E 00
2.04E 00	4.96E 00
2.56E 00	5.29E 00
2.52E 00	4.82E 00
3.06E 00	4.12E 00
3.04E 00	4.01E 00
3.54E 00	3.57E 00

6-5-25
DWORETSKY ET AL. (35)

ENERGY (KEV)	CROSS SECTION (ARB. UNITS)
4.00E-02	4.00E-02
5.05E-02	3.00E-02
5.75E-02	3.00E-02
6.35E-02	3.70E-01

6.82E-02	7.50E-01
7.15E-02	9.30E-01
7.40E-02	9.90E-01
8.10E-02	7.70E-01
8.90E-02	7.90E-01
9.45E-02	6.90E-01
1.06E-01	6.60E-01
1.18E-01	7.80E-01
1.37E-01	7.90E-01
1.54E-01	7.10E-01
1.71E-01	6.10E-01
1.99E-01	6.70E-01
2.46E-01	7.70E-01
3.00E-01	6.20E-01
3.25E-01	6.60E-01
3.78E-01	8.30E-01
4.25E-01	7.30E-01
5.45E-01	5.90E-01
8.15E-01	6.50E-01
1.07E 00	7.50E-01
1.29E 00	8.10E-01
1.79E 00	7.40E-01
2.02E 00	7.60E-01
2.55E 00	7.10E-01
3.25E 00	9.10E-01
4.50E 00	7.70E-01
5.00E 00	8.70E-01

6-5-26
DWORETSKY ET AL. (35)

ENERGY (KEV)   CROSS SECTION (ARB. UNITS)

3.95E-02	1.00E-02
4.47E-02	2.00E-02
5.50E-02	2.00E-02
5.95E-02	2.00E-02
6.45E-02	3.10E-01
7.00E-02	7.50E-01
7.45E-02	1.13E 00
8.35E-02	1.43E 00
9.30E-02	1.67E 00
1.12E-01	1.86E 00
1.32E-01	2.00E 00
1.62E-01	2.13E 00
1.80E-01	2.17E 00
2.00E-01	2.08E 00
2.40E-01	2.01E 00
3.00E-01	1.97E 00
3.25E-01	1.91E 00
3.75E-01	1.73E 00
5.00E-01	1.56E 00
6.00E-01	1.49E 00
7.10E-01	1.42E 00
8.80E-01	1.31E 00
1.08E 00	1.29E 00
1.40E 00	1.15E 00
1.60E 00	1.10E 00
2.00E 00	1.09E 00
2.50E 00	1.07E 00
3.25E 00	8.90E-01
3.80E 00	8.30E-01
4.30E 00	8.70E-01
5.10E 00	9.70E-01

6-6   EXCITATION OF HELIUM BY VARIOUS HEAVY PROJECTILES

Investigations have been made of the excitation of helium by the impact of Li^+,[202] Ne^+,[71, 218] and Ar^+.[71] The measurements are all of rather restricted scope involving only the determination of one- or two-line emission cross sections for each collision combination. The reader is referred to Section 6-1 for a general discussion of considerations applicable to the study of the collisional excitation of helium.

Van Eck et al.,[202] using the apparatus described by Sluyters et al.[164] [see Section 7-3-A(2)], measured the cross section for the emission of the He I $3^1P \to 2^1S$ line induced by the impact of Li^+ ions on a helium target. These data were taken at a target pressure of 10^{-2} torr. At such high pressures, the population of the 3^1P state will be greatly influenced by the absorption of resonance photons. Also, the projectile beam will contain a substantial proportion of neutral particles. Although linearity of emission with pressure was observed, this certainly did not indicate single-collision conditions, and a wider range of pressures would have shown that the secondary populating mechanisms were important. This determination was carried out under conditions that are in gross violation of the established criteria; therefore the results are not repeated here.

Tolk and White[218] have studied the emission of the 3888-Å line of He I ($3^3P^0 \to 2^3S$) induced by the impact of Ne^+ on a target of He. An absolute magnitude was assigned to these data by a procedure that is invalid [see Section 6-1-B(9)]. To prevent confusion, these data are shown here as a relative variation of emission cross section with projectile energy; the absolute values assigned by the authors have been omitted. These emission measurements, it should be noted, include a cascade contribution; therefore some of the structure seen on these curves might, in fact, be due to structure

in the cross sections for excitation of the cascade level, rather than to the structure in the cross section for the formation of the 3^3P^0 state. No estimates of limits of accuracy were assigned by the authors to this data. The results are shown in Data Figs. 6-6-1.

Thomas and Gilbody[71] studied the excitation of helium by Ne^+ and Ar^+ impact at energies in the 150—385-keV range. The experimental procedures were reviewed in Section 6-1-B(7). The influence of polarization of emission on these data was neglected. No tests were made of the influence of excited states in the projectile beam. The measurements are shown in Data Tables 6-6-2 and 6-6-3. Random errors were estimated as $\pm 15\%$ and systematic errors, as $\pm 12\%$.

It is not possible to make an objective assessment as to whether or not the data on collisional excitation of He by Ne^+ and Ar^+ are accurate. These measurements by Tolk and White[218] and by Thomas and Gilbody[71] should be treated with caution.

Matveyev et al.[116] have studied the threshold energy for the excitation of the 2^1P state of helium induced by the impact of K^+ on a helium target. It was shown that the minimum projectile impact energy for the emission of the 584.3-Å ($2^1P \to 1^1S$) transition of He I, measured in the laboratory frame of reference, was 228.3-eV. No quantitative cross section data are presented in the work of Matveyev et al.[116]

A. Data Tables

6-6-1 The Relative Cross Section in Arbitrary Units for the Emission of the 3888-Å Line of He I ($3^3P \to 2^3S$) Induced by the Impact of Ne^+ on a Helium Target.

6-6-2 Cross Sections for the Emission of Lines of He I and He II Induced by the Impact of Ne^+ on a Helium Target. (A) 5875-Å Line of He I ($3^3D \to 2^3P$); (B) 4685-Å Line of He[+] ($n = 4 \to n = 3$).

6-6-3 Cross Sections for the Emission of the 5875-Å Line of He I ($3^3D \to 2^3P$) Induced by the Impact of Ar^+ on a Helium Target.

6-6-1 TOLK ET AL. (218)

ENERGY (KEV)	CROSS SECTION (ARB. UNITS)
1.01E-01	2.40E-02
1.11E-01	4.20E-02
1.22E-01	4.80E-02
1.27E-01	6.00E-02
1.43E-01	6.60E-02
1.52E-01	2.40E-01
1.63E-01	7.20E-02
1.76E-01	6.00E-03
1.83E-01	6.00E-03
1.93E-01	1.20E-01
2.04E-01	4.44E-01
2.13E-01	4.20E-01
2.24E-01	4.56E-01
2.34E-01	7.02E-01
2.45E-01	6.90E-01
2.55E-01	6.42E-01
2.64E-01	6.06E-01
2.75E-01	7.08E-01
2.88E-01	7.68E-01
2.97E-01	7.56E-01
3.08E-01	8.70E-01
3.23E-01	6.66E-01
3.45E-01	6.42E-01
3.67E-01	5.70E-01
3.90E-01	7.02E-01
4.07E-01	7.92E-01
4.36E-01	8.34E-01
4.63E-01	8.76E-01
4.84E-01	9.36E-01
5.14E-01	8.16E-01
5.66E-01	5.76E-01
6.18E-01	6.42E-01
6.75E-01	6.60E-01
7.26E-01	7.50E-01
7.78E-01	9.78E-01
8.19E-01	1.17E 00
8.86E-01	1.19E 00
9.27E-01	1.55E 00
9.80E-01	1.53E 00
1.05E 00	1.36E 00
1.08E 00	1.47E 00
1.14E 00	1.39E 00
1.20E 00	1.39E 00
1.25E 00	1.66E 00
1.35E 00	1.61E 00
1.47E 00	1.78E 00
1.58E 00	2.01E 00
1.69E 00	1.98E 00
1.78E 00	2.11E 00
1.91E 00	2.33E 00
2.11E 00	2.38E 00
2.02E 00	2.42E 00
2.23E 00	2.67E 00
2.52E 00	2.79E 00
2.33E 00	2.84E 00
2.44E 00	2.84E 00
2.65E 00	3.05E 00
2.92E 00	3.25E 00
3.19E 00	3.29E 00
2.90E 00	3.34E 00
3.45E 00	3.57E 00
3.73E 00	3.83E 00
4.01E 00	4.19E 00
4.24E 00	4.37E 00
4.56E 00	4.84E 00
4.87E 00	5.15E 00

6-6-2A THOMAS ET AL. (71)

ENERGY (KEV)	CROSS SECTION (SQ. CM)
1.05E 02	3.25E-18
1.30E 02	3.35E-18
1.55E 02	3.45E-18
1.77E 02	3.50E-18
1.77E 02	3.00E-18
2.05E 02	2.75E-18
2.10E 02	3.15E-18
2.28E 02	2.30E-18
2.50E 02	2.70E-18
2.60E 02	2.30E-18
2.75E 02	2.10E-18
2.75E 02	2.35E-18
3.05E 02	2.00E-18
3.10E 02	2.10E-18
3.30E 02	2.30E-18
3.55E 02	2.00E-18

6-6-2B THOMAS ET AL. (71)

ENERGY (KEV)	CROSS SECTION (SQ. CM)
1.50E 02	5.30E-19
1.66E 02	5.90E-19
1.75E 02	5.30E-19
2.00E 02	7.10E-19
2.25E 02	7.10E-19
2.55E 02	7.30E-19
3.00E 02	7.20E-19
3.55E 02	7.20E-19
3.86E 02	6.50E-19

6-6-3 THOMAS ET AL. (71)

ENERGY (KEV)	CROSS SECTION (SQ. CM)
1.50E 02	3.27E-18
1.77E 02	3.10E-18
1.83E 02	3.10E-18
2.13E 02	3.65E-18
2.50E 02	3.50E-18
2.75E 02	3.50E-18
3.02E 02	4.10E-18
3.30E 02	3.80E-18
3.40E 02	3.70E-18
3.80E 02	3.70E-18

CHAPTER 7

EXCITATION OF
HEAVY MONOATOMIC GASES

7-1 INTRODUCTION

This chapter considers the available data pertaining to the study of the collisional excitation of monoatomic gases by impact of various projectiles. Targets include neon, argon, krypton, xenon, and mercury. Most of the work has involved light projectiles such as protons, H_2^+ and He^+. The various rare gases exhibit certain similarities of behavior and experimental problems. In contrast, the mercury target stands apart from the other cases and has been studied by only one group.

The heavy monoatomic gases have complicated energy level schemes that in turn cause intricacies in the emitted spectra. Transitions between excited states and the ground state of the atom or ion will generally lie in the vacuum ultraviolet spectral regions. This part of the spectrum exhibits little complexity, and it is relatively simple to provide sufficient resolution to ensure that the measurement of a cross section is free of interference from other emissions. There are numerous experimental difficulties associated with this spectral region, particularly in the provision of a satisfactory standard for calibration of the detection efficiency of the apparatus. Transitions between the higher excited states will generally be in the visible and near ultraviolet spectral regions that are accessible with relatively simple equipment and techniques. There is a problem associated with the complexity of the emission. For example, Moore[66] lists some 450 lines which have been identified as emanating from argon and its ions in the spectral range 3000- to 10,000-Å. As the density of available post-collision excited states increases, the individual level-excitation cross sections will, expectably, decrease. High resolutions are required in the visible spectral regions where intensity is weak; low resolutions are tolerable in the ultraviolet regions where the intensities are strong. This is unfortunate, since the provisions of high-resolution and high-detection sensitivity are mutually exclusive in most spectrometric systems.

In general, the identifications of spectral lines are taken from the tables of Moore;[66, 67] exceptions will be specifically noted. A comment must be made on the style of term designations utilized for the spectra of neutral rare gases. Moore's visible wavelength tables[66] utilize a Paschen notation. The ultraviolet tables[67] and tables of atomic energy levels[269] utilize a $j - l$ coupling notation[270] suggested by Racah. In the present monograph, the term designations found in the visible wavelength tables are converted to correspond to the Racah notation, which is favored in the ultraviolet tables.

The outer shell of the ground state neutral atom consists of two ns electrons and six np electrons; n signifies the relevant principal quantum number. When the ion spectrum is excited as the result of the impact of a projectile, one np electron is removed and another np electron is excited. The remaining two ns electrons and four np electrons may couple in three different ways, giving rise to either a ^3P, ^1D, or ^1S state of the ion core; the excited electron is coupled to this core. The spectroscopic designation of the state includes information on the state of the core indicated by the use of primes; ^3P is given no prime; ^1D is given one prime; and ^1S, two primes. For example, the spectroscopic state of Ar^+ represented by $4p'^2F_{3\ 1/2}$ means the following: a $4p$-excited electron is coupled to a ^1D-core state; the total angular momentum of the excited state is 3; the total spin is $\frac{1}{2}$; and therefore the state is a doublet with $J = 3\frac{1}{2}$.

In view of the complexities of the spectra, particularly in the visible regions, it is necessary to dwell at length on the resolution used in the various experimental determinations and the possible interference with the measurement from other emissions. In some cases where insufficient resolution was available, it has been the practice to attempt to measure the sum of the cross sections for the emission of two or more lines simultaneously. Certain precautions, described in Section 3-4-E, are necessary in order to carry out this procedure satisfactorily. To summarize briefly the results of that discussion: whenever a monochromator is used for such measurements it is necessary that the exit slit width, which governs resolution, should be wider than the entrance slit width, which governs instrumental line width, by an amount that depends on the dispersion of the instrument and the separation of the lines that it is desired to measure simultaneously. Often this precaution is omitted, and the data are in error. In two extensive investigations of emission from rare gases the detection systems have had a band width that encompasses many spectral lines. In the work of Schlumbohn[205, 228] 40-Å wide interference filters were used; in the work of Lipeles et al.[204] integral measurements over a band of about 1000-Å were used. These data represent a sum of intensities from a number of spectral lines, each weighted according to an instrumental transmission factor. Such data do not represent a measurement on a specific state, nor even on an identified group of states; moreover the data are likely to be peculiar to the apparatus used to measure it. Consequently, the data are of little fundamental value as a cross section. The results from these sources are discussed in the relevant sections of this chapter, but the data are not presented.

In general, it is impossible to estimate the level-excitation cross sections from the measurement of emission cross sections. The procedures for doing this (Section 2-2) involve measurement or theoretical estimate of the cross sections for all emissive processes which tend to populate or depopulate a particular excited state. With the complex spectra of the rare gases and mercury, this is generally impossible. In a series of experiments on excitation of rare gases by impact of He^+, de Heer and co-workers [203] have shown that cascade population of excited states of the highly ionized target can be very large and, in some cases, greatly exceeds the population by direct collisional excitation.

The question of the polarization of the emission has generally been ignored in these studies. De Heer and van Eck[77] provide some data on the polarization of emission induced by H^+ and He^+ impact on Ne, but all other work reviewed here has completely ignored it. There is every reason to suppose that emission will, in general,

be polarized. It has been noted previously (Section 3-4-D) that polarization influences the measurement of emission cross section. Polarization causes angular anisotropy of the emission. The detection sensitivity of a system will depend on the polarization of the light it detects, giving rise to so-called instrumental polarization effects. Procedures for avoiding these effects have been discussed previously (Section 3-4-D). Most of the work reviewed here neglects these problems entirely and errors may be present. Neglect of polarization will cause errors in both the absolute magnitude and in the energy dependence of the cross section. It is not possible to estimate the magnitudes of these errors without data on the polarization.

It has been noted previously (see Section 3-2-D) that care must be taken in the measurement of target pressure of heavy gases. The use of the McLeod gauge with a cold trap in the line to the experimental region may give erroneous measurements due to pumping of mercury onto the trap itself. Methods of eradicating this problem have been previously discussed with a consideration of the magnitudes of the errors that might be expected. Some of the experiments discussed here have been in error due to neglect of this point.

It will be noted during the discussions of the available data that most of the measurements are due to three research groups only. It is valuable to remember that the relative accuracy of the series of measurements by one group is probably much higher than the corresponding comparisons among different groups. For example, if a measurement on a particular spectral line is criticized on the grounds that it may have included an indeterminate amount of emission from a neighboring line, then this deficiency will occur in all measurements on that line by the one research group, even when the projectile is changed.

The available data on the excitation of neon, argon, krypton, xenon, and mercury by impact of various projectiles are discussed in Sections 7-2 through 7-6, respectively. Each section is subdivided according to the nature of the projectile. Data are collected at the end of each section. Each section also includes a discussion of the experimental arrangements which have been used in the measurements.

7-2 EXCITATION OF NEON

Studies of neon excitation have involved experiments to detect radiation from the neutral atom and also from the singly charged ion. The excitation processes may be described by the following reaction equations:

$$X^+ + Ne \rightarrow [X^+] + Ne^*$$

$$X^+ + Ne \rightarrow [X^+ + e] + Ne^{+*}. \tag{7-1}$$

There is no information on the states of excitation nor ionization of the systems contained within the square brackets. Most of the quantitative work involves impact of H^+ and He^+ projectiles. Of the six publications from which data are drawn, five emanate from the Amsterdam group composed of de Heer, van Eck, Jaecks, and others. A brief survey of the apparatus used in the research will be given (7-2-A), followed by a discussion of the data obtained for impact of H^+, He, and He^+ (Sections 7-2-B to 7-2-D, respectively), and finally the collection of data in tabular form (Section 7-2-E).

In addition to the work using optical spectroscopy there have also been recent studies of inner-shell excitation effects utilizing Auger electron spectroscopy. In this technique a collision causes the ejection of an inner-shell electron and leaves the target in an excited ion state. The state tends to decay by autoionization, ejecting an electron. Decay by the emission of an X ray is of low probability for light targets such as argon and neon. The energy of the electron characterizes the emissive transition, and the number of electrons gives the line intensity. In every way this is analogous to the decay of an excited state in an outer shell through the emission of a photon. At the present time this study is in its infancy and no quantitative data on excitation of neon are available. Edwards and Rudd[271] have made some qualitative statements about line intensities observed in the impact of H^+, He^+, and Ne^+ on neon. These intensities do allow the estimation of relative probabilities of forming different excited states of the ion by removal of inner-shell electrons. Since no quantitative cross section data are available, this work is not discussed further here.

A. Experimental Arrangements

1. The Work of de Heer et al. The work of this group commenced in 1963 and extends to the time of this writing. A total of five papers has been published. In some cases early data have been superseded and extended by later measurements. The following is a summary of the sources of information for specific reactions.

(a) $H^+ + Ne$. The first paper by van Eck et al[76] described measurements of emission cross sections of Ne I and Ne II lines emitted in the far ultraviolet spectrum. A later paper[77] revised these data and gave new measurements on emission and polarization of visible spectral lines. Further improvements in the operation of the equipment have caused another revision. Although the visible line measurements in the second article[77] are still valid, the ultraviolet emission measurements of the first article[76] are now considered to be more accurate than the so-called revised measurements. Thus, the original data are retained for use here. This conclusion is based on recent measurements using improved methods for determination of absolute detection efficiency.[67]

(b) $He + Ne$. A very limited study has been made of this case in the most recent work,[203] and this is used in the present compilation of data.

(c) $He^+ + Ne$. A situation arose which was similar to that described under the heading of the $H^+ + Ne$ studies. The ultraviolet emission measurements of van Eck et al.[76] are used here. The visible measurements in a later paper[77] are retained although the ultraviolet data are rejected. Subsequently this series of measurements was extended.[203, 272, 124] A conference paper by de Heer et al.[272] was completely superseded by two more detailed reports, one of which handles visible emission measurements[124] and the other ultraviolet emissions.[203] Data from these two latter reports[203,124] are used here and compose the bulk of the data on the $He^+ + Ne$ case.

Three different optical systems have been used to handle the different regimes of the visible and ultraviolet spectral regions. For convenience they shall be referred to as systems I, II, and III. System I was used for all the measurements in the visible spectral regions and is involved in the work of references 77 and 124. System II is a grazing incidence vacuum monochromator which was used only for ultraviolet emission measurements in references 77 and 76. System III is a vacuum monochromator that may be used in either a grazing or normal incidence mode. This was

utilized in all the ultraviolet work described in reference 203. Different techniques are used in the calibration of the detection efficiencies of these three optical systems. Systematic errors introduced when using one system will not necessarily be found in the work that involves the other optical arrangements. The three systems will be considered separately. The arrangements used for measurements in the visible region have been described previously in the context of the study of visible emissions from excited helium target [Section 6-1-B(1)], but will be reviewed again for completeness.

SYSTEM I.   The arrangement for the detection of visible light was based on a Leiss monochromator with a Czerny—Turner grating mounting, using photomultiplier detection. Descriptions of this device are found in a number of papers.[10,165] No lens was used between collision region and monochromator. Therefore, light was detected from a length of beam path defined by the size of the diffraction grating and the dimensions of the entrance slit. For calibration purposes, a standard lamp was placed a considerable distance behind the collision chamber, and diaphragms were used to ensure that only the central part of the filament was observed. When observing the standard lamp, the field of view and the solid angle from which light was collected were both much less than when measurements were made of emission from the collision region. Simple geometrical relationships were used to allow for these differences. The published work does not include tests to prove that the response of the system was independent of the solid angle and the field of view. Recent work using the same apparatus for similar experiments, however, has shown that, over the range of conditions used in these experiments, the response is almost independent of solid angle and field of view.[165, 246,163] Inadequate attention was paid to the problem of scattered light. In a paper devoted to studies of helium excitation using the same apparatus,[165] it is stated that a scattered-light correction was measured directly at wavelengths below 3300-Å, and this was assumed to remain constant at higher wavelengths. Such an assumption is not necessarily correct because most of the scattered light comes from the zero-order diffraction by the grating via spurious reflections. Such reflections are likely to vary with the position of the grating. The brief series of measurements of polarization, found only in one publication,[77] was made using a Glan—Thomson prism as a polarization analyzer. This was placed between the collision region and monochromator in a region where the light beam was slightly divergent. Use of a polarization analyzer with a divergent light beam is not good practice, but tests were made to ensure that there was a 100% rejection of light polarized perpendicularly to the principal plane.[213]

SYSTEM II.   The earliest arrangement used for detection of ultraviolet emissions was employed in reference 76, but a good description of its features must be found elsewhere.[90] The grating monochromator is based on a Rowland circle mounting. Detection was by an open-ended photomultiplier. Changes in the wavelength setting of the monochromator were achieved by moving the detector and exit slit along the Rowland circle. No attempt was made to focus light onto the entrance slit of the monochromator, and the region in the collision chamber from which light was observed was defined by the dimensions of the entrance slit and the grating.

The calibration of detection efficiency was achieved by a complicated comparison technique.[90] The ultraviolet monochromator and a nonevacuable monochromator of the type described above as System I were arranged to be at 90° to the ion beam path in the collision chamber and to view the same part of the path through the target gas. Observations were made of the emission light resulting from the formation

of excited helium atoms by charge transfer as a beam of helium ions traversed a neon or hydrogen target. The vacuum ultraviolet monochromator was set to register the 537-Å line of He I ($3p \to 1s$) and the visible-light monochromator was arranged to register the 5016-Å line of He I ($3p \to 2s$). The ratio of these two-line intensities must be equal to the ratio of the two-transition probabilities, viz.:

$$\frac{I_{3p \to 1s}}{I_{3p \to 2s}} = \frac{A_{3p \to 1s}}{A_{3p \to 2s}}. \tag{7-2}$$

These transition probabilities are known theoretically from, for example, the work of Wiese et al.,[31] for which an accuracy of 1 % is claimed. The intensity of the visible line ($3p \to 2s$) was measured absolutely using the calibration methods described above with reference to System I. Hence, the absolute intensity of the ultraviolet transition ($3p \to 1s$) was found, and a value for the efficiency of the detection system estimated. The accuracy of this part of the procedure will be limited by the statistical accuracy of the measured ratio of the two line intensities and the calibration accuracy of the detection efficiency of the visible system. There is a puzzling discrepancy of 20 % between the results of this calibration when using helium incident on hydrogen and the results using helium ions incident on neon as the source of emission.[90] It was suggested that additional emission from the hydrogen target gas was being detected and causing an error. The calibration using neon should be more accurate. The overall accuracy of the procedure for calibration of the instrument at 537-Å was estimated to be 40 %.[76] This complicated set of procedures provides a calibration of detection efficiency only at the single wavelength of 537-Å. It was then necessary to relate measurements of emission at wavelengths from 461- to 743-Å to this single calibration at 573-Å. This was achieved using a photomultiplier with a sodium salicylate screen as a detector, and making the assumption that its detection efficiency was the same at all the wavelengths of interest. According to Samson[73] this is justified at these wavelengths. Corrections were made for changes in reflectivity of the grating and the effects of astigmatism. These corrections, based on assumptions about the behavior of the apparatus, were not confirmed by experimental test. Inadequate detail is given on the methods used for making these corrections. No attempt was made to assess the influence of scattered light. Under the circumstances it is very difficult to assess the accuracy of the final calibration. No estimate was made by van Eck et al.[76] It would seem unreasonable to claim an overall accuracy for the measurement of intensity of better than $\pm 100\%$.

SYSTEM III. This has been used for the most recent measurements on ultraviolet emission that are found in reference 203. It is described fully by Moustafa Moussa and de Heer in a paper that deals with excitation of a helium target.[163] The system is based on a monochromator that may be used either in a grazing incidence or in a normal incidence mode. In this apparatus the grating may be moved on the Rowland circle in order to change wavelength. Detection was achieved using either open-ended multipliers or photomultipliers coated with a fluorescent screen of sodium salicylate. These devices could be interchanged at will. The detection efficiency was determined directly at the wavelengths of 537- and 1215-Å. At the lower wavelength, 537-Å, the techniques described above for the calibration of System II were again used. The calibration at 1215-Å was obtained by using as a standard source the collisional excitation of Lyman-alpha radiation induced by electron impact on molecular hydrogen. An auxiliary electron gun provided the source of projectiles,

and the electron beam was arranged to occupy the same position in the collision chamber as the ion beam did during experimental measurements. The intensity of the emission was calculated by taking the cross sections for this process measured by Fite and Brackmann.[98] These cross section measurements are not in themselves absolute. Fite and Brackmann compared the emission from molecular hydrogen to the emission from atomic hydrogen in the same apparatus and normalized the latter to a theoretical calculation in the Born approximation. Therefore, the final calibration of the apparatus of de Heer et al. is related to a theoretical prediction of a cross section. These procedures gave estimates of detection efficiency at 537- and 1215-Å. Values at other wavelengths were obtained by a process of "inter- and extrapolation." No details are given on the basis for these estimates. This is a very dubious procedure because the sensitivity would probably vary in a complicated manner with wavelength, and would need to be determined by detailed experiment. It is very difficult to estimate the overall accuracy of the final result. De Heer et al. state that the overall systematic error does not exceed 100%; this would certainly not appear to be a pessimistic assessment.

In all work by this group, the spectrometric resolution was chosen to allow the spectral lines of interest to be separated from neighboring emissions. In general resolutions as good as 2-Å were employed.

Throughout this series of experiments involving all three monochromator systems in both the visible and vacuum ultraviolet spectral regions, no attention was paid to the possible influence of polarization of emission. The monochromators were arranged to view the collision region at 90° to the ion beam path, and isotropy of emission was assumed. In the event that polarization is present, there will be errors in the cross section measurements due to the neglect of anisotropy and also due to instrumental polarization effects (see Section 3-4-D).

A continuing series of improvements has been made on the ion beam collimation and detection systems progressing from the fairly simple arrangements used in the early work[76] to a very complicated and sophisticated structure for the recent measurements.[203,124] There have been no comments by the authors as to any improvement in accuracy due to these changed features. It appears that full precautions to ensure accurate monitoring of ion beam current were taken for all experiments. No attention was paid to the possible influence of excited states in the projectile beams. It is unlikely that a small proportion of excited projectiles will greatly affect the cross sections for forming excited states in the target; therefore, this problem is not expected to contribute significant errors to the measurements made by this group.

A McLeod gauge was used to measure pressure with precautions to prevent the errors associated with pumping onto the cold trap and thermal transpiration (see Section 3-2-D). Linearity of emission intensity with target pressure was assured for all lines. In the case of the resonance lines of neutral neon, it was expected that absorption of resonance photons would produce considerable nonlinearity in the emission; however this was not observed. The matter will be considered further in connection with the discussion of the data.

Jaecks et al.[124] estimate that the absolute accuracy of the measurements made on visible and near ultraviolet spectral lines is within ±30%. For the vacuum ultraviolet spectral lines it would be unrealistic to estimate that the systematic errors were any less than ±100%. Reproducibility of measurement was about ±10% at all wavelengths.

2. The Work of Dufay et al. Dufay et al.[125] have made a study of the cross sections for the emission of two lines of Ne I and one of Ne II induced by the impact of protons on a neon target. The report gives inadequate details for a review, so reference is made to a more complete report[178] describing work on other targets using the same apparatus and also to an unpublished thesis by Denis.[253]

The optical system is based on a Czerny—Turner grating monochromator with photomultiplier detection. A quartz lens forms an image of the collision region on the monochromator entrance slit. The optical axis of the system was at 90° to the beam path, and it was assumed that the emission was isotropic. In the event that the emission does exhibit polarization, the data will be in error due to the neglect of the resulting anisotropy and to any errors associated with instrumental polarization effects. It is not possible to estimate the magnitude of the errors introduced into the present experiment by neglect of polarization.

The calibration of detection sensitivity was achieved using a standard lamp placed some distance from the collision region on the axis of the monochromator. The lens was moved to allow the formation of a sharp image of the lamp filament on the monochromator entrance slit. Inevitably the field of view and solid angle used during calibration were different from that used during experimental measurements. Inadequate attention was paid to ensuring that the response was the same under these two circumstances. No attention was given to the possible influence of scattered light in the monochromator.

Spectrometric resolution is not quoted in the reports of this work. Reports of other measurements made with the same apparatus suggest that resolutions as good as 2 to 4-Å were possible, and it may be presumed that this resolution was also used in the study of excitation of rare gases.

Target pressures were monitored using a McLeod gauge. No attention was directed to removing the thermal transpiration and "mercury pumping" error associated with the use of this type of gauge (see Section 3-2-D).

Details available in the publication concerning rare gas excitation are insufficient to allow any assessment of whether or not the measurements were done under conditions where light intensity was proportional to target pressure and projectile beam current.

The authors' quoted limits of accuracy in this work are up to $\pm 50\%$ possible systematic error and $\pm 10\%$ random error.

It should be noted that the published data for this work are in the form of a line of best fit drawn with the experimental points themselves being omitted from the diagrams. In the present data compilations, the original data points obtained by private communication with the authors are shown. Some of these points are revised from the original published data. The data used here are the corrected values.

3. The Work of Lipeles et al. Lipeles and co-workers[204] have made a brief study of the emissions induced in the collision $He^+ + Ne$ at rather low energies, 5- to 500-eV. The optical system consisted solely of unfiltered photon counters. A windowless multiplier was used to detect all light of wavelengths from 200- to 1200-Å and a photomultiplier detected light in the region 1050- to 3500-Å. No attempt was made to resolve individual spectral lines. Cross sections for emission of light into the wavelength regions accepted by the detectors were presented. In the absence of any attempt to measure the variation with wavelength of the detector efficiency or to make analysis of the spectra being detected, these data cannot be considered as useful cross section

information. The data will represent a sum of line emission cross sections weighted according to the efficiency of detection at each relevant wavelength. In the absence of any attempt to determine the weighting function or to analyze the emission, these data must be considered as peculiar to the apparatus used by Lipeles et al. and not as fundamental information. Consequently, these data are excluded from further discussion.

Despite the exclusion of that data from this monograph, the work might be very valuable in assisting in the understanding of collisional excitation mechanisms. The range of energies extends down to threshold. Considerable care was taken in providing a projectile beam with good energy resolution. Excited projectile states were eliminated by using an electron impact ion source with electron energies below the first excitation potential of helium ions.

B. Proton Impact

The excitation of neon by proton impact has been considered by the group of de Heer, van Eck, et al.[77, 76] and also by Dufay et al.[125] For convenience, the data will be grouped according to the term designations of the upper level of the transition. Polarization of emission has only been studied for three lines in Ne I and is considered at the end of this section.

It is emphasized that the absolute accuracy of the data may be rather poor, and that systematic errors will differ from one experiment to another.

1. Emission of Lines of Ne I. Van Eck et al.[76] have measured the sum of the cross sections for emission of the 744- and 736-Å lines ($3s[1\frac{1}{2}]_1^0 \to 2p^6\,^1S_0$ and $3s'[\frac{1}{2}]_1^0 \to 2p^6\,^1S_0$) simultaneously. Resonance reabsorption by the target gas will lead, in principle, to nonlinearity of intensity with pressure. Van Eck et al. claim that these measurements were made at about 10^{-3} torr under pressure-independent conditions. This is a surprising observation since it would be expected that nonlinearity would commence at much lower pressures. De Heer et al.[203] have also shown detailed plots of intensity against pressure for emission induced by He$^+$ impact, and again linearity was observed in the pressure range 10^{-4} to 2×10^{-3} torr. It was suggested[203] that the absorption effect is being balanced by some other secondary populating mechanism with a different pressure dependence, so that the net effect is apparent linearity of emission; however, there is no direct evidence for this. It must be concluded that the data are taken under conditions that appear to be in accordance with the basic operational definition of emission cross section, and they are therefore valid. They should be, however, treated with caution since there is a suspicion that the experimental observations are misleading. It should also be noted that a simultaneous measurement on two lines of slightly different wavelengths requires certain precautions regarding the monochromator entrance and exit slit widths (Section 3-4-E). These were omitted in the work of van Eck et al., a fault that will lead to errors. The measured sum of the cross sections for the 744- and 736-Å lines is shown in Data Table 7-2-1. As mentioned previously, a further measurement of this line by de Heer and van Eck[77] is regarded as inaccurate and is omitted.

In Data Tables 7-2-2 through 7-2-7 are shown the remaining measurements by de Heer and van Eck[77] as well as a single data set by Dufay et al.[125] The work of the two groups is in reasonable agreement for the one case where a comparison is possible.

2. Emission of Lines of Ne II. Data Table 7-2-8 shows the sum of the cross

sections for emission of the 461-Å ($2p^6\ ^2S_{1/2} \to 2p^5\ ^2P^0_{1\ 1/2}$) and 462-Å ($2p^6\ ^2S_{1/2} \to 2p^5\ ^2P^0_{1/2}$) lines as determined by van Eck et al.[76] Precautions required for the simultaneous measurement of the sum of two spectral lines (Section 3-4-E) were omitted; the consequent error is likely to be small compared with the estimated limits of accuracy of the experiment. De Heer and van Eck[77] have also measured the sum of the cross sections for emission of the 406- and 407-Å lines of Ne II, but the published values are now regarded as inaccurate.[213] To prevent confusion the data on the 406- and 407-Å emissions are not reproduced here.

De Heer and van Eck[77] have measured cross sections for the emission of the 3766-Å ($3p\ ^4P^0_{2\ 1/2} \to 3s\ ^4P_{1\ 1/2}$) and 3830-Å ($3d\ ^2D_{2\ 1/2} \to 3p\ ^2P^0_{1\ 1/2}$) lines of Ne II, shown in Data Tables 7-2-9 and 7-2-10, respectively. Dufay et al.[125] present a measurement of a line which is claimed to be a 3345-Å transition in Ne II (Data Table 7-2-11); it is likely that other lines of Ne II might have been included in the detection band and the data may therefore be unreliable.

3. Polarization of Emission. De Heer and van Eck have carried out some measurements on the polarization of Ne I spectral lines[77] (Data Tables 7-2-12 through 7-2-14). No estimates of accuracy limitations were made by the original authors.

C. He Impact

Measurements of the 5882-Å ($3p'[\frac{1}{2}]_1 \to 3s[1\frac{1}{2}]^0_2$) Ne I line by de Heer et al.[203] shown in Data Table 7-2-15, represent the only data available for this collision combination. No details are given on the method used for preparing the neutral helium beam; therefore it is impossible to assess what proportion of the atoms might be in an excited state. It is unlikely however that a small proportion of excited atoms in the projectile beam will cause appreciable errors in the measurement of the cross section for excitation of the target system. The detection system was presumably the Czerny—Turner mounted grating monochromator reviewed in Section 7-2-A(1) as System I. Also omitted from the discussion is any mention of how the neutral-atom flux was monitored. With the lack of experimental details, it is not possible to carry out a good assessment as to the validity of this measurement.

D. He$^+$ Impact

The excitation of neon has been considered only by the group of de Heer, van Eck, Jaecks, and others.[203, 77, 76, 124] Polarization has been considered for three of the lines.[77] It is emphasized that the absolute accuracy of the data, as stated by the authors, is rather poor. Work on visible spectral lines is believed to be $\pm 30\%$ accurate, and work on vacuum ultraviolet lines, only $\pm 100\%$ accurate. Reproducibility was better than $\pm 10\%$.

Throughout this work no attempt was made to assess the possible influence of excited states in the projectile beam. It is likely that the major excited component will be the metastable state of He$^+$. The population of this state will be reduced by field-quenching effects that occur in the acceleration region of the apparatus; the electric field causes mixing of the $2s$ and $2p$ excited states with a consequential rapid decay to ground. It is not possible to estimate the effect that the excited states will produce, but it is likely to be rather small.

In addition to the measurements of cross sections for the emission of discrete spectral lines, there have also been measurements on the intensity of emission into broad wavelength bands made by Lipeles et al.[204] This work is excluded from the present discussion because it does not give information on a specific transition but

rather involves a sum of many line intensities, each weighted by an apparatus function [see Section 7-2-A(3)].

   1. Emission of Lines of Ne I.   De Heer et al.[203] have measured the cross sections for the 744-Å ($3s[1\frac{1}{2}]_1^0 \to 2p^6\ {}^1S_0$) and 736-Å ($3s'[\frac{1}{2}]_1^0 \to 2p^6\ {}^1S_0$) lines of Ne I induced by the impact of He$^+$ on neon. The 736-Å line was measured at impact energies from 0.3- to 6-keV, while the 744-Å line was measured only at 6-keV. Van Eck et al[76] have measured the sum of the cross sections for emission of the 736- and 744-Å lines. These data are shown together in Data Table 7-2-16 and appear to be in good agreement. It should be noted that, although both sets of measurements were carried out by the same research group, they did constitute separate determinations and the systematic errors may be different. These two lines will be affected by resonance reabsorption of photons by the target gas. Van Eck et al. claim that their measurements were done under conditions where intensity varies linearly with target density, and no such absorption was observed. De Heer et al.[203] show detailed plots of intensity against pressure in the range 10^{-4} to 3×10^{-3} torr, these are in agreement with this observation. When nonlinearity does eventually start at high pressures, it is in the opposite direction from that which would be expected for resonance photon absorption. These are very surprising results since strong absorption effects would be expected to occur at quite low pressures. De Heer et al.[203] suggest that there is some other secondary populating mechanism that tends to cancel out the absorption effect, giving rise to apparent linearity of emission. This could probably be tested by studying the pressure dependence at even lower pressures. Based on the available evidence, it must be concluded that the data are in accordance with the fundamental operational definition of emission cross section and therefore are valid. In view of the unexpected behavior, these data should be treated with caution, and further tests are necessary. As mentioned previously, further measurement on this line by de Heer and van Eck[77] is regarded as inaccurate and is omitted.

   In Data Tables 7-2-17 and 7-2-18 are shown measurements on the 5882-Å and 5945-Å lines of Ne I; measurements by de Heer and van Eck[77] agree well with independent determinations by de Heer et al.[203] Both lines are $3p' \to 3s$ transitions. The remaining data is by de Heer and van Eck; it includes $3p' \to 3s'$ and $4d \to 3p$ transitions (Data Tables 7-2-19 through 7-2-21).

   2. Emission of Lines of Ne II.   In Data Tables 7-2-22 through 7-2-26 are shown measurements on a number of lines by the group of de Heer and co-workers (see table captions for identification of wavelength and levels). In cases where two or more lines were measured simultaneously the required precautions (Section 3-4-E) were omitted.

   Data on the 406- and 407-Å lines,[77] has been repudiated by the authors[213] and is not included here. Also omitted for the same reason is a second measurement on the 461- and 462-Å lines.

   3. Polarization of Emission.   De Heer and van Eck have carried out some measurements on the polarization of Ne I spectral lines[77] (Data Tables 7-2-27 through 7-2-29). No estimates of accuracy limitations were made by these authors.

E. Data Tables

1. Cross Sections for the Emission of Ne I Lines Induced by the Impact of H$^+$ on a Ne Target

7-2-1   744- and 736-Å Lines ($3s[1\frac{1}{2}]_1^0 \to 2p^6\ {}^1S_0$ and $3s'[\frac{1}{2}]_1^0 \to 2p^6\ {}^1S_0$) measured together.

7-2-2   5882-Å Line ($3p'[\frac{1}{2}]_1 \to 3s[1\frac{1}{2}]_2^0$).

7-2-3   5945-Å Line $(3p'[1\frac{1}{2}]_2 \to 3s[1\frac{1}{2}]_2^0)$.

7-2-4   5852-Å Line $(3p'[\frac{1}{2}]_0 \to 3s'[\frac{1}{2}]_1^0)$.

7-2-5   6074-Å Line $(3p[\frac{1}{2}]_0 \to 3s[1\frac{1}{2}]_1^0)$.

7-2-6   5764-Å Line $(4d[3\frac{1}{2}]_4^0 \to 2p[2\frac{1}{2}]_3)$.

7-2-7   5820-Å Line $(4d[3\frac{1}{2}]_3^0 \to 3p[2\frac{1}{2}]_2)$.

2.   Cross Sections for the Emission of Ne II Lines Induced by the Impact of H⁺ on a Ne Target

7-2-8   461- and 462-Å Lines $(2p^6\ ^2S_{1/2} \to 2p^5\ ^2P^0_{1\ 1/2}$ and $2p^6\ ^2S_{1/2} \to 2p^{\ 5}\ ^2P^0_{1/2})$ measured together.

7-2-9   3766-Å Line $(3p\ ^4P^0_{2\ 1/2} \to 3s\ ^4P_{1\ 1/2})$.

7-2-10   3830-Å Line $(3d\ ^2D_{2\ 1/2} \to 3p\ ^2P^0_{1\ 1/2})$.

7-2-11   3345-Å Line (Not unambiguously identified).

3.   Polarization Fractions of Ne I Lines Induced by the Impact of H⁺ on a Ne Target

7-2-12   5852-Å Line $(3p'[\frac{1}{2}]_0 \to 3s'[\frac{1}{2}]_1^0)$.

7-2-13   5882-Å Line $(3p'[\frac{1}{2}]_1 \to 3s[1\frac{1}{2}]_2^0)$.

7-2-14   5945-Å Line $(3p'[1\frac{1}{2}]_2 \to 3s[1\frac{1}{2}]_2^0)$.

4.   Cross Sections for the Emission of Ne I Lines Induced by the Impact of He on a Ne Target

7-2-15   5882-Å Line $(3p'[\frac{1}{2}]_1 \to 3s[1\frac{1}{2}]_2^0)$.

5.   Cross Sections for the Emission of Ne I Lines Induced by the Impact of He⁺ on a Ne Target

7-2-16   744- and 736-Å Lines $(3s[1\frac{1}{2}]_1^0 \to 2p^6\ ^1S_0$ and $3s'[\frac{1}{2}]_1^0 \to 2p^6\ ^1S_0)$. (A) 736-Å line alone; (B) 744-Å line alone; (C) 736- and 744-Å lines measured together.

7-2-17   5882-Å Line $(3p'[\frac{1}{2}]_1 \to 3s[1\frac{1}{2}]_2^0)$.

7-2-18   5945-Å Line $(3p'\ [1\frac{1}{2}]_2 \to 3s[1\frac{1}{2}]_2^0)$.

7-2-19   5852-Å Line $(3p'\ [\frac{1}{2}]_0 \to 3s'\ [\frac{1}{2}]_1^0)$.

7-2-20   5764-Å Line $(4d[3\frac{1}{2}]_4^0 \to 3p[2\frac{1}{2}]_3)$.

7-2-21   5820-Å Line $(4d[3\frac{1}{2}]_3^0 \to 3p[2\frac{1}{2}]_2)$.

6.   Cross Sections for the Emission of Ne II Lines Induced by the Impact of He⁺ on a Ne Target

7-2-22   461- and 462-Å Lines $(2p^6\ ^2S_{1/2} \to 2p^5\ ^2P^0_{1\ 1/2}$ and $2p^6\ ^2S_{1/2} \to 2p^5\ ^2P^0_{1/2})$ measured together.

7-2-23   (A) 3727-Å Line $(3p\ ^2D^0_{1\ 1/2} \to 3s\ ^2P_{1/2})$; (B) 3830-Å Line $(3d\ ^2D_{2\ 1/2} \to 3p\ ^2P^0_{1\ 1/2})$.

7-2-24   (A) 3568-Å Line $(3p'\ ^2F^0_{3\ 1/2} \to 3s'\ ^2D_{2\ 1/2})$; (B) 3574- and 3575-Å Lines $(3p'\ ^2F^0_{2\ 1/2} \to 3s'\ ^2D_{2\ 1/2}$ and $3p'\ ^2F^0_{2\ 1/2} \to 3s'\ ^2D_{1\ 1/2})$ measured together; (C) 3405- and 3407-Å Lines $(3d'\ ^2D_{2\ 1/2} \to 3p'\ ^2D^0_{1\ 1/2}$ and $3d'\ ^2D_{2\ 1/2} \to 3p'\ ^2D^0_{2\ 1/2})$ measured simultaneously.

7-2-25   (A) 3335-Å Line $(3p\ ^4D^0_{3\ 1/2} \to 3s\ ^4P_{2\ 1/2})$; (B) 3298-Å Line $(3p\ ^4D^0_{2\ 1/2} \to 3s\ ^4P_{2\ 1/2})$; (C) 3694-Å Line $(3p\ ^4P^0_{2\ 1/2} \to 3s\ ^4P_{2\ 1/2})$; (D) 3664-Å Line $(3p\ ^4P^0_{1\ 1/2} \to 3s\ ^4P_{2\ 1/2})$; (E) 3199-Å Line $(3d\ ^4F_{3\ 1/2} \to 3p\ ^4D^0_{2\ 1/2})$.

7-2-26   3766-Å Line $(3p\ ^4P^0_{2\ 1/2} \to 3s\ ^4P_{1\ 1/2})$.

7.   Polarization Fractions of Ne I Lines Induced by the Impact of He⁺ on a Ne Target

7-2-27   5852-Å Line $(3p'[\frac{1}{2}]_0 \to 3s'[\frac{1}{2}]_1^0)$.

7-2-28   5882-Å Line $(3p'[\frac{1}{2}]_1 \to 3s[1\frac{1}{2}]_2^0)$.

7-2-29   5945-Å Line $(3p'[1\frac{1}{2}]_2 - 3s[1\frac{1}{2}]_2^0)$.

7-2-1
VAN ECK ET AL. (76)

ENERGY (KEV)	CROSS SECTION (SQ. CM)
1.30E 01	7.80E-19
1.50E 01	6.90E-19
2.00E 01	6.30E-19
2.50E 01	5.90E-19
3.00E 01	5.80E-19
3.50E 01	6.30E-19

7-2-2
DE HEER ET AL. (77)

ENERGY (KEV)	CROSS SECTION (SQ. CM)
5.00E 00	7.20E-20
7.50E 00	1.23E-19
1.00E 01	1.32E-19
1.25E 01	1.08E-19
1.50E 01	8.30E-20
2.00E 01	6.10E-20
2.50E 01	5.20E-20
3.00E 01	6.10E-20
3.50E 01	5.70E-20

7-2-3
DE HEER ET AL. (77)

ENERGY (KEV)	CROSS SECTION (SQ. CM)
1.00E 01	1.38E-19
1.50E 01	1.38E-19
2.00E 01	1.64E-19
2.50E 01	2.00E-19
3.00E 01	2.24E-19
3.50E 01	2.25E-19

7-2-4A
DE HEER ET AL. (77)

ENERGY (KEV)	CROSS SECTION (SQ. CM)
5.00E 00	9.90E-20
7.75E 00	6.93E-19
1.00E 01	1.08E-18
1.50E 01	1.37E-18
2.00E 01	1.55E-18
2.50E 01	2.11E-18
3.00E 01	2.46E-18
3.50E 01	2.77E-18

7-2-4B
DUFAY ET AL. (125)

ENERGY (KEV)	CROSS SECTION (SQ. CM)
2.50E 01	1.87E-18
3.00E 01	1.91E-18
4.00E 01	2.12E-18
5.00E 01	2.05E-18
6.00E 01	2.06E-18
7.00E 01	1.88E-18
8.00E 01	1.80E-18
9.00E 01	1.75E-18
1.00E 02	1.62E-18
1.20E 02	1.43E-18
1.40E 02	1.21E-18
2.00E 02	9.00E-19
3.00E 02	7.30E-19
4.00E 02	5.30E-19
5.00E 02	4.65E-19
6.00E 02	4.00E-19

7-2-5
DUFAY ET AL. (125)

ENERGY (KEV)	CROSS SECTION (SQ. CM)
2.50E 01	2.40E-19
4.00E 01	3.20E-19
5.00E 01	3.00E-19
6.00E 01	3.00E-19
7.00E 01	2.80E-19
8.00E 01	2.70E-19
9.00E 01	2.80E-19
1.00E 02	2.70E-19
1.20E 02	2.80E-19
1.40E 02	2.70E-19
2.00E 02	2.50E-19
3.00E 02	2.00E-19
4.00E 02	1.70E-19
5.00E 02	1.60E-19
6.00E 02	1.45E-19

7-2-6
DE HEER ET AL. (77)

ENERGY (KEV)	CROSS SECTION (SQ. CM)
1.00E 01	1.48E-20
1.50E 01	1.54E-20
2.00E 01	1.45E-20
2.50E 01	1.59E-20
3.00E 01	1.49E-20
3.50E 01	1.46E-20

7-2-7
DE HEER ET AL. (77)

ENERGY (KEV)	CROSS SECTION (SQ. CM)
7.50E 00	8.50E-20
1.00E 01	1.15E-19
1.50E 01	1.29E-19
2.00E 01	1.41E-19
2.50E 01	1.66E-19
3.00E 01	1.73E-19
3.50E 01	1.76E-19

7-2-8
VAN ECK ET AL. (76)

ENERGY (KEV)	CROSS SECTION (SQ. CM)
1.00E 01	3.85E-18
1.50E 01	4.53E-18
2.00E 01	5.61E-18
2.50E 01	7.47E-18
3.00E 01	9.00E-18
3.50E 01	1.13E-17

7-2-9
DE HEER ET AL. (77)

ENERGY (KEV)	CROSS SECTION (SQ. CM)
1.50E 01	6.31E-21
2.00E 01	7.94E-21
2.50E 01	7.85E-21
3.00E 01	8.91E-21
3.50E 01	9.23E-21

7-2-10
DE HEER ET AL. (77)

ENERGY (KEV)	CROSS SECTION (SQ. CM)
1.00E 01	6.68E-21
1.50E 01	8.61E-21
2.00E 01	1.22E-20
2.50E 01	1.20E-20
3.00E 01	1.48E-20
3.50E 01	1.58E-20

7-2-11
DUFAY ET AL. (125)

ENERGY (KEV)	CROSS SECTION (SQ. CM)

ENERGY	CROSS SECTION
3.00E 01	9.50E-20
4.00E 01	1.25E-19
5.00E 01	1.39E-19
6.00E 01	1.44E-19
7.00E 01	1.47E-19
8.00E 01	1.61E-19
9.00E 01	1.42E-19
1.00E 02	1.35E-19
1.20E 02	1.23E-19
1.60E 02	9.40E-20
2.00E 02	8.20E-20
3.00E 02	5.90E-20
4.00E 02	4.70E-20
5.00E 02	4.30E-20

7-2-12
DE HEER ET AL. (77)

ENERGY (KEV)	POLARIZATION (PERCENT)
1.00E 01	-3.00E 00
1.50E 01	-1.30E 00
2.00E 01	2.00E 00
2.50E 01	5.00E 00
3.00E 01	4.00E 00
3.50E 01	3.00E 00

7-2-13
DE HEER ET AL. (77)

ENERGY (KEV)	POLARIZATION (PERCENT)
1.00E 01	-6.00E 00
1.50E 01	-7.00E 00
2.25E 01	-6.00E 00
3.00E 01	-6.00E 00

7-2-14
DE HEER ET AL. (77)

ENERGY (KEV)	POLARIZATION (PERCENT)
1.00E 01	-2.60E 01
1.50E 01	-2.60E 01
2.00E 01	-3.50E 01
2.50E 01	-3.60E 01
3.00E 01	-3.40E 01
3.50E 01	-3.30E 01

7-2-15
DE HEER ET AL. (203)

ENERGY (KEV)	CROSS SECTION (SQ. CM)
1.00E 01	1.04E-18
1.25E 01	9.65E-19
1.50E 01	9.00E-19
1.75E 01	6.39E-19
2.00E 01	5.77E-19
2.50E 01	5.00E-19
3.00E 01	5.00E-19
3.50E 01	5.03E-19

7-2-16A
DE HEER ET AL. (203)

ENERGY (KEV)	CROSS SECTION (SQ. CM)
3.00E-01	6.00E-18
4.00E-01	5.80E-18
5.00E-01	4.80E-18
6.00E-01	5.20E-18
7.00E-01	5.80E-18
8.00E-01	6.00E-18
9.00E-01	6.00E-18
1.00E 00	5.20E-18
2.00E 00	4.70E-18
3.00E 00	5.80E-18
4.00E 00	6.30E-18
5.00E 00	4.90E-18
6.00E 00	4.45E-18

7-2-16B
DE HEER ET AL. (203)

ENERGY (KEV)	CROSS SECTION (SQ. CM)
6.00E 00	4.12E-18

7-2-16C
VAN ECK ET AL. (76)

ENERGY (KEV)	CROSS SECTION (SQ. CM)
5.00E 00	9.00E-18
7.50E 00	5.81E-18
1.00E 01	3.62E-18
1.50E 01	2.34E-18
2.00E 01	1.87E-18
2.50E 01	1.54E-18
3.00E 01	1.25E-18
3.50E 01	1.12E-18

7-2-17A
DE HEER ET AL. (77)

ENERGY (KEV)	CROSS SECTION (SQ. CM)
5.00E 00	4.30E-19
7.50E 00	3.00E-19
1.00E 01	2.73E-19
1.50E 01	2.51E-19
2.00E 01	2.22E-19
2.50E 01	1.66E-19
3.00E 01	1.45E-19
3.50E 01	1.30E-19

7-2-17B
DE HEER ET AL. (203)

ENERGY (KEV)	CROSS SECTION (SQ. CM)
5.00E 00	4.41E-19
1.00E 01	3.09E-19
1.50E 01	2.79E-19
2.00E 01	2.24E-19
2.50E 01	1.85E-19
3.00E 01	1.76E-19
3.50E 01	1.47E-19

7-2-18A
DE HEER ET AL. (77)

ENERGY (KEV)	CROSS SECTION (SQ. CM)
5.00E 00	4.49E-19
7.50E 00	3.62E-19
1.00E 01	2.91E-19
1.50E 01	2.56E-19
2.00E 01	2.17E-19
2.50E 01	1.89E-19
3.00E 01	1.56E-19
3.50E 01	1.36E-19

7-2-18B
JAECKS ET AL. (124)

ENERGY (KEV)	CROSS SECTION (SQ. CM)
4.00E-01	3.29E-19
6.00E-01	2.75E-19
8.00E-01	3.30E-19
9.00E-01	3.70E-19
1.00E 00	3.70E-19
1.50E 00	3.92E-19
2.00E 00	3.80E-19
2.50E 00	3.20E-19
3.00E 00	3.50E-19
3.50E 00	3.82E-19
4.00E 00	4.09E-19
4.50E 00	3.70E-19
5.00E 00	3.80E-19
6.00E 00	3.82E-19
8.00E 00	3.30E-19
1.00E 01	3.02E-19
1.50E 01	2.70E-19
2.00E 01	2.28E-19
2.50E 01	2.00E-19
3.00E 01	1.68E-19

7-2-19
DE HEER ET AL. (77)

ENERGY (KEV)	CROSS SECTICN (SQ. CM)
5.0CE 00	3.92E-19
7.50E 00	2.84E-19
1.00E 01	2.16E-19
1.25E 01	1.91E-19
1.50E 01	1.70E-19
2.00E 01	1.62E-19
2.50E 01	1.61E-19
3.00E 01	1.35E-19
3.50E 01	1.25E-19

7-2-20
DE HEER ET AL. (77)

ENERGY (KEV)	CROSS SECTICN (SQ. CM)
5.00E 00	9.70E-20
1.00E 01	9.70E-20
1.50E 01	1.06E-19
2.00E 01	9.70E-20
2.50E 01	1.00E-19
3.00E 01	9.10E-20
3.50E 01	1.00E-19

7-2-21
DE HEER ET AL. (77)

ENERGY (KEV)	CROSS SECTICN (SQ. CM)
7.50E 00	9.20E-20
1.00E 01	9.00E-20
1.50E 01	7.50E-20
2.00E 01	8.10E-20
2.50E 01	6.80E-20
3.00E 01	6.50E-20
3.50E 01	7.50E-20

7-2-22
VAN ECK ET AL. (76)

ENERGY (KEV)	CROSS SECTICN (SQ. CM)
1.00E 01	6.40E-18
1.50E 01	1.14E-17
2.00E 01	1.66E-17
2.50E 01	2.01E-17
3.00E 01	2.15E-17
3.50E 01	2.26E-17

7-2-23A
JAECKS ET AL. (124)

ENERGY (KEV)	CROSS SECTION (SQ. CM)
5.00E-01	1.46E-19
7.00E-01	2.22E-19
1.00E 00	2.62F-19
2.00E 00	3.22E-19
4.00E 00	2.80E-19
6.00E 00	2.99E-19
8.00E 00	3.31E-19
1.00E 01	3.58E-19

7-2-23B
DE HEER ET AL. (77)

ENERGY (KEV)	CROSS SECTICN (SQ. CM)
7.75E 00	2.98E-20
1.00E 01	3.35E-20
1.50E 01	3.89E-20
2.00E 01	4.68E-20
2.50E 01	5.37E-20
3.00E 01	5.62E-20
3.50E 01	5.75E-20

7-2-24A
JAECKS ET AL. (124)

ENERGY (KEV)	CROSS SECTION (SQ. CM)
3.00E-01	3.60E-19
5.00E-01	6.01E-19
7.00E-01	7.43E-19
1.00E 00	1.27F-18
2.00E 00	1.97E-18
4.00E 00	2.03E-18
6.00E 00	2.00E-18
8.00E 00	2.01E-18
1.00E 01	2.02E-18

7-2-24B
JAECKS ET AL. (124)

ENERGY (KEV)	CROSS SECTICN (SQ. CM)
5.00E-01	3.31E-19
7.00E-01	4.50E-19
1.00E 00	6.63E-19
2.00E 00	1.00E-18
4.00E 00	1.20E-18
6.00E 00	1.07E-18
8.00E 00	1.03E-18
1.00E 01	9.78E-19
1.50E 01	1.03E-18
2.00E 01	1.02E-18
2.50E 01	1.05E-18
3.00E 01	1.12E-18

7-2-24C
JAECKS ET AL. (124)

ENERGY (KEV)	CROSS SECTION (SQ. CM)
7.00E-01	1.20E-19
1.00E 00	1.92E-19
2.00E 00	2.60E-19
4.00E 00	3.01E-19
6.00E 00	2.99E-19
8.00E 00	2.68E-19
1.00E 01	2.68E-19

7-2-25A
JAECKS ET AL. (124)

ENERGY (KEV)	CROSS SECTION (SQ. CM)
7.00E-01	1.56E-19
1.00E 00	2.65E-19
2.00E 00	3.20E-19
4.00E 00	3.48E-19
6.00E 00	3.79E-19
8.00E 00	4.36E-19
1.00E 01	5.54E-19
1.50E 01	7.75E-19
2.00E 01	1.02E-18
2.50E 01	1.04E-18
3.00E 01	1.08E-18

7-2-25B
JAECKS ET AL. (124)

ENERGY (KEV)	CROSS SECTICN (SQ. CM)
5.00E-01	2.60E-20
7.00E-01	4.61E-20
1.00E 00	5.87E-20
2.00E 00	8.91E-20
4.00E 00	1.01E-19
6.00E 00	1.02E-19
8.00E 00	1.07E-19
1.00E 01	1.34E-19
1.50E 01	1.97E-19
2.00E 01	2.77E-19
2.50E 01	2.77E-19
3.00E 01	2.83E-19

7-2-25C
JAECKS ET AL. (124)

ENERGY (KEV)	CROSS SECTICN (SQ. CM)
3.00E-01	8.50E-20
5.00E-01	1.80E-19
7.00E-01	3.18E-19
1.00E 00	3.38E-19
2.00E 00	3.59E-19
4.00E 00	3.00E-19
6.00E 00	3.07E-19
8.00E 00	3.81E-19
1.00E 01	4.65E-19
1.50E 01	6.53E-19
2.00E 01	7.98E-19
2.50E 01	9.35E-19
3.00E 01	9.60E-19

7-2-25D
JAECKS ET AL. (124)

ENERGY (KEV)	CROSS SECTION (SQ. CM)
7.00E-01	1.32E-19
1.00E 00	1.43E-19
2.00E 00	1.78E-19
4.00E 00	1.36E-19
6.00E 00	1.29E-19
8.00E 00	1.60E-19
1.00E 01	2.02E-19
1.50E 01	2.82E-19
2.00E 01	3.40E-19
2.50E 01	4.10E-19
3.00E 01	3.87E-19

7-2-25E
JAECKS ET AL. (124)

ENERGY (KEV)	CROSS SECTION (SQ. CM)
1.00E 00	3.72E-20
2.00E 00	6.12E-20
4.00E 00	7.65E-20
6.00E 00	7.88E-20
8.00E 00	9.38E-20
1.00E 01	1.13E-19

7-2-26
DE HEER ET AL. (77)

ENERGY (KEV)	CROSS SECTION (SQ. CM)
5.00E 00	1.62E-19
1.00E 01	1.78E-19
1.50E 01	2.85E-19
2.00E 01	3.71E-19
2.50E 01	4.73E-19
3.00E 01	4.17E-19
3.50E 01	4.36E-19

7-2-27
DE HEER ET AL. (77)

ENERGY (KEV)	POLARIZATION (PERCENT)
5.00E 00	-4.00E 00
1.00E 01	-7.00E 00
1.50E 01	-5.00E 00
2.00E 01	-6.00E 00
2.50E 01	-5.00E 00
3.00E 01	-8.00E 00
3.50E 01	-8.00E 00

7-2-28
DE HEER ET AL. (77)

ENERGY (KEV)	POLARIZATION (PERCENT)
7.50E 00	-4.00E 00
1.00E 01	-5.00E 00
1.50E 01	-4.00E 00
2.00E 01	-8.00E 00
2.50E 01	-7.00E 00
3.00E 01	-7.00E 00
3.50E 01	-7.00E 00

7-2-29
DE HEER ET AL. (77)

ENERGY (KEV)	POLARIZATION (PERCENT)
5.00E 00	-5.00E 00
1.00E 01	-1.20E 01
1.50E 01	-1.10E 01
2.00E 01	-1.00E 01
2.50E 01	-1.20E 01
3.00E 01	-1.80E 01
3.50E 01	-2.20E 01

7-3 EXCITATION OF ARGON

Studies of excitation of argon have involved experiments to detect radiation from neutral atoms and also from the single ionized state. The excitation processes may be described by the reaction equations:

$$X^+ + Ar \rightarrow [X^+] + Ar^* \tag{7-3}$$

$$X^+ + Ar \rightarrow [X^+ + e] + Ar^{+*}. \tag{7-4}$$

Here there is no information on the states of excitation nor ionization of the systems contained within the square brackets.

All the available data are in the form of emission cross sections. De Heer et al.[203] have made comments about cascade contributions in the Ar II spectrum induced by impact of He$^+$ on Ar. These allow some rough estimates of level-excitation cross sections. This work is reviewed later in the appropriate part of this section.

In addition to the direct measurements of collisionally induced emission cross sections, there have recently been a number of other studies of excitation mechanisms in argon. Kessel et al.[273] have detected X rays resulting from K-shell vacancies induced by the impact of protons at 1.5- and 3.0-MeV energy. No attempt has been made to determine cross sections for specific processes. There have been a number of studies of ultraviolet continuum emission induced by impact of protons or alpha particles on a high pressure argon target. Recent examples of such studies with high-pressure targets

are the work of Hurst et al.[274] and that of Bennet and Collinson.[275] The x-ray studies do not yet give any information on cross sections, and the continuum emission studies represent the result of multiple-body collision processes. Therefore, neither area is considered further in this monograph.

In the following section (7-3-A) a brief review is given of the various experimental systems that have been used in the emission studies. This is followed by a discussion of the data obtained for excitation by impact of protons (Section 7-3-B), H_2^+ (Section 7-3-C), He^+ (Section 7-3-D), Ne^+ (Section 7-3-E), Ar^+ (Section 7-3-F), and Cs^+ (Section 7-3-G). This is followed by a collection of the available data in graphical and tabular form (Section 7-3-H).

A. Experimental Arrangements

1. The Work of de Heer et al. The work of de Heer, van Eck, Jaecks and co-workers has extended from 1963 to the time of writing and involves a total of four relevant papers. Three different optical systems have been used in their work. These were discussed in detail in Section 7-2-A(1) in the context of their studies of excitation of neon. For convenience, the three optical systems were there labeled System I, II, and III, respectively. For a detailed discussion of procedures in the measurement of emission, pressure, and beam current the reader is referred to that section.

The data and the sources from which the cross sections are drawn are summarized as follows.

$H^+ + Ar$. Ultraviolet emissions induced by projectiles with energies from 5- to 35-keV were studied by van Eck et al.[76] using the optical System II. Later, Jaecks et al.[124] carried out some measurements of visible emissions induced by protons in the energy range 2- to 10-keV using the System I optical arrangement.

$H_2^+ + Ar$. Visible emissions induced by 1- to 10-keV projectiles were studied by Jaecks et al.[124] using the optical arrangement described as System I.

$He^+ + Ar$. Ultraviolet emissions induced by 5- to 35-keV projectiles were studied by van Eck et al.[76] using optical System II. These ultraviolet data were extended down to energies of 0.3-keV by the work of de Heer et al[203] using the apparatus defined as System III. These two sets of measurements were carried out independently and are in good agreement. Studies of visible emissions induced by 0.3- to 30-keV projectiles were made by Jaecks et al.[124] using the apparatus designated as System I. There is an additional conference paper by de Heer et al.[272] on visible and ultraviolet emissions that may be disregarded because the data are repeated in the two most recent papers[203,124] with some minor revisions.

Throughout this work the monochromators were set to observe emission at 90° to the beam direction. The polarization of the emission and its possible influence on the accuracy of cross section measurements was ignored. The absolute accuracy of the data for visible spectral lines was estimated as $\pm 30\%$ while that for vacuum ultraviolet lines was about $\pm 100\%$. Reproducibility of data was better than $\pm 10\%$ in both cases.

2. The Work of Sluyters et al. Sluyters and co-workers made a study of the emissions induced by the impact of Ar^+ ions on an argon target. No attempt was made to separate target and projectile emission, so the data are a sum of contributions from both sources. This work of Sluyters was the first attempt at excitation and emission

function measurements by the Amsterdam group that was later to be headed by F. J. de Heer; nevertheless the techniques and apparatus are so different from those later employed by de Heer, Van den Bos, Jaecks, and other co-workers that it is most practical to treat the work of Sluyters as a separate series of experiments. The full report of Sluyter's work[103] was preceded by a number of brief preliminary reports.[276, 277, 278] With the single exception of a measurement on an Ar I Line,[277] all the data are to be found in the last report.[103]

The projectile beam was at energies from 5- to 24-keV. A conventional optical system was used; photometric calibration was carried out with a tungsten strip filament lamp.

Severe criticisms may be leveled as a result of the very high target pressures that were used. At the typical operating pressure of 5×10^{-3} torr up to 63 % of the incident Ar^+ beam was neutralized.[103, 279] Sluyters et al. made the arbitrary assumption that the observed emission was induced by the remaining charged component; this unsubstantiated assumption is likely to be wrong. Pressure plots showing linearity of emission with target density all exhibited finite intercepts; these pressure plots do not indicate the existence of single collision conditions.

It must be concluded that the work of Sluyters and Kistemaker is carried out under conditions that are in gross violation of the criteria for valid measurements of cross sections. These data are wrong and misleading; all data from this source are omitted.

3. The Work of Dufay et al.   Dufay and co-workers have made measurements on the emission cross sections of one line each of Ar I and Ar II induced by the impact of H^+ on an argon target.[125] The apparatus used for this work has been assessed in Section 7-1-A(2) in the context of studies of excitation of neon by impact of protons. The reader is referred to that section, as all comments are equally valid for the work using an argon target. The maximum systematic error was estimated as $\pm 50\%$ and the reproducibility of the data was $\pm 10\%$.

4. The Work of Thomas and Gilbody.   The work of Thomas and Gilbody was concerned with emissions induced by the impact of He^+, Ne^+, and Ar^+ on an argon target. The first publication was a conference paper[280] that was later superseded with some changes by a more complete report.[71]

The projectile beam was furnished by a Van de Graaff accelerator fitted with an oscillating electron source. No precautions were taken to assess the exited state content of the projectile beam. Beam current was monitored on a deep Faraday cup. Target pressures were adequately low to ensure single collision conditions and were measured with a McLeod gauge. No precautions were taken to eliminate the effects of cold-trap pumping and thermal transpiration.

The detection system was based on a large aperture autocollimating prism monochromator fitted with photomultiplier detection. The system was set to detect light emitted at 90° to the ion beam path. No account was taken of the possible influence of polarization through anisotropy in the emission and instrumental polarization effects (see Section 3-4-D). Optical resolution was never better than 7-Å.

The detection efficiency of the system was determined using a standard tungsten strip filament lamp placed at the point normally occupied by the ion beam. In this manner the optical system was the same during the experiment as for the calibration, and it was unnecessary to determine precisely the positions and apertures of the various parts of the optical system. The linearity of the detection system's response to the intense standard lamp was checked. The lamp temperature was measured using an

optical pyrometer, and the emissive power obtained from the measurements by de Vos.[83]

Neglecting the problem of inaccurate pressure measurement using the McLeod gauge, the systematic error was estimated as $\pm 12\%$, and the random error as less than $\pm 15\%$.

In addition to the measurements carried out with the apparatus described above, a small amount of data taken with the apparatus of van Eck and de Heer in the Amsterdam laboratory are also presented. These are at a lower energy, from 30- to 100-keV. The optical system was the visible arrangement utilized by that group and described previously in Section 7-2-A(1) as "System I." This device was set at an angle of $60°$ to the ion beam path, and the influence of anisotropy associated with polarization could therefore be neglected. No correction was made for the effect of instrumental polarization. Precautions were taken to eliminate errors associated with the use of a cold trap on a McLeod gauge. All spectrometric resolutions were set to be the same as those employed in the apparatus of Thomas and Gilbody. It is estimated that the systematic errors in these measurements do not exceed $\pm 30\%$ and the reproducibility is better than $\pm 10\%$.

5. The Work of Thomas. A measurement of the emission cross sections of an Ar I and Ar II line induced by the impact of protons on an argon target is given by Thomas.[127] The apparatus used for the work was the same as that described for the study of excitation of helium[161] and was in no way related to the earlier work of Thomas and Gilbody,[280, 71] described in the previous section [7-3-A(4)].

The projectile beam for this work was provided by a Van de Graaff accelerator that had an energy range of 150- to 1000-keV. Target pressures were typically 5×10^{-4} torr and were monitored with a capacitance manometer type of pressure gauge which has a response that is independent of the nature of the target gas.

The detection system was based on a $\frac{1}{2}$ meter Ebert-Fastie monochromator fitted with photomultiplier detection and placed at an angle of $90°$ to the ion beam path. No attempt was made to determine whether or not the emission was polarized. If the emission was polarized, then the data are in error through the neglect of anisotropy in the radiation and the influence of instrumental polarization. The calibration of detection efficiency was carried out using a standard tungsten strip filament lamp as a known source of emission. Filament temperature was measured with an optical pyrometer, and the relevant values of emissive power were obtained from tables.[83] During calibration, particular care was taken to eliminate the effects of stray light within the monochromator. When calibrating the blue end of the spectrum, a band pass filter was placed in front of the standard lamp to block intense emissions in the red end of the spectrum. This precaution was found to be necessary for the 4200-Å line of Ar I.

The authors estimate that the systematic error in the data does not exceed $\pm 25\%$ and the relative error is about $\pm 5\%$.

6. The Work of Robinson and Gilbody. These authors have published a short letter[105] describing measurements of the emission cross section of the 4431-Å line of Ar^+ induced by the impact of Ar^+ on an Ar target. Doppler shift of the emission from the projectile was used to allow a separate measurement of emission from the target and the fast particle. The apparatus used for this study is identical with the system described in detail in a previous publication.[162]

The projectile beam was provided by a Van de Graaff accelerator fitted with an

oscillating electron ion source. No attempt was made to assess the possible influence of excited states in the projectile beam. Target pressures were monitored with a Pirani gauge that was calibrated frequently against a cold-trapped McLeod gauge. Satisfactory precautions were taken to eliminate the errors due to mercury pumping of the cold trap and thermal transpiration.

A large autocollimating prism monochromator fitted with photomultiplier detection was placed at an angle of 54.5° to the projectile's path. At this angle the measurement of emission cross section should be independent of anisotropy due to polarization of emission. Tests were made[252] to ensure that the instrumental polarization was small and produced a negligible error. Placing the monochromator at this angle also facilitates the separate measurement of emission from the target and projectile. Both systems will give rise to the same spectra, but the emission from the projectile will exhibit a Doppler shift in wavelength.

A lens was used to form an image of the collision region on the entrance slit of the monochromator. Consequently, only a small part of the ion beam path is observed. Calibration of detection sensitivity was carried out using a standard lamp placed in the position normally occupied by the ion beam. This procedure ensures that calibration and experiment employed precisely the same optical arrangements. No mention was made of checks for calibration errors that might be caused by scattered light in the monochromator. Linearity of response of the detection system was stated to be very good.[252]

Wavelength resolution was not quoted but was of the order of a few angstroms. The authors quote[162] an estimated systematic error of less than ± 20% and a random uncertainty up to ± 10% due to reading errors.

7. The Work of Neff. Emission of Ar II lines induced by the impact of Ar+ ions on an Ar target is recorded in two publications by Neff; the original publication[281] was superseded by a more complete report[104] with no change to the values of the data.

The projectiles were produced in a radio frequency ion source, accelerated to energies between 500- and 3000-eV and mass analyzed. No attempt was made to assess the possible importance of excited states in the projectile beam. Energy resolution of the projectile beam was not specified. Target pressure was measured with a Pirani gauge that had been calibrated against a cold-trapped McLeod gauge. No precautions were taken to eliminate the effects of cold-trap pumping and thermal transpiration. Neglect of these problems would tend to make the cross section measurements too high (see Section 3-2-D).

The detection system was based on a ½ meter Ebert-mounted grating monochromator fitted with a photomultiplier. The monochromator was placed at an angle of 90° to the ion beam path. No attempt was made to correct for anisotropy of emission that might be associated with polarization of the emission nor to eliminate the influence of instrumental polarization. The output data were in the form of a relative emission cross section. To derive an absolute scale, Neff normalized to the data of Sluyters and Kistemaker[103] on the cross section for emission of Ar II lines induced by impact of Ar+ on an Ar target. Sluyters made measurements down to only 5-keV impact energy, and it was necessary to extrapolate the data to 3-keV, the maximum energy used by Neff. There is no basis for the straight-line extrapolation of Sluyter's results. Moreover, it has been suggested in Section 7-3-A(2) that the work of Sluyters is of poor accuracy and has been omitted from this monograph. Consequently, the absolute magnitudes of the data by Neff must be considered as being inaccurate.

The resolution of the monochromator used in this work was only 30-Å. Consequently, the data may include contributions from all spectral lines within 30-Å of the nominal quoted wavelength.

The authors estimate that systematic error did not exceed $\pm 50\%$ and that the relative accuracy of measurements on lines at different wavelengths was $\pm 10\%$.

8. The Work of Lipeles et al. Lipeles and co-workers[204] have made a brief study of the emission induced in the collision of He^+ with Ar at rather low energies, 5- to 500-eV. Measurements were made of emission into rather broad wavelength bands and also of the emission of a single line in the Ar^+ spectrum.

The broad band emission measurements have been previously discussed in Section 7-2-A(3), where it was concluded that the data were of no fundamental value. The work also included a measurement of the relative variation with energy of the cross section for the emission of the 4765-Å line of Ar II; recently Lipeles et al.[282] have repeated these measurements and shown, quite conclusively, that they are in error. None of the work by Lipeles et al. is reproduced here.

9. The Work of Andersen et al. Andersen and co-workers[114] have provided a brief study of emission induced by the impact of Cs^+ on an argon target. All data were obtained with a very poor spectrometric resolution of 30-Å. The argon spectrum is so rich in lines that such an inadequate resolution cannot be said to provide an unambiguous identification of the measured lines. The data probably represent a sum of emission from various lines, each weighted according to an apparatus function. These data are not satisfactory and are excluded from further consideration here.

10. The Work of Schlumbohm. Schlumbohm [205, 228] has concentrated on the study of collisional excitation mechanisms at very low impact energies, ranging from 300-eV down to threshold.

All measurements were taken using interference filters with very broad bands. Consequently, it is impossible to determine what line, or lines, the emission represented. The absolute values of cross sections were assigned by measuring emission induced by electron impact on N_2 and normalizing to previous data by Stewart,[283] this is likely to be an incorrect procedure since the work of Stewart is in substantial disagreement with four more recent independent determinations[284, 285, 286, 287] all of which are in complete agreement with each other. It must be concluded that the work of Schlumbohm is inaccurate and that the source of the measured emission is not properly defined. Therefore these data are not considered further here.

It should be noted that this series of measurements represent one of the few attempts to study collisionally induced emission at threshold. Despite the poor quality of the data they may nevertheless provide important information on the nature of collisional excitation processes at low energies.

11. The Work of Volz and Rudd. A study of the energy distribution of electrons ejected by the impact of protons on argon has been made by Volz and Rudd.[126] The distribution consists of a continuum, with a large number of discrete peaks superimposed upon it, that represents electrons removed from the atom by ionization. The discrete peaks are related to autoionization transitions that can be used to determine cross sections for the formation of inner-shell vacancies in the atom. Most ionization events occur through the removal of electrons from an outer shell resulting, in general, in the formation of an ion and one or more free electrons. For some events the projectile will eject an electron from an inner shell. The electron will go to the continuum and the target system is left in an excited state of the singly charged ion. The excess

energy may be liberated either by the ejection of an electron through an autoionizing transition or through the emission of an X ray by normal radiative decay. The ratio of the probability of photon emission to electron ejection is called the *fluorescent yield*. The ejected electron will have a discrete energy. The formation of the excited state by collisional removal of an inner-shell electron may be detected either through monitoring X ray emission or through monitoring ejected electrons of a definite energy corresponding to autoionization transitions into the vacancy.

Quantitative measurements of the flux of autoionization electrons have been made by Volz and Rudd. Their apparatus used a Cockroft—Walton accelerator providing protons in the energy range 120- to 300-keV. Target pressures in the range 1.2 to 1.6×10^{-3} torr were measured with a capacitance manometer whose response is independent of the nature of the target medium. Ejected electrons were allowed to drift through a field-free region into an energy analyzer where the spectrum was recorded.

The ejected electron energy spectrum has the form of a continuum distribution with a large number of superimposed peaks. In principle, if the continuum distribution is subtracted from the observed spectrum, then the area under an individual peak represents the cross section for the decay of the excited state by the emission of the autoionization electron that is being studied. A practical detection system only receives electrons ejected into a rather narrow solid angle. Therefore, a measurement is made of a differential cross section for emission that should be integrated over all angles to achieve a total cross section. This total cross section for the emission of an autoionizing electron is directly analogous to the total cross section for the emission of a photon. Both are measurements of a cross section for a decay process that results from a collision excitation event. Proceeding with the same philosophy as for photon emission studies, one may now attempt to assess the various branches by which the excited state decays and also the cascade processes that tend to populate the excited state. Finally, one should, in principle, be able to derive from the measurements of electron emission a cross section for level excitation just as photon emission may be used to relate to level excitation cross sections.

Volz and Rudd assess the shape of the continuum by interpolating that part of the spectrum which is free of peaks through the region where peaks are observed; the difference is the spectrum of autoionizing electrons. The area under each line is measured and related to line intensity with corrections for the inherent width of the line and the resolution function of the detection system. This measurement is repeated at a number of angles to the projectile beam direction, and a differential cross section for ejection of electrons is produced which appears to be isotropic within the accuracy of the measurements. Integration of this differential cross section gives a total emission cross section for a given line. The nature of the transitions is determined by accurate measurement of the energy of the electrons and reference to the standard tables of atomic energy levels. All transitions leading to the decay of a particular excited state are added together to give a total cross section for the decay of that state by electron emission. The states considered involve vacancies in the $L_{2,3}$ shell and simultaneous vacancies in the $L_{2,3}$ and $M_{2,3}$ shells. It remains only to assess the contribution to decay by X ray emission and the contribution to formation by cascade processes. X ray emission resulting from L- and M-shell vacancies is negligible since the observed fluorescent yields are almost zero.[288] Cascade must involve the formation of $2s$- or $1s$-shell vacancies that are subsequently filled by an electron transition leaving a vacancy in the $L_{2,3}$ or $M_{2,3}$ shells. No autoionizing electrons resulting from such

transitions are found therefore, cascade is assessed as negligible. It follows that the total cross section for ejection of electrons by decay of an excited state is equal to the cross section for the formation of that state.

The data by Volz and Rudd provide measurements of the cross section for the formation of the $1s^2 \, 2s^2 \, 2p^5 \, 3s^2 \, 3p^6$ state of Ar^+ and also the $1s^2 \, 2s^2 \, 2p^5 \, 3s^2 \, 3p^5$ state of Ar^{2+}. The ground state configuration of Ar is $1s^2 \, 2s^2 \, 2p^6 \, 3s^2 \, 3p^6$.

No attempt will be made here to carry out a detailed assessment of the procedures utilized in this experiment. This is the only work of its type available at the present time and does not warrant a detailed discussion of all of the experimental features. The brief conference report by Volz and Rudd is inadequate to allow a detailed assessment of procedures. Consequently, the only conclusion that can be drawn concerning the validity of these measurements is that there is no obvious inadequacy in procedures; however, complexity of the method and absence of confirmatory evidence from independent experiments would suggest that these data should be treated with caution. Volz estimates that the data are accurate only to within an order of magnitude.[289]

B. Proton Impact

The group of de Heer, van Eck, and Jaecks, has made measurements on emission cross sections of a number of Ar I and Ar II lines.[124] Dufay et al.,[125] and Thomas[127] have each measured a single emission line in the Ar I and Ar II spectra. Also, there are measurements of the formation of excited states of Ar^+ and Ar^{2+} by Volz and Rudd using the technique of measuring flux of electrons ejected by autoionization.

Data Tables 7-3-1 and 7-3-2 show measurements of 7635-Å and 4200-Å emission carried out by Dufay et al. and by Thomas et al. It is possible that the 4200-Å emission was not properly resolved from neighboring lines of Ar I. Data Tables 7-7-3 and 7-7-4 show measurements by van Eck et al.[76] of the 932- and 920-Å line emission; these two lines both have a common upper level, $3p^6 \, {}^2S_{1/2}$. The errors in the absolute magnitudes of these two ultraviolet lines are likely to be rather large; as much as 100%.

Cross sections for emission of visible Ar II lines are shown in Data Tables 7-3-5 through 7-3-9; the measurements on the 4277-Å line by Dufay et al. are a revision of the original publication.[290] The 4431-Å line measured by Thomas[127] is identified as the $4p \, {}^4P^0_{2 \, 1/2} \rightarrow 3d \, {}^4D_{2 \, 1/2}$ transition; this may contain an unknown contribution from the adjacent 4430-Å ($4p \, {}^4D^0_{1 \, 1/2} \rightarrow 4s \, {}^4P_{1/2}$) line that was not separately resolved.

A further series of measurements, describing the formation of excited states which are degenerate with the continuum, is found in Volz and Rudd's work[126] that is based on the measurement of the flux of ejected autoionization electrons. Because assessing these data qualitatively as stated previously [Section 7-3-A(10)] is very difficult, they should be treated with some caution. In Data Table 7-3-10 are shown the cross sections for the formation of the $1s^2 \, 2s^2 \, 2p^5 \, 3s^2 \, 3p^6$ state of Ar^+ by collisionally induced removal of $2p$ electron from ground state Ar. The second series of measurements (Data Table 7-3-11) show the cross section for the simultaneous removal of a $2p$ and $3p$ electron from Ar to form the $1s^2 \, 2s^2 \, 2p^5 \, 3s^2 \, 3p^5$ excited state of Ar^{2+}. These measurements are level-excitation cross section while the results of light emission measurements (reviewed above) are all line emission data. The data in the original publication have been revised in absolute magnitude, and the corrected values are shown here. Volz estimates that the data are accurate only within an order of magnitude.[289]

C. H_2^+ Impact

Measurements on the emission cross sections of three Ar II lines induced by impact of H_2^+ ions on argon by Jaecks et al.[127] are shown in Data Tables 7-3-12 through 7-3-14. The data are estimated to have systematic errors less than $\pm 30\%$ and a reproducibility of $\pm 10\%$. These measurements were carried out in precisely the same manner as the work on the same lines under H^+ impact, therefore the relative accuracy between data for the two projectiles should be very good.

D. He^+ Impact

The majority of the information comes from the group of de Heer, van Eck, Jaecks and co-workers.[203, 76, 272, 124] Thomas and Gilbody[71] also provide a measurement on a single line of Ar II. The very low energy work by Lipeles[204] and by Schlumbohm[205] is not reproduced here. [See Sections 7-2-A(8) and 7-2-A(10)].

1. **Emission of Lines of Ar I.** De Heer, Jaecks, and co-workers have studied the emission of two resonance lines of Ar I,[203, 272] 1067- and 1048-Å. In principle, the observed intensity of the emission should be greatly affected by absorption of resonance photons. De Heer et al.,[203] upon investigating intensity as a function of pressure in the range 10^{-4} to 2×10^{-3} torr, found only a slight departure from linearity at 10^{-3} torr; this increased with higher pressures. Thus the situation up to 10^{-3} torr would appear to be in accordance with the operational emission cross section definition (see Section 2-1) that requires that such linearity be demonstrated. When the departure from linearity occurs, however, it is in the opposite direction from that expected for an absorption phenomenon. Moreover, it becomes significant at much higher pressures than was expected. This leads to the suggestion that there is some additional secondary populating mechanism and that the apparent linearity of the emission with pressure is due to a canceling-out effect between the two processes. Govertsen and Anderson[291] present evidence that absorption of resonance photons in argon may become significant at pressures as low as 2×10^{-6} torr; also at pressures above 4×10^{-5} torr there are indications of linear behavior that must represent cancellation of two or more nonlinear processes. It is not clear whether or not emission induced by heavy particle impact should behave in the same manner as the electron impact experiments of Govertsen and Anderson: the mean velocity with which the excited targets recoil will be different for these two cases, and the behavior of absorption will not necessarily be the same. Undoubtedly, further studies should be made of this problem.

De Heer et al. present an "apparent" emission cross section measured at a target pressure of 10^{-3} torr. Since the available evidence indicates that the operating conditions of the experiment are in accordance with the operational definition of emission cross section, the data is retained here in Data Table 7-3-15; nevertheless these should be regarded with caution since there is the suspicion that there are two or more competing secondary processes taking place. In the event that such processes are present, they will probably cause changes in both the absolute magnitudes of the cross sections and in the relative dependence on energy. The nominal systematic error in these data does not exceed $\pm 100\%$, and the reproducibility is estimated at $\pm 10\%$.

2. **Emission of Lines of Ar II.** The available data include ten measurements on vacuum ultraviolet lines and fifteen on visible emissions; the latter emissions tend to feed the upper levels of the vacuum ultraviolet transitions, therefore constituting the cascade contribution to these lines. The question of cascade has an important

bearing on the understanding of the fundamental excitation processes and will be considered later. It will be convenient to divide the discussion into two topics: vacuum ultraviolet lines and visible lines.

De Heer et al.[203] provide measurements on eight vacuum ultraviolet lines, five of them over a range of energies from 0.3- to 10-keV and three at the single energy of 2-keV. All of these data are included in the tables in the regular manner, and all of the measurements at the single energy of 2-keV are repeated in Table 7-1 in connection with a discussion on cascade. Van Eck et al.[76] also provide measurements on two ultraviolet lines; these are in agreement with the separate measurements by de Heer et al.[203] on these same lines. Some of the measurements reproduced here represent the simultaneous measurement of two or more lines; in all such cases the precautions for dealing with the simultaneous measurements of two or more different lines (Section 3-4-E) were omitted. All these ultraviolet measurements may have a systematic error of up to 100%.

The majority of the visible light emission measurements have been provided by Jaecks et al.[124] with some small contributions by Thomas and Gilbody.[71] Jaecks et al. have shown that there are considerable similarities of behavior for lines emanating from parent levels that have the same multiplicities. In presenting these measurements in Data Tables 7-3-22 through 7-3-28, the data are arranged in groups of lines that exhibit similar behaviors. The origin of the emissions are unambiguous except in the case of the 4431-Å line measured by Thomas and Gilbody[71] which probably includes components of the 4431-Å ($4p\,^4P^0_{2\,1/2} \rightarrow 3d\,^4D_{2\,1/2}$) and 4430-Å ($4p\,^4D^0_{1\,1/2} \rightarrow 4s\,^4P_{1/2}$) lines.

In order to come to some understanding of the mechanism by which the excited states are created, it is necessary to examine the emission measurements and attempt to derive a cross section for the formation of specific excited states. To do this one must assess both the branching in decay and also the possible contributions of cascade. It would be very difficult to do this for the case of the Ar II emissions due to the complexity of the spectrum, the large numbers of decay paths, and also the absence of any accurate information on branching ratios. Jaecks et al.[124] have made an approximate estimate of cascade for a number of levels. The cascade contribution to the lines that appear in the vacuum ultraviolet will generally give rise to emission lines in the visible spectrum. By measuring the cross sections for emission of visible lines, one may estimate the cross section for cascade population of an excited state. By measurement of a vacuum ultraviolet emission, one arrives at the total cross section for the formation of an excited state, and hence by subtraction of cascade, at the cross section for the formation of an excited level by direct collisional excitation. The results of the comparison are shown in Table 7-1 that is drawn from the work of de Heer et al.[203] This shows ultraviolet line-emission cross sections with estimates of cascade based on measurements of visible line emission. It should be borne in mind that the absolute accuracy of the visible measurements is of the order of $\pm 30\%$ and of the ultraviolet data, $\pm 100\%$. Any attempt to determine level-excitation cross sections from these data would be very inaccurate. Nevertheless, the data in Table 7-1 show clearly that some ultraviolet lines are dominated by cascade, and the cross section for direct excitation of the level by collision is small compared to that for formation by cascade from higher levels. Based on this limited series of observations, it would appear that $4s'\,^2D$ and $3d'\,^2D$ levels are populated by cascade while the $3p^6\,^2S$, $3d\,^2D$, and $3d'\,^2D$ levels are populated primarily by direct collisional excitation.

TABLE 7-1   Ar II Emission Cross Sections in Units 10^{-17} cm^2 at 2-keV He$^+$ Impact Energy Compared to Cascade Contribution (de Heer et al.)[203]

Wave-length (Å)	Transition	Cross Section	Wavelength (Å) of Lines Contributing to Cascade				Cascade Cross Section
679	$4s'\ ^2D_{1\,1/2}-3p^5\ ^2P^0_{1/2}$	2.02	4131,	4237,	4043		0.67
673	$4s'\ ^2D_{1\,1/2}-3p^5\ ^2P^0_{1\ 1/2}$	0.85^a	4035,	4589			
672	$4s'\ ^2D_{1\,1/2}-3p^5\ ^2P^0_{1\ 1/2}$	2.90	4277,	4072,	4080,	4609	2.32
671	$3d\ ^2D_{1\,1/2}-3p^5\ ^2P^0_{1/2}$		4637,	4610,	4277		
723	$4s\ ^2P_{1/2}-3p^5\ ^2P^0_{1/2}$						
731	$4s\ ^2P_{1\,1/2}-3p^5\ ^2P^0_{1\ 1/2}$	1.08^b	4545,	4658,	4880,	4727	0.74
726	$4s\ ^2P_{1/2}-3p^5\ ^2P^0_{1/2}$		4376,	4765,	4889,	4965	
718	$4s\ ^2P_{1/2}-3p^5\ ^2P^0_{1\ 1/2}$		4579				
662	$3d\ ^2D_{1\,1/2}-3p^5\ ^2P^0_{1\ 1/2}$	5.11					small
579 580 583	$3d'\ ^2D-3p^5\ ^2P^0$	7.16	5141,	5176,	4732,	4481	0.15
920	$3p^6\ ^2S_{1/2}-3p^5\ ^2P^0_{1\ 1/2}$	3.52					small
932	$3p^6\ ^2S_{1/2}-3p^5\ ^2P^0_{1/2}$	1.90					small

^{a} Estimated.
^{b} Correction of originally published value.[213]

E. Ne$^+$ Impact

Thomas and Gilbody[71] have carried out measurements on two lines of Ar II that are shown in Data Tables 7-3-29 and 7-3-30. The 4610-Å line is identified as the $4p'\ ^2F^0_{3\ 1/2} \rightarrow 4s'\ ^2D_{2\ 1/2}$, transition. The 4431-Å line must be an unresolved mixture of the 4431-Å ($4p\ ^4P^0_{2\ 1/2} \rightarrow 3d\ ^4D_{2\ 1/2}$) and 4430-Å ($4p\ ^4D^0_{1\ 1/2} \rightarrow 4s\ ^4P_{1/2}$) emissions.
Data by Schlumbhom[205, 228] is not reproduced here due to the poor spectroscopic resolution [see Section 7-3-A(10)].

F. Ar$^+$ Impact

In this case, both projectile and target emit the same spectra. Most of the available data represent the total cross sections for emission of a line with no determination of which particle is excited. In general, it must be supposed that emission of an Ar II line represents the sum of the following two processes:

$$Ar^+ + Ar \rightarrow Ar^{+*} + Ar \qquad (7-5)$$

$$\rightarrow [Ar^+ + e] + Ar^{+*}. \qquad (7-6)$$

The only exception to this rule is in the work of Robinson and Gilbody[105] in which the Doppler shift of wavelength of emission from the projectile was used to facilitate the separate measurement of target and projectile emissions. Where a measurement may include contributions from both the projectile and the target, it is necessary to assess the experimental procedures against both the criteria for the validity of measurements of emission from fast particles as well as those for target emission.

The emission from a projectile traversing a target gas will vary with the penetration due to the finite lifetime of the excited state (see Section 2-2-B). As penetration increases, the emission tends asymptotically to a constant intensity. A measurement of an emission cross section is valid only when made at a point where the emission has become invariant with distance. Emission from direct excitation of a state reaches two-thirds of its maximum value at a distance equal to the product of the excited-state lifetime and projectile velocity. Lifetimes for the states from which emission measurements have been made are of the order 10 nsec.[292] The maximum value of the product of velocity and lifetime applicable to any of the experiments considered here is 1.4 cm. It would appear that all experiments have been made at sufficiently large penetrations so that variation of emission from direct excitation was negligible. It is possible that the cascade population of the excited states will be significant. This also will vary with penetration, but in a more complicated manner. It is impossible to assess whether or not the experiments reviewed here have been made under conditions where the cascade was invariant with penetration.

In Section 3-4-F it was noted that the emission from a projectile will be Doppler-broadened by an amount which depends on the projectile velocity and the solid angle of the detection system. A 5000-Å line emitted by a 400-keV Ar^+ ion will have a Doppler width of 5-Å when detected by a monochromator with a 10° acceptance angle. The pass band of the monochromator may be less than this causing loss of signal. Moreover, the efficiency of transmission varies with wavelengths over the pass band of a monochromator. This problem of Doppler broadening has been neglected in all the work reviewed in this section. The magnitude of the resulting error cannot be predicted in retrospect, but it may be appreciable for high energy measurements and probably insignificant for low energy data.

The first work in this area was done by Sluyters et al.[103, 277] in the energy range 5- to 24-keV. Measurements included emission in the neutral and singly ionized spectra. In Section 7-3-A(2) these data are severely criticized and the conclusion is drawn that the data are of such poor quality that they should be omitted from the monograph altogether.

Data Table 7-3-31, showing measurements on the emission of the 4431-Å line of Ar II, includes measurements by Robinson and Gilbody[105] on the separate cross sections for emission of the line from projectile and target as well as the total cross section for emission and measurements by Thomas and Gilbody[71] on only the total cross section for emission. The identification of the 4431-Å line is ambiguous. Both sets of authors claim that it is the 4431-Å $(4p\ ^4P_{2\ 1/2} \to 3d\ ^4D_{2\ 1/2})$ transition although neither employed sufficient resolution to exclude the 4430-Å $(4p\ ^4D^0_{1\ 1/2} \to 4s\ ^4P_{1/2})$ line. Possibly, the measurement is a mixture of both.

Data Tables 7-3-32 through 7-3-34 show measurements on the emission of visible Ar II lines by Thomas and Gilbody[71] and also by Neff[104] The work by Neff involves very poor spectral resolution, and the data may not represent the emission of a single line. Also the absolute magnitudes of the data by Neff may be very inaccurate.

G. Cs^+ Impact

The data on this case by Andersen et al.[114] was measured with very poor spectral resolution and is likely to be inaccurate; that data is not reproduced here [see Section 7-3A(9)].

H. Data Tables

1. Cross Sections for the Emission of Ar I Lines Induced by Impact of H$^+$ on an Ar Target

7-3-1 7635-Å Line ($4p[1\frac{1}{2}]_2 \rightarrow 4s[1\frac{1}{2}]_2^0$).

7-3-2 4201-Å Line ($5p[\frac{1}{2}]_1 \rightarrow 4s[1\frac{1}{2}]_2^0$).

2. Cross Sections for the Emission of Ar II Lines Induced by Impact of H$^+$ on an Ar Target

7-3-3 920-Å Line ($3p^6\ ^2S_{1/2} \rightarrow 3p^5\ ^2P_{1\ 1/2}^0$).

7-3-4 932-Å Line ($3p^6\ ^2S_{1/2} \rightarrow 3p^5\ ^2P_{1/2}^0$).

7-3-5 4278-Å Line ($4p'\ ^2P_{1\ 1/2}^0 \rightarrow 4s'\ ^2D_{2\ 1/2}$).

7-3-6 4431-Å Line ($4p\ ^4P_{2\ 1/2}^0 \rightarrow 3d\ ^4D_{2\ 1/2}$).

7-3-7 4610-Å Line ($4p'\ ^2F_{3\ 1/2}^0 \rightarrow 4s'\ ^2D_{2\ 1/2}$).

7-3-8 4765-Å Line ($4p\ ^2P_{1\ 1/2}^0 \rightarrow 4s\ ^2P_{1/2}$).

7-3-9 4806-Å Line ($4p\ ^4P_{2\ 1/2}^0 \rightarrow 4s\ ^4P_{2\ 1/2}$).

3. Cross Sections for Excitation of Ar$^+$ and Ar^{2+} States Induced by the Impact of H$^+$ on an Ar Target

7-3-10 $1s^2\ 2s^2\ 2p^5\ 3s^2\ 3p^6$ State of Ar$^+$

7-3-11 $1s^2\ 2s^2\ 2p^5\ 3s^2\ 3p^5$ State of Ar^{2+}

4. Cross Sections for the Emission of Ar II Lines Induced by Impact of H$_2^+$ on an Ar Target

7-3-12 4610-Å Line ($4p'\ ^2F_{3\ 1/2}^0 \rightarrow 4s'\ ^2D_{2\ 1/2}$).

7-3-13 4765-Å Line ($4p\ ^2P_{1\ 1/2}^0 \rightarrow 4s\ ^2P_{1/2}$).

7-3-14 4806-Å Line ($4p\ ^4P_{2\ 1/2}^0 \rightarrow 4s\ ^4P_{2\ 1/2}$).

5. Cross Sections for the Emission of Ar I Lines Induced by Impact of He$^+$ on an Ar Target

7-3-15 (A) 1067-Å Line ($4s[1\frac{1}{2}]_1^0 \rightarrow 3p^6\ ^1S_0$); (B) 1048-Å ($4s'[\frac{1}{2}]_1^0 \rightarrow 3p^6\ ^1S_0$).

6. Cross Sections for Emission of Ar II Lines Induced by Impact of He$^+$ on an Ar Target

7-3-16 920-Å Line ($3p^6\ ^2S_{1/2} \rightarrow 3p^5\ ^2P_{1\ 1/2}^0$).

7-3-17 932-Å Line ($3p^6\ ^2S_{1/2} \rightarrow 3p^5\ ^2P_{1/2}^0$).

7-3-18 (A) 679-Å Line ($4s'\ ^2D_{1\ 1/2} \rightarrow 3p^5\ ^2P_{1/2}^0$); (B) 673-Å Line ($4s'\ ^2D_{1\ 1/2}^0 \rightarrow 3p^5\ ^2P_{1/2}^0$); (C) 672- and 671-Å Lines ($4s'\ ^2D_{2\ 1/2} \rightarrow 3p^5\ ^2P_{1\ 1/2}$ and $3d\ ^2D_{1\ 1/2} \rightarrow 3p^5\ ^2P_{1/2}^0$) measured together; (D) sum of 679-, 673-, 672- and 671-Å lines measured together.

7-3-19 718-, 723-, 726-, 731-Å Lines (Comprising all four components of the $4s\ ^2P \rightarrow 3p^5\ ^2P^0$ multiplet) measured together.

7-3-20 579-, 580-, 583-Å Lines (comprising all three components of the $3d'\ ^2D \rightarrow 3p^5\ ^2P^0$ multiplet) measured together.

7-3-21 622-Å Line ($3d\ ^2D_{2\ 1/2} \rightarrow 3p^5\ ^2P_{1\ 1/2}^0$).

7-3-22 (A) 4765-Å Line ($4p\ ^2P_{1\ 1/2}^0 \rightarrow 4s\ ^2P_{1/2}$); (B) 4727-Å Line ($4p\ ^2D_{1\ 1/2}^0 \rightarrow 4s\ ^2P_{1\ 1/2}$); (C) 4658-Å Line ($4p\ ^2P_{1/2}^0 \rightarrow 4s\ ^2P_{1\ 1/2}$); (D) 4965-Å Line ($4p\ ^2D_{1\ 1/2}^0 \rightarrow 4s\ ^2P_{1/2}$).

7-3-23 (A) 4348-Å Line ($4p\ ^4D_{3\ 1/2} \rightarrow 4s\ ^4P_{2\ 1/2}$); (B) 4806-Å Line ($4p\ ^2P_{1/2}^0 \rightarrow 4s\ ^4P_{1\ 1/2}$); (C) 4848-Å Line ($4p\ ^4P_{2\ 1/2}^0 \rightarrow 4s\ ^4P_{2\ 1/2}$).

7-3-24 4053-Å Line ($4p''\ ^2P_{1\ 1/2}^0 \rightarrow 4s''\ ^2S_{1/2}$).

7-3-25 (A) 4610-Å Line ($4p'\ ^2F_{3\ 1/2}^0 \rightarrow 4s'\ ^2D_{2\ 1/2}$); (B) 4610-Å Line ($4p'\ ^2F_{3\ 1/2}^0 \rightarrow 4s'\ ^2D_{2\ 1/2}$); (C) 4278-Å Line ($4p'\ ^2P_{1\ 1/2}^0 \rightarrow 4s'\ ^2D_{2\ 1/2}$).

7-3-1
DUFAY ET AL. (125)

ENERGY (KEV)	CROSS SECTION (SQ. CM)
4.00E 01	8.80E-18
5.00E 01	7.60E-18
6.00E 01	6.25E-18
7.00E 01	6.15E-18
8.00E 01	5.30E-18
9.00E 01	4.90E-18
1.10E 02	4.40E-18
1.50E 02	3.20E-18
2.00E 02	2.55E-18
3.00E 02	1.70E-18
4.00E 02	1.20E-18
5.00E 02	1.05E-18
6.00E 02	9.50E-19

7-3-2
THOMAS ET AL. (127)

ENERGY (KEV)	CROSS SECTION (SQ. CM)
1.50E 02	1.07E-19
2.00E 02	9.54E-20
2.50E 02	8.44E-20
3.00E 02	7.41E-20
3.50E 02	6.66E-20
4.00E 02	5.98E-20
4.50E 02	5.36E-20
5.00E 02	4.94E-20
6.00E 02	4.36E-20
7.00E 02	3.92E-20
8.00E 02	3.44E-20
9.00E 02	3.08E-20
1.00E 03	2.76E-20

7-3-3
VAN ECK ET AL. (76)

ENERGY (KEV)	CROSS SECTION (SQ. CM)
1.00E 01	9.76E-18
1.50E 01	1.22E-17
2.00E 01	1.67E-17
2.50E 01	1.95E-17
3.00E 01	1.88E-17
3.50E 01	1.90E-17

7-3-4
VAN ECK ET AL. (76)

ENERGY (KEV)	CROSS SECTION (SQ. CM)
1.50E 01	5.46E-18
2.00E 01	7.42E-18
2.50E 01	8.30E-18
3.00E 01	8.88E-18
3.50E 01	9.10E-18

7-3-5
DUFAY ET AL. (125)

ENERGY (KEV)	CROSS SECTION (SQ. CM)
2.50E 01	3.70E-20
4.00E 01	4.25E-20
5.00E 01	4.65E-20
6.00E 01	4.50E-20
7.00E 01	4.25E-20
8.00E 01	3.40E-20
1.00E 02	3.30E-20
1.20E 02	2.40E-20
1.60E 02	2.20E-20
2.00E 02	1.10E-20
3.00E 02	1.00E-20
4.00E 02	6.70E-21
5.00E 02	5.00E-21

7-3-6
THOMAS ET AL. (127)

ENERGY (KEV)	CROSS SECTION (SQ. CM)
1.50E 02	8.01E-20
2.00E 02	6.59E-20
2.50E 02	5.22E-20
3.00E 02	4.26E-20
3.50E 02	3.74E-20
4.00E 02	3.26E-20
4.50E 02	2.88E-20
5.00E 02	2.62E-20
6.00E 02	2.16E-20
7.00E 02	1.91E-20
8.00E 02	1.65E-20
9.00E 02	1.45E-20
1.00E 03	1.30E-20

7-3-7
JAECKS ET AL. (124)

ENERGY (KEV)	CROSS SECTION (SQ. CM)
2.00E 00	2.27E-19
2.50E 00	7.52E-19
3.00E 00	1.80E-18
3.50E 00	2.58E-18
4.00E 00	2.92E-18
4.50E 00	2.67E-18
5.00E 00	2.81E-18
5.50E 00	2.56E-18
6.00E 00	2.32E-18
7.00E 00	1.76E-18
8.00E 00	1.39E-18
8.50E 00	1.30E-18
9.00E 00	1.13E-18
1.00E 01	9.80E-19

7-3-8
JAECKS ET AL. (124)

ENERGY (KEV)	CROSS SECTION (SQ. CM)
2.00E 00	9.40E-20
2.50E 00	2.60E-19
3.00E 00	3.21E-19
3.50E 00	3.32E-19
4.00E 00	3.19E-19
4.50E 00	3.04E-19
5.00E 00	3.25E-19
6.00E 00	3.71E-19
7.00E 00	4.12E-19
8.00E 00	4.87E-19
9.00E 00	4.78E-19
1.00E 01	5.11E-19

7-3-9
JAECKS ET AL. (124)

ENERGY (KEV)	CROSS SECTION (SQ. CM)
4.00E 00	2.58E-20
6.00E 00	4.20E-20
7.00E 00	4.00E-20
8.00E 00	5.36E-20
1.00E 01	5.53E-20

7-3-10
VOLZ ET AL. (126)

ENERGY (KEV)	CROSS SECTION (SQ. CM)
1.25E 02	2.83E-19
1.50E 02	3.55E-19
1.63E 02	3.91E-19
1.75E 02	4.56E-19
2.00E 02	4.51E-19
2.25E 02	5.66E-19
2.50E 02	6.44E-19
2.75E 02	7.40E-19
3.00E 02	7.87E-19

7-3-11
VOLZ ET AL. (126)

ENERGY (KEV)	CROSS SECTION (SQ. CM)
1.25E 02	1.45E-19
1.50E 02	1.77E-19
1.63E 02	1.90E-19
1.75E 02	2.16E-19
2.00E 02	2.16E-19
2.25E 02	2.74E-19
2.50E 02	2.64E-19
2.75E 02	2.83E-19
3.00E 02	3.36E-19

7-3-12
JAECKS ET AL. (124)

ENERGY (KEV)	CROSS SECTION (SQ. CM)
1.00E 00	7.40E-20
2.00E 00	2.07E-19
4.00E 00	4.64E-19
6.00E 00	9.90E-19
7.00E 00	1.54E-18
8.00E 00	1.74E-18
1.00E 01	1.95E-18

7-3-13
JAECKS ET AL. (124)

ENERGY (KEV)	CROSS SECTION (SQ. CM)
2.00E 00	9.90E-20
4.00E 00	1.92E-19
6.00E 00	3.44E-19
8.00E 00	4.32E-19
1.00E 01	4.92E-19

7-3-14
JAECKS ET AL. (124)

ENERGY (KEV)	CROSS SECTION (SQ. CM)
2.00E 00	6.82E-21
4.00E 00	1.64E-20
6.00E 00	2.95E-20
8.00E 00	3.87E-20
1.00E 01	4.51E-20

7-3-15A
DE HEER ET AL. (203)

ENERGY (KEV)	CROSS SECTION (SQ. CM)
3.00E-01	2.45E-18
4.00E-01	2.71E-18
5.00E-01	2.98E-18
6.00E-01	2.67E-18
8.00E-01	2.35E-18
1.00E 00	2.41E-18
2.00E 00	2.08E-18
3.00E 00	2.15E-18
4.00E 00	1.95E-18
5.00E 00	1.95E-18
6.00E 00	2.00E-18
8.00E 00	2.12E-18
9.00E 00	1.92E-18

7-3-15B
DE HEER ET AL. (203)

ENERGY (KEV)	CROSS SECTION (SQ. CM)
6.00E 00	1.75E-18

7-3-16A
VAN ECK ET AL. (76)

ENERGY (KEV)	CROSS SECTION (SQ. CM)
5.00E 00	7.30E-17
7.50E 00	9.43E-17
1.00E 01	9.50E-17
1.50E 01	8.50E-17
2.00E 01	7.30E-17
2.50E 01	5.70E-17
3.00E 01	5.20E-17
3.50E 01	4.40E-17

7-3-16B
DE HEER ET AL. (203)

ENERGY (KEV)	CROSS SECTION (SQ. CM)
3.00E-01	9.75E-18
5.00E-01	1.00E-17
7.00E-01	1.16E-17
8.00E-01	2.05E-17
1.00E 00	1.84E-17
2.00E 00	3.52E-17
4.00E 00	6.94E-17
6.00E 00	9.50E-17
8.00E 00	1.04E-16
1.00E 01	9.86E-17

7-3-17A
VAN ECK ET AL. (76)

ENERGY (KEV)	CROSS SECTION (SQ. CM)
5.00E 00	4.90E-17
7.5CE 00	5.06E-17
1.00E 01	5.10E-17
1.50E 01	4.52E-17
2.00E 01	3.50E-17
2.50E 01	2.96E-17
3.00E 01	2.49E-17
3.50E 01	2.29E-17

7-3-17B
DE HEER ET AL. (203)

ENERGY (KEV)	CROSS SECTION (SQ. CM)
2.00E 00	1.90E-17

7-3-18A
DE HEER ET AL. (203)

ENERGY (KEV)	CROSS SECTION (SQ. CM)
2.00E 00	2.02E-17

7-3-18B
DE HEER ET AL. (203)

ENERGY (KEV)	CROSS SECTION (SQ. CM)
2.00E 00	8.50E-18

7-3-18C
DE HEER ET AL. (203)

ENERGY (KEV)	CROSS SECTION (SQ. CM)
2.00E 00	2.90E-17

7-3-18D
DE HEER ET AL. (203)

ENERGY (KEV)	CROSS SECTION (SQ. CM)
3.00E-01	8.20E-17
4.00E-01	1.07E-16
5.00E-01	1.17E-16
6.00E-01	1.18E-16
7.00E-01	1.45E-16
8.00E-01	1.17E-16
9.00E-01	1.15E-16
1.00E 00	9.80E-17
2.00E 00	6.06E-17
4.00E 00	4.65E-17
6.00E 00	4.28E-17
8.00E 00	3.34E-17
1.00E 01	3.10E-17

7-3-19
DE HEER ET AL. (203)

ENERGY (KEV)	CROSS SECTION (SQ. CM)
3.00E-01	1.86E-17
4.00E-01	1.85E-17
5.00E-01	1.53E-17
6.00E-01	1.53E-17
8.00E-01	1.28E-17
1.00E 00	1.28E-17
2.00E 00	7.74E-18
3.00E 00	8.25E-18
4.00E 00	8.05E-18
5.00E 00	7.16E-18

ENERGY (KEV)	CROSS SECTION (SQ. CM)
6.CCE 00	6.69E-18
7.00E 00	6.14E-18
8.0CE 00	5.21E-18
9.00E 00	4.96E-18

7-3-20
DE HEER ET AL. (203)

ENERGY (KEV)	CROSS SECTION (SQ. CM)
3.00E-01	2.44E-17
4.00E-01	3.12E-17
5.00E-01	3.38E-17
6.00E-01	4.06E-17
7.00E-01	4.86E-17
8.00E-01	5.50E-17
9.00E-01	5.98E-17
1.0CE 00	6.84E-17
2.00E 00	7.16E-17
4.00E 00	7.76E-17
6.00E 00	6.63E-17
8.00E 00	6.38E-17
1.00E 01	5.75E-17

7-3-21
DE HEER ET AL. (203)

ENERGY (KEV)	CROSS SECTION (SQ. CM)
3.00E-01	7.05E-17
4.0CE-01	7.37E-17
5.00E-01	7.20E-17
6.00E-01	7.24E-17
8.00E-01	6.82E-17
1.00E 00	5.54E-17
2.00E 00	5.11E-17
4.0CE 00	5.47E-17
6.00E 00	4.11E-17
7.00E 00	3.54E-17
8.00E 00	2.54E-17
9.CCE 00	2.11E-17

7-3-22A
JAECKS ET AL. (124)

ENERGY (KEV)	CROSS SECTION (SQ. CM)
3.00E-01	5.40E-18
5.00E-01	5.40E-18
7.00E-01	4.42E-18
1.00E 00	3.18E-18
1.50E 00	1.98E-18
2.00E 00	1.45E-18
4.0CE 00	9.30E-19
6.00E 00	9.60E-19
8.00E 00	8.10E-19
1.00E 01	8.30E-19
2.00E 01	7.00E-19
3.0CE 01	6.25E-19

7-3-22B
JAECKS ET AL. (124)

ENERGY (KEV)	CROSS SECTION (SQ. CM)
3.00E-01	2.22E-18
5.0CE-01	1.88E-18
7.00E-01	1.54E-18
1.0CE 00	1.30E-18
2.0CE 00	8.46E-19
4.00E 00	7.10E-19
6.00E 00	5.97E-19
8.00E 00	5.46E-19
1.00E 01	6.10E-19
1.50E 01	6.17E-19
2.00E 01	5.62E-19
2.50E 01	5.17E-19
3.00E 01	5.05E-19
3.50E 01	5.55E-19

7-3-22C
JAECKS ET AL. (124)

ENERGY (KEV)	CROSS SECTION (SQ. CM)
3.00E-01	2.19E-18
5.00E-01	2.21E-18
7.00E-01	1.88E-18
1.00E 00	1.38E-18
2.0CE 00	6.84E-19
4.00E 00	5.60E-19
6.00E 00	5.10E-19
7.00E 00	4.60E-19
8.00E 00	4.58E-19
9.00E 00	4.71E-19
1.00E 01	4.83E-19
1.50E 01	4.20E-19
2.00E 01	3.91E-19
2.50E 01	3.80E-19
3.00E 01	3.77E-19

7-3-22D
JAECKS ET AL. (124)

ENERGY (KEV)	CROSS SECTION (SQ. CM)
3.00E-01	1.20E-18
5.00E-01	1.05E-18
8.00E-01	7.62E-19
1.00E 00	5.42E-19
2.00E 00	3.74E-19
4.00E 00	3.14E-19
6.0CE 00	2.74E-19
8.00E 00	2.74E-19
1.00E 01	2.60E-19

7-3-23A
JAECKS ET AL. (124)

ENERGY (KEV)	CROSS SECTION (SQ. CM)
1.00E 00	2.08E-19
2.CCE 00	3.54E-19
4.00E 00	4.52E-19
6.CCE 00	4.15E-19
8.0CE 00	4.48E-19
1.00E 01	4.85E-19
1.75E 01	5.48E-19
2.00E 01	6.85E-19
3.00E 01	8.00E-19

7-3-23B
JAECKS ET AL. (124)

ENERGY (KEV)	CROSS SECTION (SQ. CM)
7.00E-01	1.08E-19
1.00E 00	1.25E-19
2.00E 00	2.10E-19
4.00E 00	2.60E-19
6.CCE 00	2.78E-19
8.00E 00	2.94E-19
1.00E 01	3.00E-19
1.50E 01	3.97E-19
2.00E 01	4.72E-19
2.50E 01	4.68E-19
3.00E 01	4.99E-19

7-3-23C
JAECKS ET AL. (124)

ENERGY (KEV)	CROSS SECTION (SQ. CM)
1.0CE 00	8.70E-20
2.00E 00	1.14E-19
4.00E 00	1.48E-19
6.00E 00	1.63E-19
8.00E 00	1.88E-19
1.00E 01	2.10E-19
1.50E 01	2.98E-19
3.0CE 01	3.70E-19

7-3-24
JAECKS ET AL. (124)

ENERGY (KEV)	CROSS SECTION (SQ. CM)
1.00E 00	9.05E-20
2.00E 00	1.22E-19
4.00E 00	2.28E-19
6.0CE 00	2.70E-19
8.00E 00	2.40E-19
1.00E 01	2.24E-19

7-3-25A
JAECKS ET AL. (124)

ENERGY (KEV)	CROSS SECTION (SQ. CM)

ENERGY (KEV)	CROSS SECTION (SQ. CM)
3.00E-01	9.45E-18
5.00E-01	9.11E-18
7.0CE-01	9.11E-18
1.00E 00	8.17E-18
2.0CE 00	6.62E-18
4.00E 00	6.23E-18
6.C0E 00	3.79E-18
8.0CE 00	2.93E-18
1.00E 01	2.88E-18
2.00E 01	2.10E-18
3.00E 01	1.50E-18

7-3-25B
THOMAS ET AL. (71)

ENERGY (KEV)	CROSS SECTICN (SQ. CM)
1.50E 02	1.44E-18
1.75E 02	1.35E-18
2.10E 02	1.19E-18
2.50E 02	1.09E-18
2.75E 02	1.21E-18
3.0CE 02	7.90E-19
3.25E 02	1.07E-18

7-3-25C
JAECKS ET AL. (124)

ENERGY (KEV)	CROSS SECTION (SQ. CM)
5.00E-01	2.16E-18
7.00E-01	2.78E-18
1.00E 00	3.40E-18
2.00E 00	2.80E-18
4.0CE 00	2.46E-18
6.C0E 00	2.37E-18
8.00E 00	2.11E-18
1.00E 01	2.07E-18
1.50E 01	1.60E-18
2.00E 01	1.36E-18
2.50E 01	1.17E-18
3.00E 01	1.15E-18
3.5CE 01	1.03E-18

7-3-26
JAECKS ET AL. (124)

ENERGY (KEV)	CROSS SECTION (SQ. CM)
1.00E 00	7.75E-20
2.0CE 00	1.23E-19
4.00E 00	2.36E-19
6.C0E 00	2.86E-19
8.0CE 00	2.44E-19
1.00E 01	2.07E-19

7-3-27
JAECKS ET AL. (124)

ENERGY (KEV)	CROSS SECTION (SQ. CM)
2.0CE 00	1.21E-20
4.00E 00	4.15E-20
6.0CE 00	5.95E-20
8.00E 00	7.95E-20
1.00E 01	9.13E-20
1.5CE 01	1.24E-19
2.00E 01	1.64F-19
2.50E 01	1.82E-19
3.00E 01	1.95E-19

7-3-28
THOMAS ET AL. (71)

ENERGY (KEV)	CROSS SECTICN (SQ. CM)
2.00E 02	9.60E-19
2.61E 02	9.00E-19
3.00E 02	7.80E-19
3.25E 02	6.50E-19
3.52E 02	8.30E-19

7-3-29
THOMAS ET AL. (71)

ENERGY (KEV)	CROSS SECTION (SQ. CM)
1.05E 02	4.77E-18
1.30E 02	5.40E-18
1.55E 02	5.32E-18
1.80E 02	5.07E-18
2.12E 02	5.18E-18
2.50E 02	4.68E-18
2.80E 02	3.92E-18
3.00E 02	3.72E-18
3.27E 02	3.35E-18
3.60E 02	3.83E-18
3.80E 02	3.19E-18

7-3-30
THOMAS ET AL. (71)

ENERGY (KEV)	CROSS SECTION (SQ. CM)
1.00E 02	2.11E-18
1.29E 02	2.14E-18
1.54E 02	2.32E-18
1.78E 02	2.25E-18
2.05E 02	2.17E-18
2.17E 02	2.18E-18
2.56E 02	2.06E-18
2.75E 02	2.22E-18
3.00E 02	1.94E-18
3.25E 02	2.01E-18
3.50E 02	1.72E-18
3.8CE 02	1.88E-18

7-3-31A
THOMAS ET AL. (71)

ENERGY (KEV)	CROSS SECTION (SQ. CM)
1.10E 02	6.15E-17
1.30E 02	7.51E-17
1.50F 02	6.73E-17
2.05E 02	9.16E-17
2.35E 02	8.02E-17
2.65E 02	8.16E-17
3.00E 02	7.87E-17
3.25E 02	8.59E-17

7-3-31B
ROBINSON ET AL. (105)

ENERGY (KEV)	CROSS SECTICN (SQ. CM)
1.30E 02	5.72E-18
1.70E 02	7.00E-18
2.00E 02	6.70E-18
2.50E 02	7.39E-18
3.00E 02	7.60E-18
3.50E 02	7.43E-18
3.60E 02	7.47E-18
4.00F 02	7.13E-18
4.20E 02	6.92E-18

7-3-31C
ROBINSON ET AL. (105)

ENERGY (KEV)	CROSS SECTICN (SQ. CM)
1.30E 02	3.58E-18
1.70E 02	4.40E-18
2.00E 02	4.35E-18
2.50E 02	4.61E-18
3.00E 02	4.61E-18
3.50E 02	4.35E-18
3.60E 02	4.44E-18
4.0CE 02	4.35E-18
4.20E 02	4.27E-18

7-3-31D
ROBINSON ET AL. (105)

ENERGY (KEV)	CROSS SECTICN (SQ. CM)
1.30E 02	2.14E-18
1.70E 02	2.60E-18
2.00E 02	2.35E-18
2.50E 02	2.78E-18
3.00E 02	2.99E-18
3.50E 02	3.08E-18
3.60E 02	3.03E-18
4.00E 02	2.78E-18
4.20E 02	2.65E-18

7-3-32A
THOMAS ET AL. (71)

ENERGY (KEV)	CROSS SECTION (SQ. CM)
1.15E 02	7.30E-18
1.55E 02	1.06E-17
1.80E 02	1.14E-17
2.00E 02	1.25E-17
2.40E 02	1.76E-17
2.60E 02	1.80E-17
2.80E 02	1.94E-17
3.12E 02	1.97E-17
3.33E 02	1.95E-17
3.50E 02	1.24E-17
3.60E 02	1.10E-17

7-3-32B
NEFF ET AL. (104)

ENERGY (KEV)	CROSS SECTION (SQ. CM)
4.00E-01	4.00E-19
6.00E-01	4.30E-19
8.00E-01	5.20E-19
1.00E 00	6.20E-19
1.50E 00	7.90E-19
2.00E 00	9.40E-19
2.50E 00	1.05E-18
3.0CE 00	1.15E-18

7-3-33A
NEFF ET AL. (104)

ENERGY (KEV)	CROSS SECTION (SQ. CM)
4.00E-01	1.40E-19
1.00E 00	2.60E-19
2.00E 00	3.20E-19

7-3-33B
NEFF ET AL. (104)

ENERGY (KEV)	CROSS SECTION (SQ. CM)
4.00E-01	2.60E-19
1.00E 00	4.10E-19
2.00E 00	6.30E-19
3.00E 00	8.00E-19

7-3-33C
NEFF ET AL. (104)

ENERGY (KEV)	CROSS SECTION (SQ. CM)
4.00E-01	3.20E-19
1.00E 00	4.60E-19
2.00E 00	8.00E-19
3.00E 00	1.05E-18

7-3-34A
THOMAS ET AL. (71)

ENERGY (KEV)	CROSS SECTION (SQ. CM)
3.00E 01	8.30E-19
4.00E 01	8.35E-19
5.00E 01	8.85E-19
6.00E 01	9.21E-19
7.00E 01	9.94E-19
8.00E 01	9.72E-19
9.00E 01	9.35E-19
9.50E 01	9.45E-19
1.00E 02	9.40E-19
1.08E 02	1.45E-18
1.35E 02	1.15E-18
1.80E 02	1.48E-18
2.05E 02	1.64E-18
2.54E 02	1.80E-18
2.57E 02	1.82E-18
2.80E 02	1.85E-18
3.10E 02	1.46E-18
3.33E 02	1.36E-18
3.45E 02	1.33E-18
3.45E 02	1.47E-18

7-3-34B
THOMAS ET AL. (71)

ENERGY (KEV)	CROSS SECTION (SQ. CM)
1.20E 02	2.42E-18
1.35E 02	1.90E-18
2.18E 02	3.00E-18
2.33E 02	2.77E-18
2.60E 02	3.05E-18
3.00E 02	2.67E-18
3.88E 02	2.37E-18

7-4 EXCITATION OF KRYPTON

Studies of excitation of krypton have involved both experiments to detect radiation from the neutral atom and also from the singly charged ion. The processes may be described by the reaction equations:

$$X^+ + Kr \rightarrow [X^+] + Kr^* \tag{7-7}$$

$$X^+ + Kr \rightarrow [X^+ + e] + Kr^{+*}. \tag{7-8}$$

Here there is no information on the states of excitation or ionization of the systems contained within the square brackets. The quantitative work involves impact of H^+ and He^+ projectiles and measurements on emission of lines in the visible and vacuum ultraviolet spectral regions. All the quantitative work is due to the research group of de Heer and co-workers in Amsterdam.

For the Kr I and Kr II spectra, identification of lines and assignment of term designations is based on the "Ultraviolet Multiplet Table" by Moore,[67] the listing of atomic lines by Dieke,[293] and the original analysis of ion lines by de Bruin et al.[294]

In the following section (7-4-A), a brief review is given of the various experimental systems that have been used in these studies. This is followed by a discussion of the data obtained for excitation by impact of protons (Section 7-4-B), He$^+$ (Section 7-4-C), Ne$^+$ (Section 7-4-D), and finally the collection of the available data in tabular form (Section 7-4-E).

A. Experimental Arrangements

1. The Work of de Heer et al. The work of this group has extended from 1963 to the time of writing and has generated a total of four relevant papers. Three different optical arrangements used for these studies are labeled Systems I, II and III, respectively. For a detailed discussion of these systems and the other operating procedures of these experiments, the reader is referred to the review in Section 7-2-A(1), where they are discussed in detail in the context of studies of neon excitation.

For convenience, the coverage of the data and also the sources from which the information used in this monograph are drawn may be summarized as follows.

H$^+$ + Kr. Ultraviolet emissions induced in this reaction have been studied by van Eck et al.[76] using the apparatus referred to as System II.

He$^+$ + Kr. This study has been reported in four separate papers. The first, by van Eck et al.[76] covers the ultraviolet Kr II emission using the apparatus known as System II. A conference paper by de Heer et al.[272] relating measurements on vacuum ultraviolet and visible lines was later completely superseded by two full reports, one of which covered visible emissions observed using System I and the other vacuum ultraviolet emissions utilizing System III.[203]

Throughout this work, the monochromators were set to observe emission at 90° to the beam direction. The polarization of the emission and its possible influence on the accuracy of cross section measurements was ignored. The absolute accuracy of data on emission of visible spectral lines was estimated as ± 30% while that for vacuum ultraviolet was ± 100%. Reproducibility was within ± 10%.

2. The Work of Lipeles et al. Lipeles and co-workers[204] have made a brief study of the emission induced in the collision He$^+$ + Kr at rather low energies, 5- to 500-eV. Measurements were made of emission in rather broad wavelength intervals, and no attempt was made to determine emission cross sections for resolved spectral lines. This work has been discussed in detail in the context of excitation of neon, and it was suggested there that the data were peculiar to the apparatus utilized and, therefore, of no fundamental value here [see Section 7-2-A(3)]. Consequently, this work is excluded from further discussion.

3. The Work of Schlumbohm. Schlumbohm[205] has carried out some measurements of emission induced in the impact of He$^+$ and Ne$^+$ on krypton in the energy range 5- to 300-eV. The apparatus has been considered in detail in Section 7-3-A(10). Considerable attention was directed at ensuring that the projectile beam was in its ground state and that the energy distribution of the projectiles should be small. The detection system relied on rather broad-band filters so that the measured cross sections were essentially a weighted sum over a number of spectral lines with the weighting factor being provided by the instrumental transmission factor. Consequently, the data do not provide information on the behavior of distinct transitions and are excluded from the present compilations.

B. Proton Impact

The study of this case is restricted to measurements of two lines of Kr II in the vacuum ultraviolet carried out by van Eck et al.[76] The emission cross sections for the 917-Å $(4p^6\ ^2S_{1/2} \rightarrow 4p^5\ ^2P^0_{1\ 1/2})$ and 965-Å $(4p^6\ ^2S_{1/2} \rightarrow 4p^5\ ^2P^0_{1/2})$ lines are shown in Data Tables 7-4-1 and 7-4-2, respectively. It is observed that these two lines exhibit a somewhat different dependence on energy of impact. Since both lines have a common parent level $(4p^6\ ^2S_{1/2})$, this observation must be in error. No explanation is given by the authors for this difference. It was estimated that the absolute accuracy of these determinations was $\pm 100\%$, while the reproducibility is about $\pm 10\%$.

C. He$^+$ Impact

A number of quantitative investigations of krypton emission have been produced by the group of de Heer and co-workers in Amsterdam. These will be discussed in detail. In addition, Lipeles et al.[204] have measured integrated intensities over broad wavelength intervals, and Schlumbohm[205] has made measurements on the emissions in a 40-Å wide interval. Since neither of these latter investigations provides information on specific transitions, they are not considered further here. The reader should note that these two investigations may be of considerable assistance in the interpretation of the collision mechanism, since they extend down to energies of 5-eV; this energy is below the threshold for onset of the emission.

The work of de Heer et al. did not include an assessment of the influence of excited states in the projectile beam. It is expected that the excited state content of the projectile beam will be quite small at entry to the collision region. Most excited states will decay radiatively. The metastable state will be Stark mixed with the 2p state while it traverses the electric field in the accelerator. Radiative decay will take place from the 2p state. The remaining excited state component of the beam will be very small and probably will have an insignificant effect on the measured cross sections.

1. Emission of Lines of Kr I. The only work on this system is measurements of the 1236-Å $(5s[1\frac{1}{2}]^\circ \rightarrow 4p^6\ ^1S)$ and 1165-Å $(5s'[\frac{1}{2}]^\circ \rightarrow 4p^6\ ^1S)$ lines by de Heer et al.[203] Both of these lines should show strong resonance absorption effects since they are direct transitions to the ground state of Kr II. De Heer et al. have carried out extensive measurements of the dependence of light intensity on target pressure in the range 10^{-4} to 2×10^{-3} torr. If no second order effects are present, then the light intensity should be proportional to target pressure. If absorption of resonance radiation is occurring, the intensity should increase less rapidly than this. De Heer et al. find the surprising result that the intensity increases more rapidly than a linear relationship would allow. They suggest that absorption is probably taking place, but that some other second-order mechanism with a different pressure dependence is greatly exceeding the absorption effect. Further, it is suggested that this might be a process whereby helium atoms are formed by charge transfer and then undergo a second collision in which a target atom is excited. No quantitative arguments are provided to back up this suggestion.

De Heer et al. provide an apparent emission cross section measured at a target pressure of 10^{-3} torr. It is not suggested that this represents the true cross section for these lines. In the event that the suggested secondary mechanism involving the

excitation by neutral helium atoms proves correct, the error on the cross section measurement will be energy dependent. Then the measurements will not even represent the relative variation of cross section with impact energy. Consequently, to avoid confusion, these data on the apparent emission cross sections are omitted from the present data compilation.

A previous measurement of emission of the 1236-Å line by de Heer et al.[272] was superseded by the work discussed above.

2. Emission of Lines of Kr II. Measurements are available for thirteen lines in the far ultraviolet spectral regions and seven in the visible spectrum. In general, the ultraviolet lines provide transitions directly to the ground state, whereas the visible lines tend to feed the upper levels of these ultraviolet transitions by cascade. The influence of cascade has great importance in the understanding of the origin of the emission induced in this collision process.

De Heer et al.[203] have provided the majority of the measurements on ultraviolet lines; two independent measurements by van Eck et al.[76] are in good agreement with the work of de Heer where comparisons are possible. All the data are shown in the Data Table (7-4-3 through 7-4-9) and all measurements at 2-keV energy are also shown in Table 7-2 as part of a discussion on cascade. In cases where two or more lines were measured simultaneously the precautions required for such measurements (Section 3-4-E) were omitted.

The work on visible lines is due to Jaecks et al.[124] and is shown in Data Tables 7-4-9 through 7-4-11. The data are grouped according to term designation of the upper state of the transition.

TABLE 7-2   Kr II Emission Cross Sections in Units 10^{-17} cm^2 at 2-keV He$^+$ Impact Energy, Compared to Contribution from Cascade (de Heer et al.[203])

Wave-length (Å)	Transition	Cross Section	Wavelength (Å) of Measured Lines Contributing to Cascade				Cascade Cross Section
844	$5s\ ^2P_{3/2}-s^2p^5\ ^2P^0_{3/2}$	0.96	4615,	4846,	4619,	4251	
865	$5s\ ^2P_{1/2}-s^2p^5\ ^2P^0_{1/2}$	0.47	4185,	6472,	5683,	5024	
			4610,	4825,	3151,	3275	1.02
			5218,	3200,	5523,	4762	
			4680,	3420,	3205		
869	$5s\ ^4P_{3/2}-s^2p^5\ ^2P^0_{3/2}$	1.10	4766,	4293,	3988,	5309	1.80
911	$5s\ ^4P_{3/2}-s^2p^5\ ^2P^0_{1/2}$	0.22[a]	5208,	4832,	3754,	3995	
891	$5s\ ^4P_{1/2}-s^2p^5\ ^2P^0_{1/2}$	0.57	4099,	4651,	4437		0.32
850	$5s\ ^4P_{1/2}-s^2p^5\ ^2P^0_{3/2}$	0.12[a]	4812,	4432,	5500,	4145	
782	$5s'\ ^2D_{3/2}-s^2p^5\ ^2P^0_{3/2}$	} 0.90	4088,	4109,	4577,	4691	2.36
784	$5s'\ ^2D_{3/2}-s^2p^5\ ^2P^0_{3/2}$		4057,	4422,	4065	4045	1.49
818	$5s'\ ^2D_{3/2}-s^2p^5\ ^2P^0_{1/2}$	Small	4633				
730	$4d\ ^2F_{5/2}-s^2p^5\ ^2P^0_{3/2}$	0.55					
669	$4d'\ ^2D_{5/2}-s^2p^5\ ^2P^0_{3/2}$	1.16					Small
965	$4p^6\ ^2S_{1/2}-s^2p^5\ ^2P^0_{1/2}$	5.0					Small
917	$4p^6\ ^2S_{1/2}-s^2p^5\ ^2P^0_{3/2}$	3.9					

[a] Estimated.

It has been repeatedly emphasized that the data presented for excitation of rare gases are in the form of emission cross sections. These cannot be readily converted into level-excitation cross sections because there is no accurate information on branching ratios. De Heer et al.[203] have made an attempt to assess the cascade contribution to the ultraviolet transitions and hence to estimate level-excitation cross sections. Many of the cascade transitions lie in the visible spectral regions and can be measured directly. In Table 7-2 the results of this investigation are summarized. The emission cross sections of the vacuum ultraviolet lines are given with an estimate of the cascade population of the upper state obtained by measuring directly the emission cross sections of the relevant cascade transitions. It must be emphasized that the absolute magnitudes of the emission cross sections are not of very high accuracy. An error of $\pm 100\%$ in the vacuum ultraviolet data and $\pm 30\%$ in the visible emission data is to be expected. This explains why some of the cascade populations have the appearance of being greater than the emission cross sections they are supposed to be influencing. Nevertheless, the comparisons do show that the cascade populations of the upper states represent an appreciable part of the total for many cases. It is clear that, in general, the emission cross sections for transitions to the ground state may be greatly affected by cascade and, therefore, do not represent the cross sections for exciting the parent level of the transition.

D.   Ne$^+$ Impact

A measurement has been made by Schlumbohm of the 3635-Å emission in a 40-Å interval for Ne$^+$ impact on Kr in the 5- to 220-eV energy range. These data do not provide information on the behavior of discrete transitions and are omitted from further discussion in this monograph.

E.   Data Tables

1.   Cross Sections for Emission of Kr II Lines Induced by Impact of H$^+$ on a Kr Target

    7-4-1   917-Å Line $(4p^6\ ^2S_{1/2} \to 4p^5\ ^2P^0_{1\ 1/2})$.

    7-4-2   965-Å Line $(4p^6\ ^2S_{1/2} \to 4p^4\ ^2P^0_{1/2})$.

2.   Cross Sections for Emission of Kr II Lines Induced by Impact of He$^+$ on a Kr Target

    7-4-3   (A) 844-Å Line $(5s\ ^2P_{1\ 1/2} \to 4p^5\ ^2P^0_{1\ 1/2})$; (B) 865-Å Line $(5s\ ^2P_{1/2} \to 4p^5\ ^2P^0_{1/2})$.

    7-4-4   (A) 891-Å Line $(5s\ ^4P_{1/2} \to 4p^5\ ^2P^0_{1/2})$; (B) 869-Å Line $(5s\ ^4P_{1\ 1/2} \to 4p^5\ ^2P^0_{1\ 1/2})$; (C) 911-Å Line $(5s\ ^4P_{1\ 1/2} \to 4p^5\ ^2P^0_{1/2})$; (D) 850-Å Line $(5s\ ^4P_{1/2} \to 4p^5\ ^2P^0_{1\ 1/2})$.

    7-4-5   784-Å and 782-Å Lines $(5s'\ ^2D_{1\ 1/2} \to 4p^5\ ^2P^0_{1\ 1/2}$ and $5s'\ ^2D_{2\ 1/2} \to 4p^5\ ^2P^0_{1\ 1/2})$ Measured Together.

    7-4-6   730-Å Line $(4d\ ^2F_{2\ 1/2} \to 4p^5\ ^2P^0_{1\ 1/2})$.

    7-4-7   669-Å Line $(4d'\ ^2D_{2\ 1/2} \to 4p^5\ ^2P^0_{1/2})$.

    7-4-8   965-Å Line $(4p^6\ ^2S_{1/2} \to 4p^5\ ^2P^0_{1/2})$.

    7-4-9   917-Å Line $(4p^6\ ^2S_{1/2} \to 4p^5\ ^2P^0_{1\ 1/2})$.

    7-4-10   (A) 4619-Å Line $(5p\ ^2D^0_{2\ 1/2} \to 5s\ ^2P_{1\ 1/2})$; (B) 4846-Å Line $(5p\ ^2P^0_{1/2} \to 5s\ ^2P_{1\ 1/2})$.

    7-4-11   (A) 4633-Å Line $(5p'\ ^2F^0_{2\ 1/2} \to 5s'\ ^2D_{1\ 1/2})$; (B) 4088-Å Line $(5p'\ ^2D^0_{2\ 1/2} \to 5s'\ ^2D_{2\ 1/2})$; (C) 4057-Å Line $(5p'\ ^2P^0_{1/2} - 5s'\ ^2D_{1\ 1/2})$.

    7-4-12   (A) 4355-Å Line $(5p\ ^4D^0_{3\ 1/2} \to 5s\ ^4P_{2\ 1/2})$; (B) 4832-Å Line $(5p\ ^4P^0_{1/2} \to 5s\ ^4P_{1\ 1/2})$.

7-4-1
VAN ECK ET AL. (76)

ENERGY (KEV)	CROSS SECTION (SQ. CM)
1.00E 01	3.72E-18
1.25E 01	5.76E-18
1.50E 01	8.66E-18
2.50E 01	1.49E-17
3.00E 01	1.59E-17
3.50E 01	1.67E-17

7-4-2
VAN ECK ET AL. (76)

ENERGY (KEV)	CROSS SECTION (SQ. CM)
1.00E 01	3.74E-18
1.25E 01	6.38E-18
1.50E 01	9.22E-18
1.75E 01	1.23E-17
2.00E 01	1.56E-17
2.25E 01	1.65E-17
2.50E 01	1.75E-17
2.75E 01	1.67E-17
3.00E 01	1.72E-17
3.25E 01	1.82E-17
3.50E 01	1.82E-17

7-4-3A
DE HEER ET AL. (203)

ENERGY (KEV)	CROSS SECTION (SQ. CM)
3.00E-01	1.87E-17
4.00E-01	1.64E-17
5.00E-01	1.43E-17
6.00E-01	1.32E-17
7.00E-01	1.26E-17
8.00E-01	1.19E-17
1.00E 00	1.07E-17
2.00E 00	9.61E-18
3.0CE 00	8.82E-18
4.00E 00	8.32E-18
5.00E 00	7.28E-18
6.00E 00	6.85E-18
7.00E 00	6.46E-18
8.0CE 00	5.99E-18
9.00E 00	5.73E-18

7-4-3B
DE HEER ET AL. (203)

ENERGY (KEV)	CROSS SECTION (SQ. CM)
2.00E 00	4.70E-18

7-4-4A
DE HEER ET AL. (203)

ENERGY (KEV)	CROSS SECTION (SQ. CM)
3.00E-01	1.07E-17
4.00E-01	8.50E-18
5.00E-01	7.21E-18
8.00E-01	7.12E-18
1.00E 00	7.12E-18
2.00E 00	5.65E-18
3.00E 00	6.03E-18
4.00E 00	5.78E-18
5.00E 00	6.07E-18
6.00E 00	5.86E-18
7.00E 00	6.28E-18

7-4-4B
DE HEER ET AL. (203)

ENERGY (KEV)	CROSS SECTION (SQ. CM)
2.00E 00	1.10E-17

7-4-4C
DE HEER ET AL. (203)

ENERGY (KEV)	CROSS SECTION (SQ. CM)
2.00E 00	2.20E-18

7-4-4D
DE HEER ET AL. (203)

ENERGY (KEV)	CROSS SECTION (SQ. CM)
2.00E 00	1.20E-18

7-4-5
DE HEER ET AL. (203)

ENERGY (KEV)	CROSS SECTION (SQ. CM)
3.00E-01	1.07E-17
4.00E-01	1.20E-17
5.00E-01	1.18E-17
6.00E-01	1.27E-17
8.00E-01	1.04E-17
1.00E 00	1.25E-17
2.00E 00	9.58E-18
3.00E 00	7.61E-18
4.00E 00	6.62E-18
5.00E 00	6.35E-18
6.00E 00	4.92E-18

7-4-6
DE HEER ET AL. (203)

ENERGY (KEV)	CROSS SECTION (SQ. CM)
3.00E-01	1.05E-17
4.00E-01	1.20E-17
5.00E-01	1.04E-17
6.00E-01	9.79E-18
8.00E-01	8.70E-18
1.00E 00	7.61E-18
2.00E 00	5.50E-18
4.0CE 00	5.76E-18
6.00E 00	4.58E-18
7.00E 00	4.41E-18

7-4-7
DE HEER ET AL. (203)

ENERGY (KEV)	CROSS SECTION (SQ. CM)
3.00E-01	7.80E-18
4.00E-01	9.53E-18
5.00E-01	1.09E-17
6.00E-01	1.33E-17
8.00E-01	1.33E-17
1.00E 00	1.30E-17
2.0CE 00	1.16E-17
3.00E 00	1.05E-17
4.00E 00	8.65E-18
5.00E 00	7.00E-18
6.00E 00	5.56E-18
7.00E 00	4.76E-18
8.00E 00	4.23E-18

7-4-8A
VAN ECK ET AL. (76)

ENERGY (KEV)	CROSS SECTION (SQ. CM)
5.0CE 00	1.12E-16
7.50E 00	1.16E-16
1.00E 01	1.04E-16
1.50E 01	8.65E-17
2.00E 01	6.91E-17
2.50E 01	5.60E-17
3.00E 01	4.60E-17
3.50E 01	4.02E-17

7-4-8B
DE HEER ET AL. (203)

ENERGY (KEV)	CROSS SECTION (SQ. CM)
3.00E-01	1.21E-17
4.00E-01	1.30E-17
5.00E-01	1.91E-17
6.00E-01	2.19E-17
8.00E-01	2.56E-17

ENERGY (KEV)	CROSS SECTION (SQ. CM)
1.00E 00	3.30E-17
2.00E 00	5.03E-17
3.00E 00	8.50E-17
4.00E 00	1.16E-16
5.00E 00	1.07E-16
6.00E 00	1.18E-16
8.00E 00	1.12E-16
1.00E 01	1.05E-16

7-4-9A
VAN ECK ET AL. (76)

ENERGY (KEV)	CROSS SECTION (SQ. CM)
5.00E 00	1.01E-16
7.50E 00	1.03E-16
1.00E 01	1.01E-16
1.50E 01	8.21E-17
2.00E 01	6.25E-17
2.50E 01	5.06E-17
3.00E 01	4.06E-17
3.50E 01	3.55E-17

7-4-9B
DE HEER ET AL. (203)

ENERGY (KEV)	CROSS SECTION (SQ. CM)
2.00E 00	3.89E-17

7-4-10A
JAECKS ET AL. (124)

ENERGY (KEV)	CROSS SECTION (SQ. CM)
3.00E-01	3.04E-18
5.00E-01	2.41E-18
7.00E-01	2.31E-18
1.00E 00	2.55E-18
2.00E 00	2.48E-18
4.00E 00	1.44E-18
6.00E 00	1.06E-18
8.00E 00	9.40E-19
1.00E 01	1.03E-18
1.50E 01	7.85E-19
2.00E 01	6.63E-19
2.50E 01	5.94E-19
3.00E 01	6.10E-19

7-4-10B
JAECKS ET AL. (124)

ENERGY (KEV)	CROSS SECTION (SQ. CM)
3.00E-01	1.80E-18
5.00E-01	1.15E-18
7.00E-01	9.80E-19
1.00E 00	9.50E-19
2.00E 00	6.10E-19
4.00E 00	4.43E-19
6.00E 00	4.08E-19
8.00E 00	3.67E-19
1.00E 01	4.00E-19
1.50E 01	3.86E-19
2.00E 01	3.28E-19
2.50E 01	2.92E-19
3.00E 01	2.87E-19

7-4-11A
JAECKS ET AL. (124)

ENERGY (KEV)	CROSS SECTION (SQ. CM)
3.00E-01	4.93E-18
5.00E-01	6.16E-18
7.00E-01	6.37E-18
1.00E 00	7.46E-18
2.00E 00	6.56E-18
4.00E 00	4.17E-18
6.00E 00	2.62E-18
8.00E 00	1.90E-18
1.00E 01	1.76E-18
1.50E 01	1.67E-18
2.00E 01	1.56E-18
2.50E 01	1.46E-18
3.00E 01	1.19E-18

7-4-11B
JAECKS ET AL. (124)

ENERGY (KEV)	CROSS SECTION (SQ. CM)
3.00E-01	6.28E-19
5.00E-01	2.52E-18
7.00E-01	2.96E-18
1.00E 00	3.25E-18
2.00E 00	2.09E-18
4.00E 00	1.48E-18
6.00E 00	1.25E-18
8.00E 00	9.20E-19
1.00E 01	8.30E-19
1.50E 01	6.23E-19
2.00E 01	5.57E-19
2.50E 01	5.60E-19
3.00E 01	5.16E-19

7-4-11C
JAECKS ET AL. (124)

ENERGY (KEV)	CROSS SECTION (SQ. CM)
3.00E-01	1.50E-18
5.00E-01	1.68E-18
7.00E-01	2.36E-18
1.00E 00	2.48E-18
2.00E 00	1.43E-18
4.00E 00	1.17E-18
6.00E 00	9.60E-19
8.00E 00	8.10E-19
1.00E 01	7.16E-19
1.50E 01	5.82E-19
2.00E 01	4.93E-19
2.50E 01	4.32E-19
3.00E 01	3.67E-19

7-4-12A
JAECKS ET AL. (124)

ENERGY (KEV)	CROSS SECTION (SQ. CM)
3.00E-01	5.05E-18
5.00E-01	3.22E-18
7.00E-01	2.63E-18
1.00E 00	2.90E-18
2.00E 00	2.76E-18
4.00E 00	2.00E-18
6.00E 00	1.37E-18
8.00E 00	1.15E-18
1.00E 01	1.18E-18
1.50E 01	1.16E-18
2.00E 01	1.16E-18
2.50E 01	1.19E-18
3.00E 01	1.19E-18

7-4-12B
JAECKS ET AL. (124)

ENERGY (KEV)	CROSS SECTION (SQ. CM)
3.00E-01	1.28E-18
5.00E-01	1.04E-18
7.00E-01	9.05E-19
1.00E 00	9.60E-19
2.00E 00	6.72E-19
4.00E 00	5.58E-19
6.00E 00	5.23E-19
8.00E 00	5.23E-19
1.00E 01	5.30E-19
2.00E 01	4.84E-19
2.50E 01	4.67E-19
3.00E 01	4.54E-19

7-5 EXCITATION OF XENON

Studies of the excitation of xenon have pursued the same basic course as those for the other rare gases. Studies have been made on the emission of radiation from atoms and ions excited by processes that may be described by the following reaction equations:

$$X^+ + Xe \rightarrow [X^+] + Xe^* \tag{7-9}$$

$$X^+ + Xe \rightarrow [X^+ + e] + Xe^{+*}. \tag{7-10}$$

Here there is no information on the states of excitation or ionization of the systems contained within the square brackets. The quantitative work is primarily devoted to the impact of He^+ ions, although there have been some studies of excitation by H^+, H_2^+, and Ar^+.

For the Xe I and Xe II spectra, the identification of lines and the assignment of term designations is based on the "Ultraviolet Multiplet Table" by Moore,[67] and the original analysis of atomic ion lines by Humphreys.[295, 296]

Most of the quantitative measurements have come from the work of de Heer, Jaecks, and co-workers.[203, 272, 124] Also, there have been a few quantitative measurements by Dufay et al.;[125] Lipeles et al.[204] have carried out measurements of emission in rather broad wavelength bands induced by the impact of He^+ on Xe. These measurements will all be discussed in detail in the following sections. It is of interest to note that some preliminary studies of X ray emission induced by the impact of protons, I^{2+}, I^{3+}, and I^{4+} on xenon have been performed by Kessel et al.[273] Impact energies were from 3- to 10-MeV. At the time of writing, the X ray spectrum has been detected and appears to result from the production of an L-shell vacancy in the xenon and possibly in the iodine. No attempt has yet been made to determine quantitative cross sections for the production of these X rays nor to investigate the probabilities of forming inner-shell vacancies. There will be no further discussion of these preliminary observations of X rays.

In the following section (7-5-A) a brief review is given of the various experimental systems that have been used in these studies. This is followed by a discussion of the data obtained for excitation by impact of protons (Section 7-5-B), H_2^+ (Section 7-5-C), He^+ (Section 7-5-D), and Ne^+ (Section 7-5-E), followed by the collection of available data in tabular form (Section 7-5-F).

A. Experimental Arrangement

1. The Work of de Heer et al. This group has produced three papers on the subject of excitation of xenon.[203,272,124] The various experimental arrangements utilized for these studies have previously been fully discussed in Section 7-2-A(1) and will not be repeated here. In that discussion, three different optical systems were identified and labeled Systems I, II, and III, respectively. Optical System I has been used for visible emissions, and System III for work in the vacuum ultraviolet.

The sources from which data are drawn may be summarized as follows:

$H^+ + Xe$. Emission cross sections for two visible lines were measured by Jaecks, de Heer, and Salop.[124]

$H_2^+ + Xe$. Emission cross sections were measured by Jaecks et al.[124] for the same two visible lines as were measured with H^+ impact.

He$^+$ + Xe. This case has been considered in three papers.[203, 272, 124] The first was a conference abstract,[272] which gave only a measurement for one vacuum ultraviolet line of Xe I. This was later superseded by a more complete paper that includes a large number of vacuum ultraviolet emission measurements determined using System III.[203] Visible emissions are found in a separate publication[124] and were all investigated with System I.

Throughout this work the monochromators were arranged to observe emission at 90° to the beam direction. The polarization of the emission and its possible influence on the accuracy of the cross section measurements was ignored. The absolute accuracy of the data on emission of visible spectral lines was estimated as ±30% while that for vacuum ultraviolet was ±100%. Reproducibility was better than ±10%.

2. The Work of Dufay et al. Dufay and co-workers have made measurements on the emission cross sections of one line each of Xe I and Xe II induced by the impact of H$^+$ on a xenon target.[125] The apparatus used for this work has been assessed in Section 7-2-A(2) in the context of studies of excitation of neon by impact of protons. The reader is referred to that section, as all comments are equally valid for work using a xenon target. No attention was paid to the problems associated with the polarization of the emission, and the pressure measurement system was in error due to the mercury pumping onto a cold trap. Accuracy of absolute cross sections was quoted as ±50% and scatter in the measurements as less than ±10%.

3. The Work of Lipeles et al. Lipeles and co-workers have made a brief study of the emission induced in the collision He$^+$ + Xe at rather low energies, 5- to 500-eV.[125] Measurements were made of emission in broad wavelength intervals, and no attempt was made to determine emission cross section for resolved spectral lines. This work has been discussed in detail in the context of excitation of neon, and it was suggested there that the data were peculiar to the apparatus utilized and therefore not relevant here [see Section 7-2-A(3)].

4. The Work of Sluyters et al. Sluyters et al.[103] have carried out a measurement on a single line of Xe II induced by the impact of Ar$^+$ on a target of xenon. It has been pointed out previously [Section 7-3-A(2)] that the target pressures utilized in this experimental program are far too high and that a substantial proportion of the projectiles was neutralized before entering the observation region. It was concluded that data from this work are both incorrect and also misleading; therefore they are excluded from any detailed discussion.

B. Proton Impact

The data on the excitation of the neutral xenon lines are restricted to the measurement on the 8232-Å line $(6p[1\frac{1}{2}]_2 \to 6s[1\frac{1}{2}]_2^0)$ by Dufay et al.[125] The published data are in the form of a line of best fit with no individual data points. That line of best fit is reproduced in Data Table 7-5-1.

Both Jaecks et al.[124] and also Dufay et al.[125] have carried out measurements on the 4844-Å line of Xe II $(6p\ ^4D_{3\ 1/2}^0 \to 6s\ ^4P_{2\ 1/2})$, shown in Data Table 7-5-2. The energy ranges of the data do not overlap. Nevertheless, it is possible to extrapolate over the intervening energy gap and surmise that the two sets of data are consistent with each other. Jaecks et al.[124] have also produced a measurement on the 4877-Å line $(6p'\ ^4F_{3\ 1/2}^0 \to 6s'\ ^2D_{2\ 1/2})$, which is shown in Data Table 7-5-3.

C. H_2^+ Impact

Jaecks et al.[124] have measured the emission cross sections of the 4844-Å ($6p\ ^4D_{3\ 1/2}^0$ $\rightarrow 6s\ ^4P_{2\ 1/2}$) and 4877-Å ($6p'\ ^4F_{3\ 1/2}^0 \rightarrow 6s'\ ^2D_{2\ 1/2}$) lines of Xe II. The data are shown in Data Tables 7-5-4 and 7-5-5. These are determined under precisely the same conditions as were used for the proton impact measurement of the same lines by the same authors. No attempt was made to assess the influence of excited states in the projectile beam. This is not expected to be very important. Accuracy is estimated at $\pm 30\%$ with $\pm 10\%$ reproducibility.

D. He^+ Impact

Quantitative measurements on the cross sections for the emission of Xe I and Xe II lines have been made by de Heer, Jaecks and co-workers in Amsterdam.[203, 272, 124] In addition, Lipeles et al.[204] have measured integrated intensities over broad wavelength intervals. This does not provide information on specific transitions and therefore the measurements by Lipeles et al.[204] are not considered further in this monograph [see Section 7-2-A(3)].

No attempt has been made to assess the possible influence on the measurements of excited states in the projectile beam; it is not expected that excited states will produce significant errors in the measurement of cross sections.

1. Emission of Lines of Xe I. De Heer et al.[203] have made measurements on two resonance lines of Xe I, namely 1470-Å ($6s[1\frac{1}{2}]_1^0 \rightarrow 5p^6\ ^1S$) and 1296-Å ($6s'[\frac{1}{2}]_1^0 \rightarrow 5p^6\ ^1S$). An early report of this work in a conference paper[272] was superseded by a more complete publication.[203] In the absence of any secondary effects, the emitted light intensity should vary linearly with target density. It is to be expected that the resonance lines will be strongly affected by self absorption in the target gas, and their intensity will increase at a rate slower than a linear relationship would allow. It was suggested that absorption might indeed be taking place but that its influence was being swamped completely by some other secondary effect with a completely different pressure dependence. No quantitative explanation of the observation was advanced. De Heer et al. provided a measurement of an apparent emission cross section measured at a target pressure of 10^{-3} torr; however, this is certainly not determined under conditions which are in accord with the operational definition of emission cross section established in Section 2-1-A. It is possible that the secondary processes are energy dependent, causing the data not to represent the relative variation of emission cross section with impact energy. These data are obtained for conditions that are not in accordance with the basic operational definition of emission cross section. In order to prevent confusion, they are omitted from the present data compilation.

2. Emission of Lines of Xe II. Measurements are available for some eleven lines in the vacuum ultraviolet spectrum and twelve in the visible regions. In general, the visible lines represent the cascade contribution to ultraviolet transitions. It will be noted later that cascade is very large in some cases and must be considered when attempting to understand the collision process that gives rise to the formation of an excited state.

De Heer et al.[203] have made measurements on a total of eleven ultraviolet lines. All data is given in the relevant Data Tables (7-5-6 to 7-5-9) and measurements at the single energy of 2-keV are repeated in Table 7-3 to illustrate comments on cascade. For some of these lines, notably the 973- and 1051.9-Å transitions, it is possible

that the resolution was inadequate to properly resolve the line, and therefore the measurement may include a small contribution from transitions at adjacent wavelengths.

The measurements on visible lines were obtained by Jaecks et al.[124] and are given in Data Tables 7-5-10 through 7-5-15. There are considerable similarities in the behavior of transitions from upper states of a particular multiplicity; the data are grouped together to illustrate these similarities of behavior. The Xe I and Xe II spectra are very complicated, and it is not clear that sufficient resolution was utilized to guarantee that the observations were free of interference from transitions at adjacent wavelengths; this is particularly true of the 5265-Å line which lies close to another Xe II line and also for the 3634-, 5419-, and 4845-Å lines that are adjacent to Xe I lines.

All the data presented here are for emission of spectral lines; these data cannot readily be converted into level-excitation cross sections because there is no accurate information on branching ratios. By comparing the intensities of visible and ultraviolet lines it is possible to make rough assessments of the contribution of cascade to the measured emission functions.

Table 7-3 by de Heer shows the measured cross sections for emission of all the vacuum ultraviolet lines at the single energy of 2-keV with estimates, based on measurements of visible lines, of the cascade contribution to these cross sections. Inevitably the accuracy of a comparison is poor, since the ultraviolet data may involve systematic errors up to 100% and those in the visible region up to 30%; consequently, some of the cascade contributions actually appear to exceed the emission cross sections to which they are supposed to contribute. It is very clear that the cascade population of some states is comparable with the total cross section for direct excitation by collision. In the case of the $6s$ and $6s'$ states, the direct excitation cross section must be smaller than the cross section for formation via cascade from higher states. Conversely, the $5d$ and $5p^6$ excited states appear to be populated predominantly by direct collisional excitation.

TABLE 7-3   Xe II Emission Cross Sections in Units 10^{-17} cm² at 2-keV He⁺ Impact Energy, Compared to Contribution from Cascade (de Heer et al.[203])

Wavelength (Å)	Transition	Cross Section	Wavelength (Å) of Measured Lines Contributing to Cascade				Cascade Cross Section
973	$6s\ ^2P_{3/2}-s^2p^5\ ^2P^0_{3/2}$	0.85	5309,	3933,	3509,	4921	0.74
1084	$6s\ ^2P_{5/2}-s^2p^5\ ^2P^0_{1/2}$	0.17	4887,	4920			
926	$6s'\ ^2D_{3/2}-s^2p^5\ ^2P^0_{3/2}$	1.40	4617,	4415,	4877		1.60
885	$6s'\ ^2D_{3/2}-s^2p^5\ ^2P^0_{3/2}$	0.50	5046,	5262,	5184,	5971	1.06
977	$6s'\ ^2D_{3/2}-s^2p^5\ ^2P^0_{1/2}$	1.27	6270				
1075	$6s\ ^4P_{5/2}-s^2p^5\ ^2P^0_{3/2}$	2.11	4845, 5339	4890,	4216,	5293	2.77
1052	$6s\ ^4P_{3/2}-s^2p^5\ ^2P^0_{3/2}$	2.0	5419,	4603,	3944,	5372	2.18
1183	$6s\ ^4P_{3/2}-s^2p^5\ ^2P^0_{1/2}$	0.25	3762,	5976,	5372		
912	$5d\ ^2D_{3/2}-s^2p^5\ ^2P^0_{3/2}$	1.40					Small
1100	$5p^6\ ^2S_{1/2}-s^2p^5\ ^2P^0_{3/2}$	28.0					Small
1244	$5p^6\ ^2S_{1/2}-s^2p^5\ ^2P^0_{1/2}$	6.3					

E. Ar$^+$ Impact

Sluyters et al.[103] have carried out a measurement on the cross section for emission of the 2475-Å line of Xe II induced by the impact of Ar$^+$. This line is listed by Moore[67] as being unclassified. It has previously been mentioned [Section 7-3-A(2)] that the target pressures for this experiment were far too high resulting in a substantial neutralization of the projectile beam. It was concluded that data from this work are both inaccurate and misleading. Therefore, the data are not considered further.

F. Data Tables

1. Cross Sections for Emission of Xe I Lines Induced by Impact of H$^+$ on an Xe Target

7-5-1 8232-Å Line $(6p[1\frac{1}{2}]_2 \rightarrow 6s[1\frac{1}{2}]_2^0)$.

2. Cross Sections for Emission of Xe II Lines Induced by Impact of H$^+$ on an Xe Target

7-5-2 4844-Å Line $(6p\ ^4D^0_{3\ 1/2} \rightarrow 6s\ ^4P_{2\ 1/2})$.

7-5-3 4877-Å Line $(6'p\ ^4F^0_{3\ 1/2} \rightarrow 6s'\ ^2D_{2\ 1/2})$.

3. Cross Sections for Emission of Xe II Lines Induced by Impact of H$_2^+$ on an Xe Target

7-5-4 4844-Å Line $(6p\ ^4D^0_{3\ 1/2} \rightarrow 6s\ ^4P_{2\ 1/2})$.

7-5-5 4877-Å Line $(6p'\ ^4F^0_{3\ 1/2} \rightarrow 6s'\ ^2D_{2\ 1/2})$.

4. Cross Sections for Emission of Xe II Lines Induced by Impact of He$^+$ on an Xe Target

7-5-6 (A) 973-Å Line $(6s\ ^2P_{1\ 1/2} \rightarrow 5p^5\ ^2P^0_{1\ 1/2})$; (B) 1084-Å Line $(6s\ ^2P_{1\ 1/2} \rightarrow 5p^5\ ^2P^0_{1\ 1/2})$.

7-5-7 (A) 977-Å Line $(6s'\ ^2D_{1\ 1/2} \rightarrow 5p^5\ ^2P^0_{1/2})$; (B) 926-Å Line $(6s'\ ^2D_{2\ 1/2} \rightarrow 5p^5\ ^2P^0_{1\ 1/2})$; (C) 885-Å Line $(6s'\ ^2D_{1\ 1/2} \rightarrow 5p^5\ ^2P^0_{1\ 1/2})$.

7-5-8 (A) 1075-Å Line $(6s\ ^4P_{2\ 1/2} \rightarrow 5p^5\ ^2P^0_{1\ 1/2})$; (B) 1052-Å Line $(6s\ ^4P_{1\ 1/2} \rightarrow 5p^5\ ^2P^0_{1\ 1/2})$; (C) 1183-Å Line $(6s\ ^4P_{1\ 1/2} \rightarrow 5p^5\ ^2P^0_{1/2})$.

7-5-9 (A) 1100-Å Line $(5p^6\ ^2S_{1/2} \rightarrow 5p^5\ ^2P^0_{1\ 1/2})$; (B) 1245-Å Line $(5p^6\ ^2S_{1/2} \rightarrow 5p^5\ ^2P^0_{1/2})$; (C) 912-Å Line $(5d\ ^2D_{2\ 1/2} \rightarrow 5p^5\ ^2P^0_{1\ 1/2})$.

7-5-10 (A) 4877-Å Line $(6p'\ ^2F^0_{3\ 1/2} \rightarrow 6s'\ ^2D_{2\ 1/2})$; (B) 5045-Å Line $(6p'\ ^2P^0_{1/2} \rightarrow 6s'\ ^2D_{1\ 1/2})$; (C) 4486-Å Line $(6p'\ ^2P^0_{1\ 1/2} \rightarrow 5d\ ^4P_{1\ 1/2})$; (D) 5262- and 5260-Å Lines $(6p'\ ^2D^0_{1\ 1/2} \rightarrow 6s'\ ^2D_{1\ 1/2}$ and $6p\ ^2P^0_{1\ 1/2} \rightarrow 6s\ ^2P_{1/2})$ Measured Together.

7-5-11 (A) 5292-Å Line $(6p\ ^4P^0_{2\ 1/2} \rightarrow 6s\ ^4P_{2\ 1/2})$; (B) 5419-Å Line $(6p\ ^4D^0_{2\ 1/2} \rightarrow 6s\ ^4P_{1\ 1/2})$; (C) 4844-Å Line $(6p\ ^4D^0_{3\ 1/2} \rightarrow 6s\ ^4P_{2\ 1/2})$.

7-5-12 3634-Å Line $(6p''\ ^2P^0_{1\ 1/2} \rightarrow 5d'\ ^2D_{2\ 1/2})$.

7-5-13 4330-Å Line $(6d\ ^4F_{3\ 1/2} \rightarrow 6p\ ^4D^0_{2\ 1/2})$.

7-5-14 4296-Å Line $(7s\ ^4P_{1/2} \rightarrow 6p\ ^4P^0_{1\ 1/2})$.

7-5-15 4310-Å Line $(7s'\ ^2D_{2\ 1/2} \rightarrow 6p\ ^2D^0_{2\ 1/2})$.

7-5-1
DUFAY ET AL. (125)

ENERGY (KEV)	CROSS SECTION (SQ. CM)
2.37E 01	1.66E-17
2.80E 01	1.90E-17
3.34E 01	2.09E-17
3.71E 01	2.23E-17
4.19E 01	2.38E-17
5.03E 01	2.58E-17
6.52E 01	2.88E-17
8.67E 01	3.08E-17
1.15E 02	3.13E-17
1.62E 02	3.10E-17
2.21E 02	2.94E-17
3.06E 02	2.65E-17
3.78E 02	2.49E-17
4.72E 02	2.33E-17
6.11E 02	2.09E-17

7-5-2A
JAECKS ET AL. (124)

ENERGY (KEV)	CROSS SECTION (SQ. CM)
1.00E 00	1.16E-18
2.00E 00	2.49E-18
2.50E 00	3.38E-18
3.00E 00	3.52E-18
3.50E 00	3.49E-18
4.0CE 00	3.16E-18
5.00E 00	2.58E-18
6.0CE 00	2.20E-18
7.00E 00	1.99E-18
8.00E 00	1.94E-18
1.00E 01	1.58E-18

7-5-2B
DUFAY ET AL. (125)

ENERGY (KEV)	CROSS SECTION (SQ. CM)
3.05E 01	1.06E-18
3.16E 01	9.72E-19
4.02E 01	9.12E-19
4.79E 01	8.39E-19
5.75E 01	7.59E-19
7.41E 01	6.61E-19
9.25E 01	5.86F-19
1.16E 02	5.10E-19
1.56E 02	4.25E-19
1.99E 02	3.66E-19
2.61E 02	3.08E-19
3.22E 02	2.63E-19
3.98E 02	2.29E-19
5.06E 02	1.90E-19

7-5-3
JAECKS ET AL. (124)

ENERGY (KEV)	CROSS SECTION (SQ. CM)
1.00E 00	8.90E-19
2.00E 00	1.69E-18
2.50E 00	3.07E-18
3.00E 00	4.50E-18
3.5E 00	5.87E-18
4.00E 00	5.70E-18
5.00E 00	4.53E-18
6.00E 00	3.56E-18
8.00E 00	1.69E-18
1.00E 01	1.32E-18

7-5-4
JAECKS ET AL. (124)

ENERGY (KEV)	CROSS SECTION (SQ. CM)
1.50E 00	1.96E-18
2.00E 00	2.04E-18
4.00E 00	3.13E-18
5.0CE 00	3.62E-18
6.00E 00	4.62E-18
7.00E 00	4.50E-18
8.0CE 00	4.80E-18
1.00E 01	4.22E-18

7-5-5
JAECKS ET AL. (124)

ENERGY (KEV)	CROSS SECTION (SQ. CM)
1.00E 00	1.46E-19
2.00E 00	6.05E-19
4.00E 00	1.51E-18
6.00E 00	2.98E-18
8.00E 00	3.39E-18
1.00E 01	3.00E-18

7-5-6A
DE HEER ET AL. (203)

ENERGY (KEV)	CROSS SECTION (SQ. CM)
3.00E-01	2.27E-17
4.00E-01	1.84E-17
5.00E-01	1.87E-17
6.00E-01	1.63E-17
8.00E-01	1.36E-17
1.00E 00	1.23E-17
2.00E 00	8.49E-18
3.0CE 00	8.41E-18
4.0CE 00	6.95E-18
5.00E 00	5.94E-18
6.0CE 00	6.38E-18

7-5-6B
DE HEER ET AL. (203)

ENERGY (KEV)	CROSS SECTION (SQ. CM)
2.00E 00	1.70E-18

7-5-7A
DE HEER ET AL. (203)

ENERGY (KEV)	CROSS SECTION (SQ. CM)
4.00E-01	1.98E-17
5.00E-01	1.63E-17
6.00E-01	1.82E-17
8.00E-01	1.90E-17
1.00E 00	1.66E-17
2.00E 00	1.27E-17
4.0CE 00	9.25E-18
5.0CE 00	7.52E-18
6.00E 00	7.40E-18
7.00E 00	6.89E-18
8.00E 00	5.86E-18
9.00E 00	5.10E-18

7-5-7B
DE HEER ET AL. (203)

ENERGY (KEV)	CROSS SECTION (SQ. CM)
2.00E 00	1.40E-17

7-5-7C
DE HEER ET AL. (203)

ENERGY (KEV)	CROSS SECTION (SQ. CM)
2.00E 00	5.00E-18

7-5-8A
DE HEER ET AL. (203)

ENERGY (KEV)	CROSS SECTION (SQ. CM)
3.00E-01	4.18E-17
4.00E-01	3.62E-17
5.00E-01	3.28E-17
6.0CE-01	2.86E-17
8.00E-01	2.60E-17
1.00E 00	2.48E-17
2.00E 00	2.11E-17
3.0CE 00	1.66E-17
4.00E 00	1.32E-17
5.00E 00	9.70E-18
6.0CE 00	8.43E-18
8.00E 00	7.93E-18
9.0CE 00	6.70E-18
1.00E 01	7.52E-18

7-5-8B
DE HEER ET AL. (203)

ENERGY (KEV)	CROSS SECTION (SQ. CM)
2.00E 00	2.00E-17

7-5-8C
DE HEER ET AL. (203)

ENERGY (KEV)	CROSS SECTION (SQ. CM)
2.00E 00	2.50E-18

7-5-9A
DE HEER ET AL. (203)

ENERGY (KEV)	CROSS SECTION (SQ. CM)
3.00E-01	1.59E-16
4.00E-01	1.60E-16
5.00E-01	1.86E-16
6.00E-01	2.32E-16
7.00E-01	2.42E-16
8.00E-01	2.53E-16
1.00E 00	2.43E-16
2.00E 00	2.80E-16
3.00E 00	2.71E-16
4.00E 00	2.92E-16
5.00E 00	2.88E-16
6.00E 00	2.70E-16
8.00E 00	2.61E-16
9.00E 00	2.60E-16
1.00E 01	2.48E-16

7-5-9B
DE HEER ET AL. (203)

ENERGY (KEV)	CROSS SECTION (SQ. CM)
2.00E 00	6.30E-17

7-5-9C
DE HEER ET AL. (203)

ENERGY (KEV)	CROSS SECTION (SQ. CM)
2.00E 00	1.40E-17

7-5-10A
JAECKS ET AL. (124)

ENERGY (KEV)	CROSS SECTION (SQ. CM)
3.00E-01	6.92E-18

5.00E-01	6.56E-18
7.00E-01	8.14E-18
1.00E 00	1.06E-17
2.00E 00	9.78E-18
4.00E 00	5.85E-18
6.00E 00	2.91E-18
8.00E 00	1.93E-18
1.00E 01	1.41E-18
1.50E 01	1.24E-18
2.00E 01	1.17E-18
2.50E 01	1.07E-18
3.00E 01	1.00E-18

7-5-10B
JAECKS ET AL. (124)

ENERGY (KEV)	CROSS SECTION (SQ. CM)
7.00E-01	8.00E-19
1.00E 00	9.36E-19
2.00E 00	1.57E-18
4.00E 00	1.38E-18
6.00E 00	1.32E-18
8.00E 00	1.20E-18
1.00E 01	1.11E-18
1.50E 01	7.20E-19
2.00E 01	6.68E-19
3.00E 01	4.53E-19

7-5-10C
JAECKS ET AL. (124)

ENERGY (KEV)	CROSS SECTION (SQ. CM)
7.00E-01	1.76E-19
1.00E 00	1.87E-19
2.00E 00	2.33E-19
4.00E 00	2.10E-19
6.00E 00	1.85E-19
8.00E 00	1.58E-19
1.00E 01	1.39E-19

7-5-10D
JAECKS ET AL. (124)

ENERGY (KEV)	CROSS SECTION (SQ. CM)
3.00E-01	3.96E-18
5.00E-01	3.77E-18
7.00E-01	3.86E-18
1.00E 00	3.63E-18
2.00E 00	2.82E-18
4.00E 00	1.90E-18
6.00E 00	1.43E-18
8.00E 00	1.09E-18
1.00E 01	8.50E-19
1.50E 01	7.90E-19
2.00E 01	7.62E-19
2.50E 01	7.05E-19
3.00E 01	6.63E-19

7-5-11A
JAECKS ET AL. (124)

ENERGY (KEV)	CROSS SECTION (SQ. CM)
3.00E-01	2.50E-17
5.00E-01	2.00E-17
7.00E-01	1.84E-17
1.00E 00	1.62E-17
2.00E 00	1.19E-17
4.00E 00	5.50E-18
6.00E 00	3.52E-18
8.00E 00	2.95E-18
1.00E 01	2.49E-18
1.50E 01	2.28E-18
2.00E 01	2.50E-18
2.50E 01	2.46E-18
3.00E 01	2.27E-18

7-5-11B
JAECKS ET AL. (124)

ENERGY (KEV)	CROSS SECTION (SQ. CM)
3.00E-01	2.68E-17
5.00E-01	2.21E-17
7.00E-01	2.04E-17
1.00E 00	1.71E-17
2.00E 00	1.10E-17
4.00E 00	5.02E-18
6.00E 00	2.92E-18
8.00E 00	2.49E-18
1.00E 01	2.12E-18
1.50E 01	1.86E-18
2.00E 01	1.86E-18
2.50E 01	1.91E-18
3.00E 01	1.78E-18

7-5-11C
JAECKS ET AL. (124)

ENERGY (KEV)	CROSS SECTION (SQ. CM)
3.00E-01	1.54E-17
5.00E-01	1.15E-17
7.00E-01	1.11E-17
1.00E 00	1.01E-17
2.00E 00	6.00E-18
4.00E 00	2.84E-18
6.00E 00	1.83E-18
8.00E 00	1.32E-18
1.00E 01	1.12E-18
1.50E 01	1.11E-18
2.00E 01	1.17E-18
2.50E 01	1.10E-18
3.00E 01	1.10E-18

7-5-12
JAECKS ET AL. (124)

ENERGY (KEV)	CROSS SECTION (SQ. CM)

1.00E 00	1.03E-19	8.00E 00	3.96E-19	1.00E 01	5.38E-19
2.00E 00	1.82E-19	1.00E 01	3.54E-19	1.50E 01	4.39E-19
4.00E 00	2.25E-19	1.50E 01	4.14E-19	2.00E 01	3.34E-19
6.0CE 00	2.90E-19	2.00E 01	4.64E-19	2.50E 01	2.88E-19
8.00E 00	3.06E-19	2.50E 01	5.36E-19	3.00E 01	2.18E-19
1.00E 01	3.60E-19	3.00E 01	5.41E-19		

7-5-13
JAECKS ET AL. (124)

7-5-14
JAECKS ET AL. (124)

7-5-15
JAECKS ET AL. (124)

ENERGY (KEV)	CROSS SECTION (SQ. CM)	ENERGY (KEV)	CROSS SECTION (SQ. CM)	ENERGY (KEV)	CROSS SECTION (SQ. CM)
3.00E-01	4.10E-18	7.00E-01	1.82E-19	5.00E-01	3.87E-19
5.00E-01	3.80E-18	1.00E 00	2.11E-19	7.00E-01	3.46E-19
7.00E-01	4.10E-18	2.0CE 00	3.55E-19	1.0CE 00	3.14E-19
1.00E 00	4.23E-18	3.00E 00	5.34E-19	2.00E 00	2.64E-19
2.00E 00	2.32E-18	4.00E 00	6.92E-19	4.00E 00	1.95E-19
4.00E 00	1.12E-18	6.00E 00	5.81E-19	6.0CE 00	1.55E-19
6.CCE CO	5.52E-19	8.00E 00	5.48E-19	8.00E 00	1.34E-19
				1.00E 01	1.17E-19

7-6 EXCITATION OF MERCURY

Studies of the excitation of mercury have not received much attention, the only systematic work being a paper by Bobashev and Pop.[168] Studies have been made on the emission of Hg I and Hg III spectral lines induced by the impact of H^+ and H_2^+, a process that can be described by the following reaction equations:

$$X^+ + Hg \rightarrow [X^+] + Hg^* \tag{7-11}$$

$$X^+ + Hg \rightarrow [X^+ + 2e] + Hg^{2+*} \tag{7-12}$$

Here there is no information on the state of excitation or ionization of the system contained within the square brackets.

The identification of Hg I spectra and their term designations are based on the tables presented by Dieke.[293]

In addition to the work of Bobashev and Pop, there were some early measurements by Clark[297] that have remained unpublished. These determined the emission functions of certain mercury lines induced by the impact of Li^+ and Na^+ on a Hg target at energies in the range 0.2- to 1.0-keV. The detection was with a photographic spectrometer. It is doubtful whether these data of Clark are of high accuracy, and they are not considered further here.

A description will be given of the apparatus used by Bobashev and Pop (Section 7-6-A), followed by a discussion of the data (Section 7-6-B), and the presentation of data tables (Section 7-6-C).

A. Experimental Arrangement

The apparatus used by Bobashev and Pop[168] is basically the same as that described for studies of emission from projectile induced by impact of H^+ and H_2^+ on rare gas target.[158]

The ion source for this work was an arc discharge within a magnetic field. Ions were extracted, accelerated to the required energy, and mass analyzed in a magnetic

field. The basic design of the projectile source and accelerator was in the form of a 180° mass spectrometer. The target cell and observation region were actually located within the magnetic field of the analyzer. A large Faraday cup was placed behind the observation region to monitor the ion beam. Secondary electrons were prevented from leaving the monitor by the analyzer magnetic field. One must inquire whether the existence of the magnetic field perpendicular to the beam direction in the observation region will in any way alter the collision process or the decay mechanism. The authors make no comment on this question.

The collision chamber was maintained at a temperature of 30°C and a side arm connected it to a source of mercury at a temperature of 22°C. The pressure of the target was taken as 1.2×10^{-3} torr, the vapor pressure of mercury at 22°C. Apparently no attempt was made to vary the pressure of the target in order to prove that the emission varied linearly with target density. It should be noted that the value for vapor pressure at 22°C is not in accordance with the value adopted in standard tables,[298] which is quoted as 1.426×10^{-3} torr. All of the internal surfaces of the apparatus were coated with graphite in order to prevent absorption of the mercury vapor by the copper walls.

A lens was used to focus the light emitted from the collision region onto the entrance slit of a prism spectrometer. Detection was by a photomultiplier. Calibration of detection sensitivity was by comparison of the collisionally induced emission with the output of a standard tungsten filament lamp. The optical system in calibration duplicated that used for the experimental measurement so that precisely the same geometry was used in both operations. No mention is made of tests to ensure that light scattered within the monochromator did not cause errors in the calibration. The monochromator was placed at 90° to the ion beam path. It was assumed that the emission was unpolarized, and no corrections were made for possible anisotropy in the emission nor instrumental polarization effects. Spectrometric resolution was not quoted.

According to the authors, the systematic error in their measurements should not exceed $\pm 50\%$ with a relative error of less than $\pm 10\%$.

B. H^+ and H_2^+ Impact

Bobashev and Pop[168] have made measurements on seven visible lines, six of them in the neutral mercury spectrum and one in the doubly ionized spectrum.† Projectiles included H^+ and H_2^+.

In Data Table 7-6-1 are shown the cross sections for emission of the 4047-, 4358-, and 5461-Å lines that compose the $7s\ ^3S_1 \to 6p\ ^3P^0_{0,1,2}$ multiplet. Since the emission from a state is directly proportional to the population of that state, these three lines with a common upper level should exhibit precisely the same energy dependence. It is surprising that this is not the case. It must be concluded that there is some error in the procedures. Possibly the measured intensities include some components from lines other than the one indicated. Inevitably, the identification of this problem for the three lines under consideration must suggest that the same error may have influenced other lines in this series of measurements.

---

† The legends for the figures of the original reference have a number of errors[299] which have been corrected in the present review.

Data Table 7-6-2 includes the emission cross section for the 4078-Å ($7s\,{}^1S_0 \to 6p\,{}^3P_1^0$) and 4916-Å ($8s\,{}^1S_0 \to 6p\,{}^1P_1^0$) lines of Hg I induced by H^+ impact. Data Table 7-6-3 shows the line identified as the 5676-Å ($9p\,{}^1P_1^0 \to 7s\,{}^3S_1$) transition of Hg I. There may be interference with this measurement from the 5677-Å transition of Hg II identified by Paschen.[300] In Data Table 7-6-4 measurements on a line identified[301] as the 4797-Å ($5d^8\,6s^2 \to 5d^9({}^2D_{2\,1/2})6P_{1/2}$) transition of Hg III are shown. Paschen[300] identifies a line of Hg II at this same wavelength. It is not clear whether the measurement represents the emission of the Hg II line, the Hg III line, or a mixture of both.

The emission cross sections for lines induced by the impact of H_2^+ on Hg are shown in Data Tables 7-6-5 to 7-6-8. These include the same lines as for H^+ impact and are arranged in the same manner. Again one notices that the three lines having the $7s\,{}^3S_1$ as a common upper level exhibit obvious difference in energy dependence, leading one to suppose that there is some error involved in the procedures utilized for this experiment. The influence of excited states in the projectile beam was not investigated, but it is expected to be of little importance.

For all of this work, the systematic error is estimated by the authors to be less than $\pm 50\%$ and the relative error as less than $\pm 10\%$.

Bobashev and Pop make no estimates of cascade contributions to the observed emission. Experimental investigations of electron impact excitation of mercury have shown that cascade is a very important contribution to some lines. As an example, the work of Anderson et al.[302] showed that in electron impact excitation the $7s\,{}^3S_1$ state is populated almost entirely by cascade, while the $7s\,{}^1S_0$ receives about 40% cascade population and the $7p\,{}^3P_2$ a neglible amount. It is to be expected that similar results will be applicable to the emission observed as a result of excitation by H^+ and H_2^+.

C.　Data Tables

1.　Cross Sections for Emission of Hg I Lines Induced by Impact of H^+ on an Hg Target

7-6-1　(A) 4047-Å Line ($7s\,{}^3S_1 \to 6p\,{}^3P_0^0$); (B) 4358-Å Line ($7s\,{}^3S_1 \to 6p\,{}^3P_1^0$); (C) 5461-Å Line ($7s\,{}^3S_1 \to 6p\,{}^3P_2^0$).

7-6-2　(A) 4078-Å Line ($7s\,{}^1S_0 \to 6p\,{}^3P_1^0$); (B) 4916-Å Line ($8s\,{}^1S_0 \to 6p\,{}^1P_1^0$).

7-6-3　5676-Å Line ($9p\,{}^1P_1^0 \to 7s\,{}^3S_1$).

7-6-4　4797-Å Line (see text for identification).

2.　Cross Sections for Emission of Hg I Lines Induced by Impact of H^+ on an Hg Target

7-6-5　(A) 4047-Å Line ($7s\,{}^3S_1 \to 6p\,{}^3P_0^0$); (B) 4358-Å Line ($7s\,{}^3S_1 \to 6p\,{}^3P_1^0$); (C) 5461-Å Line ($7s\,{}^3S_1 \to 6p\,{}^3P_2^0$).

7-6-6　(A) 4078-Å Line ($7s\,{}^1S_0 \to 6p\,{}^3P_1^0$); (B) 4961-Å Line ($8s\,{}^1S_0 \to 6p\,{}^1P_1^0$).

7-6-7　5676-Å Line ($9p\,{}^1P_1^0 \to 7s\,{}^3S_1$).

7-6-8　4797-Å Line (see text for identification).

7-6-1A
BOBASHEV ET AL. (168)

ENERGY (KEV)	CROSS SECTION (SQ. CM)
1.10E 01	1.01E-18
1.15E 01	1.08E-18
1.30E 01	1.27E-18
1.40E 01	1.37E-18
1.50E 01	1.58E-18
1.60E 01	1.68E-18
1.70E 01	1.77E-18
1.85E 01	1.89E-18
2.00E 01	1.91E-18
2.20E 01	1.92E-18
2.40E 01	2.03E-18
2.60E 01	2.23E-18
2.80E 01	2.23E-18
3.00E 01	2.34E-18
3.50E 01	2.26E-18

7-6-1B
BOBASHEV ET AL. (168)

ENERGY (KEV)	CROSS SECTION (SQ. CM)
1.10E 01	4.24E-18
1.30E 01	4.73E-18
1.50E 01	5.70E-18
1.70E 01	6.37E-18
2.00E 01	6.82E-18
2.20E 01	7.16E-18
2.40E 01	6.90E-18
2.60E 01	6.88E-18
2.80E 01	6.39E-18
3.00E 01	6.15E-18
3.40E 01	5.79E-18
3.60E 01	5.22E-18

7-6-1C
BOBASHEV ET AL. (168)

ENERGY (KEV)	CROSS SECTION (SQ. CM)
1.10E 01	4.49E-18
1.30E 01	5.25E-18
1.50E 01	6.39E-18
1.7CE 01	6.89E-18
2.00E 01	8.12E-18
2.20E 01	8.71E-18
2.60E 01	8.83E-18
2.80E 01	8.85E-18
3.00E 01	8.32E-18
3.20E 01	8.38E-18
3.40E 01	7.47E-18
3.60E 01	7.17E-18

7-6-2A
BOBASHEV ET AL. (168)

ENERGY (KEV)	CROSS SECTION (SQ. CM)
1.10E 01	1.25E-18
1.30E 01	1.56E-18
1.70E 01	2.20E-18
2.00E 01	2.83E-18
2.20E 01	2.91E-18
2.40E 01	2.80E-18
2.60E 01	2.80E-18
3.00E 01	2.53E-18
3.40E 01	2.37E-18
3.60E 01	2.13E-18

7-6-2B
BOBASHEV ET AL. (168)

ENERGY (KEV)	CROSS SECTION (SQ. CM)
1.10E 01	1.63E-18
1.30E 01	2.17E-18
1.50E 01	2.70E-18
1.70E 01	2.98E-18
2.00E 01	3.64E-18
2.20E 01	3.59E-18
2.40E 01	3.40E-18
2.60E 01	3.40E-18
2.80E 01	3.27E-18
3.00E 01	2.93E-18
3.40E 01	1.87E-18
3.60E 01	1.66E-18

7-6-3
BOBASHEV ET AL. (168)

ENERGY (KEV)	CROSS SECTION (SQ. CM)
1.10E 01	5.50E-18
1.30E 01	4.94E-18
2.00E 01	3.69E-18
2.40E 01	3.25E-18
2.60E 01	3.17E-18
2.80E 01	2.99E-18
3.30E 01	2.88E-18
3.60E 01	2.89E-18

7-6-4
BOBASHEV ET AL. (168)

ENERGY (KEV)	CROSS SECTION (SQ. CM)
1.10E 01	3.68E-18
1.30E 01	4.14E-18
1.50E 01	4.16E-18
1.70E 01	4.16E-18
2.00E 01	4.39E-18
2.20E 01	4.18E-18
2.40E 01	4.33E-18
2.60E 01	4.20E-18
2.80E 01	4.31E-18
3.00E 01	4.13E-18
3.20E 01	4.28E-18
3.40E 01	4.14E-18
3.60E 01	4.20E-18

7-6-5A
BOBASHEV ET AL. (168)

ENERGY (KEV)	POLARIZATION (PERCENT)
5.00E 00	3.47E-18
7.00E 00	3.77E-18
9.00E 00	3.70E-18
1.10E 01	3.00E-18
1.30E 01	3.40E-18
1.50E 01	3.30E-18
1.70E 01	3.37E-18
2.00E 01	3.60E-18
2.20E 01	3.47E-18
2.40E 01	3.50E-18
2.60E 01	3.30E-18
2.80E 01	3.60E-18
3.00E 01	3.80E-18
3.20E 01	3.90E-18
3.40E 01	3.70E-18
3.60E 01	4.00E-18

7-6-5B
BOBASHEV ET AL. (168)

ENERGY (KEV)	POLARIZATION (PERCENT)
5.00E 00	9.83E-18
7.00E 00	9.79E-18
9.00E 00	9.60E-18
1.10E 01	9.20E-18
1.30E 01	8.70E-18
1.50E 01	8.70E-18
1.70E 01	8.91E-18
2.00E 01	8.40E-18
2.20E 01	8.40E-18
2.40E 01	8.60E-18
2.60E 01	8.40E-18
2.80E 01	8.60E-18
3.00E 01	8.70E-18
3.20E 01	9.00E-18
3.40E 01	8.50E-18
3.60E 01	9.00E-18

7-6-5C
BOBASHEV ET AL. (168)

ENERGY (KEV)	POLARIZATION (PERCENT)
5.00E 00	1.22E-17
7.00E 00	1.23E-17
9.00E 00	1.12E-17
1.10E 01	1.05E-17
1.30E 01	1.03E-17
1.50E 01	1.02E-17
1.70E 01	1.01E-17
2.00E 01	1.00E-17
2.20E 01	9.80E-18
2.40E 01	9.80E-18
2.60E 01	9.80E-18
2.80E 01	1.02E-17
3.00E 01	1.00E-17
3.20E 01	1.00E-17
3.40E 01	1.06E-17
3.60E 01	1.07E-17

7-6-6A
BOBASHEV ET AL. (168)

ENERGY (KEV)	CROSS SECTION (SQ. CM)
5.0CE 00	7.40E-19
7.00E 00	7.30E-19
9.0CE 00	8.00E-19
1.10E 01	9.00E-19
1.30E 01	1.08E-18
1.5CE 01	1.30E-18
1.70E 01	1.55E-18
2.00E 01	1.70E-18
2.20E 01	1.90E-18
2.40E 01	2.10E-18
2.60E C1	2.24E-18
2.80E 01	2.40E-18
3.00E 01	2.40E-18
3.20E 01	2.60E-18
3.40E 01	2.60E-18
3.60E 01	2.70E-18

7-6-6B
BOBASHEV ET AL. (168)

ENERGY (KEV)	CROSS SECTION (SQ. CM)
5.00E 00	5.40E-19
7.00E 00	6.00E-19
9.00E 00	6.90E-19
1.10E 01	8.10E-19
1.30E 01	1.02E-18
1.50E 01	1.15E-18
1.70E 01	1.34E-18
2.00E 01	1.50E-18
2.20E 01	1.59E-18
2.40E 01	1.70E-18
2.60E 01	1.75E-18
2.80E 01	1.90E-18
3.00E 01	1.90E-18
3.20E 01	2.00E-18
3.40E 01	2.00E-18
3.60E 01	2.10E-18

7-6-7
BOBASHEV ET AL. (168)

ENERGY (KEV)	CROSS SECTION (SQ. CM)
5.00E 00	6.90E-18
7.00E 00	8.91E-18
9.00E 00	1.03E-17
1.10E 01	1.17E-17
1.30E 01	1.19E-17
1.50E 01	1.16E-17
1.70E 01	1.17E-17
2.00E 01	1.16E-17
2.20E 01	1.12E-17
2.40E 01	1.07E-17
2.60E 01	1.04E-17
2.80E 01	9.50E-18
3.00E 01	9.30E-18
3.20E 01	8.50E-18
3.40E 01	7.80E-18
3.60E 01	8.10E-18

7-6-8
BOBASHEV ET AL. (168)

ENERGY (KEV)	CROSS SECTION (SQ. CM)
5.00E 00	8.30E-19
7.00E 00	1.75E-18
9.0CE 00	2.80E-18
1.10E 01	3.60E-18
1.3CE 01	4.20E-18
1.50E 01	5.02E-18
1.70E 01	5.63E-18
2.00E 01	6.70E-18
2.20E 01	7.00E-18
2.40E 01	7.70E-18
2.60E 01	7.74E-18
2.80E 01	8.00E-18
3.00E 01	8.00E-18
3.20E 01	7.90E-18
3.40E 01	7.30E-18
3.60E 01	8.20E-18

CHAPTER 8

EXCITATION OF MOLECULAR TARGETS

8-1 INTRODUCTION

Considerable effort has been directed to the study of the excitation of H_2, N_2, and O_2 targets. Some limited measurements have been made for other targets, principally oxides of carbon, nitrogen, and some light hydrocarbons. For convenience these are all considered together in the present chapter, although there are only a few features that are common to all cases. The various experiments have included measurements of emission from dissociation fragments as well as those from excitation of the molecule.

Major difficulties occur with molecular emissions because of the complexity and extent of the molecular spectra. Every electronically excited state will produce emissions from various vibrational levels, each with rotational structure. It is most satisfactory to attempt a determination of a cross section for a transition between specific vibrational states. If the rotational structure extends beyond the pass band of the instrument, then the spectrum must be recorded over a suitable wavelength interval and the integrated intensity of the structure determined. General precautions for dealing with spectral structure extending over a wide wavelength interval are documented in Section 3-4-E(2). Measurements on atomic line radiation from dissociation fragments must also be carefully examined to ensure that there is no interference between lines. A detailed high resolution spectral analysis of the emission and identification of all significant features is an important requirement for experiments on the excitation of molecular targets.

The experiments for measuring emission from the target are generally designed on the assumption that an excited system emits a photon at essentially the point where it was excited. This may not be a good assumption, however, in the case of excited dissociation fragments ejected with an appreciable kinetic energy; for example, an $H(3s)$ atom, ejected with 5-eV energy by the dissociation of a molecule, will travel some 5 mm during its natural lifetime and might escape from the observation region before radiative decay occurs. When dealing with emission from dissociation fragments, it is necessary to ensure that the measured cross section is not dependent on the field of view of the optical system. Unfortunately, tests for this problem were not made when deriving any of the data reviewed here. The recoil of the dissociation fragment may be significant when measuring emission from hydrogen atoms; heavy dissociation fragments will recoil with low velocity and will generally not be subject to this problem.

In general, it is impossible to derive level-excitation cross sections from the measurements of molecular emission because branching ratios are not known with high accuracy. Franck-Condon factors may be estimated for many transitions, using simple

Morse potentials and information derived from spectroscopic analysis. The electronic transition moment cannot be readily calculated. In some cases, notably emission in the N_2^+ spectrum, it is possible to make a direct measurement of all transitions that populate and depopulate a given excited state; then Eq. 2-16 may be applied to give a direct estimate of a level excitation cross section without recourse to theoretical estimates of transition probability.

Polarization of molecular emission has recently been observed in electron impact experiments[303] and has been predicted theoretically.[29] Undoubtedly, polarization of molecular emission may occur in heavy-particle collision experiments and may influence the accuracy of cross section measurements. There is no information on whether or not the line radiation produced from dissociation fragments might exhibit polarization. Under the circumstances, it is necessary in all experiments that precautions be taken either to assess the influence of polarization or to carry out the measurements in such a manner that polarization does not influence the accuracy of the cross-section measurement (see Section 3-4-D). None of the experiments on heavy-particle impact excitation has attempted a measurement of polarization; in fact, in most cases the experimental configuration was such that if polarization had been present the quantitative cross section measurements are incorrect.

It should be noted that the target will not be in its lowest vibrational and rotational states before the collision, but will exhibit a Boltzmann population of the states appropriate to the temperature of the target region. Therefore, the measured cross sections are a weighted mean over the various prior-collision states of the molecule. Only in the case of the formation of excited N_2^+ in an N_2 target has the influence of post- and prior-collision rotational states on the excitation cross-section been studied in a systematic manner (see Section 8-3-F). For high velocity projectile impact it appears that the rotational state of the molecule is not changed by the collision, and that the probability of excitation is independent of the prior-collision rotational state; however, at low velocity impact this is not true. A measurement of target temperature is necessary in order to establish the excited state condition of the target. All experimental data have been examined for this information, and generally target temperature measurements have not been specified.

Excitation of hydrogen, nitrogen, and oxygen are considered in Sections 8-2, 8-3, and 8-4, respectively, with all other cases grouped into Section 8-5. All cross sections are presented in cm^2 per molecule of the target.

8-2 EXCITATION OF HYDROGEN

Although hydrogen is the simplest possible molecular structure, the experimental studies of collisionally induced molecular emission are more complicated than for any of the other cases surveyed in this chapter. There are many known bound electronic states of H_2, each of which contributes rotational and vibrational structure to emission spectra. At least several thousand molecular hydrogen lines have been identified.[70] The rotational structure is so open that there are no band heads nor close groups of lines to form readily identifiable characteristic features. In practical collision experiments, where wavelength resolution must be sacrificed to provide acceptable signal levels, it is difficult to unambiguously resolve emission from a specific molecular

transition because the molecular spectral lines tend to be very numerous and weak; the atomic hydrogen lines resulting from dissociation are few in number and correspondingly stronger in intensity. The majority of experiments have concentrated on the formation of excited atomic hydrogen through dissociation processes. Only two rather abbreviated studies have been made on the H_2 molecular emission. It should be noted that no bound electronically excited states of H_2^+ have been identified and neither the excitation of vibrational states nor the excitation of rotational states have been studied experimentally.

Problems common to all experiments on excitation of hydrogen will be discussed first (Section 8-2-A), followed by a survey of the various experimental arrangements used in the work (Section 8-2-B). The available data, including excitation by impact of protons, hydrogen atoms, molecular hydrogen ions, He^+, and heavier projectiles, are discussed in Sections 8-2-C to 8-2-G, respectively; the data are tabulated in Section 8-2-H.

A. General Considerations

The complexities of the H_2 emission spectrum requires that particular attention be paid to the adequacy of spectral resolution. In practice it is almost impossible to provide sufficient resolution to guarantee no interference between spectral features. The atomic hydrogen lines are usually far more intense than the molecular spectrum, and interference can then legitimately be neglected. The molecular hydrogen target, in thermal equilibrium with its surroundings at normal room temperature, will exhibit a small population of excited vibrational states and considerable population of rotational states. The population distribution among the prior-collision states of the target may be completely specified by the determination of target temperature. This has not been done, so one must presume that the nominal temperatures of experiments were in the region of $300°K$. There has been no systematic investigation of how the rotational and vibrational state populations might influence the cross section measurements.

It has been pointed out (Section 2-2-A) that most experimental arrangements are operated on the assumption that the excited atom emits at the same point where it was collisionally excited. This may not be true if a light dissociation fragment in a long-lived state is ejected with a few electron volts of kinetic energy. Where long-lived states may be involved [e.g., $H(ns)$], it is necessary to demonstrate that the measured cross section is independent of the field of view in the detection system. Most detection systems employing filter-photomultiplier combinations have a sufficiently large field of view, so that this problem might legitimately be ignored; however, for spectrometer-based systems which have small fields of view, the influence of this problem should be determined. In none of the experiments described here, were tests made to show that the measured cross section was independent of the field of view. Without definitive tests, it is impossible to assess the quantitative influence of the effect.

The multiplicity of processes that populate and depopulate the excited molecular states precludes derivation of level excitation cross sections from the line emission measurements; consequently there are no data on the cross sections for forming specific molecular states.

The analysis of emission from atomic hydrogen is complicated by the degeneracy of energy levels. For example, the wavelength of Balmer transitions that depopulate the

ns, *np*, and *nd* excited states cannot be spectroscopically resolved. There is a single experiment by Ankudinov et al.[304, 211] that introduces an indirect method for the separate determination of 3*s*, 3*p*, and 3*d*-state populations from measurement of Balmer-alpha emission. The origin of the Lyman transitions (*np* → 1*s*) is unambiguous; if cascade population is neglected, the level excitation cross sections may be calculated from the line emission measurements through the use of theoretical values of transition probabilities. Many experiments are studies of the formation of the 2*s* and 2*p* states of H; the 2*p* state gives rise to spontaneous emission of Lyman-alpha, while the metastable 2*s* state may be detected by the Lyman-alpha emitted as a result of Stark mixing of the 2*s* and 2*p* levels (see Section 2-2-C). From an experimental point of view, the quantitative detection of Lyman transitions in the vacuum uv presents definite practical problems, particularly in the absolute determination of detector sensitivity.

It is to be expected that the spontaneous emission from excited hydrogen atoms and molecules might exhibit polarization. Recently it has also been shown that field-induced emission from the metastable 2*s* state is polarized[240, 20] The available experiments ignore the anisotropy of emission that is related to polarization; inevitably the cross section measurements will be in error by an amount that depends on the polarization fraction (see Section 3-4-D).

B. Experimental Arrangements

1. The Work of Dunn, Van Zyl, and Geballe. This work has been primarily concerned with the spontaneously emitted Lyman-alpha radiation induced by the impact of hydrogen ions on H_2. There are also limited studies of Lyman-alpha emission induced by He^+ and Ar^{2+} impact on H_2. The first important paper in this series by Dunn et al.[99] includes a basic description of the apparatus; all later measurements by this group are normalized in magnitude to this work of Dunn et al. From this early paper we retain for inclusion here only the data concerning H_3^+ impact. Excitation by impact of H^+, D^+, H_2^+, D_2^+, and He^+ is dealt with most extensively in a paper by Van Zyl et al.;[149] it provides the source of the data displayed in this monograph. A brief conference report containing preliminary measurements is disregarded.[305] This same series of investigations includes extensive studies of the formation of fast excited H atoms as a result of the impact of H^+ and H_2^+ on various targets; these are considered in detail in Chapter 9. The projectile beam was produced in an oscillating electron source, mass analyzed, focused, and directed through the interaction region. In an allied experiment the half-width of the ion energy distribution was quoted as 100-eV.[194] This is important because one of the collision processes exhibited rapidly varying structure in an energy interval of about $\frac{1}{2}$-keV; in this region the cross sections were sometimes measured at energy intervals of 50-eV, which is less than the energy resolution. No details are published about the methods used to monitor the ion beam current.

Target pressure was measured with an ionization gauge. Corrections were applied to take into account the different sensitivity of the gauge for the various target gases employed. No discussion has been published about the source nor the adequacy of these gauge correction factors.

The detection system was based on an oxygen-filtered Geiger counter, similar to that described by Brackmann et al.[79] The Geiger counter responded at wavelengths

between the cutoff in the transmission of the LiF window (1080-Å) and the photo-ionization threshold of I_2 (1317-Å). The oxygen has a complicated absorption spectrum with a number of high transmission windows, one of which coincides with the Lyman-alpha wavelength.

All measurements were made on a relative basis. Dunn et al.[99] assigned absolute magnitudes to the data by a normalization procedure. An electron beam was provided in the apparatus to cross the interaction region at the same orientation with respect to the photon detection system, as the ion beam; in this manner, electron-impact excitation cross sections could be measured under conditions identical to those used in the ion-impact work. Fite and Brackmann[98] earlier had made measurements on the emission of Lyman-alpha induced by electrons on H and the "countable ultraviolet" induced by electrons on H_2; this latter quantity includes not only Lyman-alpha emission but possibly a contribution from molecular hydrogen emission that is transmitted by the filter of the detection system. Fite and Brackmann normalized their data for electrons on H to a Born approximation calculation, thereby deriving a cross section for the "countable ultraviolet." Dunn et al.[99] measured the "countable ultraviolet" emitted in the $e + H_2$ reaction, normalized their data to the work of Fite and Brackmann,[98] and thus arrived at a calibration of their apparatus. In the extensive work by Van Zyl et al.[149] the apparatus is the same as that used by Dunn et al.;[99] therefore, the cross sections are referred to the same normalization.

The apparatus used by Dunn et al. differs from that used by Fite and Brackmann in two important respects that might influence the validity of the normalization procedure. First the thickness of the oxygen filter in front of the detector was different in two experiments. A correction for thickness evaluated by Dunn et al. may be criticized on the grounds that it assumes all components of the "countable ultraviolet" emission to be absorbed by the same amount; Dunn et al. do not attempt to substantiate this assumption. Young et al.[9] present evidence that a significant amount of the "countable ultraviolet" is molecular emission and, moreover, that this component is absorbed less in oxygen than is the Lyman-alpha; this is in contradiction to the assumption by Dunn et al. The other difference arises from the use of a weak magnetic field (70 Gs) to collimate the electron beam in the apparatus of Dunn et al.; Young et al.[9] have suggested that the field might sensitize any $H(2s)$ atoms present so that stray electric fields would cause mixing with the $2p$ state and consequently a spurious contribution to the Lyman-alpha signal. It is impossible to assess in retrospect whether either of these problems would give rise to appreciable systematic errors in the normalization procedure used by Dunn et al.

All these experiments utilize a detector at an angle of 90° to the projectile beam. In the event that polarization is present the data will be in error due to neglect of the anisotropy of the emission. The data presented here represent total cross sections only if polarization is absent.

In much of the work by this group the signal comes from both target and projectile; for example, the impact of protons on molecular hydrogen produces Lyman-alpha emission from three possible reactions,

$$H^+ + H_2 \rightarrow H^* + [H + H^+],$$
$$\rightarrow [H^+ + H] + H^*,$$
$$\rightarrow [H + H^+] + H^*,$$

where the order of participants on the left has been preserved. The experiment gives no information on the state of ionization nor excitation of the products shown in brackets. No attempt was made to separate target and projectile emissions; therefore, the data are not due to the excitation of the target alone but will also include projectile contributions. When assessing these data against the criteria for accuracy established in Chapters 2 and 3 it is necessary to expect the criteria for projectile emission to be fulfilled as accurately as those for target emission. A particular problem is that the emission intensity from the projectile will vary with penetration through the target in a manner dependent on the projectile velocity and excited-state lifetime (see Section 2-2-B). The measurement is valid only if the data is obtained in a region where the emission exhibits invariance with penetration. No experimental tests were made to ensure that this criterion was met. Due to the construction of the apparatus the emission resulting from direct excitation will be invariant with penetration at the point of observation; however, there is no knowledge of whether emission resulting from cascade population of the $2p$ state will exhibit this characteristic.

In none of these experiments was attention paid to the possible influence of excited states in the projectile beam. The influence of excited projectiles is expected to be greatest for the emission induced by impact of the molecular hydrogen ions; the emission from the dissociation of the projectile molecule is included in the measurement and may be dependent on the state of vibrational excitation of the molecule. Therefore, the data presented here may be peculiar to the operating conditions of the source used in this experiment.

No attempt was made to assess cascade. As shown previously (Section 2-2-B), cascade in the excited projectile system will vary with penetration through the target. The design of the apparatus was such that the observation region was situated close to the entrance to the collision chamber, and cascade was not at its asymptotic value. Thus, the apparatus had a built-in geometrical discrimination against cascade from states in the projectile system whose lifetime is longer than that of the $2p$ level. The error involved in the neglect of cascade will be considered during the discussion of the data.

The O_2 filter has a very sharp transmission band, and any appreciable Doppler shift of emission from the projectile will cause a decrease in the efficiency with which the Lyman-alpha line is transmitted through the filter. In a related experiment using the same apparatus to measure emission induced by the collisional dissociation of H_2^+ ions, this effect was shown to produce an error up to 10%. No correction was made for this problem.

The wavelength of the Lyman-alpha from deuterium is about 0.3-Å lower than that of hydrogen. The transmission of the filter is lower at this wavelength, and deuterium radiation will be absorbed 1.25 times more than that from the hydrogen atoms.[149] No correction was made for this effect. The apparatus, which Dunn et al. used, was calibrated for H emission and, therefore, only the measurements involving Lyman-alpha emission from this light isotope will be of accurate absolute magnitude. It is expected that the measurement of the cross section for the emission of Lyman-alpha from the reaction $D^+ + D_2$ will be approximately 25% too low, whereas the emission from the similar $H^+ + H_2$ reaction is in accordance with the calibration technique and is therefore correct.

Throughout these experiments there was always the danger of detecting spurious transmissions for excited systems other than neutral hydrogen, particularly the Lyman

bands of H_2. The work of Dahlberg et al.[151] indicates [see Section 8-2-B(4)] that such emission is present for excitation of H_2 by protons. The extent that it might interfere with the Dunn et al. measurements has not been assessed in full; comments by Dunn et al.[99] that such interference is negligible for the case of He^+ on H_2 has been disputed by Young et al.;[9] the problem requires better evaluation.

The data are quoted as being accurate to $\pm 55\%$. Of this figure some $\pm 40\%$ is due to the normalization procedure. The accuracy of the data relative to the calibration point is $\pm 15\%$. In view of the above comments, there may be some additional sources of error that are not taken into account. In particular, all measurements of emission from deuterium are too low by approximately 25%.

2. The Work of Andreev et al. Andreev et al. have studied the emission of Balmer and Lyman lines induced by the impact of H_2^+, He^+, and Ne^+ on a molecular hydrogen target. [211, 148, 210] Although some features of the procedures are common to all three experiments in this series, there are major differences between the detection systems used for the study of the Lyman emissions and those used for the study of the Balmer emission.

Two optical detection systems were used in this work. The first was designed for observation of visible emissions and utilized a glass prism monochromator; the second system was used for measurements in the vacuum ultraviolet and was based on an evacuable grating monochromator. The detection sensitivity of the first system was established by direct absolute calibration; the data from the second system were measured on a relative basis, and absolute magnitudes were assigned by normalization.

All the experiments utilized a $180°$ mass spectrometer with arc discharge source to produce the projectile beam. No attention was directed to the possible influence of excited states in the projectile beam. A McLeod gauge was used to measure target pressure, but no precautions were taken to remove the mercury pumping and thermal transpiration errors associated with the cold trap between the gauge and experiment (see Section 3-2-D). Neglect of these precautions will produce an error of no more than 1 or 2% for these experiments that involve a hydrogen target.

For convenience, the work of this group will be considered in three parts. The first series of measurements[210] were of Balmer series emissions induced by He^+ and Ne^+ impact on H_2 target; it will be shown that these data are of poor quality and should be ignored. The second study[304, 211] was an attempt to derive separate cross sections for the formation of the $3s$, $3p$, and $3d$ states of H resulting from the dissociation of H_2 by He^+; this work was based on the measurement of Balmer-alpha emission and required a study of the variation of emission intensity and polarization with the application of an electric field in the observation region. The final experiments[148] are on the formation of the $2p$ and $3p$ states of H resulting from the dissociation of H_2 by impact of various projectiles; these data were based on measurements of Lyman-alpha and -beta emission.

(a) BALMER SERIES MEASUREMENTS BY ANDREEV ET AL. In this series of experiments, a glass prism monochromator was used to measure the emission of Balmer α, β, γ, δ, and ε lines induced by the impact of He^+ and Ne^+ on an H_2 target. It is stated that no molecular emission was observed; therefore, interference from this source could legitimately be ignored.

The optical system was based on a glass prism monochromator with a photomultiplier detector.[210] A lens was used to form an image of the excitation region on the monochromator entrance slit. The detection sensitivity was calibrated by using a

tungsten ribbon standard lamp, which occupied the same position with respect to the monochromator as the ion beam occupied. Precautions were taken to ensure linearity of response in the detection system through the full dynamic range employed for both experimental measurements and for calibration. Precautions to reduce or to assess the effect of scattered light in the monochromator are not discussed in the report.[210]

These measurements must be criticized for the use of very high target pressures, up to 10^{-2} torr. At this density the projectile beams will have become appreciably neutralized before entering the optical system field of view. A second problem, which might cause errors, is the presence of the collision region within the magnetic field of the mass spectrometer projectile source. Magnetic field strengths were not quoted but did reach 2000 Gs in a related experiment.[157] Due to the magnetic field, an excited hydrogen atom ejected from the dissociation event with an energy of 5-eV will experience a motional electric field up to 60-V/cm, which is quite sufficient to Stark-mix all states of a particular principal quantum number.

In view of the above comments, it is likely that the Balmer series measurements by Andreev et al.[210] are inaccurate. One may compare these data for the emission of the Balmer-alpha line with more recent measurements carried out by Ankudinov et al.[211] in an improved version of the same apparatus. For the impact of He^+ on H_2 at 20-keV energy, the early experiments measure the cross section for the emission of the Balmer-alpha line as 1.43×10^{-17} cm^2; for the same collision event the more recent data by Ankudinov et al.[211] on the separate $3s$, $3p$, and $3d$ cross sections would predict a Balmer-alpha emission cross section of 4.41×10^{-18} cm^2. The discrepancy between these two figures is substantial. Andreev[299] suggests that the original measurements of the Balmer series are in error, either through the use of high target densities or through the problems associated with the magnetic field. Consequently, these data are omitted from further consideration.

(b) $3s$, $3p$, AND $3d$ STATE MEASUREMENTS BY ANKUDINOV ET AL. The second series of experiments[304, 211] was an attempt to measure separately the cross sections for the formation of the $3s$, $3p$, and $3d$ states in the dissociation of H_2 by impact of He^+. The apparatus used for this experiment was basically the same as that described above in connection with measurements on the Balmer series of lines. The major differences were that the observation region was taken outside the magnetic field of the ion source, and proper attention was directed to the establishment of single-collision conditions. Thus, the experimental inadequacies that contribute to the invalidation of the previous experiments are not present here.

Quantitative measurements of Balmer-alpha emission provide only a weighted sum of the cross sections for exciting the $3s$, $3p$, and $3d$ states and does not allow the derivation of individual level-excitation cross sections. Ankudinov et al. develop the strong field treatment of the Stark effect and show that the field dependence of emission intensity and polarization are different functions of the three cross sections. With the three independent measurements—light intensity without field, light intensity with a strong field, and polarization in a strong field—it is possible to elucidate the three separate cross sections, $\sigma(3s)$, $\sigma(3p)$, and $\sigma(3d)$. Two important assumptions are required. First, the population of the magnetic sublevels in the field-free case should be equal; this was justified by the observation that emission was unpolarized. Second, the cross sections for the collisional excitation process should be the same in the

applied fields as in the field-free case; this was not justified either through experiment or through recourse to theory.

Ankudinov et al.[304, 211] measured the polarization using a polaroid film analyzer. Precautions were taken to ensure equal detection sensitivity to light polarized in different planes. The detection sensitivity of the optical system was determined by using a standard tungsten strip filament lamp as a calibration source.

Ankudinov et al. show that a check of the validity of these procedures may be obtained by considering the variation of the Lyman-beta $(3p \rightarrow 1s)$ intensity as a function of field strength. It is possible to predict the percentage of increase in intensity of this line in terms of the cross sections for the $3s$, $3p$, and $3d$ states that were derived from the measurements of Balmer emission. This predicted ratio agrees well with the experimental measurement.

It was estimated that the random error in the determination of the $3p$ and $3d$ cross sections was of the order $\pm 10\%$ while the error for the $3s$ state, which exhibits the smallest cross section, is of the order $\pm 100\%$. In addition, there is a possible systematic error, up to 20%, due primarily to uncertainties in the calibration of detection sensitivity. These experiments were originally published in a conference paper[304] and have been superseded by a more complete description.[211]

(c)   ULTRAVIOLET EMISSION MEASUREMENTS BY ANDREEV ET AL.   The final group of measurements[148] involve a study of the formation of the $2s$, $2p$, and $3p$ states induced by the impact of H^+, H_2^+, and He^+ on H_2. The results are based on measurements of the emission of the Lyman-alpha and -beta lines in the vacuum ultraviolet.

The optical system employed an evacuable Seya—Namioka type monochromator; a photomultiplier with a sodium salicylate screen was used as a detector. (Details of the system are in a charge transfer paper by the same authors.[75]) The monochromator was connected directly to the collision chamber, the entrance slit forming a partial barrier against the flow of target gas into the optical system. The monochromator was placed at an angle (unspecified) to the ion beam to allow separation of target and projectile emissions by the Doppler shift of wavelength.

The experiments involved measurements of spontaneous emission of Lyman-alpha $(2p \rightarrow 1s)$, spontaneous emission of Lyman-beta $(3p \rightarrow 1s)$, and field induced emission of Lyman-alpha $(2s \rightarrow 2p \rightarrow 1s)$. This experiment measured only the relative variation of emission cross section with impact energy; absolute magnitudes of cross sections were assigned by normalization to other data determined previously by the same group of authors. The measurements of Lyman-alpha emission were normalized to an absolute determination of the cross section for emission of this line induced by the impact of H^+ on Ne^{75} [see Section 9-2-A(2)]; the cross section for this process at 16-keV impact energy was taken as 8.9×10^{-18} cm^2. The cross sections for the Lyman-beta emission were obtained by normalization to the cross sections for the formation of the $H(3p)$ state in the dissociation of H_2 by He^+ at 16-keV energy; this cross section, obtained by using the visible light monochromator as described in Section (b) above, was taken as 2.1×10^{-18} cm^2.

In all of this work the possibility that the emission was polarized, and therefore anisotropic, was ignored. Furthermore, the polarization of emission was also ignored in the experiments that provided the basis for normalization of the Lyman-alpha cross section data. There is also the possibility of error due to the inherent instrumental polarization of the monochromator. Polarization of spontaneously emitted radiation

may vary as a function of impact energy; polarization of the field-induced emission[240, 20, 306] will be invariant with projectile energy. It is impossible to assess, in retrospect, the magnitudes of the errors that might have been introduced by the neglect of polarization.

In all of these experiments the influence of cascade population of the excited states was ignored. The influence of this will be considered in the discussion of the data in later sections.

The authors of this work estimate that the systematic errors in these measurements will not exceed $\pm 30\%$. This estimate does not include the possible errors due to neglect of polarization and cascade. No figure was quoted for reproducibility of the data.

3. The Work of Dahlberg, Anderson, and Dayton. Dahlberg, Anderson, and Dayton have studied Lyman-alpha emission from atomic hydrogen and Lyman-band emission from molecular hydrogen induced by the impact of H^+ and H on H_2.[151] The apparatus was described in an earlier paper devoted to the study of the collisional excitation of nitrogen.[182]

The mass-analyzed projectile beam was provided by an rf ion source and Cockcroft-Walton accelerator (10- to 130-keV). The neutral atomic hydrogen beam was formed by charge transfer neutralization of protons in an exchange cell that preceded the collision chamber; electrostatic deflection plates removed the remaining charged components in the beam. Inevitably, the neutral projectile beam will contain a small excited state component; there were no tests to determine whether this will significantly influence the measured cross sections.

The current of H^+ projectiles was monitored on a Faraday cup in a conventional manner. The flux of neutral atoms was monitored using a secondary emission technique; the efficiency of detection was calibrated with an H^+ beam, and the assumption was made that the ratio of secondary emission coefficients for H and H^+ impact were the same as those determined by Stier et al.[41] The measured H current may be in error because secondary emission coefficients will depend on the nature and condition of the material and the angle of incidence; the results of Stier et al. are not necessarily applicable to the apparatus utilized by Dahlberg et al.

Target pressures were monitored with a capacitance manometer type of pressure gauge whose response is independent of the nature of the target gas. The temperature of the target was not quoted.

The experiment was confined to the study of vacuum uv lines, and the optical system was based on an evacuated monochromator with Czerny–Turner optics placed at a 90° angle to the beam path. No assessment was made of emission isotropy nor polarization. A photomultiplier with a sodium salicylate screen was used as a detector. The measurements of Lyman-alpha emission included contributions from both excited projectiles and excited target fragments. It is therefore necessary that the experiments be examined against both the criteria for target and projectile emission measurement. The influence of cascade on these measurements was ignored.

All the measurements by Dahlberg et al. are relative and were assigned absolute values by normalization to other work. It was assumed that the detection efficiency at the wavelength of the Lyman molecular band (1606-Å) was the same as the efficiency at the Lyman-alpha wavelength (1216-Å); this unsubstantiated assumption cannot be correct. The cross section magnitudes were established by normalizing to a measurement of Lyman-alpha emission induced by H^+ impact on N_2. It will be shown [see

Section 9-2-A(6)] that the magnitude of the $H^+ + N_2$ cross section was established by a further normalization procedure that was both invalid and erroneous.[†] It follows that the absolute magnitudes of the data by Dahlberg et al. are incorrect; the magnitude of the error cannot readily be estimated.

No estimates of the reliability of this data are offered by Dahlberg and his co-workers. From the above comments, it is clear that the absolute magnitudes are wrong.

4. The Work of Edwards and Thomas. Edwards and Thomas present measurements of cross sections for atomic and molecular emissions induced by the impact of H^+ on a target of H_2. The work was originally published in a conference paper[307] and later superseded and extended by a formal publication.[150]

The projectile beam was provided by a Van de Graaff accelerator with an energy range from 150- to 1000-keV. The optical system was based on a $\frac{1}{2}$ m Ebert-Fastie mounted grating monochromator fitted with a photomultiplier. The monochromator was placed at an angle of 90° to the direction of the projectile beam for measurements of molecular emission and a 60° angle for atomic emission in order to separate target and projectile contributions by means of the Doppler shift of wavelength. For calibration a standard tungsten strip filament lamp was placed at the point normally occupied by the ion beam; the temperature of the lamp filament was measured directly with an optical pyrometer and the relevant values of emissive power were obtained from tables.[83] Target pressures were monitored with a capacitance manometer type of pressure gauge whose response is independent of the nature of the target gas.

The polarization of the emission was ignored, so errors might have been introduced through neglect of emission anisotropy and instrumental polarization effects. Target pressures were maintained sufficiently low to ensure single-collision conditions throughout. Particular care was taken to eliminate the effects of scattered light within the monochromator when carrying out the absolute calibration. The resolution was poor, and the molecular line emissions may include some interference from neighboring lines.

The authors estimate that the random scatter in the data was less than $\pm 8\%$, and the absolute accuracy was better than $\pm 25\%$.

5. The Work of Young et al. Young et al.[9] utilized a crossed beam apparatus to measure the excitation of Lyman-alpha line by impact of He^+ on H_2. The apparatus was designed primarily for studies of the interaction of ion beams with an atomic hydrogen target beam; molecular hydrogen could be thermally dissociated in a furnace and formed into a beam by effusion through an orifice. For the purposes of this experiment the furnace was kept at room temperature; therefore, the target beam was of hydrogen molecules. The projectiles were provided by an rf source and the two beams intersected at an angle of 90°. Considerable care was taken to minimize collisions of the molecular hydrogen beam on residual gases and also to prevent Stark mixing of the $2s$ and $2p$ states by stray fields in the interaction region. The molecular hydrogen beam was modulated, and true signals were identified in the presence of background noise by means of their specific frequency and phase. Throughout these experiments no attempt was made to determine the form factor

---

[†] In the discussion[151] of the normalization procedure for the $H^+ + H_2$ experiment the impression is given that the data are normalized directly to the measurements of Van Zyl et al.[149] for that collision combination. Actually the normalization used involves the less direct technique described here.[308]

describing the density distributions of the two beams (see Section 2-1-C), and no arguments were presented that would justify neglecting it.

Measurements were made only of relative variations of cross section with impact energy. Therefore, the efficiencies for detecting the Lyman-alpha emission and the H_2 beam flux were not determined. No attempt was made to assess the influence of the excited states in the projectile beam, but it is expected that most excited ions will decay before entering the observation region and that the influence of the remaining excited ions will be quite negligible

The photon detector was an iodine-filled Geiger counter with a gaseous oxygen filter. The detector was at an angle of 54.5° to the ion beam; therefore, the signal was independent of emission anisotropy associated with polarization. Provision was made for crossing the hydrogen beam with an electron beam and viewing the emission with the same photon detector. All ion impact measurements were made relative to the signal induced at 10-keV impact energy. This signal was then compared with the emission from the $e + H$ reaction at 300-eV energy, and the ion impact data were normalized to the cross section for electron impact excitation of Lyman-alpha emission determined previously by Fite et al.[98] These electron measurements had, in turn, been normalized to the theoretical predictions of the first Born approximation at energies from 200- to 700-eV. It should be noted that there will be a small error in these procedures because the measurement includes cascade contributions while the theory does not. The electron and ion beams did not intersect the molecular beam at the same point. A correction was evaluated for the different neutral beam thicknesses and densities at the two points on the assumption that gas flow from the furnace was effusive. This requires that the slit in the furnace wall be very thin and the dimensions be very much smaller than the mean free path of the atoms. No experimental evidence was presented to support the validity of this procedure.

The data from this experiment are expressed as emission cross sections, and no corrections are evaluated for cascade population. The total probable error was estimated as 35 to 40%, of which some 30% arises in the normalization to Fite's data.

6. The Work of Hughes et al. Hughes et al.[147, 198] have measured Balmer emission from an H_2 target induced by impact of H^+, H_2^+, and H_3^+. The early report of this work,[197] being completely superseded by separate reports of the measurements with H^{+}[147] and H_2^+[198] projectiles,† should be disregarded.

The optical system was based on a $\frac{1}{2}$-m grating monochromator fitted with photomultiplier detection. Separation of emission from target and projectile was facilitated by arranging the monochromator axis at an angle of 30° to the beam path; projectile emission will be Doppler-shifted in wavelength.

An unusual feature of this apparatus was the arrangement of the monochromator slit to be parallel to the ion beam path. An image of the collision region was formed on the entrance slit by a lens, and the slit width was maintained sufficiently large to accept all light emitted from the geometrical width of the ion beam. A disadvantage of this arrangement is that long-lived dissociation fragments may be ejected with sufficient velocity to cause an appreciable fraction of the emissions to occur outside the limited field of view of the optical system; this might cause a significant loss of

---

† The measurements of emission from fast H atoms formed by charge transfer and dissociation of the projectiles in these experiments[147, 198] are incorrect[175] and should be completely disregarded (see Chapter 9).

signal. A further complication is that an Ebert—Fastie grating mount will exhibit considerable curvature of the spectral line in the exit plane.[247] In this experiment the full length of straight, equal-width, entrance and exit slits were employed, thereby causing a loss of signal through the image curvature. It is not possible to assess the magnitude of any errors that might have been introduced through the use of slits parallel to the ion beam path.

The absolute sensitivity of the optical detection system was calibrated with a standard lamp in a geometrical arrangement that involved completely different solid angles and fields of view from the situation when viewing the collision region. Hughes et al. made some tests[248] to show the response of the system to be constant over the fields of view for both experimental and calibration situations.

The collision chamber was insulated from ground, and the chamber itself served as the Faraday cup. Some tests were made[248] to ensure that electrons and ions ejected from the end of the chamber and returning to the observation region did not produce an appreciable fraction of the observed radiation. Target pressure was monitored with a cold-trapped McLeod gauge, but no precautions were taken to prevent errors due to mercury pumping onto the cold trap and thermal transpiration (see Section 3-2-D).

The authors state that the systematic errors do not exceed $\pm 40\%$. Random errors, influencing the relative accuracy of measurements, are stated to be less than $\pm 5\%$. These estimated limits of accuracy are conservative.

7. The Work of Gusev et al. Gusev and co-workers have studied emission of the Balmer-alpha and -beta lines of atomic hydrogen induced by the impact of He^+, He, Ne^+, Ne, Ar^+, and Ar projectiles on an H_2 target. The report is in a brief conference abstract.[109] Details of experimental procedures are the same as those published earlier by Gusev,[111] Polyakova,[112] and co-workers.

Ions were produced in an arc source, accelerated to energies between 0.1- and 30-keV, and directed into the observation region. An achromatic lens system was used to focus the emission onto the entrance slit of a monochromator, and the light was detected by a photomultiplier. The optical system and the detection system were the same as those described in a previous paper by Polyakova et al.[309] Neutral projectile beams were formed by charge transfer neutralization of ions in a gas cell, condensor plates were used to sweep out any residual flux of ions.

Ion beam current was measured on a Faraday cup. Neutral beams were monitored with a thermal detector that had previously been calibrated with an ion beam of known current. There was no mass spectrometer included in the experimental arrangement,[112] so there is a possibility that the beams of charged projectiles will exhibit some contamination with multiple charged species. No attempt was made to examine the influence on the cross section measurements of excited atoms and ions that undoubtedly exist in the projectile beams. It is possible that at the lowest impact energies used in this experiment (100-eV) a small proportion of excited projectiles might significantly influence the measured cross section. No indication was given of the energy resolution of the projectile beam; at the lowest impact velocities the energy distribution of the projectiles emanating from the ion source might be a substantial fraction of the mean energy.

No information was given on the methods used to measure target pressure. It was stated in previous work[112] that target pressures were kept below 10^{-3} torr and that linearity of signal with target density was confirmed.

No attempt was made to calibrate the detection efficiency of the optical system; absolute magnitudes were assigned by normalization to previous measurements by Andreev et al.[210] of the Balmer emission induced by He$^+$ impact on H_2. In the discussion of this previous work by Andreev et al. [Section 8-2-B(2)] it was concluded that the absolute magnitudes of this data are unreliable; the target pressures were too high (10^{-2} torr) for single-collision conditions to be operative, and the experiment was carried out in a high magnetic field that might seriously influence both the collisional excitation and the spontaneous decay processes. Since the data of Andreev et al. are not reliable, no confidence may be placed in the absolute magnitudes of the work by Gusev et al. These comments do not, however, affect the validity of the measured variation of cross section with projectile energy, and even the ratios of cross sections for emission of a given line in different collision combinations should still be reliable.

It has previously been noted (Section 8-2-A) that when studying dissociation of H_2 it is necessary to confirm by experimental tests that the measured cross section is independent of the optical system field of view; no such tests are reported in this work. The work of Gusev et al. also ignores the possibility that the emission may be polarized, and the consequent effects on the accuracy of cross section measurements (see Section 3-4-D).

The data are expressed as total emission cross sections. No estimates were given for the accuracy limits of these measurements.

C. Proton Impact

1. Emission of Lyman-Alpha and Lyman-Beta. In Data Tables 8-2-1 and 8-2-2 are shown the measurements of Lyman-alpha emission induced by the impact of protons on an H_2 target. Data Table 8-2-1 includes measurements where emission from both target and projectile are detected together. Data Table 8-2-2 involves a single measurement on the excitation of the target alone.

Included in Data Table 8-2-1 are measurements by Van Zyl et al.[149] for excitation by D$^+$ and for targets of D_2 as well as H_2. The H$^+$ + D_2, D$^+$ + H_2, and D$^+$ + D_2 cross sections were not calibrated separately but were normalized to the cross section for H$^+$ + H_2 at 6-keV, 4.5-keV, and 3-kev energy respectively. The measurements for D$^+$ impact are plotted on the graphs and shown in the tables as though they were protons of equal velocity. Thus, the true energy of the D$^+$ projectiles is twice that shown on the data tables and graphs.

There is a substantial discrepancy in Data Table 8-2-1 between the measurements by Van Zyl et al.[149] and by Dahlberg et al.[151] It was previously concluded that the absolute values of the data by Dahlberg et al. are unreliable [Section 8-2-B(3)] due to faulty normalization procedures. Consequently, the measurements by Van Zyl et al. must be considered as the most accurate data that are available. There is no way in which the cascade contribution to the population of the $2p$ state can be estimated. Without an estimate of cascade, it is not possible to determine the relationship between the cross section for the formation of the $2p$ state and these measurements of the $2p \rightarrow 1s$ emission.

The cross sections for emission from the target alone, measured by Andreev et al.[148] and shown in Data Table 8-2-2, lie considerably below the measurements of combined target and projectile emission as shown in Data Table 8-2-1. Again in this case, there is no estimate of cascade population of the excited state.

Data Table 8-2-3 shows a measurement of the cross section for exciting the $3p$ state in the target alone, as determined by Andreev et al.[148] In this case the emission measurements were converted into a level excitation cross section on the assumption that cascade was negligible. No evidence was presented to support this assumption. Andreev et al. have also provided measurements on the $3p$ state population as a function of an electric field applied in the collision chamber; that data is not reproduced here.

2. **Excitation of the 2s State.** Andreev et al[148] have studied the formation of the $2s$ metastable excited state using the field quenching technique. The measurements shown in Data Table 8-2-4 are cross sections for field-induced emission and include cascade population of the $2s$ state from higher np levels. One may estimate cascade by taking the cross section for the excitation of the $3p$ state (Data Table 8-2-3) and multiplying by the branching ratio of 0.1185 to get the population of the $2s$ state by the transition $3p \rightarrow 2s$. The cascade is appreciable, amounting to 4.2% of the field induced emission at 18-keV and 15% at 31-keV. Obviously when cascade from higher np states is also included, the correction will be larger.

3. **Emission of the Balmer-Lines.** Hughes et al.[147] and also Edwards and Thomas[150] have made measurements of the emission in the first three lines of the Balmer series shown in Data Tables 8-2-5 to 8-2-7. The absolute magnitudes appear to differ by a factor of about 2.4. This difference between data from these two groups has been noted previously in the study of excitation of helium and was there attributed to differences in absolute calibration (Section 6-2-A).

In addition, Dieterich[259] studies Balmer-alpha emissions induced by H^+ on H_2 at 2-keV; however, the target pressures were so high that the data was unreliable; they are not reproduced here.

4. **Molecular Emissions.** The study of molecular emissions is difficult due to the great complexity of the spectrum. Two studies have been made, giving results shown in Data Tables 8-2-8 to 8-2-10. It is likely that the absolute accuracy of these measurements will be poor due to the complexity of the line structure. In neither experiment was any attempt made to ensure that the measured cross section was independent of spectral resolution.

Dahlberg et al.[151] have measured a cross section for the emission of the (4, 11), (5, 12), and (6, 13) vibrational transitions of the ultraviolet Lyman band ($B\ ^1\Sigma_u^+ \rightarrow X\ ^1\Sigma_g^+$); the measurements are shown in Data Table 8-2-8. It has been noted [Section 8-2-B(3)] that the absolute magnitudes of this data are unreliable. In Data Tables 8-2-9 and 8-2-10 are shown emission measurements by Edwards and Thomas[150] for the (1, 0) transition of the $3d\ ^2\Pi_g \rightarrow 2p\ ^1\Sigma_u^+$ decay and the (0, 0) transition of the $3d\ ^1\Sigma_g^+ \rightarrow 2p\ ^1\Sigma_u^+$ decay, both in the visible spectrum. There is no way of estimating level-excitation cross sections from this data.

Moore and Doering[152] have studied the energy loss of 150-eV H^+ that has traversed a target of H_2. Qualitative inferences concerning the excitation of vibrational states may be drawn from these data; however, there is no cross section information in this work.

D. Hydrogen Atom Impact

Dahlberg et al.[151] have provided a measurement of the cross section for emission of Lyman-alpha induced by the impact of atomic H on H_2, as shown in Data Table 8-2-11. The magnitudes of the cross sections are unreliable [see Section 8-2-B(3)].

This emission cross section includes contributions from both target and projectile excitation. Dahlberg et al. made a single measurement of molecular emission induced by atomic hydrogen impact. They arrived at a value of 1.25×10^{-18} cm^2 at 50-keV impact energy for the sum of the emission from the (4, 11), (5, 12), and (6, 13) vibrational transitions of the $B\,^1\Sigma_u^+ \to X\,^1\Sigma_g^+$ Lyman band in the vacuum uv.

In the discussion of Dahlberg's apparatus [Section 8-2-B(3)]; it was noted that inadequate attention was paid to accurate monitoring of the neutral projectile flux and that the projectiles will include a proportion of atoms in excited states. It is impossible to assess what effect, if any, will be produced on the data by these inadequacies.

E. H$_2^+$ and H$_3^+$ Impact

There are extensive measurements of the excitation of hydrogen by the impact of the molecular ions H$_2^+$ and H$_3^+$. In none of the experiments for which data are reviewed here has any study been made of how the vibrational excitation state of the projectile influences the measured cross section. It is likely that this will produce little effect on the cross section for excitation of the target, but it may be of some importance in experiments where both target and projectile excitation are measured together.

In Data Table 8-2-12 are shown the measurements by Van Zyl et al.[149] for the collisionally induced emission of Lyman-alpha for H$_2^+$ and D$_2^+$ impact on H$_2$ and D$_2$. The D$_2^+$ projectile determinations are shown on the data tables and graphs as though they were H$_2^+$ projectiles of equal velocity, that is, at D$_2^+$ energy $\div$ 2. These data include emission from both the projectile and target systems. In this experiment the detection efficiency for emission from D was lower than that for emission from H, but no correction was made for this difference. Since the calibration was made for atomic H, the measurement involving H$_2^+$ + H$_2$ must be considered as nominally correct. The D$_2^+$ + D$_2$ data will be low by about 25%. The other two cases involving H$_2^+$ + D$_2$ and D$_2^+$ + H$_2$ will also be low, but by an indeterminate amount that depends on the relative intensities of the emission from D and H. When corrections are estimated for these differences, all four sets of data would tend to approach the H$_2^+$ + H$_2$ measurements. There is no way of estimating cascade contributions to these measurements.

In Data Table 8-2-13 are shown the measurements by Andreev et al.[148] for the excitation of the 2p state in the target alone. This lies substantially below the data by Van Zyl et al. that includes both projectile and target emission. Again no correction can be made for cascade.

The cross sections for formation of the 3p level, shown in Data Table 8-2-14, include cascade population of the 3p level.[299] Andreev et al.[299] also present information concerning the influence of electric field on the emission cross section; these data are not presented here.

In Data Table 8-2-15 is given the cross section for field-induced emission from the metastable 2s excited state. One may estimate a correction for cascade by taking the cross section for the excitation of the 3p state shown in Data Table 8-2-16, multiplying by the branching ratio of 0.1185 to get the population of the 2s state by the transition $3p \to 2s$ and subtracting this from the measured 2s state measurements shown in Data Table 8-2-15. The correction amounts to about 3.7% at 18-keV and 3.2%

at 30-keV. Obviously when cascade from higher states is included the correction will be larger.

Hatfield and Hughes[259] have presented measurements of the cross sections for emission of the Balmer-alpha, -beta, and -gamma lines from the target, induced by the impact of H_2^+; these are shown in Data Table 8-2-16. There is no accurate method whereby level-excitation cross sections may be obtained from these data.

Data Table 8-2-17 shows the cross section for emission of "countable ultraviolet" induced by the impact of H_3^+ on H_2 (as measured by Dunn et al.[99]). This signal will contain emission from both the projectile and target. Although molecular emissions may be present, the predominant contribution is likely to be Lyman-alpha emission. No attempt was made to estimate the relative importance of molecular emissions; therefore, the result is given the ambiguous designation as a cross section for the emission of "countable ultraviolet." No other data are available for comparison.

Cross sections for emission of Balmer-alpha and -beta lines induced by impact of H_3^+ ions on an H_2 target are shown in Data Table 8-2-18. No attempt is made to derive level-excitation cross sections from this.

Dieterich[259] studied Balmer-alpha emission induced by the impact of H_2^+ on H_2 at 2-keV impact energy. The measurements involved high target pressures up to 2.5×10^{-2} torr, and undoubtedly considerable projectile neutralization took place. These data are likely to be both wrong and misleading; therefore they are not reproduced here.

F. He⁺ and He Impact

An appreciable amount of work has been done on the excitation of hydrogen by impact of He^+. In this case there is no problem of interference from projectile emissions. It is likely that the projectile beam emanating from the ion source will contain some ions in excited states. The presence of electric and magnetic fields in the acceleration and mass analysis regions of the apparatus will cause Stark mixing among the various excited states; therefore, longer-lived ns and nd states will tend to decay with the same order of lifetime as the np states. It is probable that a small proportion of long-lived He^+ ions in high n states will survive passage to the collision chamber. It is unlikely that a small proportion of such excited atoms will influence the cross section measurements presented here. Also included here are data on emission of Balmer lines induced by impact of neutral helium atoms.

Data Table 8-2-19 shows information on Lyman-alpha emission by three different authors. The work of Van Zyl et al.[149] is normalized to the same theoretical basis as that by Young et al.;[9] however, the experimental arrangements are very different, the former experiment employing a beam static gas configuration and the latter a crossed-beam apparatus. The work of Andreev et al.[148] is absolute and entirely independent of that from the other two workers. Agreement between the three determinations is quite good; discrepancies occur primarily in the absolute magnitudes of the data. Van Zyl et al. claim to establish the threshold for the excitation mechanism at 30- to 50-eV. The energy resolution in the experiment by Van Zyl et al. was quoted as 100-eV while the reports of the other experiments do not give a figure for this quantity. There is a systematic difference between the energy dependence of the data by Van Zyl et al. and that by Young et al. at energies below 1-keV. It has been suggested[9]

that in the apparatus used by Van Zyl et al. the ion beam exhibits considerable divergence at low impact energies; as a consequence some emission may occur in a region of the detector's field of view where the efficiency is poor, giving rise to low cross section values.

Included in Data Table 8-2-19 is the cross section for emission of Lyman-alpha induced by the impact of He^+ on D_2 measured by Van Zyl et al.[149] The detection efficiency of the apparatus for Lyman-alpha from D was stated to be less than that for emission from H by about 25%. Since the apparatus is calibrated for H, the emission from D is in error by this amount. It is interesting to note, however, that the ratio of the emission from the H_2 target to that from the D_2 target varies from approximately 1.3 at energies from 15- to 25-keV to a ratio of 1.5 at 1-keV. This suggests that there is a small difference between the behavior of the excitation process for D_2 and that for H_2.

All the data derived from the Lyman-alpha measurements are presented as emission cross sections. They are equal to the sum of cross sections for populating the $2p$ state by collisional excitation and by cascade. One may estimate the cascade population, using the cross section for exciting the $3s$ and $3d$ states of the atom measured by Ankudinov et al.[211] as displayed in Data Tables 8-2-21 and 8-2-23. These two states will decay only into the $2p$ state and produce a 12% cascade population at an impact energy of 20-keV. Cascade population must be larger since higher ns and nd states must also be taken into account. Consequently it would be inaccurate to regard the Lyman-alpha emission cross sections as being equal to the cross section for formation of the $H(2p)$ state by direct collisional dissociation of the H_2 molecule.

The data on formation of the metastable state by Andreev et al.[149] are shown in Data Table 8-2-20. The technique determines the effective cross section for populating the $2s$ state by direct excitation and by cascade. An estimate of cascade population may be determined from the measurements of the $3p$ cross section carried out in the same experiment;[299] it amounts to 11% at 20-keV. Clearly, there is appreciable population of the $2s$ state by cascade and the data do not represent the cross section for direct excitation of the $2s$ state.

In Data Tables 8-2-21 through 8-2-23 are given the cross sections for forming the $3s$, $3p$, and $3d$ states obtained from the Balmer-alpha emission by Andreev et al., using the indirect technique described in Section 8-2-A(2). Data Table 8-2-24 gives the sum of the $3s$, $3p$, and $3d$ state cross sections, that is, the total population of the $n = 3$ state. It should be noted that none of these measurements includes assessment of cascade population of the excited states; therefore, the data are the sum of cross sections for populating the relevant state by direct collisional excitation and by cascade from higher levels.

Included in Data Table 8-2-22 is the cross section for the $3p$ state derived from measurements of the Lyman-beta line.[148] This measurement is normalized to the $3p$ state cross section derived from the Balmer-alpha measurements. There is good agreement between the energy variations of these two measurements except at the lowest energy where there is an unexplained discrepancy that lies outside the random error of the experiments.

Measurements by Andreev et al.[211, 148] on the variation of the intensity of the Lyman-beta and Balmer-alpha emissions with applied electric field are not reproduced here. The variations of intensity were ascribed to mixing of the $3s$, $3p$, and $3d$ states as a result of the electric field and not to changes in the collisional excitation cross section.

Gusev et al.[109] have studied the emission of the Balmer-alpha and -beta lines induced by the impact of He^+ and He on a target of H_2; the results are shown in Data Tables 8-2-25 and 8-2-26. It has been concluded [see Section 8-2-B(7)] that the absolute magnitudes of these data are not reliable. Relative variation of cross section with impact energy and the relative magnitudes of either the H_α or H_β lines for the two different projectiles do, nevertheless, represent accurate data; the measurements are in the form of emission cross sections.

Andreev et al.[210] have carried out measurements of the emission of Balmer series lines induced by He^+ impact. In our discussion of the experimental technique [Section 8-2-B(2)] it was concluded that these data may be in error because of the high target pressures utilized and the existence of a magnetic field in the collision region. The magnitudes of these errors cannot be estimated, and the data are not reproduced here on the grounds that they might be misleading.

Hanle and Voss[199] carried out a series of relative measurements of Balmer emission induced by He^+ impact on H_2 and D_2. They used such high target pressures that up to 27% of the projectile beam was neutralized. These data, not being accurate, are omitted from the present compilation.

G. Impact of Other Projectiles

Gusev et al.[109] have studied the emission of Balmer-alpha and -beta lines induced by the impact of Ne^+, Ne, Ar^+, and Ar on a target of H_2. The results are shown in Data Tables 8-2-27 through 8-2-30. As concluded in Section 8-2-B(7), the absolute magnitudes of these data are not reliable. Nevertheless, the relative variation of cross section with impact energy and the relative magnitudes of either the H_α or H_β lines for the different projectiles do represent accurate data.

Dunn et al.[99] have provided a brief series of measurements for the emission induced by Ar^{2+} on H_2. The statistical accuracy was poor, and error bars of $\pm 20\%$ were drawn on the points. Dunn et al.[151] regard these data as being for the emission of "countable ultraviolet" on the grounds that no tests were made to determine whether emissions other than Lyman-alpha were being detected. In view of the uncertainty as to the origin of the emission and its poor accuracy, this measurement is not included in the data compilation.

Andreev et al.[210] have studied the emission of the Balmer series lines induced by impact of Ne^+ on a target of H_2. In the discussion of this experiment [Section 8-2-B(2)] it was concluded that the data are probably inaccurate due to use of high target pressures (10^{-2} torr) and to the presence of a magnetic field in the observation region. These data are excluded from the compilation on the grounds that they are inaccurate and might be misleading.

Van Eck et al.[222] have studied emission of the Balmer-beta line induced by the impact of Li^+ on H_2; however the target pressures were very high (10^{-2} torr), and an appreciable fraction of the projectile beam was neutralized. These data do not permit estimation of cross sections, and are therefore not presented here.

There have been two experiments to study excitation of vibrational transitions in H_2, induced by impact of very low energy projectiles. Schöttler and Toennies[223] have studied excitation by impact of Li^+ ions at energies up to 30-eV. The experimental method involves studying the energy loss of the projectiles. Some information is presented on the relative probability of exciting different vibrational states. Moran and Cosby[108] carried out a similar study of H_2 excitation by Ar^+ impact; data is

presented on relative probabilities of exciting the lower vibrational states of N_2. Neither the work of Schöttler and Toennies nor that of Moran and Cosby provide any data on cross sections; therefore both are excluded from further consideration.

H. Data Tables

In all the data tables the cross sections are expressed in cm^2 per molecule of the target. The reader should take particular note that data for D^+ and D_2^+ impact are shown on the tables as though they were H^+ and H_2^+ projectiles of equal velocity. Therefore, the quoted D^+ and D_2^+ energies are the actual energies in the experiment divided by two.

1. Cross Sections for Emission and Excitation Induced by Impact of H^+ on an H_2 Target

8-2-1 Lyman Alpha $(2p \rightarrow 1s)$ Emission† [(A) $H^+ + H_2$; (B) $H^+ + D_2$; (C) $D^+ + H_2$; (D) $D^+ + D_2$; (E) $H^+ + H_2$].

8-2-2 Lyman-alpha Emission‡ $(2p \rightarrow 1s)$.

8-2-3 Formation of the $3p$ state of H.‡§

8-2-4 Field Induced Emission of Lyman-alpha $(2s \rightarrow 2p \rightarrow 1s)$.‡

8-2-5 Emission of the 6563-Å (Balmer-alpha) Line $(n = 3 \rightarrow n = 2)$.‡

8-2-6 Emission of the 4861-Å (Balmer-beta) Line $(n = 4 \rightarrow n = 2)$.‡

8-2-7 Emission of the 4340-Å (Balmer-gamma) Line $(n = 5 \rightarrow n = 2)$.‡

8-2-8 Emission of the 1606-Å Lyman Band of H_2 [$B^1\Sigma_u^+ \rightarrow X^1\Sigma_g^+$ (4, 11), (5, 12), and (6, 13)].

8-2-9 Emission of the 4180-Å Band of H_2 [$3d\ ^1\Pi_g \rightarrow 2p\ ^1\Sigma_u^+$ (1, 0)].

8-2-10 Emission of the 4632-Å Band of H_2 [$3d\ ^1\Sigma_g \rightarrow 2p\ ^1\Sigma_u^+$ (0, 0)].

2. Cross Sections for Emission and Excitation Induced by the Impact of H on an H_2 Target

8-2-11 Lyman-alpha Emission.†

3. Cross Sections for Emission and Excitation Induced by Impact of H_2^+ on an H_2 Target

8-2-12 Lyman-alpha Emission† $(2p \rightarrow 1s)$. [(A) $H_2^+ + H_2$; (B) $H_2^+ + D_2$; (C) $D_2^+ + H_2$; (D) $D_2^+ + D_2$].

8-2-13 Lyman-alpha Emission $(2p \rightarrow 1s)$.‡

8-2-14 Formation of the $3p$ state of H.‡§

8-2-15 Field Induced Emission of Lyman-alpha $(2s \rightarrow 2p \rightarrow 1s)$.‡

8-2-16 Emission of Balmer Lines.‡ (A) 6563-Å (Balmer-alpha) $n = 3 \rightarrow n = 2$; (B) 4861-Å (Balmer-beta) $n = 4 \rightarrow n = 2$; (C) 4340-Å (Balmer-gamma) $n = 3 \rightarrow n = 2$.

4. Cross Sections for Emission and Excitation Induced by Impact of H_3^+ on an H_2 Target

8-2-17 Cross Section for the Emission of "Countable Ultraviolet"† (see text).

8-2-18 Emission of Balmer Lines.‡ (A) 6563-Å (Balmer-alpha) $n = 3 \rightarrow n = 2$; (B) 4861-Å (Balmer-beta) $n = 4 \rightarrow n = 2$.

5. Cross Sections for Emission and Excitation Induced by Impact of He^+ on an H_2 Target

8-2-19 Lyman Alpha Emission [(A), (C), and (D) are He^+ on H_2; (B) is He^+ on D_2].

8-2-20 Field Induced Emission of Lyman Alpha. $(2s \rightarrow 2p \rightarrow 1s)$.

8-2-21 Formation of the $3s$ State of H.§

8-2-22 Formation of the $3p$ State of H.§

8-2-23 Formation of the $3d$ State of H.§

8-2-24 Formation of the $n = 3$ State of H.§ (Sum of cross sections shown in Data Tables 8-2-21, 8-2-22, and 8-2-23.)

8-2-25 Emission of Balmer Lines. (A) 6563-Å Balmer-alpha (n $= 3 \rightarrow n = 2$); (B) 4861-Å Balmer-beta ($n = 4 \rightarrow n = 2$).

6. Cross Sections for Emission Induced by Impact of He on a Target of H_2

8-2-26 Emission of Balmer Lines. (A) 6563-Å Balmer-alpha ($n = 3 \rightarrow n = 2$); (B) 4861-Å Balmer-beta ($n = 4 \rightarrow n = 2$).

7. Cross Sections for Emission Induced by Impact of Heavy Particles on a Target of H_2

8-2-27 Emission of Balmer Lines Induced by Impact of Ne^+. (A) 6563-Å Balmer-alpha ($n = 3 \rightarrow n = 2$); (B) 4861-Å Balmer-beta ($n = 4 \rightarrow n = 2$).

8-2-28 Emission of Balmer Lines Induced by Impact of Ne. (A) 6563-Å Balmer-alpha ($n = 3 \rightarrow n = 2$); (B) 4861-Å Balmer-beta ($n = 4 \rightarrow n = 2$).

8-2-29 Emission of Balmer Lines Induced by Impact of Ar^+. (A) 6563-Å Balmer-alpha ($n = 3 \rightarrow n = 2$); (B) 4861-Å Balmer-beta ($n = 4 \rightarrow n = 2$).

8-2-30 Emission of Balmer Lines Induced by Impact of Ar. (A) 6563-Å Balmer alpha ($n = 3 \rightarrow n = 2$); (B) 4861-Å Balmer-beta ($n = 4 \rightarrow n = 2$).

† These data include both target and projectile excitation.

‡ These data include target excitation only.

§ Cross section for this level evaluated on the assumption that cascade may be neglected.

8-2-1A
VAN ZYL ET AL. (149)

ENERGY (KEV)	CROSS SECTION (SQ. CM)
5.00E-01	1.17E-17
7.50E-01	1.87E-17
1.00E 00	2.18E-17
1.50E 00	2.46E-17
2.00E 00	2.46E-17
2.50E 00	2.56E-17
3.00E 00	2.90E-17
3.50E 00	3.12E-17
4.00E 00	3.41E-17
4.50E 00	3.59E-17
5.00E 00	4.02E-17
5.50E 00	4.48E-17
6.00E 00	4.57E-17
6.50E 00	5.26E-17
7.00E 00	5.00E-17
8.00E 00	5.43E-17
9.00E 00	5.87E-17
1.00E 01	6.26E-17
1.10E 01	6.69E-17
1.20E 01	7.16E-17
1.30E 01	7.34E-17
1.40E 01	7.35E-17
1.50E 01	7.08E-17
1.60E 01	7.13E-17
1.70E 01	6.77E-17
1.80E 01	7.02E-17
1.90E 01	6.82E-17
2.00E 01	6.61E-17
2.10E 01	6.68E-17
2.20E 01	6.41E-17
2.30E 01	6.86E-17
2.40E 01	5.95E-17
2.50E 01	5.82E-17

8-2-1B
VAN ZYL ET AL. (149)

ENERGY (KEV)	CROSS SECTION (SQ. CM)
5.00E-01	8.00E-18
6.00E-01	1.03E-17
7.00E-01	1.29E-17
8.00E-01	1.81E-17
9.00E-01	2.08E-17
1.00E 00	2.24E-17
1.50E 00	2.34E-17
2.00E 00	2.37E-17
2.50E 00	2.53E-17
3.00E 00	2.81E-17
4.00E 00	3.54E-17
5.00E 00	3.95E-17
6.00E 00	4.57E-17
7.00E 00	4.97E-17
8.00E 00	5.44E-17
9.00E 00	6.06E-17
1.00E 01	6.42E-17
1.10E 01	6.59E-17
1.20E 01	6.91E-17
1.30E 01	6.67E-17
1.40E 01	6.96E-17
1.50E 01	6.90E-17
1.60E 01	6.38E-17

(continuation of 8-2-1A)

ENERGY (KEV)	CROSS SECTION (SQ. CM)
1.70E 01	6.07E-17
1.80E 01	6.10E-17
1.90E 01	5.92E-17
2.00E 01	6.43E-17
2.10E 01	6.20E-17
2.20E 01	5.60E-17
2.30E 01	5.68E-17
2.40E 01	5.84E-17
2.50E 01	5.17E-17

8-2-1C
VAN ZYL ET AL. (149)

ENERGY (KEV)	CROSS SECTION (SQ. CM)
5.00E-01	1.77E-17
1.00E 00	2.31E-17
1.50E 00	2.42E-17
2.00E 00	2.23E-17
2.50E 00	2.44E-17
3.00E 00	2.66E-17
3.50E 00	2.95E-17
4.00E 00	3.41E-17
4.50E 00	3.59E-17
5.00E 00	4.04E-17
5.50E 00	4.48E-17
6.00E 00	4.75E-17
6.50E 00	4.78E-17
7.00E 00	4.87E-17
7.50E 00	5.16E-17
8.00E 00	5.28E-17
8.50E 00	5.61E-17
9.00E 00	5.84E-17
9.50E 00	6.14E-17
1.00E 01	6.17E-17
1.05E 01	6.29E-17
1.10E 01	6.62E-17
1.15E 01	6.56E-17
1.20E 01	6.65E-17
1.25E 01	6.32E-17

8-2-1D
VAN ZYL ET AL. (149)

ENERGY (KEV)	CROSS SECTION (SQ. CM)
2.00E-01	4.00E-18
3.00E-01	5.00E-18
3.50E-01	5.10E-18
4.00E-01	8.80E-18
4.50E-01	8.10E-18
5.00E-01	9.20E-18
1.00E 00	1.96E-17
1.25E 00	2.26E-17
1.50E 00	2.38E-17
1.75E 00	2.41E-17
2.00E 00	2.37E-17
2.25E 00	2.30E-17
2.50E 00	2.63E-17
3.00E 00	2.90E-17
3.50E 00	3.37E-17
4.00E 00	3.60E-17
4.50E 00	3.91E-17
4.75E 00	3.90E-17
5.00E 00	3.92E-17
5.25E 00	4.02E-17

(continuation of 8-2-1D)

ENERGY (KEV)	CROSS SECTION (SQ. CM)
5.50E 00	4.28E-17
6.00E 00	4.85E-17
6.50E 00	5.23E-17
7.00E 00	5.49E-17
7.50E 00	5.89E-17
8.00E 00	5.86E-17
8.50E 00	6.54E-17
9.00E 00	6.88E-17
9.50E 00	7.10E-17
1.00E 01	7.40E-17
1.05E 01	7.39E-17
1.10E 01	7.61E-17
1.15E 01	7.84E-17
1.20E 01	7.56E-17
1.25E 01	7.47E-17

8-2-1E
DAHLBERG ET AL. (151)

ENERGY (KEV)	CROSS SECTION (SQ. CM)
1.30E 01	4.90E-17
1.60E 01	4.80E-17
2.00E 01	4.50E-17
2.40E 01	4.20E-17
3.00E 01	3.50E-17
3.60E 01	3.10E-17
4.00E 01	2.90E-17
4.50E 01	2.55E-17
5.00E 01	2.30E-17
5.50E 01	2.15E-17
6.00E 01	2.05E-17
7.00E 01	1.75E-17
8.00E 01	1.65E-17
9.00E 01	1.50E-17
1.00E 02	1.40E-17
1.10E 02	1.35E-17
1.20E 02	1.25E-17
1.30E 02	1.20E-17

8-2-2
ANDREEV ET AL. (148)

ENERGY (KEV)	CROSS SECTION (SQ. CM)
1.00E 01	2.60E-17
1.25E 01	3.21E-17
1.43E 01	3.35E-17
1.62E 01	3.72E-17
1.83E 01	3.80E-17
2.05E 01	3.56E-17
2.25E 01	3.41E-17
2.44E 01	3.27E-17
2.64E 01	3.16E-17
2.88E 01	2.96E-17
3.09E 01	2.85E-17
3.30E 01	2.49E-17
3.51E 01	2.17E-17

8-2-3
ANDREEV ET AL. (148)

ENERGY (KEV)	CROSS SECTION (SQ. CM)

1.22E 01	1.80E-18
1.40E 01	1.90E-18
1.83E 01	2.36E-18
2.03E 01	2.33E-18
2.24E 01	2.04E-18
2.43E 01	1.90E-18
2.65E 01	1.56E-18
2.85E 01	1.39E-18
3.07E 01	1.07E-18

8-2-4
ANDREEV ET AL. (148)

ENERGY (KEV)	CROSS SECTION (SQ. CM)
1.02E 01	4.90E-18
1.43E 01	4.85E-18
1.64E 01	5.95E-18
1.85E 01	6.65E-18
2.05E 01	7.40E-18
2.25E 01	7.75E-18
2.43E 01	6.50E-18
2.66E 01	7.90E-18
2.87E 01	6.25E-18
3.09E 01	8.40E-18
3.28E 01	8.80E-18
3.51E 01	7.50E-18

8-2-5A
HUGHES ET AL. (147)

ENERGY (KEV)	CROSS SECTION (SQ. CM)
5.0CE 00	2.12E-18
1.00E 01	4.43E-18
1.50E 01	5.01E-18
2.00E 01	4.60E-18
2.50E 01	3.85E-18
3.50E 01	2.96E-18
4.00E 01	2.75E-18
5.00E 01	2.45E-18
6.5CE 01	1.93E-18
8.50E 01	1.54E-18
1.00E 02	1.33E-18
1.30E 02	1.11F-18

8-2-5B
EDWARDS ET AL. (150)

ENERGY (KEV)	CROSS SECTION (SQ. CM)
1.50E 02	4.14E-19
2.00E 02	3.49E-19
2.50E 02	2.79E-19
3.00E 02	2.40E-19
3.50E 02	2.13E-19
4.00E 02	1.89E-19
4.50E 02	1.75E-19
5.00E 02	1.58E-19
6.00E 02	1.35E-19
7.00E 02	1.25E-19
8.00E 02	1.09E-19
9.00E 02	9.90E-20

8-2-6A
HUGHES ET AL. (147)

ENERGY (KEV)	CROSS SECTION (SQ. CM)
5.0CE 00	3.10E-19
1.00E 01	8.50E-19
1.50E 01	9.10E-19
2.00E 01	8.50F-19
2.50E 01	6.70E-19
3.50E 01	5.10E-19
4.00E 01	4.60E-19
5.00E 01	3.90F-19
6.50E 01	3.00E-19
8.50E 01	2.30E-19
1.00E 02	2.00E-19
1.30E 02	1.50E-19

8-2-6B
EDWARDS ET AL. (150)

ENERGY (KEV)	CROSS SECTION (SQ. CM)
1.50E 02	6.39E-20
2.00E 02	5.46E-20
2.50E 02	4.46E-20
3.00E 02	3.60E-20
3.50E 02	3.22E-20
4.00E 02	2.86E-20
4.50E 02	2.71E-20
5.00E 02	2.60E-20
6.00E 02	2.12E-20
7.00E 02	1.85E-20
8.00E 02	1.62E-20
9.00E 02	1.39E-20
1.00E 03	1.29E-20

8-2-7A
HUGHES ET AL. (147)

ENERGY (KEV)	CROSS SECTION (SQ. CM)
5.00E 00	7.00E-20
1.00E 01	2.00E-19
1.50E 01	2.20E-19
2.00E 01	2.10E-19
2.5CE 01	1.70E-19
3.50E 01	1.30E-19
4.00E 01	1.20E-19
5.00E 01	9.00E-20
6.50E 01	8.00E-20
8.50E 01	6.00E-20
1.00E 02	5.00E-20
1.30E 02	4.00E-20

8-2-7B
EDWARDS ET AL. (150)

ENERGY (KEV)	CROSS SECTION (SQ. CM)
1.50E 02	1.91E-20

2.00E 02	1.54E-20
2.50E 02	1.25E-20
3.00E 02	1.02E-20
3.50E 02	9.70E-21
4.00E 02	8.70E-21
4.50E 02	7.90E-21
5.00E 02	7.10E-21
6.00E 02	6.10E-21
7.00E 02	5.40E-21
8.00E 02	5.00E-21
9.00E 02	4.40E-21
1.00E 03	3.90E-21

8-2-8
DAHLBERG ET AL. (151)

ENERGY (KEV)	CROSS SECTION (SQ. CM)
2.00E 01	3.30E-18
2.50E 01	4.30E-18
3.00E 01	4.50E-18
3.50E 01	4.90E-18
4.00E 01	4.90E-18
4.50E 01	5.10E-18
5.00E 01	4.80E-18
5.50E 01	5.10E-18
6.00E 01	5.15E-18
7.00E 01	5.05E-18
8.00E 01	4.95E-18
9.00E 01	4.70E-18
1.00E 02	4.70E-18
1.10E 02	4.55E-18
1.20E 02	4.40E-18
1.30E 02	4.20E-18

8-2-9
EDWARDS ET AL. (150)

ENERGY (KEV)	CROSS SECTION (SQ. CM)
1.50E 02	7.90E-21
2.00E 02	6.40E-21
3.00E 02	4.40E-21
4.00E 02	3.10E-21
5.00E 02	2.60E-21
6.00E 02	2.35E-21
7.00E 02	1.79E-21
8.00E 02	1.63E-21
9.00E 02	1.40E-21

8-2-10
EDWARDS ET AL. (150)

ENERGY (KEV)	CROSS SECTION (SQ. CM)
1.50E 02	1.37E-20
2.00E 02	1.08E-20
2.50E 02	8.90E-21
3.00E 02	7.30E-21
3.50E 02	6.30E-21
4.00E 02	5.40E-21
4.50E 02	4.90E-21
5.00E 02	4.40E-21

ENERGY (KEV)	CROSS SECTION (SQ. CM)
6.00E 02	3.80E-21
7.00E 02	3.30E-21
8.00E 02	2.90E-21
9.00E 02	2.50E-21
1.00E 03	2.20E-21

8-2-11
DAHLBERG ET AL. (151)

ENERGY (KEV)	CROSS SECTION (SQ. CM)
1.30E 01	3.90E-17
1.60E 01	3.70E-17
2.00E 01	3.40E-17
2.40E 01	3.10E-17
2.50E 01	2.95E-17
3.00E 01	2.75E-17
3.60E 01	2.60E-17
4.00E 01	2.50E-17
4.50E 01	2.40E-17
5.00E 01	2.30E-17
5.50E 01	2.10E-17
6.00E 01	2.20E-17
7.00E 01	2.05E-17
8.00E 01	2.05E-17
9.00E 01	1.90E-17
1.00E 02	1.90E-17
1.10E 02	1.75E-17
1.20E 02	1.95E-17

8-2-12A
VAN ZYL ET AL. (149)

ENERGY (KEV)	CROSS SECTION (SQ. CM)
1.00E 00	3.24E-17
2.00E 00	4.82E-17
3.00E 00	5.19E-17
4.00E 00	5.63E-17
5.00E 00	5.94E-17
6.00E 00	6.16E-17
7.00E 00	6.58E-17
8.00E 00	6.90E-17
9.00E 00	7.08E-17
1.00E 01	7.35E-17
1.10E 01	7.46E-17
1.20E 01	7.63E-17
1.30E 01	8.27E-17
1.40E 01	8.29E-17
1.50E 01	8.54E-17
1.60E 01	8.64E-17
1.70E 01	8.73E-17
1.80E 01	9.12E-17
1.90E 01	9.32E-17
2.00E 01	9.06E-17
2.10E 01	9.28E-17
2.20E 01	9.48E-17
2.30E 01	9.42E-17
2.40E 01	9.57E-17
2.50E 01	9.73E-17

8-2-12B
VAN ZYL ET AL. (149)

ENERGY (KEV)	CROSS SECTION (SQ. CM)
1.00E 00	3.78E-17
2.00E 00	4.44E-17
3.00E 00	4.81E-17
4.00E 00	5.12E-17
5.00E 00	5.42E-17
6.00E 00	5.60E-17
7.00E 00	5.96E-17
8.00E 00	6.22E-17
9.00E 00	6.50E-17
1.00E 01	6.72E-17
1.10E 01	6.91E-17
1.20E 01	6.90E-17
1.30E 01	7.11E-17
1.40E 01	7.31E-17
1.50E 01	7.70E-17
1.60E 01	7.79E-17
1.70E 01	7.87E-17
1.80E 01	7.92E-17
1.90E 01	7.85E-17
2.00E 01	8.06E-17
2.10E 01	8.30E-17
2.20E 01	8.15E-17
2.30E 01	8.28E-17
2.40E 01	8.36E-17
2.50E 01	8.26E-17

8-2-12C
VAN ZYL ET AL. (149)

ENERGY (KEV)	CROSS SECTION (SQ. CM)
5.00E-01	2.08E-17
1.00E 00	2.67E-17
1.50E 00	3.20E-17
2.00E 00	3.65E-17
2.50E 00	4.05E-17
3.00E 00	4.30E-17
3.50E 00	4.58E-17
4.00E 00	4.77E-17
4.50E 00	4.82E-17
5.00E 00	4.99E-17
5.50E 00	5.11E-17
6.00E 00	5.34E-17
6.50E 00	5.61E-17
7.00E 00	5.75E-17
7.50E 00	5.80E-17
8.00E 00	6.11E-17
8.50E 00	6.24E-17
9.00E 00	6.43E-17
9.50E 00	6.68E-17
1.00E 01	6.81E-17
1.05E 01	6.98E-17
1.10E 01	7.22E-17
1.15E 01	7.25E-17
1.20E 01	7.23E-17
1.25E 01	7.44E-17

8-2-12D
VAN ZYL ET AL. (149)

ENERGY (KEV)	CROSS SECTION (SQ. CM)
5.00E-01	1.69E-17
1.00E 00	2.30E-17
1.50E 00	3.01E-17
2.00E 00	3.49E-17
2.50E 00	3.71E-17
3.00E 00	3.93E-17
3.50E 00	4.10E-17
4.00E 00	4.36E-17
4.50E 00	4.43E-17
5.00E 00	4.46E-17
5.50E 00	4.56E-17
6.00E 00	4.76E-17
6.50E 00	4.82E-17
7.00E 00	5.10E-17
7.50E 00	5.24E-17
8.00E 00	5.29E-17
8.50E 00	5.38E-17
9.00E 00	5.52E-17
9.50E 00	5.73E-17
1.00E 01	5.68E-17
1.05E 01	5.79E-17
1.10E 01	5.78E-17
1.15E 01	5.97E-17
1.20E 01	5.86E-17
1.25E 01	6.10E-17

8-2-13
ANDREEV ET AL. (148)

ENERGY (KEV)	CROSS SECTION (SQ. CM)
1.21E 01	1.37E-17
1.38E 01	2.11E-17
1.60E 01	2.55E-17
1.81E 01	3.04E-17
2.01E 01	2.96E-17
2.19E 01	3.22E-17
2.43E 01	3.72E-17
2.61E 01	4.08E-17
2.87E 01	3.75E-17

8-2-14
ANDREEV ET AL. (148)

ENERGY (KEV)	CROSS SECTION (SQ. CM)
1.20E 01	1.52E-18
1.39E 01	1.74E-18
1.59E 01	2.06E-18
1.79E 01	2.35E-18
1.99E 01	2.54E-18
2.19E 01	2.54E-18
2.41E 01	2.51E-18
2.62E 01	2.39E-18
2.84E 01	2.23E-18
3.06E 01	2.05E-18
3.21E 01	2.02E-18

8-2-15
ANDREEV ET AL. (148)

ENERGY (KEV)	CROSS SECTION (SQ. CM)
1.22E 01	4.15E-18
1.41E 01	7.65E-18
1.62E 01	5.40E-18
1.84E 01	7.50E-18
2.00E 01	6.80E-18
2.21E 01	6.80E-18
2.44E 01	7.15E-18
2.66E 01	7.75E-18
2.94E 01	7.65E-18

8-2-16A
FATFIELD ET AL. (198)

ENERGY (KEV)	CROSS SECTION (SQ. CM)
1.00E 01	2.55E-18
1.50E 01	3.79E-18
2.00E 01	5.03E-18
2.50E 01	5.90E-18
3.00E 01	6.34E-18
3.50E 01	6.40E-18
4.00E 01	6.37E-18
4.50E 01	6.22E-18
5.00E 01	6.03E-18
6.00E 01	5.55E-18
7.00E 01	4.91E-18
8.00E 01	4.65E-18
9.00E 01	4.12E-18
1.00E 02	3.87E-18
1.10E 02	3.49E-18
1.20E 02	3.29E-18
1.30E 02	3.12E-18

8-2-16B
FATFIELD ET AL. (198)

ENERGY (KEV)	CROSS SECTION (SQ. CM)
1.00E 01	3.94E-19
1.50E 01	6.82E-19
2.00E 01	9.22E-19
2.50E 01	1.07E-18
3.00E 01	1.18E-18
3.50E 01	1.20E-18
4.00E 01	1.19E-18
4.50E 01	1.16E-18
5.00E 01	1.15E-18
6.00E 01	1.04E-18
7.00E 01	9.64E-19
8.00E 01	8.72E-19
9.00E 01	7.82E-19
1.00E 02	7.21E-19
1.10E 02	6.60E-19
1.20E 02	6.18E-19
1.30E 02	6.00E-19

8-2-16C
HARFIELD ET AL. (198)

ENERGY (KEV)	CROSS SECTION (SQ. CM)
1.50E 01	1.70E-19
2.00E 01	2.11E-19
2.50E 01	2.55E-19
3.00E 01	2.95E-19
3.50E 01	3.15E-19
4.00E 01	3.13E-19
4.50E 01	3.11E-19
5.00E 01	3.01E-19
6.00E 01	2.68E-19
7.00E 01	2.54E-19
8.00E 01	2.25E-19
9.00E 01	1.93E-19
1.00E 02	1.90E-19
1.10E 02	1.64E-19
1.20E 02	1.60E-19
1.30E 02	1.58E-19

8-2-17
DUNN ET AL. (99)

ENERGY (KEV)	CROSS SECTION (SQ. CM)
6.00E-01	1.93E-17
7.00E-01	2.27E-17
8.00E-01	2.49E-17
9.00E-01	2.46E-17
1.00E 00	2.54E-17
1.10E 00	2.44E-17
1.20E 00	2.55E-17
1.30E 00	2.58E-17
1.40E 00	2.65E-17
1.50E 00	2.68E-17
1.60E 00	2.67E-17
1.70E 00	2.69E-17
1.80E 00	2.73E-17
1.90E 00	2.66E-17
2.00E 00	2.68E-17
2.10E 00	2.74E-17
2.20E 00	2.77E-17
2.30E 00	2.76E-17
2.40E 00	2.77E-17
2.50E 00	2.77E-17
2.60E 00	2.68E-17
2.70E 00	2.85E-17
2.80E 00	2.82E-17
2.90E 00	2.88E-17
3.00E 00	2.82E-17
3.20E 00	2.77E-17
3.30E 00	3.28E-17
3.40E 00	2.87E-17
3.50E 00	3.24E-17
3.60E 00	3.01E-17
3.80E 00	3.12E-17
4.00E 00	2.99E-17
4.10E 00	3.40E-17
4.20E 00	3.02E-17
4.30E 00	3.39E-17
4.40E 00	3.26E-17
4.50E 00	3.31E-17
4.60E 00	3.24E-17
4.80E 00	3.35E-17
5.00E 00	3.14E-17
5.20E 00	3.42E-17
5.40E 00	3.11E-17
5.50E 00	3.37E-17

8-2-18A
HATFIELD ET AL. (198)

ENERGY (KEV)	CROSS SECTION (SQ. CM)
2.00E 01	3.74E-18
2.50E 01	4.40E-18
3.00E 01	5.12E-18
3.50E 01	5.33E-18
4.00E 01	5.85E-18
4.50E 01	6.38E-18
5.00E 01	6.74E-18
5.50E 01	6.90E-18
6.00E 01	7.10E-18
6.50E 01	7.22E-18
7.00E 01	7.30E-18
7.50E 01	7.30E-18
8.00E 01	7.35E-18
9.00E 01	7.33E-18
1.00E 02	7.12E-18
1.10E 02	7.10E-18
1.20E 02	7.07E-18
1.30E 02	6.91E-18

8-2-18B
FATFIELD ET AL. (198)

ENERGY (KEV)	CROSS SECTION (SQ. CM)
2.00E 01	6.19E-19
2.50E 01	7.25E-19
3.00E 01	8.40E-19
3.50E 01	9.33E-19
4.00E 01	1.00E-18
4.50E 01	1.08E-18
5.00E 01	1.14E-18
5.50E 01	1.20E-18
6.00E 01	1.22E-18
6.50E 01	1.26E-18
7.00E 01	1.26E-18
7.50E 01	1.27E-18
8.00E 01	1.25E-18
9.00E 01	1.25E-18
1.00E 02	1.26E-18
1.10E 02	1.26E-18
1.20E 02	1.21E-18
1.30E 02	1.19E-18

8-2-19A
VAN ZYL ET AL. (149)

ENERGY (KEV)	CROSS SECTION (SQ. CM)
3.00E-02	0.0
5.00E-02	2.20E-17
1.00E-01	4.49E-17
1.50E-01	5.45E-17
2.00E-01	5.89E-17
2.50E-01	6.16E-17

ENERGY (KEV)	CROSS SECTION (SQ. CM)
3.00E-01	6.24E-17
3.50E-01	6.16E-17
4.00E-01	6.07E-17
4.50E-01	5.89E-17
5.00E-01	5.64E-17
6.00E-01	5.19E-16
7.00E-01	4.92E-17
8.00E-01	4.75E-17
9.00E-01	4.57E-17
1.00E 00	4.40E-17
1.25E 00	4.04E-17
1.50E 00	3.78E-17
1.75E 00	3.61E-17
2.00E 00	3.52E-17
3.00E 00	3.35E-17
4.00E 00	3.31E-17
5.00E 00	3.29E-17
6.00E 00	3.32E-17
7.00E 00	3.36E-17
8.00E 00	3.44E-17
9.00E 00	3.46E-17
1.00E 01	3.46E-17
1.10E 01	3.51E-17
1.20E 01	3.60E-17
1.30E 01	3.69E-17
1.40E 01	3.70E-17
1.50E 01	3.74E-17
1.60E 01	3.85E-17
1.70E 01	3.86E-17
1.80E 01	3.83E-17
1.90E 01	3.79E-17
2.00E 01	3.78E-17
2.10E 01	3.76E-17
2.20E 01	3.67E-17
2.30E 01	3.70E-17
2.40E 01	3.58E-17
2.50E 01	3.66E-17

8-2-19B
VAN ZYL ET ALL (149)

ENERGY (KEV)	CROSS SECTION (SQ. CM)
1.00E 00	2.95E-17
1.50E 00	2.56E-17
2.00E 00	2.30E-17
3.00E 00	2.23E-17
4.00E 00	2.25E-17
5.00E 00	2.29E-17
6.00E 00	2.35E-17
7.00E 00	2.48E-17
8.00E 00	2.57E-17
9.00E 00	2.60E-17
1.00E 01	2.55E-17
1.10E 01	2.70E-17
1.20E 01	2.66E-17
1.30E 01	2.68E-17
1.40E 01	2.79E-17
1.50E 01	2.89E-17
1.60E 01	2.88E-17
1.70E 01	2.91E-17
1.80E 01	2.95E-17
1.90E 01	2.98E-17
2.00E 01	3.10E-17
2.10E 01	3.09E-17
2.20E 01	3.12E-17
2.30E 01	2.95E-17
2.40E 01	2.95E-17
2.50E 01	2.81E-17

8-2-19C
YOUNG ET AL. (9)

ENERGY (KEV)	CROSS SECTION (SQ. CM)
5.00E-01	8.15E-17
7.50E-01	5.60E-17
1.00E 00	4.70E-17
1.25E 00	4.30E-17
2.00E 00	3.95E-17
3.00E 00	3.70E-17
5.00E 00	3.95E-17
7.00E 00	3.75E-17
1.00E 01	4.00E-17
1.50E 01	4.25E-17
2.00E 01	4.35E-17
2.50E 01	4.25E-17
3.00E 01	4.25E-17

8-2-19D
ANDREEV ET AL. (148)

ENERGY (KEV)	CROSS SECTION (SQ. CM)
8.20E 00	2.56E-17
1.01E 01	2.82E-17
1.24E 01	3.06E-17
1.42E 01	3.19E-17
1.61E 01	3.27E-17
1.86E 01	3.27E-17
2.08E 01	3.29E-17
2.22E 01	3.17E-17
2.42E 01	3.14E-17
2.64E 01	3.08E-17
2.84E 01	2.85E-17
3.04E 01	2.53E-17

8-2-20
ANDREEV ET AL. (148)

ENERGY (KEV)	CROSS SECTION (SQ. CM)
8.10E 00	3.00E-18
1.01E 01	3.85E-18
1.23E 01	4.85E-18
1.44E 01	4.60E-18
1.60E 01	3.85E-18
1.84E 01	3.50E-18
2.08E 01	2.75E-18
2.25E 01	3.85E-18
2.45E 01	2.40E-18
2.64E 01	2.15E-18
2.80E 01	3.50E-18
3.05E 01	3.95E-18

8-2-21
ANKUDINOV ET AL. (211)

ENERGY (KEV)	CROSS SECTION (SQ. CM)
1.00E 01	2.17E-19
1.20E 01	2.24E-19
1.40E 01	2.31E-19
1.60E 01	2.45E-19
1.80E 01	2.38E-19
2.00E 01	2.38E-19
2.20E 01	1.75E-19
2.40E 01	2.17E-19
2.60E 01	2.59E-19
2.80E 01	2.17E-19
3.00E 01	1.33E-19

8-2-22A
ANKUDINOV ET AL. (211)

ENERGY (KEV)	CROSS SECTION (SQ. CM)
1.00E 01	2.36E-18
1.20E 01	2.36E-18
1.40E 01	2.30E-18
1.60E 01	2.36E-18
1.80E 01	2.56E-18
2.00E 01	2.71E-18
2.20E 01	2.78E-18
2.40E 01	2.75E-18
2.60E 01	2.66E-18
2.80E 01	2.58E-18
3.00E 01	2.37E-18

8-2-22B
ANDREEV ET AL. (148)

ENERGY (KEV)	CROSS SECTION (SQ. CM)
9.90E 00	2.10E-18
1.21E 01	2.29E-18
1.41E 01	2.30E-18
1.59E 01	2.39E-18
1.79E 01	2.47E-18
1.98E 01	2.57E-18
2.15E 01	2.39E-18
2.37E 01	2.32E-18
2.58E 01	2.24E-18

8-2-23
ANKUDINOV ET AL. (211)

ENERGY (KEV)	CROSS SECTION (SQ. CM)
1.00E 01	3.25E-18
1.20E 01	3.26E-18
1.40E 01	3.29E-18
1.60E 01	3.47E-18
1.80E 01	3.65E-18
2.00E 01	3.85E-18
2.20E 01	3.98E-18
2.40E 01	3.88E-18
2.60E 01	3.78E-18
2.80E 01	3.65E-18
3.00E 01	3.42E-18

8-2-24
ANKUDINOV ET AL. (211)

ENERGY (KEV)	CROSS SECTION (SQ. CM)
1.00E 01	5.78E-18
1.2CE 01	5.83E-18
1.40E 01	5.76E-18
1.60E 01	6.00E-18
1.80E 01	6.38E-18
2.00E 01	6.75E-18
2.20E 01	6.97E-18
2.40E 01	6.84E-18
2.60E 01	6.63E-18
2.80E 01	6.38E-18
3.00E 01	6.00E-18

8-2-25A
GUSEV ET AL. (109)

ENERGY (KEV)	CROSS SECTION (SQ. CM)
1.00E-01	4.61E-18
1.60E-01	5.57E-18
2.00E-01	7.29E-18
2.60E-01	6.43E-18
4.60E-01	4.38E-18
6.60E-01	3.61E-18
1.00E 00	2.56E-18
2.00E 00	2.12E-18
3.00E 00	1.82E-18
5.00E 00	2.00E-18
7.00E 00	2.12E-18
1.00E 01	2.26E-18
1.50E 01	2.66E-18
2.00E 01	3.10E-18
2.50E 01	3.33E-18
3.00E 01	3.47E-18

8-2-25B
GUSEV ET AL. (1C9)

ENERGY (KEV)	CROSS SECTION (SQ. CM)
8.00E-02	3.90E-19
1.20E-01	7.50E-19
1.6CE-01	1.07E-18
2.00E-01	9.00E-19
2.20E-01	1.22E-18
2.4CE-01	1.22E-18
2.60E-01	1.21E-18
3.60E-01	1.04E-18
4.60E-01	9.14E-19
6.6CE-01	8.23E-19
1.00E 00	5.14E-19
2.00E 00	4.03E-19
3.0CE 00	3.83E-19
5.00E 00	3.94E-19
7.0CE C0	3.78E-19
1.00E 01	4.21E-19
1.50E 01	5.18E-19
2.00E 01	5.67E-19
2.50E 01	5.50E-19
3.00F 01	4.89E-19

8-2-26A
GUSEV ET AL. (109)

ENERGY (KEV)	CROSS SECTION (SQ. CM)
6.60E-01	6.40E-19
1.0CE C0	8.80E-19
2.00E 00	1.14E-18
3.00E 00	1.45E-18
5.0CE 00	1.56E-18
7.00E 00	1.77E-18
1.00E 01	1.92E-18
1.50E 01	1.85E-18
2.00E 01	1.75E-18
2.50E 01	1.74E-18
3.00E 01	1.64E-18

8-2-26B
GUSEV ET AL. (109)

ENERGY (KEV)	CROSS SECTION (SQ. CM)
2.60E-01	6.70E-20
3.60E-01	9.10E-20
4.60E-01	7.20E-20
6.60E-01	1.72E-19
1.0CE 00	2.68E-19
2.00E 00	4.39E-19
3.0CE 00	5.25E-19
5.00E 00	5.45E-19
7.00E 00	5.61E-19
1.00E 01	5.52E-19
1.50E 01	4.80E-19
2.00E 01	4.62E-19
2.50E 01	4.42E-19
3.00E 01	4.01E-19

8-2-27A
GUSEV ET AL. (109)

ENERGY (KEV)	CROSS SECTION (SQ. CM)
1.60E-01	5.40E-19
2.60E-01	9.30E-19
4.60E-01	1.17E-18
6.60E-01	1.01E-18
1.0CE 00	8.90E-19
2.00E 00	5.20E-19
3.0CE 00	5.40E-19
5.00E 00	4.50E-19
7.00E 00	5.60E-19
1.00E 01	6.90E-19
1.50E 01	9.90E-19
2.00E 01	1.21E-18
2.50E 01	1.42E-18
3.00E 01	1.41E-18

8-2-27B
GUSEV ET AL. (109)

ENERGY (KEV)	CROSS SECTION (SQ. CM)
2.00E-01	1.3CE-20
2.20E-01	5.80E-20
2.40E-01	6.10E-20
2.60E-01	1.44E-19
3.60E-01	1.44E-19
4.60E-01	1.56E-19
6.6CE-01	1.52E-19
1.00E 00	1.29E-19
2.00E 00	1.23E-19
3.0CE 00	9.30E-20
5.00E 00	1.00E-19
7.0CE 00	1.14E-19
1.00E 01	1.38E-19
1.50E 01	1.85E-19
2.0CE 01	2.23E-19
2.50E 01	2.80E-19
3.0CE 01	3.03E-19

8-2-28A
GUSEV ET AL. (109)

ENERGY (KEV)	CROSS SECTION (SQ. CM)
2.00E 00	4.20E-19
3.00E 00	3.30E-19
4.00E 00	4.20E-19
5.00E 00	5.40E-19
1.00E 01	7.20E-19
1.50E 01	1.23E-18
2.00E 01	1.40E-18
2.50E 01	1.35E-18
3.0CE 01	1.40E-18

8-2-28B
GUSEV ET AL. (109)

ENERGY (KEV)	CROSS SECTION (SQ. CM)
1.00E 00	7.00E-21
2.00E 00	5.40E-20
3.00E 00	7.70E-20
5.0CE 00	1.09E-19
7.00E 00	1.36E-19
1.00E 01	1.90E-19
1.50E 01	2.44E-19
2.00E 01	2.95E-19
2.50E 01	3.23E-19
3.00E 01	3.13E-19

8-2-29A
GUSEV ET AL. (109)

ENERGY (KEV)	CROSS SECTION (SQ. CM)
4.60E-01	8.90E-19
6.60E-01	1.93E-18
1.00E 00	4.69E-18
2.00E 00	6.09E-18
3.00E 00	6.94E-18
5.0CE 00	5.87E-18
7.00E 00	5.06E-18
1.00E 01	4.91E-18
1.50E 01	3.87E-18

2.00E 01	3.90E-18
2.50E 01	3.90E-18
3.00E 01	3.94E-18

8-2-29B
GUSEV ET AL. (109)

ENERGY (KEV)	CROSS SECTION (SQ. CM)
3.60E-01	5.10E-20
4.60E-01	9.60E-20
6.60E-01	3.20E-19
1.00E 00	6.30E-19
2.00E 00	7.90E-19
3.00E 00	1.01E-18
5.00E 00	1.07E-18
7.00E 00	9.93E-19
1.00E 01	9.59E-19
1.50E 01	7.46E-19
2.00E 01	6.59E-19
2.50E 01	6.38E-19
3.00E 01	6.38E-19

8-2-30A
GUSEV ET AL. (109)

ENERGY (KEV)	CROSS SECTION (SQ. CM)
1.00E 00	1.44E-18
2.00E 00	4.40E-18
3.00E 00	5.55E-18
5.00E 00	6.06E-18
7.00E 00	5.31E-18
1.00E 01	5.03E-18
1.50E 01	3.73E-18
2.00E 01	3.54E-18
2.50E 01	4.15E-18
3.00E 01	4.01E-18

8-2-30B
GUSEV ET AL. (109)

ENERGY (KEV)	CROSS SECTION (SQ. CM)
6.60E-01	4.50E-20
1.00E 00	2.18E-19
2.00E 00	7.46E-19
3.00E 00	1.22E-18
5.00E 00	1.36E-18
7.00E 00	1.18E-18
1.00E 01	1.15E-18
1.50E 01	8.20E-19
2.00E 01	5.80E-19
2.50E 01	5.16E-19
3.00E 01	4.80E-19

8-3   EXCITATION OF NITROGEN

The collisional formation of excited atoms and molecules in nitrogen has been more extensively studied than any other molecular target, and a rather wide variety of projectiles has been employed. The analysis of the emission is reasonably simple, the spectral lines being well separated; and the rotational structure of the molecular emissions is quite compact. The most prominent features in the collisionally induced emission spectra are the N_2 and N_2^+ molecular systems. Also observed are weaker contributions of N and N^+, resulting from collisional dissociation.

Certain problems and considerations that are common to all studies of nitrogen excitation by heavy particle impact are considered in Section 8-3-A. In Section 8-3-B the various experimental arrangements used in this work are reviewed and assessed. Experimental data are discussed in Sections 8-3-C through 8-3-F and tabulated in Section 8-3-G.

A.   General Considerations

A molecular nitrogen target, in thermal equilibrium with its surroundings at normal room temperatures, will exhibit negligible population of electronically or vibrationally excited states; however there will be a significant population of the excited rotational levels. There have been no investigations of how changes in rotational state population affect measurements of cross sections for the collisional excitation or removal of electrons. Investigations have been made on the changes in rotational population that occur simultaneously with collisionally induced changes in electronic and vibrational states (see Section 8-3-F).

It has been shown in Section 3-2-D that a McLeod gauge, or a secondary device calibrated against a McLeod gauge, may be in serious error when measuring nitrogen pressures. Those experiments that have not involved precautions against such errors are specifically noted in the following discussions.

A number of measurements have been made of the emission from excited N and N^+ formed through the dissociation of the molecular target. No accurate theoretical nor experimental values of transition probabilities are available for such cases, and the data can only be expressed as an emission cross section. The spectra of N and N^+ are very complex, and, in general, insufficient care was taken to definitely identify the spectral lines that were detected and to ensure no contribution from neighboring lines. It has been shown in Section 3-4-E that when two or more lines of different wavelengths are measured simultaneously some precautions are necessary to assess the efficiency with which each line is detected. These precautions were not taken in experiments that claim to measure the emission of two or more lines simultaneously.

The studies of the molecular spectrum of N_2 and N_2^+ involve the measurement of the emission intensity for transitions between known vibrational states of different electronic configurations. Each transition has considerable rotational structure, extending over a wavelength range of some angstroms. In general, the precautions for dealing with an extended band of emission (see Section 3-4-E) have not been fulfilled. At high impact velocities, the distribution of excited molecules among the various rotational states is close to that for room temperature thermal equilibrium (see Section 8-3-F), and the rotational structure will rapidly diminish in intensity away from the band head. Thus, the error involved in neglecting contributions from high rotational states may be small. At very low impact velocities an appreciable enhancement of the higher rotational states is observed (see Section 8-3-F), and this generalization may not be true. It is impossible to estimate in retrospect what error has been introduced into a determination by neglect of the band width of the rotational structure.

The investigation of the formation of excited N_2 has been restricted to the triplet states that give rise to the second positive system. Since the ground state is a singlet, the direct excitation process is formally forbidden, except through the mechanism of exchange. High resolution is often necessary to separate the second positive system of N_2 from adjacent lines of the first negative system of N_2^+.

The formation of excited N_2^+ has been well treated, particularly the very prominent first negative system that originates from the decay of $B^2\Sigma_u^+$ excited state to the $X^2\Sigma_g^+$ ground state. According to Gilmore[310] and Wallace,[311] there have been no reports of transitions that populate the $B^2\Sigma_u^+$ state by cascade from higher levels, and the decay to the $X^2\Sigma_g^+$ ground state is the only radiative decay mechanism. Limited studies have also been made of the formation of the $A^2\Pi_u$ state by measurement of the Meinel bands ($A \rightarrow X$). These measurements may include some cascade contributions through the Janin d'Incan bands, but in the existing measurements this cascade has not been assessed.

A valuable check on the consistency of experimental measurements of the N_2^+ emission may be achieved by evaluating the ratios of the emission cross sections for transitions from the same vibrational level. Since the ratio is equal to the branching ratio for the transition, it should be the same for all experimental determinations and independent of impact energy. A variation from one experiment to another would indicate that a wavelength-dependent error exists in the calibrations of efficiency. This test will be applied to the first negative system emissions by taking the ratios of emission cross sections in the $0 \rightarrow v''$ and the $1 \rightarrow v''$ sequences to the emission cross sections of the $0 \rightarrow 1$ and $1 \rightarrow 2$ transitions, respectively. The $0 \rightarrow 1$ and $1 \rightarrow 2$ transitions are chosen as references since they lie in the visible spectral region where it is

expected that the determination of detection efficiency should be quite accurate.

No experimental tests have been made to ensure that the measured cross sections are independent of any excited state content that may exist in the particle beam. It appears unlikely, however, that a small fraction of excited projectiles will appreciably affect the cross section for rearrangement of the target structure, except perhaps at very low velocities. Accordingly, this problem is not considered further in the present context.

B. Experimental Arrangements

1. The Work of Carleton et al. Carleton and Lawrence,[172] and Sheridan, Oldenberg, and Carleton,[173] have been involved in making measurements of excitation by impact of protons. Both sets of experiments utilized essentially the same experimental arrangement. A photomultiplier with interference filters was used as the spectrometric analysis and detection system. Colored glass filters were also included to block spurious transmission bands of the interference filters. It would appear that a different area of the photomultiplier cathode may have been used for calibration and for the experimental measurement, with no check that the sensitivity was the same in both cases. Proper precautions were taken to allow for changes in detection sensitivity over the whole rotational band structure of the transition. No attention was paid to the influence of polarization, and the measurements were all made at an angle of 90° to the ion beam path.

2. The Work of Dahlberg, Anderson, and Dayton. Dahlberg et al. have been concerned with the excitation of nitrogen by impact of H^+ and H at energies from 10- to 130-keV. The first publication[312] was a brief conference report that was later superseded by a complete discussion of the procedures and data.[182]

For visible spectral regions detection was achieved by a grating monochromator fitted with a photomultiplier. In the vacuum uv region a Czerny—Turner vacuum monochromator was attached directly to the collision chamber. The detection systems were placed at a 90° angle to the ion beam path. The influence of polarization on the accuracy and validity of the measurements was completely ignored. A complete description of the apparatus, including the methods of producing and detecting neutral atomic hydrogen beam, is given in Section 8-2-B(3).

No absolute calibration was carried out in this work, and the data were normalized to other published measurements. For the visible and near uv regions, a relative calibration of the apparatus was made using a standard lamp, and the whole set of data was normalized to the measurements of the 3914-Å emission by Philpot and Hughes[175] at energies between 40- and 70-keV. The relative cross sections at different wavelengths are independent of the normalization. A more tenuous procedure was used in the vacuum uv region where measurements were made on the N I 1493- and 1743-Å lines as well as the Lyman-alpha emission from the projectile. The H I measurements were normalized to the Lyman-alpha emission data of Van Zyl et al.[149] for the same collision combination at 20-keV energy. Van Zyl et al. state that in the case of an N_2 target the detected signal included a substantial contribution from the excited nitrogen and did not represent Lyman-alpha emission alone. In the experiment by Dahlberg et al. a completely different spectrometric system from that due to Van Zyl et al. was used; therefore the signal included a different proportion of the spurious nitrogen emission. It follows that the two experiments measure different quantities,

and the normalization procedure is invalid. In order to arrive at absolute cross sections for the N I lines at 1493- and 1743-Å, it was assumed that the detection sensitivity was the same as at the Lyman-alpha wavelength of 1216-Å, a most unlikely situation. It must be concluded that the magnitudes of these data may be greatly in error. The problems do not affect the validity of the measurements on relative cross section as a function of energy.

Dahlberg et al. note that changes took place in the charge state composition of the beam as it traversed the collision region. Small corrections were measured directly and applied to the experimental measurements of emission cross section.

3. The Work of Baker, Gardiner, and Merrill. The single report on the excitation induced by H^+ impact contains insufficient information on the experimental procedures to allow an accurate assessment to be made.[181] A photomultiplier-monochromator combination was used for spectroscopic analysis and detection. A standard blackbody source was used to calibrate the detection sensitivity of the system. Target pressures were measured with a thermocouple gauge that had been calibrated against a McLeod gauge with no precautions to prevent mercury pumping errors (Section 3-2-D).

Inadequate details were given on the techniques used for the calibration of the absolute detection efficiency, and it must be presumed that the criteria listed in Section 3-4-H were not fulfilled.

All measurements were made at an angle of 90° to the ion beam path, and no attention was paid to possible anisotropy of the emission nor to the effects of instrumental polarization.

4. The Work of Sheridan and Clark. Sheridan and Clark have been involved in work related to excitation by impact of various projectiles on nitrogen. The early report[313] is published in conference proceedings and is completely superseded by the later publication.[176] Photon detection was accomplished using a photographic spectrometer of very large aperture and employing exposure times of between one and two hours.

Most of the data was presented in the form of the ratios of cross sections for the emission of different bands. For the 3914-Å band of N_2^+ alone a cross section was presented with the absolute magnitude assigned by normalization to measurements by previous workers.[175, 173] The single cross section measurement exhibits a completely different variation with impact energy from any of the other investigations.[175, 173, 181, 178, 162, 183] Clearly a normalization under such circumstances is completely unjustified and the discrepancy casts some doubt on the validity of the whole program of measurements. Although there is no objective reason for distrusting the work of Sheridan and Clark it does seem likely that some serious error exists in the experimental procedures; consequently no further consideration is given to this work.

5. The Work of Neff and Carleton. Neff and Carleton have carried out investigations with a diverse range of projectiles including rare gas, molecular, and metallic ions. The original publication of this work in the proceedings of a conference[281] is superseded by the more recent report.[109]

The ion beam was produced in a discharge source and mass analyzed before entering the collision chamber. No attention was paid to the possible influence of excited states in the projectile beam. Light emission was detected using a monochromator photomultiplier system at an angle of 90° to the ion beam path. The possibility of emission polarization was ignored.

The absolute magnitudes were assigned by normalization to the emission of Ar II lines induced by the impact of Ar^+ on Ar, as measured previously by Sluyters and Kistemaker.[103] These data by Sluyters and Kistemaker are unreliable [see Section 7-3-A(2)]; therefore the absolute magnitudes of Neff's data are probably inaccurate. The relative accuracy of measurements on different spectral lines is governed by the photometric calibration of the system and was estimated at $\pm 10\%$ in the range of 3900- to 5200-Å and $\pm 20\%$ elsewhere.

For this study the monochromator slit was parallel to the ion beam path and was sufficiently wide to accommodate the whole image of the ion beam. Use of an exit slit that was three times wider than the entrance slit ensured a trapezoidal shape to the transmission characteristic of the monochromator; therefore the optical system satisfies the criteria required for accurate measurement of the sum of cross sections for two or more lines transmitted simultaneously by the monochromator (see Section 3-4-E). Resolution of the optical system was 30-Å.

6. The Work of Doering. Doering's investigations[227] are confined to the emission resulting from the impact of N_2^+ ions on a nitrogen target. Measurements include emission from both projectile and target. A number of unusual experimental techniques were utilized, consequently making the measurements indirect; however, no serious systematic errors are apparent.

Production of the N_2^+ ion beam was carried out in a conventional manner using an rf source and magnetic mass spectrometer. No absolute measurements were made of emission cross sections, but the data were normalized to a cross section for the emission of the 3914-Å band of N_2^+ induced by electron impact on a nitrogen target. An electron gun was built in for this purpose. The cross section for electron impact excitation at 180-eV was taken to be 5.3×10^{-18} cm^2, which is essentially the value measured by Stewart[283] and is also in close agreement with the measurements of Sheridan et al.[173] The sensitivity at other wavelengths was determined through a relative photometric calibration against a standard lamp. Precautions were taken to eliminate errors due to scattered light. It was stated that the monochromator slits were opened sufficiently so that all parts of the rotational structure of the molecular spectral lines were transmitted; however, it was not stated whether precautions were observed to ensure that the whole structure was detected with the same efficiency (see Section 3-4-E).

In the light of more recent experiments, the normalization used by Doering may be criticized. There are four more recent independent determinations of the cross section for emission of the 3914-Å band induced by electron impact;[284, 285, 286, 287] these are in complete agreement with each other and exceed Stewart's[283] values by approximately a factor of 2.6. There are good reasons to doubt the validity of Stewart's measurements; Doering's data could be normalized to the more recent electron impact data[284, 285, 286, 287] by raising it through a factor of 2.6. For the purposes of this monograph, Doering's measurements are left in their original form, normalized to the data of Stewart.

The ion beam traversed a target region of 15-cm length along the axis of a magnesium-oxide coated cylinder that acted as a photometric integrating sphere. A monochromator-photomultiplier combination viewed the interior of the sphere through a small orifice arranged so that the ion beam path was not seen directly. In this manner the detected signal was proportional to the light emitted into all directions from an extended beam path through the target. Projectile beam composition changes extensively in the long target cell; consequently it is impossible to establish observation

conditions that are consistent with the operational definition of emission cross section (Eq. 2-8). Doering measured photon emission in a region of high target density where intensity varied linearly with pressure; this regime corresponds to an equilibrium distribution of population among the various ion states in the beam. The measured intensity includes not only excitation by N_2^+ impact but also by N_2 and the dissociation fragments of the projectile. The elucidation of separate cross sections under these conditions clearly requires knowledge of the cross sections for all charge-changing processes affecting the projectile beam composition. This is a serious weak point in the approach since the measurement is dependent on the data from another experiment; also the approach is very indirect. The analysis used by Doering requires that the equilibrated beam has only a two state composition, N_2 and N_2^+. Doering's spectral analysis does itself show that some fast N^+ is formed by dissociation of the projectiles, but this factor is neglected. A further complication is that at sufficiently high target pressures an appreciable fraction of excited N_2^+ projectiles may be collisionally deactivated before emission takes place. Subsidiary experiments, using the Doppler shift to separate target and projectile emission, suggested that collisional deactivation of N_2^+ did take place; however, the emission from the fast particle accounted for less than 20% of the total emission, and it was considered that the error due to this cause could be neglected.

The approach used by Doering is indirect and undoubtedly inferior to the use of a beam of defined composition under single collision conditions. The accuracy must inevitably be poor due to the use of data from other experiments. In practice considerable scatter occurs in the data points. It is possible that these measurements contain substantial unexpected systematic errors, and should be treated with caution. The procedure as a whole does have the advantage of allowing experiments at relatively low impact energies where neutralization of the projectiles by charge transfer is substantial.

Moore and Doering have also carried out measurements on the rotational and vibrational population of the B $^2\Sigma_u^+$ state of N_2^+ using beams of H^+, H_2^+, He^+, N^+, Ne^+, and Ar^+.[184, 314, 315] The same ion accelerator system was employed but with a different optical system. Since these experiments do not give information on cross sections, they are not considered further at this point. In Section 8-3-F the available data on rotational excitation are briefly surveyed.

7. **The Work of Kurzweg, Lo, Brackmann, and Fite.** Kurzweg et al.[113] have been concerned with a study of the emission of the 3914-Å band of the first negative system $[B\ ^2\Sigma_u^+ \to X\ ^2\Sigma_g^+\ (0,0)]$ induced by the impact of Ba^+, Ba^{2+}, Xe^+, N^+, O^+, and N_2^+ on a molecular nitrogen target. It should be noted that some of the data were originally published[113] incorrectly and were corrected in a short note,[316] the source used in this monograph.

A Van de Graaff accelerator with an rf ion source was used to provide the projectile beam. No attempt was made to assess the effect of excited projectiles that might be produced in the source. Unusually detailed attention was directed to ensuring that the projectile beam did not become contaminated by charge transfer neutralization on residual gases in the flight tube between source and experimental region.

The ion beam was directed through a short collision cell, and the collisionally induced light emission was collected by a lens system and detected with a photomultiplier. An interference filter was used to isolate the 3914-Å transition. Precautions were taken to ensure that the ion beam was traveling down the center of the collision

chamber and that beam divergence was immaterial. No attempt was made to determine the polarization of the emission nor to confirm isotropy of the emission. All measurements were made at an angle of 90° to the projectile beam.

All cross section measurements were relative and were normalized to the electron-impact excitation cross section at an energy of 300-eV; an electron gun could be inserted into the position normally occupied by the projectile beam when required. Tests were made to ensure that the electrons traversed exactly the same geometrical path as the heavy projectiles. The electron-impact excitation cross section at 300-eV was taken as 1.3×10^{-17} cm^2. This figure is consistent with a number of recent independent determinations[284, 285, 286, 287] and is probably quite accurate.

No attention was directed at the possible influence of the extensive rotational structure on the measurement of the cross section. The signal detected by the photomultiplier is the sum of all rotational lines in the band, weighted according to the filter transmission at the relevant wavelength. The signal is proportional to the cross section for excitation of the band associated with the (0, 0) vibrational transition, provided that the relative intensities of the various rotational lines remain constant. In the event that these relative intensities change, through what is effectively a change in the apparent rotational temperature of the excited molecular ions, the signal is no longer proportional to the emission cross section in the (0, 0) transition. Moreover, if there is a difference between apparent rotational temperature of emission induced by electron impact and by heavy particle impact, then the calibration is also invalid. Constancy of the apparent rotational temperature is required for the validity of the procedures utilized, but it was not confirmed in any way. Moore and Doering[315] show that for projectile velocities below 10^8 cm/sec there is collisional enhancement of the population of the higher rotational states independent of the nature of the projectile. Thus, there is reason to believe that the distribution among rotational states does vary with projectile energy at the lower velocities used by Kurzweg et al.; consequently, there will be an error in their measurement of cross sections.

It is also noted that the interference filter used in the experiments of Kurzweg et al. will transmit a small fraction of the intensity of the (1, 1) transition in the first negative system of N_2^+. Moore and Doering[184] suggest that the relative intensities of the (0, 0) and (1, 1) bands vary with projectiles velocities below 10^8 cm/sec, irrespective of the nature of the projectile. This limit encompasses some of the data by Kurzweg et al. It must be concluded that the cross section measurements by Kurzweg et al. may have included a small component of signal from the (1, 1) transition and that the resulting error might be dependent on the projectile velocity. There is no mention of tests to ensure that the interference filter did not transmit appreciable signals from lines at other wavelengths outside the nominal pass band of the filter.

With the apparatus available for this experiment, there was no convenient way of ensuring that the measurements were not effected by radiation from excited projectiles; however, a test was made by using Ar, Ne, H$_2$, and O$_2$ in the collision chamber and measuring the signal received when bombarding with the various heavy projectiles. The signal was less than 2% of the total received with N$_2$ in the collision chamber, suggesting strongly that the emission from the various projectile systems could be ignored to within this level of accuracy. It should be noted that with N$_2^+$ projectiles the measured emission includes contributions from both target and projectile excitation.

The accuracy of the data was described through the standard deviation from the

mean value. It ranged from 5% for most of the work to 7.6 and 9.6%, respectively, for O^+ and Xe^+ impact, where the beam currents were small and the signals therefore low.

8. The Work of Young, Murray, and Sheridan. Young, Murray, and Sheridan[7] have carried out a very interesting experiment that involves the detection in coincidence of a photon emitted from an excited projectile and a photon emitted from an excited target atom. The reaction can be described by the following equation:

$$H^+ + N_2 = H \, (3p, 3d) + N_2^+ \, (B \, {}^2\Sigma_u^+ \, v' = 0)$$

$$H(3p, 3d) \rightarrow H(2s, 2p) + \text{Balmer-alpha photon, 6563-Å}$$

$$N_2^+ \, (B \, {}^2\Sigma_u^+ \, v' = 0) \rightarrow N_2^+ \, (X \, {}^2\Sigma_g^+ \, v'' = 0) + \text{3914-Å photon.}$$

The 3914-Å and 6563-Å photons are detected in coincidence. This is the only coincidence experiment that has been attempted in collisional excitation studies.

The ion source system was the same as that used by Sheridan and Clark and described in Section 8-3-B(4);[176] the optical systems and experimental procedure were completely different. The optical system was based on two photomultipliers, each fitted with an interference filter and identical lens systems viewing the same length of beam path at a point 17.8 cm from the entry of the beam into the chamber. Spectral analyses were made to ensure that the rather broad pass bands of these filters did not transmit any appreciable amount of radiation other than the two lines of interest.

The theory of the experimental operation is very complex, and the reader is referred to the original paper for full details.[7] Because atoms and molecules have finite radiative lifetimes, one must take into account those cases in which excitation of the desired states occurs in a single electron capture collision but where the two emissions occur at times separated by more than the coincidence resolving time. It is also necessary to account for the fast excited hydrogen atoms that pass out of the observation region before emitting. An expression is derived for the probability that an event occurring within the region of observation will be detected by the system. It is shown that the probability of detecting the long-lived $3s$ states is very much less than that for the short-lived $3p$ and $3d$ states since, in general, $3s$ states will move out of the field of view before emitting. It is possible to measure separately the emission of the $3s \rightarrow 2p$, $3p \rightarrow 2s$, and $3d \rightarrow 2p$ transitions in coincidence with the emission of the 3914-Å band by varying the length of the observation region and its distance from the point where the beam enters the target; however this proved impractical. It is argued that the contribution of the $3s$ state to the total signal was less than 10% and could be ignored to within that accuracy. Using the recent data by Hughes et al.[180] on the formation of the $3s$, $3p$, and $3d$ states, it would appear that this is an underestimate and that the contribution to the measured signal is no less than 13% at 15-keV and higher at greater impact energies. It is then argued by Young et al. that the probability of detecting the $3p$ and $3d$ states was similar and that a mean value of probability could be used for both lines. This introduced an error that was estimated to be as much as 20% at high velocities and 10% at low velocities. With the more recent work of Hughes et al.[180] it would have been possible to utilize the measured ratios of $3p$ and $3d$ state excitation cross sections to derive a weighted mean value of probability and so reduce this error. With these approximations a cross section was estimated for the emission of the $3p \rightarrow 2s$ and $3d \rightarrow 2p$ lines in coincidence with the 3914-Å band.

Target pressure was measured with an ionization gauge that had been calibrated

against a McLeod gauge in such a manner as to minimize gauge pumping effects. Linearity of signals with pressure and beam current were confirmed. No details are reported on the method by which the optical detection efficiencies were calibrated. This is a serious omission. Quantitative measurements of the 3914-Å emission function are also presented; there is good agreement with data by other workers. Tests were made to show that the emission from both systems was isotropic.

The estimated systematic error ranges from 35% at low velocities to 45% at high velocities. Random error was large due to the low coincidence signal and the consequential poor counting statistics. This error was estimated to range from 25% at the peak of the cross section to 100% where it was small. In view of the remarks made above, the estimate of systematic error may be too low.

The coincidence data for this experiment are shown in Data Table 8-3-21. There are no other comparable data with which they may be correlated.

9. The Work of Thomas, Bent, and Edwards. The report of this group's work was originally published in a brief conference report[317] that was later superseded by a formal publication.[183] The apparatus was the same as that described in Section 8-2-B(4). The detection system was placed at an angle of 90° to the ion beam path, and no attention was given to the errors that would be introduced if the emission were polarized. Entrance and exit slit widths of the monochromator were kept equal with the result that the intensity of the extensive rotational structure of the molecular bands was not determined accurately [see Section 3-4-E(2)].

The authors estimate the reproducibility of the data to be better than $\pm 5\%$ and the absolute accuracy to be better than $\pm 25\%$.

10. The Work of Philpot and Hughes. The measurements by Philpot and Hughes[175] of nitrogen emission induced by proton impact on an N_2 target were carried out using the apparatus described previously in Section 8-2-B(6). A McLeod gauge was used for pressure measurements, and no precautions were taken to prevent errors due to mercury pumping and thermal transpiration effects (see Section 3-2-D). Polarization of emission was ignored. The entrance and exit slits of the monochromator were kept of equal width, with the result that the intensity of the extensive rotational structure of the molecular bands was not determined accurately [see Section 3-4-E(2)].

The absolute accuracy of this data was quoted as $\pm 40\%$ with $\pm 5\%$ relative accuracy. An early publication by this group[174] contained inaccurate data and should be disregarded entirely in favor of the more recent publication[175] from which the data used here are drawn. Note also that an "erratum" was published[318] that has some bearing on the interpretation of these data.

11. The Work of Robinson and Gilbody. Robinson and Gilbody[162, 251] have been involved with work on the excitation of nitrogen by impact of protons and neutral hydrogen atoms. The original publication of this data in conference proceedings[251] has been superseded by a more complete report.[162]

A large prism monochromator fitted with photomultiplier detection was the primary component of the optical system. It was situated with its axis at an angle of 54.5° to the direction of the primary beam; at this angle, the measurements should be independent of anisotropy due to polarization of the emission. Tests were made[252] to ensure that the monochromator did not exhibit any appreciable degree of instrumental polarization.

A lens was used to form an image of the collision region on the entrance slit of the monochromator. Emission is, therefore, detected from only a short length of the

projectile beam path defined by the entrance slit of the monochromator. Calibration was carried out using a standard lamp placed in the position normally occupied by the ion beam. This procedure ensures that calibration and experiment employed precisely the same optical arrangement. In the report[162] no mention is made of checks for the influence of scattered light.

The entrance slits of the monochromator were kept of equal width, with the result that the intensity of the extensive rotational structure of the molecular bands was not determined accurately [see Section 3-4-E(2)]. Target densities were determined with a cold-trapped McLeod gauge, and precautions were taken to prevent errors due to mercury pumping and thermal transpiration effects.

The authors quote possible systematic error of up to $\pm 20 \%$ with an additional random uncertainty of up to $\pm 10 \%$ due to reading errors.

12. The Work of Dufay et al. Excitation by impact of H^+, as studied by Dufay, Desesquelles, Druetta, and Eidelsberg is contained in a total of three papers.[178, 202, 179] The results first published by Dufay et al.[202] were later superseded by a more complete report[178] in which numerous changes were made to the magnitudes of the cross sections. In the third report[179] a few additional measurements were presented for a single energy of 1-MeV. Inadvertently, some errors were made in plotting the data for this last publication; the corrected values that are utilized here were obtained by private communication.[290]

The experimental system utilized as the major optical element a Czerny—Turner grating monochromator with photomultiplier detection. A quartz lens formed an image of the collision region on the monochromator entrance slit. The axis of the system was at a 90° angle to the projectile beam path. No account was taken of polarization either as regards anisotropy of emission or instrumental polarization (Section 3-4-D). Inadequate attention was given to the efficient detection of the entire rotational structure associated with the molecular transitions [see Section 3-4-E(2)].

The calibration of detection sensitivity was achieved by using a standard lamp placed some distance from the collision region. Inevitably, the field of view and solid angle employed during calibration were different from that used during the experimental measurement. There were no tests to ensure that response was the same under these two circumstances. No attention was paid to the possible influence of scattered light in the monochromator.

Target pressures were monitored with a cold-trapped McLeod gauge, and no attention was given to eliminating errors due to thermal transpiration and cold-trap pumping (see Section 3-2-D).

The authors suggest that the systematic error in this work does not exceed $\pm 50 \%$; most of this uncertainty arises in the calibration of detection sensitivity. Random errors are stated to be less than $\pm 10 \%$.

13. The Work of Schlumbohm. Schlumbohm[229] has studied the emission of certain N_2^+ first negative bands induced by the impact of Ne^+ ions on N_2 at impact energies ranging from 200-eV down to threshold.

The optical system for this experiment consisted of a photomultiplier interference filter combination. The measurements are for the emission of a broad band of vibrational transitions centered on 3550- and 3845-Å. No attempt is made to detect individual transitions. Undoubtedly, there will be some radiation from neutral N_2 and dissociation fragments included in the measurements. Precautions necessary for accurate measurement of two or more lines simultaneously (see Section 3-4-E) were not followed

in the experiment by Schlumbohm. As a consequence, if the variation with energy of the cross sections for the various transitions included in the pass bands of the filter were to be different, then the data from this experiment might be very misleading. It must be concluded that these data do not represent a definite process nor a sum of definite processes; for this reason, they are excluded from the present compilation.

In Section 7-3-A(10) of this monograph the apparatus of Schlumbohm is discussed in further detail. Although the data do not represent valid cross section information, they may be of considerable value in the understanding of the collision process since they do extend down to threshold.

C. Proton Impact

Early work by Meinel and Fan[319, 320, 321] was performed with the objective of elucidating auroral phenomena but employed such poor conditions of beam purity and target density that no quantitative information may be drawn from it. There have continued to be publications of emission spectra obtained under relatively high pressure conditions. Although such experiments might provide useful diagnostic information, they do not give any insight into basic processes and will not be considered here. Much of the earlier work so excluded was reviewed by Ghosh and Srivastava.[3, 4]

The data presented here include information on the following reactions:

The formation of excited N $H^+ + N_2 \rightarrow [H^+ + N] + N^*$ [Section 8-3-C(1)]

The formation of excited N^+ $H^+ + N_2 \rightarrow [H^+ + N + e] + N^{+*}$ [Section 8-3-C(2)]

The formation of excited N_2 $H^+ + N_2 \rightarrow H^+ + N_2^*$ [Section 8-3-C(3)]

The formation of excited N_2^+ $H^+ + N_2 \rightarrow [H^+ + e] + N_2^{+*}$ [Sections 8-3-C(4) and 8-3-C(5)]

In all cases, the measurements give no information of the states of charge nor excitation of the atomic species that are placed within the square brackets in these reaction equations. The data compilations are in Section 8-3-G.

1. Formation of Excited States of N. The work of Carleton and Lawrence[172] and that of Sheridan, Oldenberg, and Carleton[173] was concerned with the measurement of the total cross section for the emission of the four lines of the $3p\ ^4P^0 \rightarrow 3s\ ^4P$ multiplet that lie between 8188- and 8216-Å. If the relative intensities listed by Moore[66] for the emission of these lines in discharge tube spectra are also applicable in the present situation, these four lines account for about one half of the total emission in this multiplet. Precautions were taken to determine sensitivity at all the wavelengths detected by the optical system and thereby arrive at an accurate sum of the emission cross sections of these four lines. There would appear to be no other atomic nor molecular emissions that might have interfered with this measurement of the sum cross section for the emission of the 8188-, 8200-, 8211-, and 8216-Å lines of N I. It should be noted that the work of Sheridan, Oldenberg, and Carleton was only published as a smoothed line of best fit and not as individual datum points. That line is reproduced here.

The 10,113-Å line measured by Dufay et al.[178] (Data Table 8-3-2) is an unresolvable blend of four components in the $3d\ ^4F \rightarrow 3p\ ^4D^0$ multiplet. The 9392-Å line measured by the same authors (Data Table 8-3-3) is the $3p\ ^2D^0_{2\ 1/2} \rightarrow 3s\ ^2P_{1\ 1/2}$ transition. Both

the 10,113-, and 9392-Å lines will be properly resolved, and there should be no interference from neighboring transitions. The measurement on the 8680-Å line by Dufay et al.[178] (Data Table 8-3-4) may have included some unknown component of other transitions in the same multiplet ($3p\ ^4D^0 \rightarrow 3s\ ^4P$) at adjacent wavelengths.

Dahlberg et al.[182] present measurements on two N I lines in the vacuum ultraviolet spectral region (Data Tables 8-3-5 and 8-3-6). The absolute magnitudes of these data were assigned by an unreliable normalization procedure; therefore no reliance should be placed on the absolute nor relative magnitudes of the 1493- and 1743-Å emission measurements. This criticism does not effect the validity of the functional dependence of cross section on projectile impact energy. It should be noted that both of these lines are unresolved doublets.

It is not possible to come to any objective conclusion as to the validity of any of these measurements. In general, criticism may be made of all the experiments, but the resulting errors cannot be properly assessed.

2. Formation of Excited States of N^+. In all but one of the determinations of collisionally induced N^+ emission, there is a possibility that the cross section included some contribution from neighboring lines. The reader is referred to Section 3-4-E for the precautions necessary when measuring a group of lines of slightly different wavelength simultaneously. No such precautions were taken in any of the experiments considered here.

The discrepancies between determinations are considerable; this is probably due primarily to error in the calibration of the optical detection efficiency. Only in the work of Thomas et al.[51] and that of Robinson and Gilbody[162] were precautions taken to eliminate thermal transpiration and trap pumping errors in the measurement of target pressure.

The measurements of the 5005-Å emission cross section shown in Data Table 8-3-7 have all been carried out using monochromators with a resolution of about 10- to 16-Å. The 5005-Å line is a blend of two emissions, $3d\ ^3F^0 \rightarrow 3p\ ^3D$ and $3p\ ^5P^0 \rightarrow 3s\ ^5P$.† Furthermore, there are a total of five other N^+ spectral lines lying within ± 7-Å of the 5005-Å emission. It seems highly unlikely that the 5005-Å line was properly resolved in any of the measurements reproduced here. Thus the identification of the 5005-Å line is in doubt. The difference between the energy dependence of the measurements by Thomas et al.[183] and those of other workers may be due to the inclusion of different proportions of the neighboring spectral lines.

Data for the 5680- and 6482-Å lines are likely to be in error due to inclusion of unknown amounts of neighboring lines in either the N II or N I systems: the 3995-Å line should be free of such interference.

The cross section for the 5680- and 5005-Å lines are found to be of almost the same magnitude. Philpot and Hughes[175] as well as Thomas et al.[183] plotted a mean of these two cross sections in their publications. Here the original measurements of the separate cross sections by Thomas et al. are used. The separated numbers were not available for the work of Philpot and Hughes;[178] therefore their "mean value" is used for both the 5005- and 5680-Å lines.

3. Formation of Excited States of N_2. Dahlberg et al.[182] and Thomas et al.[183] have measured a cross section for the emission of the 3371-Å (0, 0) transition of the second positive system ($C\ ^3\Pi_u \rightarrow B\ ^3\Pi_g$) of the N_2 molecule. Excitation of the $C\ ^3\Pi_u$

† In Fig. 11 of Ref.[182], Dahlberg et al. identify this line incorrectly.

state from the singlet ground state by H^+ impact is formally forbidden according to the Wigner spin conservation rule. Therefore, the cross section for the process will be very small, and precautions are necessary to ensure that a measurement is independent of secondary mechanisms. Neutralized projectiles or electrons, collisionally removed from N_2 molecules, may both cause excitation of the triplet state by an exchange process. Such secondary processes are dependent on two collisions, and the resulting intensity of emission of the 3371-Å transition would vary quadratically with pressure. The primary direct excitation process will give an intensity that depends linearly on pressure. Consequently, an apparent emission cross section resulting from the sum of these two contributions will vary linearly with pressure, p, in the manner: $\sigma_{ij}^* = A + Bp$. Thomas et al.[183] confirm the linear dependence of σ_{ij}^* on pressure and evaluate a cross section from the intercept of a graph of σ_{ij}^* against pressure, p. Dahlberg et al.[182] display an apparent emission cross section that is constant at low pressure and rises nonlinearly at high pressure. The observations of Dahlberg et al. are not consistent with the secondary mechanisms they claim are present, and there is inevitably some doubt that the secondary contributions were successfully eliminated from the cross section measurement.

Data Table 8-3-11 shows the measured emission cross section of the 3371-Å transition. Neither Herzberg[33] nor Gilmore[310] list any transitions that might populate the $C\,^3\Pi_u$ level by cascade; therefore the data presented here are proportional to the cross section for the formation of the $C\,^3\Pi_u$ state. It is observed that there are severe random errors in the data shown here. Thomas et al.[183] show error bars of $\pm 25\%$. Dahlberg et al.[182] do not state error limits, but their data points scatter by as much as 40% from the line of best fit.

The data by Thomas et al.[183] are an absolute measurement, but those by Dahlberg et al. involve a normalization procedure [see Section 8-3-B(2)]. Calibration of detection sensitivity at 3371 Å is likely to be of poor accuracy due to light scattering in the spectrometer [Section 3-4-B(1)]. Dahlberg et al. do not indicate that precautions were taken to eliminate this problem. It seems likely that the discrepancy between the two determinations may be attributed in part to errors in calibration.

An additional report of interest in the present context is the study by Moore and Doering[152] of the energy-loss spectrum exhibited by protons that had traversed a target of N_2. It provides qualitative information on the excitation of the lower vibrational levels of N_2; the data however are not in the form of cross sections and will not be presented here.

4. Emission of the First Negative Band System of N_2^+. The emission of the first negative band, $B\,^2\Sigma_u^+ \to X\,^2\Sigma_u^+$, may involve ionization,

$$H^+ + N_2 \to H^+ + N_2^+\,(B\,^2\Sigma_u^+) + e$$

or charge transfer,

$$H^+ + N_2 \to [H] + N_2^+\,(B\,^2\Sigma_u^+).$$

At energies above 200-keV the cross section for emission of the first positive band system is greater than the total cross section for charge transfer; therefore the ionization process predominates.

It is important to note that the rotational structure of the N_2^+ first negative bands extends over a considerable wavelength range. With the exception of the work by Carleton and Lawrence[172] and that by Sheridan et al.[173] none of the measurements

involved a calibration of the detection sensitivity over the whole of the transmission band of the detection system (see Section 3-4-E).

In Data Tables 8-3-12 through 8-3-15 are displayed the measurements of emission in the $(0, v'')$ sequence. The data by Dahlberg et al. involve a normalization to the work of Philpot and Hughes[175] [see Section 8-3-B(2)] at 40- to 70-keV and are not an independent determination. It is observed that all the determinations give a similar energy dependence, although there are some decided differences in absolute magnitude. In order to examine the agreement between these various measurements, it is helpful to normalize all the separate determinations at one energy. This is done with the 3914-Å $(0, 0)$ band since it is the strongest emission and should exhibit the best statistical accuracy. Figure 8-1 shows this comparison with all data normalized to unity at 130-keV. The data of Carleton,[172] Sheridan,[173] Baker,[181] and their respective co-workers do not extend to 100-keV and are normalized to data from other groups as indicated in the caption to Fig. 8-1. There is good agreement between all determinations, which suggests that the energy dependence of the cross section is given quite reliably by these measurements. The energy dependence of the other transitions in the $(0, v'')$ sequence must be the same as that for the $(0, 0)$ transition. Actual measurements on the $(0, 1)$, $(0, 2)$, and $(0, 3)$ lines are of poorer statistical accuracy than the 3914-Å $(0, 0)$ measurement due to smaller signal strengths.

It is not possible to make a decision as to which of the data on the $(0, v'')$ series of transitions represents the most accurate absolute measurement. There are essentially six independent determinations available for the 3914-Å line and a somewhat smaller number for the higher members of the $(0, v'')$ series. The work of Dahlberg et al.[182]

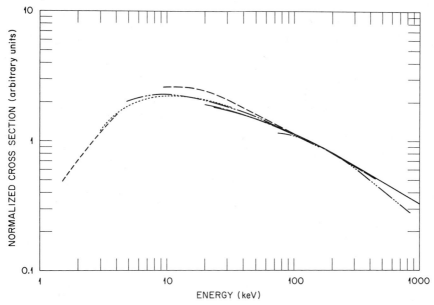

Fig. 8-1 Relative cross sections for the emission of the 3914 Å band of N_2^+ $(B\ ^2\Sigma_u^+ \rightarrow X\ ^2\Sigma_g^+$ $(0, 0))$ induced by the impact of H^+ on N_2. The data are all normalized to unity at 130 keV. . . . . . . Sheridan et al.[173] (normalized to Philpot and Hughes at 20 keV); _ _ _ _ _ Carleton and Lawrence[172] (normalized to Sheridan et al. at 4.0 keV); ___ .. ___ Philpot and Hughes[175]; ___ . ___ Baker et al.[181] (normalized to Dufay et al. at 100 keV); _ _ _ _ Dahlberg et al.[182]; _____ Dufay et al.[178]; ___ ___ Robinson and Gilbody[162]; _ ... _ Thomas et al.[183]

is normalized to that of Philpot and Hughes[175] and does not constitute an independent determination. The work of Sheridan et al.[173] and of Carleton and Lawrence[172] are both taken with the same apparatus and do not constitute independent measurements. The six independent measurements generally fall into two groups—the work of Philpot and Hughes,[175] Sheridan et al.,[173] and of Baker et al.[181] being higher than that of Dufay et al.,[178] Robinson and Gilbody,[162] and of Thomas et al.[183] The size of the discrepancy between the two groups of measurements is too large to be explained by any of the inadequacies of technique that have been identified in our discussions of the experimental procedures used by these six groups. The fact that the discrepancy remains approximately the same size for all measurements in the $(0, v'')$ series suggests that it may be attributable to an error in calibration of detection sensitivity. Taken over the whole series of $(0, v'')$ transitions, the weight of evidence tends to favor the lower group of measurements that includes the work of Dufay et al., Robinson and Gilbody, and Thomas et al.

In Data Tables 8-3-16 through 8-3-20 are shown the measurements on the $(1, v'')$ sequence. Philpot and Hughes[175] as well as Thomas et al.[183] found the $(0, 1)$ and $(0, 2)$ cross sections to be the same, within the random experimental error $(\pm 5\%)$ and both published a "mean value" for the two bands. The present work displays the mean value from the work by Philpot and Hughes, and the original separate points obtained by Thomas et al. It is observed that there are definite differences between the magnitudes of the determinations by various authors and also some differences in energy dependence. It is again useful to examine the energy dependence of the cross section by normalizing all the available data to one point. In Fig. 8-2 are shown the results for the 4236-Å $(1, 2)$ transition normalized to unity at 130-keV. This data exhibits only poor agreement of the functional dependence of cross sections

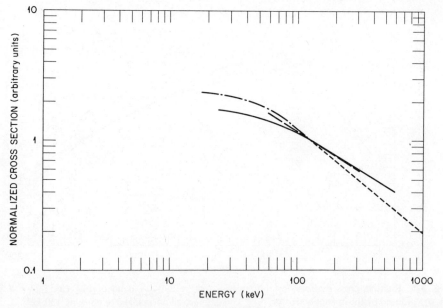

Fig. 8-2 Cross section for the emission of the 4236 Å band of N_2^+ ($B\,{}^2\Sigma_u^+ \rightarrow X\,{}^2\Sigma_g^+$ (1,2)) induced by the impact of H^+ on N_2. The data are all normalized to unity at 130 keV. ___ . ___ Philpot and Hughes[175]; ___ ___ ___ Robinson and Gilbody[162]; _____ Dufay et al.[178]; _ _ Thomas et al.[183]

on energy, and it is not possible to make any reasonable assessment as to which measurement is likely to be most accurate. There are significant discrepancies between the absolute magnitudes of the data in the various determinations. In the preceding paragraph it was concluded that the data on the $(0, v'')$ sequence by Philpot and Hughes are higher than all the other independent determinations and possibly are less accurate than those of the other groups. That same subjective conclusion is also drawn for the $(1, v'')$ sequence of emission measurements.

In addition to the measurements on the collisional excitation of the N_2^+ first negative emission, a determination has also been reported for the emission cross section of the following completely defined process:

$$H^+ + N_2 \to H(3p, 3d) + N_2^+ \ (B \ ^2\Sigma_u^+ \ v' = 0).$$

The experiment detects in coincidence the emission of a Balmer-alpha photon from the H atom and a 3914-Å photon from the N_2^+. The procedure is described in Section 8-3-B(8) and the data are shown in Data Table 8-3-21. It is interesting to note that the authors of this work deduce from these measurements that the formation of the excited H atom by charge transfer and the formation of the excited N_2^+ are completely uncorrelated events. It was pointed out in Section 8-3-B(8) that the accuracy of these data are very poor and that the published details on the methods of optical calibration are totally inadequate. Therefore, it is difficult to assess the experiment in detail. Young, Murray, and Sheridan[7] present data on the cross section for the emission of the 3914-Å band; this data was obtained at the same time as the coincidence measurements.[7] No information is given on the methods used to arrive at this cross section, and the data are shown only as a line of best fit with no individual data points. According to Sheridan[322] these data are to be published in a future report, so they are not included here.

Sheridan and Clark[176] have also carried out measurements on the emission of the 3914-Å line and the relative intensities of certain lines in the first negative system; the work is of doubtful accuracy, and the data is omitted from the present tabulations [see Section 8-3-B(4)]. Moore and Doering[184] and also Gusev et al.[111] have carried out measurements of the relative intensities of lines in the first negative system as a function of impact velocity. Since these data are not in the form of cross sections, they are not reproduced in the present data compilations.

The general consistency of the various measurements may be examined by taking the ratios $\sigma(0, v'')/\sigma(0, 1)$ and $\sigma(1, v'')/\sigma(1, 2)$, as suggested in Section 8-3-A. These ratios have been evaluated for five or more different energies in each determination, and the mean values are shown in Table 8-1. It is observed that for emission at wavelengths higher than those of the $(0, 1)$ and $(1, 2)$ transitions the agreement between the relative magnitudes of the various determinations is good and generally within the random error claimed by the various authors; however, at shorter wavelengths, and particularly for the $(0, 0)$ and $(1, 0)$ transitions, the ratios vary by sizeable amounts. It must be concluded that the measurements on the shorter wavelength lines exhibit greater inaccuracy than those at higher wavelengths.

Some authors have estimated the cross sections for populating the $v' = 0$ and $v' = 1$ levels of the $B \ ^2\Sigma_u^+$ state by summing cross sections for all transitions that depopulate these levels. They neglect contributions from transitions to v'' states greater than 2 for the $v' = 0$ estimate and transitions to states v'' greater than 3 for the $v' = 1$ level estimates; neglect of these terms contributes an error that does not exceed 5%.

TABLE 8-1

	Philpot and Hughes[175]	Dufay et al.[178]	Robinson and Gilbody[162]	Thomas et al.[183]	Dahlberg et al.[182]
The ratio $\sigma(v' = 0 \to v'')/\sigma(v' = 0 \to v'' = 1)$					
0, 0 3914-Å	3.12	3.33	4.04	2.56	3.02
0, 1 4278-Å	1	1	1	1	1
0, 2 4709-Å	0.21	0.19	0.25	0.19	0.23
0, 3 5228-Å	0.032	0.030		0.028	
The ratio $\sigma(v' = 1 \to v'')/\sigma(v' = 1 \to v'' = 2)$					
1, 0 3582-Å	1.76	1.41		1.32	
1, 1 3884-Å	1.05	1		0.93	
1, 2 4236-Å	1	1	1	1	
1, 3 4652-Å	0.44	0.31	0.48	0.46	
1, 4 5149-Å	0.088	0.078		0.105	

The data so obtained are shown in Data Tables 8-3-22 and 8-3-23. Inevitably the accuracy of these estimates is no greater than the accuracy of the emission measurements from which they are derived. It should be noted that the sums are dominated by contributions from the lower wavelength transitions; these are shown in Table 8-1 to be the cases for which the accuracy seems to be poorest. The discrepancies between the estimates of level excitation cross sections are similar to those for the band emission measurements from which they are derived. It is suggested that the most reliable estimates of the energy dependence of the $v' = 0$ and $v' = 1$ population cross sections may be obtained from the normalized emission functions shown in Figs. 8-1 and 8-2, respectively.

5. Emission of the Meinel Bands of N_2^+. The Meinel bands of N_2^+ involve the transition $A\,^2\Pi_u \to X\,^2\Sigma_g$. They lie in the infrared spectral regions and have received only limited attention. General procedures and precautions are discussed in Section 8-3-A. Special note should be taken of the high cross section for collisional deactivation,[323] which will severely restrict the pressures that may be used in an experimental investigation. The emission cross sections for the (1, 0), (2, 0), and (2, 1) transitions are shown in Data Tables 8-3-24 to 8-3-26.

Only in the case of 7826-Å (2, 0) transition is it possible to make a comparison of measurements by different research groups, and here considerable discrepancies are observed. All the experiments may be criticized for inadequate attention to the

establishment of single collision conditions. The large collisional deactivation cross section observed by Sheridan et al.[323] will produce nonlinear effects at quite low pressures. Pressure plots published by Dufay et al.[178] do not indicate that the region where emission varies linearly with pressure was definitely identified. The work of Carleton and Lawrence[172] apparently showed linear variation of emission with pressure, but the range of pressures is not indicated. It is concluded that in neither determination were single collision conditions definitely established; however it is unlikely that this fault will entirely explain the discrepancy.

D. Hydrogen Atom Impact

The general considerations that are applicable to the measurement of collisionally induced emission from a nitrogen target have been described in Section 8-3-A.

Early work by Carleton[324] and by Branscomb et al.[325] was performed with the objective of elucidating auroral phenomena. Since this work employed poor conditions of beam purity and target density, little quantitative information may be drawn from the observations.

The only work on the excitation of nitrogen by H atom impact that is considered of sufficient accuracy to be reproduced here is the work of Dahlberg et al.[182] and that of Robinson and Gilbody.[162] The experimental arrangements are described, respectively, in Sections 8-3-B(2) and 8-3-B(11). The most serious defect in both of these determinations is the lack of precautions for dealing with the wide range of wavelengths covered by the rotational structure of the molecular bands. Also, there is some ambiguity as to the identification of atomic emissions. Noticeable differences exist between the data by Dahlberg et al. and that by Robinson and Gilbody. These differences reflect the discrepancies between the data of Philpot and Hughes,[175] to which Dahlberg et al. normalize their work, and the absolute measurements by Robinson and Gilbody.

In both the work of Dahlberg et al.[182] and also that of Robinson and Gilbody[162] the neutral projectiles were detected using a secondary emission technique. The response of the detector used by Robinson and Gilbody had previously been shown[265] to be the same for protons and neutral H atoms; therefore the response could be calibrated using the proton beam. Dahlberg et al.[182] assumed that the secondary emission coefficients due to H^+ and H impact were in the same ratio as those determined by Stier et al.[41] In the discussion of the experimental procedures [Section 8-2-B(3)], it was suggested that this assumption might be erroneous. It is impossible to determine in retrospect what the magnitude of such an error might be.

The data presented here include information on the following reactions:

The formation of excited N	$H + N_2 \rightarrow [H + N] + N^*$	[Section 8-3-D(1)]
The formation of excited N^+	$H + N_2 \rightarrow [H + N + e] + N^{+*}$	[Section 8-3-D(2)]
The formation of excited N_2	$H + N_2 \rightarrow H + N_2^*$	[Section 8-3-D(3)]
The formation of excited N_2^+	$H + N_2 \rightarrow [H + e] + N_2^{+*}$	[Section 8-3-D(4)]

In all cases the measurements give no information on the states of charge nor excitation of the atomic species that are placed within the square brackets of the reaction equations. The data compilations are collected in Section 8-3-G.

1. Formation of Excited States of N. The only data on this case are the measurements of Dahlberg et al.[182] on the emission functions of the 1493-Å ($3s\ {}^2P \to 2p^3\ {}^2D^0$) and 1743-Å ($3s\ {}^2P \to 2p^3\ {}^2P^0$) transitions shown in Data Tables 8-3-27 and 8-3-28. The method by which these measurements were placed on an absolute basis is incorrect [see Section 8-3-B(2)]. The absolute magnitudes of the cross sections are not accurate, and reliance should be placed only on the relative variation of the cross section with impact energy. It should be noted that both of these lines are unresolved doublets.

2. Formation of Excited States of N^+. Dahlberg et al.[182] and Robinson and Gilbody[162] have measured the emission cross section of the 5005-Å line, which is a blend of two emissions, $3d\ {}^3F^0 \to 3p\ {}^3D$ and $3p\ {}^5P^0 \to 3s\ {}^5P$ (Data Table 8-3-29). There are a total of five other N^+ spectral lines lying within ± 7-Å of the 5005-Å emission, and it seems quite unlikely that the 5005-Å lines were properly resolved in the data presented here. Moreover, the precautions required when measuring a group of lines of adjacent wavelengths (Section 3-4-E) were not observed.

3. The Formation of Excited State of N_2. In Data Table 8-3-30 is shown the cross section for the emission of the 3371-Å [$C\ {}^3\Pi_u \to B\ {}^3\Pi_g$ (0,0)] transition as measured by Dahlberg et al.[182] There are no known transitions that could cause a cascade contribution; therefore, the cross section presented here is proportional to the cross section for forming the $C\ {}^3\Pi_u\ (v' = 0)$ state.

Carleton and Lawrence[172] detected the emission of the first positive band of N_2 under proton bombardment. It was assumed that direct excitation by proton impact was forbidden by the Wigner spin conservation rule and that the excitation of the triplet state of the molecule occurred by the impact of an H atom formed in a previous charge transfer collision. The pressure dependence of the emission was consistent with this mechanism. These data are not reproduced here since the measurements were of very poor accuracy, exhibiting considerable scatter; moreover, the experimental approach is indirect and may be subject to many sources of error.

4. Formation of Excited States of N_2^+. The measurements of this excitation mechanism are shown in Data Tables 8-3-31 to 8-3-35. In neither of the two experiments[182, 162] was sufficient care taken to include all of the rotational structure of the first negative band emission in the manner indicated in Section 3-4-E.

In so far as is possible with these limited series of measurements, the consistency of the data has been examined using the techniques suggested in Section 8-3-A. The mean values of the ratios $\sigma(0, v'') \div \sigma(0, 1)$ and $\sigma(1, v'') \div \sigma(1, 2)$ determined at a series of different energies are shown in Table 8-2. Some differences are observed between the ratios evaluated here for H impact data and those evaluated previously in Section 8-3-C(4) for H^+ impact. It is noted that the 3914-Å transition measured by Robinson and Gilbody[162] exhibits a ratio with the 4278-Å emission that is some 19% higher than for H^+ impact; this must indicate a lack of consistency. The ratios determined for the data presented by Dahlberg et al.[182] agree with the H^+ impact determinations within estimated random experimental error.

Both experimental determinations have included estimates of the cross sections for populating the $v' = 0$ vibrational level of the $B\ {}^2\Sigma_u^+$ state, obtained by summing together the emission cross sections for all the transitions that depopulate this level. This is shown in Data Table 8-3-36.

Gusev et al.[111] have studied the first negative emission induced by H impact on N_2. The data is in the form of ratios of cross sections for the formation of the $v' = 1$

TABLE 8-2

	Robinson and Gilbody[162]	Dahlberg et al.[182]
The ratio $\dfrac{\sigma(v' = 0 \to v'')}{\sigma(v'' = 0 \to v'' = 1)}$		
0, 0 3914-Å	4.79	3.25
0, 1 4278-Å	1	1
0, 2 4709-Å	0.28	0.21
The ratio $\dfrac{\sigma(v' = 1 \to v'')}{\sigma(v' = 1 \to v'' = 2)}$		
1,2 4236-Å	1	
1, 3 4652-Å	0.40	

and $v' = 2$ states of $B\ ^2\Sigma_u^+$. The data give no direct measurement of a cross section and is not presented here.

E. Impact of Other Projectiles

Studies of collisional excitation of nitrogen by projectiles other than protons and atomic hydrogen have not been pursued very deeply at the present time. The available data is fragmentary, and generally there is little or no confirmation of values by independent measurements due to separate research groups. Most of the information is for excitation by impact of molecular hydrogen ions, rare gas ions, alkali metal ions, and N_2^+ molecular ions. The accuracy of this data is likely to be rather poor.

1. Excitation of Nitrogen by H_2^+ and H_3^+ Impact. Only fragmentary data are available on the collisional excitation of nitrogen by H_2^+ and H_3^+.

The work by Thomas and Bent[201] gives a ratio of the cross sections for emission of the 3914-Å band of N_2^+ induced by H^+, H_2^+, and H_3^+, respectively, as $1:1.30:1.44$, at the single impact velocity of 6.2×10^8 cm sec^{-1}.

Dufay et al.[202] measured certain cross sections for H_2^+ impact during the course of an investigation that was primarily devoted to the use of protons as projectiles. The work with protons was later revised as a result of improvements in calibration techniques,[178] but no new cross sections for H_2^+ impact were presented. It is considered that the ratio of the H_2^+ and H^+ impact cross sections was quite accurate in the first experiments; therefore, revised values for H_2^+ impact may be obtained by normalizing the H^+ data to the more recent measurements. Revised data for the emission of the 5005-Å transition obtained in this manner are shown in Table 8-3-37. It should be recalled that this line is a blend of two transitions ($3d\ ^3F^0 \to 3p\ ^3D$ and $3p\ ^5p^0 \to 3s\ ^5P$) and that, in practice, the measurement by Dufay[178, 202] probably includes an unknown contribution from certain other transitions at neighboring wavelengths.

Measurements were also presented for the population of the $v' = 0$ and $v' = 1$ levels of the $B\ ^2\Sigma_u^+$ state of N_2^+ induced by H_2^+ impact. These cross sections were obtained by the summation of the emission cross sections for the relevant transitions in the first negative system of N_2^+; the emission cross sections themselves were not published. The measurements for H_2^+ impact were found to be a factor two higher than those for H^+ at the same energy. This suggests that an estimate of emission cross sections for H_2^+ impact may be obtained by taking the revised data for H^+ impact shown in Data Tables 8-3-12 through 8-3-20 and multiplying by a factor of two.

   2. Excitation of Nitrogen by Impact of Rare Gas Ions.   The collisional excitation by impact of Ne^+ and Ar^+ is due to the work of Neff[104] using the apparatus described in Section 8-3-B(5). The work with Xe^+ impact is due to Kurzweg et al.,[113] who used the apparatus described in Section 8-3-B(7).

   Neff presents data on the emission of the 3995-, 4640-, 5005-, and 5675-Å lines of N II (Data Table 8-3-38). Due to the poor resolution used the identification of all of these transitions is in doubt.

   Measurements on the emission of the 3914-Å band induced by impact of Ne^+, Ar^+, and Xe^+ (Data Tables 8-3-38 through 8-3-40) are unambiguous. The cross section for formation of the $B\ ^2\Sigma_u^+$ state may be estimated by multiplying these data by a factor of 1.37, the branching ratio evaluated from theoretical calculations of Franck-Condon factors.[326]

   Data by Schlumbohm[229] on collisionally induced emission by Ne^+ impact and data by Sheridan and Clark on emission induced by He^+ and Ne^+ impact is omitted for the reasons discussed previously [see Section 8-3-B(13) and 8-3-B(4)]. Polyakova et al.[111, 110, 334] and also Moore and Doering[184] have studied the ratios of populations of the $v' = 0$ and $v' = 1$ levels; these data do not give direct information on cross sections and are not reproduced here. Early measurements by Fan[321] on emission induced by He^+ impact were carried out under very poorly defined conditions and are likely to be of little quantitative value; these also are omitted.

   3. Excitation of Nitrogen by Impact of Alkali Metal Ions.   Data for this case are available only from a limited series of measurements by Neff[104] whose experimental technique is described in Section 8-3-B(5). The data on the emission of the 3914-Å band by Na^+ and K^+ (Data Tables 8-3-41 and 8-3-42) is unambiguous. There is also data for emission of two N II lines induced by Na^+ impact (Data Table 8-3-41); the 3995-Å line is unambiguously identified as the $3p\ ^1D_2 \rightarrow 3s\ ^1P_1^0$ transition; however, the origin of the 5005-Å line is in doubt because of the poor resolution employed.

   Polyakova et al.[110] have studied the relative populations of the $v' = 0$ and $v' = 1$ levels of the $B\ ^2\Sigma_u^+$ state induced by impact of Li^+, Na^+, and K^+; however, these give no direct information on cross sections and the data is not reproduced here.

   4. Excitation of Nitrogen by Impact of N_2^+ Ions.   Contributions to the study of this case have been made by Neff[104] [see Section 8-3-B(5)], Doering[227] [see Section 8-3-B(6)], and Kurzweg et al.[113] [see Section 8-3-B(7)]. In all three cases the measurements included emission from both the projectile and target. It is therefore necessary that the results be shown to be independent of penetration through the target (see Section 2-2-B). The lifetime of the $B\ ^2\Sigma_u^+$ state is 6.6×10^{-8} sec.[327] At the velocities used by both Doering and by Neff the decay length will be very much less than the size of the apparatus so that the error is negligible; however, in the work of Kurzweg et al. the decay length at the highest velocities will exceed the size of the apparatus and neglect of this fact will have caused appreciable errors. Also, in the work of

Kurzweg the neglect of Doppler shift of emission will have caused energy-dependent errors (see Section 3-4-F).

The discrepancies between data for emission of the first negative transition (Data Table 8-3-43) may reflect inadequacies in the determination of absolute cross sections.

In Data Tables 8-3-44 and 8-3-45 are shown the emission cross sections of the 4278- and 4236-Å bands of the N_2^+ first negative system as measured by Doering. The scatter of the data points is considerable. The average value of the ratio of 3914-Å to 4278-Å emission cross sections measured by Doering is quoted as 3.33 ± 0.45, which is consistent with the majority of the values of that same ratio evaluated for proton impact excitation [see Section 8-3-C(4)].

In Data Table 8-3-46 is shown the emission cross section of the 5005-Å line of N II. As mentioned previously in the context of H^+ impact excitation, the origin of that line cannot be positively established with the poor resolutions employed in these experiments.

5. Excitation of Nitrogen by Other Projectiles. Kurzweg et al.[113] have provided measurements on the cross sections for emission of the 3914-Å band of the first negative system of N_2^+ induced by the impact of Ba^+, Ba^{2+}, N^+, and 0^+ that are shown in Data Tables 8-3-47 to 8-3-50, respectively. It is possible to estimate the cross sections for forming the $B\ ^2\Sigma_u^+$ excited state from these data through multiplication by a factor of 1.37, the branching ratio for emissions from this state.[326] The data was placed on an absolute basis by a normalization to electron impact data.

Moore and Doering[184] have published ratios of the cross sections for exciting the $v' = 1$ and $v' = 0$ levels of the $B\ ^2\Sigma_u^+$ state of N_2^+ induced by the impact of N^+ ions. These data are not reproduced here since they provide no direct cross section information.

Sheridan and Clark[176] have published measurements on the excitation of nitrogen by N^+ impact; however, there are grounds for doubting the accuracy of these measurements [see Section 8-3-B(4)], so the data are not reproduced here.

Neff[104] has investigated the excitation of nitrogen by the alkali earth ions Mg^+ and Ca^+ in the energy range 400-eV to 2-keV. No nitrogen emissions were observed, suggesting that a very approximate upper limit of 2×10^{-19} cm^2 may be placed on the emission cross sections for nitrogen lines produced by this collision combination.

As part of an extensive series of measurements on the excitation of nitrogen by the impact of N_2^+, Doering[227] produced an estimate of the cross section for the emission of the 3914-Å first negative band of N_2^+ induced by the impact of N_2 on molecular nitrogen. The value of 3×10^{-19} cm^2 at 1-keV includes contributions from both target and projectile systems and is of very limited accuracy.

F. Rotational Structure of N_2^+ Emission

A molecular system in thermal equilibrium with its surroundings will exhibit a Boltzmann population distribution among the rotational states. Collisionally induced changes in the rotation of the molecule may be investigated by comparing the population distribution among the rotational states of the excited molecules with the distribution among the unexcited target molecules. In principle, the changes in the distribution are directly related to collision cross sections, although in practice, due to the complexity of the situation, cross sections have not been derived from such information.

In this section the present status of these investigations will be reviewed, and the various available sources of data will be listed. No attempt is made to reproduce the data, since its form varies from one publication to another.

1. Interpretation of Measurements. The distribution among rotational states may be investigated by measuring the relative intensities of the rotational fine structure of a given transition. Such investigations have been limited to the $0 \rightarrow 0$ and $1 \rightarrow 0$ transitions of the first negative system of N_2^+ giving information on the rotational population of the $v' = 0$ and $v' = 1$ vibrational levels of the $B\ ^2\Sigma_g^+$ state of N_2^+. The usual procedure is, first, to assume that the population distribution is Boltzmann, and then to attempt a determination of an apparent rotational temperature. The apparent temperature or, alternatively, the degree of deviation from the distribution for a specific temperature will characterize the rotational population distribution.

For a Boltzmann distribution appropriate to a temperature $T\ °K$, the intensity $I_{K'K''}$ for a transition $K' \rightarrow K''$ for lines of the R branch (i.e., $K' - K'' = 1$) in the N_2^+ first negative system is given by

$$I_{K'K''} = C\ v_{K'K''}^4\ K' \exp - \frac{1.44\ B\ K'\ (K' + 1)}{T} \tag{8-1}$$

The factors C and B are constants for a given band. For small segments of the overall band system, the frequency of the transition $v_{K'K''}$ may be considered to be approximately constant. It has been the practice to plot $\log (I_{K'K''}/K')$ against $K'\ (K' + 1)$. If a straight line is obtained, then an apparent temperature is evaluated from the slope of the graph using a value of B obtained from spectroscopic measurements. There is an intensity alternation that arises from the alternating statistical weights of the rotational levels of both the upper and lower states;[33] usually one multiplies the intensities of the K'-even lines by a factor of two, to eliminate this and to produce a single curve. It was shown first by Lowe and Ferguson[328] and later by Moore and Doering[314] that it is necessary to include the effect of the P-branch overlap. The K' line of the R-branch in the $(0, 0)$ transition is closely overlapped by the P-branch line that originates in the $K' + 25$ upper rotational state; for sufficiently high temperatures, this overlap will cause the plot to become nonlinear. In Fig. 8-3 is shown a computed plot by Moore and Doering[314] for three different temperatures in which the effect of the overlap is included by adding to the R-branch intensity the intensity for the overlapping P-branch line. It can be seen that for low temperatures the P-branch contribution is negligible, and the plot for the R-branch remains linear; for higher temperatures, the plot becomes nonlinear. At the highest temperatures the line for K'-even separates from that line for the odd states since the weaker K'-even lines from the R-branch are overlapped by the stronger K'-odd lines from the P-branch. At $3000°K$ the intensity alternation in the spectrum will have almost disappeared for K' up to 23. From this analysis, it is clear that for high temperatures it is necessary to allow for the P-branch overlap before attempting any analysis of the significance of data.

It should be noted that in the 4278-Å $(0, 1)$ band of the first negative system, the R-branch line from the K' state is overlapped by the $(K' + 22)$ line of the P-branch. In this case an even line in the R-branch is overlapped by an even line in the P-branch, the factor of two intensity alternations is maintained, and a graph corresponding to Fig. 8-3 for this case will show only a single line at all temperatures.

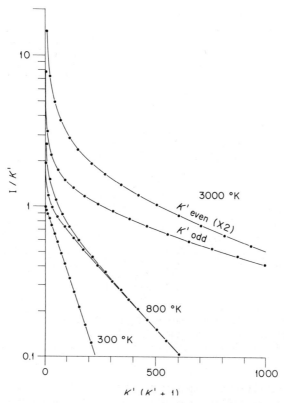

Fig. 8-3   Computed plot of I/K' versus $K'(K'+1)$ at various rotational temperatures for the N_2^+ first negative (0, 0) band. Intensities of points corresponding to K' even lines have been multiplied by 2. From Moore and Doering.[314]

It is clear that in order to determine an effective rotational temperature it is necessary to allow for the P-branch overlap. Moore and Doering[314] have produced a set of correction factors that may be applied to the measured line intensities to derive the intensity of the R-branch alone. For conditions at or below room temperatures the correction is negligible. These corrections are based on the assumption of a Boltzmann distribution among the rotational states; in the event of some other distribution a correction for the effect of overlap cannot readily be evaluated.

Figure 8-4, taken from Moore and Doering,[314] shows a plot of measured intensities for the (0, 0) transition induced by the impact of H_2^+ on N_2 at 400-eV impact energy. Care was taken to eliminate the interference of the $A\ ^2\Pi_u$ state that perturbs the P and R branches at $K' = 39$; for this reason, data in the region of $K' = 39$ in the R-branch and also for $K' = 14$ (which is overlapped by the $K' = 39$ line in the P-branch), are omitted. The graph shows the separation of the even and odd states. At high K' the limiting slope of the graph is appropriate to a temperature of 3500°K. Moore and Doering calculate the expected intensities for such a temperature and plot them as the dotted line; there is clear disagreement with experiment at low values of K'. Attempts to fit other temperatures, taking due account of overlap, are equally unsuccessful. It must therefore be concluded that the observed intensity variation is not appropriate to a Boltzmann distribution. Moore and Doering point out that in most previous

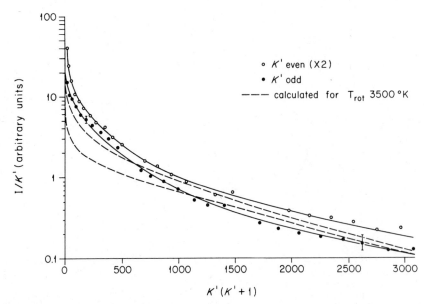

Fig. 8-4 Plot of I/K' versus $K'(K' + 1)$ taken from spectra of the first negative (0, 0) band excited by 0.4 keV H_2^+. The intensities of the lines of even K' have been multiplied by 2. The calculated curves for $I_{rot} = 3500°K$ is included for comparison. The reproducibility of the data is indicated by the error bars at each end of the plot. From Moore and Doering.[314]

experiments the data have extended only out to $K' = 10$ or $K' = 20$, and with some poor statistical accuracy of the data points it would be possible to draw a straight line through the data and infer a rotational temperature. Most of the data in the literature are of this type.

It is clear that, to make a meaningful analysis of the rotational emission, it is necessary to carry out an accurate measurement of relative line intensities, which should extend out to values of K' of at least 30. Corrections must be applied for over-lapping of branches. Only then is it justifiable to examine the data to see whether it indicates an effective temperature or a deviation from a Boltzmann distribution.

2. Experimental Procedures. From an experimental point of view most of the precautions enumerated in Chapters 2 and 3, which are applicable to cross section measurements, are not appropriate here. It is necessary that the light emission be shown to be due to single collision mechanisms; that is, that the intensity is propor-tional to target pressure and projectile beam current. Moore and Doering[314] point out that a possible interpretation of a non-Boltzmann distribution is that there are two or more collision mechanisms contributing to the excitation of the N_2^+ state, and each involves a different apparent rotational temperature; strict adherance to single collision conditions obviates this possibility. It is also necessary that the detection system have a uniform response over the whole of the wavelength interval covered by the rotational bands or alternatively that a correction be made for any variation.

It is very necessary, particularly when measuring intensity at states of high K', that attention be paid to removing any interfering spectral lines of other systems. The intensities will be very small, and the interpretation of the data depends critically on these measurements. Moore and Doering[314] point out that scattered light in the

TABLE 8-3   ROTATIONAL TEMPERATURE OF THE B $^2\Sigma_u^+$ STATE OF N_2 EXCITED BY THE IMPACT OF HEAVY PARTICLES ON A NITROGEN TARGET

Projectile	References	Energy (keV)	Apparent Temp (°K)	Violated Criteria
H^+	Carleton[324]	3	300 ($\pm$10)[a]	4, 5
H^+	Roesler et al.[329]	10–30	305 ($\pm$8)[a]	2, 4, 5
H^+	Sheridan, Clark[176]	15–65	300 ($\pm$60)[a]	2, 4, 5
D^+	Sheridan, Clark[176]	15–65	300 ($\pm$60)[a]	2, 4, 5
H^+	Polyakova et al.[330]	30, 10, 5	[b]	2, 4
H^+	Polyakova et al.[309]	30	[b]	2, 4
H^+	Dufay et al.[178]	20–600	305[a]	4, 5
H^+	Thomas et al.[183]	150–1000	310 ($\pm$30)[a]	4, 5
H^+	Reeves, Nicholls[331]	500	301 ($\pm$80)[a]	2, 4, 5
H^+	Reeves, Nicholls[331]	1000	276 ($\pm$10)[a]	2, 4, 5
H^+	Reeves et al.[332]	1000	276 ($\pm$10)[a]	2, 4, 5
H^+	Moore, Doering[315]	0.6–10	[b]	2
D^+	Moore, Doering[315]	0.6–10	[b]	2
H_2^+	Moore, Doering[314]	0.4–3	[b]	2
H_2^+	Polyakova et al.[330]	26	[b]	2, 4
H_2^+	Reeves et al.[332]	1000	276 ($\pm$10)[a]	2, 4, 5
H_2^+	Moore, Doering[315]	0.6–10	[b]	2
D_2^+	Moore, Doering[315]	0.6–10	[b]	2
H_3^+	Reeves, Nicholls[331]	1500	301 ($\pm$80)[a]	2, 4, 5
H_3^+	Reeves et al.[332]	1500	301 ($\pm$80)[a]	2, 4, 5
He^+	Doering[333]	2–17	300 $\rightarrow$ 650 (see text)	2, 4, 5
He^+	Polyakova et al.[330]	30, 20, 10	[b]	2, 4
He^+	Sheridan, Clark[176]	20–65	280 ($\pm$60)[a]	2, 4, 5
He^+	Polyakova et al.[110]	2–30	[b]	2, 4
He^+	Moore, Doering[315]	0.6–10	[b]	2
He^+	Polyakova et al.[334]	5–30	[b]	2, 4
He	Polyakova et al.[334]	5–30	[b]	2, 4
Li^+	Reeves, Nicholls[331]	2	[b]	2, 4, 5
Li^+	Lowe, Ferguson[328]	3	3500	2
Li^+	Lowe, Ferguson[335]	6	1425 (1350–1525)	2
Li^+	Lowe, Ferguson[335]	8	1160 (1080–1280)	2
Li^+	Lowe, Ferguson[335]	10	980 (930–1060)	2
Li^+	Polyakova et al.[110]	4.5, 10	[b]	1, 2, 4
C^+	Polyakova et al.[330]	10	[b]	2, 4
N^+	Polyakova et al.[330]	10	[b]	2, 4
N_2^+	Polyakova et al.[330]	10	[b]	2, 4
N_2^+	Doering[227]	10	475	2, 4, 5
N_2^+	Doering[333]	2–17	300 $\rightarrow$ 650 (see text)	2, 4, 5
CO^+	Polyakova et al.[330]	10	[b]	2, 4
O^+	Polyakova et al.[330]	10	[b]	2, 4
Ne^+	Polyakova et al.[330]	16, 5	[b]	2, 4

Table 8-3—*continued*

Projectile	References	Energy (keV)	Apparent Temp (°K)	Violated Criteria
Ne⁺	Sheridan, Clark[176]	25	280 (±60)[a]	2, 4, 5
Ne⁺	Polyakova et al.[110]	2–15	[b]	2, 4
Ne⁺	Moore, Doering[315]	0.6–10	[b]	2
Ne⁺	Polyakova et al.[334]	5–30	[b]	2, 4
Ne	Polyakova et al.[334]	5–30	[b]	2, 4
Na⁺	Lowe, Ferguson[335]	6	3050 (±350)	2
Na⁺	Lowe, Ferguson[335]	8	3120 (2850–3500)	2
Na⁺	Lowe, Ferguson[335]	10	2270 (2020–2720)	2
Na⁺	Polyakova et al.[110]	5, 10	[b]	1, 2, 4
Ar⁺	Polyakova et al.[330]	7	[b]	2, 4
Ar⁺	Doering[333]	2–17	300–650 (see text)	2, 4, 5
Ar⁺	Polyakova et al.[110]	7	[b]	2, 4
Ar⁺	Polyakova et al.[334]	5–30	[b]	2, 4
Ar	Polyakova et al.[334]	5–30	[b]	2, 4
K⁺	Lowe, Ferguson[335]	5–10	450 (420–500)	1, 2
K⁺	Polyakova et al.[110]	5–10	5	1, 2, 4
Kr⁺	Polyakova et al.[334]	5–30	[b]	2, 4
Kr	Polyakova et al.[334]	5–30	[b]	2, 4
Rb⁺	Lowe, Ferguson[335]	6–10	400 (350–450)	1, 2
Xe⁺	Polyakova et al.[334]	5–30	[b]	2, 4
Xe	Polyakova et al.[334]	5–30	[b]	2, 4
Cs⁺	Lowe, Ferguson[335]	6–10	400 (375–425)	1, 2

[a] Within stated experimental error the same as the nominal temperature of the target region.
[b] Non-Boltzmann population distribution.

spectrometer coming from systems at other wavelengths might provide sufficient intensity to be troublesome.

3. Survey of Data from Rotational Excitation Experiments. The data available from such studies varies greatly in its form. In many experiments only the apparent Boltzmann temperature was quoted, and the individual data points were not reproduced. For the present purposes the references to published data are presented in Table 8-3.

The following list of basic criteria is provided to which a study of rotational excitation must adhere in order that the results should have meaning.

(1) Single collision conditions must be demonstrated (i.e., intensity proportional to pressure and beam current).

(2) Variation of sensitivity over the relevant wavelength interval must be assessed and corrections applied.

(3) Interfering spectral lines from other systems must be removed.

(4) Overlapping of the *P*-branch with the *R*-branch must be allowed for.

(5) Measurements should include upper rotational states through $K' = 20$.

In the event that criterion (1) is violated, then the nature of the process is ambiguous. For situations where the apparent rotational temperature is approximately 300°K or less, then the violation of criteria (4) and (5) will not cause large errors in the interpretation of the data. The result of neglecting criterion (2) will depend on the particular experimental system that is used and under some circumstances might lead to erroneous interpretations of the data.

The available sources of data are presented in Table 8-3 with an indication of the criteria that were violated in the experiment. No uniform presentation of results is found in the literature, and no attempt is made to reproduce the data here. In general a report will either conclude that Boltzmann distribution is appropriate and quote a rotational temperature or, alternatively, will conclude that the distribution is not Boltzmann and will provide some indication of the actual distribution among the rotational states. In Table 8-3 it is noted whether the distribution appeared to be Boltzmann or non-Boltzmann; in the former case the apparent temperature quoted by the authors is listed.

One of the most useful recent papers is that by Moore and Doering[315] in which they observe the rotational population of excited N_2^+ formed through impact of H^+, D^+, H_2^+, D_2^+, He^+, and Ne^+. For impact velocities less than 10^8 cm/sec the rotational state population is not appropriate to a Boltzmann distribution. They suggest that this conclusion was applicable to any collision combination, irrespective of the nature of the projectile.

G. Data Tables

Some of the data for excitation by H^+ impact include studies with D^+. In such cases the D^+ data is shown in the tables as though it were H^+ of the same velocity; that is, at the true D^+ energy divided by two.

1. Cross Sections for Emission of N I and N II Lines Induced by H^+ Impact on an N_2 Target

8-3-1 8188-, 8200-, 8211-, 8216-Å Lines of N I ($3p\ ^4P^0 \rightarrow 3s\ ^4P$ multiplet).

8-3-2 10,113-Å Line of N I ($3d\ ^4F \rightarrow 3p\ ^4D^0$ multiplet).

8-3-3 9392-Å Line of N I ($3p\ ^2D_{5/2} \rightarrow 3s\ ^2P_{3/2}$).

8-3-4 8680-Å Line of N I ($3p\ ^4D^0 \rightarrow 3s\ ^4P$ multiplet).

8-3-5 1743-Å Line of N I ($3s\ ^2P \rightarrow 2p^3\ ^2P^0$; unresolved doublet).

8-3-6 1493-Å Line of N I ($3s\ ^2P \rightarrow 2p^3\ ^2D^0$; unresolved doublet).

8-3-7 5005-Å Line of N II (identification uncertain; 8-3-7D is data for D^+ impact).

8-3-8 5680-Å Line of N II ($3p\ ^3D \rightarrow 3s\ ^3P^0$).

8-3-9 3995-Å Line of N II ($3p\ ^1D_2 \rightarrow 3s\ ^1P_1^0$).

8-3-10 6482-Å Line of N II (identification uncertain).

2. Cross Sections for Emission of N_2 Bands Induced by Impact of H^+ on an N_2 Target

8-3-11 3371-Å Band [$C\ ^3\Pi_u \rightarrow B\ ^3\Pi_g$ (0, 0)].

3. Cross Sections for Emission of N_2^+ First Negative Bands Induced by Impact of H^+ on an N_2 Target

8-3-12   3914-Å Band $[B\ ^2\Sigma_u^+ \to X\ ^2\Sigma_g^+\ (0, 0)]$ [8-3-12 (F) is data for D^+ impact].

8-3-13   4278-Å Band $[B\ ^2\Sigma_u^+ \to X\ ^2\Sigma_g^+\ (0, 1)]$.

8-3-14   4709-Å Band $[B\ ^2\Sigma_u^+ \to X\ ^2\Sigma_g^+\ (0, 2)]$.

8-3-15   5228-Å Band $[B\ ^2\Sigma_u^+ \to X\ ^2\Sigma_g^+\ (0, 3)]$.

8-3-16   3582-Å Band $[B\ ^2\Sigma_u^+ \to X\ ^2\Sigma_g^+\ (1, 0)]$; [8-3-16(C) is data for D^+ impact].

8-3-17   3884-Å Band $[B\ ^2\Sigma_u^+ \to X\ ^2\Sigma_g^+\ (1, 1)]$.

8-3-18   4236-Å Band $[B\ ^2\Sigma_u^+ \to X\ ^2\Sigma_g^+\ (1, 2)]$.

8-3-19   4652-Å Band $[B\ ^2\Sigma_u^+ \to X\ ^2\Sigma_g^+\ (1, 3)]$.

8-3-20   5149-Å Band $[B\ ^2\Sigma_u^+ \to X\ ^2\Sigma_g^+\ (1, 4)]$.

8-3-21   3914-Å Band $[B\ ^2\Sigma_u^+ \to X\ ^2\Sigma_g^+\ (0, 0)]$ in Coincidence with the Emission of Balmer Alpha Radiation ($3p \to 2s$ plus $3d \to 2p$).

4. Cross Sections for Excitation of $B\ ^2\Sigma_u^+$ Levels Induced by Impact of H^+ on an N_2 Target (Derived from Emission Data)

8-3-22 $B\ ^2\Sigma_u^+\ v' = 0$ State.

8-2-23 $B\ ^2\Sigma_u^+\ v' = 1$ State.

5. Cross Sections for Emission of N_2^+ Meinel Bands Induced by Impact of H^+ on an N_2 Target

8-3-24   9145-Å Band $[A\ ^2\Pi_u \to X\ ^2\Sigma_g^+\ (1, 0)]$.

8-3-25   7826-Å Band $[A\ ^2\Pi_u \to X\ ^2\Sigma_g^+\ (2, 0)]$.

8-3-26   9431-Å Band $[A\ ^2\Pi_u \to X\ ^2\Sigma_g^+\ (2, 1)]$.

6. Cross Sections for Emission of Lines and Bands Induced by Impact of H on an N_2 Target

8-3-27   1743-Å Lines of N I ($3s\ ^2P \to 2p^3\ ^2P^0$; unresolved doublet).

8-3-28   1493-Å Lines of N I ($3s\ ^2P \to 2p^3\ ^2D^0$; unresolved doublet).

8-3-29   5005-Å Line of N II (identification of line uncertain).

8-3-30   3371-Å Band of N_2 $[C\ ^3\Pi_u \to B\ ^3\Pi_g\ (0, 0)]$.

8-3-31   3914-Å Band of N_2^+ $[B\ ^2\Sigma_u^+ \to X\ ^2\Sigma_g^+\ (0, 0)]$.

8-3-32   4278-Å Band of N_2^+ $[B\ ^2\Sigma_u^+ \to X\ ^2\Sigma_g^+\ (0, 1)]$.

8-3-33   4709-Å Band of N_2^+ $[B\ ^2\Sigma_u^+ \to X\ ^2\Sigma_g^+\ (0, 2)]$.

8-3-34   4236-Å Band of N_2^+ $[B\ ^2\Sigma_u^+ \to X\ ^2\Sigma_g^+\ (1, 2)]$.

8-3-35   4652-Å Band of N_2^+ $[B\ ^2\Sigma_u^+ \to X\ ^2\Sigma_g^+\ (1, 3)]$.

7. Cross Sections for Excitation of $B\ ^2\Sigma_u^+$ Levels Induced by Impact of H on an N_2 Target (Derived from Emission Data)

8-3-36 $B\ ^2\Sigma_u^+\ v' = 0$ State.

8. Cross Sections for Emission of Lines and Bands under Impact of Various Projectiles

8-3-37   5005-Å Line of N II Induced by the Impact of H_2^+ (identification of line is uncertain).

8-3-38   Certain Lines of N II and N_2^+ Induced by the Impact of Ne^+ (A) 3995-Å ($3p\ ^1D_2 \to 3s\ ^1P_1^0$); (B) 4640-Å ($3p\ ^3P \to 3s\ ^3P^0$, see text); (C) 5005-Å (identification uncertain); (D) 5675-Å ($3p\ ^3D \to 3s\ ^3P^0$, see text); (E) 3914-Å $[B\ ^2\Sigma_u^+ \to X\ ^2\Sigma_g^+\ (0, 0)]$.

8-3-39   3914-Å Band of N_2^+ $[B\ ^2\Sigma_u^+ \to X\ ^2\Sigma_g^+\ (0, 0)]$ Induced by the Impact of Ar^+.

8-3-40   3914-Å Band of N_2^+ $[B\ ^2\Sigma_u^+ \to X\ ^2\Sigma_g^+\ (0,0)]$ Induced by the Impact of Xe^+.

8-3-41 Emission of Certain Lines of N II and N_2^+ Induced by the Impact of Na^+ (A) 3995-Å ($3p\ ^1D_2 \rightarrow 3s\ ^1P_1^0$); (B) 5005-Å (identification uncertain); (C) 3914-Å [$B\ ^2\Sigma_u^+ \rightarrow X\ ^2\Sigma_g^+$ (0, 0)].

8-3-42 3914-Å Band of N_2^+ [$B\ ^2\Sigma_u^+ \rightarrow X\ ^2\Sigma_g^+$ (0, 0)] Induced by the Impact of K^+.

8-3-43 3914-Å Band of N_2^+ [$B\ ^2\Sigma_u^+ \rightarrow X\ ^2\Sigma_g^+$ (0, 0)] Induced by the Impact of N_2^+ (this measurement is the sum of contributions from the excited projectile and the excited target).

8-3-44 4278-Å Band of N_2^+ [$B\ ^2\Sigma_u^+ \rightarrow X\ ^2\Sigma_g^+$ (0, 1)] Induced by the Impact of N_2^+ (this measurement is the sum of contributions from the excited projectile and the excited target).

8-3-45 4236-Å Band of N_2^+ [$B\ ^2\Sigma_u^+ \rightarrow X\ ^2\Sigma_g^+$ (1, 2)] Induced by the Impact of N_2^+ (this measurement is the sum of contributions from the excited projectile and the excited target).

8-3-46 5005-Å Line of N II Induced by the Impact of N_2^+ (this measurement is the sum of contributions from the excited projectile and the excited target; identification of line is uncertain).

8-3-47 3914-Å Band of N_2^+ [$B\ ^2\Sigma_u^+ \rightarrow X\ ^2\Sigma_g^+$ (0, 0)] Induced by the Impact of Ba^+.

8-3-48 3914-Å Band of N_2^+ [$B\ ^2\Sigma_u^+ \rightarrow X\ ^2\Sigma_g^+$ (0, 0)] Induced by the Impact of Ba^{2+}.

8-3-49 3914-Å Band of N_2^+ [$B\ ^2\Sigma_u^+ \rightarrow X\ ^2\Sigma_g^+$ (0, 0)] Induced by the Impact of N^+.

8-3-50 3914-Å Band of N_2^+ [$B\ ^2\Sigma_u^+ \rightarrow X\ ^2\Sigma_g^+$ (0, 0)] Induced by the Impact of O^+.

8-3-1A
SHERIDAN ET AL. (173)

ENERGY (KEV)	CROSS SECTION (SQ. CM)
3.10E 00	5.36E-18
3.70E 00	5.75E-18
5.00E 00	6.08E-18
6.00E 00	6.20E-18
7.40E 00	6.14E-18
8.90E 00	6.00E-18
1.03E 01	5.75E-18
1.17E 01	5.62E-18
1.28E 01	5.41E-18
1.41E 01	5.30E-18
1.51E 01	5.13E-18
1.60E 01	5.03E-18
1.71E 01	4.90E-18
1.82E 01	4.80E-18
1.93E 01	4.64E-18
2.03E 01	4.49E-18
2.13E 01	4.40E-18
2.23E 01	4.28E-18
2.36E 01	4.19E-18
2.47E 01	4.15E-18
2.60E 01	4.07E-18
2.70E 01	4.00E-18
2.85E 01	3.89E-18
2.98E 01	3.89E-18

8-3-1B
CARLETON ET AL. (172)

ENERGY (KEV)	CROSS SECTION (SQ. CM)
2.00E 00	4.56E-18
2.50E 00	5.64E-18
3.00E 00	5.31E-18
3.50E 00	5.73E-18
4.00E 00	5.62E-18
4.50E 00	6.04E-18

8-3-2
DUFAY ET AL. (178)

ENERGY (KEV)	CROSS SECTION (SQ. CM)
4.00E 01	4.00E-18
5.00E 01	3.60E-18
6.00E 01	3.40E-18
7.00E 01	3.00E-18
8.00E 01	3.20E-18
1.00E 02	3.00E-18
1.20E 02	2.40E-18

8-3-3
DUFAY ET AL. (178)

ENERGY (KEV)	CROSS SECTION (SQ. CM)
4.00E 01	2.70E-18

5.00E 01	2.70E-18
6.00E 01	2.20E-18
7.00E 01	2.00E-18
8.00E 01	1.95E-18
9.00E 01	1.65E-18
1.20E 02	1.55E-18

8-3-4
DUFAY ET AL. (178)

ENERGY (KEV)	CROSS SECTION (SQ. CM)
4.00E 01	4.60E-18
5.00E 01	4.00E-18
8.00E 01	3.30E-18
1.00E 02	3.20E-18
1.20E 02	2.60E-18

8-3-5
DAHLBERG ET AL. (182)

ENERGY (KEV)	CROSS SECTION (SQ. CM)
2.00E 01	4.90E-18
2.60E 01	5.10E-18
3.00E 01	4.80E-18
3.20E 01	5.00E-18
3.50E 01	4.90E-18
4.00E 01	4.40E-18
4.00E 01	4.20E-18
4.50E 01	4.40E-18
5.00E 01	4.40E-18
5.20E 01	4.30E-18
5.50E 01	4.00E-18
6.00E 01	3.80E-18
6.20E 01	3.80E-18
6.80E 01	3.50E-18
7.00E 01	3.50E-18
8.00E 01	3.20E-18
9.00E 01	3.00E-18
1.00E 02	2.70E-18
1.10E 02	2.70E-18
1.20E 02	2.50E-18
1.30E 02	2.40E-18

8-3-6
DAHLBERG ET AL. (182)

ENERGY (KEV)	CROSS SECTION (SQ. CM)
1.50E 01	9.50E-18
1.90E 01	9.30E-18
2.50E 01	9.90E-18
3.00E 01	9.90E-18
3.50E 01	9.00E-18
4.00E 01	9.20E-18
4.50E 01	8.90E-18
5.00E 01	8.20E-18
5.50E 01	7.80E-18
6.00E 01	7.40E-18
7.00E 01	6.90E-18
8.00E 01	6.40E-18
9.00E 01	5.80E-18

1.00E 02	5.90E-18
1.10E 02	5.00E-18
1.30E 02	4.70E-18

8-3-7A
PHILPOT ET AL. (175)

ENERGY (KEV)	CROSS SECTION (SQ. CM)
1.00E 01	9.10E-19
1.50E 01	1.14E-18
2.00E 01	1.30E-18
3.00E 01	1.53E-18
4.00E 01	1.56E-18
5.00E 01	1.46E-18
6.00E 01	1.33E-18
7.00E 01	1.15E-18
8.00E 01	1.02E-18
9.00E 01	9.00E-19
1.00E 02	8.20E-19
1.10E 02	7.50E-19
1.20E 02	7.00E-19
1.30E 02	6.50E-19

8-3-7B
DUFAY ET AL. (178)

ENERGY (KEV)	CROSS SECTION (SQ. CM)
2.50E 01	8.50E-19
3.00E 01	9.00E-19
4.00E 01	8.80E-19
5.00E 01	8.50E-19
6.00E 01	7.40E-19
7.00E 01	6.20E-19
8.00E 01	5.50E-19
9.00E 01	5.00E-19
1.00E 02	4.60E-19
1.20E 02	3.80E-19
1.50E 02	3.40E-19
2.00E 02	2.40E-19
3.00E 02	1.70E-19
4.00E 02	1.30E-19
5.00E 02	1.10E-19
6.00E 02	9.50E-20
1.00E 03	5.50E-20

8-3-7C
ROBINSON ET AL. (162)

ENERGY (KEV)	CROSS SECTION (SQ. CM)
6.00E 01	7.60E-19
7.00E 01	7.00E-19
8.00E 01	6.00E-19
9.00E 01	5.80E-19
1.00E 02	4.90E-19
1.30E 02	4.10E-19
1.50E 02	3.70E-19
2.00E 02	3.00E-19
2.50E 02	2.50E-19
3.00E 02	2.00E-19

8-3-7D
THOMAS ET AL. (183)

ENERGY (KEV)	CROSS SECTION (SQ. CM)
7.50E 01	6.04E-19
1.00E 02	4.29E-19
1.25E 02	3.25E-19
1.5CE 02	2.47E-19
1.75E 02	2.08E-19
2.00E 02	1.61E-19

8-3-7E
THOMAS ET AL. (183)

ENERGY (KEV)	CROSS SECTION (SQ. CM)
1.50E 02	2.26E-19
2.00E 02	1.74E-19
2.50E 02	1.28E-19
3.00E 02	1.01E-19
3.50E 02	8.38E-20
4.00E 02	6.67E-20
5.00E 02	5.12E-20
6.00E 02	4.36E-20
7.00E 02	3.73E-20
8.00E 02	3.29E-20
9.00E 02	2.92E-20

8-3-7F
CAHLBERG ET AL. (182)

ENERGY (KEV)	CROSS SECTION (SQ. CM)
1.50E 01	1.40E-18
2.00E 01	1.55E-18
2.00E 01	1.50E-18
2.50E 01	1.65E-18
3.00E 01	1.75E-18
3.50E 01	1.80E-18
4.0CE 01	1.65E-18
4.50E 01	1.55E-18
4.50E 01	1.60E-18
5.00E 01	1.55E-18
5.50E 01	1.33E-18
5.50E 01	1.48E-18
5.80E 01	1.40E-18
6.00E 01	1.35E-18
7.00E 01	1.10E-18
7.00E 01	1.15E-18
8.00E 01	9.50E-19
9.00E 01	8.20E-19
9.00E 01	9.00E-19
1.00E 02	7.50E-19
1.00E 02	7.80E-19
1.20E 02	6.20E-19
1.20E 02	6.50E-19

8-3-8A
PHILPOT ET AL. (175)

ENERGY (KEV)	CROSS SECTION (SQ. CM)
1.00E 01	9.10E-19
1.50E 01	1.14E-18
2.00E 01	1.30E-18
3.00E 01	1.53E-18
4.00E 01	1.56E-18
5.00E 01	1.46E-18
6.00E 01	1.33E-18
7.00E 01	1.15E-18
8.00E 01	1.02E-18
9.00E 01	9.00E-19
1.00E 02	8.20E-19
1.10E 02	7.50E-19
1.20E 02	7.00E-19
1.30E 02	6.50E-19

8-3-8B
DUFAY ET AL. (178)

ENERGY (KEV)	CROSS SECTION (SQ. CM)
2.50E 01	8.00E-19
3.00E 01	8.00E-19
4.00E 01	7.80E-19
5.00E 01	7.50E-19
6.00E 01	6.70E-19
7.00E 01	6.20E-19
8.00E 01	5.40E-19
9.00E 01	4.60E-19
1.00E 02	4.20E-19
1.20E 02	3.40E-19

8-3-8C
THOMAS ET AL. (183)

ENERGY (KEV)	CROSS SECTION (SQ. CM)
1.50E 02	1.94E-19
2.00E 02	1.64E-19
2.50E 02	1.33E-19
3.00E 02	8.61E-20
3.50E 02	8.13E-20
4.00E 02	6.50E-20
4.50E 02	5.51E-20
5.00E 02	5.00E-20
6.00E 02	4.14E-20
7.00E 02	3.54E-20
8.00E 02	2.99E-20
9.00E 02	2.57E-20

8-3-8D
BAKER ET AL. (181)

ENERGY (KEV)	CROSS SECTION (SQ. CM)
2.00E 01	1.11E-18
3.00E 01	1.41E-18

4.00E 01	1.65E-18
6.00E 01	1.27E-18
8.00E 01	9.34E-19
1.00E 02	6.93E-19

8-3-9
DUFAY ET AL. (178)

ENERGY (KEV)	CROSS SECTION (SQ. CM)
2.50E 01	1.90E-19
3.00E 01	2.00E-19
4.00E 01	2.10E-19
5.00E 01	1.80E-19
6.00E 01	1.50E-19
7.00E 01	1.38E-19
8.00E 01	1.36E-19
1.00E 02	1.00E-19
1.20E 02	8.50E-20

8-3-10
DUFAY ET AL. (178)

ENERGY (KEV)	CROSS SECTION (SQ. CM)
4.00E 01	4.90E-19
5.00E 01	4.20E-19
6.00E 01	3.80E-19
7.00E 01	3.40E-19
8.00E 01	3.00E-19
9.00E 01	3.00E-19
1.00E 02	2.60E-19
1.20E 02	2.40E-19

8-3-11A
THOMAS ET AL. (183)

ENERGY (KEV)	CROSS SECTION (SQ. CM)
1.50E 02	1.44E-20
2.00E 02	1.11E-20
2.50E 02	1.02E-20
3.00E 02	7.10E-21
4.00E 02	5.50E-21
5.00E 02	4.40E-21
6.00E 02	2.65E-21
8.00E 02	1.33E-21

8-3-11B
CAHLBERG ET AL. (182)

ENERGY (KEV)	CROSS SECTION (SQ. CM)
1.40E 01	1.65E-19
2.00E 01	1.20E-19
2.00E 01	1.70E-19
2.50E 01	1.25E-19
2.50E 01	1.95E-19
3.00E 01	1.60E-19
3.00E 01	2.25E-19

ENERGY (KEV)	CROSS SECTION (SQ. CM)
3.50E 01	2.20E-19
4.00E 01	1.90E-19
4.50E 01	1.80E-19
4.70E 01	1.75E-19
5.00E 01	1.90E-19
5.50E 01	1.75E-19
5.50E 01	1.80E-19
6.00E 01	1.50E-19
7.00E 01	1.53E-19
7.00E 01	1.55E-19
8.00E 01	1.60E-19
8.50E 01	1.35E-19
9.00E 01	1.30E-19
1.00E 02	1.28E-19
1.00E 02	1.48E-19

8-3-12A
SHERIDAN ET AL. (173)

ENERGY (KEV)	CROSS SECTION (SQ. CM)
3.10E 00	5.67E-17
3.70E 00	6.85E-17
4.50E 00	7.80E-17
5.50E 00	8.80E-17
6.70E 00	9.60E-17
8.00E 00	1.00E-16
9.00E 00	1.00E-16
1.06E 01	9.90E-17
1.20E 01	9.85E-17
1.35E 01	9.62E-17
1.52E 01	9.50E-17
1.67E 01	9.40E-17
1.87E 01	9.30E-17
2.09E 01	9.20E-17
2.28E 01	9.03E-17
2.47E 01	8.80E-17
2.64E 01	8.70E-17
2.80E 01	8.50E-17
2.90E 01	8.44E-17
3.10E 01	8.45E-17

8-3-12B
CARLETON ET AL. (172)

ENERGY (KEV)	CROSS SECTION (SQ. CM)
1.50E 00	2.34E-17
2.00E 00	3.52E-17
2.50E 00	4.62E-17
3.00E 00	5.73E-17
3.50E 00	6.82E-17
4.00E 00	7.78E-17

8-3-12C
PHILPOT ET AL. (175)

ENERGY (KEV)	CROSS SECTION (SQ. CM)
5.00E 00	8.20E-17
1.00E 01	9.30E-17
1.50E 01	8.70E-17
2.00E 01	8.20E-17

ENERGY (KEV)	CROSS SECTION (SQ. CM)
3.00E 01	7.40E-17
4.00E 01	7.00E-17
5.00E 01	6.20E-17
6.00E 01	5.60E-17
7.00E 01	5.20E-17
8.00E 01	5.00E-17
9.00E 01	4.70E-17
1.00E 02	4.40E-17
1.10E 02	4.20E-17
1.20E 02	4.10E-17
1.30E 02	4.00E-17

8-3-12D
DUFAY ET AL. (178)

ENERGY (KEV)	CROSS SECTION (SQ. CM)
2.50E 01	4.00E-17
3.00E 01	3.80E-17
4.00E 01	3.50E-17
5.00E 01	3.20E-17
6.00E 01	3.00E-17
7.00E 01	2.90E-17
8.00E 01	2.70E-17
1.00E 02	2.50E-17
1.20E 02	2.20E-17
1.50E 02	2.00E-17
2.00E 02	1.80E-17
3.00E 02	1.40E-17
4.00E 02	1.25E-17
5.00E 02	1.10E-17
6.00E 02	9.60E-18
1.00E 03	7.00E-18

8-3-12E
ROBINSON ET AL. (162)

ENERGY (KEV)	CROSS SECTION (SQ. CM)
6.00E 01	3.35E-17
7.00E 01	2.90E-17
8.00E 01	3.00E-17
9.00E 01	2.70E-17
1.00E 02	2.65E-17
1.50E 02	2.30E-17
1.80E 02	1.95E-17
2.00E 02	1.80E-17
2.30E 02	1.75E-17
2.50E 02	1.65E-17
3.00E 02	1.50E-17
4.00E 02	1.30E-17

8-3-12F
THOMAS ET AL. (183)

ENERGY (KEV)	CROSS SECTION (SQ. CM)
7.50E 01	1.97E-17
1.00E 02	1.67E-17
1.25E 02	1.55E-17
1.50E 02	1.57E-17
1.75E 02	1.26E-17
2.00E 02	1.05E-17

8-3-12G
THOMAS ET AL. (183)

ENERGY (KEV)	CROSS SECTION (SQ. CM)
1.50E 02	1.32E-17
2.00E 02	1.22E-17
2.50E 02	1.13E-17
3.00E 02	9.77E-18
3.50E 02	8.86E-18
4.00E 02	8.08E-18
4.50E 02	7.15E-18
5.00E 02	6.65E-18
6.00E 02	5.65E-18
7.00E 02	5.02E-18
8.00E 02	4.50E-18

8-3-12H
BAKER ET AL. (181)

ENERGY (KEV)	CROSS SECTION (SQ. CM)
2.00E 01	7.87E-17
3.00E 01	6.86E-17
4.00E 01	6.33E-17
6.00E 01	5.80E-17
8.00E 01	4.87E-17
1.00E 02	4.57E-17

8-3-12I
DAHLBERG ET AL. (182)

ENERGY (KEV)	CROSS SECTION (SQ. CM)
1.00E 01	1.01E-16
1.40E 01	9.90E-17
1.50E 01	9.40E-17
1.80E 01	9.70E-17
1.90E 01	9.10E-17
2.50E 01	8.45E-17
3.00E 01	8.00E-17
3.50E 01	7.50E-17
4.00E 01	7.00E-17
4.50E 01	6.60E-17
5.00E 01	6.30E-17
5.50E 01	6.00E-17
6.00E 01	5.60E-17
6.50E 01	5.35E-17
7.00E 01	5.20E-17
8.00E 01	4.90E-17
9.00E 01	4.60E-17
1.00E 02	4.30E-17
1.10E 02	4.20E-17
1.20E 02	3.90E-17
1.30E 02	3.80E-17

8-3-13A
PHILPOT ET AL. (175)

ENERGY (KEV)	CROSS SECTION (SQ. CM)

ENERGY (KEV)	CROSS SECTION (SQ. CM)
1.00E 01	2.90E-17
1.50E 01	2.74E-17
2.00E 01	2.60E-17
3.00E 01	2.36E-17
4.00E 01	2.20E-17
5.00E 01	2.03E-17
6.00E 01	1.86E-17
7.00E 01	1.73E-17
8.00E 01	1.63E-17
9.00E 01	1.55E-17
1.00E 02	1.47E-17
1.10E 02	1.40E-17
1.20E 02	1.34E-17
1.30E 02	1.30E-17

8-3-13B
DUFAY ET AL. (178)

ENERGY (KEV)	CROSS SECTION (SQ. CM)
2.50E 01	1.20E-17
3.00E 01	1.18E-17
4.00E 01	1.00E-17
5.00E 01	1.00E-17
6.00E 01	9.50E-18
7.00E 01	8.80E-18
8.00E 01	8.20E-18
1.00E 02	7.40E-18
1.20E 02	6.80E-18
1.50E 02	6.00E-18
2.00E 02	5.20E-18
3.00E 02	4.20E-18
4.00E 02	3.60E-18
5.00E 02	3.25E-18
6.00E 02	2.95E-18

8-3-13C
ROBINSON ET AL. (162)

ENERGY (KEV)	CROSS SECTION (SQ. CM)
6.00E 01	8.20E-18
7.00E 01	7.80E-18
8.00E 01	7.60E-18
9.00E 01	7.40E-18
1.00E 02	6.50E-18
1.30E 02	6.05E-18
1.50E 02	5.80E-18
1.75E 02	5.20E-18
2.00E 02	4.75E-18
2.50E 02	4.30E-18
3.00E 02	3.80E-18
3.50E 02	3.33E-18

8-3-13D
THOMAS ET AL. (183)

ENERGY (KEV)	CROSS SECTION (SQ. CM)
1.50E 02	5.41E-18
2.00E 02	4.69E-18
2.50E 02	3.94E-18
3.00E 02	3.60E-18

ENERGY (KEV)	CROSS SECTION (SQ. CM)
3.50E 02	3.21E-18
4.00E 02	3.08E-18
4.50E 02	2.86E-18
5.00E 02	2.58E-18
6.00E 02	2.34E-18
7.00E 02	2.05E-18
8.00E 02	1.64E-18
9.00E 02	1.49E-18
1.00E 03	1.25E-18

8-3-13E
DAHLBERG ET AL. (182)

ENERGY (KEV)	CROSS SECTION (SQ. CM)
1.50E 01	3.20E-17
1.50E 01	3.30E-17
1.90E 01	3.00E-17
2.00E 01	3.00E-17
2.50E 01	2.75E-17
2.50E 01	2.80E-17
3.00E 01	2.60E-17
3.50E 01	2.45E-17
4.00E 01	2.30E-17
4.50E 01	2.28E-17
5.00E 01	2.02E-17
6.00E 01	1.85E-17
6.30E 01	1.85E-17
7.00E 01	1.75E-17
8.00E 01	1.60E-17
9.00E 01	1.50E-17
1.00E 02	1.40E-17
1.10E 02	1.35E-17
1.20E 02	1.25E-17
1.30E 02	1.20E-17

8-3-14A
PHILPOT ET AL. (175)

ENERGY (KEV)	CROSS SECTION (SQ. CM)
1.00E 01	5.80E-18
1.50E 01	5.70E-18
2.00E 01	5.40E-18
3.00E 01	4.90E-18
4.00E 01	4.60E-18
5.00E 01	4.20E-18
6.00E 01	3.90E-18
7.00E 01	3.70E-18
8.00E 01	3.30E-18
9.00E 01	3.20E-18
1.00E 02	3.00E-18
1.10E 02	2.90E-18
1.20E 02	2.75E-18
1.30E 02	2.65E-18

8-3-14B
DUFAY ET AL. (178)

ENERGY (KEV)	CROSS SECTION (SQ. CM)
2.50E 01	2.25E-18
3.00E 01	2.20E-18

ENERGY (KEV)	CROSS SECTION (SQ. CM)
4.00E 01	2.00E-18
5.00E 01	1.90E-18
6.00E 01	1.80E-18
7.00E 01	1.70E-18
8.00E 01	1.60E-18
1.00E 02	1.45E-18
1.20E 02	1.30E-18
1.50E 02	1.25E-18
2.00E 02	1.10E-18
3.00E 02	9.00E-19
4.00E 02	7.20E-19
5.00E 02	6.20E-19
6.00E 02	5.60E-19

8-3-14C
ROBINSON ET AL. (162)

ENERGY (KEV)	CROSS SECTION (SQ. CM)
6.00E 01	2.10E-18
7.00E 01	2.15E-18
8.00E 01	2.10E-18
9.00E 01	1.80E-18
1.00E 02	1.76E-18
1.50E 02	1.50E-18
2.00E 02	1.21E-18
2.10E 02	1.18E-18
2.30E 02	1.13E-18
3.00E 02	9.20E-19
3.50E 02	7.90E-19

8-3-14D
THOMAS ET AL. (183)

ENERGY (KEV)	CROSS SECTION (SQ. CM)
1.50E 02	1.02E-18
2.00E 02	8.11E-19
2.50E 02	7.18E-19
3.00E 02	6.66E-19
3.50E 02	6.51E-19
4.00E 02	5.88E-19
4.50E 02	5.46E-19
5.00E 02	5.06E-19
6.00E 02	4.76E-19
7.00E 02	3.98E-19
8.00E 02	3.47E-19
9.00E 02	3.45E-19
1.00E 03	2.61E-19

8-3-14E
DAHLBERG ET AL. (182)

ENERGY (KEV)	CROSS SECTION (SQ. CM)
2.00E 01	6.50E-18
2.50E 01	6.10E-18
3.00E 01	6.10E-18
3.50E 01	5.70E-18
4.00E 01	5.40E-18
4.50E 01	5.10E-18
5.00E 01	4.80E-18
5.50E 01	4.60E-18

6.00E 01	4.40E-18
7.00E 01	3.90E-18
8.00E 01	3.50E-18
9.00E 01	3.20E-18
1.00E 02	3.00E-18
1.20E 02	2.65E-18

8-3-15A
PHILPOT ET AL. (175)

ENERGY (KEV)	CROSS SECTION (SQ. CM)
1.00E 01	9.10E-19
1.50E 01	9.00E-19
2.00E 01	8.50E-19
3.00E 01	7.70E-19
4.00E 01	7.10E-19
5.00E 01	6.60E-19
6.00E 01	6.00E-19
7.00E 01	5.70E-19
8.00E 01	5.20E-19
9.00E 01	5.00E-19
1.00E 02	4.70E-19
1.10E 02	4.60E-19
1.20E 02	4.40E-19
1.30E 02	4.30E-19

8-3-15B
DUFAY ET AL. (178)

ENERGY (KEV)	CROSS SECTION (SQ. CM)
2.50E 01	3.50E-19
3.00E 01	3.40E-19
4.00E 01	3.20E-19
5.00E 01	2.80E-19
6.00E 01	2.70E-19
7.00E 01	2.40E-19
1.20E 02	1.95E-19
1.50E 02	1.85E-19
2.00E 02	1.65E-19
3.00E 02	1.30E-19
4.00E 02	1.10E-19
5.00E 02	1.00E-19

8-3-15C
THOMAS ET AL. (183)

ENERGY (KEV)	CROSS SECTION (SQ. CM)
1.50E 02	1.48E-19
2.00E 02	1.34E-19
2.50E 02	1.22E-19
3.00E 02	1.05E-19
3.50E 02	9.94E-20
4.00E 02	8.70E-20
4.50E 01	8.11E-20
5.00E 02	7.37E-20
6.00E 02	6.19E-20
7.00E 02	5.54E-20
8.00E 02	5.06E-20
9.00E 02	4.59E-20

8-3-16A
PHILPOT ET AL. (175)

ENERGY (KEV)	CROSS SECTION (SQ. CM)
5.00E 00	7.80E-18
1.00E 01	8.50E-18
1.50E 01	8.00E-18
2.00E 01	7.50E-18
3.00E 01	6.20E-18
4.00E 01	5.40E-18
5.00E 01	4.90E-18
6.00E 01	4.50E-18
7.00E 01	4.20E-18
8.00E 01	3.90E-18
9.00E 01	3.70E-18
1.00E 02	3.50E-18
1.10E 02	3.30E-18
1.20E 02	3.10E-18
1.30E 02	3.00E-18

8-3-16B
DUFAY ET AL. (178)

ENERGY (KEV)	CROSS SECTION (SQ. CM)
2.50E 01	2.60E-18
3.00E 01	2.50E-18
4.00E 01	2.40E-18
5.00E 01	2.20E-18
6.00E 01	2.10E-18
7.00E 01	2.00E-18
8.00E 01	1.80E-18
1.00E 02	1.65E-18
1.20E 02	1.50E-18
1.50E 02	1.40E-18
2.00E 02	1.20E-18
3.00E 02	9.40E-19
4.00E 02	7.60E-19
5.00E 02	6.75E-19
6.00E 02	6.00E-19

8-3-16C
THOMAS ET AL. (183)

ENERGY (KEV)	CROSS SECTION (SQ. CM)
7.50E 01	1.33E-18
1.00E 02	1.05E-18
1.25E 02	8.94E-19
1.50E 02	7.69E-19
1.75E 02	6.99E-19
2.00E 02	6.06E-19

8-3-16D
THOMAS ET AL. (183)

ENERGY (KEV)	CROSS SECTION (SQ. CM)
1.50E 02	7.34E-19
2.00E 02	6.32E-19

2.50E 02	5.51E-19
3.00E 02	4.78E-19
3.50E 02	4.13E-19
4.00E 02	3.94E-19
4.50E 02	3.62E-19
5.00E 02	3.41E-19
6.00E 02	2.86E-19
7.00E 02	2.63E-19
8.00E 02	2.47E-19
9.00E 02	2.09E-19

8-3-17A
PHILPOT ET AL. (175)

ENERGY (KEV)	CROSS SECTION (SQ. CM)
2.00E 01	3.90E-18
3.00E 01	3.70E-18
4.00E 01	3.40E-18
5.00E 01	3.10E-18
6.00E 01	2.90E-18
7.00E 01	2.70E-18
8.00E 01	2.40E-18
9.00E 01	2.30E-18
1.00E 02	2.00E-18
1.10E 02	1.87E-18
1.20E 02	1.75E-18
1.30E 02	1.65E-18

8-3-17B
DUFAY ET AL. (178)

ENERGY (KEV)	CROSS SECTION (SQ. CM)
2.50E 01	1.85E-18
3.00E 01	1.80E-18
4.00E 01	1.65E-18
5.00E 01	1.55E-18
6.00E 01	1.45E-18
7.00E 01	1.40E-18
8.00E 01	1.30E-18
1.00E 02	1.20E-18
1.20E 02	1.10E-18
1.50E 02	9.60E-19
2.00E 02	8.00E-19
3.00E 02	6.50E-19
4.00E 02	5.20E-19
5.00E 02	4.70E-19
6.00E 02	4.40E-19

8-3-17C
THOMAS ET AL. (183)

ENERGY (KEV)	CROSS SECTION (SQ. CM)
1.50E 02	6.19E-19
2.00E 02	4.59E-19
2.50E 02	3.91E-19
3.00E 02	3.33E-19
3.50E 02	3.09E-19
4.00E 02	2.69E-19
4.50E 02	2.46E-19
5.00E 02	2.26E-19

ENERGY (KEV)	CROSS SECTION (SQ. CM)
6.00E 02	1.96E-19
7.00E 02	1.76E-19
8.00E 02	1.67E-19
9.00E 02	1.52E-19

8-3-18A
PHILPOT ET AL. (175)

ENERGY (KEV)	CROSS SECTION (SQ. CM)
2.00E 01	3.90E-18
3.00E 01	3.70E-18
4.00E 01	3.40E-18
5.00E 01	3.10E-18
6.00E 01	2.90E-18
7.00E 01	2.70E-18
8.00E 01	2.40E-18
9.00E 01	2.30E-18
1.00E 02	2.00E-18
1.10E 02	1.87E-18
1.20E 02	1.75E-18
1.30E 02	1.65E-18

8-3-18B
DUFAY ET AL. (178)

ENERGY (KEV)	CROSS SECTION (SQ. CM)
2.50E 01	1.85E-18
3.00E 01	1.80E-18
4.00E 01	1.65E-18
5.00E 01	1.55E-18
6.00E 01	1.45E-18
7.00E 01	1.40E-18
8.00E 01	1.30E-18
1.00E 02	1.20E-18
1.20E 02	1.10E-18
1.50E 02	9.60E-19
2.00E 02	8.00E-19
3.00E 02	6.50E-19
4.00E 02	5.20E-19
5.00E 02	4.70E-19
6.00E 02	4.40E-19

8-3-18C
ROBINSON ET AL. (162)

ENERGY (KEV)	CROSS SECTION (SQ. CM)
6.00E 01	1.60E-18
7.00E 01	1.60E-18
8.00E 01	1.35E-18
9.00E 01	1.28E-18
1.00E 02	1.19E-18
1.30E 02	1.04E-18
1.50E 02	9.60E-19
1.80E 02	9.00E-19
2.00E 02	8.00E-19
2.50E 02	7.00E-19
3.00E 02	6.00E-19

8-3-18D
THOMAS ET AL. (183)

ENERGY (KEV)	CROSS SECTION (SQ. CM)
1.50E 02	5.86E-19
2.00E 02	5.02E-19
2.50E 02	4.36E-19
3.00E 02	3.76E-19
3.50E 02	3.38E-19
4.00E 02	2.90E-19
4.50E 02	2.71E-19
5.00E 02	2.41E-19
6.00E 02	2.09E-19
7.00E 02	1.89E-19
8.00E 02	1.70E-19
9.00E 02	1.55E-19

8-3-19A
PHILPOT ET AL. (175)

ENERGY (KEV)	CROSS SECTION (SQ. CM)
1.00E 01	1.60E-18
1.50E 01	1.58E-18
2.00E 01	1.55E-18
3.00E 01	1.45E-18
4.00E 01	1.33E-18
5.00E 01	1.24E-18
6.00E 01	1.14E-18
7.00E 01	1.07E-18
8.00E 01	9.90E-19
9.00E 01	9.30E-19
1.00E 02	8.70E-19
1.10E 02	8.30E-19
1.20E 02	7.90E-19
1.30E 02	7.60E-19

8-3-19B
DUFAY ET AL. (178)

ENERGY (KEV)	CROSS SECTION (SQ. CM)
2.50E 01	6.00E-19
3.00E 01	5.30E-19
4.00E 01	4.90E-19
5.00E 01	4.50E-19
6.00E 01	4.20E-19
7.00E 01	4.00E-19
1.00E 02	3.50E-19
1.20E 02	3.20E-19
2.00E 02	2.50E-19
3.00E 02	2.00E-19
4.00E 02	1.80E-19
6.00E 02	1.40E-19

8-3-19C
ROBINSON ET AL. (162)

ENERGY (KEV)	CROSS SECTION (SQ. CM)

ENERGY (KEV)	CROSS SECTION (SQ. CM)
6.00E 01	7.40E-19
7.00E 01	7.40E-19
8.00E 01	7.00E-19
9.00E 01	6.20E-19
1.00E 02	5.80E-19
1.20E 02	5.10E-19
1.50E 02	4.60E-19
1.80E 02	4.30E-19
2.00E 02	4.10E-19
2.50E 02	3.20E-19
3.00E 02	2.80E-19

8-3-19D
THOMAS ET AL. (183)

ENERGY (KEV)	CROSS SECTION (SQ. CM)
1.50E 02	2.46E-19
2.00E 02	2.20E-19
2.50E 02	1.81E-19
3.00E 02	1.61E-19
3.50E 02	1.47E-19
4.00E 02	1.42E-19
4.50E 02	1.28E-19
5.00E 02	1.12E-19
6.00E 02	1.05E-19
7.00E 02	9.70E-20
8.00E 02	9.01E-20

8-3-19E
DAHLBERG ET AL. (182)

ENERGY (KEV)	CROSS SECTION (SQ. CM)
1.98E 01	2.17E-18
2.38E 01	2.05E-18
2.95E 01	2.03E-18
4.00E 01	1.73E-18
4.53E 01	1.61E-18
4.95E 01	1.54E-18
5.55E 01	1.45E-18
5.95E 01	1.41E-18
6.60E 01	1.34E-18
6.95E 01	1.31E-18
7.50E 01	1.27E-18
8.05E 01	1.21E-18
9.04E 01	1.13E-18
1.00E 02	1.04E-18
1.16E 02	9.82E-19
1.26E 02	9.20E-19

8-3-20A
PHILPOT ET AL. (175)

ENERGY (KEV)	CROSS SECTION (SQ. CM)
1.00E 01	3.50E-19
1.50E 01	3.40E-19
2.00E 01	3.20E-19
3.00E 01	2.90E-19
4.00E 01	2.60E-19
5.00E 01	2.40E-19
6.00E 01	2.20E-19

ENERGY (KEV)	CROSS SECTION (SQ. CM)
7.00E 01	2.10E-19
8.00E 01	1.94E-19
9.00E 01	1.94E-19
1.00E 02	1.83E-19
1.10E 02	1.70E-19
1.20E 02	1.70E-19
1.30E 02	1.60E-19

8-3-20B
DUFAY ET AL. (178)

ENERGY (KEV)	CROSS SECTION (SQ. CM)
4.00E 01	1.30E-19
6.00E 01	1.10E-19
1.00E 02	8.60E-20
2.00E 02	6.60E-20

8-3-20C
THOMAS ET AL. (183)

ENERGY (KEV)	CROSS SECTION (SQ. CM)
1.50E 02	6.16E-20
2.00E 02	5.36E-20
2.50E 02	4.68E-20
3.00E 02	3.99E-20
3.50E 02	3.40E-20
4.00E 02	3.13E-20
4.50E 02	3.14E-20
5.00E 02	2.72E-20
6.00E 02	2.40E-20
7.00E 02	2.22E-20
8.00E 02	1.99E-20
9.00E 02	1.82E-20

8-3-21
YOUNG ET AL. (7)

ENERGY (KEV)	CROSS SECTION (SQ. CM)
1.57E 00	1.14E-19
2.14E 00	2.40E-19
2.90E 00	6.38E-19
4.65E 00	1.01E-18
5.88E 00	1.11E-18
7.36E 00	1.51E-18
9.50E 00	9.02E-19
1.17E 01	1.27E-18
1.47E 01	5.54E-19
1.78E 01	5.18E-19
2.48E 01	2.64E-19
3.01E 01	5.16E-19

8-3-22A
PHILPOT ET AL. (175)

ENERGY (KEV)	CROSS SECTION (SQ. CM)
5.00E 00	1.13E-16
1.00E 01	1.23E-16
1.50E 01	1.19E-16
2.00E 01	1.13E-16
3.00E 01	1.05E-16
4.00E 01	9.60E-17
5.00E 01	8.80E-17
6.00E 01	8.10E-17
7.00E 01	7.50E-17
8.00E 01	6.90E-17
9.00E 01	6.70E-17
1.00E 02	6.20E-17
1.10E 02	6.00E-17
1.20E 02	5.80E-17
1.30E 02	5.60E-17

8-3-22B
DUFAY ET AL. (178)

ENERGY (KEV)	CROSS SECTION (SQ. CM)
2.50E 01	5.80E-17
3.00E 01	5.50E-17
4.00E 01	5.00E-17
5.00E 01	4.70E-17
6.00E 01	4.40E-17
7.00E 01	4.00E-17
8.00E 01	3.80E-17
1.00E 02	3.60E-17
1.20E 02	3.20E-17
1.50E 02	2.80E-17
2.00E 02	2.50E-17
3.00E 02	2.00E-17
4.00E 02	1.65E-17
5.00E 02	1.45E-17
6.00E 02	1.30E-17

8-3-22C
ROBINSON ET AL. (162)

ENERGY (KEV)	CROSS SECTION (SQ. CM)
6.00E 01	4.38E-17
7.00E 01	3.89E-17
8.00E 01	3.97E-17
9.00E 01	3.62E-17
1.00E 02	3.48E-17
1.50E 02	3.03E-17
2.00E 02	2.40E-17
3.00E 02	1.98E-17

8-3-22D
THOMAS ET AL. (183)

ENERGY (KEV)	CROSS SECTION (SQ. CM)
1.50E 02	1.98E-17
2.00E 02	1.78E-17
2.50E 02	1.61E-17
3.00E 02	1.41E-17
3.50E 02	1.28E-17
4.00E 02	1.18E-17
4.50E 02	1.06E-17
5.00E 02	9.81E-18

ENERGY (KEV)	CROSS SECTION (SQ. CM)
6.00E 02	8.53E-18
7.00E 02	7.52E-18
8.00E 02	6.54E-18

8-3-22E
DAHLBERG ET AL. (182)

ENERGY (KEV)	CROSS SECTION (SQ. CM)
2.50E 01	1.19E-16
3.00E 01	1.12E-16
3.50E 01	1.05E-16
4.00E 01	9.84E-17
4.50E 01	9.39E-17
5.00E 01	8.80E-17
6.00E 01	7.89E-17
7.00E 01	7.34E-17
8.00E 01	6.85E-17
9.00E 01	6.42E-17
1.00E 02	6.00E-17
1.20E 02	5.41E-17

8-3-23A
PHILPOT ET AL. (175)

ENERGY (KEV)	CROSS SECTION (SQ. CM)
5.00E 00	1.90E-17
1.00E 01	2.00E-17
1.50E 01	1.85E-17
2.00E 01	1.70E-17
3.00E 01	1.53E-17
4.00E 01	1.37E-17
5.00E 01	1.27E-17
6.00E 01	1.16E-17
7.00E 01	1.08E-17
8.00E 01	9.90E-18
9.00E 01	9.30E-18
1.00E 02	8.60E-18
1.10E 02	8.10E-18
1.20E 02	7.60E-18
1.30E 02	7.20E-18

8-3-23B
DUFAY ET AL. (178)

ENERGY (KEV)	CROSS SECTION (SQ. CM)
2.50E 01	6.80E-18
3.00E 01	6.50E-18
4.00E 01	6.00E-18
5.00E 01	5.60E-18
6.00E 01	5.40E-18
7.00E 01	5.00E-18
8.00E 01	4.80E-18
1.00E 02	4.40E-18
1.20E 02	4.10E-18
1.50E 02	3.60E-18
2.00E 02	3.00E-18
3.00E 02	2.50E-18
4.00E 02	2.15E-18
5.00E 02	1.90E-18
6.00E 02	1.70E-18

8-3-23C
THOMAS ET AL. (183)

ENERGY (KEV)	CROSS SECTION (SQ. CM)
1.50E 02	2.25E-18
2.00E 02	1.87E-18
2.50E 02	1.61E-18
3.00E 02	1.39E-18
3.50E 02	1.24E-18
4.00E 02	1.13E-18
4.50E 02	1.04E-18
5.00E 02	9.50E-19
6.00E 02	8.20E-19
7.00E 02	7.50E-19
8.00E 02	6.90E-19

8-3-24
DUFAY ET AL. (178)

ENERGY (KEV)	CROSS SECTION (SQ. CM)
4.00E 01	3.35E-17
5.00E 01	3.40E-17
6.00E 01	3.30E-17
7.00E 01	3.10E-17
8.00E 01	3.00E-17
1.00E 02	2.60E-17
1.20E 02	2.40E-17

8-3-25A
CARLETON ET AL. (172)

ENERGY (KEV)	CROSS SECTION (SQ. CM)
1.50E 00	1.40E-17
1.75E 00	1.48E-17
2.00E 00	1.48E-17
2.50E 00	1.55E-17
3.00E 00	1.53E-17
3.50E 00	1.56E-17
4.00E 00	1.62E-17
4.50E 00	1.60E-17

8-3-25B
SHERIDAN ET AL. (173)

ENERGY (KEV)	CROSS SECTION (SQ. CM)
2.90E 00	1.51E-17
3.80E 00	1.64E-17
4.40E 00	1.67E-17
5.10E 00	1.67E-17
5.80E 00	1.66E-17
6.90E 00	1.62E-17
8.10E 00	1.57E-17
9.50E 00	1.52E-17
1.10E 01	1.49E-17
1.20E 01	1.45E-17
1.30E 01	1.41E-17
1.42E 01	1.37E-17

1.51E 01	1.36E-17
1.60E 01	1.33E-17
1.69E 01	1.30E-17
1.81E 01	1.27E-17
1.91E 01	1.25E-17
2.03E 01	1.23E-17
2.14E 01	1.21E-17
2.26E 01	1.19E-17
2.38E 01	1.18E-17
2.46E 01	1.17E-17
2.55E 01	1.15E-17
2.66E 01	1.15E-17
2.76E 01	1.14E-17
2.85E 01	1.14E-17
2.94E 01	1.12E-17
3.02E 01	1.12E-17
3.12E 01	1.11E-17
3.24E 01	1.11E-17

8-3-25C
DUFAY ET AL. (178)

ENERGY (KEV)	CROSS SECTION (SQ. CM)
4.00E 01	1.80E-17
5.00E 01	1.85E-17
6.00E 01	1.70E-17
7.00E 01	1.72E-17
8.00E 01	1.60E-17
1.00E 02	1.50E-17
1.20E 02	1.40E-17

8-3-26
DUFAY ET AL. (178)

ENERGY (KEV)	CROSS SECTION (SQ. CM)
4.00E 01	1.40E-17
5.00E 01	1.40E-17
6.00E 01	1.42E-17
7.00E 01	1.30E-17
8.00E 01	1.32E-17
1.00E 02	1.20E-17
1.20E 02	1.10E-17

8-3-27
DAHLBERG ET AL. (182)

ENERGY (KEV)	CROSS SECTION (SQ. CM)
2.60E 01	3.70E-18
3.20E 01	3.70E-18
3.50E 01	3.70E-18
4.00E 01	3.80E-18
5.00E 01	3.30E-18
5.50E 01	3.00E-18
6.00E 01	3.10E-18
7.00E 01	2.90E-18
8.00E 01	2.70E-18
9.00E 01	2.80E-18

8-3-28
DAHLBERG ET AL. (182)

ENERGY (KEV)	CROSS SECTION (SQ. CM)
1.50E 01	5.30E-18
1.90E 01	8.50E-18
2.50E 01	7.30E-18
3.00E 01	7.50E-18
3.50E 01	7.10E-18
4.00E 01	6.70E-18
4.50E 01	6.30E-18
5.00E 01	6.20E-18
5.50E 01	5.90E-18
6.00E 01	5.60E-18
7.00E 01	5.70E-18
8.00E 01	5.30E-18
9.00E 01	5.30E-18

8-3-29A
ROBINSON ET AL. (162)

ENERGY (KEV)	CROSS SECTION (SQ. CM)
6.00E 01	5.30E-19
7.00E 01	5.15E-19
8.00E 01	4.20E-19
9.00E 01	4.00E-19
1.00E 02	3.70E-19
1.10E 02	3.60E-19
1.20E 02	3.53E-19

8-3-29B
DAHLBERG ET AL. (182)

ENERGY (KEV)	CROSS SECTION (SQ. CM)
1.50E 01	9.20E-19
2.00E 01	9.90E-19
2.00E 01	1.00E-18
2.50E 01	1.05E-18
3.00E 01	1.05E-18
3.00E 01	1.15E-18
3.50E 01	1.16E-18
4.00E 01	1.10E-18
4.30E 01	1.05E-18
4.50E 01	1.20E-18
5.00E 01	1.05E-18
5.50E 01	9.60E-19
5.50E 01	1.08E-18
5.70E 01	1.00E-18
6.00E 01	9.50E-19
7.00E 01	8.20E-19
7.00E 01	8.40E-19
8.00E 01	8.10E-19
9.00E 01	6.80E-19
9.00E 01	7.50E-19
1.00E 02	6.20E-19
1.20E 02	7.70E-19

8-3-30
DAHLBERG ET AL. (182)

ENERGY (KEV)	CROSS SECTION (SQ. CM)
1.40E 01	4.80E-18
2.00E 01	3.70E-18
2.50E 01	2.90E-18
2.50E 01	3.10E-18
3.00E 01	2.80E-18
3.10E 01	2.25E-18
3.50E 01	2.10E-18
4.00E 01	1.95E-18
4.50E 01	1.98E-18
4.70E 01	1.98E-18
5.00E 01	1.96E-18
5.50E 01	1.15E-18
5.50E 01	1.35E-18
6.00E 01	1.10E-18
7.00E 01	9.00E-19
7.00E 01	1.02E-18
8.00E 01	7.60E-19
8.50E 01	7.10E-19
9.00E 01	6.90E-19
1.00E 02	5.80E-19
1.00E 02	6.10E-19

8-3-31A
ROBINSON ET AL. (162)

ENERGY (KEV)	CROSS SECTION (SQ. CM)
6.00E 01	2.60E-17
7.00E 01	2.70E-17
8.00E 01	2.30E-17
9.00E 01	2.10E-17
1.00E 02	1.95E-17
1.10E 02	1.85E-17
1.20E 02	1.82E-17

8-3-31B
DAHLBERG ET AL. (22)

ENERGY (KEV)	CROSS SECTION (SQ. CM)
1.00E 01	3.40E-17
1.40E 01	3.40E-17
1.80E 01	3.40E-17
1.90E 01	3.30E-17
2.50E 01	3.50E-17
3.00E 01	3.60E-17
3.50E 01	3.60E-17
4.00E 01	3.50E-17
4.50E 01	3.60E-17
5.00E 01	3.60E-17
5.50E 01	3.50E-17
6.00E 01	3.50E-17
6.50E 01	3.30E-17
7.00E 01	3.30E-17
8.00E 01	3.20E-17
9.00E 01	3.10E-17
1.00E 02	3.10E-17
1.10E 02	3.05E-17
1.20E 02	2.90E-17

1.30E 02	2.90E-17

8-3-32A
ROBINSON ET AL. (162)

ENERGY (KEV)	CROSS SECTION (SQ. CM)
6.00E 01	5.40E-18
7.00E 01	4.80E-18
8.00E 01	4.70E-18
9.00E 01	4.45E-18
1.00E 02	4.30E-18
1.10E 02	4.10E-18
1.20E 02	4.10E-18

8-3-32B
DAHLBERG ET AL. (182)

ENERGY (KEV)	CROSS SECTION (SQ. CM)
1.50E 01	1.03E-17
1.50E 01	1.02E-17
1.90E 01	1.08E-17
2.50E 01	1.10E-17
2.50E 01	1.10E-17
3.00E 01	1.10E-17
3.50E 01	1.10E-17
4.00E 01	1.15E-17
4.50E 01	1.10E-17
5.00E 01	1.10E-17
6.30E 01	1.05E-17
7.00E 01	1.00E-17
8.00E 01	1.00E-17
9.00E 01	9.80E-18
1.00E 02	9.70E-18
1.10E 02	9.10E-18
1.20E 02	1.00E-17
1.30E 02	9.00E-18

8-3-33A
ROBINSON ET AL. (162)

ENERGY (KEV)	CROSS SECTION (SQ. CM)
6.00E 01	1.50E-18
7.00E 01	1.55E-18
8.00E 01	1.32E-18
9.00E 01	1.11E-18
1.00E 02	1.20E-18
1.10E 02	1.10E-18
1.20E 02	1.08E-18

8-3-33B
DAHLBERG ET AL. (182)

ENERGY (KEV)	CROSS SECTION (SQ. CM)
2.00E 01	2.30E-18
2.50E 01	2.25E-18
3.00E 01	2.15E-18
3.50E 01	2.40E-18
4.00E 01	2.35E-18
4.50E 01	2.40E-18
5.00E 01	2.20E-18
5.50E 01	2.20E-18
6.00E 01	2.25E-18
7.00E 01	2.20E-18
8.00E 01	2.05E-18
9.00E 01	2.10E-18
1.00E 02	2.10E-18
1.20E 02	2.20E-18

8-3-34
ROBINSON ET AL. (162)

ENERGY (KEV)	CROSS SECTION (SQ. CM)
6.00E 01	1.20E-18
7.00E 01	1.05E-18
8.00E 01	1.15E-18
9.00E 01	1.02E-18
1.00E 02	1.00E-18
1.10E 02	9.00E-19
1.20E 02	9.60E-19

8-3-35A
ROBINSON ET AL. (162)

ENERGY (KEV)	CROSS SECTION (SQ. CM)
7.00E 01	4.30E-19
8.00E 01	4.30E-19
9.00E 01	3.90E-19
1.00E 02	3.90E-19
1.10E 02	3.90E-19
1.20E 02	4.00E-19

8-3-35B
DAHLBERG ET AL. (182)

ENERGY (KEV)	CROSS SECTION (SQ. CM)
1.95E 01	1.14E-18
2.39E 01	1.13E-18
2.94E 01	1.13E-18
3.38E 01	1.11E-18
3.90E 01	1.10E-18
4.35E 01	1.06E-18
4.87E 01	1.04E-18
5.32E 01	1.01E-18
5.78E 01	9.91E-19
6.30E 01	9.55E-19
6.75E 01	9.29E-19
7.90E 01	8.63E-19
8.75E 01	8.24E-19
9.85E 01	7.73E-19

8-3-36A
ROBINSON ET AL. (162)

ENERGY (KEV)	CROSS SECTION (SQ. CM)
6.00E 01	3.29E-17
7.00E 01	3.34E-17
8.00E 01	2.91E-17
9.00E 01	2.65E-17
1.00E 02	2.50E-17
1.10E 02	2.37E-17
1.20E 02	2.34E-17

8-3-36B
DAHLBERG ET AL. (182)

ENERGY (KEV)	CROSS SECTION (SQ. CM)
2.50E 01	4.80E-17
3.00E 01	4.90E-17
3.50E 01	4.90E-17
4.00E 01	4.90E-17
4.50E 01	4.90E-17
5.00E 01	4.90E-17
7.00E 01	4.50E-17
8.00E 01	4.40E-17
9.00E 01	4.30E-17
1.00E 02	4.30E-17
1.20E 02	4.12E-17

8-3-37
DUFAY ET AL. (202)

ENERGY (KEV)	CROSS SECTION (SQ. CM)
8.00E 01	1.10E-18
1.00E 02	1.01E-18
2.00E 02	5.28E-19
3.00E 02	4.15E-19
4.00E 02	2.82E-19
5.00E 02	2.39E-19
6.00E 02	2.00E-19

8-3-38A
NEFF (104)

ENERGY (KEV)	CROSS SECTION (SQ. CM)
4.00E-01	1.10E-19
6.00E-01	1.60E-19
8.00E-01	2.30E-19
1.20E 00	3.10E-19
1.50E 00	3.60E-19
2.00E 00	4.20E-19

8-3-38B
NEFF (104)

ENERGY (KEV)	CROSS SECTION (SQ. CM)
8.00E-01	1.60E-19
1.20E 00	2.10E-19

8-3-38C
NEFF (104)

ENERGY (KEV)	CROSS SECTION (SQ. CM)
8.00E-01	4.00E-19
1.20E 00	5.40E-19

8-3-38D
NEFF (104)

ENERGY (KEV)	CROSS SECTION (SQ. CM)
8.00E-01	5.00E-19
1.20E 00	7.30E-19

8-3-38E
NEFF (104)

ENERGY (KEV)	CROSS SECTION (SQ. CM)
4.00E-01	2.10E-19
6.00E-01	2.70E-19
8.00E-01	3.70E-19
1.20E 00	6.00E-19
1.50E 00	6.60E-19
2.00E 00	9.70E-19

8-3-39
NEFF (104)

ENERGY (KEV)	CROSS SECTION (SQ. CM)
6.00E-01	4.20E-19
8.00E-01	5.90E-19
1.00E 00	7.60E-19
1.50E 00	1.20E-18
2.00E 00	1.50E-18

8-3-40
KURZWEG (113)

ENERGY (KEV)	CROSS SECTION (SQ. CM)
2.74E 02	5.00E-17
3.80E 02	5.60E-17
4.85E 02	6.33E-17
5.92E 02	6.72E-17
7.50E 02	7.12E-17
9.62E 02	8.20E-17
1.18E 03	7.70E-17
1.44E 03	8.19E-17
1.65E 03	8.64E-17
1.81E 03	8.59E-17

8-3-41A
NEFF (104)

ENERGY (KEV)	CROSS SECTION (SQ. CM)
1.00E 00	2.50E-20
2.00E 00	5.90E-20

8-3-41B
NEFF (104)

ENERGY (KEV)	CROSS SECTION (SQ. CM)
1.00E 00	3.70E-20
2.00E 00	8.70E-20

8-3-41C
NEFF (104)

ENERGY (KEV)	CROSS SECTION (SQ. CM)
1.00E 00	2.50E-20
2.00E 00	3.70E-20

8-3-42
NEFF (104)

ENERGY (KEV)	CROSS SECTION (SQ. CM)
4.00E-01	2.10E-20
1.00E 00	5.50E-20
1.50E 00	1.05E-19
2.00E 00	2.20E-19

8-3-43A
DOERING (227)

ENERGY (KEV)	CROSS SECTION (SQ. CM)
1.00E 00	1.72E-18
2.00E 00	3.10E-18
3.00E 00	5.40E-18
3.00E 00	5.60E-18
4.00E 00	6.40E-18
5.00E 00	9.60E-18
6.00E 00	9.90E-18
7.00E 00	1.22E-17
8.00E 00	1.19E-17

ENERGY (KEV)	CROSS SECTION (SQ. CM)
9.0CE 00	1.11E-17
1.00E 01	7.90E-18
1.04E 01	9.90E-18

8-3-43B
NEFF (104)

ENERGY (KEV)	CROSS SECTION (SQ. CM)
4.00E-01	1.20E-18
6.00E-01	1.50E-18
8.00E-01	2.00E-18
1.00E 00	2.50E-18
1.2CE 00	2.70E-18
1.50E 00	3.10E-18
2.00E 00	3.60E-18
2.50E 00	3.90E-18
3.00E 00	4.30E-18

8-3-43C
KURZWEG (113)

ENERGY (KEV)	CROSS SECTION (SQ. CM)
2.80E 01	6.37E-17
6.00E 01	7.80E-17
9.00E 01	9.22E-17
1.26E 02	1.04E-16
1.68E 02	1.04E-16
2.50E 02	1.09E-16
2.74E 02	1.14E-16
3.06E 02	1.10E-16
3.60E 02	1.11E-16
4.23E 02	1.13E-16
5.18E 02	1.03E-16
6.55E 02	9.24E-17
7.83E 02	8.80E-17
9.84E 02	8.14E-17
1.23E 03	8.40E-17
1.43E 03	7.60E-17
1.74E 03	7.50E-17
1.97E 03	6.66E-17

8-3-44
DOERING (227)

ENERGY (KEV)	CROSS SECTION (SQ. CM)
2.00E 00	9.40E-19
3.00E 00	1.10E-18
4.00E 00	1.90E-18
5.0CE 00	1.80E-18
5.00E 00	2.80E-18
6.CCE 00	3.30E-18
7.00E 00	4.00E-18
8.00E 00	3.60E-18
8.0CE 00	3.50E-18
9.00E 00	2.50E-18
9.00E 00	4.00E-18
1.00E 01	3.80E-18
1.00E 01	2.20E-18
1.04E 01	2.60E-18

8-3-45
DOERING (227)

ENERGY (KEV)	CROSS SECTION (SQ. CM)
2.0CE 00	8.20E-19
2.50E 00	8.30E-19
3.00E 00	7.00E-19
3.50E 00	6.40E-19
4.00E 00	1.12E-18
4.50E 00	9.90E-19
5.00E 00	9.50E-19
5.00E 00	8.60E-19
5.5CE 00	1.53E-18
6.00E 00	1.41E-18
6.5CE 00	1.59E-18
7.00E 00	1.60E-18
7.00E 00	2.15E-18
7.00E 00	1.91E-18
7.50E 00	1.41E-18
7.5CE 00	1.68E-18
8.00E 00	2.72E-18
8.0CE 00	1.91E-18
8.50E 00	1.31E-18
8.50E 00	2.01E-18
9.00E 00	1.72E-18
9.50E 00	1.52E-18
9.50E 00	2.10E-18
1.00E 01	1.40E-18
1.00E 01	1.87E-18
1.05E 01	2.09E-18
1.05E 01	1.73E-18
1.10E 01	2.41E-18
1.13E 01	2.59E-18
1.15E 01	2.34E-18

8-3-46
DOERING (227)

ENERGY (KEV)	CROSS SECTION (SQ. CM)
3.00E 00	1.30E-19
4.00E 00	3.20E-19
5.00E 00	4.40E-19
6.00E 00	6.40E-19
7.00E 00	7.20E-19
8.00E 00	1.01E-18
9.00E 00	1.01E-18
1.00E 01	1.27E-18
1.10E 01	1.57E-18

8-3-47
KURZWEG (113)

ENERGY (KEV)	CROSS SECTICN (SQ. CM)
3.28E 02	3.82E-17
4.02E 02	3.59E-17
4.34E 02	3.59E-17
5.28E 02	3.59E-17
7.08E 02	4.20E-17
7.52E 02	3.74E-17
8.73E 02	4.95E-17
1.08E 03	5.45E-17

ENERGY (KEV)	CROSS SECTION (SQ. CM)
1.23E 03	6.15E-17
1.32E 03	6.25E-17
1.43E 03	7.15E-17
1.58E 03	7.80E-17
1.75E 03	8.94E-17
1.95E 03	1.05E-16
2.05E 03	1.15E-16
2.13E 03	1.10E-16

8-3-48
KURZWEG (113)

ENERGY (KEV)	CROSS SECTION (SQ. CM)
8.04E 02	6.15E-17
9.52E 02	6.65E-17
1.08E 03	8.40E-17
1.29E 03	1.00E-16
1.50E 03	1.29E-16
1.67E 03	1.37E-16
2.10E 03	1.66E-16
2.42E 03	1.72E-16
2.44E 03	1.66E-16
2.67E 03	1.82E-16
2.98E 03	1.88E-16
3.41E 03	1.80E-16
3.79E 03	1.90E-16
3.83E 03	1.74E-16
4.16E 03	1.83E-16

8-3-49
KURZWEG (113)

ENERGY (KEV)	CROSS SECTION (SQ. CM)
3.20E 01	1.21E-16
6.80E 01	1.23E-16
9.10E 01	1.25E-16
1.29E 02	1.24E-16
1.48E 02	1.25E-16
1.81E 02	1.21E-16
2.19E 02	1.22E-16
2.68E 02	1.21E-16
2.74E 02	1.06E-16
3.44E 02	1.24E-16
3.54E 02	1.15E-16
3.70E 02	1.11E-16
4.33E 02	9.39E-17
5.44E 02	9.12E-17
6.71E 02	8.80E-17
8.25E 02	7.80E-17
1.01E 03	7.95E-17
1.23E 03	7.10E-17
1.49E 03	6.10E-17
1.76E 03	6.39E-17
1.97E 03	5.85E-17

8-3-50
KURZWEG (113)

ENERGY (KEV)	CROSS SECTICN (SQ. CM)
3.00E 01	1.17E-16

5.90E 01	1.40E-16	2.75E 02	1.10E-16	9.65E 02	8.95E-17
8.40E 01	1.38E-16	3.06E 02	1.12E-16	1.18E 03	8.69E-17
1.14E 02	1.40E-16	3.82E 02	1.10E-16	1.39E 03	8.35E-17
1.66E 02	1.36E-16	4.87E 02	1.07E-16	1.55E 03	8.45E-17
2.14E 02	1.25E-16	5.93E 02	1.12E-16	1.65E 03	8.80E-17
2.60E 02	1.19E-16	7.52E 02	9.52E-17	1.81E 03	8.53E-17
				1.92E 03	8.00E-17

8-4   EXCITATION OF OXYGEN

The collisionally induced oxygen spectrum tends to be dominated by the O_2^+ first negative bands with small contributions from O I, O II, and the O_2^+ second negative system. The discussion will concentrate first on the general features common to any study of excitation of oxygen, particularly the assessment of overlap in the molecular structure (Section 8-4-A). The only reliable data is for excitation by proton impact (Section 8-4-B); fragmentary information on excitation by H_2^+ and Ne^+ is unreliable (Section 8-4-C). The data tables and figures are presented in Section 8-4-D.

In addition to the quantitative measurement of cross sections, some effort has been expended on the qualitative identification of collisionally induced emission lines. Such studies were generally carried out at high target pressures, sometimes as high as one atmosphere. Metal films were often placed between the ion source and collision experiment to inhibit the flow of gas, a procedure which will produce spurious effects. Projectiles included H^+, H, Li^+, and He^{2+}. It is doubtful whether any useful information can be drawn from such studies. The reader is referred to the original papers for further details.[325, 336, 337, 338, 339, 340]

A.   General Considerations

A molecular oxygen target in thermal equilibrium with its surroundings at normal room temperature will exhibit negligible population of excited electronic or vibrational states. There will be a significant population of excited rotational states. In order to specify the population distribution it is necessary that the temperature of the target region be measured. There have been no systematic observations of how the initial rotational state population might effect measurements of cross sections for the collisional excitation or removal of electrons.

In Section 3-2-D it was shown that pressure measurements with a McLeod gauge might be in serious error due to the effects of thermal transpiration and mercury streaming onto the gauge cold trap. Experiments where precautions have not been taken to prevent this error will be specifically mentioned.

Some measurements have been made on the emission from excited O and O^+, formed through the dissociation of the molecular target. Weise et al.[31] have presented calculated transition probabilities that might permit estimates of level-excitation cross sections. The accuracy of these calculations is poor and no attempt is made here to derive level-excitation cross sections from the measurements of emission cross sections. The atomic spectra are very complex and considerable care is necessary to ensure that the lines are unambiguously identified and that a single line emission is measured free of interference from neighboring transitions. In all cases where two or more lines have been detected simultaneously, the precautions outlined in Section 3-4-E for measurement of the sum of the cross sections were ignored.

Most measurements of molecular emission have concentrated on the O_2^+ first negative system. Quantitative measurements on the first negative system are complicated by the width of the rotational structure, as much as 100-Å in some cases, and by overlapping of bands. Figure 8-5 illustrates the problem. The cross section for a particular vibrational transition $v' \to v''$ must be obtained by integrating the area under the rotational structure and subtracting any contribution from overlapped bands.

All the experiments described here, ranging in projectile energy from 5- to 1000-keV, demonstrated that the shape of the rotational structure did not change as a function of projectile energy. Therefore, the variation with impact energy of the intensity of one part of the structure will represent the relative variation of the cross section for the vibrational transition. All published observations indicated that the population of the rotational states of the excited molecule is appropriate to a room-temperature Boltzmann distribution, similar to the observations for nitrogen (see Section 8-3-F). It cannot necessarily be assumed that such behavior will be observed in other situations; it is likely that changes in rotational population will occur for low-velocity projectile impact.

For most of the vibrational transitions in the O_2^+ first negative system it is necessary to assess the overlap of bands. The position of the various rotational lines are known, and the population distribution among the rotational states may be inferred from analysis of bands where little or no overlap occurs. With this information and knowledge of the resolution of the detection system, it is possible to synthesize a rotational structure for a given line and normalize this to some feature of the band system where no overlap occurs, as shown in Fig. 8-5. Measurement of the areas under the two curves allows an assessment of the relative contributions of the two lines to the overlapped bands.

The final step in the measurement of the band emission cross sections is to determine the integrated intensity of the band system. This has customarily been done by setting a monochromator with resolution β at a series of wavelengths $\lambda_1, \lambda_2 \ldots \lambda_n \ldots \lambda_m$ and measuring the signal S_n at each setting. The wavelength range λ_1 to λ_m must encompass the complete rotational structure of interest. The settings are separated in wavelength by β (i.e., $\lambda_{n+1} = \lambda_n + \beta$). The sum of the signals $\sum_n S_n$ is taken as the total signal from the band. A mean value of detector sensitivity $K(\lambda)$ at some wavelength in the band is determined by the normal calibration procedures. These two quantities, $\sum_n S_n$ and $K(\lambda)$, are inserted into Eq. 3-1 to arrive at a determination of the total emission J_{ij} in the band. This procedure with its possible limitations has been discussed at length in Section 3-4-E(2); there it is shown that necessary precautions

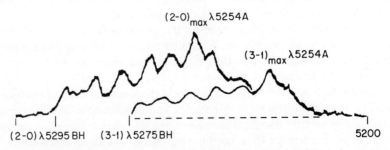

Fig. 8-5 The overlapping (2, 0) and (3, 1) bands of the O_2^+ first negative system induced by proton impact on O_2 at 20-keV energy. This scan was taken by Hughes and Ng[185] with 3.2-Å resolution.

include demonstrating that the detection sensitivity of the system is invariant with wavelength over the band system, and that the final result is independent of spectral resolution. Tests of these requirements have been inadequate.

According to Herzberg,[33] there are no transitions that populate the $b\ ^4\Sigma_g^-$ or the $A^2\Pi_u$ states of O_2^+. Therefore, the first and second negative band systems ($b\ ^4\Sigma_g^- \to a\ ^4\Pi_u$ and $A\ ^2\Pi_u \to X\ ^2\Pi_g$, respectively) are unaffected by cascade. It is possible to calculate Franck—Condon factors and branching ratios for these cases. Thus, an estimate of a level excitation cross section may be arrived at by use of Eq. 2-15.

B. Proton Impact

Absolute measurements on the excitation of oxygen by protons have been provided by four independent studies. For all experiments, the optical system was based on a monochromator with the photomultiplier detector arranged at a fixed angle with respect to the direction of the projectile beam. All four studies neglected the influence of polarization on the cross section data. With the exception of the work by Thomas et al.[230] the experiments measured target pressure using a McLeod gauge without attention to the possible influence of errors introduced by use of the cold trap (see Section 3-2-D).

Dufay et al. are responsible for the most extensive consideration of this collision combination. They produced a total of four published papers. The first was concerned solely with qualitative spectral analysis and provides no cross section information.[341] A second paper[202] provides preliminary data that was completely changed and superseded by a final publication from which most of the present data were obtained.[178] A brief conference paper[179] repeated these data and provided additional measurements at an energy of 1-MeV. Errors were made in the presentation of data in this last publication; the data utilized in this monograph have been corrected.[290] The apparatus described in Section 8-3-B(12) was employed for all these measurements. Target pressures were maintained at 4×10^{-3} torr above 150-keV impact energy and 2×10^{-3} torr below it. At energies below 30-keV appreciable corrections were needed for neutralization of the projectile beam. During cross section measurements resolution was set at 4-Å for wavelengths from 2000- to 5000-Å, and 10- or 20-Å above 5000-Å. In general, the descriptions are very abbreviated and do not allow a detailed assessment of their adequacy. Absolute accuracy is quoted as $\pm 50\%$ with relative accuracy between data points of $\pm 10\%$.

Hughes and Ng[185] used the apparatus described in Section 8-3-B(10). Target pressures were typically 10^{-3} to 5×10^{-3} torr. Their published spectra indicated appreciable contamination of the target by N_2. Systematic errors were assessed to be less than $\pm 40\%$ while reproducibility was $\pm 5\%$.

Baker et al.[181] used the apparatus described in Section 8-3-B(3) for measurements on one O_2^+ band and one O I line. Published details are insufficient to allow a proper assessment of the procedures employed. The limits of accuracy were not stated.

Thomas et al.[230] employed the apparatus described in Section 8-3-B(9). Relative accuracy was assessed at $\pm 10\%$. The poor limits of absolute accuracy, quoted as $\pm 50\%$, were due, in large measure, to the difficulties of handling the extensive molecular structure.

1. Formation of Excited States of O. Dufay et al.[178, 179] measured cross sections for emission of three infrared lines of O I. Their results are shown in Data Table 8-4-1.

Spectral resolutions of 4- to 10-Å were used for these measurements; this resolution is insufficient to allow unambiguous resolution of spectral lines. The 7772-Å line ($3\,^5P \to 3\,^5S^0$) lies within 2-Å of another line in the same multiplet and within 1.5-Å of three lines in another multiplet ($7\,^5D^0 \to 3\,^5P$). The 8446-Å emission must be a blend of three lines in the same multiplet ($3\,^3P \to 3\,^3S^0$). The 9266-Å line lies within 3-Å of another line in the same multiplet ($3\,^5D^0 \to 3\,^5P$). It is concluded that in the case of the 7772-Å and 9266-Å lines the data may represent the sum of two or more cross sections in unknown proportions determined without the precautions detailed in Section 3-4-E. The 8446-Å line is free of interference and represents the total emission in one multiplet. Clearly, the origin of the emissions is inadequately defined, and the accuracy of the cross section measurements may be poor.

2. Formation of Excited States of O^+. All four groups have carried out measurements on a line of O II centered at about 4416-Å (Data Table 8-4-2). In view of the poor resolution used in all experiments, this probably represents the sum of the 4415- and 4417-Å lines in the $3p\,^2D^0 \to 3s\,^2P$ multiplet. There is also one other line (4417.37-Å) within one angstrom; this line is listed as weak in discharge tube spectra.[66] The neglect of precautions for dealing with the simultaneous measurement of the intensity of two different lines (see Section 3-4-E) will probably cause some error. The measurements by Baker et al.[181] agree with those by Hughes and Ng;[185] both appear to lie almost a factor of three above the work of Thomas et al.[230] Discrepancies of this magnitude between data from these research groups have been identified in studies of helium excitation (see Section 6-2-A) and are ascribed to differences in calibration of detection sensitivity. The data by Dufay et al.[178, 179] show a somewhat different energy dependence at extreme energies from those by Thomas et al.[230] but agree in magnitude.

Dufay et al.[178, 179] have also considered three other lines of O II (Data Table 8-4-3). With 4-Å resolution, the 4185-Å line ($3d'\,^2G \to 3p'\,^2F^0$) should be resolved from a neighboring transition in the same multiplet at 4190-Å. Similarly the 4591-Å line ($3p'\,^2F^0 \to 3s'\,^2D$) would be properly resolved. The 4650-Å line would be a sum of two transitions in the $3p\,^4D^0 \to 3s\,^4P$ multiplet.

Hughes and Ng[185] quote a relative measurement on three O II lines in the $3p\,^4D^0 \to 3s\,^4P$ multiplet at impact energies of 20- and 100-keV. Relative to the 4416-Å line the cross sections for the 4649-, 4674-, and 4662-Å transitions are 1.7, 0.2_6, and 0.3_5, respectively. It is stated that these transitions were completely resolved from the neighboring lines.

3. Formation of Excited States of O_2^+. There are available a number of measurements on the O_2^+ first negative system ranging in energy from 5- to 1000-keV. Dufay et al.[178] have also performed a brief study of the second negative system. These molecular emissions exhibit considerable rotational structure and it is to be expected that a report of an experiment will contain full details on the methods utilized to assess overlap of structures and to sum all significant rotational contributions to a line.

All four experiments consist of two distinct steps. First, a measurement is made of the relative variation of the intensity of one segment of a band as a function of projectile energy. Second, at a single energy the band is integrated, the band overlap is assessed, and an absolute cross section evaluated. All experiments have shown that the shape of the rotational structure is invariant with projectile impact energy. Thus, the first step provides a measure of the relative variation of the band emission cross

section with impact energy; this data will probably be quite accurate. The second step assigns the absolute magnitude to the cross section through a series of rather complicated procedures; there are many opportunities for error and consequently the absolute magnitudes may not be reliable.

The work of Dufay et al.,[178] Hughes and Ng,[185] and that of Thomas et al.[230] have included scans of the rotational bands to assess overlap. Baker et al.[181] present a measurement of the sum of emission in the 5175- and 5300-Å bands without separating the relative contributions. In all experiments the integrated intensity over the whole band was determined in the manner described in Section 3-4-E(2). A necessary test for the validity of these procedures is to show the result to be independent of the wavelength resolution of the system [see Section 3-4-E(2)]. Only in the work of Dufay et al.[178] was this done, resolution being varied from 1- to 10-Å. Thomas et al.[230] and also Hughes and Ng[185] neglected the problem. Baker et al.[181] supplied so few experimental details that it is impossible to determine what procedure was used.

The results are shown in Data Tables 8-4-4 to 8-4-11. The data of Hughes and Ng agree well with that by Dufay et al. This is surprising, since measurements on helium excitation by these two groups differ consistently from each other (see Section 6-2-A). There the discrepancy was ascribed to errors in calibration that should logically occur here also. The work by Thomas et al. shows a different energy dependence from that by other workers. There is no obvious explanation for this serious discrepancy. The work by Baker et al. on the sum of the 5175- and 5300-Å bands is shown in Data Table 8-4-11 compared with the same sum deduced from the data by Hughes and Ng; agreement is good. On a subjective basis the weight of evidence favors the measurements by Dufay, Hughes, Baker, and their respective co-workers.

The O_2^+ first negative system represents the $b\ ^4\Sigma_g^- \to a\ ^4\Pi_u$ electronic transition. There are no other transitions that might depopulate the $b\ ^4\Sigma_g^-$ state or populate it by cascade.[33] Therefore, the line emission cross sections are proportional to level excitation cross sections. Dufay et al.[178] measured relative intensities of all lines that provided a significant depopulation of the upper state. Cross sections were given for only a few of these emissions, the remainder were presented as relative intensities in tabular form only. These measurements allowed a direct estimate of level excitation cross section by adding together the emission cross sections for all relevant depopulating transitions. This approach is completely experimental and makes no recourse to theory. Neither Hughes and Ng nor Thomas et al. carried out measurements on sufficient lines to allow such a direct estimate to be made, but these workers used theoretical estimates of branching ratios to calculate level excitation cross sections by Eq. 2-15. Franck—Condon factors were obtained from the work of Jarmain et al.[342] The estimated level excitation cross sections presented by the authors are shown in Data Tables 8-4-13 to 8-4-15. Both approaches should give the same result. Inevitably these may be less accurate than the emission measurements from which they are derived.

In addition to the extensive work on the first negative system, Dufay et al.[178] have carried out a limited series of measurements on two lines in the weak O_2^+ second negative system. This system has a very complex double-headed structure. The measurements are shown in Data Tables 8-4-16 and 8-4-17. No attempt has been made to derive level-excitation cross sections. Relative intensities of other transitions are also given in tabular form by Dufay et al.[178] but are not reproduced here.

C. Impact of Other Projectiles

Studies on excitation by projectiles other than protons are confined to fragmentary information on H_2^+ and Ne^+ impact; they are of poor accuracy and will not be reproduced.

The work on H_2^+ impact is found in an early paper by Dufay et al.[202] which is devoted primarily to H^+ impact excitation. The data is confined to the 4416-Å line of the O II spectrum; this consists presumably of a blend of the 4415- and 4417-Å transitions in the $3p\ ^2D^0 \to 3s\ ^2P$ multiplet. Other measurements in this early paper have been subsequently revised with large changes to cross section magnitudes; the H_2^+ data has not been remeasured. It is concluded that these data may be unreliable.

Schlumbohm[229] has studied excitation induced by the impact of Ne^+ using the apparatus described in Section 7-3-A(10). Emission into a band of ± 40-Å centered at 5250-Å was measured using an interference filter. This data must include components of the (2, 0) and (3, 1) transitions in the O_2^+ first negative system. In view of the poor resolution it is difficult to determine exactly what these measurements represent.

D. Data Tables

1. Cross Sections for Emission of Lines (and Bands) and Excitation of Levels Induced by Impact of H^+ on an O_2 Target

8-4-1   O I Lines (A) 7772-Å $(3\ ^5P \to 3\ ^5S^0)$; (B) 8446-Å $(3\ ^3P \to 3\ ^3S^0)$; (C) 9266-Å $(3\ ^5D^0 \to 3\ ^5P)$.

8-4-2   4416-Å Line of O II $(3p\ ^2D^0 \to 3s\ ^2P)$.

8-4-3   O II Lines (A) 4185-Å $(3d'\ ^2G \to 3p'\ ^2F^0)$; (B) 4591-Å $(3p'\ ^2F^0 \to 3s'\ ^2D)$; (C) 4650-Å $(3p\ ^4D^0 \to 3s\ ^4P)$.

8-4-4   6026-Å Band of O_2^+ $[b\ ^4\Sigma_g^- \to a\ ^4\Pi_u\ (0, 0)]$.

8-4-5   6419-Å Band of O_2^+ $[b\ ^4\Sigma_g^- \to a\ ^4\Pi_u\ (0, 1)]$.

8-4-6   5632-Å Band of O_2^+ $[b\ ^4\Sigma_g^- \to a\ ^4\Pi_u\ (1, 0)]$.

8-4-7   5290-Å Band of O_2^+ $[b\ ^4\Sigma_g^- \to a\ ^4\Pi_u\ (2, 0)]$.

8-4-8   5597-Å Band of O_2^+ $[b\ ^4\Sigma_g^- \to a\ ^4\Pi_u\ (2, 1)]$.

8-4-9   5006-Å Band of O_2^+ $[b\ ^4\Sigma_g^- \to a\ ^4\Pi_u\ (3, 0)]$.

8-4-10   5275-Å Band of O_2^+ $[b\ ^4\Sigma_g^- \to a\ ^4\Pi_u\ (3, 1)]$.

8-4-11   5275- and 5290-Å Bands of O_2^+ $[b\ ^4\Sigma_g^- \to a\ ^4\Pi_{uu}\ (2, 0)$ and $(3, 1)]$ Measured Together.

8-4-12 O_2^+ $(b\ ^4\Sigma_g^-\ v' = 0)$ Level.

8-4-13 O_2^+ $(b\ ^4\Sigma_g^-\ v' = 1)$ Level.

8-4-14 O_2^+ $(b\ ^4\Sigma_g^-\ v' = 2)$ Level.

8-4-15 O_2^+ $(b\ ^4\Sigma_g^-\ v' = 3)$ Level.

8-4-16   3398-Å Band of O_2^+ $[a\ ^2\Pi_u \to X\ ^2\Pi_g\ (0, 5)]$.

8-4-17   3830-Å Band of O_2^+ $[a\ ^2\Pi_u \to X\ ^2\Pi_g\ (0, 7)]$.

8-4-1A
DUFAY ET AL. (178,179)

ENERGY (KEV)	CROSS SECTION (SQ. CM)
3.00E 01	8.00E-18
4.00E 01	7.30E-18
5.00E 01	7.50E-18
6.00E 01	7.50E-18
7.00E 01	7.00E-18
8.00E 01	7.60E-18
1.00E 02	6.50E-18
1.20E 02	6.00E-18
1.00E 03	2.60E-18

8-4-1B
DUFAY ET AL. (178)

ENERGY (KEV)	CROSS SECTION (SQ. CM)
3.00E 01	3.90E-18
4.00E 01	3.10E-18
5.00E 01	3.50E-18
6.00E 01	2.90E-18
7.00E 01	2.80E-18
8.00E 01	3.10E-18
1.00E 02	2.80E-18
1.20E 02	2.70E-18

8-4-1C
DUFAY ET AL. (178)

ENERGY (KEV)	CROSS SECTION (SQ. CM)
3.00E 01	2.65E-18
4.00E 01	2.40E-18
5.00E 01	2.40E-18
6.00E 01	2.20E-18
7.00E 01	2.05E-18
8.00E 01	1.95E-18
1.00E 02	1.85E-18
1.20E 02	1.80E-18

8-4-2A
DUFAY ET AL. (178)

ENERGY (KEV)	CROSS SECTION (SQ. CM)
4.00E 01	5.20E-19
5.00E 01	5.10E-19
6.00E 01	5.10E-19
7.00E 01	4.40E-19
8.00E 01	4.30E-19
1.00E 02	3.40E-19
1.20E 02	3.80E-19
1.50E 02	2.30E-19
2.00E 02	2.00E-19
3.00E 02	1.40E-19
4.00E 02	1.15E-19
5.00E 02	1.00E-19

8-4-2B
HUGHES ET AL. (185)

ENERGY (KEV)	CROSS SECTION (SQ. CM)
1.00E 01	3.88E-19
1.50E 01	5.46E-19
2.00E 01	7.53E-19
3.00E 01	9.27E-19
4.00E 01	1.10E-18
4.50E 01	1.09E-18
5.00E 01	1.07E-18
5.50E 01	1.05E-18
6.00E 01	1.07E-18
6.50E 01	9.89E-19
7.00E 01	9.25E-19
8.00E 01	9.18E-19
9.00E 01	8.80E-19
1.00E 02	8.09E-19
1.10E 02	7.50E-19
1.20E 02	7.01E-19
1.30E 02	6.37E-19

8-4-2C
THOMAS ET AL. (230)

ENERGY (KEV)	CROSS SECTION (SQ. CM)
1.50E 02	2.09E-19
2.00E 02	1.76E-19
2.50E 02	1.27E-19
3.00E 02	1.00E-19
3.50E 02	8.20E-20
4.00E 02	7.00E-20
4.50E 02	6.00E-20
5.00E 02	5.30E-20
6.00E 02	4.40E-20
7.00E 02	3.70E-20
8.00E 02	3.20E-20
9.00E 02	2.80E-20
1.00E 03	2.50E-20

8-4-2D
BAKER ET AL. (181)

ENERGY (KEV)	CROSS SECTION (SQ. CM)
2.00E 01	8.45E-19
3.00E 01	9.27E-19
4.00E 01	1.02E-18
6.00E 01	9.57E-19
8.00E 01	8.65E-19
1.00E 02	7.25E-19

8-4-3A
DUFAY ET AL. (178)

ENERGY (KEV)	CROSS SECTION (SQ. CM)
3.00E 01	2.00E-19
4.00E 01	2.20E-19
5.00E 01	2.00E-19
6.00E 01	2.20E-19
7.00E 01	1.60E-19
8.00E 01	1.80E-19
1.00E 02	1.30E-19
1.20E 02	1.05E-19

8-4-3B
DUFAY ET AL. (178)

ENERGY (KEV)	CROSS SECTION (SQ. CM)
3.00E 01	4.20E-19
4.00E 01	4.40E-19
5.00E 01	4.10E-19
6.00E 01	3.70E-19
7.00E 01	3.60E-19
8.00E 01	3.40E-19
1.00E 02	2.60E-19
1.20E 02	2.20E-19

8-4-3C
DUFAY ET AL. (178,179)

ENERGY (KEV)	CROSS SECTION (SQ. CM)
3.00E 01	1.00E-18
4.00E 01	1.10E-18
5.00E 01	9.80E-19
6.00E 01	9.00E-19
7.00E 01	8.50E-19
8.00E 01	8.00E-19
1.00E 02	7.20E-19
1.20E 02	5.40E-19
1.50E 02	4.50E-19
2.00E 02	3.40E-19
3.00E 02	2.60E-19
4.00E 02	2.20E-19
5.00E 02	1.70E-19
1.00E 03	1.00E-19

8-4-4A
DUFAY ET AL. (178)

ENERGY (KEV)	CROSS SECTION (SQ. CM)
2.00E 01	1.20E-17
3.00E 01	9.50E-18
5.00E 01	8.10E-18
8.00E 01	7.15E-18
1.50E 02	6.45E-18
3.00E 02	5.80E-18
5.00E 02	5.40E-18
1.00E 03	4.30E-18

8-4-4B
HUGHES ET AL. (185)

ENERGY (KEV)	CROSS SECTION (SQ. CM)

ENERGY (KEV)	CROSS SECTION (SQ. CM)
5.00E 00	9.60E-18
6.00E 00	1.06E-17
8.00E 00	1.20E-17
1.00E 01	1.27E-17
1.10E 01	1.30E-17
1.20E 01	1.27E-17
1.30E 01	1.26E-17
1.40E 01	1.23E-17
1.50E 01	1.20E-17
1.60E 01	1.15E-17
1.80E 01	1.10E-17
2.00E 01	1.06E-17
3.00E 01	8.88E-18
4.00E 01	8.08E-18
5.00E 01	7.64E-18
6.00E 01	7.36E-18
7.00E 01	7.09E-18
8.00E 01	6.88E-18
9.00E 01	6.68E-18
1.00E 02	6.63E-18
1.10E 02	6.50E-18
1.20E 02	6.40E-18
1.30E 02	6.19E-18

8-4-4C
THOMAS ET AL. (230)

ENERGY (KEV)	CROSS SECTION (SQ. CM)
1.50E 02	4.58E-18
2.00E 02	3.96E-18
2.50E 02	3.42E-18
3.00E 02	2.97E-18
3.50E 02	2.77E-18
4.00E 02	2.47E-18
4.50E 02	2.37E-18
5.00E 02	2.18E-18
6.00E 02	1.97E-18
7.00E 02	1.80E-18
8.00E 02	1.69E-18
9.00E 02	1.69E-18

8-4-5
HUGHES ET AL. (185)

ENERGY (KEV)	CROSS SECTION (SQ. CM)
5.00E 00	1.03E-17
6.00E 00	1.12E-17
8.00E 00	1.26E-17
1.00E 01	1.34E-17
1.10E 01	1.36E-17
1.20E 01	1.34E-17
1.30E 01	1.32E-17
1.40E 01	1.29E-17
1.50E 01	1.26E-17
1.60E 01	1.20E-17
1.80E 01	1.16E-17
2.00E 01	1.12E-17
3.00E 01	9.30E-18
4.00E 01	8.46E-18
5.00E 01	8.00E-18
6.00E 01	7.73E-18
7.00E 01	7.32E-18
8.00E 01	7.22E-18
9.00E 01	7.00E-18

ENERGY (KEV)	CROSS SECTION (SQ. CM)
1.00E 02	6.95E-18
1.10E 02	6.85E-18
1.20E 02	6.72E-18
1.30E 02	6.50E-18

8-4-6A
DUFAY ET AL. (178)

ENERGY (KEV)	CROSS SECTION (SQ. CM)
2.00E 01	1.60E-17
7.00E 01	1.30E-17
5.00E 01	1.10E-17
8.00E 01	9.70E-18
1.50E 02	8.75E-18
3.00E 02	7.85E-18
5.00E 02	7.30E-18

8-4-6B
HUGHES ET AL. (185)

ENERGY (KEV)	CROSS SECTION (SQ. CM)
5.00E 00	1.45E-17
6.00E 00	1.57E-17
8.00E 00	1.77E-17
1.00E 01	1.88E-17
1.10E 01	1.92E-17
1.20E 01	1.88E-17
1.30E 01	1.86E-17
1.40E 01	1.82E-17
1.50E 01	1.77E-17
1.60E 01	1.70E-17
1.80E 01	1.63E-17
2.00E 01	1.57E-17
3.00E 01	1.35E-17
4.00E 01	1.21E-17
5.00E 01	1.13E-17
6.00E 01	1.09E-17
7.00E 01	1.04E-17
8.00E 01	1.02E-17
9.00E 01	9.85E-18
1.00E 02	9.76E-18
1.10E 02	9.60E-18
1.20E 02	9.43E-18
1.30E 02	9.14E-18

8-4-6C
THOMAS ET AL. (230)

ENERGY (KEV)	CROSS SECTION (SQ. CM)
1.50E 02	5.94E-18
2.00E 02	5.04E-18
2.50E 02	4.10E-18
3.00E 02	3.91E-18
3.50E 02	3.50E-18
4.00E 02	3.25E-18
4.50E 02	3.23E-18
5.00E 02	2.83E-18
6.00E 02	2.74E-18
7.00E 02	2.51E-18
8.00E 02	2.34E-18

ENERGY (KEV)	CROSS SECTION (SQ. CM)
9.00E 02	2.16E-18

8-4-7A
DUFAY ET AL. (178)

ENERGY (KEV)	CROSS SECTION (SQ. CM)
2.00E 01	7.00E-18
3.00E 01	5.85E-18
5.00E 01	5.00E-18
8.00E 01	4.40E-18
1.50E 02	4.00E-18
3.00E 02	3.60E-18
5.00E 02	3.30E-18

8-4-7B
HUGHES ET AL. (185)

ENERGY (KEV)	CROSS SECTION (SQ. CM)
5.00E 00	5.15E-18
6.00E 00	5.59E-18
8.00E 00	6.30E-18
1.00E 01	6.79E-18
1.10E 01	6.82E-18
1.20E 01	6.70E-18
1.30E 01	6.63E-18
1.40E 01	6.48E-18
1.50E 01	6.32E-18
1.60E 01	6.00E-18
1.80E 01	5.82E-18
2.00E 01	5.59E-18
3.00E 01	4.66E-18
4.00E 01	4.24E-18
5.00E 01	4.00E-18
6.00E 01	3.88E-18
7.00E 01	3.72E-18
8.00E 01	3.61E-18
9.00E 01	3.51E-18
1.00E 02	3.48E-18
1.10E 02	3.43E-18
1.20E 02	3.37E-18
1.30E 02	3.25E-18

8-4-8A
DUFAY ET AL. (178)

ENERGY (KEV)	CROSS SECTION (SQ. CM)
2.00E 01	4.00E-18
3.00E 01	3.40E-18
5.00E 01	2.90E-18
8.00E 01	2.50E-18
1.50E 02	2.30E-18
3.00E 02	2.05E-18
5.00E 02	1.90E-18

8-4-8B
HUGHES ET AL. (185)

ENERGY (KEV)	CROSS SECTION (SQ. CM)
5.00E 00	2.97E-18
6.00E 00	3.22E-18
8.00E 00	3.63E-18
1.00E 01	3.86E-18
1.10E 01	3.94E-18
1.20E 01	3.86E-18
1.30E 01	3.82E-18
1.40E 01	3.73E-18
1.50E 01	3.63E-18
1.60E 01	3.49E-18
1.80E 01	3.34E-18
2.00E 01	3.22E-18
3.00E 01	2.77E-18
4.00E 01	2.48E-18
5.00E 01	2.32E-18
6.00E 01	2.24E-18
7.00E 01	2.13E-18
8.00E 01	2.09E-18
9.00E 01	2.02E-18
1.00E 02	2.00E-18
1.10E 02	1.97E-18
1.20E 02	1.93E-18
1.30E 02	1.87E-18

8-4-9
DUFAY ET AL. (178)

ENERGY (KEV)	CROSS SECTION (SQ. CM)
2.00E 01	1.50E-18
3.00E 01	1.20E-18
5.00E 01	1.00E-18
8.00E 01	9.00E-19
1.50E 02	8.00E-19
3.00E 02	7.00E-19
5.00E 02	6.50E-19

8-4-10
HUGHES ET AL. (185)

ENERGY (KEV)	CROSS SECTION (SQ. CM)
5.00E 00	2.76E-18
6.00E 00	3.03E-18
8.00E 00	3.38E-18
1.00E 01	3.64E-18
1.10E 01	3.66E-18
1.20E 01	3.60E-18
1.30E 01	3.56E-18
1.40E 01	3.48E-18
1.50E 01	3.39E-18
1.60E 01	3.22E-18
1.80E 01	3.12E-18
2.00E 01	3.00E-18
3.00E 01	2.50E-18
4.00E 01	2.28E-18
5.00E 01	2.15E-18
6.00E 01	2.08E-18
7.00E 01	2.00E-18
8.00E 01	1.94E-18
9.00E 01	1.88E-18
1.00E 02	1.87E-18
1.10E 02	1.84E-18
1.20E 02	1.81E-18
1.30E 02	1.74E-18

8-4-11A
BAKER ET AL. (181)

ENERGY (KEV)	CROSS SECTION (SQ. CM)
2.00E 01	9.25E-18
3.00E 01	8.50E-18
4.00E 01	7.90E-18
6.00E 01	7.30E-18
8.00E 01	6.60E-18
1.00E 02	6.35E-18

8-4-11B
HUGHES ET AL. (185)

ENERGY (KEV)	CROSS SECTION (SQ. CM)
5.00E 00	7.91E-18
6.00E 00	8.59E-18
8.00E 00	9.68E-18
1.00E 01	1.04E-17
1.10E 01	1.05E-17
1.30E 01	1.02E-17
1.40E 01	9.96E-18
1.50E 01	9.71E-18
1.60E 01	9.22E-18
1.80E 01	8.94E-18
2.00E 01	8.59E-18
3.00E 01	7.16E-18
4.00E 01	6.52E-18
5.00E 01	6.15E-18
6.00E 01	5.96E-18
7.00E 01	5.72E-18
8.00E 01	5.55E-18
9.00E 01	5.39E-18
1.00E 02	5.35E-18
1.10E 02	5.27E-18
1.20E 02	5.18E-18
1.30E 02	4.99E-18

8-4-12A
DUFAY ET AL. (178,179)

ENERGY (KEV)	CROSS SECTION (SQ. CM)
2.00E 01	3.60E-17
3.00E 01	2.60E-17
5.00E 01	2.20E-17
8.00E 01	2.00E-17
1.50E 02	1.80E-17
3.00E 02	1.60E-17
5.00E 02	1.50E-17
1.00E 03	1.30E-17

8-4-12B
HUGHES ET AL. (185)

ENERGY (KEV)	CROSS SECTION (SQ. CM)
5.00E 00	2.87E-17
6.00E 00	3.10E-17
8.00E 00	3.55E-17
1.00E 01	3.76E-17
1.10E 01	3.84E-17
1.20E 01	3.77E-17
1.30E 01	3.75E-17
1.40E 01	3.64E-17
1.50E 01	3.56E-17
1.60E 01	3.42E-17
1.80E 01	3.28E-17
2.00E 01	3.15E-17
3.00E 01	2.60E-17
4.00E 01	2.39E-17
5.00E 01	2.26E-17
6.00E 01	2.18E-17
7.00E 01	2.08E-17
8.00E 01	2.04E-17
9.00E 01	1.98E-17
1.00E 02	1.96E-17
1.10E 02	1.93E-17
1.20E 02	1.90E-17
1.30E 02	1.84E-17

8-4-12C
THOMAS ET AL. (230)

ENERGY (KEV)	CROSS SECTION (SQ. CM)
1.50E 02	1.33E-17
2.00E 02	1.15E-17
2.50E 02	9.90E-18
3.00E 02	8.60E-18
3.50E 02	8.00E-18
4.00E 02	7.20E-18
4.50E 02	6.90E-18
5.00E 02	6.30E-18
6.00E 02	5.70E-18
7.00E 02	5.20E-18
8.00E 02	4.90E-18
9.00E 02	4.90E-18

8-4-13A
DUFAY ET AL. (178)

ENERGY (KEV)	CROSS SECTION (SQ. CM)
2.00E 01	2.70E-17
3.00E 01	2.20E-17
5.00E 01	1.80E-17
8.00E 01	1.60E-17
1.50E 02	1.45E-17
3.00E 02	1.30E-17
5.00E 02	1.20E-17

8-4-13B
HUGHES ET AL. (185)

ENERGY (KEV)	CROSS SECTION (SQ. CM)
5.00E 00	2.21E-17
6.00E 00	2.40E-17
8.00E 00	2.71E-17
1.00E 01	2.86E-17
1.10E 01	2.94E-17
1.20E 01	2.86E-17
1.30E 01	2.84E-17
1.40E 01	2.79E-17
1.50E 01	2.71E-17
1.60E 01	2.60E-17
1.80E 01	2.50E-17
2.00E 01	2.40E-17
3.00E 01	2.06E-17
4.00E 01	1.84E-17
5.00E 01	1.72E-17
6.00E 01	1.66E-17
7.00E 01	1.59E-17
8.00E 01	1.55E-17
9.00E 01	1.51E-17
1.00E 02	1.49E-17
1.10E 02	1.47E-17
1.20E 02	1.44E-17
1.30E 02	1.39E-17

8-4-13C
THOMAS ET AL. (230)

ENERGY (KEV)	CROSS SECTION (SQ. CM)
1.50E 02	9.50E-18
2.00E 02	8.10E-18
2.50E 02	6.60E-18
3.00E 02	6.30E-18
3.50E 02	5.60E-18
4.00E 02	5.20E-18
4.50E 02	5.20E-18
5.00E 02	4.50E-18
6.00E 02	4.40E-18
7.00E 02	4.00E-18
8.00E 02	3.70E-18
9.00E 02	3.50E-18

8-4-14A
CUFAY ET AL. (178)

ENERGY (KEV)	CROSS SECTION (SQ. CM)

2.00E 01	1.90E-17
3.00E 01	1.45E-17
5.00E 01	1.20E-17
8.00E 01	1.10E-17
1.50E 02	1.00E-17
3.00E 02	8.70E-18
5.00E 02	8.00E-18

8-4-14B
HUGHES ET AL. (185)

ENERGY (KEV)	CROSS SECTION (SQ. CM)
5.00E 00	1.09E-17
6.00E 00	1.18E-17
8.00E 00	1.33E-17
1.00E 01	1.43E-17
1.10E 01	1.44E-17
1.20E 01	1.42E-17
1.30E 01	1.40E-17
1.40E 01	1.37E-17
1.50E 01	1.34E-17
1.60E 01	1.27E-17
1.80E 01	1.23E-17
2.00E 01	1.18E-17
3.00E 01	9.87E-18
4.00E 01	9.00E-18
5.00E 01	8.46E-18
6.00E 01	8.22E-18
7.00E 01	7.86E-18
8.00E 01	7.64E-18
9.00E 01	7.42E-18
1.00E 02	7.36E-18
1.10E 02	7.26E-18
1.20E 02	7.13E-18
1.30E 02	6.87E-18

8-4-15A
CUFAY ET AL. (178)

ENERGY (KEV)	CROSS SECTION (SQ. CM)
2.00E 01	1.40E-17
3.00E 01	1.10E-17
5.00E 01	9.00E-18
8.00E 01	8.00E-18
1.50E 02	7.40E-18
3.00E 02	6.50E-18
5.00E 02	6.00E-18

8-4-15B
HUGHES ET AL. (185)

ENERGY (KEV)	CROSS SECTION (SQ. CM)
5.00E 00	4.79E-18
6.00E 00	5.20E-18
8.00E 00	5.86E-18
1.00E 01	6.32E-18
1.10E 01	6.35E-18
1.20E 01	6.25E-18
1.30E 01	6.16E-18
1.40E 01	6.03E-18
1.50E 01	5.86E-18
1.60E 01	5.59E-18
1.80E 01	5.30E-18
2.00E 01	5.20E-18
3.00E 01	4.33E-18
4.00E 01	3.94E-18
5.00E 01	3.72E-18
6.00E 01	3.60E-18
7.00E 01	3.45E-18
8.00E 01	3.36E-18
9.00E 01	3.26E-18
1.00E 02	3.24E-18
1.10E 02	3.19E-18
1.20E 02	3.13E-18
1.30E 02	3.02E-18

8-4-16
CUFAY ET AL. (178)

ENERGY (KEV)	CROSS SECTION (SQ. CM)
3.00E 01	8.20E-19
4.00E 01	7.50E-19
5.00E 01	6.60E-19
6.00E 01	7.00E-19
8.00E 01	6.00E-19
1.00E 02	5.00E-19
1.20E 02	5.00E-19

8-4-17
CUFAY ET AL. (178)

ENERGY (KEV)	CROSS SECTION (SQ. CM)
3.00E 01	6.00E-19
4.00E 01	5.50E-19
5.00E 01	5.00E-19
6.00E 01	4.80E-19
1.00E 02	4.00E-19

8-5 EXCITATION OF OTHER MOLECULAR TARGETS

Some measurements are available on the excitation of CO, NO, CH_4, and C_2H_2 by H^+ impact and also CO by N_2^+. These are considered in detail in Sections 8-5-A, 8-5-B, and 8-5-C. The various data tables and graphs are collected in Section 8-5-D. There are also a few qualitative observations that may be mentioned here. Bayes,

Haugh, and co-workers[343, 107] studied spectra emitted on bombardment of N_2O, CS_2, CO, HBr, and COS by Ar^+ and Kr^+ ions at 2- and 2.5-keV impact energy. The spectra were recorded using a photographic spectrometer. No mention was made of the target density employed. The major spectral features are noted and some comments made on the relative intensities of emission in different collision combinations. Polyakova et al.[340] have made similar observations of emission from CO, CO_2, H_2, H_2O, NH_3, CH_4, N_2, and O_2 bombarded by a mixed beam of protons and hydrogen atoms at 38-keV impact energy. No quantitative information can be obtained from this work.

The data presented in this section are not of high accuracy. They are all drawn from reports that provide inadequate explanation of the techniques adopted. The emission spectra are complicated, with many opportunites for overlapping bands and lines. These data should all be treated with caution.

A. Oxides of Carbon

The principal work has been an extensive study of excitation of CO by H^+ impact carried out by Poulizac et al.[179, 135, 344, 345] Utterback and Broida have made relative measurements on excitation of CO by N_2^+ at low energies.[226] Other work has been done on identification of spectral lines, but this provides little data of relevance to the study of cross sections.

1. Excitation of CO by H^+ Impact. Poulizac et al. have carried out a detailed study of the emission induced by H^+ impact on CO. Preliminary results,[344, 345] were revised and completely superseded by a later article[135] from which all the data used here are obtained. Data on this case presented in a conference paper[179] are the same as in a published article.†[135]

The experiments were carried out using the apparatus described in Section 8-3-B(12). The major part of the estimated error limits of $\pm 50\%$ were incurred in the calibration of detection sensitivity using a standard tungsten strip filament lamp at high wavelengths and a hydrogen discharge at low wavelengths. No precautions were mentioned to assess the possible influence of stray light in the spectrometer; this might prove serious, particularly at lower wavelengths.

Poulizac et al.[135] show high resolution spectra that contain lines of CO^+, CO, C I, C II, O I, and O II. The resolution was adequate to provide definite identification of these features. Cross Section measurements were made with 2- to 4-Å resolution.

Emission of the CO bands was weak and no systematic cross section measurements were made on them. The Angstrom ($B\ ^1\Sigma^+ \rightarrow A\ ^1\Pi$) and Herzberg ($C\ ^1\Sigma^+ \rightarrow A\ ^1\Pi$) bands were stated to have cross sections of less than 10^{-19} cm^2 at 100-keV. The third positive system was observed to have a square law dependence on pressure indicating that excitation occurred via a secondary process. The Asundi bands were completely absent.

The majority of the cross section measurements were devoted to emissions of the CO^+ ion. The Baldet—Johnson system ($B\ ^2\Sigma^+ \rightarrow A\ ^2\Pi$) was very weak, and it was estimated that the (1, 0) (0, 0), and (0, 1) bands had cross sections of 2 to 5 × 10^{-19} cm^2 at 100-keV. Extensive measurements on the first negative ($B\ ^2\Sigma^+ \rightarrow X\ ^2\Sigma^+$)

† Figure 1 of Reference 179 has an incorrect caption. The data are for the population of the $v' = 0$ level not $v' = 1$. The data are identical to that presented in Reference 135.

and comet-tail systems ($A\ ^2\Pi \to X\ ^2\Sigma^+$) are reproduced in Data Tables 8-5-1 and 8-5-2. The relative intensity of certain other weak features of these bands was presented in tabular form[135] but are not reproduced here. It is to be noted that the CO^+ comet-tail systems are quite complex, being composed of double-headed bands with two Q and two R branches degraded towards the red. The band heads are separated by up to 50-Å. Further complications arise from the extensive rotational structure. A measured cross section must include all emissions from a rather wide wavelength interval. In an unpublished thesis[346] Poulizac notes briefly that the cross sections were measured by integrating the area under a spectral scan for the whole of the relevant wavelength interval. The published reports contain inadequate details to permit an assessment of the accuracy with which these procedures were carried out.

Poulizac et al. estimated the cross sections for populating the various vibrational levels of the $A\ ^2\Pi$ and $B\ ^2\Sigma$ states. There are no known transitions which might populate the $A\ ^2\Pi$ state by cascade. The Baldet—Johnson bands depopulate the $A\ ^2\Pi$ state and populate the $B\ ^2\Sigma$ state by cascade; however the intensity of these lines is weak; they were ignored, introducing an error that was estimated as less than 10%. The branching ratios required to relate emission data to level-excitation cross sections by Eq. 2-15 were calculated, using Franck—Condon factors and neglecting any variation in electronic transition moment. The level-excitation cross sections so obtained are shown in Data Tables 8-5-3 and 8-5-4. Inevitably the level-excitation cross sections are less accurate than the line-emission cross sections from which they are derived.

Measurements of the cross sections for emission of certain lines in the O I, C I, and C II systems formed by collisional dissociation of the target are shown in Data Tables 8-5-5 to 8-5-7.† The atomic lines measured by Poulizac et al.[135] are all components of closely spaced multiplets. In some cases it is unclear as to which line or lines from the multiplets were measured. The 8446-Å line of O I consists of transitions at 8446.35- and 8446.76-Å and includes all components of the $3\ ^3P \to 3\ ^3S^0$ multiplet. The 9262.73-Å O I line lies close to the 9260.88 line in the same multiplet; the observed signal must include some indeterminate contribution from both lines. The C I lines are probably free of interference from neighboring transitions. The C II line at 4267-Å was presumably the sum of the 4267.27- and 4267.02-Å lines that compose all the transitions in the $4\ ^2F^0 \to 3\ ^2D$ multiplet. Similarly, the 2837.6-Å line was not resolved from a second component of the same doublet ($3p\ ^2P^0 \to 2p\ ^2S$). Some detailed discussion of the resolution used for each case would be necessary in order to decide precisely what transition, or transitions, were measured. Also, it appears that, in cases where two lines were being detected simultaneously, the precautions for such measurements, outlined in Section 3-4-3, were ignored; therefore the data do not accurately represent a sum of the two line emission cross sections.

Poulizac et al.[135] also measured the apparent rotational temperature of the excited CO^+, using the rotational structure of the (0, 0) and (0, 1) band of the first negative system. It was found to be equal to room temperature, a similar result to the many studies on excitation of nitrogen (see Section 8-3-F).

Lipeles[208] has studied the emission of the 4267-Å doublet of C^+ ($4\ ^2F^0_{3\ 1/2,\ 2\ 1/2}$ $3\ ^2D_{2\ 1/2,\ 1\ 1/2}$) induced by the impact of He^+ on a target of CO. The data are in the

---

† Figure 14 of Reference 135, which displays these data, exhibits incorrect labeling of the wavelength for two transitions.

form of the relative variation of intensity with energy of the projectile. Inadequate details are given to permit a proper assessment of the experimental procedures. It appears from the spectra published by Lipeles that the C^+ line was not properly resolved from the comet-tail system of CO^+. Thus, the measurements do not unambiguously represent the behavior of the C^+ line. It must be concluded that the data do not represent the behavior of an identified transition; they are, therefore, excluded from further consideration.

2. Excitation of CO by N_2^+ Impact. Utterback and Broida[226] made a relative measurement of the energy dependence for the collisionally induced emission of the CO^+ comet-tail bands $(A\ ^2\Pi \rightarrow X\ ^2\Sigma^+)$ induced by the impact of N_2^+ on CO at energies from 11- to 300-eV. N_2^+ ions were produced in an electron bombardment source. Target pressure was 5×10^{-3} torr. Secondary electron emission from the ion beam Faraday cup was not suppressed. The collisionally induced spectrum was analyzed using a scanning monochromator with 100-Å resolution. The spectrum showed the comet-tail system clearly. Resolution was insufficient to definitely exclude emission from other spectral systems of target and projectile. The analysis showed that the relative intensity of the various lines in the comet-tail system were independent of projectile energy. A photomultiplier was used to provide an integrated measurement of the intensity of emission at all wavelengths over an undetermined part of the visible spectrum. This measurement, shown in Data Table 8-5-8 was expected to represent the relative variation of the comet-tail emission cross section with impact energy. In view of the poor resolution, it is likely that emissions from other systems are included in this integrated measurement. Tests showed that excited state population in the projectile beam did not appreciably influence the measured cross sections. The data on the relative cross section were obtained with the electron energy in the source at 24-eV; therefore, the excited-state content of the projectile beam may readily be reproduced in some other experiment.

3. Other Work. Schlumbohm[229] has studied the emission of CO_2^+ and CO^+ bands induced by the impact of 2- to 250-eV Ne^+ ions on CO_2 using the apparatus described in Section 7-3-A(10). The quantitative measurements represent the integrated emission intensity over broad regions of the spectrum. In view of the poor resolution it is impossible to relate the data to the occurrence of particular transitions; therefore the data are not reproduced here.

Surveys of spectra emitted as a result of H^+ and H impact on CO and CO_2 have been presented by Polyakova.[340] Similar information for Ne^+ and Ar^+ impact on CO are presented by Haugh et al.[343] Neither of these studies provides any cross section information. Dufay and Poulizac[345] discuss spectra emitted by bombardment of CO_2 with H^+ ions at 300- and 400-keV impact energy. Cross sections were measured but are available only in an unpublished thesis by Poulizac.[346] The thesis by Poulizac also contains unpublished cross sections on the excitation of CO by Li^+. None of these data are reproduced here.

B. Oxides of Nitrogen

In Data Tables 8-5-9 and 8-5-10 are shown measurements of the cross sections for the emission of the 4649-Å O II and 5005-Å N II lines induced by the impact of protons on NO. These measurements were carried out by Robinson and Gilbody[162, 251] with the apparatus described in Section 8-3-B (11). An early conference

report[251] was superseded by a more complete publication[162] from which the data are obtained.

Unambiguous identification of lines in the complex O^+ and N^+ spectra requires the use of high spectral resolution; the published reports give no information on the band width of the monochromator used for these studies. It is impossible to identify the transitions which contribute to the measured intensities. Moore[66] lists some ten lines with wavelengths within ± 7-Å of 5005-Å including two ($3d\ ^3F^0 \rightarrow 3p\ ^3D$ and $3p\ ^5P^0 \rightarrow 3s\ ^5P$) that form a blend at 5005-Å. Similarly, there are three lines that lie within ± 7-Å of the 4649-Å wavelength. It must be concluded that the published emission cross sections may include contributions from a number of different transitions.

C. Light Hydrocarbons

Carré and Dufay[133] have made a study of emission induced by proton impact on light hydrocarbons. Cross section measurements were presented for C_2H_2 and CH_4 targets; also some qualitative remarks were made about excitation of C_2H_4 and C_2H_6. An unpublished thesis by Carré[347] contains additional measurements on these cases and also considers excitation by impact of Li^+ ions. The published cross sections for C_2H_2 and CH_4 targets are reviewed here, the measurements are shown in Data Tables 8-5-11 to 8-5-14.

It is interesting to note that Carré has published detailed analyses of the spectra induced by H^+ impact on CH_4 and C_2H_2.[348, 349] Two new band systems were discovered and tentatively identified as $^1\Delta \rightarrow {}^1\Pi$ and $^3\Sigma \rightarrow {}^3\Pi$ transitions of CH^+.[348] These band systems were not previously observed in conventional high pressure discharge sources. Carré definitely ascribes these emissions to excitation by primary particle impact (i.e., protons) and not to a secondary collision mechanism. These studies also provided new identifications of components in the $A\ ^1\Pi \rightarrow X\ ^1\Sigma^+$ systems that had previously not been observed. These studies[348, 349] provide no cross section information and are not considered further.

The cross section measurements were carried out using the experimental arrangement described in Section 6-1-A(5). Pressures were measured using a cold-trapped McLeod gauge with no precautions against errors due to thermal transpiration and mercury pumping effects (see Section 3-2-D). The possible influences of emission polarization were ignored. Target pressures were quoted[347] as being below 10^{-3} torr, and emission varied linearly with pressure. The authors quote a possible systematic error of $\pm 50\%$, primarily ascribed to uncertainties in calibration, and a possible random error of $\pm 10\%$.

Measurements were made of the emission cross sections of the 3900-Å and 4300-Å bands of CH and the Balmer lines of H induced by the impact of H^+ on CH_4 and C_2H_2. The emitted spectrum is very complex.[347] It is doubtful whether overlapping of spectral lines was completely eliminated from the measurements. The Balmer alpha and beta lines appear to be free of interference but certain spectrograms by Carré[347] show the Balmer-gamma line to be overlapped by a broad-band system, possibly C_2. It is likely that neglect of the interference from the molecular emission causes a small error in the Balmer-gamma measurement. The data presented here for Balmer emission are for the target alone; Balmer emission from the projectile was separated by using the Doppler shift of the emission from the fast particle.

It has previously been noted (Section 2-2-A) that the operational definition of emission cross section makes a tacit assumption that the excited atom emits at the point where it was excited. A light dissociation fragment may recoil with an appreciable kinetic energy and emit at some distance from the ion beam path. Tests to ensure that the measured cross section is independent of the optical field of view are necessary for the measurements of Balmer emission; however, they were omitted in the present case.

No attempt was made to derive level-excitation cross sections from the line emission measurements. It might be possible to estimate level-population cross sections for the CH excited states by using calculated values of branching ratios; however, the authors did not do this, and the data are left in the original form. It is not possible to estimate level-excitation cross sections from Balmer-emission measurements because of the degeneracy between the emitting levels.

Quantitative results for the molecular emission were restricted to measurements on the CH bands. The sum of emissions at 4315- and 4312-Å in the "4300-Å System" $[A\ ^2\Delta \to X\ ^2\Pi\ (0,\ 0)$ and $(1,\ 1)]$ was determined and also the 3889-Å band of the "3900-Å system" $[B\ ^2\Sigma^- \to X\ ^2\Pi\ (0,\ 0)]$. In an unpublished thesis[347] Carré states that the cross sections were evaluated by making a spectral scan and integrating the structure in order to get the total cross section. No allowance was made for variation of detection sensitivity over the wavelength range of the structure, and there were no tests to ensure that the final result was independent of the resolution utilized when making the spectrogram [see Section 3-4-E(2)]. In the case of the 3900-Å system, some interference from the 3914-Å band of N_2^+ was observed,[347] presumably from excitation of the residual background gas. Since the 3900-Å CH system is degraded towards the red and the N_2^+ 3914-Å system is degraded towards the blue, it is difficult to ensure complete separation of the signals from these two sources. Better attention to system cleanliness would presumably reduce the 5×10^{-6} torr background pressure and remove the interference from N_2^+.

D.   Data Tables

1.   Cross Sections for Emission and Excitation Induced by the Impact of H^+ on CO

8-5-1   First Negative Band System of CO^+ $(B\ ^2\Sigma^+ \to X\ ^2\Sigma^+)$; (A) 2190-Å Band (0, 0); (B) 2300-Å Band (0, 1); (C) 2112-Å Band (1, 0).

8-5-2   Comet-Tail Bands of CO^+ $(A\ ^2\Pi \to X\ ^2\Sigma^+)$; (A) 4911- and 4879-Å Band (0, 0); (B) 4566- and 4539-Å Band (1, 0); (C) 4249- and 4274-Å Band (2, 0); (D) 3997- and 4020-Å Band (3, 0); (E) 3778- and 3796-Å Band (4, 0); (F) 3584- and 3601-Å Band (5, 0).

8-5-3 CO^+ $(B\ ^2\Sigma^+)$ Level; (A) $v' = 0$; (B) $v' = 1$.

8-5-4 CO^+ $(A\ ^2\Pi)$ Level; (A) $v' = 0$; (B) $v' = 1$; (C) $v' = 2$; (D) $v' = 3$; (E) $v' = 4$; (F) $v' = 5$.

8-5-5   O I Lines; (A) 7771-Å $(3\ ^5P \to 3\ ^5S^0)$; (B) 8446-Å $(3\ ^3P \to 3\ ^3S^0)$; (C) 9263-Å $(3\ ^5D^0 \to 3\ ^5P)$.

8-5-6   C I Lines; (A) 9094-Å $(3p\ ^3P \to 3s\ ^3P^0)$; (B) 9406-Å $(3p\ ^1D \to 3s\ ^1P^0)$.

8-5-7   C II Lines; (A) 6578-Å $(3\ ^2P^0 \to 3\ ^2S)$; (B) 4267-Å $(4\ ^2F^0 \to 3\ ^2D)$; (C) 2838-Å $(3p\ ^2P^0 \to 2p\ ^2S)$.

2. Cross Section for Emission Induced by Impact of N_2^+ on a CO Target

8-5-8 Relative Cross Section for the comet-tail System of CO^+ ($A\ ^2\Pi \to X\ ^2\Sigma^+$). The data represent an integrated intensity over all bands in the visible spectrum (see text).

3. Cross Sections for Emission Induced by Impact of H^+ on a NO Target

8-5-9 4649-Å Line of O II.

8-5-10 5005-Å Line of N I.

4. Cross Sections for Emission Induced by Impact of H^+ on CH_4 and C_2H_2 Targets

8-5-11 Bands of CH Induced in a CH_4 Target. (A) 3889-Å Band [$B\ ^2\Sigma^- \to X\ ^2\Pi$ (0, 0)]; (B) 4315- and 4312-Å Bands [$A\ ^2\Delta \to X\ ^2\Pi$ (0, 0) and (1, 1)].

8-5-12 Bands of CH Induced in a C_2H_2 Target. (A) 3889-Å Band [$B\ ^2\Sigma^- \to X\ ^2\Pi$ (0, 0)]; (B) 4315- and 4312-Å Bands [$A\ ^2\Delta \to X\ ^2\Pi$ (0, 0) and (1, 1)].

8-5-13 Balmer Lines of H I Induced in a CH_4 Target (Target Emission Only). (A) 6563-Å Balmer—alpha ($n = 3 \to n = 2$); (B) 4861-Å Balmer—beta ($n = 4 \to n = 2$); (C) 4340-Å Balmer—gamma ($n = 5 \to n = 2$).

8-5-14 Balmer Lines of H I Induced in a C_2H_2 Target (Target Emission Only). (A) 6563-Å Balmer—alpha ($n = 3 \to n = 2$); (B) 4861-Å Balmer—beta ($n = 4 \to n = 2$).

8-5-1A
POULIZAC ET AL. (135)

ENERGY (KEV)	CROSS SECTION (SQ. CM)
3.00E 01	9.30E-18
4.00E 01	8.65E-18
5.00E 01	7.35E-18
6.00E 01	7.70E-18
7.00E 01	7.16E-18
8.00E 01	6.40E-18
9.00E 01	6.70E-18
1.00E 02	6.65E-18
1.20E 02	6.00E-18
1.40E 02	5.50E-18
2.00E 02	4.80E-18
3.00E 02	4.15E-18
4.00E 02	3.50E-18
5.00E 02	3.00E-18
6.00E 02	2.74E-18
1.00E 03	2.00E-18

8-5-1B
POULIZAC ET AL. (135)

ENERGY (KEV)	CROSS SECTION (SQ. CM)
3.00E 01	5.25E-18
4.00E 01	4.50E-18
5.00E 01	3.70E-18
6.00E 01	3.90E-18
7.00E 01	3.70E-18
8.00E 01	3.80E-18
9.00E 01	3.44E-18
1.00E 02	3.70E-18
1.20E 02	3.30E-18
1.40E 02	3.00E-18
2.00E 02	2.54E-18
3.00E 02	2.20E-18
4.00E 02	1.85E-18
5.00E 02	1.60E-18
6.00E 02	1.50E-18

8-5-1C
POULIZAC ET AL. (135)

ENERGY (KEV)	CROSS SECTION (SQ. CM)
3.00E 01	1.23E-18
4.00E 01	1.13E-18
5.00E 01	9.60E-19
6.00E 01	1.00E-18
7.00E 01	9.60E-19
8.00E 01	8.40E-19
9.00E 01	8.80E-19
1.00E 02	8.80E-19
1.20E 02	7.70E-19
1.40E 02	7.00E-19
2.00E 02	6.10E-19
3.00E 02	5.20E-19
4.00E 02	4.40E-19
5.00E 02	3.80E-19
6.00E 02	3.40E-19

8-5-2A
POULIZAC ET AL. (135)

ENERGY (KEV)	CROSS SECTION (SQ. CM)
3.00E 01	3.10E-18
4.00E 01	2.70E-18
5.00E 01	2.30E-18
6.00E 01	2.30E-18
7.00E 01	2.40E-18
8.00E 01	2.25E-18
9.00E 01	2.30E-18
1.00E 02	2.20E-18
1.20E 02	2.10E-18
1.40E 02	1.85E-18
2.00E 02	1.48E-18
3.00E 02	1.10E-18
4.00E 02	8.50E-19
5.00E 02	7.00E-19
6.00E 02	5.70E-19

8-5-2B
POULIZAC ET AL. (135)

ENERGY (KEV)	CROSS SECTION (SQ. CM)
3.00E 01	1.00E-17
4.00E 01	9.00E-18
5.00E 01	7.00E-18
6.00E 01	7.50E-18
7.00E 01	7.80E-18
8.00E 01	7.30E-18
9.00E 01	7.80E-18
1.00E 02	7.30E-18
1.20E 02	7.30E-18
1.40E 02	6.00E-18
2.00E 02	4.80E-18
3.00E 02	3.65E-18
4.00E 02	2.80E-18
5.00E 02	2.30E-18
6.00E 02	1.85E-18

8-5-2C
POULIZAC ET AL. (135)

ENERGY (KEV)	CROSS SECTION (SQ. CM)
3.00E 01	1.65E-17
4.00E 01	1.47E-17
5.00E 01	1.21E-17
6.00E 01	1.20E-17
7.00E 01	1.25E-17
8.00E 01	1.15E-17
9.00E 01	1.21E-17
1.00E 02	1.15E-17
1.20E 02	1.08E-17
1.40E 02	9.70E-18
2.00E 02	7.80E-18
3.00E 02	5.40E-18
4.00E 02	4.55E-18
5.00E 02	3.66E-18
6.00E 02	2.96E-18
1.00E 03	2.00E-18

8-5-2D
POULIZAC ET AL. (135)

ENERGY (KEV)	CROSS SECTION (SQ. CM)
3.00E 01	1.20E-17
4.00E 01	1.07E-17
5.00E 01	8.90E-18
6.00E 01	8.70E-18
7.00E 01	9.30E-18
8.00E 01	8.60E-18
9.00E 01	8.90E-18
1.00E 02	8.60E-18
1.20E 02	8.20E-18
1.40E 02	7.40E-18
2.00E 02	5.70E-18
3.00E 02	4.30E-18
4.00E 02	3.40E-18
5.00E 02	2.75E-18
6.00E 02	2.20E-18

8-5-2E
POULIZAC ET AL. (135)

ENERGY (KEV)	CROSS SECTION (SQ. CM)
3.00E 01	8.40E-18
4.00E 01	7.40E-18
5.00E 01	6.00E-18
6.00E 01	6.20E-18
7.00E 01	6.40E-18
8.00E 01	6.00E-18
9.00E 01	6.00E-18
1.00E 02	5.80E-18
1.20E 02	5.50E-18
1.40E 02	4.85E-18
2.00E 02	3.80E-18
3.00E 02	2.85E-18
4.00E 02	2.25E-18
5.00E 02	1.75E-18
6.00E 02	1.50E-18

8-5-2F
POULIZAC ET AL. (135)

ENERGY (KEV)	CROSS SECTION (SQ. CM)
3.00E 01	3.00E-18
4.00E 01	2.80E-18
5.00E 01	2.40E-18
6.00E 01	2.10E-18
7.00E 01	2.30E-18
8.00E 01	2.20E-18
9.00E 01	2.10E-18
1.00E 02	2.00E-18
1.20E 02	2.00E-18
1.40E 02	1.75E-18
2.00E 02	1.40E-18
3.00E 02	1.05E-18
4.00E 02	8.00E-19
5.00E 02	6.60E-19
6.00E 02	5.50E-19

8-5-3A
POULIZAC ET AL. (135)

ENERGY (KEV)	CROSS SECTION (SQ. CM)
3.00E 01	1.65E-17
5.00E 01	1.28E-17
7.00E 01	1.23E-17
1.00E 02	1.15E-17
2.00E 02	9.40E-18
3.00E 02	7.70E-18
4.00E 02	7.00E-18
6.00E 02	5.40E-18

8-5-3B
POULIZAC ET AL. (135)

ENERGY (KEV)	CROSS SECTION (SQ. CM)
3.00E 01	4.80E-18
5.00E 01	3.90E-18
7.00E 01	3.70E-18
1.00E 02	3.40E-18
2.00E 02	2.40E-18
3.00E 02	2.00E-18
4.00E 02	1.70E-18
6.00E 02	1.30E-18

8-5-4A
POULIZAC ET AL. (135)

ENERGY (KEV)	CROSS SECTION (SQ. CM)
3.00E 01	3.20E-17
5.00E 01	2.35E-17
7.00E 01	2.40E-17
1.00E 02	2.20E-17
2.00E 02	1.50E-17
3.00E 02	1.10E-17
4.00E 02	8.40E-18
6.00E 02	6.00E-18

8-5-4B
POULIZAC ET AL. (135)

ENERGY (KEV)	CROSS SECTION (SQ. CM)
3.00E 01	2.80E-17
5.00E 01	2.00E-17
1.00E 02	1.90E-17
2.00E 02	1.25E-17
3.00E 02	1.00E-17
4.00E 02	7.70E-18
6.00E 02	5.60E-18

8-5-4C
POULIZAC ET AL. (135)

ENERGY (KEV)	CROSS SECTION (SQ. CM)
3.00E 01	3.40E-17
5.00E 01	2.50E-17
7.00E 01	2.50E-17
1.00E 02	2.30E-17
2.00E 02	1.60E-17
3.00E 02	1.15E-17
4.00E 02	9.20E-18
6.00E 02	6.50E-18

8-5-4D
POULIZAC ET AL. (135)

ENERGY (KEV)	CROSS SECTION (SQ. CM)
3.00E 01	2.20E-17
5.00E 01	1.60E-17
1.00E 02	1.50E-17
2.00E 02	9.50E-18
3.00E 02	7.30E-18
4.00E 02	5.70E-18
6.00E 02	4.00E-18

8-5-4E
POULIZAC ET AL. (135)

ENERGY (KEV)	CROSS SECTION (SQ. CM)
3.00E 01	1.30E-17
5.00E 01	9.00E-18
1.00E 02	8.50E-18
2.00E 02	5.50E-18
3.00E 02	4.25E-18
4.00E 02	3.30E-18
6.00E 02	2.40E-18

8-5-4F
POULIZAC ET AL. (135)

ENERGY (KEV)	CROSS SECTION (SQ. CM)
3.00E 01	3.08E-18
5.00E 01	2.17E-18
1.00E 02	2.11E-18
2.00E 02	1.47E-18
3.00E 02	1.08E-18
4.00E 02	8.50E-19
6.00E 02	6.10E-19

8-5-5A
POULIZAC ET AL. (135)

ENERGY (KEV)	CROSS SECTION (SQ. CM)
3.00E 01	4.30E-19
4.00E 01	4.70E-19
5.00E 01	4.30E-19
6.00E 01	3.90E-19
8.00E 01	3.60E-19
1.00E 02	3.30E-19
1.20E 02	2.80E-19

8-5-5B
POULIZAC ET AL. (135)

ENERGY (KEV)	CROSS SECTION (SQ. CM)
3.00E 01	8.20E-19
4.00E 01	1.00E-18
5.00E 01	9.00E-19
6.00E 01	9.00E-19
8.00E 01	8.10E-19
1.00E 02	7.70E-19
1.20E 02	6.60E-19

8-5-5C
POULIZAC ET AL. (135)

ENERGY (KEV)	CROSS SECTION (SQ. CM)
3.00E 01	2.10E-19
4.00E 01	2.80E-19
5.00E 01	2.40E-19
6.00E 01	2.25E-19
8.00E 01	2.10E-19
1.00E 02	1.80E-19
1.20E 02	1.50E-19
1.40E 02	1.45E-19

8-5-6A
POULIZAC ET AL. (135)

ENERGY (KEV)	CROSS SECTION (SQ. CM)
3.00E 01	3.50E-19
4.00E 01	4.15E-19
5.00E 01	3.60E-19
6.00E 01	3.60E-19
8.00E 01	3.30E-19
1.00E 02	3.40E-19
1.20E 02	2.90E-19
1.40E 02	2.80E-19

8-5-6B
POULIZAC ET AL. (75)

ENERGY (KEV)	CROSS SECTION (SQ. CM)
3.00E 01	3.10E-19
4.00E 01	3.40E-19
5.00E 01	2.85E-19
6.00E 01	2.85E-19
8.00E 01	2.50E-19
1.00E 02	2.30E-19

ENERGY (KEV)	CROSS SECTION (SQ. CM)
1.20E 02	2.00E-19
1.40E 02	2.00E-19

8-5-7A
POULIZAC ET AL. (135)

ENERGY (KEV)	CROSS SECTION (SQ. CM)
3.00E 01	6.70E-19
4.00E 01	8.20E-19
5.00E 01	7.20E-19
8.00E 01	5.10E-19
1.00E 02	4.00E-19
1.20E 02	3.60E-19

8-5-7B
POULIZAC ET AL. (135)

ENERGY (KEV)	CROSS SECTION (SQ. CM)
2.00E 01	2.76E-19
3.00E 01	3.60E-19
4.00E 01	4.65E-19
5.00E 01	3.87E-19
6.00E 01	3.60E-19
7.00E 01	2.96E-19
8.00E 01	2.76E-19
1.00E 02	1.95E-19
1.20E 02	1.80E-19
1.40E 02	1.48E-19

8-5-7C
POULIZAC ET AL. (135)

ENERGY (KEV)	CROSS SECTION (SQ. CM)
2.00E 01	9.00E-20
3.00E 01	1.37E-19
4.00E 01	1.62E-19
5.00E 01	1.41E-19
6.00E 01	1.08E-19
7.00E 01	1.02E-19
8.00E 01	9.00E-20
1.00E 02	7.30E-20
1.20E 02	6.00E-20
1.40E 02	5.50E-20

8-5-8
UTTERBACK ET AL. (226)

ENERGY (KEV)	CROSS SECTION (ARB. UNITS)
1.10E-02	2.00E 00
1.20E-02	2.15E 00
1.30E-02	2.55E 00
1.40E-02	2.70E 00
1.60E-02	3.50E 00
1.80E-02	4.20E 00
2.30E-02	4.90E 00
2.80E-02	5.30E 00
3.80E-02	6.30E 00
5.80E-02	5.60E 00
7.80E-02	5.80E 00
9.60E-02	6.10E 00
9.60E-02	6.70E 00
1.48E-01	7.20E 00
2.00E-01	7.20E 00
3.00E-01	7.10E 00

8-5-9
ROBINSON ET AL. (162)

ENERGY (KEV)	CROSS SECTION (SQ. CM)
8.00E 01	3.75E-19
1.00E 02	2.60E-19
1.50E 02	1.30E-19
2.00E 02	8.00E-20
2.50E 02	5.50E-20
3.00E 02	4.00E-20

8-5-10
ROBINSON ET AL. (162)

ENERGY (KEV)	CROSS SECTION (SQ. CM)
6.00E 01	7.50E-19
7.00E 01	5.55E-19
8.00E 01	5.05E-19
1.00E 02	4.25E-19
1.20E 02	3.20E-19
1.50E 02	2.45E-19
1.70E 02	2.28E-19
2.00E 02	1.50E-19
2.50E 02	1.10E-19
3.00E 02	8.40E-20

8-5-11A
CARRE ET AL. (133)

ENERGY (KEV)	CROSS SECTION (SQ. CM)
3.00E 01	1.37E-18
4.00E 01	1.08E-18
5.00E 01	8.80E-19
6.00E 01	7.30E-19
7.00E 01	6.40E-19
8.00E 01	5.80E-19
9.00E 01	5.30E-19
1.00E 02	4.70E-19
1.20E 02	4.20E-19

8-5-11B
CARRE ET AL. (133)

ENERGY (KEV)	CROSS SECTION (SQ. CM)
3.00E 01	5.42E-18
4.00E 01	4.83E-18
5.00E 01	4.15E-18

ENERGY (KEV)	CROSS SECTION (SQ. CM)
6.00E 01	3.90E-18
7.00E 01	3.80E-18
8.00E 01	3.49E-18
9.00E 01	3.10E-18
1.00E 02	2.86E-18
1.20E 02	2.38E-18
1.50E 02	1.80E-18
2.00E 02	1.50E-18
3.00E 02	1.06E-18
4.00E 02	7.70E-19
5.00E 02	6.40E-19

8-5-12A
CARRE ET AL. (133)

ENERGY (KEV)	CROSS SECTION (SQ. CM)
3.00E 01	3.73E-18
4.00E 01	2.56E-18
6.00E 01	2.01E-18
7.00E 01	1.63E-18
8.00E 01	1.47E-18
9.00E 01	1.28E-18
1.00E 02	1.07E-18
1.20E 02	8.80E-19

8-5-12B
CARRE ET AL. (133)

ENERGY (KEV)	CROSS SECTION (SQ. CM)
3.00E 01	8.80E-18
4.00E 01	7.60E-18
5.00E 01	6.44E-18
6.00E 01	5.93E-18
7.00E 01	5.28E-18
8.00E 01	5.20E-18
9.00E 01	4.45E-18
1.00E 02	4.10E-18
1.20E 02	3.50E-18
1.50E 02	2.72E-18
2.00E 02	2.25E-18
3.00E 02	1.55E-18
4.00E 02	1.40E-18
5.00E 02	1.17E-18
6.00E 02	9.30E-19

8-5-13A
CARRE ET AL. (133)

ENERGY (KEV)	CROSS SECTION (SQ. CM)
3.00E 01	1.19E-17
5.00E 01	9.90E-18
7.00E 01	5.90E-18
9.00E 01	5.40E-18
1.20E 02	4.20E-18

8-5-13B
CARRE ET AL. (133)

ENERGY (KEV)	CROSS SECTICN (SQ. CM)
3.00E 01	2.80E-18
4.00E 01	2.36E-18
5.00E 01	1.71E-18
6.00E 01	1.48E-18
7.00E 01	1.42E-18
8.00E 01	1.26E-18
9.00E 01	1.04E-18
1.00E 02	9.20E-19
1.20E 02	7.20E-19
1.50E 02	5.70E-19
2.00E 02	4.10E-19
3.00E 02	2.80E-19
4.00E 02	1.70E-19
5.00E 02	1.40E-19

8-5-13C
CARRE ET AL. (133)

ENERGY (KEV)	CROSS SECTICN (SQ. CM)
3.00E 01	9.60E-19
4.00E 01	7.30E-19
5.00E 01	6.40E-19
6.00E 01	5.50E-19
7.00E 01	4.80E-19
8.00E 01	4.00E-19
9.00E 01	3.60E-19
1.00E 02	3.50E-19
1.20E 02	2.60E-19

8-5-14A
CARRE ET AL. (133)

ENERGY (KEV)	CROSS SECTICN (SQ. CM)
3.00E 01	2.90E-18
4.00E 01	2.48E-18
5.00E 01	2.09E-18
6.00E 01	1.57E-18
7.00E 01	1.50E-18
8.00E 01	1.25E-18
9.00E 01	1.02E-18
1.00E 02	8.90E-19
1.20E 02	7.80E-19
1.50E 02	6.40E-19
2.00E 02	5.20E-19
3.00E 02	3.40E-19
4.00E 02	2.60E-19
5.00E 02	2.00E-19
6.00E 02	1.60E-19

8-5-14B
CARRE ET AL. (133)

ENERGY (KEV)	CROSS SECTICN (SQ. CM)
3.00E 01	1.16E-18
4.00E 01	8.30E-19
5.00E 01	7.00E-19
6.00E 01	5.80E-19
7.00E 01	4.60E-19
8.00E 01	4.30E-19
9.00E 01	3.70E-19
1.00E 02	3.30E-19
1.20E 02	2.80E-19

CHAPTER 9

FORMATION OF EXCITED ATOMIC
HYDROGEN PROJECTILES

This chapter considers mechanisms by which fast hydrogen atoms are formed when hydrogenic projectiles traverse a target. The experiments detect only excited hydrogen atoms resulting from collisional rearrangement of the projectile; generally there is no information on the target particles's postcollision state of excitation nor ionization.† Experiments on the production of fast excited hydrogen atoms have included such diverse processes as charge transfer neutralization of protons, stripping of H^-, dissociation of H_2^+ and H_3^+, as well as direct excitation of hydrogen atoms. Although there is little fundamental similarity between these mechanisms, the experimental procedures used for their study are almost identical, differing only in the source of the projectiles. In practice, most research groups active in this area have used a single apparatus to study two or more of the relevant mechanisms.

Most studies have been carried out with targets of H_2, N_2, O_2, or the rare gases; data for other molecular targets and metallic vapors are very limited. There is a rather specialized group of experiments designed to study formation of excited atoms by charge transfer of protons on an atomic hydrogen target; these experiments require special techniques for the production of the unstable target medium; they are considered separately in Chapter 5.

It is the objective of this monograph to discuss only experiments for which the data can, in principle, be related to single collisions between atomic systems. It is recognized, however, that there are important practical applications for information concerning the excited state content of beams that have traversed these targets. There are a number of reliable experiments that provide quantitative measurements under multiple collision conditions; the sources of such information are listed, without critical assessment, in Appendix II.

The discussion will concentrate on the problems that are common to all studies of excited hydrogen atom formation (Section 9-1) and then will continue with a review of the experimental systems utilized for research on this subject (Section 9-2). The available data are discussed according to the nature of the collision mechanism—charge transfer, excitation of neutrals, stripping of H^-, collisional dissociation of H_2^+ and H_3^+ (Sections 9-3 to 9-7, respectively). Data tables are presented at the end of each appropriate section.

† An exception to this generality is a single study of the simultaneous formation of an excited target molecule and excited projectile induced by the impact of H^+ on N_2. This experiment—the one example of how coincidence techniques may be applied to excitation studies—is discussed in Section 9-3-G.

9-1 GENERAL CONSIDERATIONS

A measurement of emission from an excited particle moving with a high velocity is complicated by the finite lifetime of the excited state. In general, an excited projectile or dissociation fragment will move an appreciable distance before emitting, causing a spatial variation in intensity. In Section 2-2-B it is suggested that emission from the projectile may be measured either as it penetrates the target gas or, alternatively, after the beam has completely traversed the target region and emerges into an evacuated flight tube. In Section 2-2-B the simple relationships between emission intensity and excitation cross section are formulated.

A simple measurement of emission at some arbitrary penetration through a target gas does not give useful cross section information. The data will be dependent on the point of observation in the apparatus. Two techniques for valid measurement are available. If the observation point is located within the gas target itself, then at a sufficient distance from the point where the beam enters this region the emission will become invariant with penetration. Under these circumstances a measurement of emissions gives a valid emission cross section, related to level excitation cross sections through Eq. 2-22. The second technique is to observe the variation of emission intensity with distance either as the beam traverses the target or after it exits from the target and proceeds through an evacuated flight tube; the excitation cross section is then obtained by fitting the emission intensity to either Eq. 2-21 or 2-23. The decay length (product of velocity and excited state lifetime) for the excited states may range from some tens of centimeters to one meter, depending on the circumstances of the experiment. It is difficult to provide sufficiently long target chambers so that the emission is invariant with further penetration. Therefore, the first technique being rarely appropriate, most accurate experiments employ the second approach.

Measurements of cross sections by detection of spontaneous emission are directed to spectral lines in either the Lyman series or the Balmer series. The parent level of Lyman transition is unambiguous; the line originates from excited levels with angular momentum equal to unity. Unfortunately, these lines all lie in the vacuum ultraviolet; therefore, there is some experimental complexity in working with this series. The Balmer series can be detected more simply since it lies in the visible and near ultraviolet regions. The Balmer line from the level with a principal quantum number n includes contributions from the $ns \rightarrow 2p$, $np \rightarrow 2s$, and $nd \rightarrow 2p$ transitions that cannot be resolved spectroscopically. It is therefore impossible to obtain a cross section for a specific transition from the measurement of the emission of a Balmer line.

The emission of a line in the Lyman series may be related directly to a level excitation cross section. Branching ratios for the decay of a particular np excited state may be obtained from accurate theoretical estimates of transition probabilities, such as those by Weiss et al.[31] Cascade contributions to Lyman emission cannot readily be assessed. Cascade from higher states can be studied only through Balmer emission; as explained previously, there is no direct method by which Balmer emission may be separated into contributions from the ns, np, and nd states. In practice all the experiments evaluate cross sections on the assumption that cascade may be ignored; in a few cases, the discussion will include order-of-magnitude estimates of the error that this assumption introduces.

Recently a technique has been pioneered by Hughes and co-workers[120] that utilizes the lifetimes of the excited states to allow separation of the contributions

to a Balmer line. A Balmer line includes the $ns \rightarrow 2p$, $np \rightarrow 2s$, and $nd \rightarrow 2p$ emissions. The ns, np, and nd states will have decidedly different lifetimes. Thus, the emission observed in the target region will be given by a sum of three Eqs. such as 2-21, and emission in an evacuated flight tube will be given by a sum of three Eqs. such as 2-23. Accurate measurement of intensity, as a function of distance, and suitable fitting of the theoretical expressions will, in principle, allow the separation of cross sections for the formation of the ns, np, and nd states. This procedure makes no allowance for cascade that must be assessed by some subsidiary measurement.

There have been some experiments in which the emission in a Lyman series transition has been measured at a single distance of penetration through a target. Excitation cross sections have been deduced by taking a theoretical value for the excited state lifetime and inserting it into Eq. 2-21, which relates emission penetration through the target to cross section for the formation of the state. This procedure is satisfactory since the parent level for the transition is an np state with a definite lifetime; however, it does neglect cascade that must be assessed by some subsidiary technique.

In some experiments measurements have been made of the Balmer line intensity at a single penetration through a target; this emission cannot be related to a level-excitation cross section using the technique that was described above for a Lyman-series line. The parent levels of a Balmer line are the three ns, np, and nd states; it is impossible to arrive at an "effective" lifetime for the parent level of the transition without knowledge of the relative population of these three levels. Measurements of Balmer emission intensity made at a single distance of penetration into the target cannot be related directly to the cross section for the formation of a specific excited state; data of this type are excluded from this monograph.

The metastable $2s$ state of hydrogen has been subject to considerable attention. It may be detected quite conveniently by application of an electric field that causes Stark mixing of the $2s$ and $2p$ states. As a result, the excited atom will decay with the emission of a Lyman-alpha photon ($2p \rightarrow 1s$). The difference between Lyman-alpha emission with and without quench field will be related to the density of $2s$ excited states. This technique is discussed more fully in Section 2-2-C. It should be recalled that the effect of the field is to shorten the lifetime of the metastable state. The minimum lifetime is twice that of the $2p$ level. It is still necessary to take account of the variation of emission with penetration through the quenching region. There are two distinct experimental configurations in which the field-quenching approach is applied. The field may be applied either inside the collision region itself; or alternatively the beam may be allowed to traverse the target and emerge into an evacuated region where the field is applied. The former technique requires the important assumption that the mechanism of the collision process is unaffected by the presence of an ambient electric field; there is at present no experimental test of whether or not charge transfer, dissociation, and excitation processes are affected when the excited system is in a field. The second approach requires an experimental test to ensure that no appreciable fraction of the flux of excited atoms is intercepted by the exit aperture from the target region; an advantage of this second method is that there can be no possibility of interference by emission from the target gas.

Many of the experiments that measure emission of the Lyman and Balmer series claim to deduce level-excitation cross sections from the data, but omit any consideration of possible cascade population of the excited states. Cascade might be appreciable, particularly into the $2s$ and $2p$ states, and cannot be arbitrarily neglected.

Throughout this chapter, the spontaneous emission of Lyman alpha ($2p \to 1s$) and field-induced emission of Lyman-alpha ($2s \to 2p \to 1s$) will be regarded as line-emission cross sections. If one chooses to ignore cascade, then these will be equal to the cross sections for exciting the $2p$ and $2s$ levels, respectively. Often there is no basis on which cascade can be estimated, and the error involved in neglecting cascade is not known. Measurements of emission from higher states, such as $n = 3$ and $n = 4$, are also quoted in many papers as being level-excitation cross sections. Often there is some basis for assuming that cascade into these states is small; such cases will be discussed as they occur. It should be noted that cascade will vary with penetration of the beam through a target in a manner that depends on both the lifetime of the cascade level and the lifetime of the parent level of the transition under observation. For example, consider observation of emission from an np state at a penetration through the target that is equal to five times the decay length of the np state. The emission resulting from direct excitation of the np state will be within 2% of its maximum value. The emission due to the cascade from the higher ns levels, which have lifetimes of more than an order of magnitude greater than that of the np level, will have reached only 10 to 20% of its maximum intensity at that point. Thus, the geometrical construction of the apparatus, particularly the distance of the detector from the point of entry of the beam into the target or exit of the beam from the target, has a very direct bearing on the contribution of cascade to the observations. It is to be expected that every experimental investigation will include a detailed assessment of how cascade might influence the measurements. In practice, no such systematic studies are presented in any of the papers reviewed here.

It is vitally important to ensure that the excited state of the hydrogen atom does not change before spontaneous emission occurs. Small stray electric fields may cause Stark mixing of the various angular momentum states of a given principal quantum number. Complete mixing of $np_{3/2}$ and $nd_{3/2}$ states will occur for fields of $\frac{1}{2}$ and 2-V/cm parallel to the axis of quantization for $n = 4$ and $n = 3$, respectively. The critical fields for mixing $ns_{1/2}$ and $np_{1/2}$ are 12 and 58-V/cm, respectively, for $n = 4$ and $n = 3$. The field for complete mixing of the $2s_{1/2}$ and $2p_{1/2}$ states is in excess of 1-kV/cm; however, a field as small as 1-V/cm will produce a lifetime of 10^{-3} sec which is greatly different from the field-free value of 0.14 sec. States of higher principal quantum number will be mixed by smaller electric fields. In order to measure the cross section for the formation of different n states by detection of emission, it is clearly necessary to prevent any appreciable mixing before decay takes place. Stray electric fields of sufficient magnitude to cause mixing can be produced accidentally. Sources of such spurious fields might include the space charge of the ion beam, and stray-charge build-up on insulators or on insulating films. The motion of the atom through stray magnetic fields may also produce an appreciable motional electric field. It follows: an experiment must include tests to ensure that there is only a negligible amount of state mixing by stray electric and magnetic fields.

A further mechanism by which the excited atoms might be lost before emission takes place is collisional destruction of the excited state. For example, stripping of the excited electron into the continuum has a large cross section. The existence of such a loss mechanism would be detectable through nonlinear variation of emission intensity with target pressure. It has already been established that linearity is a basic criterion for the validity of the operational definition of cross section [see Section 2-1-B(10)].

Spontaneous emission from ns states will be inherently unpolarized; all other transitions may exhibit polarization. In general, the precautions for assessing the influence of polarization on the measurement of cross section (see Section 3-4-D) have not been fully carried out. If polarization is present, then account must be taken of anisotropy of emission and also the possible effects of instrumental polarization. Recent experiments have shown that emission resulting from the field-induced decay of the metastable $2s$ state is polarized and that this polarization changes with electric field.[239, 240, 20] Again, account must be taken of anisotropy of emission and also of the possible effects of instrumental polarization. Polarization of field-induced Lyman-alpha emission is dependent only on the electric field; therefore, if an experiment employs the same electric field for all particle impact energies then the errors inherent in the neglect of polarization will influence only the magnitude of the cross section and will not affect the behavior as a function of impact energy. In contrast, polarization of spontaneous emission depends on the population of magnetic quantum number sublevels; this population varies with projectile impact energy; therefore, an error inherent in the neglect of polarization influences both the absolute magnitudes of the data as well as the behavior as a function of impact energy.

Emission from a fast moving particle will exhibit a transverse Doppler shift and also a Doppler-broadening that is related to the acceptance angle of the optical system. As an example, the Balmer-alpha emission from a 30-keV H atom observed perpendicular to the direction of motion by an optical system with a 0.1-rad acceptance angle will be shifted by 2.1-Å and broadened by 5.3-Å. Apart from ensuring that a monochromator is adjusted to the correct wavelength to accommodate the transverse shift, there is also a necessity to consider the influence of the Doppler broadening on the calibration of detection sensitivity. These problems have been discussed in full in Section 3-4-F, and the required procedures are documented. In practice most of the experiments reviewed in this chapter have ignored entirely problems of Doppler shift and broadening. It is not possible to determine exactly what the resulting errors might be; inevitably, neglect of shift and broadening would cause the measured cross sections to decrease below the true value by an amount that will increase with projectile energy.

Many experiments have been carried out on the formation of hydrogen atoms in high states of excitation using detection procedures based on field ionization. It is noted (Section 3-3-B) that the technique is designed only to measure total population of a state of a given principal quantum number; under most conditions, signals produced by the n level cannot be readily resolved from $n + 1$ and $n - 1$ level signals. The inherent resolution of such measurements is substantially poorer than resolution for optical emission measurements. Therefore, data obtained by the field-ionization technique are not included in this monograph; the available experiments are listed in Chapter 4 and discussed briefly in the relevant sections of the present chapter, but the data are not reproduced. This rule is consistent with the consideration of optical emission experiments in which data taken with poor optical resolution are excluded from detailed consideration.

For a well defined collision experiment, it is necessary to ensure that the excitation state of a projectile beam is known. This is most important when it is the projectile beam that is being excited. Protons, of course, have no electronic structure, and there have been no reports of identification of excited states of the H^- ion.[306] Any experiments employing H, H_2^+, and H_3^+ projectiles should include an assessment of

the excited state content of the projectile beam and its possible influence on the measured cross section. In general, this precaution is neglected.

Targets used in these experiments have included both light and heavy permanent gases and a few cases of metallic vapors. Many of the data for rare gas and heavy molecular targets reviewed here have involved the measurement of target pressure using a McLeod gauge. It has previously been shown (see Section 3-2-D) that the use of this gauge may involve considerable errors unless attention is applied to the problems of thermal transpiration and mercury pumping onto the cold trap. The measurement of the density of metallic vapor targets is generally deduced from the temperature of the target region and the known vapor pressure of the material.

9-2   EXPERIMENTAL ARRANGEMENTS

1.   The Work of Geballe et al.   This group has produced a total of eight reports on studies of the formation of the $2s$ and $2p$ states of H as beams of H^+, D^+, H_2^+, D_2^+, and H_3^+ traverse gaseous targets. The first major report provides a description of the basic apparatus used for all the studies.[99] This early work on formation of the $2p$ state by charge transfer and dissociation has been extended by further work on charge transfer in rare gases[117] and dissociation of H_2^+ on various targets.[194] The formation of the metastable $2s$ state has been studied for processes of charge transfer[16] and dissociation.[195] Finally, this group has begun studies on the polarization of the spontaneously emitted Lyman-alpha radiation induced by charge transfer and dissociation processes.[28] There have been three additional papers that are superseded by the reports listed above. (References 305, 169, and 350 have been superseded by references 99, 117, and 28, respectively.)

This group has also studied emission from a molecular hydrogen target and simultaneous emission from projectile and target.[99, 305, 149] These results are reviewed in Section 8-2.

Ions were produced in an oscillating electron source accelerated to energies between 0.2 and 25-keV, mass analyzed, and directed into a gaseous target. In studies with the molecular ion beams of H_2^+ and H_3^+, no attempt was made to assess the possible influence on the cross section measurement of the excited state content of the projectile beam. Ion beam current was monitored on a Faraday cup in the conventional manner. The half-width of the ion energy distribution was 100-V.[194] Any structure occurring in intervals smaller than 100-eV will not be resolved. Target pressures were measured with an ion gauge. No details are published on the basis for the relative calibration of this gauge for different gases. It is not possible to assess whether or not the various problems of pressure measurement, identified in Section 3-2-D, contribute errors to this work.

The detection system was based on an oxygen-filtered helium-iodine Geiger counter similar to that described by Brackmann et al.[79] The counter responded at wavelengths between the cutoff in the transmission of the LiF window (1080-Å) and the photoionization threshold of iodine (1317-Å). The oxygen filter has a complicated absorption spectrum with a number of high transmission windows, one of which coincides with the Lyman-alpha wavelength. This group always took precautions

to determine whether or not other emission from target or projectile were being transmitted by one of the other oxygen transmission bands in the range 1080 to 1317-Å. It was calculated that the oxygen filter should absorb about 50% of the Lyman-alpha emission. Therefore, if no emissions were occurring at other wavelengths in the 1080 to 1317-Å interval, the evacuation of the filter should cause a 100% rise in signal. The presence of other lines in that interval was indicated when the rise in signal was greater than this. In any case where appreciable emission from systems other than the Lyman-alpha line was indicated, then the data taken with the filter were designated as cross sections for the emission of "countable uv", to indicate that the origin of the emission was indeterminate. In most situations such data will be due primarily to the Lyman-alpha line although there is uncertainty in the extent of interference from other systems. Interference was definitely observed for measurements using targets of N_2 and Kr. In order to avoid confusion, all measurements involving N_2 and Kr targets have been excluded from this data compilation.

All measurements of cross sections by this group were relative, and absolute values were assigned by a normalization procedure. Fite and Brackmann[98] have made measurements on the emission of Lyman-alpha induced by electrons on H and also the "countable uv" induced by electrons on H_2. The $e + H$ data were normalized to the theoretical predictions of the Born approximation, and a cross section was estimated for the "countable uv" from the $e + H_2$ reaction. In the first experiments by Dunn, Geballe, and co-workers,[99] an electron gun was provided in the apparatus so that the reaction $e + H_2$ could be observed with exactly the same optical conditions as those used for the collisions between heavy particles. The energy dependence of the "countable uv" emission measured by Fite and Brackmann was reproduced exactly. All data were normalized to this previously determined cross section. Subsequent studies by this group were in turn normalized to the work of Dunn. The experiments of Fite and Brackmann and those of Dunn et al. exhibit important differences that might cause errors in the normalization procedures; these possible sources of error were both pointed out by Young et al.[9] First, although both experiments employed similar detectors, the filter used by Dunn et al. was thicker than that used by Fite and Brackmann; a correction for this difference was evaluated on the assumption that the absorption coefficient for "countable uv" was the same as that for emission at the Lyman-alpha wavelength. This assumption was challenged by Young et al. who noted that the "countable uv" includes an appreciable amount of molecular H_2 emission and, moreover, that the absorption coefficient of H_2 emission is different from that for the Lyman-alpha line; as a consequence, the correction for filter thickness evaluated by Dunn et al. is incorrect. A second difference is that Dunn et al. used a weak magnetic field (70 Gs) to collimate the electrons during the normalization procedure. It was suggested by Young et al. that the magnetic field might sensitize the $H(2s)$ atoms to quenching by stray electric fields and thereby cause a spurious enhancement of the Lyman-alpha emission. It is not possible to determine the manner in which the differences between experiments might influence the validity of the normalization used by Dunn et al. The accuracy of the data by Fite and Brackmann, which form the basis for this normalization, is estimated at $\pm 30\%$.

In many of these experiments beams of D^+ and D_2^+ were employed in order to extend the velocity range of the observations. The wavelength of the Lyman-alpha

from deuterium is about 0.3-Å lower than that of hydrogen. The transmission of the filter at this wavelength was estimated to be 25% lower than the transmission at the Lyman-alpha wavelength from the lighter isotope; no attempt was made to correct for this effect. The magnitudes of cross sections measured for H emission are nominally correct since the calibration procedure is for this case; however, the cross sections for the D emission will be about 25% low due to the neglect of the additional absorption. Most cross sections for D^+ and D_2^+ impact appear to be about 25% lower than the corresponding data for H^+ and H_2^+. A difference of this magnitude indicates that the cross sections are in fact the same at equal impact velocities.

Measurements of $2p$ state emission were made using the counter to observe the collision region. For measurements of the $2s$ state a quench field was arranged to alternate spatially, transverse to the projectile beam in the collision region.[16] In this way, the metastables could be quenched without appreciable net deflection of the projectile ions. Care was taken to eliminate spurious emission induced by impact of ions and electrons on surfaces within the detector's field of view; this problem was particularly severe when charged particles were accelerated by the quench field. A study of emission intensity as a function of quench field confirmed that the correct saturation conditions were obtained.

Emission from the excited fast particles will exhibit Doppler shift and broadening. To study the influence of these effects on cross section measurements a comparison was made of the signals observed when using the full aperture of the detection system and those signals observed when the detector was baffled to allow detection only of light emitted at 90° to the beam. The difference between signals was found to be negligible at 5-keV but about 20% at 30-keV.[99] It was also measured to be 10% for collisional dissociation of H_2^+ at 25-keV.[194] No correction was ever applied for this effect.

The influence of cascade was neglected in all of these studies. The authors contend that since cascading transitions have considerably longer lifetimes than the $2p$ state and the emission due to cascade builds up over a greater target penetration distance, then, by making observations close to the entry into the target, the measurements discriminate against contributions from the long-lived cascading transitions. A direct test of the influence of cascade was made in the study of the collisional dissociation of H_2^+ and D_2^+ to produce the excited $2p$ state;[194] it was shown that tripling the distance from the point of entry of the beam into the collision chamber produced changes in the data that were less than the statistical accuracy of the measurements. It appears that the geometrical construction of this experiment provides quite effective discrimination against cascade from long-lived states. In general, cascade contributions to the data measured by this group are quite small. All observations on spontaneous emission were made at a sufficient penetration into the gas cell to cause emission resulting from direct excitation of the $2s$ and $2p$ state to be invariant with further increase of penetration through the target.

Recently, direct studies have been made of the polarization of the spontaneous emission resulting from charge transfer as protons traverse rare gases.[28, 350] Measurements of intensity were made at angles of 90° and 54° to the ion beam path, and polarization was computed from Eq. 2-27. The same type of helium-iodine Geiger counter was used, as in the earlier work. Since the radiation to be detected exhibits Doppler shift, the oxygen filter was removed so that the detector responded to all wavelengths in the range of 1050 to 1300-Å. Preliminary tests showed that none of the

target gases used for these experiments emitted a substantial amount of radiation in the wavelength range of the detector. The detectors were baffled so that the angular acceptance was small and the intensities corresponded accurately to emission at angles of 90° and 54°. Corrections were made for the different beam path lengths observed in the two positions. All measurements were at a sufficient penetration into the target gas so that the variation of emission with distance could be disregarded. Depolarization of emission by stray magnetic fields in the apparatus was shown to be small. Polarization fractions were shown to be independent of target pressure. Since the operational definition of polarization fraction (Eq. 2-26) is in terms of ratios of intensities, there was no need to determine detection efficiency. The reliability of the data will be governed primarily by the statistical accuracy of the measurement; the opportunities for systematic error are small. Polarization is related to the relative populations of the three magnetic substates of the $2p$ level. Denoting the cross section for population of the $m = 1$ and $m = -1$ states as $\sigma(1)$ and that for the $m = 0$ level as $\sigma(0)$, it is possible to show[25] that the polarization fraction P is given by

$$P = \frac{\sigma(0) - \sigma(1)}{a\sigma(0) + b\sigma(1)} \tag{9-1}$$

where a and b are constants.

The total cross section for emission of Lyman-alpha is given by

$$\sigma_{\text{total}}(2p) = \sigma(0) + 2\sigma(1). \tag{9-2}$$

From the measurement of P and the previous data on σ_{total}, it is possible to derive values of $\sigma(0)$ and $\sigma(1)$. Following Percival and Seaton,[25] the values of a and b for H atoms are 2.375 and 3.794, respectively, and those for D atoms are slightly different at 2.341 and 3.681, respectively.[28] Gaily, Jaecks, and Geballe[28] used these relationships to derive $\sigma(0)$ and $\sigma(1)$.

A study of emission resulting from dissociation of molecular ions gave zero polarization.[28] This implies that the population of the $m = 0$ and $m = \pm 1$ levels are equal.

With the exception of the publication by Gaily et al.,[28] all cross section measurements cited here were evaluated on the assumption that polarization was negligible. In the case of dissociation of the molecular ions, this appears to be correct; however, for the formation of atomic hydrogen by charge transfer this is not true, and the original data are in error by a small amount due to the neglect of the anisotropy of the emission. Gaily et al. have provided corrections for anisotropy based on their measurements of polarization fraction. The corrected data are displayed here in the data tables and graphs alongside the data as originally published. In all cases the correction was less than 7%. This is compared with the random error in the measurements which was generally about 5%. The emission from the field-induced quenching of the $2s$ excited state was assumed to be unpolarized and therefore isotropic. Recently, it has been shown that this assumption is incorrect; the emission is polarized by an amount of -33% or more depending on electric field strength.[239, 240, 20] The data for formation of the $2s$ state derived by Geballe and co-workers, therefore, contain a systematic error due to the neglect of anisotropy. It is not possible to evaluate exactly the magnitude of the error in these experiments because the quench field was not uniform over the region viewed by the detectors; a very rough estimate is that the cross sections may be too high by some 20%.

The original measurements of emission cross sections by Dunn et al. were estimated by the authors to be accurate to within $\pm 40\%$, with a 20% reproducibility.[99] Later work, which was normalized to that by Dunn et al., had an estimated absolute accuracy of $\pm 50\%$, with reproducibilities of $\pm 5\%$.[117, 194, 16, 195] The polarization measurements had a statistical accuracy of about $\pm 15\%$ (i.e., $P = X\% \pm 15\%$). No estimate was given of the possible accuracy of the measurements of the cross sections for forming the different magnetic sublevel states of H $(2p)$, but it is likely to be poor since this quantity is derived from both polarization and cross section measurements.

2. The Work of Ankudinov, Andreev, and Bobashev. The work of this group has contributed a total of ten papers that deal with the formation of excited atomic hydrogen by excitation of atoms,[191] the stripping of negative ions,[192] the charge transfer of protons,[148, 168, 75, 351, 122, 123, 158, 157] and the dissociation of molecular hydrogen ions.[168, 158, 157] Projectile energies are in the range of 5 to 40-keV.

Certain features of the apparatus are common to all the experiments by this group. The projectile ions (H^-, H^+, H_2^+, H_3^+) were formed in an arc discharge source within a magnetic field, extracted by the required acceleration potential, and then mass analyzed by a large $180°$ magnet. With the exception of the early experiments of Bobashev et al.,[158, 157] which are of doubtful accuracy, the ions were deflected through a $10°$ angle before entering the collision chamber to ensure that no neutrals were present. In none of the experiments was attention directed at the possible influence of vibrationally excited H_2^+ and H_3^+ ions on the measured cross sections. Neutral H atoms were prepared by passing an H^+ beam through a neon neutralizing cell and removing any residual ions with an electric field. Such a field also serves to remove metastable H atoms and to reduce the proportion of long-lived ns and np excited states by Stark mixing with the short-lived np levels. Systems for monitoring ion beam current were adequate. Neutral beam current was monitored by using a thermal detector that had been calibrated by a proton beam of known current.

Throughout all the experiments on permanent gas targets, the target pressure was monitored with a cold-trapped McLeod gauge without attention to errors caused by thermal transpiration and mercury pumping effects (see Section 3-2-D). For the experiments on a mercury target the collision chamber was maintained at a temperature of $30°C$; a side arm connected it to a source of mercury at a temperature of $22°C$.[168] The pressure of the target was taken as 1.2×10^{-3} torr, the vapor pressure of mercury at $22°C$. It should be noted that this value is not in accordance with the value adopted in standard tables: 1.426×10^{-3} torr.[298] It appears that all experiments included adequate attention to ensure the independence of the cross section measurement from target pressure.

It is convenient to divide the discussion of the optical part of the apparatus into three sections. First, the early work by Bobashev et al. on emission in the Balmer series; second the measurements of the $2s$, $2p$, and $3p$ state which utilize detection in the vacuum ultraviolet; third, experiments on the formation of the $3s$, $3p$, and $3d$ states utilizing a simultaneous measurement of Balmer-alpha and Lyman-beta emission.

(a) Balmer Series Emission Measurements by Bobashev et al. This series of measurements was devoted to the emission of the Balmer series line induced by the impact of H^+, H_2^+, and H_3^+ on targets of helium, neon, and mercury.[75, 158, 157] The light emission measurements were made with a prism monochromator fitted with a photomultiplier detector. The detection sensitivity was calibrated with a standard

lamp. These experiments, the first by this group, were all carried out with the collision chamber placed in the magnetic field of the mass analyzer.

Measurements were made at a single distance of penetration into the target region and a correction evaluated for the spatial variation of projectile emission associated with the lifetimes of the excited states. This correction was first made by assuming that only the np state contributed significantly to the Balmer line;[158] this was later revised using a statistical distribution of population among the various levels contributing to the Balmer line.[158]

Both methods of correcting for lifetime are based on inadequate justifications and are likely to be wrong. Moreover, the experiments were all carried out in a strong magnetic field that would have caused appreciable mixing of the contributing states and destroyed their separate identity. The data are inconsistent with later measurements by the same group on the separate formation of the $3s$, $3p$, and $3d$ states with a neon target.[123] It is concluded that these data are misleading and therefore they are omitted from further consideration.

(b) LYMAN SERIES EMISSION MEASUREMENTS BY ANDREEV, ANKUDINOV, AND BOBASHEV. This series of measurements comprises the bulk of the data on formation of excited atomic hydrogen produced by this group. It includes work with incident protons[148, 75, 351, 122] H^-[192] and neutral atoms.[191] The apparatus is fully described in only one reference.[75]

All emission measurements were made at a sufficient penetration into the target so that the emission resulting from the direct excitation of the projectile could be taken as invariant with penetration. Cascade was ignored completely, and no tests were made to ensure that its contribution to the total emission was either negligible or invariant with penetration.

The apparatus was used to detect $2p$ and $3p$ states by spontaneous emission and $2s$ states by field-induced emission. Tests were made to ensure that the field used for the quenching technique was sufficient to cause complete mixing of the $2s$ and $2p$ states. The collision chamber for these experiments was located outside the magnetic field of the mass spectrometer, so the only electric field was that applied to quench the $2s$ state.

An evacuable Seya-Namioka grating monochromator with photomultiplier detection was employed for all relative measurements of the line intensity as a function of impact energy. It was normally placed [192, 75, 351, 122] at 90° to the beam direction, but an angle of 35° was also used[148] when target and projectile emissions were to be separated utilizing the Doppler wavelength shift. No account was taken of how emission polarization might influence the cross section determinations.

The cross section for the formation of $H(2p)$ by charge transfer as protons traverse neon was determined absolutely in a separate experiment and was utilized as a normalization point for all the other measurements of Lyman-alpha emission.[75] The detector in this standardization experiment was based on a NO-filled ionization chamber fitted with a LiF window, and attached directly to the collision chamber. Spectral analysis, using the monochromator, showed that in the wavelength range of the detector only the Lyman-alpha line was produced in the $H^+ + Ne$ collision combination. The photoionization coefficient of NO was taken as 0.83. Watanabe[97] has recently revised this value to 0.81, a change that will render the cross section measurements too low by an amount of 2.5%. The transmission of the LiF plate in front of the counter was determined before and after cross section measurement in a

subsidiary experiment. The estimated accuracy of the absolute cross section measurement was $\pm 20\%$. It should be noted that the whole calibration procedure is absolute except for the information on the NO photoionization coefficient. The work of Samson[94] indicates that the thermopile used by Watanabe in the measurement of this photoionization coefficient has the same response at 1215-Å as it has at visible wavelengths, where it was calibrated; therefore, there is little doubt that this value for photoionization coefficient is substantially correct. The cross section for the emission of Lyman-alpha radiation induced by the impact of protons on neon at 16-keV was taken as 8.9×10^{-18} cm^2. Other measurements of the spontaneous and field-induced emission of Lyman alpha are normalized to this value. It should be noted that this calibration method provides a measure of the Lyman-alpha emission induced by H$^+$ on Ne at 16-keV emitted at an angle of 90° to the beam path. It is not corrected for anisotropy of emission, nor is any account taken of cascade into the $2p$ state.

The detection efficiency at 1026-Å, the wavelength of the $3p \to 1s$ line, was determined indirectly from a previous[211] absolute measurement of the separate cross sections for the emission of the $3p \to 2s$ transition induced by the collisional dissociation of H$_2^+$ by He [see Section 8-2-B(2)]; this cross section multiplied by the relevant ratio of transition probabilities gives the cross section for the emission of the $3p \to 1s$ line (2.1×10^{-18} cm^2 at 16-keV impact energy). All measurements on Lyman-beta emission are normalized to that value. The accuracy of this procedure is quoted as $\pm 30\%$. When comparing absolute magnitudes of the $2p \to 1s$ and $3p \to 1s$ lines, it should be remembered that one relies on a direct calibration in the vacuum uv and that the other is obtained indirectly from measurements on a visible emission. The systematic errors are likely to be different.

(c) MEASUREMENT OF THE $3s$, $3p$, $3d$ CROSS SECTIONS BY ANDREEV, ANKUDINOV, AND BOBASHEV. Andreev et al.[123] have attempted a measurement of the separate cross section for forming the $3s$, $3p$, and $3d$ excited states by reactions such as H$^+$ + $X = $ H* + X^+. It is based on measurements of both Lyman-beta and Balmer-alpha emissions. Quantitative measurement of the Lyman-beta line will give the cross section for formation of the $3p$ state (provided cascade is disregarded). Measurement of the Balmer-alpha emission gives a sum of the $3s$, $3p$, and $3d$ states weighted according to the appropriate branching ratios and according to the correction factors related to the lifetime of the excited state and the penetration into the target at which the observations were made. These two measurements of Balmer-alpha and Lyman-beta are not sufficient to provide the separate values of the $3s$ and $3d$ state cross sections. As a final step Andreev et al. measured the Lyman-beta and Balmer-alpha emissions when a strong electric field (600-V/cm) was applied to the observation region. An effective lifetime correction was made to provide a measure of the emission that would be observed if the observations were made at sufficient penetration so that emission was independent of the point of observation. The sum of the Lyman-beta and Balmer-alpha cross sections, measured with an applied field, must be equal to the total cross section for forming the $n = 3$ level; that is to say, the sum of the $3s$, $3p$, and $3d$ state formation cross sections. From these four measurements of Balmer-alpha and Lyman-beta with and without field, it is possible to separate out the three cross sections for forming the $3s$, $3p$, and $3d$ states.

There are two important assumptions made in this experimental procedure: first, that the sum of the cross sections for forming the three states is unaffected by the application of the field; secondly, that there is no polarization and therefore the

emission is isotropic. No information is given on how the "effective lifetime" for the Balmer and Lyman emissions in the applied electric field were determined. Cascade contributions, which may be different in the zero-field and high-field situations, were completely ignored. The validity of this whole procedure depends greatly on the accuracy of these assumptions.

The Balmer-alpha line was measured with a prism monochromator fitted with photomultiplier detection. The sensitivity was determined by calibration against a standard lamp. The Lyman-beta line was measured with the evacuable Seya-Namioka monochromator described in Section 9-2(2b); the absolute calibration of that monochromator at 1026-Å is related to a measurement of Balmer-alpha emission with the prism monochromator.

It would appear that the nature of this procedure offers many opportunities for errors. The available report is inadequate to determine whether the doubts raised in the preceding paragraphs have any foundation. The data obtained from this experiment should be treated with some caution until more detailed reports are provided or confirmatory evidence from another source is published. The data are quoted as level-excitation cross sections in the original report; this designation is continued here. In fact, cascade contributions are completely neglected.

Throughout all three groups of studies no specific reference is made to precautions to eliminate errors due to the Doppler shift and broadening of spectral lines, errors from this source are likely to be very small.

The discussion of possible errors is inadequate throughout these experiments. The limits of absolute accuracy are variously estimated as $\pm 20\%$ to $\pm 30\%$. This is conservative since the errors in measurement of pressure have not been included. Reproducibility of measurements ranged from 5 to 10%, depending on the particular case under consideration.

3. The Work of Dose et al. The work of Dose et al. involves studies of the formation of the $2s$ and $2p$ states in charge transfer as protons traverse helium,[159] and cross sections for emission of Lyman-alpha induced by hydrogen atoms incident on helium, neon and argon.[188] In the latter case, measurements were also made of the polarization of the emission.[189] All work was in the energy range of 3 to 55-keV, and in some cases deuterium atoms were used as projectiles to extend the velocity range of the experiments.

For studies of the impact of H atoms, the proton beam was partially neutralized in a gas cell. A field of 10-kV/cm extending over 10 cm was used to remove the residual flux of ions and also to quench the metastable state. All experiments with H projectiles were repeated with oxygen, nitrogen, and air as neutralizer gases, and no change to the measured cross sections was observed. This was taken to indicate that any excited atoms in the beam were exercising a neglible effect on the measurement of cross sections. Targets of helium, neon, and argon were used in the various experiments. An ion gauge was used to monitor target pressure, but no information was given on the basis of its calibration.

The target cell was located at the center of a rather large chamber that was evacuated to a background pressure of 2×10^{-6} torr. Observations of Lyman-alpha emission were made through a LiF window at one end of the cell. The detectors were helium-iodine Geiger counters that were located outside the target region but inside the large evacuated chamber. Such a detector would have responded to all emissions in the range 1080 to 1260-Å.

The first set of experiments was concerned with the formation of the $2s$ and $2p$ states of H by charge transfer as protons traversed a target of helium.[159] The photon detector was arranged at an angle of $90°$ to the beam path. A series of nine closely spaced annular apertures were placed coaxial to the beam in the collision region. Fields to quench the metastable $2s$ state were produced by applying voltage to alternate plates, the remainder being maintained at ground potential. This arrangement provided a longitudinal field which did not affect the focussing of the beam. Precautions were taken to correct for excitation by charge transfer in the residual gas of the surrounding vacuum chamber. The finite lifetime of the excited states will cause a variation of emission with penetration of the projectile through the target; the decay length of the $2p$ state at the highest impact velocities was 0.59 cm, and for the $2s$ state in a high electric field the decay length was twice this figure. This spatial variation was neglected; consequently, the cross sections may be too low. The error for the $2s$ state will be greater than for the $2p$ states, and in both cases the error will increase with projectile impact energy.

Measurements of Lyman-alpha emission without the quench field were taken as proportional to the $2p$ state population, and the difference between the emission with and without quench field was taken as proportional to the $2s$ state population. The cascade contribution to the population of the excited states was ignored. The $2p$ data at 70-keV impact energy were then normalized to a Born approximation prediction of the cross section for forming the $2p$ state. The absolute magnitudes of the data assigned by this normalization technique cannot be taken as reliable because the measured energy variation of cross section was different from that given by the theory.

That the experimental procedures do not take into account the possible anisotropy of the emission which is associated with polarization is noteworthy. Gaily et al.[28] have shown that the spontaneous emission of Lyman-alpha does exhibit polarization at low energies. Also, it has been shown that the field-induced emission will exhibit polarization.[239, 240, 20] Neglect of the polarization of field-induced emission from the $2s$ state will cause the published cross sections to be systematically too high by an amount that is independent of impact energy; an error of about 10 to 20% may have been involved. Polarization associated with the spontaneous decay of the $2p$ state will vary with projectile energy; error caused by neglect of this polarization cannot be estimated.

The second set of experiments, which were concerned with the excitation of the $2p$ state by impact of H atoms on helium, neon, and argon targets, were carried out in the same basic manner as the charge transfer experiments. No correction was made for anisotropy associated with polarization, although a separate measurement of polarization fraction was reported in a companion paper.[189] Cascade was again neglected. No attempt was made to measure the flux of incident H projectiles directly. It was assumed that the angular distribution of scattered H atoms could be predicted by the Rutherford cross section with some allowance for the classical screening effect of the electron on the projectile. The scattered H atom flux at $2.2°$ was monitored and related to the incident flux of H atoms. An apparent difference between excitation cross sections for impact of H and D at the same velocities was ascribed to inadequate allowance for electron screening in the calculation; the D impact data were arbitrarily normalized to the H impact measurements. There is no evidence that the Rutherford formulation provides a satisfactory prediction of neutral particle scattering at low energies; indeed the discrepancy between data for H and for D impact suggests

strongly that the formulation is in error. An experimental error in the determination of neutral flux is that the scattered neutral particles were detected with a multiplier that had been calibrated using a proton beam; there is no evidence that the efficiency for detection of H is the same as the efficiency for detection of protons. It must be concluded that the procedure for determination of projectile flux is inadequate. Absolute magnitudes of the emission cross sections were established by a further normalization procedure; the formation of the $2p$ state of H by charge transfer as protons traverse helium was measured in the same apparatus and normalized to measurements by Pretzer et al.[169] and by Andreev et al.[75] The absolute magnitudes of the data and the functional dependence of cross section with energy established by these complex normalization procedures may be rather inaccurate, particularly in view of the inadequate justification of the validity of the method for determining projectile beam flux.

The third aspect of the work of Dose et al. is a measurement of Lyman-alpha polarization induced in collisions of H with targets of helium, neon, and argon.[189] Three detectors were arranged to be at angles of 56°, 90°, and 126° to the beam direction. They could be rotated so that all three viewed the beam at 90° and also so that the 56° and 126° counters could be interchanged. Comparison of signals with all three at the 90° angle to the beam allowed comparison of detection efficiency; measurements at different angles allowed the deduction of the polarization fraction from the known variation of emission with angle (Eq. 2-27). Corrections were made for the variation of light emission with the penetration distance of the beam through the target gas. This caused definite differences in the response of the detectors at 56 and 126°. Corrections were also made for the difference in the transmission of the LiF window to light polarized parallel to the beam direction and to light polarized perpendicular to the beam direction. Measurements, being repeated with H and D projectiles, gave essentially the same results at equal impact velocities.

Throughout this work, the reproducibility of cross section measurements was given as from 5 to 10%. In view of the comments made above on the normalization procedures, it would be dangerous to hazard a guess as to the possible accuracy limits on the absolute magnitudes of cross sections. In the event that the use of theoretical Rutherford scattering cross sections is inaccurate, the energy variation of cross section for excitation of incident neutrals will be wrong. The error limits in the polarization measurement were such as to give the uncertainty in polarization P as $\pm 10\%$ with the random error of about the same magnitude (i.e., $P = X\% \pm 10\%$). All cross section data are quoted as being for level excitation, although there is no evidence that the cascade can be disregarded.

4. The Work of Hughes et al. This group has provided a sizable series of cross section measurements for the formation of excited atomic hydrogen by charge transfer as protons traverse the target, and also by collisional dissociation of molecular ions. Projectile energies range from 5 to 130-keV.

Ions were formed in an rf discharge source, accelerated, and mass analyzed. The possible influence of excited states of molecular ions on the measurement of dissociation cross sections was ignored. Monitoring of the ion beam was adequate. Target pressures were monitored with a cold-trapped McLeod gauge; no attention was paid to the effect of cold-trap pumping effects; therefore, the measurements of pressures of heavy gases might be in serious error (see Section 3-2-D).

Two optical arrangements have been used by this group. In the early work a

grating monochromator was employed to measure emission at an arbitrary penetration of the projectiles through the target region. No correction was made for the lifetime of the excited hydrogen atoms, so the data are in error, sometimes by as much as 300%. This work is of no value and is discarded. All the relevant references are included in the summary assessment table of Chapter 4 and mentioned briefly in the text, but the data are not reproduced.[147,174,175,197,198] It should be noted that measurements of target emissions in these same papers are not affected by this type of error. The apparatus used for this incorrect work was reviewed in Section 6-1-A(2) and will not be considered further here. Following this series of incorrect measurements Hughes and co-workers developed the technique of utilizing the lifetime of excited states to allow measurement of separate cross sections of the 3s, 3p, 3d, and 4s states. Measurements have involved charge transfer processes[120,353,180] and also dissociation.[196] The data from these experiments are fully reproduced here, and the apparatus is reviewed below.

The projectile beam traversed a low pressure target region and emerged into an evacuated flight tube. The intensity of emission from excited projectiles decays exponentially as a function of distance from the target exit. The relationship between intensity of emission and cross section for the formation of the excited state was derived as Eq. 2-23. In the case of the Balmer-alpha line the emission comes from three states of different lifetimes; therefore, the emission varies with distance like a sum of three exponential decays. Hughes and co-workers carefully measured the emission as a function of distance and analyzed the curve into three exponential decays. The intensity of each component can be related through Eq. 2-23 to the collision excitation cross section.

The target gas cells utilized for this work ranged in length from 2 to 12 cm. A pressure differential of 60:1 was maintained across the exit aperture from the gas cell. The flight tube was 35 cm long. Light intensity was monitored by using a photomultiplier fitted with an interference filter centered on the Balmer-alpha line. This multiplier could be tracked along the flight tube to measure emission as a function of distance. Light from the beam was focused onto the photomultiplier by a lens system. A diaphragm at the focal point determined the length of beam path that was observed. In the earlier experiments, the observed length was 1 cm, while for the most recent work—in which the only successful measurements of 3p and 3d cross sections were made—the observed length was restricted to 1/2 cm.[120, 353, 196] Corrections were made for the light emission caused by the interaction of the projectile beam with the background gas in the flight tube. Appreciable (10%) neutralization of the projectile beam took place under some circumstances; the mean proton current in the gas cell was taken as the effective current for data analysis.

The first experiment[120] was an attempt to measure the 3s, 3p, and 3d state cross sections in charge transfer. Reproducibility of the 3p and 3d cross sections were very poor, and it was suggested that this was due to Stark mixing caused by stray fields related to charge buildup on surfaces. It was concluded that reliable measurements of the 3p and 3d state cross sections could not be made with the available apparatus. At an observation point chosen where the 3p and 3d state contributions had decayed to negligible proportions, relative measurements of the 3s intensity were made. These were converted to cross sections for excitation of the 3s level. Data for the 3p and 3d states that had smaller cross sections than the 3s level were not published. In the second paper concerned with the Balmer-beta line originating from the $n = 4$

level, it was reported that the emission appeared to be characteristic of the 4s state lifetime with a shorter-lived (26 nsec) component of lower intensity superimposed.[353] It was suggested that the shorter-lived component was the result of Stark mixing of the 4p and 4d states. On the basis of experimental tests it was concluded that 4s state emission was not appreciably affected by fields of less than 1-V/cm and, therefore, that measured cross sections for this state are reliable. Emission from the 4p state could not be separated from the 4d state emission; consequently, no data were presented for these cases. The third experiment dealt with the formation of the 3s state in dissociation of H_2^+ and H_3^+ using the same experimental techniques as those used in the former two studies.[196] The final experiments involved a very detailed study of Balmer emission induced in the charge transfer of protons in nitrogen at energies from 10 to 35-keV only.[180] Considerable effort was made to eliminate stray fields. It proved possible to make reliable measurements of the 3p and 3d cross sections as well as of the much larger 3s emission cross section. Experimental tests were consistent with negligible Stark mixing of the states.

All experiments were repeated with gas cells of different lengths in order to test whether or not appreciable angular scattering of excited projectiles was taking place. For all charge transfer experiments the different cells gave essentially the same results, suggesting that this problem could be ignored. In the case of dissociation the shorter cells gave slightly lower cross sections, the discrepancy being 5 to 15% at the lowest energies depending on the collision combination under study. For this work, the data from the shorter cells were weighted heavily in order to arrive at the final values for publication. It is admitted that the dissociation cross sections might be underestimated by a "small amount" at lower energies due to loss of some particles.

Emission from ns states, being unpolarized, is expected to be isotropic (see Section 3-4-D). Emission from the 3p and 3d states will, in general, be polarized. During the single series of measurements involving 3p and 3d states, the polarization fraction was measured although no correction was made for the resulting anisotropy in the emission. In the case of the 3p state the correction would appear to be approximately zero. The construction of the detector system is such that instrumental polarization effects are not expected.

The absolute values of the cross sections were assigned by normalizing to previous measurements of Balmer-alpha emission from the collisional dissociation of an H_2 target into excited H atoms by the impact of 100-keV protons (see Section 8-2-C).[147] Hughes and co-workers determined the magnitudes of the Balmer-beta emission measurements by a relative calibration. The Balmer lines emitted by a hydrogen discharge tube were compared with the intensity of a standard lamp at these same wavelengths. The discharge tube was then used to calibrate the relative detection efficiency of the photomultiplier filter combination, and thus relate the Balmer-beta measurements to the alpha measurements that had been normalized to previous data. This same apparatus was also used for a check on the calibration of the Balmer-alpha data, and it was found that it agreed within 15% with the data obtained through the normalization procedure. There is a basic fault in these calibration procedures; the emissions used for normalization are line emission at definite wavelengths; the emission observed from the excitation of projectiles will be both Doppler-shifted and also broadened. Assuming an entrance aperture to the optics to be 0.1 rad, the broadening of the alpha line at 100-keV would be 9.6-Å, and the shift, 0.72-Å. The precautions for dealing with this situation (see Section 3-4-F) were omitted; therefore,

it is expected that the cross section measurements will be low by an amount that increases with projectile velocity; the magnitude of this error cannot be estimated.

The experiments by Hughes and co-workers did not account for cascade contributions to excited state population. The importance of cascade will vary greatly between the 4s, 3s, 3p, and 3d states. Cascade population of the 3s and 4s levels from higher np states is likely to be small because branching ratios favor direct decay of the np state to the 2s or 1s levels; for example, only 4.2% of atoms in the 4p state will decay to the 3s level. In contrast, the cascade population of the 3p level may be large because some 41% of the 4s excited atoms decay to the 3p level. Apart from the inherent error in cross section that is introduced by neglect of cascade, there is also some danger that the analysis of the experiment is in error; a decay sequence such as 4s → 3p → 2s is characteristic of the lifetimes of both the 4s and 3p states and does not produce a simple exponentially decaying contribution to the spatial variation of light intensity. Hughes et al. ignore this problem completely in all their work. Data are presented as "cross sections" for forming the 4s, 3s, 3p, and 3d states. This designation is not correct if cascade is appreciable. It is not possible to estimate the extent of any errors that might have been incurred through this cause.

In this work the reproducibility of data varied from 5 to 15% depending on circumstances. Absolute accuracy was not quoted, but inevitably the error limits must be no less than the ±50% that is assigned to the data to which these measurements were normalized.

5. The Work of Sellin and Granoff. Sellin and Granoff have measured the cross sections for the formation of metastable H atoms as a beam of protons traversed targets of H_2, K, Cs, and Rb. Two papers have been published. In the first, Sellin[17] directed protons through a target of H_2 and determined the cross sections for formation of the metastable atom by charge transfer and also by excitation of neutral H atoms; the second paper, by Sellin and Granoff,[139] considers the formation of metastables by charge transfer as protons traversed targets of potassium, cesium, and rubidium.

(a) CHARGE TRANSFER ON H_2 BY SELLIN. An H^+ beam was directed through a target cell containing hydrogen and emerged into an evacuated region where both the metastable component and projectile flux could be monitored. The metastables were detected by field quenching with the subsequent detection of the emitted Lyman-alpha photon. The detector was sensitive to all wavelengths in the region 1050 to 1300-Å. Particular attention was paid to ensuring that the lifetime of the metastable atom in the electric field was in accordance with theoretical predictions. Corrections were made for light emitted outside the field of view of the detector.

Detailed measurements were made of the metastable flux and neutral beam current for target thicknesses up to 300 torr-cm. The pressure dependent variation of H^+, H*(2s) and H(1s) fluxes was assumed to be governed by the six cross sections for transitions between these three components; by careful fitting of the predicted pressure variations to the measurements, it was possible to deduce cross sections for the formation of H(2s) by charge transfer (10/0*1) and also by excitation of neutrals (00/0*0). The charge transfer cross section comes in fact from the low pressure behavior where single collision conditions obtain; the excitation of the neutral atom is related to multiple collision conditions. This whole technique is very indirect and may be much in error. It is not clear that all other excited states of H and the existence of H^- may be legitimately neglected.

The absolute magnitude of the data was established by inserting an electron gun into the apparatus and observing the signal emitted in the $e + H_2$ collisional excitation. It was assumed that the signal was the same as the "countable ultraviolet" observed by Fite and Brackmann[98] and that Sellin's data could be normalized to that previous work. This is an incorrect procedure. Fite and Brackmann utilized a wavelength selective detector that would respond to little of the collisionally induced molecular hydrogen emission; in contrast, the detector used by Sellin had a broad band response and undoubtedly detected a large amount of molecular emission. Unpublished data suggests that the error in the work of Sellin may be as much as 65%.[196]

The work by Sellin is not very reliable. The normalization may exhibit substantial errors and the technique for studying excitation of neutrals is indirect. Moreover, there will be other errors due to neglect of cascade and of the polarization of field quenched emission.[239, 240, 20] The accuracy must be very much poorer than the $\pm 30\%$ limits set by Sellin and the results of this experiment should all be treated with caution.

(b) CHARGE TRANSFER ON K, Cs, AND Rb BY SELLIN AND GRANOFF. The discussion of the experimental method for this measurement is rather brief and does not allow a detailed assessment.[139] It appears that the only major differences from the techniques used previously by Sellin for measurements with the H_2 target lie in the nature of the target, the measurement of its pressure, and the assignment of an absolute magnitude to the data.

The target cell was an oven in which the metal was heated to provide a vapor. The density of the target was deduced from the vapor pressure of the metal at the operating temperature of the target.

The normalization technique by which absolute magnitudes were assigned to the data may involve considerable inaccuracy; the $2s$ state excitation data for targets of Cs and K at the energy of 10-keV were set to be one quarter of the total electron capture cross section for those same targets taken from the work of Il'in et al.[118] It is therefore assumed that the dominant contribution to the total cross section comes from the four nearly resonant $n = 2$ states and that the populations of the states will be equal. None of these assumptions can be substantiated. The data for rubidium were normalized to that for cesium using a scaling factor given by Rapp and Francis.[354]

The relative accuracy of these measurements should be reliable. The assumptions made in the assignment of absolute magnitudes are not properly substantiated, therefore, these absolute magnitudes are not reliable. The authors make no estimate of the accuracy of their data.

6. The Work of Dahlberg et al. Dahlberg and co-workers studied the spontaneous emission of Lyman-alpha induced by the impact of H atoms and protons on a nitrogen target. A conference paper[312] on this work was completely superseded by a more complete report.[182] Work has also been done on emission as H and H^+ traverse a target of H_2; however, in this case the measurement includes contributions resulting from the excitation of the target, and these data have been previously reviewed in Section 8-2.

Ions were provided from a Cockcroft-Walton accelerator with an rf ion source operating at energies from 10 to 130-keV. Neutrals were formed by passing the proton beam through a cell containing a molecular gas (unspecified), and removing any residual ions in the emergent beam by electrostatic deflection. This electric field

also served to quench any metastable atoms that might be present. No attempt was made to determine the influence on the cross sections of any other excited states remaining in the projectile beam on entry to the observation region. Monitoring of the ion beam was carried out satisfactorily using a Faraday cup. Neutrals were monitored with a secondary emission device that was calibrated by using a proton beam, under the assumption that the ratio of secondary emission coefficients for proton and H atom impact were the same as those given by Stier et al.[41] Since Stier et al. used a different type of surface and a different angle of incidence, the use of their data to correct for response to H impact might not be accurate.

Target pressures were measured with a capacitance manometer pressure gauge for which the response is independent of the nature of the target gas.

The light emission was analyzed using a Czerny-Turner vacuum monochromator with an open ended multiplier as a detector. The monochromator was placed at an angle of 90° to the beam path. The point of observation was at a sufficient distance from the entrance to the collision chamber so that the emission induced by direct collisional excitation of the projectile would be invariant with further penetration. No attempt was made to estimate the cascade population. States contributing to the cascade population of the $2p$ level have longer lifetimes, and the cascade contribution to emission requires a large penetration to build up to its maximum value. Since the measurements of Lyman-alpha emission by Dahlberg et al. are made close to the entrance to the target region, it is likely that the cascade contribution to the data is rather small. No account was taken of the possible effects of polarization, which would have caused anisotropy of emission and influenced the calibration of detection sensitivity (see Section 3-4-D).

Absolute magnitudes were assigned to the data by normalizing to previous measurements by Van Zyl et al. of Lyman-alpha emission induced by the $H^+ + N_2$ collision; the normalization was made to preliminary unpublished data quoted by Murray et al.[177] that appears to be different from the data that Van Zyl eventually published.[149] Van Zyl et al. did not claim that the data represented only the emission of the Lyman-alpha line; indeed there is ample evidence in the report that an unknown proportion of the signal is from excitation of the nitrogen target. Dahlberg claims to utilize sufficient resolution to separate Lyman-alpha alone. It is clearly unjustifiable to normalize the work of Dahlberg to that of Van Zyl.

Dahlberg et al. completely ignored the Doppler shift of emission from the projectile. At 100-keV impact energy, an optical system with an entrance aperture of 0.1 rad will accept radiation having a Doppler width of 9.6-Å. The entrance and exit slits of the monochromator used by Dahlberg were set at 8-Å. The error involved due to the neglect of Doppler broadening (see Section 3-4-F), will cause the measured cross section to be too low by an amount that increases with impact energy.

In view of the remarks made above the accuracy of these measurements must be regarded as very poor. There are potential sources of error that will affect the magnitudes of the cross sections and also the energy dependence.

7. The Work of de Heer et al. This group has measured the emission of the Lyman-alpha and -beta lines of H induced as a result of charge transfer as protons traverse targets of He and Ne.[78] Experimental details are to be found in a preceding paper.[155]

The ions were produced in a duoplasmatron source, accelerated to energies of 5 to 35-keV, mass analyzed, and directed into the target region. Adequate attention

was paid to the monitoring of projectile beam flux. Target pressure was monitored with a cold-trapped McLeod gauge with no precautions to prevent errors due to mercury pumping and thermal transpiration (see Section 3-2-D).

Light emission was monitored using a grazing incidence vacuum monochromator placed at 90° to the ion beam path. An electron multiplier served as the detector. No additional optical components were used to focus light onto the monochromator entrance slit; therefore, the field of view was governed by the area of the grating. Spectral resolution was about 14-Å. The measurements were made at a sufficient penetration into the target so that variation of emission intensity along the beam path was small. A correction was necessary for this effect at the highest energies.

The calibration of detection sensitivity was described in a separate report.[90] The vacuum monochromator and a visible light grating monochromator were both arranged to be at 90° to the ion beam and to view the same section of its path through the chamber. A He^+ ion beam was directed through a neon target, and the visible and uv monochromator were arranged to detect the 5016-Å ($3\ ^1P \to 2\ ^1S$) and 537-Å ($3\ ^1P \to 1\ ^1S$) lines of the He I, respectively. The ratio of the two line intensities must be in the ratio of the probabilities for the $3\ ^1P \to 2\ ^1S$ and $3\ ^1P \to 1\ ^1S$ transitions, which may be obtained accurately by theoretical predictions. The visible light monochromator was calibrated absolutely with a standard tungsten strip filament lamp of known emissive power. By the use of known transition probabilities and the measured intensity of the 5016-Å line, the intensity of the 537-Å line may be calculated and the calibration of the vacuum uv monochromator established. A calibration of the monochromator at other wavelengths was obtained by an extrapolation procedure. It was assumed that the variation of response of the whole optical system with wavelength was due only to change in response of the electron multiplier, grating reflectivity, and instrumental astigmatism. The variation of multiplier response was found by comparison with a conventional photomultiplier fitted with sodium salicylate screen that was assumed to have a constant fluorescent yield with wavelength. Changes in grating reflectivity and instrumental astigmatism were evaluated by theoretical calculation. The procedure used to calibrate detection sensitivity can be criticized on a number of grounds. The effective optical aperture utilized during calibration of the visible light monochromator was different from that used for measurements of emission from the collision region. Calculations of the effect of instrumental astigmatism and changes in reflectivity of the grating should not be relied on without experimental tests. Finally, if the collisionally induced He I emission is polarized, then these results will be incorrect due to neglect of instrumental polarization effects in the two monochromators.

The experimental measurements were of the $2p \to 1s$ and $3p \to 1s$ emission cross sections. The data were published in the form of level excitation cross sections evaluated on the assumption that cascade population could be neglected. No experimental evidence was given to substantiate this assumption.

The authors of this work estimate that the systematic error in their measurements does not exceed $\pm 40\%$. In the opinion of the present writer this is a conservative estimate.

8. The Work of Cristofori et al. This group has made measurements on the formation of the metastable state of H as a result of proton impact on H_2, He, and Cs targets. The earliest publication by this group also gave some qualitative assessments of the rate at which metastable H atoms were produced, as protons traversed

various molecular targets at fairly high pressures.[355] No quantitative cross section information is available from that study, therefore it is not considered further here.

The first measurement of cross sections for the H_2 and He targets was in conference proceedings[356] and was later superseded by a more complete report.[145] The data originally published were assigned absolute values by normalization to theory. Subsequently, the data were made absolute using a novel detection system[357] and these new values (obtained by private communication) are used here.

The data for the Cs target are restricted to measurements at four energies only. The method is basically the same as that used for the He and H_2 targets and employed the new detection system that gave absolute cross sections directly. The data used here for that case are from the original publication.[138]

The ions were produced in an rf discharge, accelerated, passed through the gaseous target, and emerged into an evacuated region where the flux of metastables was determined by detection of field induced Lyman-alpha emission. It is most important to note that no mass analysis was used between the source and the experimental region; therefore, the incident beam may have been contaminated with H_2^+ and H_3^+ ions. Frigerio[356] states that a mass analysis system was sometimes used behind the collision chamber to check the relative proportions of these ions in the projectile beam, and that operating conditions were chosen to give a "negligible" flux of H_2^+ and H_3^+. It is to be noted that in an extensive review of available ion source configurations Blanc and Degeilh[358] do not list any source that produces an output having more than 92% proton content. The formation of H(2s) by dissociation of H_2^+ is two to six times more likely than the formation by neutralization of H^+;[16,195] it follows that a small percentage of molecular contaminants in the beam might cause a disproportionate error in the measured cross section. It is a reasonable conclusion that the measurements by Cristofori et al. may be in error due to contamination of the projectile beam.

Beam current was monitored in a conventional manner by using a Faraday cup. No details on how target density was measured are provided for either the study using He and H_2 targets or for that using cesium target. All measurements were carried out under single collision conditions.

The flux of metastable atoms was determined by applying an electric field to Stark mix the 2s and 2p states and then measuring the emitted Lyman-alpha radiation. The detector was an electron multiplier with a LiF window and a gaseous oxygen filter to discriminate against spurious radiation. The quenching field decreases the effective lifetime of the 2s state to twice that of the 2p state. The intensity of emission will vary as a function of the distance from the point at which the particle enters the field, and a velocity dependent correction is necessary to take into account any variation of detection sensitivity with this distance. There is no note of whether this correction was necessary nor of how it was implemented.

The original measurements for the He and H_2 targets were placed on an absolute basis by normalizing them to the Born approximation predictions of the cross section. It is not clear that the Born approximation would be valid at energies around 25 to 40-keV. Subsequently, a calibration procedure has been developed that provides direct absolute measurements.

The method for determining absolute efficiency is remarkably simple in concept and differs from the conventional techniques used by other research groups.[357] An electron beam was used to excite a target of H_2. The Lyman-alpha detector, which

was used for the cross section measurements, and a normal photomultiplier with an interference filter for the Balmer-alpha line were arranged to view the interaction region simultaneously. A field was provided in the observation region to quench the $2s$ levels so that they also radiate. The apparatus measures the rates at which Balmer-alpha and Lyman-alpha photons are emitted; also, it monitors the rate of coincidences between photons of the two wavelengths. Each Balmer photon is a cascade contribution to the Lyman-alpha line and should be in coincidence with a Lyman-alpha photon. The ratio of the coincidence rate to the Balmer-alpha emission rate may readily be shown equal to the product of the efficiency and the effective solid angle appropriate to the Lyman-alpha detector. This procedure has a very great advantage over other methods: nowhere is it necessary to use a standard of emission. The great disadvantage is that the resolving time of the coincidence circuit must be kept appreciably longer than the lifetime of the $2p$ state, and as a result the accidental coincidence rate is quite large. There will be an error in this calibration due to detection of molecular H_2 emission at wavelengths in the transmission bands of the Lyman-alpha and Balmer-alpha filters; the influence of this problem was neglected.

The authors estimate that the systematic error in these data does not exceed $\pm 50\%$. Within these broad limits, the absolute magnitudes assigned by normalization are in agreement with those obtained from the direct calibration procedures described above. The statistical accuracy of the cross section measurements is about $\pm 10\%$.

9. The Work of Thomas, Edwards, and Ford. Thomas et al.[129] have carried out preliminary measurements on the cross sections for the formation of the $3s$ and $3d$ state of H by charge transfer as protons traverse targets of He, Ar, and N_2 at the single energy of 168-keV.

Ions were produced in an rf discharge, accelerated, mass analyzed, and directed through a long gas cell. Observations were made of the Balmer-alpha emission as a function of penetration distance through the target. A photomultiplier with an interference filter was used as the detector. A lens system provided an image of the emission on the sensitive surface of the photomultiplier with a 1:1 magnification. A screen limited the observed length of the beam path to 6.0 mm. Target pressures were measured with a capacitance manometer that is insensitive to the nature of the target gas. Adequate attention was directed at accurate monitoring of ion beam flux.

The emission from a single excited state is related to the level excitation cross section by Eq. 2-21. For the Balmer-alpha line, which consists of three transitions, the observed emission is a sum of three such expressions. An equation was fitted to the penetration dependence of the emission from which the $3s$, $3p$, and $3d$ state cross sections were determined. The contribution of the $3p$ state was so small that data could not be obtained with any significant statistical accuracy. Absolute values for the cross sections were assigned by normalizing the measurements to a previous determination of the Balmer-alpha emission from the collisional dissociation of a H_2 target [see Section 8-2-C(3)].[150] The emission from the projectile exhibits considerable Doppler shift and broadening, while the target emission measurements to which these data are normalized represent emissions at a single unshifted wavelength. As a result, the detection efficiency for emission from the projectile is different from that for the target emission. A correction was evaluated for this difference.

Strictly speaking, the data obtained in this study are line emission cross sections; however, they were presented as level-excitation cross sections evaluated on the

assumption that the contribution of cascade could be neglected. Polarization-related anisotropy in the distribution of emission was ignored; because of this the $3d$ state cross section is in error, but the $3s$ state measurement is not influenced since the $3s \rightarrow 2p$ transition is inherently unpolarized.

The data originally published have been revised, and the numbers presented here are the new values.[262] The authors estimate the systematic errors do not exceed $\pm 50\%$. Of this, some $\pm 25\%$ is due to possible uncertainty in the cross sections to which the data are normalized. The reproducibility was $\pm 15\%$.

10. The Work of Ryding, Wittkower, and Gilbody. This group has made measurements on the formation of the fast metastable H atoms as protons traverse targets of H_2, and He.[142] The H^+ ions were directed through a target cell and monitored on a Faraday cup. The metastable atoms were monitored about 40 cm from the point of their exit from the target cell by detection of field quenched Lyman-alpha radiation. Precautions were taken to prevent quenching of metastables by fringe fields outside the observation region. It was assumed that the detector viewed a length of beam path equal to the width of the detector window and that the response of the detector was constant over this region. A correction was applied for the small fraction of the metastable flux that will traverse the detection region without radiating.

The experiment measured the variation of Lyman-alpha signal with projectile energy for a fixed target thickness; this gives directly the variation of the charge transfer cross section with projectile energy. No attempt was made to determine absolute values of target thickness or detector sensitivity. The relative cross section values were placed on an absolute basis by normalization to theoretical predictions. For the He target the data at energies in excess of 100-keV had the same energy dependence as the theoretical predictions of the $2s$ cross section by Mapleton;[359] the experimental data were normalized to the theory at 160-keV energy. For the H_2 target, the data were normalized at 100-keV to twice Mapleton's theoretical prediction of charge transfer on an atomic hydrogen target;[237] subsidiary experiments showed that the cross section for the molecule was approximately twice that of the atom.

There are few opportunities for systematic experimental errors because no measurements are required of target density, beam current, or detection sensitivity. The experimental method may be criticized on the grounds that the detector sensitivity might not have been independent of projectile velocity; the field-induced emission will exhibit a velocity-dependent distribution in the detector's field of view; any nonuniformity in detection efficiency will cause erroneous measurements.

The relative accuracy of the measurements was estimated to be within $\pm 25\%$. The absolute accuracy is entirely dependent on the validity of the theoretical predictions to which the data were normalized.

11. The Work of Berkner et al. Berkner et al.[170] have measured the emission in the Balmer transition resulting from the formation of excited H atoms as protons traverse targets of neon and magnesium. Ions were accelerated to energies from 5 to 70-keV, mass analyzed, passed through a target cell, and emission from the neutralized projectiles monitored in an evacuated region beyond the target.

The aim of the experiment was to determine the total population of the $n = 6$ state formed by charge transfer, from the measurement of Balmer-delta emission at a single distance from the exit from the target cell. The Balmer line includes the $6s \rightarrow 2p$, $6p \rightarrow 2s$, and $6d \rightarrow 2p$ transitions. In order to determine the populations of the $6s$, $6p$, and $6d$ states, it is necessary to correct for radiative loss of excited atoms

by spontaneous decay between the emergence from the cell and the point of observation. The assumption was made that at the emergence from the cell, the states are populated according to their statistical weight, and during the flight through the observation region there is no change to the population except through normal radiative decay.

Berkner et al.[170] present no evidence that a statistical distribution is an appropriate description of the excited state population. Indeed there are a number of experimental measurements of $3s$, $3p$, and $3d$ state formation that show the charge transfer process to preferentially populate the $3s$ state (see Section 9-3). Berkner et al. argue that whatever the individual cross sections might be, weak fields will Stark mix the various components of the $n = 6$ state and cause a statistical distribution to be appropriate in the observation region; however, no experimental evidence is provided to support this contention. Under the circumstances one must consider that these results could be considerably in error and their inclusion in this text might be misleading. Accordingly, they are omitted from detailed consideration here.

2. The Work of Bayfield. Bayfield has studied the formation of the metastable H atom by charge transfer as protons traverse targets of argon and molecular hydrogen.[130] Ions were produced in an rf source, accelerated to energies between 2 and 70-keV, mass analyzed, and finely collimated. The beam was directed through a target cell, and the determination of metastable content made as the beam emerges from the exit of the cell into a long evacuated flight tube. In the region of the flight tube weak fields were provided which could sweep away the residual proton flux without appreciably quenching the metastable content. The projectile flux was detected by a Faraday cup arrangement that could be used either to measure charged particles as a current or to measure neutral particles by secondary emission. The efficiency of the cup for neutral atoms was calibrated by measuring directly the secondary emission coefficient for protons using a beam of known flux and applying the ratio of secondary emission coefficients for neutrals and protons that was determined by Stier et al.[41] Since secondary emission coefficients will vary with surface condition and angle of impact, there is no justification for assuming that the results of Stier et al. should be applicable to the conditions used in Bayfield's experiment.

Target thickness was measured by an indirect method. Argon was introduced into the target cell; a proton beam was directed through it, and the resulting flux of fast neutral atoms determined with the secondary emission detector. A survey of previously published cross sections for charge transfer beam neutralization showed that most determinations were in very good agreement. A mean value of the cross section was taken, and from the measured neutralization of the proton beam the effective target thickness could be obtained. The effective thickness for a hydrogen target was determined by measuring the rates of gas flow into the target cell for an argon target of known thickness and a hydrogen target of unknown thickness; the ratio of gas flow rates was equal to the ratio of target thicknesses. A correction was made for the different pumping conductances for argon and hydrogen. The whole procedure for target thickness measurement is not very satisfactory. Over the range of energies used in Bayfield's experiment there are only two independent determinations of total charge transfer cross section for an argon target. Agreement between these two sets of measurements might be fortuitous. It was suggested that the final result of the measurement of target thickness was accurate to within $\pm 5\%$.

Particular care was taken to remove spurious fields that might quench the metastable atoms before they reached the detection region. A novel operating procedure eliminated the effect of a magnetic field that arose in the observation region due to the use of a magnetically focused photomultiplier. The efficiency of the detection system was determined by assessing the influence of each component part. The secondary emission coefficient for the photomultiplier cathode was taken as the mean of published values. The transmission of a grid in front of the cathode was measured. The spatial dependence of photomultiplier detection efficiency over the field of view was taken from published reports. Finally, the counts lost below the setting of the discriminator in the electronics was also allowed for. The overall detection efficiency was said to be known to an accuracy of $\pm 55\%$.

A study was made of the angular distribution of the scattered metastables with the objective of ensuring that no appreciable part of the total flux was intercepted by the defining apertures on the beam path. At energies below 5-keV, where the angular distribution was quite large, some corrections were necessary.

The detector was set to collect photons emitted at $90°$ to the ion beam path. It was also perpendicular to the quenching field. No correction was made for the anisotropy in the field-induced radiation that will occur when electric-quenching fields are employed.[239, 240, 20]

No assessment was made of the cascade contribution to these measurements. The detection region was at a considerable distance from the gas cell (258 cm); therefore, all cascading into the $2s$ state will have finished before detection. The measurement gives the total population of the $2s$ state by charge transfer and by cascade population.

It was estimated that the systematic errors in the experiment did not exceed $\pm 55\%$. Error bars were given to include all error sources that might influence the energy dependence of the cross section. These ranged from $\pm 25\%$ at 2-keV to $\pm 15\%$ at energies from 10 to 70-keV.

13. The Work of Teubner et al.   Teubner et al.[128] have presented a brief conference report on the measurement of polarization of Lyman-alpha photons emitted as a result of the impact of H^+ on an argon target.

Protons were fired through a static volume of the target gas. The polarization of the emission was studied using a Brewster-angle reflection analyzer. The analyzer was arranged to observe the radiation emitted perpendicular to the projectile beam path. A lithium fluoride plate, set at the Brewster-angle, reflected emission into an oxygen-filtered Geiger counter. The whole analyzer could be rotated around an axis perpendicular to the projectile beam, so that the components of emission polarized parallel and perpendicular to the direction of the beam could be separately recorded.

Since polarization is related to the ratios of light intensities of different polarization, there is no need to perform absolute measurements of beam flux, target density, nor photon emission. There is no record of tests to ensure that the measurements are independent of target density, beam flux, and stray fields in the experimental region.

The estimate of the possible error in these data is related to the intensities of the emission and hence to the energy of projectile impact. Typically, the uncertainty is $\pm 13\%$ in the percentage of polarization (i.e., $P = X\% \pm 13\%$). At low energies it is somewhat larger.

14. The Work of Donnally et al.   Donnally et al.[137] have studied the formation of the metastable state of hydrogen as protons traverse a target of cesium at impact energies of 0.6 to 3.0-keV.

The protons are produced in a Nier type ion source, accelerated, mass analyzed, and directed across a beam of cesium atoms. The cesium is produced by heating the metal in an oven. After traversing the target the projectiles enter an evacuated region where the metastable content is analyzed by the conventional technique of applying an electric field to Stark mix the 2s and 2p states and then detecting the resulting Lyman-alpha photon. The photon detector was an iodine-vapor-filled Geiger counter.

The cesium beam intensity was monitored on a surface ionization detector. No attention was directed to the influence of inhomogeneities in either the ion beam or the cesium beam. In the introductory discussions of crossed-beam experiments (Section 3-1-C) it is shown that these problems may be considered through the evaluation of a form factor for the density distribution in the region of interaction.

Absolute magnitudes were assigned to the data by normalization to previous measurements. Hydrogen was introduced into the detection region, and an electron gun was utilized to excite ultraviolet emission. This emission was taken to be the same as the "countable uv" emission that Fite and Brackmann[98] had measured in a previous experiment; this normalization is inaccurate. The detectors used by Fite and Brackmann were provided with an oxygen gas filter that removed much of the molecular emission in the region of the Lyman-alpha line. The detector used by Donnally[137] did not have a filter and therefore responded to both Lyman-alpha and also to all the relevant molecular emissions in the range of detector sensitivity. The signals from the two experiments do in fact refer to different quantities. The absolute magnitudes of the data by Donnally must be considered as inaccurate by an unknown amount.

All measurements were made at an angle of 90° to the beam direction, and no account was taken of the anisotropy associated with the polarization of field-induced emission.[239, 240, 20] No tests were mentioned to ensure that the loss of metastables between the crossed-beam region and the detector could be neglected. No attempt was made to assess the population of the 2s state by cascade from higher np states.

The authors estimate that the systematic error in the determination did not exceed $\pm 50\%$ and that the reproducibility was about $\pm 10\%$. In view of the criticisms discussed above, the systematic error is probably far higher than this figure.

9-3 CHARGE TRANSFER INTO THE EXCITED STATE OF THE PROJECTILE

The process of charge transfer into the excited state may be described by the reaction equation:

$$H^+ + X \rightarrow H^* + X^+. \tag{9-3}$$

In general, these experiments provide no information on the state of excitation or ionization of the postcollision products. There has been a single experiment, using a nitrogen target, where the postcollision states of excitation and ionization of both particles was defined in a coincidence experiment [see Section 9-3-G(4)].

In general, measurements are available for the formation of the lower-lying ns and np states with a few measurements on the 3d level. The most extensive coverage has been given to the case of a helium target.

The experiments described here have taken no account of cascade. The reader's attention is drawn to the remarks of Section 9-1 where the problem of cascade is discussed. In particular, it should be noted that the geometrical construction of all experimental systems provides some discrimination against the cascade contributions from long-lived state. When, in the following discussions, estimates of cascade are given they refer to the maximum possible contribution and do not take into account any attenuation due to geometrical construction; because experimental arrangements are poorly reported, it is impossible to make an accurate assessment of the actual cascade. In general, one finds that the cross sections for formation of a $n'p$ state is far less than that for an ns level (where $n' > n$); the branching ratios favor decay of $n'p$ to the $1s$ rather than the ns state. As a result the cascade population of ns states is very low, often less than the statistical accuracy of measurement, and to a good approximation may be neglected. Thus level excitation functions from ns states may be derived from measurements of $ns-2p$ emission; in particular, the many measurements of cross section for field induced emission of Lyman-alpha ($2s \rightarrow 2p \rightarrow 1s$) may be taken as equal to the $2s$ excitation cross section. Cascade into the np levels (from $n's$ and $n'd$ states, where $n' > n$) always appears to be large. Consequently, measurements of $np \rightarrow 2s$ and $np \rightarrow 1s$ emission cannot be accurately converted to level-excitation cross sections without a direct estimate of cascade; furthermore, without a measurement of cascade an emission cross section may be peculiar to the apparatus in which it was studied. For the few cases of measurements on $3d$ state formation there is no way by which cascade into the $3d$ level can be estimated.

In Sections 9-3-A through 9-3-J the data are discussed under the headings of different target systems. Finally, the data tables are collected in Section 9-3-K.

A. Helium Target

Data are available on the formation of the $2s$, $3s$, $4s$, $2p$, $3p$, and $3d$ states at energies from 0.25 to 200-keV. Also, the polarization of the $2p \rightarrow 1s$ emission has been measured, and from this are deduced the cross sections for formation of the different magnetic quantum number sublevels of the $2p$ state.

Certain papers are excluded from detailed consideration. Bobashev et al.[158, 157] have made measurements on the emission cross sections for the Balmer series lines. It was suggested previously [see Section 9-2(2)] that these data may be misleading due to the method by which the lifetime correction was determined; therefore they are excluded. Data on the formation of the high n states ($n \geq 10$) by Il'in et al.,[118, 57, 119] using field ionization for detection, will also be excluded from detailed consideration (see Section 3-3-A). An early measurement of emission of the Balmer-beta line by Hughes[154] is omitted because the data are not of high accuracy and, also, because a correction for the lifetime of the emitting state was omitted. Work by Dieterich[259] is not reviewed here since it contains no quantitative information on cross sections.

1. Cross Sections for the Formation of the ns States. (a) $2s$ STATE. The measurements on the field-induced emission of Lyman-alpha radiation are shown in Data Table 9-3-1. The data originally published by Cristofori, Colli, and co-workers[356, 145] were normalized to the Born approximation prediction at energies from 25 to 40-keV. More recently, the method of absolute measurement reviewed in Section 9-2-(8) has been applied to this experiment and the data renormalized. These new data are used

here.[146] Only the work of Andreev et al.[75] and the unpublished revisions of the data by Colli et al.[146] are independent absolute measurements. The normalizations by Dose[159] and Ryding et al.[142] to theoretical predictions of the charge transfer cross section are hazardous since the range of validity of this theory has not been properly established. The normalization by Jaecks et al.[16] is less subject to criticism since the $e + H$ cross section is quite accurately predicted by theory.

Three different sets of data are shown for the work by Jaecks and co-workers. The original publication included measurements for H^+ and D^+ impact.[16] Both of these are given, with the D^+ data shown in the tables as though they were H^+ of the same velocity (i.e., at one-half true D^+ energy). Recent measurements of polarization by that group have caused a small change in the $2p$ state cross section through a correction for anisotropy of emission.[28] The $2s$ data were referred to the $2p$ data and required a corresponding change. The correction, only a matter of 3%, is not available for all data points, but revised values are shown separately on the data tables wherever available.[360]

None of these data have been corrected for cascade population from higher np states. Andreev et al.[122] show that the cross section for the formation of the $3p$ state is only 10% of that for the $2s$ level. Since the branching ratio for decay will ensure that only 10% of the $3p$ states will decay to the $2s$ state (the remainder go to the $1s$ ground state), the problem of cascade is probably negligible. Therefore, the measurements on field induced emission of Lyman-alpha are equal to the cross section for excitation of the $2s$ state to a good approximation.

The various measurements on the formation of the $2s$ states are shown in Data Table 9-3-1. It is clear that there are significant discrepancies between the determinations in energy dependence as well as in magnitude. There appears to be excellent agreement between the measurements by Andreev et al. and those by Jaecks et al. This agreement is illusory since both experiments neglect the anisotropy of field-induced Lyman-alpha emission;[239, 240, 20] corrections for anisotropy would raise the data by Andreev et al. and would lower that by Jaecks et al., causing the results to show a disagreement of approximately 40%.

It is difficult to make a completely objective judgement as to which of these cross sections are likely to be most accurate. The work of Cristofori, Colli, and co-workers[356, 145, 146] might be eliminated on the grounds that the projectile beam was not mass analyzed and might be contaminated with molecular ions. The work of Dose[159] might be set aside on the grounds that the absolute magnitudes are established by normalization to poorly established theory and the energy dependence disagrees with two other measurements. One may then conclude that the measurements by Jaecks et al.[16] and by Andreev et al.[75] provide the most reliable determinations of the magnitude and energy dependence of the cross section; both of these experiments, however, are in error due to neglect of polarization. The work by Ryding et al.[142] might be used to extend the energy dependence of the cross section to higher energies although with less certainty due to the lack of confirmatory evidence.

(b) $3s$ AND $4s$ STATES. Measurements on the $3s$ state by Hughes et al.,[120] by Andreev et al.,[123] and also the single point by Thomas et al.[129] are shown in Data Table 9-3-2. There are no data for the formation of high np states from which cascade might be estimated. The measurements by Hughes et al. and by Andreev et al. differ in magnitude by a factor of about two. It is not possible to make an objective assessment as to which of these two sets of data is more accurate. The energy dependence

of both is essentially the same, except perhaps for a few points between 15 and 20-keV. It may be noted that in the assessment of previous work by Hughes et al. on excitation of helium (Section 6-2) and nitrogen (Section 8-3-C) their data were found to lie consistently higher than that from other groups by a factor of about two. The measurements on charge transfer rely on the same basic calibration procedure as those for target excitation and might be expected to exhibit the same errors in calibration.

Hughes et al.[353] provided a measurement for the 4s state, as shown in Data Table 9-3-3; the data were evaluated on the assumption that cascade was negligible.

2. Cross Sections for the Formation of the np States. (a) 2p STATE. There are a total of seven cross section determinations for the emission of the Lyman-alpha line shown in Data Table 9-3-4. Gaily et al.[28] have taken the measurements of cross section and the measurements of polarization [see Section 9-3-A(4)] to derive cross sections for the formation of the $m = 0$ and $m = 1$ magnetic sublevels of the 2p state. These results are shown in Data Table 9-3-5 and will also be discussed in detail below.

The group of Geballe and co-workers has produced a number of cross section measurements, all of which are shown in the data table. The early measurements by Pretzer et al.[117] for both proton and deuteron impact were not corrected for anisotropy of emission associated with polarization. Subsequently, Gaily[28] has determined polarization and applied a correction to some of the data published by Pretzer et al.[117] Here, both the corrected and uncorrected data are shown. The difference is only 3 to 5%, which is of the same order as the random error in the original measurements. Other publications by this group represent preliminary data and are not used here.[169, 350]

The data by Andreev et al.[75] are an absolute measurement. The data by Pretzer et al.[117] and the corrected values by Gaily et al.[28] are assigned an absolute magnitude by a normalization technique; this can be traced to a Born approximation prediction of the formation of H(2p) by the impact of electrons on atomic hydrogen. Although this is not absolute, the normalization is expected to be reliable. The data by Dose[159] are normalized to a Born approximation prediction of the charge transfer into the 2p state. There is insufficient evidence that this approximation is valid at such low energies; therefore, the basis for this procedure may be incorrect. The work by de Heer[79] is also an inherently absolute measurement.

One may estimate the cascade contribution by considering the population of the 3s and 3d states, both of which will decay only by transitions to the 2p level. Using the measurements of Andreev et al.,[123] the cascade population might be as high as 30 to 50%.

There are definite discrepancies between the measurements of cross section shown in Data Table 9-3-4. All the cross section determinations have a similar dependence on energy. There is excellent agreement between the data of Andreev[75] and that by Pretzer et al.[117] The data by de Heer et al.[78] and that by Dose[159] all lie considerably higher and exhibit a slightly different energy dependence from the data by Andreev et al. and that by Pretzer et al. The discrepancies cannot reasonably be attributed to differences in cascade contributions to the measurements.

It is not possible to make an objective assessment as to which set of measurements is likely to be most accurate; however, the agreement between the independent measurements by Pretzer et al.[117] and that of Andreev et al.[75] suggests that these are likely to represent the most accurate data that are available.

In Data Table 9-3-5 are displayed the cross sections for the population of the $m = 0$ and $m = 1$ magnetic sublevels of the $2p$ state, as deduced by Gaily[28] from his measurements of polarization [reviewed in Section 9-3-A(4)], and the data of Pretzer et al.[117] for the total cross section. It must be noted that the accuracy of the final result may not be very high since it depends on two separate experimental measurements, each of which has a limited statistical accuracy.

(b) $3p$ STATE. De Heer et al.[78] have measured the emission function of the $3p \to 1s$ Lyman-beta transition (1026-Å) and converted it to cross section for the excitation of the $3p$ level on the assumption that cascade may be neglected. Andreev et al.[122] have presented the measured cross section for the emission of the $3p \to 1s$ line; also shown is a value of the $3p$ state cross section derived from the emission data on the assumption that cascade might be neglected. These sets of data by de Heer et al. and also by Andreev et al. are shown in Data Table 9-3-6. Two additional conference publications by Andreev et al.[351, 123] would appear to contain the same data as the full report[122] and may therefore be disregarded.

Andreev et al.[122] have provided measurements on the variation of the Lyman-beta line intensity with applied electric field in the collision region. These data are not reproduced here because they provide no information on behavior of cross sections.

Neither of the determinations reviewed here provides an assessment of the cascade contribution to the measurements. It is possible to use the data of Hughes et al.[353] on the cross section for forming the $4s$ state (Data Table 9-3-3) to estimate that the cascade to the $3p$ level may be as high as 20% (at 10-keV) to 40% (at 40-keV).

The two determinations show approximately the same variation of cross section with impact energy. There is a difference of a factor of two in absolute magnitudes; this is very similar to the difference between cross sections for the $2p$ state determined by these two groups (Data Table 9-3-4). This probably indicates an error in absolute calibrations. There is no method by which an objective choice as to the more accurate measurement might be made; however, in the previous discussion of the formation of the $3p$ level it was suggested that the weight of evidence supported the measurements by Andreev et al. If indeed the discrepancy is due only to differences in calibration, then the same subjective choice must be made here.

3. Cross Sections for the Formation of the $3d$ State. Andreev et al.[123] have measured the cross section for the $3d \to 2p$ emission at energies from 14 to 30-keV. Thomas et al.[129] have measured the cross section at 150-keV impact energy. Neither measurement estimates the cascade contribution to this data.

4. Polarization of Emission. In Data Table 9-3-8 are shown the polarization fractions of the $2p \to 1s$ emission measured by Gaily et al.[28] for impact of protons and deuterons. The statistical accuracy of these data is estimated as no better than $\pm 15\%$ (i.e., $P = X\% \pm 15\%$).

B. Neon Target

Data are available on the formation of the $2s$, $3s$, $4s$, $2p$, $3p$, and $3d$ states. The polarization of the $2p \to 1s$ emission has been measured; from this has been deduced the cross section for the formation of the different magnetic quantum number sublevels of the $2p$ state.

Work by Bobashev et al.[158, 157] and by Berkner et al.[170] on Balmer emission is likely to be inaccurate [see Sections 9-2(2) and 9-2-(11)] and is omitted. Measurements

by Il'in[118, 119, 54] which use the field-ionization technique to measure the total population of states for which the principal quantum number is nine or greater are also excluded (see Section 3-3-A).

1. Cross Sections for the Formation of the *ns* States. (a) 2*s* STATE. Jaecks et al.[16] have provided an extensive series of measurements on the formation of this state by both H^+ and D^+ impact. The absolute magnitudes of these data were obtained by relating to the measurements of the 2*p* state. Recently, Gaily[28] has revised the 2*p* measurements by making a small correction for the anisotropy of emission associated with polarization; this requires a 4% correction to the 2*s* data, which is less than the random error in the original determinations. These revised values for a few energies are also shown here.[360]

It is possible to estimate from Andreev's[122] data (shown in Data Table 9-3-14) that cascade from the 3*p* state is the order of 1 to 2%. Even after allowance for cascade from higher states, the magnitude of this contribution is substantially less than the random error in the experimental measurements.

The cross section measurements are shown in Data Table 9-3-9. Although the two determinations are in apparent agreement, it should be recalled that both experiments neglect anisotropy of the field induced emission;[239, 240, 20] the required correction cannot be accurately estimated but will have the effect of raising the data by Jaecks et al., lowering the data by Andreev et al. and thereby introducing a discrepancy of perhaps 40% between the data from the two experiments. There is a small difference in the energy dependence of the data from these two experiments; this discrepancy is within the quoted reproducibilities of the data.

(b) 3*s* AND 4*s* STATES. Hughes et al.[120] and Andreev et al.[123] have provided the measurements for the 3*s* level shown in Data Table 9-3-10. Neither author estimates the possible influence of cascade on their measurements. There is a marked discrepancy between the data of these two groups in both absolute magnitude and also in energy dependence. It may be noted that the method used by Hughes is relatively direct, involving only the measurement of the Balmer-alpha emission. In contrast, the technique of Andreev et al. involves measurements of the Lyman-beta and Balmer-alpha lines, both with and without an electric field. The systematic errors are likely to be different for the two wavelengths; this will tend to influence both the energy dependence and absolute magnitude of the data. There is no basis on which the work of Andreev can be subject to definite criticism, but certainly the opportunities for systematic errors are more numerous.

Hughes et al.[353] have provided a measurement for the 4*s* state shown in Data Table 9-3-11. The measurements are published in the form of level-excitation cross sections, evaluated from the emission on the assumption that cascade may be neglected.

2. Cross Sections for the Formation of the *np* States. (a) 2*p* STATES. In Data Table 9-3-12 are shown measurements of the cross section for the spontaneous emission of the Lyman-alpha ($2p \rightarrow 1s$) line. Four sets of data are shown from the work of Pretzer,[117] Gaily[28] and co-workers. Pretzer's emission measurements[117] were later corrected by Gaily[28] for anisotropy of emission associated with polarization. The correction amounts to some 4% and is less than the random error in the determinations. Both the corrected and uncorrected data are shown. Other publications by this group[169, 350] represent preliminary data and are not used here. Each report includes work with both D^+ ions as well as protons.

All the measurements are emission cross sections and may include a cascade contribution arising from the ns and nd states ($n > 3$). Cascade from the $3s$ and $3d$ states, estimated by adding the cross sections for the $3s$ and $3d$ state [Data Tables 9-3-10(B) and 9-3-15, respectively] ranges from 30% at 14-keV to 56% at 30-keV. These figures show that cascade cannot be neglected.

The data of Andreev et al.[75] and those of Pretzer et al.[117] are in excellent agreement, while the work of de Heer et al.[78] lies above the other two determinations but exhibits similarity in energy dependence; a similar comparison of data for protons on helium shows the same features. The absolute calibration by de Heer et al. involved some unsubstantiated assumptions about the variation of the detection sensitivity of their system with wavelength [see Section 9-2-(7)] and may be unreliable. The good agreement between the other two sets of data by Andreev et al. and Pretzer et al. does suggest that their data should provide a reliable estimate of the true behavior of the cross section.

In Data Table 9-3-13 are displayed the cross section for the population of the $m = 0$ and $m = \pm 1$ magnetic sublevels of the $2p$ state, as deduced by Gaily et al.[28] from the measurements of emission [Data Table 9-3-12(D) and (E)] and polarization (Data Table 9-3-16). Using the method described in Section 9-2(1), the accuracy of the final result may not be very high because it depends on two separate measurements, each of which has limited statistical accuracy. Moreover, the procedure assumed that cascade can be neglected.

(b) $3p$ STATE. De Heer et al.[78] have measured the emission function of the $3p \rightarrow 1s$ Lyman-beta transition (1026-Å) and converted it to a cross section for the excitation of the $3p$ level on the assumption that cascade may be neglected. Andreev et al.[122] have presented the cross section for the emission of the $3p \rightarrow 1s$ line and also a cross section for the formation of the $3p$ state calculated on the assumption that cascade might be neglected. Two additional conference publications by Andreev et al.[351, 123] appear to contain the same data as the full report[122] and may therefore be disregarded. The data from these two sources is shown in Data Table 9-3-14.

Andreev et al.[122] have also investigated the intensity of the Lyman-beta line as a function of an electric field applied in the observation region. This work provides no information on cross sections and therefore is not considered here.

Neither of the determinations reviewed here provide an assessment of the cascade contribution to the measurements. It is possible to use the data by Hughes et al.[353] on the formation of the $4s$ state (Data Table 9-3-11) to estimate that at energies of 10-keV and 40-keV the cascade into the $3p$ state will be some 12% and 48% of the $3p$ state cross section measured by Andreev et al.[122]

The two determinations show a similar energy dependence except at the lowest energies but disagree in magnitude. This discrepancy probably represents an error in calibration. The method by which de Heer et al. calibrated their detection system involved making unsubstantiated assumptions about the variation of sensitivity with wavelength. Therefore, it seems reasonable to regard the work of Andreev et al.[122] as being closer to the true absolute values.

3. **Cross Sections for the Formation of the $3d$ State.** Andreev et al.[123] have measured the cross section for the emission of the $3d \rightarrow 2p$ line at energies from 14 to 30-keV shown in Data Table 9-3-15. No attempt was made to assess cascade.

4. **Polarization of Emission.** In Data Table 9-3-16 are shown the polarization fractions of the $2p \rightarrow 1s$ emission as measured by Gaily et al.[28] for impact of protons

and deuterons. The statistical accuracy of these data is estimated as no better than $\pm 15\%$ (i.e., $P = X\% \pm 15\%$).

C. Argon Target

Data are available for the formation of the 2s, 3s, 4s, 2p, 3p, and 3d states, with impact energies ranging from 0.6 to 168-keV. The polarization of the $2p \rightarrow 1s$ emission has been measured; from this have been deduced cross sections for the formation of the different magnetic quantum number sublevels of the 2p state.

Il'in et al.[118, 119] have presented data on the total population of high principal quantum number states measured by using the field ionization technique. In a review of this method of detection (Section 3-3-A) it is suggested that data obtained by this method should be excluded from detailed consideration in this monograph, so work of Il'in is not considered further here.

1. Cross Section for the Formation of the ns States. (a) 2s STATE. Data on this case are available from three sources and are displayed in Data Table 9-3-17. The data by Jaecks were obtained by relating to the measurements of the formation of the 2p state carried out in the same apparatus. Recently, Gaily[28] has shown that the 2p data must be revised to take into account anisotropy associated with the polarization of the emission. This requires a small correction (2 to 7%) which is of the same order as the reproducibility of the data. Data Table 9-3-17 also includes some of the corrected data.[360] The three determinations, by Bayfield, Andreev, Jaecks, and their respective co-workers are all completely independent. The work of Andreev et al. and Bayfield are inherently absolute. The absolute magnitude of the data by Jaecks et al. is related to a reliable theoretical prediction of the excitation of H by electrons.

None of the experiments reviewed here has made any assessment of cascade. An estimate of the contribution from the 3p level using the data by Andreev et al.[122] (Data Table 9-3-22) gives contributions of 2.4 to 4.0%; this is negligible.

All the experiments exhibit good agreement in the functional dependence of cross section on projectile energy. There also appears to be agreement in the absolute magnitudes; however, it should be recalled that all three experiments neglect to correct for the anisotropic distribution of emission.[239, 240, 20] A correction for anisotropy would raise the data of Bayfield and of Andreev et al. by approximately 20% while reducing the values of Jaecks et al. by a like amount; after correction, the work of Bayfield remains in good agreement with the data by Andreev et al. while the work of Jaecks et al. lies appreciably lower. It is not possible to make an objective choice of the most accurate data.

(b) 3s AND 4s STATES. Measurements of the 3s level have been provided by Hughes et al.,[120] Andreev et al.,[123] and Thomas et al.;[129] in each case the data are presented as a level excitation function with no correction for cascade. There is a decided discrepancy between the magnitudes in the work by Andreev et al.[123] and by Hughes et al.,[120] although the energy dependence of both sets of data is approximately the same. It is possible that this difference indicates an error in the determination of detection efficiency. The data by Thomas et al.[129] cannot be realistically compared with those by Hughes et al. due to the difference in energy between the two determinations. There is no basis on which one might choose one set of data to be more reliable than the other.

In Data Table 9-3-19 is shown the data for the formation of the 4s state as measured by Hughes et al.[353] The data are given as level-excitation cross sections evaluated on the assumption that cascade is negligible.

2. Cross Sections for the Formation of the np States. (a) 2p STATE. In Data Table 9-3-20 are shown measurements of the cross section for the spontaneous emission of the Lyman-alpha ($2p \to 1s$) line resulting from the impact of H^+ on an argon target. Three groups of data are shown by Pretzer,[117] Gaily,[28] and co-workers. Pretzer's work includes data for both H^+ and D^+ impact. Pretzer's original measurements of emission were later corrected by Gaily[28] for the anisotropy of the emission associated with polarization. The correction ranges from 2 to 4% and is less than the random error in the measurements. Early papers by Pretzer,[169] Gaily,[350] and co-workers were superseded and extended by the references from which the data used here are drawn.[117, 28]

The measurements of emission cross sections might include a contribution due to cascade from the lower lying ns and nd levels. Cascade from the 3s and 3d states may be estimated by taking the sum of the cross sections for the formation of the 3s and 3d states as measured by Andreev et al.[123] [Data Tables 9-3-18(B) and 9-3-23 (A)]; cascade amounts to about 26% at 10-keV and 40% at 30-keV.

The data presented by Andreev et al.[75] are in very good agreement with those of Pretzer et al.[117] The experimental procedures used by these two groups are completely independent of each other. The excellent agreement shown here, and also in the case of charge transfer on other targets, lends considerable confidence to the accuracy of these measurements.

On Data Table 9-3-21 are displayed the cross sections for the formation of the $m = 0$ and $m = \pm 1$ magnetic sublevels of the 2p state as deduced by Gaily et al.[12] [see Section 9-2(1)] from the measurements of emission [Data Table 9-3-20(C)] and polarization [Data Table 9-3-4(A)] for the $2p \to 1s$ line. The procedure assumes that cascade can be disregarded. It should be noted that there are two independent determinations of the polarization fraction (Data Table 9-3-24) which are in disagreement with each other. In view of the uncertainty as to which polarization measurement is correct, there must also be uncertainty as to the validity of the cross sections for the separate magnetic quantum number sublevels derived from Gaily's polarization measurements.

(b) 3p STATE. Andreev et al. have measured the cross section for the emission of the Lyman-beta ($3p \to 1s$) line and deduced a cross section for the formation of the 3p state on the assumption that cascade was negligible[122] [Data Table 9-3-22]. Two additional conference papers by Andreev et al.[351, 123] appear to contain the same data as the full report and may therefore be disregarded.

Andreev et al. also measure the variation of the Lyman-beta emission with applied electric field; these studies provide no cross section information and therefore are not considered further.

From the data by Hughes et al.[353] [Data Table 9-3-19] cascade from the 4s level into the 3p may be from 14 to 27%. Although Andreev's apparatus discriminates somewhat against cascade, this contribution will not be negligible.

3. Cross Sections for the Formation of the 3d State. Data on the formation of this state by Andreev et al.[123] and by Thomas et al.[129] are shown in Data Table 9-3-23. The measurement is the cross section for the emission of the $3d \to 2p$ transition; there is no way of estimating cascade.

4. Polarization of Emission. In Data Table 9-3-24 are shown measurements of the polarization of the Lyman-alpha line by Gaily et al.[28] and also by Teubner et al.[128] The statistical accuracy of the work by Gaily et al. is estimated as $\pm 15\%$ (i.e., $P = X\% \pm 15\%$), that for Teubner et al. is better than $\pm 13\%$.

There is no apparent reason for the discrepancy between these two determinations. The data due to Teubner et al.[128] were obtained by using a Brewster's angle polarization analyzer, and that due to Gaily et al.[28] were derived by employing a measurement of the variation of emission intensity with angle of observation. It is not possible to resolve the differences in the data from the two groups.

Andreev et al.[122] also present a measurement of the polarization of the Balmer-alpha line induced in this collision combination. The Balmer line will contain components from both the $3p \rightarrow 2s$ and $3d \rightarrow 2p$ transitions. It is not clear from the report whether or not the observations were made at a sufficient penetration through the target, so that the intensities of emission from the $3p$ and $3d$ states were invariant with distance. If this condition is not achieved, a measured polarization will be peculiar to the particular point where the observation was made. In view of this uncertainty the data are excluded from further consideration.

D. Krypton Target

The data available for this case is limited to work by Andreev et al. who have studied the formation of the $2s$, $2p$, and $3p$ states.[75, 122]

Pretzer and Geballe[117, 169] present some measurements on Lyman-alpha emission induced as a result of H^+ and D^+ impact on krypton. There is evidence that the detector used for this work was not completely selective to the Lyman-alpha line and that a line of the Kr I spectrum is also transmitted.[117] Apparently this line is excited with a sufficiently large cross section so that it can contribute an appreciable part of the signal. It was shown that part of the structure on the cross section curve is due to this line and not to the Lyman-alpha. The data by Pretzer et al. may be misleading due to the inclusion of an unknown fraction of the Kr I emission; therefore, it is not reproduced in the tabulations of this monograph.

The data on the $2s$, $2p$, and $3p$ emission measurements by Andreev et al.[75, 122] are shown in Data Tables 9-3-25, 9-3-26, and 9-3-27, respectively. The data are quoted as having systematic errors of less than $\pm 20\%$ and a reproducibility of better than $\pm 10\%$. There are no other data with which these numbers might be compared; however, in the preceding sections concerning He, Ne, and Ar targets, it is concluded that the data by this group are apparently quite reliable. There is no reason to doubt that this is also true for the present work.

Andreev et al. made no assessment of cascade population of the $2s$, $2p$, and $3p$ levels. The data for the $3p$ state is presented both in the form of the emission cross section for the Lyman-beta line and also in the form of a cross section for the formation of the $3p$ state evaluated on the assumption that cascade is negligible. An estimate of cascade into the $2s$ level based only on transitions from the $3p$ gives a contribution from 1.7 to 2.3% depending on energy; it seems reasonable to conclude that cascade is negligible. Cascade into the $2p$ and $3p$ states cannot be estimated on the basis of available information; however, the cascade is large for the other targets such as He, Ne, and Ar, and therefore it is likely to be large here also.

Andreev et al.[122] include measurements on the variation of the $3p \rightarrow 1s$ emission

cross section with an applied electric field. This provides no information on cross sections; therefore, is not considered further here.

E. Xenon Target

Measurements are available for the formation of the 2s, 2p, and 3p states from the work of Jaecks,[16] Pretzer,[117] Andreev,[75, 122] and their respective co-workers.

1. Cross Sections for the Formation of the 2s State. The measurements on this state, shown in Data Table 9-3-28, are not in good agreement for either absolute magnitude or energy dependence. The difference cannot be fully explained by an error in calibration nor in the measurement of projectile energy. Both groups—Jaecks et al. and Andreev et al.—neglect anisotropy of field-induced emission;[239, 240, 20] correction for this problem would tend to increase the discrepancy. Neither group attempts any assessment of the cascade contribution to the measured emission. Cascade from the 3p level [Data Table 9-3-30(B)] will not exceed 2%; it would be safe to assume that cascade is negligible. It is impossible to come to a decision as to which of the available measurements is the more accurate.

2. Cross Sections for the Formation of the np States. (a) 2p STATE. The measurements of the cross section for spontaneous emission of Lyman-alpha are shown in Data Tables 9-3-29. The work of Pretzer et al.[117] includes measurements with both H^+ and D^+ projectiles. The agreement between the data by Pretzer et al. and that of Andreev et al.[75] is rather poor. The energy dependences appear to be the same, but the magnitudes of the cross sections are different. There is no objective criterion by which one cross section may be chosen as more accurate than the other. Both experiments ignore the influence of cascade. For targets of He, Ne, and Ar cascade into the 2p state has been found to be large. It is expected that the same result would hold here. Both experimental arrangements provide some degree of discrimination against cascade contributions from states that are longer-lived than the 2p. Early measurements on the formation of the 2p state by Pretzer[169] are superseded and changed by the data that are used here.[117]

(b) 3p STATE. In Data Table 9-3-30 are shown measurements by Andreev et al.[122] of the cross section for the emission of the 1026-Å $(3p \rightarrow 1s)$ line and also of the cross section for the formation of the 3p state deduced on the assumption that cascade may be neglected. The cascade contribution to this state cannot be assessed on the basis of available evidence. For targets of He, Ne, and Ar, the cascade was large, and the same result might be expected to be true here also.

The work of Andreev et al. also includes information on the variation of the $3p \rightarrow 1s$ emission cross section with an applied electric field. This provides no cross section information and is not considered further here.

F. Molecular Hydrogen Target

Data are available on the formation of the 2s, 3s, 4s, 2p, and 3p states as well as total cross sections for a level of a given principal quantum number summed over all angular momentum substates. There have been no measurements on the polarization of collisionally induced emission. The data reviewed here all refer to the formation of excited H atoms by neutralization of the projectile. Data on the simultaneous formation of excited projectiles and excited fragments of dissociated target molecules are reviewed in Section 8-2-C.

A number of the papers listed in the summary assessment table given in Chapter 4 are not considered in detail here. An early measurement on the formation of the 2s state by Madansky and Owen[144] provided only an estimate of the lower limit to the magnitude of cross section. Since it is of poor accuracy and more accurate measurements are now available, that data is omitted. Hughes et al.[147] carried out measurements on the emission of Balmer-series lines but omitted to correct for the lifetime of the excited atoms; these data are incorrect. There have been measurements of the total population for all angular quantum number substates of levels with high principal quantum numbers carried out by using the field-quenching techniques. These are due to Riviere and Sweetman,[56] Il'in et al.,[118, 57, 119] McFarland et al.,[132] and Le Doucen and Guidini.[134] In Section 3-3-A the technique of field ionization was discussed, and it was concluded that data obtained by this technique are not comparable in definition to measurements by the optical methods. These data are also excluded from further consideration here. Dose and Meyer[153] have carried out measurements on the angular distribution of metastable H formed by charge transfer as protons are incident on H_2, but the data yield no information on total cross sections.

1. Cross Sections for the Formation of the ns States. (a) 2s STATE. The data for this case, shown in Data Table 9-3-31, are all obtained by field-quenching of the metastable state and detecting the emitted Lyman-alpha photon. All data include some contribution from cascade population.

The work of Andreev et al.[148] and of Bayfield[130] are both independent absolute determinations of the field-induced emission cross section. The work of Colli and Cristofori et al., originally published as a conference paper[356] and later superseded by a more complete report,[145] was originally normalized to a theoretical prediction of charge transfer; however, these data have recently been renormalized to a direct absolute measurement of cross section [see Section 9-2(8)]; these revised measurements, obtained by private communication,[146] are used in this monograph.

The work of Sellin,[17] at the single energy of 15-keV, is placed on an absolute basis by a normalization procedure that is ultimately traceable to the Born approximation prediction of H excitation by electrons; this normalization procedure is invalid and will involve substantial errors [see Section 9-2(5)]. The work of Ryding et al.[142] is normalized to a Born approximation prediction of a charge transfer cross section at 160-keV [see Section 9-2(10)]. There is insufficient evidence that the Born approximation prediction of charge transfer should be accurate at this energy.

None of the experiments reviewed here make any assessment of cascade. The contribution from the 3p state (Data Table 9-3-35) amounts to about 1%; it is reasonable to conclude that cascade is negligible.

Numerous discrepancies are exhibited in Data Table 9-3-31. It is necessary to attempt to arrive at an estimate of where the true value of the cross section lies. The apparatus used by Colli et al.[145] involved a mixed projectile beam containing both protons and molecular hydrogen ions. On these grounds one might regard it as considerably less accurate than the other measurements. Except for the work of Colli et al., all the remaining sets of data exhibit similar dependence on impact energy. Absolute magnitudes were determined by Andreev et al.[148] and by Bayfield;[130] the work of Ryding et al.[142] is a relative measurement normalized to theory. It is surprising that disagreement exists between the data by Andreev et al. and that by Bayfield; for other similar measurements using an argon target (see Data Table

9-3-17) there was good agreement between the data obtained by these two groups. It should be recalled that all the experiments neglected to account for the polarization-related emission anisotropy; the correction for this problem would raise about 20% both the data published by Bayfield and that published by Andreev et al. There are no objective criteria by which one might determine whether or not the data of Bayfield are any more reliable than those of Andreev et al.

(b) 3s AND 4s STATES. Hughes et al.[120,353] provide the measurement on the emission of the $3s \rightarrow 2p$ and $4s \rightarrow 2p$ lines shown in Data Tables 9-3-32 and 9-3-33. No attempt is made by the authors to assess cascade; however, since the cascade into the $2s$ state has been shown to be negligible [see Section 9-3-F(1a)], it seems reasonable to assume that cascade into the $3s$ and $4s$ states may also be neglected.

2. **Cross Sections for the Formation of the np States.** (a) $2p$ STATE. Andreev et al.[148] have provided the only measurement of $2p \rightarrow 1s$ emission (Data Table 9-3-34). Based on the measurements by Hughes et al.[120] of the cross section for the $4s$ state, (Data Table 9-3-32) the cascade contribution may be as high as 49% (30-keV).

(b) $3p$ STATE. Andreev et al.[148] have measured the emission in the $3p \rightarrow 1s$ transition and present a cross section for the formation of the $3p$ state evaluated on the assumption that cascade may be neglected (Data Table 9-3-35). Cross section data are also given for the emission of the Lyman-beta transition when high electric fields are applied to the detection region. These data are not repeated here because they give no additional information on cross sections.

Cascade population from the $4s$ level estimated using the work of Hughes et al.[353] (Data Table 9-3-33) may be as high as 50% (30-keV).

G. Molecular Nitrogen Target

Data are available on the formation of the $3s$, $4s$, $2p$, $3p$, and $3d$ states as well as on the total cross section for a level of a given principal quantum number summed over all angular momentum states. Some information is also available on polarization of emission.

For this case, there have been a large number of publications that are of little value; such publications are not considered in detail here. Measurements of Balmer emission taken by Carleton and Lawrence,[172] Sheridan et al.,[173] Philpot and Hughes,[174, 175] Sheridan and Clark,[176] and Dufay et al.[178] do not include lifetime corrections and therefore give no cross section information. Murray, Young, and Sheridan[177] attempted to measure Balmer-alpha emission at a sufficient distance of penetration through the target region so that emission would be invariant with further increase in penetration, and therefore a lifetime correction was unnecessary. A target cell of one meter in length was used with a target pressure of 10^{-4} torr. At the lower energies of this experiment (2 to 35-keV), the projectile beam will be neutralized by as much as 40% in traversing a chamber of this length. Thus the experiment was carried out with a very badly contaminated beam, and the results are likely to be both incorrect and misleading; therefore the experiment is omitted from further consideration here. Measurements of the emission of Lyman-alpha radiation have been made by Dunn et al.[99] and by Van Zyl et al.[149] using oxygen-filtered Geiger counters. Van Zyl showed quite definitely that the oxygen filter was transmitting a substantial amount of radiation from the excited molecule or dissociation fragments. Their data, not representing the spontaneous emission of the Lyman-

alpha line and including unknown contributions from other emitting systems, is excluded from detailed consideration here. Finally, there have been measurements by Il'in et al.[57] and McFarland et al.[132] of the total population of all angular momentum substates of a particular high principal quantum number. These have been carried out by using field ionization detection techniques. In Section 3-3-A it is concluded that data obtained by this technique are not comparable in definition to measurements by the optical methods and should be excluded from detailed consideration in this monograph.

After elimination of the work discussed in the preceding paragraph, only six papers remain. Three of these are by Hughes and co-workers,[120, 353, 180] dealing with the formation of the 3s, 4s, 3p, and 3d states.

A paper by Dahlberg[182] gives the cross section for Lyman-alpha emission ($2p \rightarrow 1s$); it will be suggested later that the normalization of these data to a previous experimental measurement is inadequate. Thomas et al.[129] have measured the formation of the 3s and 3d states at the single energy of 168-keV. The data originally published by Thomas et al. have been revised, and the new values are used here.[262] Young et al.[7] have carried out a coincidence experiment that detects the simultaneous formation of an excited H atom and an excited N_2^+ ion in the target. This rather specialized experimental measurement is considered separately in Section 9-3-G(4).

1. Cross Sections for the Formation of the 3s and 4s States. Cross Sections for the emission of the $3s \rightarrow 2p$ transition are shown in Data Table 9-3-36. Hughes et al. have produced two measurements of this cross section,[120, 180] both of which are shown here to indicate the excellent reproducibility of the experiment. The single measurement by Thomas et al.[129] cannot be readily compared with the data of Hughes et al., which lie at somewhat lower energies. There is no way in which the cascade contribution to these data can be assessed accurately.

The cross section for the formation of the 4s state as measured by Hughes et al.[353] is shown in Data Table 9-3-37. This cross section was evaluated from the $4s \rightarrow 2p$ line emission cross section on the assumption that cascade was negligible.

2. Cross Section for the Formation of the np States. (a) 2p STATE. In Data Table 9-3-38 is shown the cross section for spontaneous emission of the Lyman-alpha line as measured by Dahlberg et al.[182] In a review of this experiment [Section 9-2(6)] it is noted that Dahlberg normalized his data to the measurements on "Lyman-alpha" emission measured by Van Zyl et al. Van Zyl[149] points out that his data include appreciable contributions from nitrogen emissions and do not represent the emission of Lyman-alpha alone. For that reason the data is excluded from the present compilation. It is clearly unsatisfactory to normalize data to these misleading measurements. Dahlberg's data are presented here in the form in which they were originally published; however, the absolute magnitudes have no real validity.

No attempt was made to assess the cascade contribution to the measurement of Lyman-alpha emission. By adding the cross sections for the $3s \rightarrow 2p$ emission (Data Table 9-3-36) and the cross section for the $4s \rightarrow 2p$ emission evaluated from the data on the formation of the 4s state (Data Table 9-3-37), the cascade from these two states alone may be from 12 to 100% of Dahlberg's measured emission cross section.

(b) 3p STATE. Data Table 9-3-39 shows the cross section for the formation of the 3p state evaluated by Hughes et al.[180] Cascade was ignored. An estimate of cascade from the 4s state can be made using Hughes's own data on the measurement of this

level (Data Table 9-3-37).[353] Based on this estimate alone cascade may amount to 15% of the $3p \rightarrow 2s$ emission at 20-keV.

3. Cross Section for the Formation of the 3d State. Data Table 9-3-40 shows the cross section for the formation of the $3d$ state evaluated by Hughes et al.[180] and the single measurement at 169-keV by Thomas et al.[129] The data by Thomas et al. are a revised value.[262] Both determinations assume that cascade is negligible without providing any supporting evidence.

4. The Formation of an Excited H Atom and Excited Target Molecule in Coincidence. Young et al.[7] have made a measurement of the formation of an excited H atom and the formation of an excited target molecule in coincidence. The reaction may be represented by the equation,

$$H^+ + N_2 \rightarrow H(3p + 3d) + N_2^+ [B\,^2\Sigma_u^+, (v = 0)]. \tag{9-4}$$

The data are, in fact, the cross section for the subsequent emission of photons from both excited systems. The data are not in the form of cross sections for formation of the excited state.

This experiment, reviewed in Section 8-3-B(8) in the context of studies of target excitation, is not discussed again here. It should be noted that a large number of assumptions and approximations are made in arriving at the final results. These cause the accuracy to be rather poor. No details are given on the method utilized to calibrate detection sensitivity. The authors estimate the systematic error to range from 35% at low velocities to 45% at high velocities. Random error was very large due to poor counting statistics, ranging from 25% at the peak of the cross section to 100% where it was small.

The results of this experiment are shown in Data Table 9-3-41.[7] There are no other data with which this might be compared.

5. Polarization of Emission. Hughes et al.[180] have measured the polarization of the $3d \rightarrow 2p$ transition. The data are shown in Data Table 9-3-42. Accuracy is not very high. The error bars range from $\pm 5\%$ at 10-keV (i.e., $P = X\% \pm 5\%$) to $\pm 10\%$ at high energies. There are no other measurements with which these might be compared.

In the same paper Hughes et al. also attempted to measure the polarization of the $3p \rightarrow 2s$ transition. The accuracy was very poor and no detailed results are published. It is stated that "the polarization was essentially zero at the higher energies (35-keV) and dropped to negative values at the lower energies (10-keV)."[180]

H. Molecular Oxygen Target

There is a limited amount of data on the formation of excited H atoms by charge transfer as protons traverse a target of oxygen. Hughes et al.[120, 353] have presented cross sections for the formation of the $3s$ and $4s$ states. Work by Dufay et al.[178] on the emission of the Balmer-alpha and -beta lines omitted corrections for the lifetime of the excited state; these data are in error and omitted from further consideration here.

The cross sections for the emission of the $3s \rightarrow 2p$ and $4s \rightarrow 2p$ lines are shown in Data Tables 9-3-43 and 9-3-44. These data are evaluated on the assumption that

cascade may be neglected; such an assumption has been shown valid for other targets though not explicitly tested in oxygen.

I. Other Molecular Target

In addition to the data reviewed in the preceding sections for impact of H^+ on molecular targets of H_2, N_2, and O_2, there have been a number of other studies. Poulizac et al.[344, 135] has measured the Balmer-alpha emission from a H^+ projectile incident on a target of CO. No correction was made for the lifetime of the excited atom, therefore the data are wrong. Measurements on the cross section for formation of high principal quantum number states are not considered here because all involve field ionization as the detection technique (Section 3-3A). Such data included measurements with an CO target by Il'in et al.[57] with H_2, NH_3, C_8F_{16}, and Freon by McFarland et al.[132] and with C_2H_2 and C_6H_6 by Le Doucen and Guidini.[134]

J. Metallic Vapor Targets

There have been a large number of measurements using metallic vapor targets. Most of these have involved studies of the formation of highly excited H atoms with the field-ionization technique being utilized for detection; data obtained by this method are excluded from detailed consideration (see Section 3-3-A). After eliminating certain other determinations on the grounds of very poor accuracy there remain only the measurements of $2s$ formation on targets of potassium, rubidium, and cesium made by Sellin and Granoff,[139] Donnally, et al.,[137] and Cesati et al.[138] It should be recalled that only the work of Cesati et al. is inherently absolute, the work by Donnally et al. and by Sellin and Granoff both being placed on an absolute basis by questionable normalization procedures. In our assessments of the measurements for targets of H_2 and He, [see Sections 9-3-A(1) and 9-3-F(1)] it is concluded that the data by Colli et al.[145] may be incorrect. Therefore, one cannot have confidence in the absolute magnitudes of any of the data presented for metallic vapor targets. All of the measurements are essentially cross sections for the field-induced emission of Lyman-alpha radiation. The data contain unknown contributions by cascade from higher states.

 ·1. **Cross Section Measurements for Targets of Potassium, Rubidium, and Cesium.** The cross sections for the formation of the metastable state for targets of K, Rb, and Cs are shown, respectively, in Data Tables 9-3-45 through 9-3-47. The only case where comparison between data may be made is cesium. There is considerable disagreement. The general form of the energy dependence appears to be the same in all three cases, but the absolute magnitudes are greatly different. In view of the comments made above these discrepancies are not too surprising. It is suggested that no great weight can be placed on the absolute accuracy of any of these data.

There have been measurements on the formation of excited states of high principal quantum number for targets of potassium[118, 119, 132] and cesium,[118] but these data are not considered further in this monograph.

2. **Charge Transfer on Other Metallic Targets.** There are no measurements, other than those for targets of potassium, rubidium, and cesium, that are of sufficient accuracy to justify detailed discussion in this monograph. Studies of Balmer emission induced by H^+ impact on Mg and Hg have been published by Berkner et al.[170] and

by Bobashev and Pop;[168] these data are likely to be wrong due to inadequate experimental procedures [see Sections 9-2(11) and 9-2(2), respectively].

There have been a very large number of measurements of the formation of highly excited atomic hydrogen states by protons, incident on various metallic targets. The work is mainly due to the groups of Il'in et al.[118, 119, 54, 136] and McFarland et al.[132] Targets have included Li,[118, 119] Na,[118, 119, 54] Mg,[118, 54, 132, 136] Ca,[136] Zn,[136] Cd,[136] and Ba.[132] None of these data are discussed further here.

K.   Data Tables

In some of the experiments D^+ ions have been used in addition to H^+ projectiles. The data for D^+ impact are shown in the tables as though they were for D^+ of equal velocity, that is, at one-half the true D^+ energy.

1. Formation of Excited H States Induced by Impact of H⁺ on an He Target (Includes Line-Emission Cross Sections, Level-Excitation Cross Sections, and Polarization Fractions)

9-3-1   Field Induced Emission of Lyman-Alpha ($2s \to 2p \to 1s$). [9-3-1A is revised data by private communication;[146] 9-3-1C is data for D^+ impact; 9-3-1D is data from 9-3-1B corrected for polarization effects.[360]]

9-3-2   Emission of the 6563-Å Line ($3s \to 2p$).

9-3-3   Formation of the $4s$ State.†

9-3-4   Emission of the Lyman-Alpha Line ($2p \to 1s$). [9-3-4C and 9-3-4E are data for D^+ impact; 9-3-4D and 9-3-4E are respectively the data from tables 9-3-4B and 9-3-4C corrected for polarization effects.[28]]

9-3-5   Formation of the $m = 0$ and $m = 1$ Magnetic Substates of $2p$. These are derived from the measurements in Data Tables 9-3-4D, 9-3-4E, and 9-3-8; see text for details. [9-3-5A and 9-3-5C are for $m = 0$; 9-3-5B and 9-3-5D are for $m = +1$ or -1; 9-3-5C and 9-3-5D are data for D^+ impact.]

9-3-6   Formation of the $3p$ State.† [9-3-6C are the original emission cross sections ($3p \to 1s$) from which 9-3-6B was evaluated.]

9-3-7   Emission of the 6563-Å Line ($3d \to 2p$).

9-3-8   Polarization Fraction of the Lyman-Alpha Line ($2p \to 1s$). [9-3-8B are data for D^+ impact.]

2. Formation of Excited H States Induced by Impact of H⁺ on an Ne Target (Includes Line Emission Cross Sections, Level-Excitation Cross Sections and Polarization Fractions)

9-3-9   Field-Induced Emission of Lyman-Alpha ($2s \to 2p \to 1s$). [9-3-9B and 9-3-9D are data for D^+ impact; 9-3-9C and 9-3-9D are the data from 9-3-9A and 9-3-9B corrected for polarization effects.[360]]

9-3-10 Emission of the 6563-Å Line ($3s \to 2p$).

9-3-11 Formation of the $4s$ State.†

9-3-12 Emission of the Lyman-Alpha Line ($2p \to 1s$). [9-3-12C and 9-3-12E are data for D^+ impact; 9-3-12D and 9-3-12E are, respectively, the data from Tables 9-3-12B and 9-3-12C corrected for polarization effects.[28]]

9-3-13 Formation of the $m = 0$ and $m = 1$ Magnetic Substates of $2p$. These are derived from the measurements in Data Tables 9-3-12D, 9-3-12F, and 9-3-16; see text for details. [9-3-13A and 9-3-13C are for $m = 0$; 9-3-13B and 9-3-13D are for $m = +1$ or -1; 9-3-13C and 9-3-13D are data for D^+ impact.]

9-3-14   Formation of the $3p$ State.† [9-3-14C are the original emission cross sections ($3p \rightarrow$ $1s$) from which 9-3-14B was evaluated.]

9-3-15   Emission of the 6563-Å Line ($3d \rightarrow 2p$).

9-3-16   Polarization Fraction of the Lyman-Alpha Line ($2p \rightarrow 1s$). [9-3-16B are data for D^+ impact.]

3.   Formation of Excited H States Induced by Impact of H^+ on an Ar Target (Includes Line Emission Cross Sections, Level-Excitation Cross Sections, and Polarization Fractions)

9-3-17   Field-Induced Emission of Lyman-Alpha ($2s \rightarrow 2p \rightarrow 1s$). [9-3-17B is data for D^+ impact; 9-3-17C is data from 9-3-17A corrected for polarization effects.[360]]

9-3-18   Emission of the 6563-Å Line ($3s \rightarrow 2p$).

9-3-19   Formation of the $4s$ State.†

9-3-20   Emission of the Lyman-Alpha Line ($2p \rightarrow 1s$). [9-3-20B is data for D^+ impact; 9-3-20C are data from Table 9-3-20A corrected for polarization effects.[28]]

9-3-21   Formation of the $m = 0$ and $m = 1$ Magnetic Substates of $2p$. These are derived from the measurements in Data Tables 9-3-20C and 9-3-24; see text for details. [9-3-21A is for $m = 0$; 9-3-21B is for $m = +1$ or -1.]

9-3-22   Emission of the 1026-Å Line ($3p \rightarrow 1s$). [9-3-22A is the cross section for emission; 9-3-22B is the cross section for the formation of the $3p$ state derived from 9-3-22A on the assumption of negligible cascade.]

9-3-23   Emission of the 6563-Å Line ($3d \rightarrow 2p$).

9-3-24   Polarization Fraction of the Lyman-Alpha Line ($2p \rightarrow 1s$). [9-3-24C is data for D^+ impact.]

4.   Formation of Excited H States Induced by Impact of H^+ on a Kr Target

9-3-25   Field-Induced Emission of Lyman-Alpha ($2s \rightarrow 2p \rightarrow 1s$).

9-3-26   Emission of the Lyman-Alpha Line ($2p \rightarrow 1s$).

9-3-27   Emission of the 1026-Å Line ($3p \rightarrow 1s$). [9-3-27A is the cross section for emission; 9-3-27B is the cross section for the formation of the $3p$ state derived from 9-3-27A on the assumption of negligible cascade.]

5.   Formation of Excited H States Induced by Impact of H^+ on a Xe Target

9-3-28   Field-Induced Emission of Lyman-Alpha ($2s \rightarrow 2p \rightarrow 1s$). [9-3-28B are data for D^+ impact.]

9-3-29   Emission of the Lyman-Alpha Line ($2p \rightarrow 1s$). [9-3-29B is data for D^+ impact.]

9-3-30   Emission of the 1026-Å Line ($3p \rightarrow 1s$). [9-3-30A is the cross section for the formation of the $3p$ state derived from 9-3-30A on the assumption of negligible cascade.]

6.   Formation of Excited H States Induced by Impact of H^+ on a H_2 Target (Includes Cross Sections for Emission and Cross Sections for Level Excitation; all Data are for Excited States of the Projectile only)

9-3-31   Field-Induced Emission of Lyman-Alpha ($2s \rightarrow 2p \rightarrow 1s$). [9-3-31A is revised data by private communication.[146]]

9-3-32   Emission of the 6563-Å Line ($3s \rightarrow 2p$).

9-3-33   Formation of the $4s$ State.†

9-3-34   Emission of Lyman-Alpha Line ($2p \rightarrow 1s$).

9-3-35   Formation of the $3p$ State.†

7. Formation of Excited H States Induced by Impact of H⁺ on a N₂ Target (Includes Cross Sections for Emission, Cross Sections for Level Excitation, and Polarization Fractions)

9-3-36 Emission of the 6563-Å Line ($3s \rightarrow 2p$).

9-3-37 Formation of the $4s$ State.†

9-3-38 Emission of the Lyman-Alpha Line ($2p \rightarrow 1s$).

9-3-39 Formation of the $3p$ State.†

9-3-40 Emission of the 6563-Å Line ($3d \rightarrow 2p$).

9-3-41 Cross section for the Emission of the 3914-Å Line of N_2^+ $[B^2\Sigma_u^+ \rightarrow X\ ^2\Sigma_g^+\ (0,\ 0)]$ in Coincidence with the Emission of Balmer-Alpha Radiation from the Decay of the $3p$ and $3d$ States of H ($3p \rightarrow 2s$ plus $3d \rightarrow 2p$).

9-3-42 Polarization Fractions of the 6563-Å line ($3d \rightarrow 2p$).

8. Formation of Excited H States Induced by Impact of H⁺ on a O₂ Target

9-3-43 Emission of the 6563-Å Line ($3s \rightarrow 2p$).

9-3-44 Formation of the $4s$ state.†

9. Cross Sections for Field-Induced Emission of Lyman-Alpha ($2s \rightarrow 2p \rightarrow 1s$) **Induced by Impact of H⁺ on Various Metallic Targets**

9-3-45 K Target.

9-3-46 Rb Target.

9-3-47 Cs Target.

† Data evaluated on the assumption that cascade may be neglected.

CULLI ET AL. (145)

ENERGY (KEV)	CROSS SECTION (SQ. CM)
7.00E 00	3.90E-19
1.22E 01	5.50E-19
1.70E 01	6.73E-19
2.20E 01	1.17E-18
2.70E 01	1.07E-18
3.20E 01	1.04E-18
3.90E 01	7.49E-19

9-3-1B
JAECKS ET AL. (16)

ENERGY (KEV)	CROSS SECTION (SQ. CM)
6.00E 00	6.60E-19
9.00E 00	1.06E-18
1.20E 01	1.37E-18
1.60E 01	1.59E-18
1.90E 01	2.22E-18
2.10E 01	2.76E-18
2.30E 01	3.28E-18

9-3-1C
JAECKS ET AL. (16)

ENERGY (KEV)	CROSS SECTION (SQ. CM)
3.00E 01	1.75E-19
6.00E 01	5.80E-19

9-3-1D
JAECKS ET AL. (16)

ENERGY (KEV)	CROSS SECTION (SQ. CM)
6.00E 00	6.30E-19
9.00E 00	1.02E-18
1.20E 01	1.33E-18

9-3-1E
DOSE (159)

ENERGY (KEV)	CROSS SECTION (SQ. CM)
3.04E 00	1.21E-18
3.68E 00	1.12E-18
4.16E 00	1.60E-18
4.80E 00	1.83E-18
5.20E 00	2.02E-18
5.84E 00	2.35E-18
6.40E 00	2.56E-18
7.04E 00	2.71E-18
7.84E 00	2.68E-18
8.56E 00	3.09E-18

ENERGY (KEV)	CROSS SECTION (SQ. CM)
9.20E 00	2.86E-18
1.00E 01	2.94E-18
1.08E 01	3.73E-18
1.15E 01	3.30E-18
1.25E 01	3.48E-18
1.34E 01	3.52E-18
1.44E 01	3.07E-18
1.54E 01	3.76E-18
1.65E 01	3.78E-18
1.74E 01	4.21E-18
1.86E 01	4.15E-18
1.97E 01	4.99E-18
2.09E 01	5.20E-18
2.20E 01	5.71E-18
2.32E 01	6.52E-18
2.44E 01	6.90E-18
2.58E 01	7.66E-18
2.70E 01	8.53E-18
2.84E 01	9.19E-18
2.96E 01	8.73E-18
3.11E 01	1.07E-17
3.26E 01	1.17E-17
3.40E 01	1.27E-17
3.55E 01	1.33E-17
3.70E 01	1.34E-17
3.85E 01	1.35E-17
4.02E 01	1.37E-17
4.18E 01	1.34E-17
4.34E 01	1.34E-17
4.50E 01	1.32E-17
4.67E 01	1.28E-17
4.86E 01	1.19E-17
5.04E 01	1.22E-17
5.21E 01	1.16E-17
5.38E 01	1.10E-17
5.56E 01	1.05E-17
5.77E 01	1.00E-17
5.94E 01	1.00E-17
6.15E 01	9.39E-18
6.35E 01	9.12E-18
6.55E 01	8.88E-18
6.74E 01	8.26E-18
6.94E 01	7.86E-18
7.17E 01	7.47E-18

9-3-1F
ANDREEV ET AL. (75)

ENERGY (KEV)	CROSS SECTION (SQ. CM)
1.00E 01	1.01E-18
1.20E 01	1.18E-18
1.40E 01	1.53E-18
1.60E 01	1.72E-18
1.80E 01	2.14E-18
2.00E 01	2.67E-18
2.20E 01	3.43E-18
2.40E 01	4.60E-18
2.60E 01	4.91E-18
2.80E 01	5.30E-18
3.00E 01	5.90E-18
3.20E 01	6.72E-18
3.40E 01	7.22E-18
3.80E 01	7.70E-18
4.00E 01	8.14E-18

9-3-1G
RYDING ET AL. (142)

ENERGY (KEV)	CROSS SECTION (SQ. CM)
4.42E 01	4.28E-18
5.00E 01	4.49E-18
6.00E 01	4.14E-18
7.00E 01	3.91E-18
8.00E 01	3.44E-18
1.00E 02	2.49E-18
1.20E 02	1.84E-18
1.40E 02	1.14E-18
1.60E 02	7.93E-19
1.80E 02	5.46E-19
2.00E 02	4.12E-19

9-3-2A
HUGHES ET AL. (120)

ENERGY (KEV)	CROSS SECTION (SQ. CM)
5.00E 00	1.60E-19
6.00E 00	2.30E-19
7.50E 00	3.10E-19
1.00E 01	4.00E-19
1.50E 01	6.10E-19
2.00E 01	7.90E-19
2.50E 01	1.10E-18
3.00E 01	1.42E-18
3.50E 01	1.63E-18
4.00E 01	1.79E-18
4.50E 01	1.76E-18
5.00E 01	1.64E-18
5.50E 01	1.52E-18
6.00E 01	1.42E-18
6.50E 01	1.31E-18
7.00E 01	1.24E-18
7.50E 01	1.16E-18
8.00E 01	1.08E-18
8.50E 01	1.00E-18
9.00E 01	9.50E-19
9.50E 01	8.30E-19
1.00E 02	7.90E-19
1.05E 02	7.10E-19
1.10E 02	6.90E-19
1.15E 02	6.00E-19

9-3-2B
ANDREEV ET AL. (123)

ENERGY (KEV)	CROSS SECTION (SQ. CM)
1.40E 01	2.91E-19
1.60E 01	2.94E-19
1.80E 01	2.99E-19
1.96E 01	3.22E-19
2.20E 01	4.11E-19
2.40E 01	4.44E-19
2.60E 01	4.99E-19
2.80E 01	6.11E-19
3.00E 01	7.31E-19

9-3-2C
THOMAS ET AL. (129)

ENERGY (KEV)	CROSS SECTION (SQ. CM)
1.50E 02	1.88E-19

9-3-3
HUGHES ET AL. (353)

ENERGY (KEV)	CROSS SECTION (SQ. CM)
5.00E 00	4.30E-20
7.00E 00	6.10E-20
9.00E 00	8.40E-20
1.00E 01	9.30E-20
1.10E 01	1.13E-19
1.20E 01	1.19E-19
1.30E 01	1.32E-19
1.50E 01	1.47E-19
1.70E 01	1.64E-19
1.80E 01	1.72E-19
2.00E 01	2.02E-19
2.20E 01	2.21E-19
2.40E 01	2.49E-19
2.60E 01	2.67E-19
2.80E 01	3.52E-19
3.00E 01	3.37E-19
3.50E 01	4.62E-19
4.00E 01	5.61E-19
4.50E 01	6.10E-19
5.00E 01	7.00E-19
6.00E 01	6.78E-19
7.00E 01	6.14E-19
8.00E 01	5.15E-19
9.00E 01	3.85E-19
1.00E 02	3.02E-19
1.10E 02	2.50E-19
1.20E 02	1.99E-19

9-3-4A
DE HEER ET AL. (78)

ENERGY (KEV)	CROSS SECTION (SQ. CM)
1.00E 01	6.87E-18
1.25E 01	8.28E-18
1.50E 01	8.28E-18
2.00E 01	8.11E-18
2.50E 01	7.26E-18
3.00E 01	6.64E-18
3.50E 01	5.82E-18

9-3-4B
PRETZER ET AL. (117)

ENERGY (KEV)	CROSS SECTION (SQ. CM)
2.10E 00	1.11E-18
2.20E 00	1.08E-18
3.10E 00	1.32E-18
4.00E 00	1.60E-18
5.00E 00	1.71E-18
6.00E 00	1.95E-18
7.00E 00	1.92E-18
8.00E 00	2.19E-18
9.00E 00	2.31E-18
1.04E 01	2.54E-18
1.10E 01	2.78E-18
1.20E 01	2.95E-18
1.31E 01	3.10E-18
1.40E 01	3.58E-18
1.50E 01	3.51E-18
1.60E 01	3.59E-18
1.80E 01	3.69E-18
2.00E 01	3.72E-18
2.12E 01	3.63E-18
2.20E 01	3.45E-18
2.30E 01	3.46E-18
2.40E 01	3.31E-18
2.50E 01	3.35E-18
2.60E 01	3.33E-18
2.70E 01	3.27E-18
2.80E 01	3.25E-18

9-3-4C
PRETZER ET AL. (117)

ENERGY (KEV)	CROSS SECTION (SQ. CM)
2.50E-01	5.10E-19
5.00E-01	4.50E-19
7.50E-01	5.00E-19
8.60E-01	5.50E-19
1.00E 00	6.20E-19
1.25E 00	6.60E-19
1.37E 00	7.20E-19
1.50E 00	6.80E-19
1.62E 00	8.00E-19
1.75E 00	7.90E-19
2.00E 00	8.30E-19
2.12E 00	8.70E-19
2.25E 00	9.20E-19
2.50E 00	9.80E-19
2.25E 00	1.06E-18
3.00E 00	1.07E-18
3.50E 00	1.22E-18
4.00E 00	1.30E-18
4.50E 00	1.31E-18
5.00E 00	1.50E-18
5.50E 00	1.59E-18
6.00E 00	1.64E-18
6.50E 00	1.66E-18
7.00E 00	1.76E-18
7.50E 00	1.62E-18
8.00E 00	1.86E-18
8.50E 00	1.95E-18
9.00E 00	1.95E-18
9.50E 00	2.15E-18
1.00E 01	2.00E-18
1.05E 01	2.18E-18
1.10E 01	2.25E-18
1.15E 01	2.30E-18
1.20E 01	2.41E-18
1.25E 01	2.45E-18
1.30E 01	2.36E-18

9-3-4D
GAILY ET AL. (28)

ENERGY (KEV)	CROSS SECTION (SQ. CM)
6.00E-01	5.50E-19
1.00E 00	6.80E-19
2.00E 00	9.50E-19
3.00E 00	1.20E-18
4.50E 00	1.53E-18
6.00E 00	1.86E-18
7.00E 00	1.85E-18
8.00E 00	2.08E-18
9.00E 00	2.23E-18
1.00E 01	2.43E-18
1.10E 01	2.69E-18
1.20E 01	2.87E-18

9-3-4E
GAILY ET AL. (28)

ENERGY (KEV)	CROSS SECTION (SQ. CM)
6.00E-01	4.30E-19
1.00E 00	5.63E-19
2.00E 00	7.84E-19
4.00E 00	1.23E-18

9-3-4F
DOSE (159)

ENERGY (KEV)	CROSS SECTION (SQ. CM)
3.04E 00	6.04E-18
3.68E 00	6.38E-18
4.08E 00	6.98E-18
4.72E 00	7.02E-18
5.12E 00	7.62E-18
5.84E 00	7.92E-18
6.40E 00	8.10E-18
7.04E 00	8.44E-18
7.76E 00	8.80E-18
8.32E 00	9.16E-18
9.20E 00	9.66E-18
1.02E 01	1.03E-17
1.06E 01	1.05E-17
1.18E 01	1.11E-17
1.25E 01	1.20E-17
1.35E 01	1.26E-17
1.45E 01	1.27E-17
1.55E 01	1.38E-17
1.65E 01	1.48E-17
1.76E 01	1.49E-17
1.85E 01	1.53E-17
1.97E 01	1.54E-17
2.09E 01	1.54E-17
2.21E 01	1.54E-17
2.30E 01	1.52E-17
2.45E 01	1.43E-17
2.58E 01	1.41E-17
2.71E 01	1.40E-17
2.84E 01	1.34E-17
2.95E 01	1.31E-17
3.11E 01	1.26E-17

ENERGY (KEV)	CROSS SECTION (SQ. CM)
3.26E 01	1.22E-17
3.38E 01	1.10E-17
3.54E 01	1.03E-17
3.70E 01	9.76E-18
3.85E 01	9.40E-18
4.01E 01	8.88E-18
4.18E 01	8.82E-18
4.32E 01	8.24E-18
4.50E 01	7.82E-18
4.66E 01	7.44E-18
4.85E 01	6.92E-18
5.02E 01	6.18E-18
5.20E 01	5.94E-18
5.38E 01	5.58E-18
5.55E 01	5.60E-18
5.76E 01	5.10E-18
5.94E 01	4.84E-18
6.13E 01	4.84E-18
6.34E 01	4.50E-18
6.54E 01	4.24E-18
6.74E 01	4.28E-18
6.94E 01	4.00E-18
7.17E 01	4.10E-18

9-3-4G
ANDREEV ET AL. (75)

ENERGY (KEV)	CROSS SECTION (SQ. CM)
1.00E 01	2.31E-18
1.20E 01	2.67E-18
1.40E 01	3.06E-18
1.60E 01	3.52E-18
1.80E 01	3.87E-18
2.00E 01	4.05E-18
2.20E 01	4.18E-18
2.40E 01	4.18E-18
2.60E 01	3.93E-18
2.80E 01	3.83E-18
3.00E 01	3.70E-18
3.20E 01	3.40E-18
3.40E 01	3.03E-18
3.60E 01	2.73E-18
3.80E 01	2.43E-18
4.00E 01	2.16E-18

9-3-5A
GAILY ET AL. (28)

ENERGY (KEV)	CROSS SECTION (SQ. CM)
6.00E-01	4.88E-19
1.00E 00	5.36E-19
2.00E 00	6.08E-19
3.00E 00	8.51E-19
4.50E 00	9.34E-19
6.00E 00	9.98E-19
7.00E 00	9.16E-19
8.00E 00	1.19E-18
9.00E 00	1.08E-18
1.00E 01	1.13E-18
1.10E 01	1.25E-18
1.20E 01	1.26E-18

9-3-5B
GAILY ET AL. (28)

ENERGY (KEV)	CROSS SECTION (SQ. CM)
6.00E-01	3.12E-20
1.00E 00	7.18E-20
2.00E 00	1.71E-19
3.00E 00	1.74E-19
4.50E 00	2.98E-19
6.00E 00	4.24E-19
7.00E 00	4.67E-19
8.00E 00	4.46E-19
9.00E 00	5.78E-19
1.00E 01	6.55E-19
1.10E 01	7.25E-19
1.20E 01	8.06E-19

9-3-5C
GAILY ET AL. (28)

ENERGY (KEV)	CROSS SECTION (SQ. CM)
6.00E-01	3.43E-19
1.00E 00	4.20E-19
2.00E 00	4.51E-19
4.00E 00	6.62E-19

9-3-5D
GAILY ET AL. (28)

ENERGY (KEV)	CROSS SECTION (SQ. CM)
6.00E-01	4.36E-20
1.00E 00	7.14E-20
2.00E 00	1.65E-19
4.00E 00	2.84E-19

9-3-6A
DE HEER ET AL. (78)

ENERGY (KEV)	CROSS SECTION (SQ. CM)
1.50E 01	6.50E-19
2.00E 01	9.18E-19
2.50E 01	9.64E-19
3.00E 01	1.00E-18
3.50E 01	9.38E-19

9-3-6B
ANDREEV ET AL. (122)

ENERGY (KEV)	CROSS SECTION (SQ. CM)
1.00E 01	2.26E-19
1.20E 01	2.72E-19
1.40E 01	3.17E-19
1.60E 01	3.98E-19
1.80E 01	4.79E-19
2.00E 01	5.77E-19
2.20E 01	6.64E-19
2.40E 01	7.29E-19
2.60E 01	7.41E-19
2.80E 01	7.92E-19
3.00E 01	7.76E-19
3.20E 01	7.87E-19
3.40E 01	7.67E-19
3.60E 01	7.50E-19
3.80E 01	7.41E-19
4.00E 01	6.51E-19

9-3-6C
ANDREEV ET AL. (122)

ENERGY (KEV)	CROSS SECTION (SQ. CM)
1.00E 01	2.08E-19
1.20E 01	2.40E-19
1.40E 01	2.82E-19
1.60E 01	3.52E-19
1.80E 01	4.25E-19
2.00E 01	5.16E-19
2.20E 01	5.81E-19
2.40E 01	6.52E-19
2.60E 01	6.56E-19
2.80E 01	7.21E-19
3.00E 01	6.71E-19
3.20E 01	6.90E-19
3.40E 01	6.92E-19
3.60E 01	6.82E-19
3.80E 01	6.67E-19
4.00E 01	5.93E-19

9-3-7A
ANDREEV ET AL. (123)

ENERGY (KEV)	CROSS SECTION (SQ. CM)
1.40E 01	5.57E-19
1.60E 01	5.94E-19
1.80E 01	6.67E-19
2.00E 01	7.48E-19
2.20E 01	7.48E-19
2.40E 01	7.24E-19
2.60E 01	7.11E-19
2.80E 01	7.18E-19
3.00E 01	7.41E-19

9-3-7B
THOMAS ET AL. (129)

ENERGY (KEV)	CROSS SECTION (SQ. CM)
1.50E 02	2.70E-21

9-3-8A
GAILY ET AL. (28)

ENERGY (KEV)	POLARIZATION (PERCENT)
6.00E-01	3.60E 01
1.00E 00	3.01E 01
2.00E 00	2.10E 01
3.00E 00	2.20E 01
4.50E 00	1.90E 01
6.00E 00	1.44E 01
7.00E 00	1.14E 01
8.00E 00	1.42E 01
9.00E 00	1.05E 01
1.00E 01	9.00E 00
1.10E 01	9.10E 00
1.20E 01	7.40E 00

9-3-8B
GAILY ET AL. (28)

ENERGY (KEV)	POLARIZATION (PERCENT)
6.00E-01	3.08E 01
1.00E 00	2.77E 01
2.00E 00	1.68E 01
4.00E 00	1.44E 01

9-3-9A
JAECKS ET AL. (16)

ENERGY (KEV)	CROSS SECTION (SQ. CM)
3.00E 00	1.32E-18
5.00E 00	1.67E-18
6.00E 00	2.88E-18
7.00E 00	3.15E-18
1.00E 01	3.30E-18
1.10E 01	3.80E-18
1.20E 01	4.80E-18
1.30E 01	5.55E-18
1.50E 01	6.90E-18
1.70E 01	8.15E-18
1.90E 01	9.10E-18
2.10E 01	8.95E-18
2.30E 01	9.10E-18

9-3-9B
JAECKS ET AL. (16)

ENERGY (KEV)	CROSS SECTION (SQ. CM)
1.50E 00	6.38E-19
2.00E 00	6.72E-19
3.00E 00	1.43E-18
4.00E 00	1.82E-18
6.00E 00	2.79E-18
7.00E 00	3.96E-18

9-3-9C
JAECKS ET AL. (16)

ENERGY (KEV)	CROSS SECTION (SQ. CM)
3.00E 00	1.26E-18
6.00E 00	2.67E-18
7.00E 00	2.94E-18
1.00E 01	3.22E-18
1.10E 01	3.75E-18
1.20E 01	4.67E-18
1.50E 01	6.72E-18

9-3-9D
JAECKS ET AL. (16)

ENERGY (KEV)	CROSS SECTION (SQ. CM)
2.00E 00	6.32E-19
3.00E 00	1.32E-18
4.00E 00	1.66E-18
6.00E 00	2.55E-18
7.00E 00	3.71E-18

9-3-9E
ANDREEV ET AL. (75)

ENERGY (KEV)	CROSS SECTION (SQ. CM)
1.00E 01	4.20E-18
1.20E 01	5.48E-18
1.40E 01	6.92E-18
1.60E 01	7.88E-18
1.80E 01	9.98E-18
2.00E 01	1.11E-17
2.10E 01	1.20E-17
2.20E 01	1.26E-17
2.30E 01	1.27E-17
2.40E 01	1.25E-17
2.60E 01	1.20E-17
2.80E 01	1.18E-17
3.00E 01	1.12E-17
3.20E 01	1.09E-17
3.40E 01	1.05E-17
3.60E 01	9.98E-18
3.80E 01	9.64E-18

9-3-10A
HUGHES ET AL. (120)

ENERGY (KEV)	CROSS SECTION (SQ. CM)
2.00E 00	3.00E-19
3.00E 00	6.00E-19
4.00E 00	1.10E-18
5.00E 00	1.10E-18
6.00E 00	1.00E-18
7.00E 00	1.03E-18
8.00E 00	1.08E-18
9.00E 00	1.13E-18
1.00E 01	1.23E-18

1.25E 01	1.42E-18
1.40E 01	1.55E-18
1.50E 01	1.70E-18
1.75E 01	2.15E-18
2.00E 01	2.62E-18
2.25E 01	3.00E-18
2.50E 01	3.15E-18
2.75E 01	3.20E-18
3.00E 01	3.15E-18
3.50E 01	2.96E-18
4.00E 01	2.69E-18
4.50E 01	2.45E-18
5.00E 01	2.24E-18
5.50E 01	2.05E-18
6.00E 01	1.91E-18
6.50E 01	1.84E-18
7.00E 01	1.80E-18
7.50E 01	1.74E-18
8.00E 01	1.68E-18
8.50E 01	1.59E-18
9.00E 01	1.53E-18
9.50E 01	1.48E-18
1.00E 02	1.46E-18
1.05E 02	1.42E-18
1.10E 02	1.44E-18
1.15E 02	1.38E-18

9-3-10B
ANDREEV ET AL. (123)

ENERGY (KEV)	CROSS SECTION (SQ. CM)
1.20E 01	4.13E-19
1.40E 01	5.62E-19
1.60E 01	9.16E-19
1.80E 01	1.20E-18
1.96E 01	1.47E-18
2.00E 01	1.83E-18
2.20E 01	2.33E-18
2.40E 01	2.27E-18
2.60E 01	2.17E-18
2.80E 01	2.03E-18
3.00E 01	2.05E-18

9-3-11
HUGHES ET AL. (353)

ENERGY (KEV)	CROSS SECTION (SQ. CM)
2.00E 00	1.20E-19
3.00E 00	1.50E-19
4.00E 00	2.20E-19
5.00E 00	3.40E-19
6.00E 00	3.50E-19
7.00E 00	3.20E-19
8.00E 00	2.90E-19
9.00E 00	3.10E-19
1.00E 01	2.60E-19
1.10E 01	3.50E-19
1.20E 01	3.10E-19
1.30E 01	3.60E-19
1.40E 01	3.60E-19
1.50E 01	4.30E-19
2.00E 01	7.30E-19
2.50E 01	9.40E-19
3.00E 01	1.07E-18

4.00E 01	9.80E-19
5.00E 01	8.30E-19
6.00E 01	7.30E-19
7.00E 01	6.90E-19
8.00E 01	6.50E-19
9.00E 01	5.60E-19
1.00E 02	4.80E-19
1.20E 02	4.00E-19

9-3-12A
DE HEER ET AL. (78)

ENERGY (KEV)	CROSS SECTION (SQ. CM)
7.50E 00	2.19E-17
1.00E 01	3.18E-17
1.25E 01	2.86E-17
1.50E 01	2.59E-17
2.00E 01	2.21E-17
2.50E 01	1.87E-17
3.00E 01	1.67E-17
3.50E 01	1.30E-17

9-3-12B
PRETZER ET AL. (117)

ENERGY (KEV)	CROSS SECTION (SQ. CM)
1.70E 00	2.49E-18
2.20E 00	2.34E-18
3.00E 00	5.51E-18
3.50E 00	7.23E-18
4.00E 00	8.10E-18
4.60E 00	9.32E-18
5.00E 00	9.19E-18
5.50E 00	9.25E-18
6.00E 00	9.48E-18
7.00E 00	8.99E-18
8.00E 00	8.54E-18
9.00E 00	8.18E-18
1.00E 01	8.08E-18
1.10E 01	8.31E-18
1.20E 01	8.24E-18
1.30E 01	8.27E-18
1.40E 01	8.32E-18
1.50E 01	8.15E-18
1.60E 01	7.92E-18
1.70E 01	8.09E-18
1.80E 01	7.76E-18
1.90E 01	7.33E-18
2.01E 01	7.18E-18
2.05E 01	7.17E-18
2.20E 01	6.89E-18

9-3-12C
PRETZER ET AL. (117)

ENERGY (KEV)	CROSS SECTION (SQ. CM)
2.50E-01	8.30E-19
3.70E-01	7.70E-19
5.00E-01	7.60E-19
7.30E-01	1.07E-18

7.50E-01	7.40E-19
1.00E 00	1.00E-18
1.25E 00	1.15E-18
1.50E 00	1.51E-18
1.75E 00	1.72E-18
2.00E 00	2.05E-18
2.25E 00	2.62E-18
2.50E 00	3.27E-18
2.75E 00	4.30E-18
3.00E 00	5.11E-18
3.50E 00	6.83E-18
4.00E 00	7.50E-18
4.50E 00	8.01E-18
5.00E 00	8.54E-18
5.50E 00	8.53E-18
6.00E 00	8.65E-18
6.50E 00	8.38E-18
7.00E 00	8.19E-18
7.50E 00	7.86E-18
8.00E 00	8.08E-18
8.50E 00	7.64E-18
9.00E 00	7.41E-18
9.50E 00	7.78E-18
1.00E 01	7.82E-18
1.05E 01	7.65E-18
1.10E 01	7.79E-18
1.15E 01	7.91E-18
1.20E 01	8.01E-18
1.25E 01	7.20E-18

9-3-12D
GAILY ET AL. (28)

ENERGY (KEV)	CROSS SECTION (SQ. CM)
6.00E-01	6.00E-19
1.00E 00	8.30E-19
2.00E 00	2.34E-18
3.00E 00	5.03E-18
4.50E 00	8.33E-18
6.00E 00	8.78E-18
7.00E 00	8.39E-18
8.00E 00	8.01E-18
9.00E 00	7.94E-18
1.00E 01	7.89E-18
1.10E 01	8.21E-18
1.20E 01	8.01E-18
1.50E 01	7.94E-18

9-3-12E
GAILY ET AL. (28)

ENERGY (KEV)	CROSS SECTION (SQ. CM)
4.50E-01	6.60E-19
6.00E-01	7.26E-19
1.00E 00	9.31E-19
2.00E 00	1.93E-18
2.50E 00	3.03E-18
3.00E 00	4.73E-18
3.50E 00	6.30E-18
4.00E 00	6.85E-18
4.50E 00	7.38E-18
5.00E 00	7.89E-18
6.00E 00	7.92E-18
6.50E 00	7.73E-18

7.00E 00	7.67E-18
7.50E 00	7.42E-18
8.00E 00	7.69E-18

9-3-12F
ANDREEV ET AL. (75)

ENERGY (KEV)	GROSS SECTION (SQ. CM)
1.00E 01	7.88E-18
1.20E 01	8.08E-18
1.40E 01	8.26E-18
1.60E 01	8.82E-18
1.80E 01	8.40E-18
2.00E 01	7.86E-18
2.20E 01	7.48E-18
2.30E 01	7.52E-18
2.40E 01	7.06E-18
2.47E 01	6.86E-18
2.58E 01	6.66E-18
2.78E 01	6.32E-18
3.00E 01	6.04E-18
3.18E 01	5.66E-18
3.40E 01	5.22E-18
3.57E 01	4.52E-18
3.80E 01	4.18E-18

9-3-13A
GAILY ET AL. (28)

ENERGY (KEV)	CROSS SECTION (SQ. CM)
6.00E-01	5.40E-19
1.00E 00	6.37E-19
2.00E 00	1.45E-18
3.00E 00	3.63E-18
4.50E 00	5.83E-18
6.00E 00	5.75E-18
7.00E 00	5.25E-18
8.00E 00	4.84E-18
9.00E 00	3.65E-18
1.00E 01	3.43E-18
1.10E 01	3.18E-18
1.20E 01	3.61E-18
1.50E 01	3.54E-18

9-3-13B
GAILY ET AL. (28)

ENERGY (KEV)	CROSS SECTION (SQ. CM)
6.00E-01	3.02E-20
1.00E 00	9.68E-20
2.00E 00	4.47E-19
3.00E 00	7.07E-19
4.50E 00	1.25E-18
6.00E 00	1.52E-18
7.00E 00	1.57E-18
8.00E 00	1.58E-18
9.00E 00	2.15E-18
1.00E 01	2.23E-18
1.10E 01	2.51E-18
1.20E 01	2.20E-18

1.50E 01   2.19E-18

9-3-13C
GAILY ET AL. (28)

ENERGY (KEV)	CROSS SECTICN (SQ. CM)
4.50E-01	5.57E-19
6.00E-01	5.41E-19
1.00E 00	5.84E-19
2.00E CO	1.12E-18
2.50E CO	1.97E-18
3.00E 00	3.13E-18
3.50E 00	4.25E-18
4.00E 00	4.94E-18
4.50E 00	5.03E-18
5.00E 00	5.30E-18
6.00E 00	5.64E-18
6.50E 00	5.24E-18
7.00E 00	4.69E-18
7.5CE 00	4.23E-18
8.00E 00	4.18E-18

9-3-13D
GAILY ET AL. (28)

ENERGY (KEV)	CROSS SECTICN (SQ. CM)
4.50E-01	5.10E-20
6.00E-01	9.20E-20
1.00E 00	1.73E-19
2.0CE 00	4.01E-19
2.50E 00	5.32E-19
3.0CE 00	7.98E-19
3.50E 00	1.03E-18
4.0CE 00	9.58E-19
4.50E 00	1.17E-18
5.00E 00	1.30E-18
6.00E 00	1.14E-18
6.50E 00	1.25E-18
7.0CE 00	1.49E-18
7.50E 00	1.60E-18
8.00E 00	1.76E-18

9-3-14A
CE HEER ET AL. (78)

ENERGY (KEV)	CROSS SECTION (SQ. CM)
7.50E 00	2.75E-18
1.00E 01	2.49E-18
1.25E 01	2.64E-18
1.50E 01	2.65E-18
1.75E 01	2.64E-18
2.00E 01	2.64E-18
2.50E 01	2.35E-18
3.00E 01	2.26E-18
3.50E 01	2.09E-18

9-3-14B
ANDREEV ET AL. (122)

ENERGY (KEV)	CROSS SECTION (SQ. CM)
1.00E 01	7.94E-19
1.20E 01	9.95E-19
1.40E 01	1.15E-18
1.60E 01	1.29E-18
1.80E 01	1.38E-18
2.00E 01	1.44E-18
2.20E 01	1.48E-18
2.40E 01	1.47E-18
2.60E 01	1.44E-18
2.80E 01	1.39E-18
3.00E 01	1.32E-18
3.20E 01	1.25E-18
3.40E 01	1.15E-18
3.60E 01	1.04E-18
3.80E 01	9.20E-19
4.00E 01	8.32E-19

9-3-14C
ANDREEV ET AL. (122)

ENERGY (KEV)	CROSS SECTICN (SQ. CM)
1.00E 01	7.28E-19
1.20E 01	8.91E-19
1.40E 01	1.06E-18
1.60E 01	1.17E-18
1.80E 01	1.26E-18
2.00E 01	1.31E-18
2.20E 01	1.31E-18
2.40E 01	1.32E-18
2.60E 01	1.29E-18
2.80E 01	1.25E-18
3.00E 01	1.16E-18
3.20E 01	1.14E-18
3.40E 01	1.05E-18
3.60E 01	9.42E-19
3.80E 01	8.20E-19
4.00E 01	7.58E-19

9-3-15
ANDREEV ET AL. (123)

ENERGY (KEV)	CROSS SECTICN (SQ. CM)
1.20E 01	1.95E-18
1.40E 01	1.92E-18
1.60E 01	1.81E-18
1.80E 01	1.79E-18
2.00E 01	1.69E-18
2.20E 01	1.20E-18
2.40E 01	1.26E-18
2.64E 01	1.39E-18
2.80E 01	1.26E-18
3.00E 01	1.43E-18

9-3-16A
GAILY ET AL. (28)

ENERGY (KEV)	POLARIZATION (PERCENT)
6.00E-01	3.67E 01
1.00E 00	2.89E 01
2.00E 00	1.94E 01
3.00E 00	2.61E 01
4.50E 00	2.47E 01
6.00E 00	2.20E 01
7.00E 00	2.00E 01
8.00E 00	1.87E 01
9.00E 00	8.90E 00
1.00E 01	7.10E 00
1.10E 01	3.70E 00
1.20E 01	8.30E 00
1.50E 01	7.90E 00

9-3-16B
GAILY ET AL. (28)

ENERGY (KEV)	POLARIZATION (PERCENT)
4.50E-01	3.31E 01
6.00E-01	2.71E 01
1.00E 00	1.97E 01
2.00E 00	1.67E 01
2.50E 00	2.11E 01
3.00E 00	2.20E 01
3.50E 00	2.26E 01
4.00E 00	2.55E 01
4.5CE 00	2.30E 01
5.00E 00	2.23E 01
6.0CE CO	2.51E 01
6.50E CO	2.28E 01
7.00E 00	1.88E 01
7.50E 00	1.56E 01
8.00E 00	1.44E 01

9-3-17A
JAFCKS ET AL. (16)

ENERGY (KEV)	CROSS SECTION (SQ. CM)
2.0CE 00	1.19E-17
3.00E 00	1.10E-17
4.0CE 00	9.83E-18
5.00E 00	6.93E-18
6.00E 00	7.10E-18
7.0CE 00	5.68E-18
8.00E 00	7.65E-18
9.0CE 00	1.08E-17
1.00E 01	1.35E-17
1.10E 01	1.76E-17
1.20E 01	1.86E-17
1.30E 01	2.19E-17
1.40E 01	2.49E-17
1.50E 01	2.71E-17
1.60E 01	2.46E-17
1.70E 01	2.54E-17
1.80E 01	2.78E-17
1.90E 01	2.67E-17
2.00E 01	3.05E-17

2.10E 01	2.57E-17
2.20E 01	2.57E-17
2.30E 01	2.80E-17

9-3-17B
JAECKS ET AL. (16)

ENERGY (KEV)	CROSS SECTION (SQ. CM)
1.50E 00	8.60E-18
2.00E 00	9.62E-18
3.00E 00	1.04E-17
4.00E 00	8.13E-18

9-3-17C
JAECKS ET AL. (16)

ENERGY (KEV)	CROSS SECTION (SQ. CM)
2.00E 00	1.12E-17
3.00E 00	1.04E-17
6.00E 00	6.70E-18
7.00E 00	5.51E-18
8.00E 00	7.33E-18
9.00E 00	1.06E-17
1.00E 01	1.32E-17
1.10E 01	1.73E-17
1.20E 01	1.82E-17

9-3-17D
ANDREEV ET AL. (75)

ENERGY (KEV)	CROSS SECTION (SQ. CM)
8.00E 00	1.01E-17
9.00E 00	1.53E-17
1.00E 01	2.21E-17
1.10E 01	2.23E-17
1.20E 01	2.46E-17
1.30E 01	2.78E-17
1.40E 01	3.15E-17
1.60E 01	3.25E-17
1.80E 01	3.02E-17
2.00E 01	3.13E-17
2.20E 01	2.89E-17
2.40E 01	3.01E-17
2.60E 01	3.01E-17
2.80E 01	3.02E-17
3.00E 01	3.15E-17
3.20E 01	2.97E-17
3.40E 01	3.02E-17
3.60E 01	3.00E-17
3.80E 01	2.86E-17
4.20E 01	2.58E-17

9-3-17E
BAYFIELD (130)

ENERGY (KEV)	CROSS SECTION (SQ. CM)
2.00E 00	9.90E-18
3.00E 00	9.00E-18
3.12E 00	9.50E-18
3.47E 00	9.60E-18
4.00E 00	8.60E-18
4.89E 00	7.30E-18
5.47E 00	8.10E-18
5.97E 00	6.60E-18
6.95E 00	1.07E-17
7.62E 00	1.18E-17
7.91E 00	1.41E-17
9.04E 00	1.79E-17
9.64E 00	2.10E-17
1.05E 01	2.03E-17
1.12E 01	2.54E-17
1.21E 01	2.82E-17
1.25E 01	2.59E-17
1.30E 01	2.92E-17
1.37E 01	2.83E-17
1.47E 01	3.02E-17
1.60E 01	2.95E-17
1.75E 01	3.08E-17
1.83E 01	2.82E-17
1.88E 01	3.12E-17
1.95E 01	2.91E-17
2.14E 01	3.08E-17
2.22E 01	2.86E-17
2.42E 01	2.96E-17
2.59E 01	2.84E-17
2.82E 01	2.74E-17
3.03E 01	2.73E-17
3.12E 01	2.52E-17
3.44E 01	2.42E-17
3.75E 01	2.27E-17
4.05E 01	2.37E-17
4.57E 01	2.00E-17
5.10E 01	1.71E-17
5.49E 01	1.43E-17
6.03E 01	1.31E-17
6.52E 01	1.47E-17
7.01E 01	1.24E-17

9-3-18A
HUGHES ET AL. (120)

ENERGY (KEV)	CROSS SECTION (SQ. CM)
4.00E 00	3.20E-18
5.00E 00	3.70E-18
6.00E 00	3.70E-18
7.00E 00	3.48E-18
8.00E 00	3.52E-18
9.00E 00	3.73E-18
1.00E 01	3.95E-18
1.25E 01	4.90E-18
1.50E 01	5.75E-18
2.00E 01	7.40E-18
2.50E 01	7.90E-18
3.00E 01	7.75E-18
4.00E 01	6.40E-18
5.00E 01	5.28E-18
6.00E 01	4.52E-18
7.00E 01	4.03E-18
8.00E 01	3.53E-18
9.00E 01	3.05E-18
1.00E 02	2.66E-18
1.10E 02	2.33E-18
1.15E 02	2.22E-18

9-3-18B
ANDREEV ET AL. (123)

ENERGY (KEV)	CROSS SECTION (SQ. CM)
1.00E 01	1.64E-18
1.20E 01	3.06E-18
1.40E 01	2.62E-18
1.80E 01	3.89E-18
1.96E 01	4.59E-18
2.00E 01	4.70E-18
2.20E 01	4.63E-18
2.40E 01	5.81E-18
2.60E 01	4.61E-18
3.00E 01	5.27E-18

9-3-18C
THOMAS ET AL. (129)

ENERGY (KEV)	CROSS SECTION (SQ. CM)
1.68E 02	4.70E-19

9-3-19
HUGHES ET AL. (353)

ENERGY (KEV)	CROSS SECTION (SQ. CM)
5.00E 00	1.02E-18
6.00E 00	1.27E-18
7.00E 00	1.15E-18
8.00E 00	1.19E-18
9.00E 00	1.17E-18
1.00E 01	1.27E-18
1.10E 01	1.32E-18
1.30E 01	1.40E-18
1.50E 01	1.67E-18
1.70E 01	2.00E-18
2.00E 01	2.51E-18
2.60E 01	2.65E-18
3.00E 01	2.70E-18
4.00E 01	2.51E-18
5.00E 01	2.27E-18
6.00E 01	1.98E-18
7.00E 01	1.84E-18
8.00E 01	1.57E-18
9.00E 01	1.24E-18
1.00E 02	9.90E-19
1.10E 02	8.50E-19

9-3-20A
PRETZER ET AL. (117)

ENERGY (KEV)	CROSS SECTION (SQ. CM)
1.00E 00	2.43E-17
1.50E 00	2.72E-17
2.00E 00	3.28E-17
2.50E 00	3.85E-17
2.70E 00	3.96E-17
3.00E 00	4.33E-17

3.20E 00	4.35E-17
3.50E 00	4.35E-17
4.00E 00	4.29E-17
4.50E 00	4.10E-17
5.00E 00	4.13E-17
6.00E 00	3.93E-17
7.0CE 00	4.05E-17
8.00E 00	4.24E-17
9.00E 00	4.49E-17
1.00E 01	4.72E-17
1.10E 01	4.74E-17
1.20E 01	4.74E-17
1.30E 01	4.83E-17
1.40E 01	4.70E-17
1.60E 01	4.47E-17
1.80E 01	4.22E-17
1.90E 01	3.90E-17
2.00E 01	3.78E-17
2.20E 01	3.34E-17
2.43E 01	3.21E-17

9-3-20B
PRETZER ET AL. (117)

ENERGY (KEV)	CROSS SECTION (SQ. CM)
5.00E-01	1.57E-17
7.50E-01	1.92E-17
1.00E 00	2.19E-17
1.25E 00	2.49E-17
1.50E 00	2.74E-17
1.75E 00	2.95E-17
2.00E 00	3.19E-17
2.25E 00	3.60E-17
2.50E 00	3.82E-17
2.75E 00	3.98E-17
3.0CE 00	4.10E-17
3.50E 00	4.22E-17
4.0CE 00	4.17E-17
4.50E 00	3.96E-17
5.00E 00	3.76E-17
5.5CE 00	3.76E-17
6.00E 00	3.76E-17
6.50E 00	3.92E-17
7.00E 00	4.14E-17
7.50E 00	4.25E-17
8.00E 00	4.20E-17
8.50E 00	4.43E-17
9.00E 00	4.64E-17
9.50E 00	4.79E-17
1.00E 01	4.67E-17
1.05E 01	4.67E-17
1.10E 01	4.65E-17
1.15E 01	4.70E-17
1.20E 01	4.60E-17
1.25E 01	4.58E-17
1.30E 01	4.20E-17

9-3-20C
GAILY ET AL. (28)

ENERGY (KEV)	CROSS SECTION (SQ. CM)
6.00E-01	9.74E-18
1.00E 00	2.32E-17
2.00E 00	3.08E-17

3.00E 00	4.10E-17
4.50E 00	3.88E-17
6.CCE 00	3.73E-17
6.50E 00	3.74E-17
7.00E 00	3.93E-17
7.50E 00	3.97E-17
8.00E 00	4.06E-17
8.50E 00	4.16E-17
9.00E 00	4.40E-17
1.00E 01	4.63E-17
1.10E 01	4.66E-17
1.20E 01	4.65E-17

9-3-20D
ANDREEV ET AL. (75)

ENERGY (KEV)	CROSS SECTION (SQ. CM)
8.00E 00	4.08E-17
9.00E 00	4.19E-17
1.00E 01	4.67E-17
1.10E 01	4.90E-17
1.20E 01	5.23E-17
1.30E 01	5.20E-17
1.38E 01	5.04E-17
1.60E 01	4.74E-17
1.80E 01	4.40E-17
2.00E 01	4.12E-17
2.20E 01	3.73E-17
2.40E 01	3.33E-17
2.60E 01	3.21E-17
2.80E 01	2.83E-17
3.00E 01	2.47E-17
3.20E 01	2.28E-17
3.40E 01	2.08E-17
3.60E 01	1.94E-17
3.80E 01	1.77E-17
4.00E 01	1.54E-17
4.20E 01	1.40E-17

9-3-21A
GAILY ET AL. (28)

ENERGY (KEV)	CROSS SECTION (SQ. CM)
6.00E-01	6.72E-18
1.00E 00	1.24E-17
2.00E 00	1.86E-17
3.00E 00	2.30E-17
4.50E 00	2.19E-17
6.00E 00	2.05E-17
6.50E 00	2.20E-17
7.0CE 00	1.81E-17
7.50E 00	1.86E-17
8.00E 00	2.07E-17
8.50E 00	1.90E-17
9.00E 00	1.84E-17
1.00E 01	1.92E-17
1.10E 01	1.90E-17
1.20E 01	1.93E-17

9-3-21B
GAILY ET AL. (28)

ENERGY (KEV)	CROSS SECTION (SQ. CM)
6.00E-01	1.51E-18
1.00E 00	5.39E-18
2.0CE 00	6.14E-18
3.00E 00	8.97E-18
4.50E 00	8.43E-18
6.00E 00	8.43E-18
6.50E 00	7.70E-18
7.00E 00	1.06E-17
7.50E 00	1.05E-17
8.00E 00	9.94E-18
8.50E 00	1.13E-17
9.00E 00	1.28E-17
1.00E 01	1.35E-17
1.10E 01	1.38E-17
1.20E 01	1.36E-17

9-3-22A
ANDREEV ET AL. (122)

ENERGY (KEV)	CROSS SECTION (SQ. CM)
1.00E 01	3.44E-17
1.20E 01	4.22E-17
1.40E 01	5.20E-17
1.60E 01	5.43E-17
1.80E 01	5.41E-17
2.00E 01	5.27E-17
2.20E 01	5.10E-17
2.40E 01	4.81E-17
2.60E 01	4.55E-17
2.80E 01	4.47E-17
3.00E 01	4.22E-17
3.20E 01	3.98E-17
3.40E 01	3.66E-17
3.60E 01	3.49E-17
3.80E 01	3.48E-17
4.00E 01	3.19E-17

9-3-22B
ANDREEV ET AL. (122)

ENERGY (KEV)	CROSS SECTION (SQ. CM)
1.00E 01	3.87E-17
1.20E 01	4.71E-17
1.40E 01	5.64E-17
1.60E 01	6.00E-17
1.80E 01	5.98E-17
2.00E 01	5.76E-17
2.20E 01	5.59E-17
2.40E 01	5.27E-17
2.60E 01	5.04E-17
2.80E 01	4.87E-17
3.00E 01	4.59E-17
3.20E 01	4.38E-17
3.40E 01	4.07E-17
3.60E 01	3.89E-17
3.80E 01	3.84E-17
4.00E 01	3.48E-17

9-3-23A
ANDREEV ET AL. (123)

ENERGY (KEV)	CROSS SECTION (SQ. CM)
1.00E 01	1.01E-17
1.20E 01	9.90E-18
1.40E 01	8.63E-18
1.60E 01	8.31E-18
1.80E 01	7.24E-18
2.00E 01	6.85E-18
2.20E 01	6.22E-18
2.43E 01	5.49E-18
3.00E 01	4.28E-18

9-3-23B
THOMAS ET AL. (129)

ENERGY (KEV)	CROSS SECTION (SQ. CM)
1.68E 02	1.10E-19

9-3-24A
GAILY ET AL. (28)

ENERGY (KEV)	POLARIZATION (PERCENT)
6.00E-01	2.43E 01
1.00E 00	1.40E 01
2.00E 00	1.84E 01
3.00E 00	1.59E 01
4.50E 00	1.60E 01
6.00E 00	1.50E 01
6.50E 00	1.75E 01
7.00E 00	8.80E 00
7.50E 00	9.50E 00
8.00E 00	1.24E 01
8.50E 00	8.70E 00
9.00E 00	5.90E 00
1.00E 01	5.70E 00
1.10E 01	5.20E 00
1.20E 01	5.70E 00

9-3-24B
TEUBNER ET AL. (128)

ENERGY (KEV)	POLARIZATION (PERCENT)
4.00E-01	-5.20E 00
6.00E-01	-1.23E 01
1.00E 00	-1.37E 01
1.25E 00	-3.80E 00
1.50E 00	1.60E 00
2.00E 00	5.90E 00
2.55E 00	4.90E 00
3.00E 00	3.70E 00
3.55E 00	4.30E 00
4.00E 00	3.80E 00
4.50E 00	2.30E 00
5.00E 00	-1.20E 00
5.5CE 00	-1.10E 00

6.00E 00	-2.40E 00
6.50E 00	-3.00E 00
7.00E 00	-5.40E 00
7.50E 00	-5.10E 00
8.00E 00	-6.60E 00
9.00E 00	-6.00E 00
1.00E 01	-5.00E 00
1.20E 01	-3.70E 00
1.60E 01	-4.50E 00
2.00E 01	-5.00E 00

9-3-24C
TEUBNER ET AL. (128)

ENERGY (KEV)	POLARIZATION (PERCENT)
3.00E-01	-4.80E 00
4.00E-01	-1.58E 01
6.00E-01	-1.73E 01
1.00E 00	-1.02E 01
2.00E 00	9.00E-01
4.25E 00	3.10E 00

9-3-25
ANDREEV ET AL. (75)

ENERGY (KEV)	CROSS SECTION (SQ. CM)
1.00E 01	4.52E-17
1.20E 01	4.59E-17
1.40E 01	4.98E-17
1.60E 01	5.12E-17
1.80E 01	5.20E-17
2.00E 01	5.23E-17
2.20E 01	5.11E-17
2.58E 01	4.95E-17
2.80E 01	4.55E-17
3.00E 01	4.30E-17
3.20E 01	4.15E-17
3.60E 01	3.69E-17
3.80E 01	3.54E-17
4.00E 01	3.36E-17

9-3-26
ANDREEV ET AL. (75)

ENERGY (KEV)	CROSS SECTION (SQ. CM)
1.00E 01	7.07E-17
1.20E 01	7.45E-17
1.40E 01	7.31E-17
1.60E 01	7.11E-17
1.80E 01	6.65E-17
2.00E 01	6.15E-17
2.20E 01	5.70E-17
2.38E 01	5.23E-17
2.58E 01	4.64E-17
2.80E 01	4.27E-17
3.00E 01	3.84E-17
3.17E 01	3.60E-17
3.40E 01	3.36E-17
3.60E 01	3.25E-17
3.80E 01	3.07E-17

4.00E 01	2.76E-17

9-3-27A
ANDREEV ET AL. (122)

ENERGY (KEV)	CROSS SECTION (SQ. CM)
1.00E 01	5.97E-18
1.20E 01	7.69E-18
1.40E 01	9.03E-18
1.60E 01	9.81E-18
1.80E 01	9.81E-18
2.00E 01	9.24E-18
2.20E 01	9.03E-18
2.40E 01	8.59E-18
2.60E 01	8.39E-18
2.80E 01	7.72E-18
3.00E 01	7.41E-18
3.20E 01	6.76E-18
3.40E 01	6.85E-18
3.60E 01	6.36E-18
3.80E 01	5.94E-18
4.00E 01	5.39E-18

9-3-27B
ANDREEV ET AL. (122)

ENERGY (KEV)	CROSS SECTION (SQ. CM)
1.00E 01	6.51E-18
1.20E 01	8.20E-18
1.40E 01	9.59E-18
1.60E 01	1.06E-17
1.80E 01	1.04E-17
2.00E 01	1.00E-17
2.20E 01	9.55E-18
2.40E 01	9.20E-18
2.60E 01	9.03E-18
2.80E 01	8.27E-18
3.00E 01	7.87E-18
3.20E 01	7.31E-18
3.40E 01	7.31E-18
3.60E 01	6.85E-18
3.80E 01	6.31E-18
4.00E 01	5.86E-18

9-3-28A
JAECKS ET AL. (16)

ENERGY (KEV)	CROSS SECTION (SQ. CM)
8.00E 00	5.17E-17
9.00E 00	4.49E-17
1.00E 01	5.46E-17
1.10E 01	5.00E-17
1.20E 01	5.70E-17
1.30E 01	5.76E-17
1.50E 01	6.30E-17
1.60E 01	6.30E-17
1.80E 01	5.98E-17
2.00E 01	5.46E-17
2.20E 01	5.17E-17

356

9-3-28B
JAECKS ET AL. (16)

ENERGY (KEV)	CROSS SECTION (SQ. CM)
1.00E 00	6.00E-18
2.00E 00	1.50E-18
3.00E 00	1.24E-17
5.00E 00	2.73E-17
6.00E 00	3.50E-17
7.00E 00	4.00E-17
8.00E 00	4.57E-17
1.00E 01	5.40E-17

9-3-28C
ANDREEV ET AL. (75)

ENERGY (KEV)	CROSS SECTION (SQ. CM)
8.00E 00	5.74E-17
1.00E 01	6.94E-17
1.20E 01	7.80E-17
1.40E 01	8.30E-17
1.60E 01	8.56E-17
1.80E 01	8.54E-17
2.00E 01	8.28E-17
2.40E 01	7.78E-17
2.60E 01	7.50E-17
2.80E 01	7.56E-17
3.00E 01	6.88E-17
3.20E 01	6.88E-17
3.40E 01	6.66E-17
3.60E 01	6.30E-17
3.80E 01	6.10E-17
4.00E 01	5.76E-17

9-3-29A
PRETZER ET AL. (117)

ENERGY (KEV)	CROSS SECTION (SQ. CM)
8.00E-01	1.64E-16
1.00E 00	1.23E-16
1.50E 00	9.49E-17
2.00E 00	6.84E-17
2.50E 00	5.70E-17
3.00E 00	5.27E-17
3.5CE 00	5.35E-17
4.00E 00	5.59E-17
5.0CE 00	7.05E-17
6.00E 00	8.29E-17
7.0CE 00	9.30E-17
8.00E 00	1.00E-16
9.00E 00	1.03E-16
1.00E 01	1.04E-16
1.10E 01	1.03E-16
1.15E 01	1.02E-16
1.20E 01	9.78E-17
1.30E 01	9.68E-17
1.40E 01	9.15E-17
1.50E 01	8.85E-17
1.60E 01	8.26E-17
1.70E 01	7.93E-17
1.80E 01	7.14E-17
1.90E 01	6.69E-17
2.00E 01	6.77E-17
2.10E 01	6.49E-17
2.20E 01	6.07E-17

9-3-29B
PRETZER ET AL. (117)

ENERGY (KEV)	CROSS SECTION (SQ. CM)
3.00E-01	1.83E-16
3.50E-01	1.99E-16
4.00E-01	2.13E-16
4.90E-01	1.83E-16
6.25E-01	1.62E-16
7.50E-01	1.40E-16
1.00E 00	1.21E-16
1.25E 00	1.08E-16
1.50E 00	9.16E-17
1.75E 00	8.37E-17
2.00E 00	7.07E-17
2.25E 00	6.50E-17
2.50E 00	6.10E-17
2.75E 00	5.64E-17
3.00E 00	5.33E-17
3.25E 00	5.44E-17
3.50E 00	5.95E-17
4.0CE 00	6.10E-17
4.50E 00	7.31E-17
5.00E 00	7.41E-17
5.5CE 00	8.56E-17
6.00E 00	9.12E-17
6.5CE 00	1.04E-16
7.00E 00	1.06E-16
7.5CE 00	1.11E-16
8.0CE 00	1.14E-16
8.50E 00	1.22E-16
9.00E 00	1.22E-16
9.50E 00	1.21E-16
1.00E 01	1.25E-16
1.05E 01	1.24E-16
1.10E 01	1.22E-16

9-3-29C
ANDREEV ET AL. (75)

ENERGY (KEV)	CROSS SECTION (SQ. CM)
8.12E 00	1.11E-16
1.00E 01	1.15E-16
1.22E 01	1.15E-16
1.40E 01	1.09E-16
1.60E 01	1.02E-16
1.80E 01	9.56E-17
2.00E 01	8.84E-17
2.20E 01	8.26E-17
2.40E 01	7.20E-17
2.60E 01	6.76E-17
2.80E 01	6.26E-17
3.00E 01	5.80E-17
3.20E 01	5.66E-17
3.40E 01	5.34E-17
3.60E 01	5.06E-17
3.80E 01	4.66E-17
4.00E 01	4.46E-17

9-3-30A
ANDREEV ET AL. (122)

ENERGY (KEV)	CROSS SECTION (SQ. CM)
1.00E 01	1.04E-17
1.20E 01	1.15E-17
1.40E 01	1.30E-17
1.60E 01	1.39E-17
1.80E 01	1.42E-17
2.00E 01	1.37E-17
2.20E 01	1.23E-17
2.40E 01	1.16E-17
2.60E 01	1.09E-17
2.80E 01	1.01E-17
3.00E 01	9.50E-18
3.20E 01	9.20E-18
3.40E 01	8.55E-18
3.60E 01	8.32E-18
3.80E 01	8.24E-18
4.00E 01	8.02E-18

9-3-30B
ANDREEV ET AL. (122)

ENERGY (KEV)	CROSS SECTION (SQ. CM)
1.00E 01	1.09E-17
1.20E 01	1.23E-17
1.40E 01	1.37E-17
1.60E 01	1.47E-17
1.80E 01	1.50E-17
2.00E 01	1.47E-17
2.20E 01	1.35E-17
2.40E 01	1.27E-17
2.60E 01	1.17E-17
2.80E 01	1.09E-17
3.00E 01	1.02E-17
3.20E 01	9.73E-18
3.40E 01	9.16E-18
3.60E 01	8.95E-18
3.80E 01	8.67E-18
4.00E 01	8.35E-18

9-3-31A
COLLI ET AL. (145)

ENERGY (KEV)	CROSS SECTION (SQ. CM)
9.00E 00	6.10E-19
1.21E 01	7.30E-19
1.80E 01	1.50E-18
2.02E 01	1.80E-18
2.40E 01	2.45E-18
2.70E 01	2.50E-18
3.00E 01	1.80E-18
3.38E 01	1.30E-18
3.68E 01	1.00E-18
4.00E 01	8.40E-19

9-3-31B
SELLIN (17)

ENERGY (KEV)	CROSS SECTION (SQ. CM)
1.50E 01	1.00E-17

9-3-31C
RYDING ET AL. (142)

ENERGY (KEV)	CROSS SECTION (SQ. CM)
4.40E 01	9.15E-18
5.00E 01	8.03E-18
6.00E 01	5.96E-18
7.00E 01	4.87E-18
8.00E 01	3.59E-18
9.00E 01	2.60E-18
1.00E 02	2.04E-18
1.20E 02	1.17E-18
1.40E 02	5.71E-19
1.60E 02	3.35E-19
1.80E 02	2.30E-19

9-3-31D
ANDREEV ET AL. (148)

ENERGY (KEV)	CROSS SECTION (SQ. CM)
1.00E 01	1.42E-17
1.20E 01	1.77E-17
1.40E 01	2.25E-17
1.60E 01	2.88E-17
1.80E 01	3.58E-17
2.00E 01	3.74E-17
2.20E 01	3.95E-17
2.60E 01	3.93E-17
2.80E 01	4.07E-17
3.20E 01	3.64E-17
3.40E 01	3.70E-17

9-3-31E
BAYFIELD (130)

ENERGY (KEV)	CROSS SECTION (SQ. CM)
2.00E 00	3.00E-18
3.03E 00	2.66E-18
4.01E 00	5.44E-18
4.63E 00	6.88E-18
5.12E 00	7.76E-18
5.59E 00	7.83E-18
6.08E 00	7.94E-18
6.63E 00	8.16E-18
7.14E 00	8.01E-18
7.65E 00	8.27E-18
8.24E 00	8.55E-18
8.71E 00	8.39E-18
9.20E 00	9.07E-18
1.00E 01	1.05E-17
1.15E 01	1.16E-17
1.25E 01	1.35E-17
1.31E 01	1.32E-17
1.46E 01	1.54E-17
1.53E 01	1.62E-17
1.60E 01	1.81E-17
1.77E 01	2.16E-17
1.89E 01	2.06E-17
2.04E 01	2.33E-17
2.24E 01	2.43E-17
2.45E 01	2.43E-17
2.66E 01	2.59E-17
2.84E 01	2.62E-17
3.04E 01	2.68E-17
3.29E 01	2.67E-17
3.51E 01	2.62E-17
3.75E 01	2.35E-17
4.11E 01	2.04E-17
4.55E 01	1.99E-17
5.05E 01	1.80E-17
5.67E 01	1.44E-17
6.16E 01	1.25E-17
6.69E 01	1.12E-17
7.11E 01	8.83E-18

9-3-32
HUGHES ET AL. (120)

ENERGY (KEV)	CROSS SECTION (SQ. CM)
5.00E 00	9.79E-19
7.50E 00	1.66E-18
1.00E 01	2.72E-18
1.25E 01	3.27E-18
1.50E 01	3.94E-18
1.75E 01	4.85E-18
2.00E 01	5.56E-18
2.25E 01	5.13E-18
2.50E 01	6.49E-18
2.75E 01	6.89E-18
3.00E 01	7.23E-18
3.25E 01	7.19E-18
3.50E 01	6.73E-18
4.00E 01	5.68E-18
4.50E 01	4.80E-18
5.00E 01	4.27E-18
5.50E 01	3.78E-18
6.00E 01	3.11E-18
6.50E 01	2.57E-18
7.00E 01	2.22E-18
7.50E 01	1.99E-18
8.00E 01	1.66E-18
8.50E 01	1.49E-18
9.00E 01	1.26E-18
9.50E 01	1.00E-18
1.00E 02	8.93E-19
1.05E 02	7.86E-19
1.10E 02	7.28E-19
1.15E 02	6.26E-19
1.20E 02	5.90E-19

9-3-33
HUGHES ET AL. (353)

ENERGY (KEV)	CROSS SECTION (SQ. CM)
5.00E 00	2.50E-19
6.00E 00	2.80E-19
1.00E 01	4.70E-19
1.40E 01	6.90E-19
1.80E 01	9.80E-19
2.20E 01	1.27E-18
2.60E 01	1.56E-18
3.00E 01	1.80E-18
3.50E 01	1.89E-18
4.00E 01	1.82E-18
4.50E 01	1.63E-18
5.00E 01	1.50E-18
5.50E 01	1.37E-18
6.00E 01	1.18E-18
6.50E 01	1.06E-18
7.00E 01	9.40E-19
7.50E 01	8.50E-19
8.00E 01	7.50E-19
8.50E 01	6.60E-19
9.00E 01	5.60E-19
9.50E 01	5.00E-19
1.00E 02	4.40E-19
1.10E 02	3.60E-19
1.20E 02	2.90E-19

9-3-34
ANDREEV ET AL. (148)

ENERGY (KEV)	CROSS SECTION (SQ. CM)
1.00E 01	2.36E-17
1.20E 01	2.81E-17
1.40E 01	2.81E-17
1.60E 01	3.06E-17
1.80E 01	2.78E-17
2.00E 01	2.65E-17
2.20E 01	2.33E-17
2.40E 01	2.27E-17
2.60E 01	1.89E-17
2.80E 01	1.72E-17
3.00E 01	1.51E-17
3.20E 01	1.20E-17
3.40E 01	1.12E-17

9-3-35
ANDREEV ET AL. (148)

ENERGY (KEV)	CROSS SECTION (SQ. CM)
1.20E 01	1.71E-18
1.40E 01	1.99E-18
1.60E 01	2.05E-18
1.80E 01	2.72E-18
2.00E 01	3.21E-18
2.20E 01	3.20E-18
2.40E 01	2.80E-18
2.60E 01	2.27E-18
2.80E 01	2.05E-18
3.00E 01	1.48E-18

HUGHES ET AL. (120)

ENERGY (KEV)	CROSS SECTION (SQ. CM)
5.00E 00	3.26E-18
7.50E 00	3.82E-18
1.00E 01	4.81E-18
1.50E 01	6.95E-18
2.00E 01	9.25E-18
2.50E 01	1.01E-17
3.00E 01	9.64E-18
3.50E 01	8.63E-18
4.00E 01	7.86E-18
5.00E 01	6.49E-18
6.00E 01	4.94E-18
7.00E 01	4.22E-18
8.00E 01	3.62E-18
9.00E 01	3.37E-18
1.00E 02	3.02E-18
1.10E 02	3.02E-18
1.20E 02	2.97E-18

9-3-36B
HUGHES ET AL. (180)

ENERGY (KEV)	CROSS SECTION (SQ. CM)
1.00E 01	4.74E-18
1.25E 01	5.65E-18
1.50E 01	6.84E-18
1.75E 01	7.85E-18
2.00E 01	8.81E-18
2.25E 01	9.77E-18
2.50E 01	1.00E-17
2.75E 01	1.00E-17
3.00E 01	9.57E-18
3.50E 01	9.46E-18

9-3-36C
THOMAS ET AL. (129)

ENERGY (KEV)	CROSS SECTION (SQ. CM)
1.68E 02	5.69E-19

9-3-37
HUGHES ET AL. (353)

ENERGY (KEV)	CROSS SECTION (SQ. CM)
5.00E 00	1.11E-18
6.00E 00	9.40E-19
8.00E 00	1.16E-18
1.00E 01	1.18E-18
1.20E 01	1.19E-18
1.40E 01	1.71E-18
1.60E 01	2.03E-18
1.80E 01	2.49E-18
2.00E 01	2.72E-18
2.20E 01	3.09E-18
2.40E 01	3.07E-18
2.50E 01	3.18E-18
2.60E 01	3.31E-18
2.80E 01	3.28E-18
3.00E 01	3.31E-18
3.50E 01	3.14E-18
4.00E 01	2.83E-18
4.50E 01	2.72E-18
5.00E 01	2.62E-18
5.50E 01	2.39E-18
6.00E 01	2.14E-18
6.50E 01	2.04E-18
7.00E 01	1.89E-18
7.50E 01	1.74E-18
8.00E 01	1.59E-18
8.50E 01	1.37E-18
9.00E 01	1.29E-18
1.00E 02	1.20E-18
1.10E 02	1.00E-18

9-3-38
DAHLBERG ET AL. (182)

ENERGY (KEV)	CROSS SECTION (SQ. CM)
2.00E 01	4.00E-17
2.50E 01	3.05E-17
3.00E 01	2.55E-17
3.50E 01	1.95E-17
4.00E 01	1.70E-17
4.50E 01	1.25E-17
5.00E 01	1.05E-17
5.50E 01	8.80E-18
6.00E 01	8.20E-18
7.00E 01	6.20E-18
8.00E 01	4.90E-18
9.00E 01	4.30E-18
1.00E 02	3.50E-18
1.10E 02	3.10E-18
1.20E 02	2.55E-18
1.30E 02	2.65E-18

9-3-39
HUGHES ET AL. (180)

ENERGY (KEV)	CROSS SECTION (SQ. CM)
1.25E 01	3.75E-18
1.50E 01	4.80E-18
1.75E 01	4.22E-18
2.00E 01	6.93E-18
2.25E 01	7.09E-18
2.50E 01	6.41E-18
2.75E 01	5.70E-18
3.00E 01	5.28E-18
3.50E 01	4.54E-18

9-3-40A
HUGHES ET AL. (180)

ENERGY (KEV)	CROSS SECTION (SQ. CM)
1.00E 01	5.07E-18
1.25E 01	4.14E-18
1.50E 01	3.71E-18
1.75E 01	3.52E-18
2.00E 01	2.79E-18
2.25E 01	2.28E-18
2.50E 01	2.02E-18
2.75E 01	1.87E-18
3.00E 01	1.75E-18
3.50E 01	1.30E-18

9-3-40B
THOMAS ET AL. (129)

ENERGY (KEV)	CROSS SECTION (SQ. CM)
1.68E 02	1.13E-20

9-3-41
YOUNG ET AL. (7)

ENERGY (KEV)	CROSS SECTION (SQ. CM)
1.57E 00	1.14E-19
2.14E 00	2.40E-19
2.90E 00	6.38E-19
4.65E 00	1.00E-18
5.88E 00	1.11E-18
7.36E 00	1.51E-18
9.50E 00	9.02E-19
1.17E 01	1.27E-18
1.47E 01	5.54E-19
1.78E 01	5.18E-19
2.48E 01	2.46E-19
3.01E 01	5.16E-19

9-3-42
HUGHES ET AL. (180)

ENERGY (KEV)	POLARIZATION (PERCENT)
1.25E 01	1.99E 01
1.50E 01	1.91E 01
1.75E 01	1.65E 01
2.00E 01	1.90E 01
2.25E 01	1.50E 01
2.50E 01	1.38E 01
2.75E 01	1.08E 01
3.00E 01	1.26E 01
3.50E 01	9.04E 00

9-3-43
HUGHES ET AL. (120)

ENERGY (KEV)	CROSS SECTION (SQ. CM)
5.00E 00	3.69E-18
1.00E 01	6.09E-18
1.50E 01	7.81E-18
2.00E 01	8.13E-18
2.50E 01	8.70E-18

ENERGY (KEV)	CROSS SECTION (SQ. CM)
3.00E 01	8.90E-18
3.50E 01	8.51E-18
4.00E 01	7.70E-18
4.50E 01	6.52E-18
5.00E 01	5.76E-18
5.50E 01	4.92E-18
6.00E 01	4.42E-18
6.50E 01	4.13E-18
7.00E 01	3.84E-18
7.50E 01	3.62E-18
8.00E 01	3.54E-18
9.00E 01	3.23E-18
1.00E 02	3.08E-18
1.10E 02	3.01E-18
1.20E 02	2.87E-18

9-3-44
HUGHES ET AL. (353)

ENERGY (KEV)	CROSS SECTION (SQ. CM)
5.00E 00	6.00E-19
6.00E 00	5.90E-19
8.00E 00	7.60E-19
1.00E 01	7.00E-19
1.20E 01	7.60E-19
1.40E 01	9.20E-19
1.60E 01	1.13E-18
1.80E 01	1.44E-18
2.00E 01	1.55E-18
2.20E 01	1.82E-18
2.40E 01	1.98E-18
2.60E 01	2.04E-18
2.80E 01	2.33E-18
3.00E 01	2.34E-18
3.50E 01	2.26E-18
4.00E 01	2.12E-18
4.50E 01	2.08E-18
5.50E 01	1.79E-18
6.00E 01	1.64E-18
6.50E 01	1.50E-18
7.00E 01	1.54E-18
7.50E 01	1.32E-18
8.50E 01	1.20E-18
9.00E 01	1.17E-18
9.50E 01	1.15E-18
1.00E 02	9.90E-19
1.05E 02	9.80E-19
1.10E 02	7.60E-19
1.15E 02	8.00E-19
1.20E 02	6.10E-19

9-3-45
SELLIN ET AL. (139)

ENERGY (KEV)	CROSS SECTION (SQ. CM)
1.01E 01	7.76E-16
1.52E 01	3.38E-16
1.52E 01	2.57E-16
1.92E 01	2.48E-16
2.38E 01	1.84E-16
2.95E 01	1.58E-16

9-3-46
SELLIN ET AL. (139)

ENERGY (KEV)	CROSS SECTION (SQ. CM)
5.03E 00	8.22E-15
7.14E 00	3.16E-15
7.17E 00	2.66E-15
9.07E 00	1.66E-15
9.07E 00	1.46E-15
9.95E 00	1.06E-15
1.13E 01	8.81E-16
1.33E 01	5.62E-16
1.70E 01	3.71E-16
2.15E 01	1.53E-16
2.72E 01	1.00E-16
3.16E 01	1.17E-16

9-3-47A
SELLIN ET AL. (139)

ENERGY (KEV)	CROSS SECTION (SQ. CM)
2.14E 00	2.02E-14
2.52E 00	1.93E-14
2.84E 00	2.43E-14
3.22E 00	1.80E-14
3.58E 00	2.02E-14
3.83E 00	1.68E-14
4.36E 00	1.38E-14
4.78E 00	1.62E-14
4.92E 00	2.45E-14
4.96E 00	1.40E-14
5.03E 00	1.15E-14
5.44E 00	6.45E-15
5.54E 00	9.02E-15
5.67E 00	7.08E-15
6.05E 00	6.31E-15
6.42E 00	5.25E-15
7.07E 00	4.41E-15
7.17E 00	5.01E-15
7.69E 00	4.07E-15
9.07E 00	2.04E-15
9.07E 00	1.68E-15
9.46E 00	1.07E-15
1.00E 01	9.33E-16
1.07E 01	9.12E-16
1.33E 01	4.78E-16
1.60E 01	2.92E-16
1.72E 01	2.26E-16
1.92E 01	2.19E-16
2.00E 01	1.99E-16
2.10E 01	1.82E-16
2.50E 01	1.82E-16
2.51E 01	1.46E-16
3.16E 01	1.32E-16

9-3-47B
DONNALLY ET AL. (137)

ENERGY (KEV)	CROSS SECTION (SQ. CM)
1.54E-01	4.16E-15
1.54E-01	3.98E-15
1.54E-01	3.88E-15
2.03E-01	4.10E-15
2.03E-01	4.06E-15
2.03E-01	3.70E-15
2.04E-01	3.64E-15
2.04E-01	3.41E-15
2.51E-01	4.17E-15
2.52E-01	3.90E-15
2.52E-01	3.48E-15
3.53E-01	3.51E-15
3.53E-01	3.37E-15
4.01E-01	3.46E-15
4.50E-01	3.86E-15
4.55E-01	3.69E-15
4.55E-01	3.57E-15
4.55E-01	3.30E-15
5.97E-01	3.57E-15
5.97E-01	3.47E-15
5.97E-01	3.37E-15
8.05E-01	3.75E-15
8.01E-01	3.30E-15
8.01E-01	3.24E-15
1.00E 00	3.38E-15
1.52E 00	3.40E-15
2.07E 00	3.59E-15
2.08E 00	3.55E-15
2.63E 00	3.68E-15
2.61E 00	3.47E-15
3.13E 00	3.27E-15

9-3-47C
CESATI ET AL. (138)

ENERGY (KEV)	CROSS SECTION (SQ. CM)
5.00E-01	1.00E-16
1.00E 00	5.60E-17
2.00E 00	8.60E-17
5.00E 00	1.70E-17

9-4   EXCITATION OF FAST H ATOMS

The process may be described by the reaction equation,

$$H(1s) + X \rightarrow H^* + [X], \tag{9-5}$$

where the square brackets indicate that there is no information about the state of ionization or excitation of the target system after collision. The general considerations pertaining to this type of experiment are discussed at length in Section 9-1.

The most comprehensive treatment of this problem is the work of Ankudinov, Orbeli, and co-workers[191, 187] who have studied the formation of the 2s, 2p, and 3p states by impact of H on targets of He, Ne, Ar, Kr, and Xe. An early publication of some of these data in conference proceedings[191] was superseded and extended in a more complete report, the source of the data used here.[187] The apparatus used for this work and the methods employed to establish absolute cross sections have been detailed in Section 9-2(2).

Further studies of this problem are in the work of Dose[188] with fragmentary contributions by Sellin[17] and Dahlberg;[182] the experimental systems used by these workers are described in parts (3), (5), and (6), respectively, of Section 9-2. These experiments are restricted to study of 2s and 2p state formation; Dose includes measurements of polarization of the $2p \rightarrow 1s$ emission.

The measured cross sections are for the spontaneous emission of the Lyman-alpha line or for the field-induced emission of that line. Measurements of spontaneous emission have been made close to the entrance of the beam into the target region. At this point cascade into the 2p level from long-lived ns and nd states will not have reached its maximum value. One might argue that cascade into the 2p state is likely to be small, and the measured $2p \rightarrow 1s$ emission cross section is at least approximately equal to the cross section for the excitation of the 2p state. It is not possible to make similar arguments for the 2s state since the cascade will come from shortlived np levels.

In the discussion of the experimental systems, there are identified some potential sources of serious error in these experimental data. Using the published reports as a basis for assessment, there are a number of doubts about the accuracy of the absolute measurements and even about the validity of the relative variation of cross section with energy in some cases. The work to which the fewest objections have been raised is that of Ankudinov et al.; the limits of absolute accuracy were estimated at $\pm 20\%$, with a reproducibility from 5 to 10% depending on experimental conditions. The possibility of systematic errors in pressure measurement, neglect of polarization of Doppler shift and broadening were not included in these estimates. In the study of 2s state formation by Orbeli et al.[187] one may estimate that the data are too low by about 22% due to neglect of polarization-related emission anisotropy. The experimental measurements by the other groups may be considerably worse than this. The most reliable data is that of Ankudinov et al.

The measurements of polarization by Dose are not subject to most of the systematic errors associated with measurement of cross section. Dose quotes a possible systematic error of $\pm 10\%$ (i.e., $P = X\% \pm 10\%$) and a random error of about the same amount. These would appear to be reasonably accurate estimates of the validity of the measurements.

A. Cross-Section Measurements

In Data Tables 9-4-1 through 9-4-14 are shown the cross sections for the spontaneous- and field-induced emission of Lyman-alpha resulting from impact of H on rare gases. When comparisons are possible, the work of Orbeli et al.[187] tends to lie about 20% above that of Dose et al.[188] and exhibits a similar dependence on energy. There is a considerable discrepancy in the case of H impact on Ar, both in the magnitude and energy dependence of the cross sections. It has been noted [see Section 9-2(3)] that Dose et al.[188] have employed a most indirect method for determining the flux of the previous atom beam. The flux of scattered H atoms was monitored directly; from this, the incident atom flux was calculated, utilizing the Rutherford scattering formula. Dose et al. contend that it is necessary to correct the Rutherford formulation for the screening due to the electron in the H atom. A serious criticism of this procedure is that there is no direct evidence for the validity of the scattering cross sections predicted by a screened Rutherford formulation. Dose et al. present excitation cross sections evaluated using both screened and unscreened Rutherford prediction; the former is expected to give the more valid results. In the case of H + He, the results evaluated by the screened and unscreened formulations are almost identical. In the case of H + Ne, the difference is up to 30%, but the general form is the same; in contrast, the H + Ar data differ by 50% between the two formulations and an oscillatory behavior of cross section is observed only for the screened Rutherford prediction. It is quite possible that the procedure utilized to evaluate the primary atom flux from the predictions of cross section is invalid and that the apparent behavior of the H + Ar cross section is the result of this erroneous procedure. It must be concluded that there is some doubt as to the validity of the cross sections presented by Dose et al. and that the work of Orbeli et al. is therefore expected to be more accurate.

Studies of the formation of excited atomic hydrogen by impact of H on molecular targets are confined to two measurements. Sellin[17] determines the $2s$ state excitation cross section appropriate to an H_2 target (Data Table 9-4-15). Dahlberg et al. measure the $2p$ state excitation cross section for an N_2 target (Data Table 9-4-16).[182] The absolute magnitudes of both sets of data are very unreliable.

There are a number of studies of H impact on H_2, where emissions from target and projectile are detected simultaneously. These data are reviewed in Section 8-2-D.

B. Polarization of Emission

In Data Table 9-4-17 to 9-4-19 are shown the polarization fractions for the Lyman-alpha emission, collisionally induced by the impact of H on targets of helium, neon, and argon.[189] Some of the data are obtained with impact by deuterium atoms and are shown on data tables and graphs as though they were hydrogen atoms of equal velocity. There seems to be no difference between the behavior of H and D atoms.

C. Data Tables

Certain of the cross section determinations by Dose et al. have involved measurements using beams of neutral deuterium atoms. It was concluded that there was no essential difference between the cross section for H and for D impact at the same

velocities. In cross section determinations the D impact data are normalized to measurements for H impact at the same velocity. The data for impact of D are shown in the data tables and graphs as though they were H atoms of the same velocity, that is, at one-half the true D energy.

1. Cross Sections for Formation of Excited H States Induced by Impact of H on Various Targets (Includes Line Emission and Level Excitation Cross Sections)

9-4-1 Field-Induced Emission of the Lyman-Alpha Line ($2s \rightarrow 2p \rightarrow 1s$), Resulting from H Impact on He.

9-4-2 Emission of the Lyman-Alpha Line ($2p \rightarrow 1s$) Induced by H Impact on He. [9-4-2C is data for D Impact normalized to 9-4-2B at an unspecified energy.]

9-4-3 Formation of the $3p$ State Induced by H Impact on He.†

9-4-4 Field-Induced Emission of the Lyman-Alpha Line ($2s \rightarrow 2p \rightarrow 1s$), Resulting from H Impact on Ne.

9-4-5 Emission of the Lyman-Alpha Line ($2p \rightarrow 1s$) Induced by H Impact on Ne. [9-4-5C are data for D impact normalized to 9-4-5B at an unspecified energy.]

9-4-6 Formation of the $3p$ State Induced by H Impact on Ne.†

9-4-7 Field-Induced Emission of the Lyman-Alpha Line ($2s \rightarrow 2p \rightarrow 1s$), Resulting from H Impact on Ar.

9-4-8 Emission of the 1216-Å Lyman-Alpha Line ($2p \rightarrow 1s$) Induced by H Impact on Ar. [9-4-8C are data for D impact normalized to 9-4-8B at an unspecified energy.]

9-4-9 Formation of the $3p$ State Induced by H Impact on Ar.

9-4-10 Field-Induced Emission of the Lyman-Alpha Line ($2s \rightarrow 2p \rightarrow 1s$), Resulting from H Impact on Kr.

9-4-11 Emission of the Lyman-Alpha Line ($2p \rightarrow 1s$) Induced by H Impact on Kr.

9-4-12 Formation of the $3p$ State Induced by H Impact on Kr.†

9-4-13 Field-Induced Emission of the Lyman-Alpha Line ($2s \rightarrow 2p \rightarrow 1s$), Resulting from H Impact on Xe.

9-4-14 Emission of the Lyman-Alpha Line ($2p \rightarrow 1s$) Induced by H Impact on Xe.

9-4-15 Field-Induced Emission of the Lyman-Alpha Line ($2s \rightarrow 2p \rightarrow 1s$) from H Atoms that have traversed a H_2 Target.

9-4-16 Emission of the Lyman-Alpha Line ($2p \rightarrow 1s$), Induced by H on N_2.

2. Polarization Fractions of H Emission Induced by H Impact on Rare Gases

9-4-17 Lyman-Alpha Line ($2p \rightarrow 1s$) Induced by H Impact on He. [9-4-17B is data for D Impact.]

9-4-18 Lyman-Alpha Line ($2p \rightarrow 1s$) Induced by H Impact on Ne. [9-4-18B is data for D impact.]

9-4-19 Lyman-Alpha Line ($2p \rightarrow 1s$) Induced by H Impact on Ar. [9-4-19B is data for D impact.]

† Data evaluated on the assumption that cascade may be neglected.

9-4-1
ORBELI ET AL. (187)

ENERGY (KEV)	CROSS SECTION (SQ. CM)
5.00E 00	4.74E-18
6.00E 00	5.04E-18
7.00E 00	5.52E-18
8.00E 00	5.64E-18
9.00E 00	5.16E-18
1.00E 01	5.16E-18
1.10E 01	3.60E-18
1.20E 01	4.14E-18
1.40E 01	3.54E-18
1.60E 01	3.12E-18
1.80E 01	3.24E-18
2.00E 01	3.12E-18
2.20E 01	3.48E-18
2.40E 01	3.24E-18
2.60E 01	3.24E-18
2.80E 01	2.82E-18
3.00E 01	3.24E-18
3.20E 01	2.10E-18
3.40E 01	2.64E-18
3.60E 01	3.24E-18
3.80E 01	2.46E-18
4.00E 01	2.76E-18

9-4-2A
ORBELI ET AL. (187)

ENERGY (KEV)	CROSS SECTION (SQ. CM)
5.20E 00	2.40E-17
6.20E 00	2.18E-17
7.20E 00	1.92E-17
8.00E 00	1.78E-17
9.00E 00	1.68E-17
1.00E 01	1.49E-17
1.10E 01	1.26E-17
1.20E 01	1.22E-17
1.40E 01	1.00E-17
1.60E 01	9.12E-18
1.80E 01	8.72E-18
2.00E 01	7.44E-18
2.20E 01	6.88E-18
2.40E 01	6.32E-18
2.60E 01	6.16E-18
2.80E 01	4.80E-18
3.00E 01	5.20E-18
3.20E 01	4.64E-18
3.40E 01	4.80E-18
3.60E 01	4.16E-18
3.80E 01	4.40E-18
4.00E 01	4.56E-18

9-4-2B
COSE ET AL. (188)

ENERGY (KEV)	CROSS SECTION (SQ. CM)
7.10E 00	1.28E-17
7.73E 00	1.24E-17
8.44E 00	1.25E-17

ENERGY (KEV)	CROSS SECTION (SQ. CM)
9.13E 00	1.07E-17
9.99E 00	1.05E-17
1.07E 01	9.30E-18
1.15E 01	8.80E-18
1.24E 01	8.60E-18
1.33E 01	8.00E-18
1.42E 01	7.90E-18
1.52E 01	7.00E-18
1.62E 01	6.90E-18
1.71E 01	6.50E-18
1.82E 01	6.40E-18
1.92E 01	6.00E-18
2.03E 01	5.70E-18
2.16E 01	5.30E-18
2.26E 01	5.20E-18
2.38E 01	4.80E-18
2.50E 01	4.40E-18
2.64E 01	4.20E-18
2.75E 01	4.30E-18
2.89E 01	3.60E-18
3.03E 01	3.70E-18
3.17E 01	3.80E-18
3.30E 01	3.40E-18
3.45E 01	3.20E-18
3.58E 01	3.10E-18
3.74E 01	2.60E-18
3.90E 01	2.60E-18
4.07E 01	2.40E-18
4.22E 01	2.20E-18
4.39E 01	2.20E-18
4.54E 01	1.80E-18
4.72E 01	2.00E-18
4.89E 01	2.00E-18
5.06E 01	1.80E-18
5.25E 01	1.70E-18
5.41E 01	1.70E-18

9-4-2C
COSE ET AL. (188)

ENERGY (KEV)	CROSS SECTION (SQ. CM)
2.42E 00	2.10E-17
2.82E 00	2.18E-17
3.28E 00	2.13E-17
3.69E 00	1.93E-17
4.21E 00	1.80E-17
4.77E 00	1.68E-17
5.22E 00	1.62E-17
5.80E 00	1.53E-17
6.49E 00	1.40E-17

9-4-3
ORBELI ET AL. (187)

ENERGY (KEV)	CROSS SECTION (SQ. CM)
4.88E 00	2.44E-18
6.00E 00	2.48E-18
6.64E 00	2.14E-18
8.00E 00	1.96E-18
9.00E 00	1.81E-18
1.00E 01	1.81E-18
1.10E 01	1.68E-18
1.20E 01	1.61E-18
1.40E 01	1.50E-18
1.60E 01	1.34E-18
1.80E 01	1.16E-18
2.00E 01	1.11E-18
2.20E 01	1.06E-18
2.40E 01	1.15E-18
2.60E 01	1.02E-18
2.80E 01	9.90E-19
3.00E 01	9.10E-19
3.20E 01	9.10E-19
3.40E 01	9.10E-19
3.60E 01	8.00E-19
3.80E 01	7.70E-19
4.00E 01	8.40E-19

9-4-4
ORBELI ET AL. (187)

ENERGY (KEV)	CROSS SECTION (SQ. CM)
5.25E 00	2.62E-17
6.37E 00	2.20E-17
7.35E 00	1.94E-17
7.28E 00	1.84E-17
8.26E 00	1.74E-17
9.31E 00	1.46E-17
1.00E 01	1.30E-17
1.10E 01	1.21E-17
1.20E 01	1.03E-17
1.40E 01	1.02E-17
1.60E 01	9.66E-18
1.80E 01	9.42E-18
2.00E 01	7.38E-18
2.20E 01	7.62E-18
2.40E 01	6.42E-18
2.60E 01	6.66E-18
2.80E 01	6.60E-18
3.00E 01	6.00E-18
3.20E 01	6.00E-18
3.40E 01	6.42E-18
3.60E 01	5.28E-18
3.80E 01	4.80E-18
4.00E 01	5.88E-18

9-4-5A
ORBELI ET AL. (187)

ENERGY (KEV)	CROSS SECTION (SQ. CM)
5.36E 00	2.74E-17
6.24E 00	2.76E-17
7.36E 00	2.62E-17
8.40E 00	2.61E-17
9.00E 00	2.51E-17
1.00E 01	2.45E-17
1.10E 01	2.38E-17
1.20E 01	2.10E-17
1.40E 01	2.09E-17
1.60E 01	1.99E-17
1.80E 01	1.86E-17
2.00E 01	1.74E-17
2.20E 01	1.66E-17
2.40E 01	1.44E-17
2.60E 01	1.36E-17
2.80E 01	1.34E-17
3.00E 01	1.23E-17
3.20E 01	1.21E-17

ENERGY (KEV)	CROSS SECTION (SQ. CM)
3.40E 01	1.22E-17
3.64E 01	1.24E-17
3.80E 01	1.30E-17
4.00E 01	1.31E-17

9-4-5B
DOSE ET AL. (188)

ENERGY (KEV)	CROSS SECTION (SQ. CM)
7.27E 00	2.16E-17
7.99E 00	2.17E-17
8.62E 00	2.16E-17
9.31E 00	2.12E-17
1.01E 01	2.13E-17
1.09E 01	2.05E-17
1.17E 01	2.02E-17
1.26E 01	1.95E-17
1.34E 01	1.90E-17
1.44E 01	1.94E-17
1.54E 01	1.89E-17
1.64E 01	1.82E-17
1.73E 01	1.76E-17
1.84E 01	1.73E-17
1.94E 01	1.70E-17
2.05E 01	1.63E-17
2.17E 01	1.59E-17
2.29E 01	1.55E-17
2.41E 01	1.50E-17
2.54E 01	1.47E-17
2.66E 01	1.44E-17
2.79E 01	1.40E-17
2.93E 01	1.39E-17
3.05E 01	1.31E-17
3.20E 01	1.28E-17
3.33E 01	1.26E-17
3.50E 01	1.21E-17
3.64E 01	1.20E-17
3.77E 01	1.19E-17
3.93E 01	1.15E-17
4.10E 01	1.13E-17
4.26E 01	1.10E-17
4.42E 01	1.07E-17
4.59E 01	1.08E-17
4.76E 01	1.07E-17
4.93E 01	1.07E-17
5.10E 01	1.07E-17
5.29E 01	1.06E-17
5.48E 01	1.06E-17

9-4-5C
DOSE ET AL. (188)

ENERGY (KEV)	CROSS SECTION (SQ. CM)
1.8CE 00	1.66E-17
2.14E 00	1.71E-17
2.50E 00	1.80E-17
2.93E 00	1.88E-17
3.31E 00	1.96E-17
3.75E 00	2.05E-17
4.24E 00	2.15E-17
4.80E 00	2.20E-17
5.40E 00	2.21E-17
5.95E 00	2.25E-17
6.57E C0	2.25E-17

9-4-6
CRBELI ET AL. (187)

ENERGY (KEV)	CROSS SECTION (SQ. CM)
5.00E 00	4.19E-18
6.20E 00	4.48E-18
6.80E 00	4.28E-18
8.0CE 00	4.30E-18
9.30E 00	4.27E-18
1.00E 01	4.42E-18
1.10E 01	4.17E-18
1.20E 01	4.33E-18
1.40E 01	4.05E-18
1.60E 01	3.96E-18
1.80E 01	3.58E-18
2.00E 01	3.40E-18
2.20E 01	3.35E-18
2.40E 01	3.07E-18
2.60E 01	3.04E-18
2.80E 01	2.84E-18
3.00E 01	2.62E-18
3.20E 01	2.52E-18
3.4CE 01	2.44E-18
3.60E 01	2.62E-18
3.80E 01	2.52E-18
4.00E 01	2.39E-18

9-4-7
ORBELI ET AL. (187)

ENERGY (KEV)	CROSS SECTION (SQ. CM)
5.0CE 00	1.91E-17
6.23E 00	1.95E-17
7.21E 00	2.06E-17
8.26E 00	1.97E-17
9.31E 00	2.03E-17
9.87E 00	1.98E-17
1.10E 01	2.31E-17
1.20E 01	2.24E-17
1.40E 01	2.21E-17
1.80E 01	2.48E-17
2.00E 01	2.48E-17
2.20E 01	2.57E-17
2.40E 01	2.59E-17
2.60E 01	2.62E-17
2.80E 01	2.54E-17
3.00E 01	2.39E-17
3.20E 01	2.36E-17
3.40E 01	2.34E-17
3.60E 01	2.34E-17
4.00E 01	2.18E-17

9-4-8A
ORBELI ET AL. (187)

ENERGY (KEV)	CROSS SECTION (SQ. CM)
5.36E 00	6.61E-17
7.76E 00	6.54E-17
6.72E 00	6.33E-17
8.40E 00	6.19E-17
9.52E 00	5.74E-17

ENERGY (KEV)	CROSS SECTION (SQ. CM)
1.04E 01	5.47E-17
1.16E 01	5.34E-17
1.26E 01	4.73E-17
1.42E 01	4.44E-17
1.65E 01	4.60E-17
1.83E 01	4.53E-17
2.00E 01	4.48E-17
2.20E 01	4.32E-17
2.40E 01	4.45E-17
2.60E 01	4.51E-17
2.80E 01	4.55E-17
3.00E 01	4.44E-17
3.20E 01	4.45E-17
3.40E 01	4.47E-17
3.64E 01	4.26E-17
3.85E 01	4.20E-17
4.00E 01	4.11E-17

9-4-8B
DOSE ET AL. (188)

ENERGY (KEV)	CROSS SECTION (SQ. CM)
7.18E 00	2.13E-17
7.82E 00	2.04E-17
8.57E 00	2.00E-17
9.31E 00	1.98E-17
9.99E 00	1.93E-17
1.08E 01	1.92E-17
1.16E 01	1.98E-17
1.24E 01	2.03E-17
1.33E 01	1.95E-17
1.42E 01	1.97E-17
1.52E 01	1.92E-17
1.61E 01	1.90E-17
1.72E 01	1.82E-17
1.82E 01	1.90E-17
1.93E 01	1.92E-17
2.05E 01	1.90E-17
2.15E 01	1.88E-17
2.28E 01	1.87E-17
2.39E 01	1.83E-17
2.50E 01	1.84E-17
2.64E 01	1.79E-17
2.77E 01	1.86E-17
2.90E 01	1.87E-17
3.05E 01	1.84E-17
3.18E 01	1.86E-17
3.33E 01	1.84E-17
3.47E 01	1.79E-17
3.62E 01	1.82E-17
3.76E 01	1.82E-17
3.93E 01	1.80E-17
4.07E 01	1.77E-17
4.25E 01	1.75E-17
4.41E 01	1.76E-17
4.58E 01	1.74E-17
4.74E 01	1.77E-17
4.92E 01	1.71E-17
5.08E 01	1.70E-17
5.27E 01	1.71E-17
5.43E 01	1.63E-17

9-4-8C
DOSE ET AL. (188)

ENERGY (KEV)	CROSS SECTION (SQ. CM)
1.77E 00	2.46E-17
2.10E 00	2.49E-17
2.52E 00	2.40E-17
2.88E 00	2.32E-17
3.29E 00	2.24E-17
3.75E 00	2.14E-17
4.28E 00	2.08E-17
4.74E 00	2.07E-17
5.33E 00	2.10E-17
5.92E 00	2.15E-17
6.57E 00	2.16E-17

9-4-9
ORBELI ET AL. (187)

ENERGY (KEV)	CROSS SECTION (SQ. CM)
4.88E 00	5.81E-18
6.16E 00	5.73E-18
7.36E 00	5.64E-18
8.24E 00	6.44E-18
9.44E 00	6.16E-18
1.04E 01	6.41E-18
1.14E 01	5.85E-18
1.22E 01	6.26E-18
1.46E 01	6.96E-18
1.83E 01	6.34E-18
2.00E 01	6.65E-18
2.20E 01	6.31E-18
2.40E 01	6.03E-18
2.60E 01	6.31E-18
3.00E 01	6.24E-18
3.20E 01	5.70E-18
3.40E 01	6.06E-18
3.60E 01	6.34E-18
3.80E 01	5.56E-18
4.00E 01	5.95E-18

9-4-10
ORBELI ET AL. (187)

ENERGY (KEV)	CROSS SECTION (SQ. CM)
5.50E 00	1.66E-17
6.30E 00	1.85E-17
8.19E 00	1.91E-17
9.00E 00	2.16E-17
1.00E 01	2.29E-17
1.40E 01	2.66E-17
1.60E 01	2.83E-17
1.80E 01	2.80E-17
2.00E 01	2.85E-17
2.60E 01	2.92E-17
3.00E 01	2.75E-17
3.20E 01	2.54E-17
3.40E 01	2.45E-17
3.60E 01	2.52E-17
3.80E 01	2.32E-17
4.00E 01	2.48E-17

9-4-11
ORBELI ET AL. (187)

ENERGY (KEV)	CROSS SECTION (SQ. CM)
5.50E 00	5.84E-17
7.68E 00	5.99E-17
9.68E 00	6.39E-17
1.05E 01	6.70E-17
1.26E 01	6.29E-17
1.45E 01	5.86E-17
1.66E 01	5.33E-17
1.80E 01	5.21E-17
2.04E 01	5.33E-17
2.27E 01	5.51E-17
2.45E 01	5.38E-17
2.66E 01	5.15E-17
2.80E 01	5.29E-17
3.00E 01	4.98E-17
3.20E 01	4.78E-17
3.40E 01	4.77E-17
3.60E 01	4.77E-17
3.80E 01	4.48E-17
4.00E 01	4.48E-17

9-4-12
ORBELI ET AL. (187)

ENERGY (KEV)	CROSS SECTION (SQ. CM)
4.96E 00	8.46E-18
7.28E 00	9.60E-18
6.40E 00	9.84E-18
8.32E 00	1.12E-17
1.06E 01	1.15E-17
9.44E 00	1.17E-17
1.14E 01	1.20E-17
1.26E 01	1.22E-17
1.40E 01	1.17E-17
1.60E 01	1.13E-17
1.80E 01	1.05E-17
2.00E 01	1.02E-17
2.20E 01	1.08E-17
2.40E 01	1.10E-17
2.60E 01	1.07E-17
2.80E 01	1.09E-17
3.00E 01	1.13E-17
3.26E 01	1.02E-17
3.40E 01	1.08E-17
3.60E 01	1.05E-17
3.84E 01	1.02E-17
4.00E 01	1.00E-17

9-4-13
ORBELI ET AL. (187)

ENERGY (KEV)	CROSS SECTION (SQ. CM)
5.00E 00	2.14E-17
6.00E 00	2.56E-17
7.49E 00	2.57E-17
8.12E 00	2.48E-17
1.00E 01	2.75E-17
1.25E 01	2.74E-17
1.40E 01	2.95E-17
1.60E 01	3.24E-17
1.80E 01	3.34E-17
2.00E 01	3.70E-17
2.20E 01	3.68E-17
2.40E 01	3.88E-17
2.60E 01	3.71E-17
2.80E 01	3.59E-17
3.00E 01	3.68E-17
3.20E 01	3.50E-17
3.40E 01	2.98E-17
3.60E 01	2.87E-17
3.80E 01	2.95E-17
4.00E 01	3.05E-17

9-4-14
ORBELI ET AL. (187)

ENERGY (KEV)	CROSS SECTION (SQ. CM)
5.44E 00	6.75E-17
6.24E 00	6.89E-17
8.64E 00	7.06E-17
7.28E 00	7.13E-17
9.52E 00	7.41E-17
1.17E 01	6.77E-17
1.40E 01	6.00E-17
1.64E 01	5.83E-17
1.83E 01	6.10E-17
2.00E 01	6.26E-17
2.20E 01	6.65E-17
2.46E 01	6.72E-17
2.60E 01	7.05E-17
2.80E 01	7.32E-17
3.06E 01	7.34E-17
3.20E 01	7.57E-17
3.40E 01	7.47E-17
3.60E 01	7.12E-17
3.84E 01	7.30E-17

9-4-15
SELLIN ET AL. (17)

ENERGY (KEV)	CROSS SECTION (SQ. CM)
1.50E 01	1.50E-18

9-4-16
DAHLBERG ET AL. (182)

ENERGY (KEV)	CROSS SECTION (SQ. CM)
2.00E 01	4.90E-17
2.50E 01	3.90E-17
3.00E 01	3.50E-17
3.50E 01	3.20E-17
4.00E 01	2.50E-17
4.50E 01	2.20E-17
5.00E 01	2.20E-17
5.50E 01	2.20E-17
6.00E 01	2.20E-17
7.00E 01	2.20E-17
8.00E 01	2.25E-17

9.00E 01	2.30E-17	6.57E 00	1.32E 01	6.03E 00	4.34E 01
1.00E 02	2.30E-17	7.35E 00	9.04E 00	6.65E 00	3.32E 01
1.10E 02	2.10E-17	8.04E 00	1.03E 01	7.39E 00	3.89E 01
1.20E 02	2.15E-17	8.71E 00	1.23E 01	8.17E 00	3.44E 01
1.30E 02	2.30E-17	9.55E 00	2.33E 01	8.71E 00	2.86E 01
		1.11E 01	2.52E 01	9.50E 00	3.42E 01
				1.04E 01	3.40E 01
				1.11E 01	3.28E 01

9-4-17A
DOSE ET AL. (189)

9-4-18A
DOSE ET AL. (189)

9-4-19A
DOSE ET AL. (189)

ENERGY (KEV)	POLARIZATION (PERCENT)	ENERGY (KEV)	POLARIZATION (PERCENT)	ENERGY (KEV)	POLARIZATION (PERCENT)
7.99E 00	1.51E 01				
8.62E 00	1.88E 01	7.27E 00	4.18E 01		
9.41E 00	2.10E 01	7.99E 00	4.09E 01		
1.02E 01	1.03E 01	8.66E 00	4.09E 01	7.14E 00	5.05E 01
1.04E 01	1.64E 01	9.46E 00	2.95E 01	7.95E 00	4.07E 01
1.10E 01	8.30E 00	1.03E 01	3.72E 01	8.66E 00	4.19E 01
1.18E 01	2.00E 01	1.11E 01	4.00E 01	9.31E 00	4.32E 01
1.27E 01	1.79E 01	1.18E 01	3.63E 01	1.01E 01	2.60E 01
1.36E 01	1.36E 01	1.28E 01	3.51E 01	1.10E 01	3.22E 01
1.46E 01	7.60E 00	1.37E 01	3.50E 01	1.18E 01	3.32E 01
1.55E 01	1.42E 01	1.46E 01	3.14E 01	1.26E 01	3.38E 01
1.65E 01	1.76E 01	1.56E 01	3.36E 01	1.36E 01	1.68E 01
1.78E 01	1.48E 01	1.65E 01	2.72E 01	1.45E 01	2.53E 01
1.85E 01	1.92E 01	1.76E 01	3.32E 01	1.55E 01	2.11E 01
1.97E 01	1.73E 01	1.87E 01	3.12E 01	1.65E 01	1.28E 01
2.07E 01	1.79E 01	1.97E 01	2.36E 01	1.76E 01	6.30E 00
2.19E 01	2.04E 01	2.09E 01	3.39E 01	1.87E 01	2.13E 01
2.31E 01	2.13E 01	2.19E 01	3.10E 01	1.97E 01	1.42E 01
2.42E 01	1.28E 01	2.32E 01	3.48E 01	2.09E 01	1.48E 01
2.56E 01	1.22E 01	2.43E 01	2.47E 01	2.20E 01	9.36E 00
2.69E 01	1.10E 01	2.57E 01	2.40E 01	2.32E 01	7.86E 00
2.81E 01	5.50E 00	2.69E 01	3.96E 01	2.43E 01	5.28E 00
2.97E 01	1.06E 01	2.83E 01	2.36E 01	2.56E 01	1.42E 01
3.09E 01	4.60E 00	2.97E 01	3.12E 01	2.69E 01	1.15E 01
3.23E 01	1.72E 01	3.08E 01	2.39E 01	2.81E 01	8.70E 00
3.38E 01	1.96E 01	3.24E 01	2.52E 01	2.96E 01	1.06E 01
3.53E 01	1.31E 01	3.38E 01	2.57E 01	3.08E 01	2.64E 00
3.67E 01	2.56E 01	3.53E 01	2.53E 01	3.22E 01	9.00E 00
3.82E 01	6.80E 00	3.67E 01	8.80E 00	3.38E 01	8.88E 00
3.98E 01	1.66E 01	3.82E 01	2.84E 01	3.51E 01	9.00E-01
4.13E 01	6.40E 00	3.98E 01	2.30E 01	3.66E 01	6.72E 00
4.30E 01	2.02E 01	4.13E 01	1.80E 01	3.79E 01	1.79E 01
4.45E 01	1.79E 01	4.30E 01	2.32E 01	3.97E 01	7.56E 00
4.63E 01	2.63E 01	4.46E 01	2.41E 01	4.10E 01	1.26E 01
4.79E 01	1.44E 01	4.61E 01	2.20E 01	4.29E 01	1.05E 01
4.96E 01	3.47E 01	4.77E 01	2.42E 01	4.45E 01	2.16E 01
5.14E 01	1.48E 01	4.96E 01	2.64E 01	4.61E 01	8.34E 00
5.31E 01	2.36E 01	5.13E 01	2.17E 01	4.78E 01	1.14E 01
5.51E 01	6.08E 00	5.31E 01	2.45E 01	4.96E 01	2.20E 01
		5.49E 01	2.27E 01	5.12E 01	1.76E 01
				5.32E 01	1.75E 01
				5.49E 01	1.44E 01

9-4-17B
DOSE ET AL. (189)

9-4-18B
DOSE ET AL. (189)

9-4-19B
DOSE ET AL. (189)

ENERGY (KEV)	POLARIZATION (PERCENT)	ENERGY (KEV)	POLARIZATION (PERCENT)	ENERGY (KEV)	POLARIZATION (PERCENT)
1.77E 00	3.70E 01				
2.07E 00	3.08E 01	1.82E 00	7.60E 01		
2.50E 00	3.22E 01	2.17E 00	5.84E 01		
2.85E 00	2.01E 01	2.57E 00	6.08E 01	1.77E 00	4.59E 01
3.28E 00	1.91E 01	2.90E 00	4.24E 01	2.14E 00	4.16E 01
3.78E 00	2.14E 01	3.34E 00	4.60E 01	2.52E 00	3.51E 01
4.31E 00	1.92E 01	3.84E 00	3.94E 01	2.88E 00	5.13E 01
4.80E 00	1.52E 01	4.31E 00	5.02E 01	3.31E 00	5.08E 01
5.36E 00	7.84E 00	4.84E 00	3.70E 01	3.75E 00	5.29E 01
5.95E 00	1.31E 01	5.44E 00	3.29E 01	4.31E 00	4.56E 01

4.80E 00	5.57E 01	6.61E 00	3.82E 01	8.66E 00	3.95E 01
5.36E 00	4.89E 01	7.22E 00	4.53E 01	9.31E 00	2.67E 01
5.95E 00	4.47E 01	7.86E 00	3.37E 01	1.04E 01	2.25E 01
				1.10E 01	4.60E 01

9-5 SIMULTANEOUS STRIPPING AND EXCITATION
OF FAST H⁻ IONS

This process may be described by the reaction equation,

$$H^- + X \rightarrow H^* + [X + e], \tag{9-6}$$

where the square brackets indicate that there is no information about the state of
ionization or excitation of the systems. There are no known excited states of the
H^- ion; therefore, excited projectiles are not a problem. The general considerations
regarding this type of experiment have been discussed in Section 9-1.

All of the information reviewed here is from the work of Andreev et al.[192] and
involves the formation of the $2s$ and $2p$ states by H^- impact on helium, neon, argon,
krypton, and xenon. The apparatus is reviewed in Section 9-2(2). The authors estimate
that the absolute accuracy of the data is within $\pm 20\%$ and that the reproducibility
is between ± 5 and $\pm 10\%$. These estimates of accuracy limitations do not take into
account error in pressure measurement, neglect of emission anisotropy, instrumental
polarization, Doppler shift and broadening, and the influence of cascade.

Berkner et al.[193] have studied the formation of highly excited states of D formed
by the impact of D^- on H_2 at 20-MeV impact energy. These data are obtained by
the field ionization technique for detection of the excited state. It is noted in Section
3-3-A that this detection procedure does not give a fine resolution of the state under
investigation and certainly cannot provide distinction between angular momentum
substates of a given principal quantum number. It is concluded that data of this
type should be excluded from detailed consideration here.

A. Cross-Section Measurements

Data Tables 9-5-1 through 9-5-10 show measurements of the emission in the
$2p \rightarrow 1s$ and $2s \rightarrow 2p \rightarrow 1s$ transitions. It is interesting to note that, with the exception
of the xenon target, there is little variation of cross section with impact energy in
this energy range. Also, the cross sections for the formation of the $2p$ and $2s$ levels
by this reaction are considerably greater than those for formation of the same states
by the impact of H atoms or protons on these same targets.

There is no objective procedure for estimating the cascade contribution to these
data. It is possible to argue that the cascade into the $2p$ level should be small. The
observation region is close to the entrance into the chamber. The cascade from the
long-lived ns and nd states will build up slowly to its maximum value and will be
rather small close to the entrance. Even if cascade ultimately becomes important, it
seems likely that the geometrical construction of the apparatus will reduce its influence

on the $2p \to 1s$ emission data to negligible proportions. Therefore, the measurements on spontaneous emission of the Lyman-alpha line are at least approximately equal to the cross sections for the formation of the $2p$ state. It is not possible to make similar arguments concerning cascade into the $2s$ level.

The field-induced Lyman-alpha emission will be polarized and therefore exhibit anisotropy.[239, 240, 20] From the recent studies of this polarization[20] and knowledge of the constructional details applicable to the apparatus used by Andreev et al., one may estimate that neglect of polarization causes the $2s$ cross sections to be too low by approximately 22%.

B. Data Tables

1. Cross Sections for H Emission Resulting from the Impact of H⁻ on Various Rare Gas Targets

9-5-1 Lyman-Alpha Line ($2p \to 1s$), Induced by H⁻ Impact on He.

9-5-2 Field-Induced Emission of the Lyman Alpha Line ($2s \to 2p \to 1s$), Resulting from H⁻ Impact on He.

9-5-3 Lyman-Alpha Line ($2p \to 1s$), Induced by H⁻ Impact on Ne.

9-5-4 Field-Induced Emission of the Lyman-Alpha Line ($2s \to 2p \to 1s$), Resulting from H⁻ Impact on Ne.

9-5-5 Lyman-Alpha Line ($2p \to 1s$) Induced by H⁻ Impact on Ar.

9-5-6 Field-Induced Emission of the Lyman-Alpha Line ($2s \to 2p \to 1s$), Resulting from H⁻ Impact on Ar.

9-5-7 Lyman-Alpha Line ($2p \to 1s$) Induced by H⁻ Impact on Kr.

9-5-8 Field-Induced Emission of the Lyman-Alpha Line ($2s \to 2p \to 1s$), Resulting from H⁻ Impact on Kr.

9-5-9 Lyman-Alpha Line ($2p \to 1s$) Induced by H⁻ Impact on Xe.

9-5-10 Field-Induced Emission of the Lyman Alpha Line ($2s \to 2p \to 1s$), Resulting from H⁻ Impact on Xe.

9-5-1
ANDREEV ET AL. (192)

ENERGY (KEV)	CROSS SECTION (SQ. CM)
6.16E 00	5.44E-17
7.08E 00	5.11E-17
8.12E 00	4.75E-17
8.92E 00	4.99E-17
1.01E 01	3.91E-17
1.22E 01	3.31E-17
1.42E 01	3.06E-17
1.63E 01	3.06E-17
1.84E 01	3.18E-17
2.04E 01	2.86E-17
2.24E 01	2.67E-17
2.45E 01	2.68E-17
2.65E 01	2.44E-17
2.85E 01	2.65E-17
3.06E 01	2.44E-17
3.27E 01	2.47E-17
3.46E 01	2.67E-17
3.87E 01	2.52E-17

9-5-2
ANDREEV ET AL. (192)

ENERGY (KEV)	CROSS SECTION (SQ. CM)
5.04E 00	2.65E-17
6.92E 00	2.94E-17
9.08E 00	2.26E-17
1.01E 01	2.44E-17
1.21E 01	2.34E-17
1.42E 01	2.59E-17
1.61E 01	2.44E-17
1.80E 01	2.38E-17
2.02E 01	2.37E-17
2.22E 01	2.38E-17
2.41E 01	2.14E-17
2.62E 01	2.29E-17
2.82E 01	2.13E-17
3.02E 01	2.35E-17
3.23E 01	2.38E-17
3.42E 01	2.44E-17
3.82E 01	2.02E-17

9-5-3
ANDREEV ET AL. (192)

ENERGY (KEV)	CROSS SECTION (SQ. CM)
5.04E 00	6.48E-17
7.04E 00	6.01E-17
8.08E 00	6.40E-17
9.12E 00	6.55E-17
1.01E 01	5.58E-17
1.21E 01	6.13E-17
1.43E 01	5.79E-17
1.63E 01	5.38E-17
1.83E 01	6.19E-17
2.04E 01	6.00E-17
2.25E 01	5.59E-17
2.44E 01	5.41E-17

2.65E 01	5.28E-17
2.85E 01	4.92E-17
3.06E 01	5.41E-17
3.27E 01	5.58E-17

9-5-4
ANDREEV ET AL. (192)

ENERGY (KEV)	CROSS SECTION (SQ. CM)
5.00E 00	5.52E-17
6.08E 00	4.75E-17
7.04E 00	6.36E-17
8.04E 00	5.68E-17
8.92E 00	6.18E-17
1.01E 01	5.77E-17
1.20E 01	6.99E-17
1.42E 01	5.76E-17
1.61E 01	5.37E-17
2.02E 01	6.15E-17
2.22E 01	6.18E-17
2.43E 01	6.15E-17
2.62E 01	6.00E-17
2.83E 01	5.76E-17
3.03E 01	5.83E-17
3.42E 01	5.16E-17

9-5-5
ANDREEV ET AL. (192)

ENERGY (KEV)	CROSS SECTION (SQ. CM)
6.04E 00	7.77E-17
7.08E 00	8.77E-17
8.16E 00	7.54E-17
9.08E 00	8.61E-17
1.01E 01	7.98E-17
1.22E 01	8.02E-17
1.42E 01	7.99E-17
1.63E 01	7.87E-17
1.83E 01	7.81E-17
2.04E 01	7.99E-17
2.25E 01	8.20E-17
2.46E 01	7.65E-17
2.66E 01	8.08E-17
2.86E 01	8.28E-17
3.06E 01	8.59E-17
3.27E 01	8.77E-17
3.46E 01	7.80E-17
3.67E 01	7.86E-17

9-5-6
ANDREEV ET AL. (192)

ENERGY (KEV)	CROSS SECTION (SQ. CM)
4.96E 00	5.35E-17
6.08E 00	5.31E-17
6.96E 00	6.06E-17
8.04E 00	7.11E-17
9.00E 00	6.45E-17
1.00E 01	6.76E-17
1.20E 01	6.19E-17

1.42E 01	7.27E-17
1.62E 01	6.31E-17
2.02E 01	6.55E-17
2.22E 01	6.58E-17
2.62E 01	6.70E-17
2.83E 01	6.27E-17
3.02E 01	7.05E-17
3.23E 01	6.75E-17
3.44E 01	6.57E-17

9-5-7
ANDREEV ET AL. (192)

ENERGY (KEV)	CROSS SECTION (SQ. CM)
5.08E 00	6.48E-17
6.08E 00	6.21E-17
7.08E 00	6.31E-17
8.12E 00	6.63E-17
9.12E 00	6.39E-17
1.02E 01	7.12E-17
1.43E 01	7.17E-17
1.64E 01	8.20E-17
1.84E 01	7.53E-17
2.04E 01	8.32E-17
2.28E 01	8.22E-17
2.46E 01	8.86E-17
2.66E 01	8.28E-17
2.86E 01	8.07E-17
3.06E 01	8.01E-17
3.27E 01	7.96E-17
3.47E 01	7.65E-17

9-5-8
ANDREEV ET AL. (192)

ENERGY (KEV)	CROSS SECTION (SQ. CM)
6.08E 00	3.90E-17
8.04E 00	5.32E-17
9.08E 00	3.54E-17
1.00E 01	5.38E-17
1.21E 01	4.69E-17
1.41E 01	4.84E-17
1.62E 01	5.92E-17
1.82E 01	5.25E-17
2.03E 01	5.46E-17
2.22E 01	5.64E-17
2.42E 01	5.67E-17
2.62E 01	5.62E-17
2.83E 01	6.10E-17
3.03E 01	6.63E-17
3.24E 01	6.55E-17
3.44E 01	6.67E-17

9-5-9
ANDREEV ET AL. (192)

ENERGY (KEV)	CROSS SECTION (SQ. CM)
6.00E 00	1.36E-16
7.04E 00	1.22E-16
8.08E 00	1.18E-16

8.96E 00	1.21E-16	3.26E 01	1.39E-16	8.96E 00	1.00E-16
1.01E 01	1.21E-16	3.46E 01	1.44E-16	8.00E 00	8.59E-17
1.21E 01	1.05E-16			7.04E 00	7.36E-17
1.41E 01	1.03E-16			1.40E 01	6.88E-17
1.62E 01	1.10E-16	9-5-10		1.61E 01	6.58E-17
1.84E 01	1.01E-16	ANDREEV ET AL. (192)		1.80E 01	6.57E-17
2.04E 01	1.08E-16			2.01E 01	7.32E-17
2.24E 01	1.07E-16	ENERGY	CROSS	2.42E 01	7.47E-17
2.46E 01	1.12E-16	(KEV)	SECTION	2.62E 01	8.08E-17
2.66E 01	1.25E-16		(SQ. CM)	2.80E 01	7.87E-17
2.86E 01	1.39E-16			3.02E 01	8.08E-17
3.06E 01	1.39E-16	6.04E 00	1.07E-16	3.23E 01	9.28E-17
				3.42E 01	8.70E-17

9-6 DISSOCIATION OF H$_2^+$ IONS

The process may be described by the reaction equation,

$$H_2^+ + X \rightarrow H^* + [H^+ + X]; \qquad (9\text{-}7)$$

there is no information about the states of ionization or excitation of the postcollision products that are contained within the square brackets. The general considerations pertaining to this type of experiment have been discussed at length in Section 9-1. The data are for the formation of the 2s, 3s, 2p, and 3p states measured by the groups of Geballe, Andreev, Hughes, and their respective co-workers. The experimental arrangements used by these groups have been reviewed, respectively, in parts (1), (2), and (4) of Section 9-2.

It is well known that vibrational excitation of H$_2^+$ ions will influence the cross section for dissociation. None of the experimental investigations includes an assessment of whether or not this factor might have influenced the validity of the cross section measurements.

The work of Hughes et al.[196] and that of Andreev[148] involves making measurements of target pressures using McLeod gauges. For heavy target gases, this measurement will be in error due to the thermal transpiration and mercury pumping effects (Section 3-2-D).

The work of Van Zyl et al.[194] on the spontaneous emission of the Lyman-alpha line by dissociation on various rare gas targets included a direct test of the influence of cascade. No change in signal was observed when the distance between collision chamber entrance and observation region was varied. This indicated that cascade from the longer-lived ns and nd states was negligible at the point where observations were being carried out. The work of Andreev et al.[148] on the formation of the 2p and 3p state utilized a geometry similar to that of Van Zyl et al. and will also discriminate against cascade from long-lived states. It would appear that cascade contributions to all the measurements on the formation of the 2p and 3p states is likely to be rather small. Therefore, the cross sections measured for spontaneous Lyman-alpha emission may be taken as equal to the cross sections for the formation of the 2p state, and the cross section for spontaneous Lyman-beta emission can be converted to the 3p level-excitation cross section using the relevant branching ratio and neglecting cascade.

The measurements on the formation of the 3s state by Hughes et al.[196] and the formation of the 2s state by Jaecks and Tynan[195] may both have been influenced

by cascade from np levels. The branching ratios favor decay of np states to the $1s$ ground state, and only if the $3p$ and $4p$ states have substantially higher cross sections than the $2s$ and $3s$ levels, respectively, is there any likelihood of appreciable cascade. There is no evidence available on the cross sections for the formation of these states with which this criterion can be compared.

The work of Andreev et al.[148] on the spontaneous emission of the $3p \to 1s$ line is published in the form of the cross section for excitation of the $3p$ level, calculated on the assumption that cascade may be neglected; as discussed above, this may be a justifiable assumption, so the data are presented here in the form originally published. Measurements of the $2s \to 2p \to 1s$, $3s \to 2p$, and $2p \to 1s$ emissions are described in the data figure captions as line emission cross sections; if cascade is neglected, these are equal to the cross sections for forming the $2s$, $3s$, and $2p$ levels. Neglect of cascade appears to be a justifiable assumption for the $2p$ level.

Gailey et al.[28] have shown empirically that the emission from the $2p$ state is unpolarized. Emission from the $3s$ state must inherently be unpolarized because there are no sublevels of this state. Field-induced decay of the $2s$ state does exhibit polarization;[234, 240, 20] the degree of polarization depends on the strength of the quenching field. The $2s$ state data reproduced here have not been corrected for anisotropy; one can estimate that a correction would raise the data of Andreev et al.[148] by about 22% and lower the data of Jaecks and Tynan[195] by approximately 20%.

The work of Van Zyl et al.[194] and Jaecks and Tynan[195] are estimated by the authors to have an absolute accuracy of better than $\pm 50\%$ and a reproducibility of better than $\pm 5\%$. The corresponding figures for the work of Hughes et al.[196] are $\pm 50\%$ and ± 5 to $\pm 15\%$; those for the work of Andreev et al.[148] $\pm 30\%$ and ± 5 to $\pm 10\%$.

A. Rare Gas Targets

In Data Tables 9-6-1 through 9-6-12 are shown measurements on the formation of the $2s$, $3s$, and $2p$ states by impact of H_2^+ on targets of helium, neon, argon, krypton, and xenon.

The data on the formation of the $2p$ excited state from the group of Geballe and co-workers are from the paper by Van Zyl et al.[194] These are in complete agreement with the previous work by Dunn et al.[99] An earlier conference paper by Dunn et al.[305] is disregarded.

Van Zyl et al. examined the possibility that their measurement of spontaneous emission in the $H_2^+ + Kr$ collision includes a contribution from the 1165-Å line of Kr I that lies very close to one of the transmission bands of the oxygen filter (1167-Å). Emission at this wavelength has been identified as interfering with measurements of Lyman-alpha emission in the $H^+ + Kr$ charge transfer measurements.[117] It was concluded that there is some Kr I emission contributing to the measured cross section, but it is "appreciably smaller" than the Lyman-alpha contribution. It was also shown that the Kr I emission could not be used to explain any of the prominent features of the measured cross section.

In addition to the data reproduced here, there are a number of other attempted measurements, which are excluded because the accuracy is likely to be very poor and the data therefore misleading. Bobashev et al.[158, 157] studied the emission in the

Balmer-series lines resulting from dissociation of H_2^+ on helium and neon. It is suggested in the review of this work [Section 9-2(2)] that these data may be incorrect despite attempts to correct for the lifetime of the excited state and that it should be disregarded. Hughes et al.[197] have studied the emission of Balmer-alpha induced by H_2^+ on helium but have omitted to take account of the lifetime of the excited dissociation fragments and thereby have invalidated the measurement [see Section 9-2(4)]. Hanle and Voss[199] have studied Balmer-alpha emission induced by H_2^+ on helium. Correction for lifetime was made using a "mean lifetime" for the excited state without explaining how this quantity was obtained and without justifying its use. Target pressure of 5×10^{-2} torr were used; these pressures are certainly outside the range where single collision conditions can be guaranteed, and therefore these data must be disregarded. Dieterich[259] made some early measurements of relative emission due to the 2-keV-H_2^+ incident on helium, but this provides no quantitative information.

B. Molecular Targets

In Data Tables 9-6-13 through 9-6-17 are shown the cross sections for the formation of the $3s$, $3p$, $2s$, and $2p$ states of H by impact of H_2^+ on targets of molecular hydrogen and nitrogen.

In Section 8-2-E there is a review of data on the sum of target and projectile emissions induced by the impact of H_2^+ on an H_2 target; in these cases, it is impossible to estimate what proportion of the emission was due to the target. The data reviewed here are for emission from excited fragments of the projectile only.

In addition to these data there have been a number of other attempts at such measurements; these results are excluded on the grounds that their accuracy is likely to be very poor and the data therefore misleading. Hughes et al. carried out measurements of Balmer emission induced by impact of H_2^+ on H_2[197, 198] using Doppler shift to separate target and projectile emission. Neglect of a correction for the excited state lifetime invalidates the data. Dufay et al.[202] made measurements of Balmer-beta emission induced by H_2^+ impact on N_2 and O_2. Most of the data in that paper were later superseded, but no attempt was made to remeasure the emission from the dissociated projectile fragments; moreover, the effect of the lifetime of the H atom was ignored completely. Therefore, this data should be disregarded. Van Zyl and co-workers have made measurements on the emission induced by H_2^+ impact on an N_2 target by using an oxygen-filtered Geiger counter.[99, 149] Van Zyl showed quite clearly that in addition to the Lyman-alpha emission the detectors were also responding to some emission from the target or from its dissociation fragments. It is impossible to determine from this data what proportion of the emission was from H and what was emission from the target. These data are not repeated here on the grounds that its inclusion would be misinforming. Berkner et al.[193] have studied the formation of highly excited H atoms by dissociation of H_2^+ on an H_2 target at 20-MeV impact energy. As concluded in Section 3-3-A, experiments that involve detection by the field-ionization procedure are not of the same quality as those obtained by optical methods and should therefore be excluded from detailed consideration in this monograph.

The work by Andreev regarding the impact of H_2^+ on H_2 includes a study of the change in the intensity of emission of the Lyman-beta line as the result of the

application of an electric field to the collision region. Those data, not providing information on cross sections, are not reproduced here.

C. Metallic Vapor Targets

Bobashev and Pop[168] have studied the emission of the Balmer-alpha line induced by the dissociation of H_2^+ on an Hg target. The experiment, reviewed in Section 9-2(2), did not provide an accurate method of correcting for the lifetime of the excited dissociation fragment. The data may therefore be incorrect and are omitted from further consideration.

D. Polarization of Emission

The only information on this subject is from the study by Gailey et al.[28] of the polarization of Lyman-alpha emission resulting from the impact of H_2^+ on helium at 3 and 6-keV energy. The polarization fraction was found to be zero.

E. Data Tables

In some of the experiments, D_2^+ ions have been used in addition to H_2^+ projectiles. The data for D_2^+ impact is shown in the tables as though it were for H_2^+ of equal velocity that is, at one half of the true D_2^+ energy.

1. Cross Sections for Emission of H I Lines Induced by Impact of H_2^+ on Various Targets

9-6-1   Field-Induced Emission of Lyman-Alpha ($2s \rightarrow 2p \rightarrow 1s$); He Target.

9-6-2   6563-Å Line ($3s \rightarrow 2p$); He Target.

9-6-3   Lyman-Alpha Line ($2p \rightarrow 1s$); He Target. [9-6-3B is data for D_2^+ Impact.]

9-6-4   Field-Induced Emission of Lyman-Alpha ($2s \rightarrow 2p \rightarrow 1s$); Ne Target [9-4-6B is data for D_2^+ Impact.]

9-6-5   6563-Å Line ($3s \rightarrow 2p$); Ne Target.

9-6-6   Lyman-Alpha Line ($2p \rightarrow 1s$); Ne Target. [9-3-6B is data for D_2^+ Impact.]

9-6-7   Field-Induced Emission of Lyman-Alpha ($2s \rightarrow 2p \rightarrow 1s$); Ar Target.

9-6-8   6563-Å Line ($3s \rightarrow 2p$); Ar Target.

9-6-9   Lyman-Alpha Line ($2p \rightarrow 1s$); Ar Target. [9-6-9B is data for D_2^+ Impact.]

9-6-10   Lyman-Alpha Line ($2p \rightarrow 1s$); Kr Target. [9-6-10B is data for D_2^+ Impact.]

9-6-11   Field-Induced Emission of Lyman-Alpha ($2s \rightarrow 2p \rightarrow 1s$); Xe Target.

9-6-12   Lyman-Alpha Line ($2p \rightarrow 1s$); Xe Target. [9-6-12B is data for D_2^+ Impact.]

9-6-13   Field-Induced Emission of Lyman-Alpha ($2s \rightarrow 2p \rightarrow 1s$); H_2 Target. [projectile excitation only.]

9-6-14   6563-Å Line ($2s \rightarrow 2p$); H_2 Target. [projectile excitation only.]

9-6-15   Lyman-Alpha Line ($2p \rightarrow 1s$); H_2 Target. [projectile excitation only.]

9-6-16   Formation of the $3p$ State; H_2 Target. [excitation of projectile only; data evaluated on the assumption that cascade is negligible.]

9-6-17   6563-Å Line ($3s \rightarrow 2p$); N_2 Target.

JAECKS ET AL. (195)

ENERGY (KEV)	CROSS SECTION (SQ. CM)
2.00E 00	6.20E-18
3.00E 00	5.50E-18
4.00E 00	6.10E-18
5.00E 00	5.80E-18
6.00E 00	6.20E-18
7.00E 00	5.70E-18
8.00E 00	6.00E-18
1.00E 01	6.40E-18
1.10E 01	5.20E-18
1.20E 01	5.90E-18
1.30E 01	5.70E-18
1.40E 01	5.20E-18
1.70E 01	6.00E-18
1.80E 01	5.50E-18
2.00E 01	5.30E-18
2.30E 01	4.70E-18
2.50E 01	5.10E-18

9-6-2
HUGHES ET AL. (196)

ENERGY (KEV)	CROSS SECTION (SQ. CM)
2.00E 01	1.88E-18
3.00E 01	2.05E-18
4.00E 01	2.07E-18
5.00E 01	1.97E-18
6.00E 01	1.97E-18
7.00E 01	1.91E-18
8.00E 01	1.90E-18
9.00E 01	1.87E-18
1.00E 02	1.81E-18
1.10E 02	1.73E-18
1.20E 02	1.62E-18

9-6-3A
VAN ZYL ET AL. (194)

ENERGY (KEV)	CROSS SECTION (SQ. CM)
5.00E-01	2.45E-17
7.50E-01	3.00E-17
1.00E 00	3.36E-17
1.50E 00	3.93E-17
2.00E 00	4.28E-17
2.50E 00	4.41E-17
3.00E 00	4.41E-17
3.50E 00	4.40E-17
4.00E 00	4.30E-17
4.50E 00	4.20E-17
5.00E 00	4.17E-17
5.50E 00	4.10E-17
6.00E 00	4.04E-17
6.50E 00	4.02E-17
7.00E 00	4.04E-17
7.50E 00	3.96E-17
8.00E 00	3.91E-17
9.00E 00	3.81E-17
1.00E 01	3.56E-17

1.10E 01	3.56E-17
1.20E 01	3.52E-17
1.30E 01	3.40E-17
1.40E 01	3.27E-17
1.50E 01	3.25E-17
1.60E 01	3.28E-17
1.70E 01	3.22E-17
1.80E 01	3.14E-17
1.90E 01	3.04E-17
2.00E 01	2.99E-17
2.10E 01	2.92E-17
2.20E 01	2.90E-17
2.30E 01	2.70E-17
2.40E 01	2.79E-17
2.50E 01	2.69E-17

9-6-3B
VAN ZYL ET AL. (194)

ENERGY (KEV)	CROSS SECTION (SQ. CM)
2.50E-01	1.31E-17
3.00E-01	1.67E-17
3.50E-01	1.93E-17
4.00E-01	1.99E-17
4.50E-01	2.03E-17
5.00E-01	2.01E-17
1.00E 00	2.69E-17
1.50E 00	3.15E-17
2.00E 00	3.39E-17
2.50E 00	3.58E-17
3.00E 00	3.59E-17
3.50E 00	3.63E-17
4.00E 00	3.56E-17
4.50E 00	3.55E-17
5.00E 00	3.55E-17
5.50E 00	3.49E-17
6.00E 00	3.44E-17
6.50E 00	3.36E-17
7.00E 00	3.32E-17
7.50E 00	3.29E-17
8.00E 00	3.22E-17
8.50E 00	3.29E-17
9.00E 00	3.19E-17
9.50E 00	3.18E-17
1.00E 01	3.18E-17
1.05E 01	3.25E-17
1.10E 01	3.04E-17
1.15E 01	3.08E-17
1.20E 01	3.17E-17
1.25E 01	3.19E-17

9-6-4A
JAECKS ET AL. (195)

ENERGY (KEV)	CROSS SECTION (SQ. CM)
2.00E 00	1.18E-17
3.00E 00	9.30E-18
4.00E 00	9.50E-18
5.00E 00	1.01E-17
6.00E 00	9.80E-18
7.00E 00	9.50E-18
8.00E 00	1.02E-17
9.00E 00	9.80E-18
1.00E 01	1.02E-17

1.20E 01	1.01E-17
1.40E 01	1.06E-17
1.60E 01	1.14E-17
1.80E 01	9.90E-18
2.00E 01	1.01E-17
2.30E 01	7.80E-18
2.40E 01	1.02E-17

9-6-4B
JAECKS ET AL. (195)

ENERGY (KEV)	CROSS SECTION (SQ. CM)
1.50E 00	1.00E-17
2.00E 00	8.00E-18
3.00E 00	8.10E-18
5.00E 00	6.80E-18
6.00E 00	8.30E-18
8.00E 00	8.20E-18

9-6-5
HUGHES ET AL. (196)

ENERGY (KEV)	CROSS SECTION (SQ. CM)
2.00E 01	4.65E-18
3.00E 01	4.74E-18
4.00E 01	4.55E-18
5.00E 01	4.18E-18
6.00E 01	4.28E-18
7.00E 01	4.38E-18
8.00E 01	4.21E-18
9.00E 01	3.90E-18
1.00E 02	3.76E-18
1.10E 02	3.51E-18
1.20E 02	3.23E-18

9-6-6A
VAN ZYL ET AL. (194)

ENERGY (KEV)	CROSS SECTION (SQ. CM)
5.00E-01	1.11E-17
1.00E 00	2.17E-17
1.50E 00	2.77E-17
2.00E 00	3.06E-17
3.00E 00	3.44E-17
4.00E 00	3.55E-17
5.00E 00	3.57E-17
6.00E 00	3.62E-17
7.00E 00	3.58E-17
8.00E 00	3.63E-17
9.00E 00	3.63E-17
1.00E 01	3.74E-17
1.10E 01	3.68E-17
1.20E 01	3.72E-17
1.30E 01	3.70E-17
1.40E 01	3.71E-17
1.50E 01	3.82E-17
1.60E 01	3.80E-17
1.70E 01	3.79E-17
1.80E 01	3.77E-17
1.90E 01	3.71E-17

ENERGY (KEV)	CROSS SECTION (SQ. CM)
2.00E 01	3.65E-17
2.10E 01	3.63E-17
2.20E 01	3.69E-17
2.30E 01	3.60E-17
2.40E 01	3.67E-17
2.50E 01	3.55E-17

9-6-6B
VAN ZYL ET AL. (194)

ENERGY (KEV)	CROSS SECTION (SQ. CM)
3.00E-01	5.40E-18
4.00E-01	8.30E-18
5.00E-01	9.70E-18
1.00E 00	1.74E-17
1.50E 00	2.21E-17
2.00E 00	2.43E-17
2.50E 00	2.67E-17
3.00E 00	2.77E-17
3.50E 00	2.80E-17
4.00E 00	2.85E-17
4.50E 00	2.87E-17
5.00E 00	2.89E-17
5.50E 00	2.90E-17
6.00E 00	2.93E-17
6.50E 00	2.88E-17
7.0CE 00	2.99E-17
7.50E 00	3.05E-17
8.00E 00	3.14E-17
8.50E 00	3.14E-17
9.00E 00	3.20E-17
9.50E 00	3.27E-17
1.00E 01	3.27E-17
1.05E 01	3.33E-17
1.10E 01	3.37E-17
1.15E 01	3.33E-17
1.20E 01	3.29E-17
1.25E 01	3.39E-17

9-6-7
JAECKS ET AL. (195)

ENERGY (KEV)	CROSS SECTION (SQ. CM)
2.0CE 00	1.60E-17
3.00E 00	1.78E-17
4.00E 00	1.54E-17
5.00E 00	1.65E-17
6.00E 00	1.56E-17
7.00E 00	1.36E-17
8.00E 00	1.48E-17
9.00E 00	1.32E-17
1.00E 01	1.32E-17
1.10E 01	1.45E-17
1.20E 01	1.55E-17
1.30E 01	1.50E-17
1.40E 01	1.50E-17
1.50E 01	1.59E-17
1.60E 01	1.56E-17
1.70E 01	1.83E-17
1.80E 01	1.53E-17
1.90E 01	1.97E-17
2.00E 01	1.71E-17
2.10E 01	1.93E-17
2.20E 01	2.01E-17

ENERGY (KEV)	CROSS SECTION (SQ. CM)
2.30E 01	1.71E-17

9-6-8
HUGHES ET AL. (196)

ENERGY (KEV)	CROSS SECTION (SQ. CM)
2.00E 01	6.25E-18
3.00E 01	6.71E-18
4.00E 01	7.27E-18
5.00E 01	7.17E-18
6.00E 01	7.65E-18
7.00E 01	7.46E-18
8.00E 01	7.74E-18
9.00E 01	7.62E-18
1.00E 02	7.73E-18
1.10E 02	7.63E-18
1.20E 02	7.50E-18

9-6-9A
VAN ZYL ET AL. (194)

ENERGY (KEV)	CROSS SECTION (SQ. CM)
5.00E-01	7.65E-17
6.00E-01	8.15E-17
6.50E-01	8.65E-17
7.00E-01	8.40E-17
7.50E-01	8.82E-17
8.00E-01	8.40E-17
8.50E-01	8.55E-17
9.00E-01	8.74E-17
1.00E 00	8.46E-17
1.25E 00	8.26E-17
1.50E 00	8.08E-17
2.00E 00	7.57E-17
2.50E 00	7.30E-17
3.00E 00	7.46E-17
4.00E 00	7.18E-17
5.00E 00	7.00E-17
6.00E 00	6.68E-17
7.00E 00	6.33E-17
8.00E 00	5.96E-17
9.00E 00	6.06E-17
1.00E 01	5.61E-17
1.10E 01	5.79E-17
1.20E 01	5.67E-17
1.30E 01	5.79E-17
1.40E 01	5.65E-17
1.50E 01	5.83E-17
1.60E 01	5.87E-17
1.70E 01	5.82E-17
1.80E 01	5.93E-17
1.90E 01	6.07E-17
2.00E 01	6.22E-17
2.10E 01	6.14E-17
2.20E 01	6.12E-17
2.30E 01	6.10E-17
2.40E 01	6.00E-17
2.50E 01	6.16E-17

9-6-9B
VAN ZYL ET AL. (194)

ENERGY (KEV)	CROSS SECTION (SQ. CM)
2.50E-01	5.14E-17
3.00E-01	5.33E-17
3.50E-01	5.58E-17
3.75E-01	5.55E-17
4.00E-01	5.37E-17
4.25E-01	5.44E-17
4.50E-01	5.40E-17
5.00E-01	5.56E-17
7.50E-01	5.87E-17
1.00E 00	5.74E-17
1.25E 00	5.83E-17
1.50E 00	5.60E-17
2.00E 00	5.51E-17
2.50E 00	5.42E-17
3.00E 00	5.23E-17
3.50E 00	5.19E-17
4.00E 00	5.23E-17
4.50E 00	5.17E-17
5.00E 00	5.28E-17
5.50E 00	5.22E-17
6.00E 00	5.09E-17
6.50E 00	5.05E-17
7.00E 00	4.90E-17
7.50E 00	4.99E-17
8.00E 00	4.79E-17
8.50E 00	5.03E-17
9.00E 00	4.79E-17
9.50E 00	4.79E-17
1.00E 01	4.81E-17
1.05E 01	4.87E-17
1.10E 01	4.54E-17
1.15E 01	4.61E-17
1.20E 01	4.17E-17
1.25E 01	5.21E-17

9-6-10A
VAN ZYL ET AL. (194)

ENERGY (KEV)	CROSS SECTION (SQ. CM)
4.00E-01	1.15E-16
5.00E-01	1.00E-16
6.00E-01	1.23E-16
7.00E-01	1.26E-16
8.00E-01	1.22E-16
9.00E-01	1.07E-16
1.00E 00	1.03E-16
1.25E 00	9.75E-17
1.50E 00	9.51E-17
2.00E 00	8.78E-17
2.50E 00	8.12E-17
3.00E 00	8.08E-17
4.00E 00	7.95E-17
5.00E 00	7.20E-17
6.00E 00	7.02E-17
7.00E 00	6.80E-17
8.00E 00	6.37E-17
9.00E 00	6.22E-17
1.00E 01	6.18E-17
1.10E 01	6.27E-17
1.20E 01	6.39E-17
1.30E 01	6.50E-17

ENERGY (KEV)	CROSS SECTION (SQ. CM)
1.40E 01	6.61E-17
1.50E 01	6.76E-17
1.60E 01	6.72E-17
1.70E 01	7.02E-17
1.80E 01	7.12E-17
1.90E 01	7.04E-17
2.00E 01	7.16E-17
2.10E 01	7.34E-17
2.20E 01	7.04E-17
2.30E 01	7.19E-17
2.40E 01	7.29E-17
2.50E 01	7.22E-17
2.60E 01	7.07E-17

9-6-10B
VAN ZYL ET AL. (194)

ENERGY (KEV)	CROSS SECTION (SQ. CM)
2.50E-01	6.73E-17
3.00E-01	6.98E-17
3.50E-01	6.98E-17
4.00E-01	7.91E-17
4.50E-01	7.89E-17
5.00E-01	7.20E-17
1.00E 00	6.77E-17
1.50E 00	6.20E-17
2.00E 00	5.78E-17
2.50E 00	5.38E-17
3.00E 00	5.30E-17
3.50E 00	5.17E-17
4.00E 00	5.00E-17
4.50E 00	5.23E-17
5.00E 00	5.04E-17
5.50E 00	4.95E-17
6.00E 00	4.96E-17
6.50E 00	4.85E-17
7.00E 00	4.68E-17
7.50E 00	4.77E-17
8.00E 00	4.45E-17
8.50E 00	4.49E-17
9.00E 00	4.28E-17
9.50E 00	4.58E-17
1.00E 01	4.66E-17
1.05E 01	4.75E-17
1.10E 01	4.95E-17
1.15E 01	4.96E-17
1.20E 01	4.66E-17
1.25E 01	5.42E-17

9-6-11
JAECKS ET AL. (195)

ENERGY (KEV)	CROSS SECTION (SQ. CM)
2.00E 00	1.99E-17
3.00E 00	2.01E-17
4.00E 00	1.85E-17
5.00E 00	1.99E-17
6.00E 00	1.84E-17
7.00E 00	1.66E-17
8.00E 00	1.61E-17
9.00E 00	2.01E-17
1.00E 01	2.04E-17
1.20E 01	2.31E-17
1.40E 01	2.40E-17

ENERGY (KEV)	CROSS SECTION (SQ. CM)
1.60E 01	2.48E-17
1.80E 01	2.35E-17
1.90E 01	3.12E-17
2.00E 01	3.15E-17
2.20E 01	2.59E-17

9-6-12A
VAN ZYL ET AL. (194)

ENERGY (KEV)	CROSS SECTION (SQ. CM)
4.00E-01	1.74E-16
5.00E-01	1.50E-16
5.50E-01	1.50E-16
6.00E-01	1.43E-16
6.50E-01	1.37E-16
7.00E-01	1.54E-16
8.00E-01	1.45E-16
9.00E-01	1.36E-16
1.00E 00	1.28E-16
1.50E 00	1.12E-16
2.00E 00	1.03E-16
2.50E 00	9.93E-17
3.00E 00	9.42E-17
4.00E 00	9.32E-17
5.00E 00	9.19E-17
6.00E 00	8.50E-17
7.00E 00	8.41E-17
8.00E 00	8.22E-17
9.00E 00	8.44E-17
1.00E 01	8.45E-17
1.10E 01	8.49E-17
1.20E 01	8.77E-17
1.30E 01	8.96E-17
1.40E 01	9.40E-17
1.50E 01	9.18E-17
1.60E 01	9.24E-17
1.70E 01	9.46E-17
1.80E 01	9.37E-17
1.90E 01	9.66E-17
2.00E 01	9.83E-17
2.10E 01	9.83E-17
2.20E 01	9.66E-17
2.30E 01	9.07E-17
2.40E 01	9.45E-17
2.50E 01	9.50E-17

9-6-12B
VAN ZYL ET AL. (194)

ENERGY (KEV)	CROSS SECTION (SQ. CM)
2.00E-01	1.20E-16
2.50E-01	1.16E-16
3.00E-01	1.32E-16
3.50E-01	1.29E-16
4.00E-01	1.28E-16
4.50E-01	1.17E-16
5.00E-01	1.12E-16
1.00E 00	9.09E-17
1.50E 00	7.67E-17
2.00E 00	7.20E-17
2.50E 00	6.80E-17
3.00E 00	6.73E-17
3.50E 00	7.00E-17
4.00E 00	6.80E-17

ENERGY (KEV)	CROSS SECTION (SQ. CM)
4.50E 00	6.80E-17
5.00E 00	7.00E-17
5.50E 00	6.39E-17
6.00E 00	6.21E-17
6.50E 00	6.35E-17
7.00E 00	6.37E-17
7.50E 00	6.54E-17
8.00E 00	6.49E-17
8.50E 00	6.93E-17
9.00E 00	6.73E-17
9.50E 00	6.80E-17
1.00E 01	6.62E-17
1.05E 01	6.68E-17
1.10E 01	7.07E-17
1.15E 01	6.73E-17
1.20E 01	7.00E-17
1.25E 01	6.93E-17

9-6-13
ANDREEV ET AL. (148)

ENERGY (KEV)	CROSS SECTION (SQ. CM)
1.20E 01	1.71E-17
1.40E 01	1.84E-17
1.60E 01	1.90E-17
1.80E 01	2.03E-17
2.00E 01	2.12E-17
2.40E 01	2.62E-17
2.60E 01	2.52E-17
2.80E 01	2.61E-17

9-6-14
HUGHES ET AL. (196)

ENERGY (KEV)	CROSS SECTION (SQ. CM)
2.00E 01	3.44E-18
3.00E 01	3.96E-18
4.00E 01	4.55E-18
5.00E 01	4.54E-18
6.00E 01	4.38E-18
7.00E 01	4.20E-18
8.00E 01	3.94E-18
9.00E 01	3.80E-18
1.00E 02	3.59E-18
1.10E 02	3.55E-18
1.20E 02	3.29E-18

9-6-15
ANDREEV ET AL. (148)

ENERGY (KEV)	CROSS SECTION (SQ. CM)
1.20E 01	4.46E-17
1.40E 01	4.40E-17
1.60E 01	4.86E-17
2.00E 01	5.17E-17
2.20E 01	5.36E-17
2.40E 01	5.74E-17
2.60E 01	5.45E-17
2.80E 01	4.62E-17

9-6-16		1.40E 01	3.06E-18	9-6-17	
ANDREEV ET AL. (148)		1.80E 01	3.30E-18	HUGHES ET AL. (196)	
		2.00E 01	3.62E-18		
ENERGY	CROSS	2.20E 01	3.39E-18	ENERGY	CROSS
(KEV)	SECTION	2.40E 01	3.55E-18	(KEV)	SECTION
	(SQ. CM)	2.60E 01	3.39E-18		(SQ. CM)
		2.80E 01	3.14E-18		
1.20E 01	3.00E-18			1.00E 02	8.80E-18

9-7 DISSOCIATION OF H_3^+ IONS

The process may be described by the reaction equation,

$$H_3^+ + X \rightarrow H^* + [H_2^+ + X];$$ (9-8)

there is no information about the states of ionization or excitation of the postcollision products that are contained within the square brackets. The general considerations pertaining to this type of experiment have been discussed at length in Section 9-1. The study includes the formation of the 2p state by Dunn, Geballe,[99] and co-workers, and the formation of the 3s state by Hughes and co-workers.[196] The experimental arrangements have been reviewed, respectively, in parts (1) and (4) of Section 9-2.

The state of excitation of the H_3^+ ion may influence the cross section for collisional dissociation. Neither of the experimental investigations included any assessment of the influence of excited state population on the validity of the experimental measurements.

The work of Hughes et al. involves making measurements of target pressure with a McLeod gauge. For heavy target gases, this measurement will be in error due to thermal transpiration and mercury pumping effects.

In the preceding Section (9-6), it has been argued that the apparatus used by Dunn, Geballe, and co-workers has an inherent discrimination against cascade into the 2p state from the longer-lived ns and nd levels. There is no way of assessing the cascade contributions to the measurement of the 3s → 2p transition by Hughes et al.

Dunn et al. estimated their work to have an accuracy of $\pm 40\%$ with a reproducibility of 20%. Hughes et al. estimate their absolute accuracy as $\pm 50\%$ with reproducibility of $\pm 10\%$.

A. Rare Gas Targets

In Data Tables 9-7-1 and 9-7-2 are shown measurements on the formation of the 3s and 2p state of H by impact of H_3^+ on helium. Data Tables 9-7-3 through 9-7-4 show data on the formation of the 3s state by impact of H_3^+ on neon and argon.

The data by Dunn et al.[99] were listed by the authors as cross sections for emission of "countable uv" on the grounds that no tests were made to ensure that emission from other systems was absent. It is noted that the emission in the H_2^+ + He and H^+ + He measurements[194, 16] was shown to be entirely Lyman-alpha emitted as a result of formation of the excited H (2p) state. It seems highly unlikely that the H_3^+ + He case did exhibit any appreciable emission of systems that were not observed for the other two collision combinations. Therefore, the data by Dunn are included

here and are regarded as cross sections for the emission of the Lyman-alpha line, notwithstanding Dunn's pessimistic classification.

There have been a number of other attempts at such measurements that are excluded because the accuracy is likely to be small and the data misleading. Bobashev et al.[158, 157] studied emission in the Balmer series resulting from dissociation of H_3^+ on helium and neon. It is suggested in the review of this work [Section 9-2(2)] that these data may be incorrect and should be disregarded. Hughes et al. made some early measurements of Balmer-alpha emission induced by the impact of H_3^+ on helium but omitted to take into account the lifetime of the excited dissociation fragments and thereby invalidated the measurement [see Section 9-2(4)].[197, 198] Hanle and Voss[199] have studied Balmer-alpha emission induced by H_3^+ on helium. Corrections were made using a "mean lifetime" for the excited state without explaining how this quantity was determined and without justifying its use. Furthermore, for the experiment the target pressures that were used (5×10^{-2} torr) are certainly outside the range where single-collision conditions can be guaranteed. These data must be disregarded.

B. Molecular Targets

The only information that is reproduced here are the measurements on the formation of the $3s$ state of H by impact of H_3^+ on N_2 carried out by Hughes et al.[196] It should be noted that in Section 8-2-3 there is a review of the measurements by Dunn et al. on the Lyman-alpha emission induced by H_3^+ on H_2; these include contributions from both the target and the projectile.

In addition to the work tabulated here, Hughes et al.[197, 198] have made measurements of Balmer emission induced by H_3^+ on an H_2 target. These data did not take into account the finite lifetime of the excited H atom and therefore are invalid. Dunn et al.[99] have also made a study of the emission induced by the impact of H_3^+ on N_2. The detector for this study was an oxygen-filtered Geiger counter. Subsequently, Van Zyl et al.[149] showed that the impact of H_2^+ on N_2 caused the emission of a substantial amount of radiation from the target system, some of which might be transmitted by the filter and thereby contribute to the measured signal. Undoubtedly, this same situation must be expected for the case of H_3^+ on N_2. Since the data may include an indeterminate amount of emission from the target system, their inclusion here would be misleading and therefore they are disregarded.

C. Data Tables

1. Cross Sections for Emission of H I Lines Induced by Impact of H_3^+ on Various Targets

9-7-1 6563-Å Line ($3s \rightarrow 2p$); He Target.

9-7-2 Lyman-Alpha Line ($2p \rightarrow 1s$); He Target.

9-7-3 6563-Å Line ($3s \rightarrow 2p$); Ne Target.

9-7-4 6563-Å Line ($3s \rightarrow 2p$); Ar Target.

9-7-5 6563-Å Line ($3s \rightarrow 2p$); H_2 Target. [projectile excitation only.]

9-7-1
HUGHES ET AL. (196)

ENERGY (KEV)	CROSS SECTION (SQ. CM)
2.00E 01	2.42E-18
3.00E 01	2.51E-18
4.00E 01	2.83E-18
5.00E 01	2.91E-18
6.00E 01	2.86E-18
7.00E 01	2.64E-18
8.00E 01	2.67E-18
9.00E 01	2.66E-18
1.00E 02	2.57E-18
1.10E 02	2.51E-18
1.20E 02	2.37E-18

9-7-2
DUNN ET AL. (99)

ENERGY (KEV)	CROSS SECTION (SQ. CM)
7.00E-01	2.37E-17
8.00E-01	2.38E-17
9.00E-01	2.32E-17
1.00E 00	2.38E-17
1.20E 00	2.46E-17
1.50E 00	2.70E-17
1.80E 00	2.86E-17
2.00E 00	2.97E-17
2.35E 00	3.04E-17
2.50E 00	2.95E-17
2.80E 00	3.08E-17
3.00E 00	3.22E-17
3.50E 00	3.41E-17
4.00E 00	3.25E-17

9-7-3
HUGHES ET AL. (196)

ENERGY (KEV)	CROSS SECTION (SQ. CM)
2.00E 01	4.07E-18
3.00E 01	5.04E-18
4.00E 01	6.14E-18
5.00E 01	6.33E-18
6.00E 01	6.54E-18
7.00E 01	6.46E-18
8.00E 01	6.47E-18
9.00E 01	6.19E-18
1.00E 02	6.43E-18
1.10E 02	6.24E-18
1.20E 02	5.95E-18

9-7-4
HUGHES ET AL. (196)

ENERGY (KEV)	CROSS SECTION (SQ. CM)

ENERGY (KEV)	CROSS SECTION (SQ. CM)
2.00E 01	3.93E-18
3.00E 01	5.12E-18
4.00E 01	6.76E-18
5.00E 01	7.56E-18
6.00E 01	8.45E-18
7.00E 01	8.55E-18
8.00E 01	8.71E-18
9.00E 01	8.95E-18
1.00E 02	9.24E-18
1.10E 02	9.27E-18
1.20E 02	9.44E-18

9-7-5
HUGHES ET AL. (196)

ENERGY (KEV)	CROSS SECTION (SQ. CM)
2.00E 01	2.99E-18
3.00E 01	3.58E-18
4.00E 01	4.46E-18
5.00E 01	4.86E-18
6.00E 01	5.37E-18
7.00E 01	5.52E-18
8.00E 01	5.36E-18
9.00E 01	5.32E-18
1.00E 02	5.18E-18
1.10E 02	5.01E-18
1.20E 02	4.81E-18

CHAPTER 10

EXCITATION OF HELIUM AND OTHER
HEAVY PROJECTILES

Consideration is given here to the processes by which heavy projectiles are collisionally excited. The processes of interest include direct excitation of neutrals (00/0*0)†, direct excitation of ions (10/1*0) and charge transfer into an excited state (10/0*1). This chapter will be divided into two parts considering respectively helium projectiles and all other projectiles. Only in the case of helium is the data comprehensive, reasonably reliable and free of ambiguities. In the other cases it is fragmentary and often subject to uncertainty due to experimental inadequacies. Discussion of techniques and validity of data will be restricted to a minimum, since there are few areas where a detailed objective analysis is possible.

10-1 GENERAL CONSIDERATIONS

Most of the data have been obtained by measurement of spontaneous emission, consequently the remarks of the present section are related primarily to the optical techniques.

Excluded from detailed consideration here are the studies by Solov'ov, Il'in and co-workers[54, 217] on the formation of highly excited atoms (principal quantum numbers greater than 10) by neutralization of He^+ in Ne, Na, Mg neutralization of N^+, O^+, Ne^+ in Mg and by dissociation of O_2^+ and N_2^+ in Mg. Detection of these highly excited states is by field ionization. It has been concluded (see Section 3-3-A) that data obtained by this technique are poorly defined and should not be tabulated in this monograph.

In the case of excitation of helium projectiles, there are studies by Schlachter et al.[209, 63] and Gilbody et al.[62, 61] based on other techniques that purport to measure the sum of the cross sections for the formation of a number of excited states. Later [Section 10-2-A(4)] it will be shown that the methods used by Schlachter et al. are open to criticism. This leaves the work of Gilbody et al. as the only reliable series of measurements that do not involve optical techniques.

There is one important practical difference between the procedures for study of emission from fast heavy atoms and the corresponding procedures for fast hydrogen

† See Table 1-1 for explanation of this notation.

atoms. For heavy atoms there is no degeneracy of the excited states of different angular momentum quantum number; therefore transitions may be resolved spectroscopically.

In most experiments the excited projectiles move a considerable distance before spontaneous emission takes place. The spatial distribution of emission is characterized by the decay length (product of excited state lifetime and velocity) of the projectile. Two different experimental arrangements have been employed for emission measurements. Observations of spontaneous emission from the projectile have been made within the target region itself and related to the level excitation cross section by Eq. 2-21. The alternative technique is to measure emission after the projectiles have emerged from a target cell in which case emission decreases exponentially with distance from the exit and is related to level-excitation cross sections by Eq. 2-23. It is legitimate to make a measurement of emission at a single point and relate it to the level excitation cross section using either Eq. 2-21 or 2-23. It is more satisfactory to measure part of the spatial variation of the emission to confirm that it is characterized by the theoretical lifetime, and then to fit the appropriate equation to determine cross section. This latter procedure has the important advantage of confirming that the equations from which cross sections are determined do indeed represent the operation of the experiment.

In all of the experiments there have been no studies of how the excited state content of the projectile beam will influence the validity of the cross section measurements; this may be a serious omission.

10-2 EXCITATION OF HELIUM PROJECTILES

The data on helium excitation is generally in the form of level-excitation cross sections derived from measurements of line emission. The accuracy of the final cross section data will be influenced by the accuracy of the theoretical values of lifetimes and branching ratios that are used in the analysis of the raw data. There are some significant differences among values quoted in the literature. Errors in the assumed values of lifetimes will cause serious inaccuracies in cross section when lifetimes are used to estimate the correction for the spatial variation of emission. Weise et al.[31] have made a critical evaluation of the available calculations and have assigned estimates of accuracy to their tabulated values; their data should therefore be utilized in the analysis of experiments involving helium emission. The experiments reviewed here have utilized the predictions of Gabriel and Heddle;[243] these are generally in good agreement with the values of Weise et al. None of the analyses of data makes any estimate of the accuracy limitation that would be caused by errors in the calculated values of transition probabilities; this is a serious omission.

It is important that some assessment be made of the emission resulting from cascade, because this will exhibit a different spatial dependence from that due to direct excitation. The proportion of the measured emission that results from cascade is a function of the point at which the observations are made. A measurement of cross section based on a single observation of emission and a correction for spatial variation using the lifetime of the parent level of the observed transition will be inaccurate if cascade is large at the observation point. For helium emission it is often possible to

measure the cascading transitions directly and thereby make a correction for cascade population.

Spontaneous emission from the n^1S and n^3S states will be inherently unpolarized. All other transitions will exhibit nonzero polarization if the cross sections for populating the magnetic quantum number sublevels of the excited states are unequal. Polarization will cause anisotropy of emission requiring that measurements of emission at one angle be related to total cross section through the techniques discussed in Section 3-4-D. In some of the published work this precaution is ignored.

A. Experimental Arrangements

1. The Work of de Heer et al. De Heer and co-workers have carried out two series of investigations of the formation of excited helium projectiles. The first study was directed at the formation of the 2^1P, 3^1P, 4^1D, and 4^1S states induced by the impact of He^+ ions on targets of H_2, Ne, and Kr.[78] Measurements for the 2^1P and 3^1P states were made using the vacuum uv Lyman lines, and the other levels were investigated through the emission of visible lines. The second series of measurements, published originally in the proceedings of a conference[361] and later revised with a full report,[12] were concerned with emissions of both visible and near uv lines from the projectile induced by the impact on a helium target. Subsequently, many of these published measurements from both series of experiments have been revised.[213] The revised values, although unpublished, are utilized here since they undoubtedly represent an appreciable improvement in reliability. The only data retained here from the original publications are: the measurements on the formation of the 2^1P state for targets of H_2, Ne, and Kr,[78] measurements on the formation of the 3^1P state for a target of H_2,[78] and measurements on the formation of the 6^1D state for a target of He.[12]

The vacuum uv monochromator[78] used for measurements on the emission of the $2^1P - 1^2S$ and $3^1P - 1^1S$ lines was described originally by van Eck et al.[10] with the calibration method described separately.[90] The photometric calibration[90] is likely to be inaccurate since it involved theoretical assumptions concerning monochromator grating efficiency and astigmatism that were not confirmed experimentally. The level-excitation data were derived on the assumption that cascade population could be neglected; no evidence was given to substantiate this assumption. The authors of this work estimate that the systematic error in these measurements does not exceed $\pm 40\%$. This is a conservative estimate in the opinion of the present writer.

In the visible region of the spectrum a grating monochromator was used for the original published measurements[361, 12] on excitation by impact of He^+ on a helium target and also for unpublished revisions of that data.[213] Measurements were made of the emission from the excited projectiles as they traversed the target gas. The monochromator was placed at an angle of 60° to the beam path; the target and projectile emission were resolved through the Doppler shift of the projectile wavelength. The revised data used here[213] include attention to Doppler broadening of the emission in the manners indicated previously (Section 3-4-F). Adequate allowance was made for cascade. The influence of polarization was assessed and some direct measurements of polarization presented. De Heer et al. estimate that the systematic

error in the excitation cross sections will be less than $\pm 20\%$ and that the cross section data are likely to be too high rather than low. Reproducibility is estimated at $\pm 10\%$. In the measurements of polarization fraction, the error in the measured ratio of light intensity polarized parallel to the beam and polarized perpendicular to the beam should not exceed $\pm 5\%$.

2. The Work of Head and Hughes. Head and Hughes[214] have studied the formation of excited helium as a result of the impact of He^+ on targets of He, N_2, and O_2 at energies from 20 to 120-keV.

Collisionally induced emission was monitored with a 1/2-meter Ebert mounted grating monochromator fitted with photomultiplier detection. The axis was at an angle of 90° to the beam path for studies with targets of N_2 and O_2 but at 60° for target of helium in order to facilitate the separation of target and projectile emission through the Doppler shift of wavelength. A lens was used to form an image of the beam on the entrance slit of the monochromator. Emission was observed inside the target region for all of the work with N_2 and O_2 and for most of the work with He. For some of the helium lines, the Doppler effect tended to shift the line so that it interfered with emissions from the target at certain velocities. To remove this problem, the projectile beam was allowed to traverse the collision region entirely and to enter an evacuated flight tube where the emission intensity was measured. It was shown that both modes of observation gave the same results.

Calibration of detection sensitivity of the system was described in a previous paper.[154] [See Section 6-1-B(2) for analysis]. Cascade was neglected.

3. The Work of Tolk and White. This experiment[218] involves only a relative measurement of the cross section for emission of the 3888-Å line induced by He^+ impact on Ne. No attempt was made to derive level-excitation cross sections. A normalization to previous data by de Heer et al.[12, 212] is unreliable.

4. The Work of Schlachter et al. Schlachter and co-workers have studied the formation of excited states of helium as a result of the impact of He^+ ions on targets of hydrogen, argon, xenon, and cesium. The technique is indirect and does not resolve the individual excited states. It is designed to provide measurements on the total cross sections for the formation of all triplet states and the total cross section for the formation of all singlet states. Although different from studies of definite levels carried out by optical techniques, the data would be of value if it did indeed represent the stated quantities.

The first experiment determined the cross sections for the formation and destruction of helium atoms as a beam of He^+ traversed a target of cesium.[209] In addition to the charge transfer data, this measurement also gave cross sections for the formation of the He^- and He^+ ions resulting from impact of the "singlet" and "triplet" states of neutral helium on cesium.

The second series of experiments were designed to measure charge transfer by impact of He^+ ions on various permanent gases.[63] The fraction of neutral helium atoms formed in the "singlet" and "triplet" states was determined by passing the neutral beam through a cesium cell and monitoring the He^- and He^+ flux. The ratio of the "singlet" and "triplet" populations was derived from the He^- and He^+ flux with the assistance of the cross sections determined in the first series of measurements. Clearly, the validity of this detection procedure depends critically on the accuracy with which the measurements using the cesium target were carried out.

For the cesium target experiments, an He^+ beam was directed through Cs and a

study made of how the emergent He^+, He^0 and He^- fractions varied with target density. It was found that the variation of these fractions with density could be explained only by assuming that the neutral helium be considered as two components, "singlet" and "triplet"; the beam may then be treated as consisting of four components. This improved agreement does not in fact indicate that the four component beam is the unique solution, rather that it is better than three; inevitably, the more components are introduced, the better the fit will become. A further assumption was that the cross sections for formation and stripping of the 2^1S state was the same as for the 1^1S state—a most unlikely situation. In view of these unsubstantiated assumptions it is concluded that the data by Schlachter are unreliable. It follows that the work with permanent gases, which depends on the use of cesium data, is equally unreliable. For these reasons as well as other minor considerations, it is concluded that these data are all unreliable; therefore they are omitted from further consideration.

5. The Work of Gilbody et al. Gilbody et al. provide measurements of the cross sections for formation of the metastable state of helium by impact of He^+ on targets of H_2.[62, 61] It is claimed that the data represent the sum of the cross section for formation of the 2^1S and 2^3S states.

The He^+ ions were produced in an oscillating electron source, accelerated to between 75 and 250-keV energy, momentum analyzed, and directed into the experimental area. Two gas cells separated by a long flight tube formed the experimental system; the first cell was the target chamber, the second cell functioned as an analyzer. The primary He^+ beam first traversed the hydrogen-filled target cell in which it was partially neutralized. On emerging into the flight tube, the residual ion flux was removed by electrostatic deflection, and excited atoms decayed into either the ground or metastable states. The neutral beam then traversed the second gas cell, which has the function of an analyzer; in this region, some of the neutrals were collisionally ionized and the resulting charged components were removed by electrostatic deflection. Finally, the current of the attenuated neutral beam emerging from the second cell was monitored with a secondary emission detector.

It was assumed that, at impact on the second cell, the beam consisted only of metastable atoms and ground state atoms; all other excited states had decayed to these levels. In traversing the second cell, the neutral beam was attenuated by stripping of the ground state and metastable atoms; it was assumed that, at the high energies of this experiment, the formation of negative ions by one-electron capture or doubly charged ions by two-electron loss, is negligible. Each component of the neutral beam will decrease exponentially with distance from the entrance to the cell. The attenuation of the beam was measured as a function of target thickness and deconvoluted into two exponential terms; the exponents gave the cross sections for stripping from the metastable and ground states, and the ratios of the amplitudes of the two terms gave the ratios of the metastable and ground state components in the beam emerging from the first cell. For single-collision conditions in the first cell, the ratio of the metastable to ground state neutral components was equal to the ratio of the cross sections for charge transfer into the metastable and ground states. The cross section for charge transfer into the ground state was taken from previous measurements, and hence the cross section for formation of the metastable state was determined.

It is important to examine the validity of this method of determining metastable atom flux. A number of different gases and various pressures were utilized in the first cell in order to provide different proportions of metastables in the neutral beam. Under

all conditions, the decay curve in the second cell was the sum of two exponentials, each exhibiting a decay length that was independent of the method by which the excited neutrals were formed. This adds considerable certainty to the assumption that the beam entering the detection region does, in fact, consist only of two components. The important assumption is made that the cross section for stripping from the metastable singlet state is the same as for the metastable triplet; again, the deconvolution of the decay curve into only two components tends to confirm this assumption. The observations could also be interpreted as showing that one metastable component was present to a negligible extent or even that the cross section for stripping from one of the metastable states was negligible compared with the stripping from the other; neither of these two alternative interpretations is likely.

The accuracy of the data will depend on the validity of the model used to analyze the measurements. The assumption of equal stripping cross sections from the triplet and singlet metastable states has not been unambiguously confirmed. It was estimated by the authors that the overall accuracy of the data was better than $\pm 10 \%$; this error limit allowed for inaccuracy in measurements of target pressure and also for possible error in the cross section for one-electron capture into the ground state.

B. Experimental Results

The data are principally by de Heer et al.[12, 213] and by Head and Hughes.[214] Both series of experiments are subject to only minor criticisms and the considerable discrepancies between the data in some cases is puzzling. A possible cause for the discrepancy may be due to inadequate attention to establishing single-collision conditions in the work of Head and Hughes. They used pressures of the order 3×10^{-3} torr. Taking the cross section for stripping of the triplet levels as being the same as found by Gilbody et al.[61] for a hydrogen target, it seems likely that stripping of the atom before spontaneous decay may occur under some conditions. This will be most severe for the ^1S and ^3S states, which have long lifetimes, and will probably be negligible for the shortlived ^1P, ^1D, ^3P, ^3D states. It is noted that the most serious discrepancies between data from the two sources occurs for the ^1S and ^3S states. It is concluded that the data of de Heer et al. are likely to be more reliable.

Polarization measurements are all by de Heer et al.;[78, 213] there is no reason to doubt the accuracy of these data.

It is noted that Utterback[65] has provided qualitative information on the formation of metastable helium at low energies by charge transfer on helium target. During a study of total ionization cross sections for He incident on H_2, it was noticed that the results were dependent on whether He was the target and H_2 the projectile or vice versa. It was concluded that the difference was due to metastable helium atoms formed in the neutral projectile beam when it is created by charge transfer neutralization of He^+ ions in a helium cell. The difference between the He + H_2 and H_2 + He ionization cross sections may be taken as a measure of the formation of metastables. The energy range of the study, measured in the laboratory frame, ranged from 44-eV (the apparent threshold) to 66-eV. There appeared to be a sharp cross section maximum at 56-eV. Unfortunately, it is not possible to derive from these data a quantitative measurement of the cross section for formation of metastable helium, even on a relative basis; therefore this work is not considered further.

In addition to the work mentioned above, there have been some other studies that

are excluded from detailed consideration here. Hanle and Voss[199] studied helium emission induced by the impact of He^+ on H_2. The target pressure was 6×10^{-2} torr, at which some 27% of the beam was neutralized; for this reason, these data are expected to be inaccurate and misleading.

C. Data Tables

1. Cross Sections for the Formation of He States by the Impact of He $^+$ on a He Target (All Data Includes Excitation of the Projectile only)

10-2-1 4^1S State. [10-2-1(A)†, 10-2-2(B).‡]

10-2-2 5^1S State.†

10-2-3 3^1P State. [10-2-3(A)†, 10-2-3(B).‡]

10-2-4 4^1P State.†

10-2-5 3^1D State.‡

10-2-6 4^1D State. [10-2-6(A)†, 10-2-6(B).‡]

10-2-7 5^1D State. [10-2-7(A)†, 10-2-7(P).‡]

10-2-8 6^1D State.

10-2-9 3^3S State.‡

10-2-10 4^3S State. [10-2-10(A)†, 10-2-10(B).‡]

10-2-11 5^3S State.†

10-2-12 3^3P State. [10-2-12(A)†, 10-2-12(B).‡]

10-2-13 4^3P State.†

10-2-14 3^3D State. [10-2-14(A)†, 10-2-14(B).‡]

10-2-15 4^3D State. [10-2-15(A)†, 10-2-15(B).‡]

10-2-16 5^3D State.†

2. Polarization Fractions for Lines of He I Induced by Impact of He $^+$ on a He Target (This includes Excitation of the Projectile Only)

10-2-17 5016-Å $(3^1P \to 2^1S)$.†

10-2-18 4922-Å $(4^1D \to 2^1P)$.†

10-2-19 3889-Å $(3^3P \to 2^3S)$.

10-2-20 5876-Å $(3^3D \to 2^3P)$.†

3. Cross Sections for the Formation of He States by the Impact of He $^+$ on a Ne Target

10-2-21 4^1S State.†

10-2-22 2^1P State.‡

10-2-23 3^1P State.§

10-2-24 4^1D State.†

4. Cross Sections for the Emission of He I Lines Induced by the Impact of He $^+$ on a Ne Target

10-2-25 3889-Å $(3^3P \to 2^3S)$.

5. Polarization Fractions for Lines of He I Induced by the Impact of He $^+$ on a Ne Target

10-2-26 5016-Å $(3^1P \to 2^1S)$.§

10-2-27 4922-Å $(4^1D \to 2^1P)$.§

6. Cross Sections for the Formation of He States by the Impact of He⁺ on a Kr Target
10-2-28 2^1P State.‡

7. Cross Sections for the Formation of He State by the Impact of He⁺ on a H₂ Target
10-2-29 2^1P State.‡
10-2-30 3^1P State.‡
10-2-31 Sum of Cross Sections for All Metastable States (Direct Formation of 2^1S and 2^3S plus all Cascade Contributions)

8. Cross Sections for the Formation of He States by the Impact of He⁺ on a N₂ Target
10-2-32 3^1D State.‡
10-2-33 4^1D State.‡
10-2-34 3^3D State.‡
10-2-35 4^3D State.‡

9. Cross Sections for the Formation of He States by the Impact of He⁺ on a O₂ Target
10-2-36 3^3D State.

† Revisions[213] of the data originally published.[12]
‡ Data evaluated on the assumption that cascade may be neglected.
§ Revisions[213] of the data originally published.[78]

10-2-1A
DE HEER ET AL. (213)

ENERGY (KEV)	CROSS SECTION (SQ. CM)
1.50E 01	3.60E-20
1.75E 01	3.97E-20
2.00E 01	4.58E-20
2.25E 01	4.82E-20
2.50E 01	4.98E-20
2.75E 01	5.30E-20
3.00E 01	6.10E-20
3.50E 01	6.52E-20
4.00E 01	7.76E-20
5.00E 01	1.31E-19
6.00E 01	1.68E-19
7.00E 01	2.23E-19
8.00E 01	2.43E-19
9.00E 01	2.50E-19

10-2-1B
HEAD ET AL. (214)

ENERGY (KEV)	CROSS SECTION (SQ. CM)
2.00E 01	5.00E-20
3.00E 01	1.10E-19
4.00E 01	1.80E-19
5.00E 01	3.20E-19
6.00E 01	4.50E-19
7.00E 01	5.10E-19
8.00E 01	5.60E-19
9.00E 01	5.80E-19
1.00E 02	6.20E-19
1.10E 02	6.50E-19
1.20E 02	6.50E-19

10-2-2
DE HEER ET AL. (213)

ENERGY (KEV)	CROSS SECTION (SQ. CM)
5.00E 01	5.73E-20
6.00E 01	8.48E-20
7.00E 01	1.13E-19
8.00E 01	1.19E-19
9.00E 01	1.21E-19

10-2-3A
DE HEER ET AL. (213)

ENERGY (KEV)	CROSS SECTION (SQ. CM)
1.25E 01	6.95E-19
1.50E 01	5.71E-19
1.75E 01	5.20E-19
2.00E 01	4.99E-19
2.25E 01	4.92E-19
2.50E 01	4.83E-19
2.75E 01	4.63E-19
3.00E 01	4.72E-19
3.50E 01	5.16E-19
4.00E 01	6.14E-19
5.00E 01	1.04E-18
6.00E 01	1.27E-18
7.00E 01	1.39E-18
8.00E 01	1.45E-18
9.00E 01	1.42E-18

10-2-3B
HEAD ET AL. (214)

ENERGY (KEV)	CROSS SECTION (SQ. CM)
2.00E 01	1.85E-18
3.00E 01	1.67E-18
4.00E 01	1.66E-18
5.00E 01	1.68E-18
6.00E 01	2.12E-18
7.00E 01	2.33E-18
8.00E 01	2.57E-18
9.00E 01	2.92E-18
1.00E 02	2.30E-18
1.10E 02	2.49E-18
1.20E 02	2.13E-18

10-2-4
DE HEER ET AL. (213)

ENERGY (KEV)	CROSS SECTION (SQ. CM)
5.00E 01	3.94E-19
6.00E 01	4.62E-19
7.00E 01	5.63E-19
8.00E 01	6.30E-19
9.00E 01	6.07E-19

10-2-5
HEAD ET AL. (214)

ENERGY (KEV)	CROSS SECTION (SQ. CM)
2.00E 01	8.99E-19
3.00E 01	8.24E-19
4.00E 01	7.22E-19
5.00E 01	7.49E-19
6.00E 01	7.13E-19
7.00E 01	6.44E-19
8.00E 01	5.77E-19
9.00E 01	4.79E-19
1.00E 02	4.09E-19
1.10E 02	3.27E-19
1.20E 02	2.00E-19

10-2-6A
DE HEER ET AL. (213)

ENERGY (KEV)	CROSS SECTION (SQ. CM)
7.00E 00	6.89E-20
8.00E 00	8.25E-20
9.00E 00	1.05E-19
1.00E 01	1.20E-19
1.25E 01	6.34E-20
1.50E 01	7.34E-20
1.75E 01	1.20E-19
2.00E 01	1.51E-19
2.25E 01	1.83E-19
2.50E 01	2.21E-19
2.75E 01	2.36E-19
3.00E 01	2.22E-19
3.50E 01	2.37E-19
4.00E 01	2.67E-19
5.00E 01	3.12E-19
6.00E 01	3.41E-19
7.00E 01	3.25E-19
8.00E 01	2.79E-19
9.00E 01	2.39E-19

10-2-6B
HEAD ET AL. (214)

ENERGY (KEV)	CROSS SECTION (SQ. CM)
2.00E 01	1.73E-19
3.00E 01	2.83E-19
4.00E 01	3.21E-19
5.00E 01	3.82E-19
6.00E 01	4.23E-19
7.00E 01	4.20E-19
8.00E 01	3.95E-19
9.00E 01	3.58E-19
1.00E 02	2.97E-19
1.10E 02	2.46E-19
1.20E 02	2.24E-19

10-2-7A
DE HEER ET AL. (213)

ENERGY (KEV)	CROSS SECTION (SQ. CM)
1.25E 01	5.30E-20
1.50E 01	4.07E-20
1.75E 01	3.86E-20
2.00E 01	4.55E-20
2.25E 01	6.09E-20
2.50E 01	7.13E-20
2.75E 01	8.07E-20
3.00E 01	9.47E-20
3.50E 01	1.16E-19
4.00E 01	1.35E-19
5.00E 01	1.47E-19
6.00E 01	1.83E-19
7.00E 01	1.76E-19
8.00E 01	1.66E-19
9.00E 01	1.37E-19

10-2-7B
HEAD ET AL. (214)

ENERGY (KEV)	CROSS SECTION (SQ. CM)
2.50E 01	8.50E-20
3.00E 01	1.10E-19

4.00E 01   1.57E-19
5.00E 01   1.77E-19
6.00E 01   2.10E-19
7.00E 01   2.34E-19
8.00E 01   2.20E-19
9.00E 01   1.96E-19
1.00E 02   1.97E-19
1.10E 02   1.67E-19
1.20E 02   1.46E-19

10-2-8
DE HEER ET AL. (12)

ENERGY   CROSS
(KEV)   SECTION
         (SQ. CM)

2.50E 01   3.45E-20
3.00E 01   3.14E-20
3.50E 01   3.30E-20
5.00E 01   6.91E-20
5.50E 01   7.85E-20
6.00E 01   8.31E-20
6.50E 01   8.40E-20
7.00E 01   8.40E-20
7.50E 01   7.55E-20
8.00E 01   7.17E-20
9.00E 01   7.14E-20

10-2-9
HEAD ET AL. (214)

ENERGY   CROSS
(KEV)   SECTION
         (SQ. CM)

2.00E 01   4.60E-19
3.00E 01   5.10E-19
4.00E 01   6.70E-19
5.00E 01   9.50E-19
6.00E 01   1.20E-18
7.00E 01   1.60E-18
8.00E 01   2.00E-18
9.00E 01   2.30E-18
1.00E 02   2.70E-18
1.10E 02   3.00E-18
1.20E 02   3.20E-18

10-2-10A
DE HEER ET AL. (213)

ENERGY   CROSS
(KEV)   SECTION
         (SQ. CM)

1.50E 01   6.41E-20
1.75E 01   6.69E-20
2.00E 01   8.44E-20
2.25E 01   9.45E-20
2.50E 01   1.02E-19
2.75E 01   9.89E-20
3.00E 01   1.10E-19
3.50E 01   1.24E-19
4.00E 01   1.51E-19
5.00E 01   2.06E-19
6.00E 01   2.77E-19
7.00E 01   2.92E-19
8.00E 01   4.53E-19
9.00E 01   5.52E-19

10-2-10B
HEAD ET AL. (214)

ENERGY   CROSS
(KEV)   SECTION
         (SQ. CM)

2.00E 01   1.00E-19
3.00E 01   2.30E-19
4.00E 01   3.90E-19
5.00E 01   6.00E-19
6.00E 01   6.70E-19
7.00E 01   9.60E-19
8.00E 01   1.20E-18
9.00E 01   1.40E-18
1.00E 02   1.70E-18
1.10E 02   1.90E-18
1.20E 02   2.00E-18

10-2-11
DE HEER ET AL. (213)

ENERGY   CROSS
(KEV)   SECTION
         (SQ. CM)

4.00E 01   5.11E-20
5.00E 01   7.33E-20
6.00E 01   9.92E-20
7.00E 01   1.36E-19
8.00E 01   1.62E-19
9.00E 01   2.34E-19

10-2-12A
DE HEER ET AL. (213)

ENERGY   CROSS
(KEV)   SECTION
         (SQ. CM)

5.00E 00   3.29E-19
6.00E 00   5.05E-19
7.00E 00   7.68E-19
8.00E 00   8.12E-19
9.00E 00   7.52E-19
1.00E 01   6.40E-19
1.25E 01   6.47E-19
1.50E 01   6.75E-19
1.75E 01   7.10E-19
2.00E 01   7.02E-19
2.25E 01   6.98E-19
2.50E 01   6.74E-19
2.75E 01   7.12E-19
3.00E 01   7.08E-19
3.50E 01   7.40E-19
4.00E 01   8.29E-19
5.00E 01   9.47E-19
6.00E 01   1.05E-18
7.00E 01   1.10E-18
8.00E 01   1.15E-18
9.00E 01   8.76E-19

10-2-12B
HEAD ET AL. (214)

ENERGY   CROSS
(KEV)   SECTION
         (SQ. CM)

2.00E 01   1.22E-18
2.50E 01   1.18E-18
3.00E 01   1.27E-18
4.00E 01   1.59E-18
5.00E 01   1.74E-18
6.00E 01   1.85E-18
7.00E 01   2.04E-18
8.00E 01   2.17E-18
9.00E 01   2.36E-18
1.00E 02   2.47E-18
1.10E 02   2.50E-18
1.20E 02   2.67E-18

10-2-13
DE HEER ET AL. (213)

ENERGY   CROSS
(KEV)   SECTION
         (SQ. CM)

1.00E 01   1.10E-19
1.25E 01   1.00E-19
1.50E 01   9.00E-20
1.75E 01   7.56E-20
2.00E 01   7.90E-20
2.25E 01   9.00E-20
2.50E 01   9.50E-20
2.75E 01   1.05E-19
3.00E 01   1.12E-19
3.50E 01   1.19E-19
4.00E 01   1.29E-19
5.00E 01   1.79E-19
6.00E 01   2.12E-19
7.00E 01   2.25E-19
8.00E 01   2.48E-19
9.00E 01   2.16E-19

10-2-14A
DE HEER ET AL. (213)

ENERGY   CROSS
(KEV)   SECTION
         (SQ. CM)

7.00E 00   9.85E-19
8.00E 00   6.98E-19
9.00E 00   6.65E-19
1.00E 01   6.63E-19
1.25E 01   8.05E-19
1.50E 01   8.49E-19
1.75E 01   8.58E-19
2.00E 01   7.81E-19
2.25E 01   6.83E-19
2.50E 01   7.22E-19
2.75E 01   7.93E-19
3.00E 01   9.55E-19
3.50E 01   1.09E-18
4.00E 01   9.60E-19
5.00E 01   7.82E-19
6.00E 01   6.41E-19
7.00E 01   4.75E-19
8.00E 01   4.16E-19
9.00E 01   3.47E-19

1.00E 00	6.70E-18
1.5CE 00	9.40E-18
2.00E 00	1.10E-17

10-3-11A
MATVEYEV ET AL. (220)

ENERGY (KEV)	CROSS SECTION (RELATIVE UNITS)
7.38E-01	0.0
8.73E-01	3.98E 00
9.96E-01	5.88E 00
9.63E-01	7.00E 00
1.10E 00	1.02E 01
1.22E 00	1.32E 01
1.3CE 00	1.84E 01
1.35E 00	2.25E 01
1.42E 00	2.66E 01
1.56E C0	3.28E 01
1.64E 00	3.73E 01
1.84E 00	4.22E 01
2.11E 00	5.38E 01
2.29E 00	6.07E 01
2.46E 00	6.50E 01
2.68E 00	7.04E 01
2.95E 00	7.55E 01
3.12E 00	7.78E 01
3.35E 00	8.09E 01
3.94E 00	8.60E 01
4.09E 00	8.75E 01
4.36E 00	8.86E 01
4.65E 00	8.99E 01
4.95E 00	9.07E 01
5.42E 00	9.11E 01
5.98E 00	9.07E 01
6.48E 00	8.86E 01
6.94E 00	8.64E 01
7.51E 00	8.47E 01
7.97E 00	8.17E 01
8.5CE 00	8.09E 01
8.99E 00	7.94E 01

10-3-11B
MATVEYEV ET AL. (220)

ENERGY (KEV)	CROSS SECTION (RELATIVE UNITS)
1.20E 00	7.40E 00
1.42E 00	9.50E 00
1.62E 00	1.20E 01
2.00E 00	1.49E 01
2.18E 00	1.69E 01
2.52E 00	1.94E 01
2.99E 00	2.24E 01
3.37E 00	2.41E 01
3.77E C0	2.52E 01
4.13E 00	2.59E 01
4.58E 00	2.66E 01
4.95E 00	2.58E 01
5.46E 00	2.55E 01
5.98E 00	2.48E 01
6.47E 00	2.38E 01
6.50E 00	2.46E 01
6.98E 00	2.27E 01
7.53E 00	2.18E 01
8.02E 00	2.14E 01

10-3-11C
MATVEYEV ET AL. (220)

ENERGY (KEV)	CROSS SECTION (RELATIVE UNITS)
1.05E 00	3.50E 00
1.20E 00	5.10E 00
1.39E 00	7.20E 00
1.57E 00	9.10E 00
1.83E 00	1.08E 01
2.00E 00	1.25E 01
2.21E 00	1.47E 01
2.57E 00	1.73E 01
3.00E 00	1.99E 01
3.40E 00	2.15E 01
3.79E 00	2.27E 01
4.13E 00	2.37E 01
4.54E 00	2.43E 01
4.94E 00	2.43E 01
5.45E 00	2.40E 01
5.97E 00	2.31E 01
6.93E 00	2.13E 01
7.52E 00	2.01E 01
7.97E 00	1.94E 01

10-3-12
MATVEYEV ET AL. (115)

ENERGY (KEV)	CROSS SECTION (ARB. UNITS)
1.55E 00	4.77E 00
1.75E 00	1.03E 01
1.98E 00	1.29E 01
2.18E 00	1.57E 01
2.47E 00	2.48E 01
2.65E 00	2.95E 01
2.95E 00	3.41E 01
3.37E 00	3.41E 01
3.91E 00	3.08E 01
4.87E 00	2.71E 01
5.87E 00	2.87E 01

10-3-13
NEFF (104)

ENERGY (KEV)	CROSS SECTION (SQ. CM)
4.00E-01	5.50E-20
6.00E-01	7.90E-20
8.00E-01	1.20E-19
1.50E 00	1.80E-19
2.00E 00	2.70E-19

10-3-14
BOBASHEV ET AL. (362)

ENERGY (KEV)	CROSS SECTION (RELATIVE UNITS)
3.90E 00	4.00E-01
4.34E 00	3.30E 00
4.89E 00	5.10E 00
5.37E 00	8.70E 00
5.90E 00	1.04E 01
6.35E 00	1.36E 01
6.83E 00	1.64E 01
7.43E 00	2.18E 01
7.89E 00	2.61E 01
8.40E 00	3.37E 01
8.86E 00	4.24E 01
9.38E 00	5.20E 01
9.94E 00	5.94E 01
1.04E 01	7.52E 01
1.09E 01	7.90E 01
1.14E 01	9.38E 01
1.19E 01	1.00E 02

10-3-15A
ANDERSEN ET AL. (114)

ENERGY (KEV)	CROSS SECTION (RELATIVE UNITS)
1.00E 01	1.11E 02
1.25E 01	2.18E 02
1.50E 01	3.93E 02
1.75E 01	5.63E 02
2.00E 01	7.62E 02
2.25E 01	8.45E 02
2.50E 01	9.50E 02
2.75E 01	1.06E 03
3.00E 01	1.18E 03
3.00E 01	1.15E 03
3.25E 01	1.13E 03
3.50E 01	1.20E 03
3.75E 01	1.20E 03
4.00E 01	1.23E 03
4.25E 01	1.17E 03
4.50E 01	1.23E 03
4.75E 01	1.20E 03
5.00E 01	1.15E 03
5.25E 01	1.18E 03
5.50E 01	1.26E 03
5.75E 01	1.17E 03
6.00E 01	1.18E 03
6.25E 01	1.25E 03
6.50E 01	1.25E 03
6.75E 01	1.24E 03
7.00E 01	1.21E 03
7.25E 01	1.15E 03
7.50E 01	1.15E 03

10-3-15B
ANDERSEN ET AL. (114)

ENERGY (KEV)	CROSS SECTION (RELATIVE UNITS)
4.00E 00	3.25E 02
6.00E 00	4.62E 02
8.00E 00	5.11E 02
1.00E 01	6.00E 02
1.20E 01	1.10E 03
1.40E 01	1.49E 03
1.50E 01	2.08E 03
1.60E 01	2.06E 03
1.80E 01	2.75E 03
2.00E 01	3.07E 03
2.20E 01	3.41E 03
2.40E 01	3.43E 03

ENERGY (KEV)	CROSS SECTION (SQ. CM)
2.75E 01	1.13E-18
3.00E 01	1.21E-18
3.50E 01	1.98E-18

10-2-24
DE HEER ET AL. (213)

ENERGY (KEV)	CROSS SECTICN (SQ. CM)
5.00E 00	5.51E-19
6.00E 00	4.33E-19
7.00E 00	3.45E-19
8.00E 00	2.98E-19
9.00E 00	3.07E-19
1.00E 01	3.73E-19
1.25E 01	4.63E-19
1.50E 01	3.73E-19
1.75E 01	3.30E-19
2.00E 01	3.73E-19
2.25E 01	4.40E-19
2.50E 01	5.24E-19
2.75E 01	3.74E-19
3.00E 01	5.82E-19
3.50E 01	6.12E-19

10-2-25
TOLK ET AL. (218)

ENERGY (KEV)	CROSS SECTION (SQ. CM)
2.39E-02	8.00E-21
2.56E-02	9.00E-22
2.82E-02	3.00E-21
2.92E-02	2.40E-20
2.94E-02	4.60E-20
2.98E-02	6.10E-20
3.02E-02	7.80E-20
3.13E-02	1.02E-19
3.13E-02	1.30E-19
3.24E-02	1.37E-19
3.47E-02	1.66E-19
3.64E-02	1.66E-19
3.67E-02	2.06E-19
3.69E-02	2.27E-19
3.82E-02	2.50E-19
4.15E-02	2.51E-19
4.36E-02	2.72E-19
4.48E-02	2.81E-19
4.61E-02	2.74E-19
4.70E-02	2.59E-19
4.74E-02	2.40E-19
4.84E-02	2.23E-19
4.97E-02	2.02E-19
5.11E-02	1.56E-19
5.22E-02	1.71E-19
5.33E-02	1.89E-19
5.44E-02	2.21E-19
5.51E-02	2.39E-19
5.75E-02	2.52E-19
5.87E-02	2.36E-19
5.95E-02	2.21E-19
6.12E-02	1.84E-19
6.24E-02	1.71E-19
6.51E-02	1.84E-19
6.69E-02	2.34E-19
6.78E-02	2.48E-19
7.07E-02	2.43E-19
7.27E-02	2.27E-19
7.42E-02	2.04E-19
7.58E-02	1.71E-19
7.79E-02	1.61E-19
7.79E-02	1.39E-19
8.35E-02	1.39E-19
8.46E-02	1.53E-19
8.88E-02	1.70E-19
9.13E-02	1.92E-19
9.52E-02	2.10E-19
1.02E-01	1.97E-19
1.08E-01	1.71E-19
1.12E-01	1.46E-19
1.20E-01	1.53E-19
1.23E-01	1.71E-19
1.28E-01	1.90E-19
1.33E-01	2.45E-19
1.36E-01	2.97E-19
1.43E-01	3.24E-19
1.47E-01	3.50E-19
1.53E-01	3.57E-19
1.58E-01	4.00E-19
1.60E-01	4.33E-19
1.67E-01	4.82E-19
1.67E-01	5.33E-19
1.86E-01	5.37E-19
1.83E-01	5.15E-19
2.05E-01	5.05E-19
2.22E-01	4.82E-19
2.26E-01	4.88E-19
2.37E-01	5.04E-19
2.42E-01	5.20E-19
2.53E-01	5.20E-19
2.60E-01	5.37E-19
2.74E-01	5.39E-19
2.82E-01	5.22E-19
2.92E-01	4.89E-19
3.15E-01	5.04E-19
3.33E-01	5.13E-19
3.57E-01	5.04E-19
3.72E-01	4.86E-19
3.99E-01	4.61E-19
4.21E-01	3.99E-19
4.42E-01	3.96E-19
4.64E-01	3.84E-19
4.90E-01	4.05E-19
5.22E-01	4.18E-19
5.40E-01	4.08E-19
5.67E-01	4.26E-19
5.95E-01	4.44E-19
6.51E-01	5.04E-19
6.93E-01	5.75E-19
7.51E-01	6.25E-19
7.95E-01	6.57E-19
8.40E-01	6.93E-19
9.07E-01	7.27E-19
8.76E-01	7.36E-19
9.78E-01	7.40E-19
1.10E 00	7.56E-19
1.18E 00	7.17E-19
1.3CE 00	7.17E-19
1.49E 00	7.25E-19
1.59E 00	6.91E-19
1.38E 00	6.68E-19
1.69E 00	6.49E-19
1.78E 00	6.16E-19
1.88E 00	5.67E-19
1.98E 00	5.42E-19
2.17E 00	4.95E-19
2.41E 00	4.77E-19
2.62E 00	4.82F-19
3.02E 00	6.15E-19
3.28E 00	6.50E-19
3.82E 00	6.79E-19
3.49E 00	6.93E-19
4.27E 00	8.28E-19
4.67E 00	8.55E-19
4.04E 00	8.89E-19

10-2-26
CE HEER ET AL. (213)

ENERGY (KEV)	PCLARIZATION (PERCENT)
5.00E 00	1.80E 01
6.00E 00	-4.16E CO
8.0CE 00	2.43E 00
1.00E 01	-5.65E 00
2.50E 01	4.30E 00
3.00E 01	2.03E 01
3.50E 01	1.80E 01

10-2-27
DE HEER ET AL. (213)

ENERGY (KEV)	PCLARIZATION (PERCENT)
4.00E 00	2.43E 00
6.00E 00	1.67E 01
8.00E 00	1.90E 01
1.00E 01	2.81E 01
2.00E 01	2.88E 01
2.50E 01	2.88E 01
3.00E 01	3.53E C1
3.50E 01	3.53E 01

10-2-28
DE HEER ET AL. (78)

ENERGY (KEV)	CROSS SECTION (SQ. CM)
1.00E 01	4.69E-17
1.25E 01	4.76E-17
1.50E 01	4.89E-17
2.00E 01	5.70E-17
2.50E 01	6.32E-17
3.00E 01	6.50E-17
3.50E 01	7.01E-17

10-2-29
DE HEER ET AL. (78)

ENERGY (KEV)	CROSS SECTICN (SQ. CM)
5.0CE 00	2.99E-17
7.00E 00	3.12E-17
1.00E 01	3.81E-17
1.50E 01	3.95E-17
2.00E 01	4.20E-17
2.50E 01	4.03E-17
3.00E 01	4.20E-17

ENERGY (KEV)	CROSS SECTION (SQ. CM)
3.50E 01	3.87E-17

10-2-30
DE HEER ET AL. (78)

ENERGY (KEV)	CROSS SECTION (SQ. CM)
1.00E 01	2.06E-18
1.50E 01	1.55E-18
1.75E 01	1.48E-18
2.00E 01	1.18E-18
2.25E 01	1.36E-18
2.50E 01	1.30E-18
3.00E 01	1.58E-18
3.50E 01	1.90E-18

10-2-31
GILBODY ET AL. (61)

ENERGY (KEV)	CROSS SECTION (SQ. CM)
8.00E 01	6.70E-17
1.00E 02	6.50E-17
1.25E 02	5.70E-17
1.50E 02	5.20E-17
1.75E 02	4.40E-17
2.00E 02	3.50E-17
2.50E 02	2.70E-17

10-2-32
HEAD ET AL. (214)

ENERGY (KEV)	CROSS SECTION (SQ. CM)
2.00E 01	4.03E-18
3.00E 01	3.89E-18
4.00E 01	4.11E-18
5.00E 01	3.87E-18
6.00E 01	3.49E-18
7.00E 01	3.37E-18
8.00E 01	3.19E-18
9.00E 01	2.77E-18
1.00E 02	2.38E-18
1.10E 02	2.28E-18
1.20E 02	2.07E-18

10-2-33
HEAD ET AL. (214)

ENERGY (KEV)	CROSS SECTION (SQ. CM)
1.50E 01	9.20E-19
2.00E 01	1.03E-18
3.00E 01	1.24E-18
4.00E 01	1.47E-18
5.00E 01	1.66E-18
6.00E 01	1.68E-18
7.00E 01	1.64E-18
8.00E 01	1.55E-18
9.00E 01	1.41E-18
1.00E 02	1.31E-18
1.10E 02	1.23E-18
1.20E 02	1.12E-18

10-2-34
HEAD ET AL. (214)

ENERGY (KEV)	CROSS SECTION (SQ. CM)
1.00E 01	1.37E-17
2.00E 01	1.08E-17
3.00E 01	1.00E-17
3.50E 01	9.84E-18
4.00E 01	9.71E-18
4.50E 01	9.96E-18
5.00E 01	9.53E-18
5.50E 01	9.57E-18
6.00E 01	9.27E-18
7.00E 01	8.65E-18
8.00E 01	8.05E-18
9.00E 01	7.58E-18
1.00E 02	6.38E-18
1.10E 02	5.86E-18
1.20E 02	5.18E-18

10-2-35
HEAD ET AL. (214)

ENERGY (KEV)	CROSS SECTION (SQ. CM)
2.00E 01	2.29E-18
3.00E 01	2.71E-18
4.00E 01	3.22E-18
5.00E 01	3.71E-18
6.00E 01	4.12E-18
7.00E 01	4.34E-18
8.00E 01	3.99E-18
9.00E 01	4.01E-18
1.00E 02	3.71E-18
1.10E 02	3.12E-18
1.20E 02	3.08E-18

10-2-36
HEAD ET AL. (214)

ENERGY (KEV)	CROSS SECTION (SQ. CM)
2.00E 01	9.47E-18
2.50E 01	8.48E-18
3.00E 01	8.04E-18
4.00E 01	7.82E-18
5.00E 01	7.41E-18
6.00E 01	6.53E-18
7.00E 01	5.85E-18
8.00E 01	5.53E-18
9.00E 01	5.13E-18
1.00E 02	4.75E-18
1.10E 02	4.34E-18
1.20E 02	4.08E-18

10-3 EXCITATION OF HEAVY PROJECTILES

The data on these cases are fragmentary and of poor quality. Little reliance should be placed on any of the measurements. There are no measurements of cross sections for the formation of levels; all are line emission data.

A. Experimental Arrangements

In most experiments Doppler effects and polarization were ignored. In some cases the invariance of emission with distance of penetration into the target was not confirmed. In many studies there is ambiguity as to the identification of lines.

Sluyters and Kistemaker[103, 276, 277, 278] carried out a number of studies of Ar^+ impact on targets of He, Ne, and Kr. In all cases, the target pressure was far too high (5×10^{-3} mm), single-collision conditions were not established and a large fraction

of the projectile beam was neutralized. These data are unreliable and are not repeated here.

Thomas and Gilbody studied excitation of Ar^+ and Kr^+ resulting from impact on He, Ne, Ar, and Kr. Their first publication was a conference paper[280] that was later superseded by a more complete report.[71] The apparatus was previously reviewed in Section 6-1-B(7). Some of the data was taken using the apparatus of van Eck and de Heer; descriptions of this apparatus may be found in a number of publications.[10, 165] In all experiments there must be some uncertainty as to the identification of the measured lines. This is particularly true for the 4431-Å transition ($4p\ ^4P^0_{1\ 1/2} \rightarrow 3d\ ^4D_{2\ 1/2}$) that was not resolved from the neighboring 4430-Å ($4p\ ^4D^0_{1\ 1/2} \rightarrow 4s\ ^4P_{1/2}$) line.

Robinson and Gilbody[105] studied Ar^+ impact on Ar and measured the separate emission of the 4431-Å Ar^+ line from the target and from the projectile. This line is unresolved from the 4430-Å transition, similar to the situation in the work of Thomas and Gilbody.

Neff[281, 104] has studied a number of reactions involving excitation of Ne^+, Na^+, K^+, and Ca^+ projectiles by impact on N_2, and excitation of N_2^+ by impact on Ar. Spectrometric resolution was poor (30-Å) and identification of lines is not adequate. The absolute magnitudes of cross sections were assigned by normalization to data of Sluyters and Kistemaker.[103] It was concluded above that the work of Sluyters and Kistemaker is unreliable, consequently the absolute magnitudes of Neff's data must also be considered as being unreliable.

There are some studies by Haugsjaa et al.[64] on the formation of metastable argon atoms by charge transfer as Ar^+ traverses Ar. It was noticed that the cross section for ionization of C_2H_2 by Ar atoms depended on whether the atomic projectiles were produced by charge transfer of Ar^+ ions in a cell of H_2 or of Ar. In the case of Ar^+ + H_2 the available energy in the center-of-mass system was insufficient to allow the formation of excited Ar; whereas in the $Ar^+ + Ar$ case metastable argon atoms could be formed. A determination was made of the difference between the ionization cross section of C_2H_2, measured by using a beam of Ar prepared in H_2, and the same cross section by using a beam of Ar prepared in Ar; this difference in cross section was taken as a measure of the formation of metastable Ar atoms by charge transfer of Ar^+ in Ar. The energy range of the study was from 22-eV, the apparent threshold, to 48-eV. Sharp peaks were exhibited at 30 and 42-eV impact energy. Unfortunately, it is not possible to derive any quantitative information about the cross sections for the formation of the metastables, even on a relative basis. Therefore, these observations are not considered further here.

Matveyev, Bobashev, and co-workers[220, 115, 362, 116] have studied Na^+, K^+, Rb^+, and Cs^+ impact on rare gases. In many cases the energy range encompasses the threshold of the excitation function. The study of emission induced by Cs^+ impact on He was discussed first in a conference paper[362] and later in a formal publication;[116] both reports contain identical data. A description of the apparatus is to be found in the earliest paper.[220] Projectiles were all produced by thermionic emission, therefore they do not contain any excited states. Many of the studies were of the thresholds for specific transitions; these thresholds[115] are not repeated here. The cross section data is in the form of relative variation of cross section with energy; no absolute data is given.

H. Anderson et al.[114] have studied Cs II lines induced by Cs^+ impact on He and Ar. The existing report is inadequate for a detailed analysis of these data.

R. W. Anderson et al.[221] have studied excitation of K by impact on H_2, O_2, N_2, Cl_2, He, Ne, Ar, Kr, and Xe at energies from 25 to 600-keV. The optical system consisted of a photomultiplier with a broad band interference filter ($\pm$ 70-Å) centered at a wavelength of 7680-Å. It was intended that the optical system should detect, simultaneously, both components of the red K I doublet at 7665 and 7699-Å ($4\,^2P_{1/2,\,3/2} - 4\,^2S_{1/2}$). Approximate absolute values were assigned to the cross sections; however, no information was given on how the detection efficiency of the optical system was evaluated. Due to the poor optical resolution there is no guarantee that the measured emission was only from the K I doublet; the data may include contributions from other projectile states or target emission. It is concluded that the origin of the emission is ambiguous; therefore the data is not considered further here.

B.　Data Tables

The reader is cautioned that much of the data listed here may be unreliable. The listing contains, first, monoatomic rare gas projectiles, second, the data for molecular projectiles, and, finally, data for metallic projectiles.

1.　Cross Sections for Emission Resulting from Excitation of Rare Gas Projectiles

10-3-1　6402-Å Line of Ne I ($3p[2\frac{1}{2}]_3 \rightarrow 3s[1\frac{1}{2}]_2^0$) Induced by Ne^+ Impact on N_2.

10-3-2　Lines of Ar II Induced by Ar^+ Impact on He. (A) 4431-Å ($4p\,^4P_{2\,1/2}^0 \rightarrow 3d\,^4D_{2\,1/2}$); (B) 4610-Å ($4p'\,^2F_{3\,1/2}^0 \rightarrow 4s'\,^2D_{2\,1/2}$); (C) 4658-Å ($4p\,^2P_{1/2}^0 \rightarrow 4s\,^2P_{1\,1/2}^0$); (D) 4765-Å ($4p\,^2P_{1\,1/2}^0 \rightarrow 4s\,^2P_{1/2}$).

10-3-3　Lines of Ar II Induced by Ar^+ Impact on Ne. (A) 4431-Å ($4p\,^4P_{2\,1/2}^0 \rightarrow 3d\,^4D_{2\,1/2}$); (B) 4610-Å ($4p'\,^2F_{3\,1/2}^0 \rightarrow 4s'\,^2D_{2\,1/2}$); (C) 4658-Å ($4p\,^2P_{1/2}^0 \rightarrow 4s\,^2P_{1\,1/2}$); (D) 4765-Å ($4p\,^2P_{1\,1/2}^0 \rightarrow 4s\,^2P_{1/2}$).

10-3-4　4431-Å Line of Ar II ($4p\,^4P_{2\,1/2}^0 \rightarrow 3d\,^4D_{2\,1/2}$) Induced by Ar^+ Impact on Ar. [Excitation of the projectile only.]

10-3-5　Lines of Ar II Induced by Ar^+ Impact on Kr. (A) 4431-Å ($4p\,^4P_{2\,1/2}^0 \rightarrow 3d\,^4D_{2\,1/2}$); (B) 4610-Å ($4p'\,^2F_{3\,1/2}^0 \rightarrow 4s'\,^2D_{2\,1/2}$); (C) 4658-Å ($4p\,^2P_{1/2}^0 \rightarrow 4s\,^2P_{1\,1/2}$).

10-3-6　Lines of Kr II Induced by Kr^+ Impact on He. (A) 4577-Å ($5p'\,^2F_{3\,1/2}^0 \rightarrow 5s'\,^2D_{2\,1/2}$); (B) 4738-Å ($5p\,^4P_{2\,1/2}^0 \rightarrow 5s\,^4P_{1\,1/2}$); (C) 4765-Å ($5p\,^4D_{2\,1/2}^0 \rightarrow 5s\,^4P_{1\,1/2}$).

10-3-7　Lines of Kr II Induced by Kr^+ Impact on Ne. (A) 4577-Å ($5p'\,^2F_{3\,1/2}^0 \rightarrow 5s'\,^2D_{2\,1/2}$); (B) 4738-Å ($5p\,^4P_{2\,1/2}^0 \rightarrow 5s\,^4P_{1\,1/2}$); (C) 4765-Å ($5p\,^4D_{2\,1/2}^0 \rightarrow 5s\,^4P_{1\,1/2}$).

10-3-8　Lines of Kr II Induced by Kr^+ Impact on Ar. (A) 4577-Å ($5p'\,^2F_{3\,1/2}^0 \rightarrow 5s'\,^2D_{2\,1/2}$); (B) 4738-Å ($5p\,^4P_{2\,1/2}^0 \rightarrow 5s\,^4P_{1\,1/2}$); (C) 4765-Å ($5p\,^4D_{2\,1/2}^0 \rightarrow 5s\,^4P_{1\,1/2}$).

2.　Cross Sections for Emission Resulting from Excitation of Molecular Projectiles

10-3-9　3914-Å Line of N_2^+ [($B\,^2\Sigma_u^+ \rightarrow X\,^2\Sigma_g^+$ (0, 0)] Induced by N_2^+ Impact on Ar.

3.　Cross Sections for Emission Resulting from Excitation of Metallic Projectiles

10-3-10　Na I Lines Induced by Na^+ Impact on N_2. (A) Group of three lines—5682.6, 5688.20, and 5688.19-Å that comprises the $4d\,^2D_{2\,1/2,\,1\,1/2} \rightarrow 3p\,^2P_{1\,1/2,\,1/2}^0$ multiplet; (B) Group of two lines—5890 and 5895-Å that comprises the $3p\,^2P_{1\,1/2,\,1/2}^0 \rightarrow 3s\,^2S_{1/2}$ multiplet.

10-3-11 K II Lines Induced by K^+ Impact on He. (A) 600.7-Å $(4s'[\tfrac{1}{2}]_1^0 \to 3p^6\ {}^1S_0)$; (B) 607.9-Å $(3d[\tfrac{1}{2}]_1^0 \to 3p^6\ {}^1S^0)$; (C) 612.6-Å $(4s[1\tfrac{1}{2}]_1^0 \to 3p^6\ {}^1S_0)$. The cross sections are given in arbitrary units; the relative magnitudes for the three lines are valid.

10-3-12 600.7-Å Line of K II $(4s'[\tfrac{1}{2}]_1^0 \to 3p^6\ {}^1S_0)$ Induced by K^+ Impact on Ne.

10-3-13 4044 and 4047-Å Lines of K I $(5p\ {}^2P_{1\ 1/2\ ,\ 1/2} \to 4s\ {}^2S_{1/2})$ Induced by K^+ Impact on N_2. [Two lines measured together.]

10-3-14 926.75-Å Line of Cs II $(6s[1\tfrac{1}{2}]_1^0 \to 5p^6\ {}^1S_0)$ Induced by Cs^+ Impact on He.

10-3-15 Lines of Cs II Induced by Cs^+ Impact on He. (A) 4527-Å $(6p[1\tfrac{1}{2}]_1 \to 6s[1\tfrac{1}{2}]_1^0)$; (B) 4604-Å $(6p[2\tfrac{1}{2}]_3 \to 6s[1\tfrac{1}{2}]_2^0)$.

10-3-16 901.3-Å Line of Cs II $(5d[1\tfrac{1}{2}]_1^0 \to 5p^6\ {}^1S_0)$ Induced by Cs^+ Impact on Ne.

10-3-17 4604-Å Line of Cs II $(6p[1\tfrac{1}{2}]_3 \to 6s[1\tfrac{1}{2}]_2^0)$ Induced by Cs^+ Impact on Ar.

10-3-18 Lines of Mg I Induced by Mg^+ Impact on N_2. (A) 3835-Å, a blend of the five lines—3829, 3832 (two lines), and 3838 (two lines)—that composes all of the components of the $3d\ {}^3D \to 3p\ {}^2P$ multiplet; (B) 5176-Å, a blend of the 5184-, 5173-, 5167-Å lines—that comprises all three components of the $4s\ {}^3S_1 \to 3p\ {}^3P^0$ multiplet.

10-3-19 4481-Å Line of Mg II $(4f\ {}^2F_{2\ 1/2}^0 \to 3d\ {}^2D_{2\ 1/2\ ,\ 1\ 1/2})$ Induced by Mg^+ Impact on N_2.

10-3-20 Lines of Ca I Induced by Ca^+ Impact on N_2. (A) 4227-Å $(4p\ {}^1P_1^0 \to 4s\ {}^1S_0)$; (B) 4455-Å, a blend of the 4455-, 4456-, and 4457-Å lines—that comprises three of the six components in the $4d\ {}^3D \to 4p\ {}^3P^0$ multiplet; (C) 4878-Å $(4f\ {}^1F_3^0 \to 3d\ {}^1D_2)$; (D) 5588-Å, a blend of all six components of the $4p'\ {}^3D^0 \to 3d\ {}^3D$ multiplet lying between 5582 and 5598-Å.

10-3-21 Lines of Ca II Induced by Ca^+ Impact on N_2. (A) 3737-Å $(5s\ {}^2S_{1/2} \to 4p\ {}^2P_{1\ 1/2}^0)$; (B) 3934-Å $(4p\ {}^2P_{1\ 1/2}^0 \to 4s\ {}^2S_{1/2})$.

10-3-1
NEFF (104)

ENERGY (KEV)	CROSS SECTION (SQ. CM)
4.00E-01	1.30E-18
6.00E-01	2.10E-18
8.00E-01	2.50E-18
1.20E 00	2.60E-18
1.50E 00	2.60E-18
2.00E 00	2.70E-18

10-3-2A
THOMAS ET AL. (71)

ENERGY (KEV)	CROSS SECTION (SQ. CM)
1.10E 02	1.35E-18
1.30E 02	1.35E-18
1.50E 02	1.35E-18
1.77E 02	1.08E-18
2.00E 02	9.40E-19
2.10E 02	1.04E-18
2.40E 02	9.00E-19
2.65E 02	8.60E-19
2.85E 02	9.20E-19
3.05E 02	1.04E-18
3.30E 02	8.80E-19

10-3-2B
THOMAS ET AL. (71)

ENERGY (KEV)	CROSS SECTION (SQ. CM)
1.25E 02	2.60E-18
1.25E 02	2.90E-18
1.55E 02	3.13E-18
1.75E 02	3.13E-18
2.00E 02	2.80E-18
2.05E 02	3.18E-18
2.35E 02	3.24E-18
2.50E 02	2.64E-18
2.75E 02	3.02E-18
3.05E 02	2.91E-18
3.35E 02	2.80E-18

10-3-2C
THOMAS ET AL. (71)

ENERGY (KEV)	CROSS SECTION (SQ. CM)
3.00E 01	4.42E-19
4.00E 01	4.96E-19
5.00E 01	5.06E-19
5.50E 01	5.61E-19
6.00E 01	5.22E-19
6.50E 01	5.51E-19
7.00E 01	5.02E-19
8.00E 01	5.06E-19
9.00E 01	4.77E-19
1.00E 02	4.37E-19

1.30E 02	5.40E-19
1.55E 02	4.30E-19
1.80E 02	4.20E-19
2.05E 02	3.60E-19
2.40E 02	4.20E-19
2.40E 02	3.80E-19
2.80E 02	3.30E-19
2.80E 02	2.80E-19
3.00E 02	3.60E-19
3.30E 02	4.40E-19
3.50E 02	3.10E-19

10-3-2D
THOMAS ET AL. (71)

ENERGY (KEV)	CROSS SECTION (SQ. CM)
1.30E 02	1.12E-18
1.50E 02	1.05E-18
1.75E 02	9.70E-19
2.00E 02	8.20E-19
2.50E 02	6.80E-19
2.75E 02	5.80E-19
3.04E 02	5.50E-19
3.28E 02	7.50E-19
3.45E 02	4.60E-19

10-3-3A
THOMAS ET AL. (71)

ENERGY (KEV)	CROSS SECTION (SQ. CM)
1.10E 02	3.13E-18
1.30E 02	2.64E-18
1.50E 02	3.10E-18
1.80E 02	3.52E-18
2.05E 02	3.24E-18
2.40E 02	3.16E-18
2.55E 02	3.27E-18
2.80E 02	2.53E-18
3.00E 02	1.98E-18

10-3-3B
THOMAS ET AL. (71)

ENERGY (KEV)	CROSS SECTION (SQ. CM)
1.10E 02	5.67E-18
1.30E 02	5.95E-18
1.50E 02	7.82E-18
1.75E 02	7.15E-18
2.00E 02	7.75E-18
2.40E 02	7.90E-18
2.55E 02	7.57E-18
3.05E 02	7.70E-18

10-3-3C
THOMAS ET AL. (71)

ENERGY (KEV)	CROSS SECTION (SQ. CM)
3.00E 01	5.30E-19
4.00E 01	6.11E-19
5.00E 01	6.45E-19
6.00E 01	6.36E-19
7.00E 01	6.62E-19
8.00E 01	6.75E-19
9.00E 01	6.80E-19
1.00E 02	6.91E-19
1.53E 02	7.92E-19
1.80E 02	6.05E-19
2.00E 02	6.56E-19
2.33E 02	7.46E-19
2.50E 02	6.51E-19
2.75E 02	6.42E-19
3.00E 02	4.90E-19

10-3-3D
THOMAS ET AL. (71)

ENERGY (KEV)	CROSS SECTION (SQ. CM)
1.30E 02	1.33E-18
1.80E 02	1.40E-18
2.02E 02	1.56E-18
2.50E 02	1.68E-18
2.80E 02	1.34E-18
3.00E 02	1.37E-18

10-3-4
ROBINSON ET AL. (105)

ENERGY (KEV)	CROSS SECTION (SQ. CM)
1.30E 02	2.14E-18
1.70E 02	2.60E-18
2.00E 02	2.35E-18
2.50E 02	2.78E-18
3.00E 02	2.99E-18
3.50E 02	3.08E-18
3.60E 02	3.03E-18
4.00E 02	2.78E-18
4.20E 02	2.65E-18

10-3-5A
THOMAS ET AL. (71)

ENERGY (KEV)	CROSS SECTION (SQ. CM)
1.50E 02	3.48E-18
1.80E 02	4.04E-18
2.13E 02	4.74E-18
2.50E 02	5.66E-18
2.75E 02	5.12E-18
3.00E 02	6.24E-18
3.25E 02	6.00E-18
3.50E 02	4.94E-18
3.70E 02	4.35E-18

10-3-5B
THOMAS ET AL. (71)

ENERGY (KEV)	CROSS SECTION (SQ. CM)
1.29E 02	4.50E-18
1.50E 02	3.82E-18
1.75E 02	5.31E-18
2.10E 02	5.59E-18
2.50E 02	6.52E-18
2.70E 02	6.90E-18
3.00E 02	6.77E-18
3.20E 02	6.51E-18
3.43E 02	5.78E-18

10-3-5C
THOMAS ET AL. (71)

ENERGY (KEV)	CROSS SECTION (SQ. CM)
1.50E 02	2.14E-18
1.70E 02	2.76E-18
2.14E 02	2.62E-18
2.50E 02	2.82E-18
2.80E 02	2.93E-18
3.00E 02	3.11E-18
3.22E 02	2.84E-18
3.50E 02	2.44E-18
3.75E 02	2.72E-18
4.00E 02	2.63E-18

10-3-6A
THOMAS ET AL. (71)

ENERGY (KEV)	CROSS SECTION (SQ. CM)
1.17E 02	3.18E-18
1.38E 02	3.65E-18
1.60E 02	4.67E-18
1.75E 02	5.40E-18
1.95E 02	5.39E-18
2.00E 02	5.30E-18

10-3-6B
THOMAS ET AL. (71)

ENERGY (KEV)	CROSS SECTION (SQ. CM)
1.33E 02	1.80E-18
1.50E 02	1.62E-18
1.65E 02	1.68E-18
1.90E 02	1.71E-18
1.97E 02	1.58E-18
2.12E 02	1.59E-18

10-3-6C
THOMAS ET AL. (71)

ENERGY (KEV)	CROSS SECTION (SQ. CM)
1.15E 02	4.02E-18
1.30E 02	4.33E-18
1.40E 02	4.26E-18
1.50E 02	4.24E-18
1.55E 02	3.85E-18
1.69E 02	3.98E-18
1.82E 02	4.15E-18
1.89E 02	4.04E-18
2.10E 02	2.93E-18
2.10E 02	3.59E-18

1C-3-7A
THOMAS ET AL. (71)

ENERGY (KEV)	CROSS SECTION (SQ. CM)
1.40E 02	5.35E-18
1.50E 02	5.42E-18
1.70E 02	6.05E-18
1.90E 02	6.72E-18
2.10E 02	7.00E-18

10-3-7B
THOMAS ET AL. (71)

ENERGY (KEV)	CROSS SECTION (SQ. CM)
1.30E 02	1.78E-18
1.55E 02	1.82E-18
1.75E 02	1.89E-18
1.95E 02	2.02E-18
1.95E 02	2.11E-18
2.13E 02	2.31E-18

10-3-7C
THOMAS ET AL. (71)

ENERGY (KEV)	CROSS SECTION (SQ. CM)
1.15E 02	3.83E-18
1.35E 02	4.32E-18
1.55E 02	4.77E-18
1.75E 02	5.21E-18
1.93E 02	5.79E-18
2.12E 02	6.23E-18

10-3-8A
THOMAS ET AL. (71)

ENERGY (KEV)	CROSS SECTION (SQ. CM)
1.13E 02	3.18E-18
1.43E 02	3.94E-18
1.67E 02	4.85E-18
1.86E 02	4.95E-18
2.12E 02	4.80E-18
2.12E 02	5.26E-18

10-3-8B
THOMAS ET AL. (71)

ENERGY (KEV)	CROSS SECTION (SQ. CM)
1.13E 02	2.62E-18
1.30E 02	2.84E-18
1.47E 02	3.24E-18
1.67E 02	3.18E-18
1.90E 02	4.37E-18
2.12E 02	4.06E-18

10-3-8C
THOMAS ET AL. (71)

ENERGY (KEV)	CROSS SECTION (SQ. CM)
1.30E 02	3.36E-18
1.70E 02	3.94E-18
1.90E 02	4.21E-18
2.13E 02	5.45E-18

10-3-9
NEFF (104)

ENERGY (KEV)	CROSS SECTION (SQ. CM)
4.00E-01	5.90E-19
6.00E-01	7.60E-19
8.00E-01	8.00E-19
1.00E 00	8.90E-19
1.20E 00	9.30E-19
1.50E 00	1.10E-18
2.00E 00	1.20E-18

10-3-10A
NEFF (104)

ENERGY (KEV)	CROSS SECTION (SQ. CM)
4.00E-01	5.00E-20
1.00E 00	1.80E-19
2.00E 00	4.10E-19

10-3-10B
NEFF (104)

ENERGY (KEV)	CROSS SECTION (SQ. CM)
4.00E-01	2.70E-18
6.00E-01	4.70E-18
8.00E-01	6.10E-18

10-2-14B
HEAD ET AL. (213)

ENERGY (KEV)	CROSS SECTION (SQ. CM)
2.00E 01	8.98E-19
3.00E 01	1.11E-18
4.00E 01	1.09E-18
5.00E 01	8.50E-19
6.00E 01	6.26E-19
7.00E 01	5.21E-19
8.00E 01	4.44E-19
9.00E 01	3.79E-19
1.00E 02	3.26E-19
1.10E 02	3.00E-19
1.20E 02	2.76E-19

10-2-15A
DE HEER ET AL. (213)

ENERGY (KEV)	CROSS SECTION (SQ. CM)
7.00E 00	1.19E-19
8.00E 00	1.45E-19
9.00E 00	1.63E-19
1.00E 01	1.48E-19
1.25E 01	1.17E-19
1.50E 01	1.37E-19
1.75E 01	1.90E-19
2.00E 01	2.04E-19
2.25E 01	2.01E-19
2.50E 01	2.32E-19
2.75E 01	2.55E-19
3.00E 01	2.88E-19
3.50E 01	3.35E-19
4.00E 01	3.35E-19
5.00E 01	2.94E-19
6.00E 01	2.64E-19
7.00E 01	2.15E-19
8.00E 01	1.66E-19
9.00E 01	1.52E-19

10-2-15B
HEAD ET AL. (214)

ENERGY (KEV)	CROSS SECTION (SQ. CM)
2.00E 01	2.94E-19
2.50E 01	3.00E-19
3.00E 01	3.71E-19
4.00E 01	4.66E-19
5.00E 01	4.38E-19
6.00E 01	3.60E-19
7.00E 01	3.01E-19
8.00E 01	2.48E-19
9.00E 01	2.54E-19
1.00E 02	2.23E-19

10-2-16
DE HEER ET AL. (213)

ENERGY (KEV)	CROSS SECTION (SQ. CM)
1.25E 01	7.86E-20
1.50E 01	7.46E-20
1.75E 01	8.75E-20
2.00E 01	1.00E-19
2.25E 01	9.60E-20
2.50E 01	1.03E-19
2.75E 01	1.12E-19
3.00E 01	1.17E-19
3.50E 01	1.42E-19
4.00E 01	1.45E-19
5.00E 01	1.53E-19
6.00E 01	1.32E-19
7.00E 01	1.21E-19
8.00E 01	9.82E-20
9.00E 01	9.32E-20

10-2-17
DE HEER ET AL. (213)

ENERGY (KEV)	POLARIZATION (PERCENT)
2.50E 01	4.30E 00
3.00E 01	2.03E 01
3.50E 01	1.80E 01
4.00E 01	1.27E 01
5.00E 01	3.62E 00
6.00E 01	5.00E-01

10-2-18
DE HEER ET AL. (213)

ENERGY (KEV)	POLARIZATION (PERCENT)
2.00E 01	2.88E 01
2.50E 01	2.88E 01
3.00E 01	3.53E 01
3.50E 01	3.53E 01
4.00E 01	3.59E 01
5.00E 01	3.90E 01
6.00E 01	3.46E 01

10-2-19
DE HEER ET AL. (12)

ENERGY (KEV)	POLARIZATION (PERCENT)
1.50E 01	-5.30E 00
2.00E 01	-5.80E 00
2.50E 01	-4.70E 00
3.00E 01	-4.70E 00
3.50E 01	-4.70E 00
4.00E 01	-3.60E 00
4.50E 01	5.70E 00
5.00E 01	0.0
5.50E 01	5.00E-01
6.00E 01	1.52E 00

10-2-20
DE HEER ET AL. (213)

ENERGY (KEV)	POLARIZATION (PERCENT)
2.50E 01	1.84E 01
3.00E 01	1.63E 01
3.50E 01	1.74E 01
4.00E 01	2.00E 01
6.00E 01	1.70E 01

10-2-21
DE HEER ET AL. (213)

ENERGY (KEV)	CROSS SECTION (SQ. CM)
5.00E 00	8.60E-20
6.00E 00	9.34E-20
7.00E 00	1.04E-19
8.00E 00	1.95E-19
9.00E 00	2.49E-19
1.00E 01	2.48E-19
1.25E 01	1.42E-19
1.50E 01	1.88E-19
1.75E 01	3.09E-19
2.00E 01	4.02E-19
2.25E 01	4.18E-19
2.50E 01	3.71E-19
2.75E 01	3.16E-19
3.00E 01	2.49E-19
3.50E 01	1.82E-19

10-2-22
DE HEER ET AL. (78)

ENERGY (KEV)	CROSS SECTION (SQ. CM)
7.50E 00	4.47E-17
1.00E 01	4.96E-17
1.25E 01	5.50E-17
1.50E 01	5.32E-17
2.00E 01	4.93E-17
2.50E 01	4.47E-17
3.00E 01	4.17E-17
3.25E 01	3.79E-17

10-2-23
DE HEER ET AL. (213)

ENERGY (KEV)	CROSS SECTION (SQ. CM)
7.00E 00	1.18E-18
8.00E 00	1.53E-18
9.00E 00	1.89E-18
1.00E 01	1.93E-18
1.25E 01	1.61E-18
1.50E 01	8.51E-19
1.75E 01	6.63E-19
2.00E 01	7.34E-19
2.25E 01	8.07E-19
2.50E 01	9.25E-19

2.60E 01	3.62E 03	3.75E 01	2.30E 03	4.00E-01	1.80E-19
2.80E 01	3.68E 03	4.00E 01	2.19E 03	6.00E-01	4.70E-19
3.00E 01	3.65E 03	4.00E 01	2.24E 03	1.00E 00	1.10E-18
3.00E 01	3.53E 03	4.25E 01	2.25E 03	1.50E 00	1.90E-18
3.20E 01	3.60E 03	4.50E 01	2.32E 03	2.00E 00	2.00E-18
3.40E 01	3.58E 03	4.75E 01	2.36E 03		
3.60E 01	3.68E 03	5.00E 01	2.26E 03		
3.80E 01	3.57E 03	5.25E 01	2.20E 03		
4.00E 01	3.72E 03	5.50E 01	2.18E 03		
4.25E 01	3.93E 03	5.75E 01	2.14E 03		
4.50E 01	3.92E 03	6.00E 01	2.16E 03		
4.75E 01	3.93E 03	6.25E 01	2.08E 03		
5.00E 01	3.88E 03	6.50E 01	2.10E 03		
5.25E 01	4.08E 03	6.75E 01	2.04E 03		
5.50E 01	4.20E 03	7.00E 01	2.00E 03		
5.75E 01	4.29E 03	7.25E 01	1.97E 03		
6.00E 01	4.29E 03	7.50E 01	1.89E 03		
6.25E 01	4.72E 03	7.75E 01	1.85E 03		
6.50E 01	4.30E 03	8.00E 01	1.96E 03		
6.75E 01	4.47E 03	8.00E 01	1.98E 03		
7.00E 01	4.50E C3	8.25E 01	1.88E 03		
7.25E 01	4.61E 03	8.50E 01	2.02E 03		
7.50E 01	4.53E C3				

10-3-20B
NEFF (104)

ENERGY (KEV)	CROSS SECTION (SQ. CM)
1.00E 00	2.20E-19
2.00E 00	5.50E-19

10-3-20C
NEFF (104)

ENERGY (KEV)	CROSS SECTION (SQ. CM)
2.00E 00	1.50E-19

10-3-20D
NEFF (104)

ENERGY (KEV)	CROSS SECTION (SQ. CM)
1.00E 00	7.00E-19
2.00E 00	1.50E-19

10-3-21A
NEFF (104)

ENERGY (KEV)	CROSS SECTION (SQ. CM)
4.00E-01	1.10E-18
6.00E-01	2.10E-18
1.00E 00	2.70E-18
1.50E 00	3.10E-18
2.00E 00	3.70E-18

10-3-16
MATVEYEV ET AL. (115)

ENERGY (KEV)	CROSS SECTION (RELATIVE UNITS)
3.01E 00	6.00E-02
3.18E 00	3.72E 00
3.37E 00	6.72E 00
3.60E 00	1.18E 01
3.83E 00	1.50E 01
3.99E 00	1.88E 01
4.21E 00	2.83E 01
4.38E 00	3.22E 01
4.59E 00	3.82E 01
4.80E 00	4.24E 01
5.04E 00	4.42E 01
5.49E 00	4.66E 01
6.03E 00	4.37E 01
7.03E 00	4.05E 01
8.05E 00	4.00E 01
9.07E 00	4.00E 01
1.01E 01	3.99E 01
1.11E 01	3.88E 01

10-3-18A
NEFF (104)

ENERGY (KEV)	CROSS SECTION (SQ. CM)
4.00E-01	1.30E-19
6.00E-01	3.30E-19
1.00E 00	5.30E-19
1.50E 00	8.20E-19
2.00E 00	1.00E-18

10-3-18B
NEFF (104)

ENERGY (KEV)	CROSS SECTION (SQ. CM)
4.00E-01	1.40E-19
6.00E-01	2.10E-19
1.00E 00	3.30E-19
1.50E 00	5.10E-19
2.00E 00	5.50E-19

10-3-21B
NEFF (104)

ENERGY (KEV)	CROSS SECTION (SQ. CM)
4.00E-01	1.40E-17
6.00E-01	1.60E-17
1.00E 00	1.70E-17
2.00E 00	1.90E-17

10-3-17
ANDERSEN ET AL. (114)

ENERGY (KEV)	CROSS SECTION (RELATIVE UNITS)
6.00E 00	1.34E 03
8.00E 00	1.27E 03
1.00E 01	1.28E 03
1.20E 01	1.47E 03
1.40E 01	1.55E 03
1.60E 01	1.70E 03
1.80E 01	1.64E 03
2.00E 01	1.81E 03
2.40E 01	1.96E 03
2.80E 01	2.15E 03
3.00E 01	2.22E 03
3.25E 01	2.22E 03
3.50E 01	2.31E 03

10-3-19
NEFF (104)

ENERGY (KEV)	CROSS SECTION (SQ. CM)
4.00E-01	4.00E-19
6.00E-01	9.10E-19
1.00E 00	1.40E-18
1.50E 00	2.60E-18
2.00E 00	3.20E-18

10-3-20A
NEFF (104)

ENERGY (KEV)	CROSS SECTION (SQ. CM)

400

BIBLIOGRAPHY

1. W. Maurer, *Physik., Z.*, **40,** 161 (1939).
2. H. S. W. Massey and E. H. S. Burhop, *Electronic and Ionic Phenomena*, Oxford University Press, London, 1952.
3. S. N. Ghosh and B. N. Srivastava, *Z. Astrophys.*, **53,** 186 (1961).
4. S. N. Ghosh and B. N. Srivastava, *Proc. Natl. Acad. Sci., India, A*, 32, 231 (1962).
5. F. J. de Heer, *Advances in Atomic and Molecular Physics*, Vol. II, p. 328, (D. R. Bates and I. M. Estermann, eds.) Academic Press, New York, 1966.
6. J. B. Hasted, *Physics of Atomic Collisions*, page 3, Butterworths, London, 1964.
7. S. J. Young, J. S. Murray, and J. R. Sheridan, *Phys. Rev.*, **178,** 40 (1969).
8. R. F. Stebbings, R. A. Young, C. L. Oxley, and H. Ehrhardt, *Phys. Rev.*, **138,** A1312 (1965).
9. R. A. Young, R. F. Stebbings, and J. W. McGowan, *Phys. Rev.*, **171,** 85 (1968).
10. J. van Eck, F. J. de Heer, and J. Kistemaker, *Physica*, **30,** 1171 (1964).
11. D. R. Bates and J. C. G. Walker, *Planetary Space Sci.*, **14,** 1367 (1966).
12. F. J. de Heer, L. Wolterbeek Muller, and R. Geballe, *Physica*, **31,** 1745 (1965).
13. H. A. Bethe and E. E. Salpeter, "Lifetimes of Excited States in Hydrogen," Sec. 67, p. 370 in *Encyclopedia of Physics*, Vol. XXXV, (S. Flügge, ed.) Springer-Verlag, Berlin, 1957.
14. W. E. Lamb and R. C. Retherford, *Phys. Rev.*, **79,** 549 (1950).
15. G. Lüders, *Z. Naturforsch.*, **5a,** 608 (1950).
16. D. Jaecks, B. Van Zyl, and R. Geballe, *Phys. Rev.*, **137,** A340 (1965).
17. I. A. Sellin, *Phys. Rev.*, **136,** A1245 (1964).
18. D. Jaecks et al., *Phys. Rev.*, **137,** A340 (1965); I. A. Sellin, *Phys. Rev.*, **136,** A1245 (1964); W. L. Fite et al., *Phys. Rev.*, **116,** 363 (1959); J. E. Bayfield, *Phys. Rev.*, **182,** 115 (1969).
19. W. L. Fite, W. E. Kauppila, and W. R. Ott, *Phys. Rev. A*, **1,** 1089 (1970).
20. I. A. Sellin, J. A. Biggerstaff, and P. M. Griffin, *Phys. Rev. A,* **2,** 423 (1970).
21. W. Lichten, *Phys. Rev. Letters*, **6,** 12 (1961).
22. W. L. Fite, R. T. Brackmann, D. G. Hummer, and R. F. Stebbings, *Phys. Rev.*, **116,** 363 (1959).
23. J. A. Smit, *Physica*, **2,** 104 (1935).
24. R. H. McFarland and E. A. Soltysik, *Phys. Rev.*, **129,** 2581 (1963).
25. I. C. Percival and M. J. Seaton, *Phil. Trans. Roy. Soc. London, Ser. A.*, **251,** 113 (1958).
26. J. R. Oppenheimer, *Z. Physik*, **43,** 27 (1927); *Proc. Nat. Acad. Sci. U.S.*, **13,** 800 (1927); *Phys. Rev.*, **32,** 361 (1928).
27. W. G. Penney, *Proc. Natl. Acad. Sci. U.S.*, **18,** 231 (1932).
28. T. D. Gaily, D. H. Jaecks, and R. Geballe, *Phys. Rev.*, **167,** 81 (1968).
29. A. N. Jette and P. Cahill, *Phys. Rev.*, **176,** 186 (1968); *Phys. Rev. A*, **1,** 558 (1970).
30. B. M. Glennon and W. L. Weise, *Bibliography on Atomic Transition Probabilities*, National Bureau of Standards Monograph 50, U.S. Government Printing Office, Washington, D.C., 1962.
31. W. L. Weise, M. W. Smith, and B. M. Glennon, *Atomic Transition Probabilities, Hydrogen Through Neon*, National Standard Reference Data Series-National Bureau of Standards, Vol. I, U.S. Government Printing Office, Washington, D.C., 1966.
32. D. R. Bates, *Monthly Notices Roy. Astron. Soc.*, **112,** 614 (1952).

33. G. Herzberg, *Molecular Spectra and Molecular Structure: I. Spectra of Diatomic Molecules*, 2nd ed., D. Van Nostrand, Princeton, N. J., 1950.

34. R. W. Nicholls, *Ann. Geophys.*, **20**, 144 (1964).

35. S. Dworetsky, R. Novick, W. W. Smith, and N. Tolk, *Phys. Rev. Letters*, **18**, 939 (1967).

36. G. W. McClure, *Phys. Rev.*, **166**, 22 (1968).

37. C. F. Barnett and H. B. Gilbody, "Measurements of Atomic Cross Sections in Static Gases," Sec. 4.2, p. 390 in *Atomic and Electronic Physics: Atomic Interactions*, Vol. VIII, Part A, (B. Bederson and W. L. Fite, eds.) of series *Methods of Experimental Physics*, (L. Marton, ed.) Academic, New York, 1968.

38. G. Carter and J. S. Colligon, *Ion Bombardment of Solids*, American Elsevier, New York, 1968.

39. G. Carter and J. S. Colligon, *Ion Bombardment of Solids*, Chapter 3, American Elsevier, New York, 1968.

40. G. Carter and J. S. Colligon, *Ion Bombardment of Solids* p. 75, American Elsevier, New York, 1968.

41. P. M. Stier, C. F. Barnett, and G. E. Evans, *Phys. Rev.* **96**, 973 (1954).

42. C. F. Barnett and H. K. Reynolds, *Phys. Rev.*, **109**, 355 (1958).

43. G. J. Lockwood, H. F. Helbig, and E. Everhart, *J. Chem. Phys.*, **41**, 3820 (1964).

44. P. H. Carr, *Vacuum*, **14**, 37 (1964).

45. M. Rusch and O. Bunge, *Z. Tech. Physik*, **13**, 77 (1932).

46. T. Takaishi and Y. Sensui, *Trans. Faraday Soc.*, **59**, 2503 (1963).

47. S. Dushman, *Scientific Foundations of Vacuum Technique*, 2nd ed., John Wiley and Sons, New York, 1962.

48. W. Gaede, *Ann. Physik*, **46**, 357 (1915).

49. H. Ishii and K. I. Nakayama. *Transactions of the Eighth National Vacuum Symposium and Second International Congress on Vacuum Science and Technology* Washington, D.C., 1961, Vol. I, p. 519, (L. E. Preuss, ed.) Pergamon, New York, 1962.

50. T. Takaishi, *Trans. Faraday Soc.*, **61**, 840 (1965).

51. N. G. Utterback and T. Griffith, *Rev. Sci. Instr.*, **37**, 866 (1966).

52. N. V. Fedorenko, V. A. Ankudinov, and R. N. Il'in, *Zh. Tekhn., Fiz.*, **35**, 585 (1965); [*Soviet Phys.,-Tech. Phys.*, **10**, 461 (1965)].

53. C. F. Barnett, Oak Ridge National Laboratory Report ORNL-4150, p. 103, 1967.

54. R. N. Il'in, V. A. Oparin, I. T. Serenkov, E. S. Solov'ev, and N. V. Fedorenko, *Sixth Intern. Conf. on the Physics of Electronic and Atomic Collisions: Abstracts of Papers*, Cambridge, Mass., p. 446, MIT Press, 1969.

55. K. H. Berkner, S. N. Kaplan, G. A. Paulikas, and R. V. Pyle, *Phys. Rev.*, **138**, A410 (1965).

56. A. C. Riviere and D. R. Sweetman, *Atomic Collision Processes: Proc. Third Intern. Conf. on the Physics of Electronic and Atomic Collisions*, London, 1963, p. 734, (M. R. C. McDowell, ed.) North-Holland, Amsterdam, 1964.

57. R. N. Il'in, B. I. Kikiani, V. A. Oparin, E. S. Solov'ev, and N. V. Fedorenko, *Zh. Eksperim. i Theor. Fiz.*, **47**, 1235 (1964); [Soviet Phys.-JETP, 20, 835 (1965)].

58. D. C. Lorents, W. Aberth, and V. W. Hesterman, *Phys. Rev. Letters*, **17**, 849 (1966).

59. D. Coffey, D. C. Lorents, and F. T. Smith, *Sixth Intern. Conf. on the Physics of Electronic and Atomic Collisions: Abstracts of Papers*, Cambridge, Mass., p. 299, MIT Press, 1969.

60. J. T. Park and F. D. Schowengerdt, *Phys. Rev.*, **185**, 152 (1969).

61. H. B. Gilbody, R. Browning, G. Levy, A. I. McIntosh, and K. F. Dunn, *J. Phys. B*, **1**, 863 (1968).

62. H. B. Gilbody, R. Browning, and G. Levy, *J. Phys. B*, **1**, 230 (1968).

63. R. E. Miers, A. S. Schlachter, and L. W. Anderson, *Phys. Rev.*, **183**, 213 (1969).

64. P. O. Haugsjaa, R. C. Amme, and N. G. Utterback, *Phys. Rev. Letters*, **22**, 322 (1969).

65. N. G. Utterback, *Phys. Rev. Letters*, **12**, 295 (1964).

66. C. E. Moore, *A Multiplet Table of Astrophysical Interest*, rev. ed., National Bureau of Standards, Technical Note 36, U.S. Government Printing Office, Washington, D.C., 1959.

67. C. E. Moore, *An Ultraviolet Multiplet Table*, National Bureau of Standards Circular 488, Sections 1, 2, 3, 4, and 5, U.S. Government Printing Office, Washington, D.C., 1950–1962.

68. R. W. B. Pearse and A. G. Gaydon, *The Identification of Molecular Spectra*, 2nd rev. ed., John Wiley and Sons, New York, 1950.

69. H. G. Gale, G. S. Monk, and V. O. Lee, *Astrophys. J.*, **67,** 89 (1928).

70. O. W. Richardson, *Molecular Hydrogen and Its Spectrum*, Yale University Press, New Haven, Conn., 1934.

71. E. W. Thomas and H. B. Gilbody, *Proc. Phys. Soc. (London)*, **85,** 363 (1965).

72. P. Jacquinot, *J. Opt. Soc. Am.*, **44,** 761 (1954).

73. J. A. R. Samson, *Techniques of Vacuum Ultraviolet Spectroscopy*, John Wiley and Sons, New York, 1967.

74. E. Ya. Shreider, *Zh. Tekhn., Fiz.*, **34,** 2089 (1964); [*Soviet Phys.,-Tech. Phys.*, **9,** 1609 (1965)].

75. E. P. Andreev, V. A. Ankudinov, and S. V. Bobashev, *Zh. Eksperim. i Theor. Fiz.*, **50,** 565 (1966); [*Soviet Phys.-JETP*, **23,** 375 (1966)].

76. J. van Eck, F. J. de Heer, and J. Kistemaker, *Phys. Rev.*, **130,** 656 (1963).

77. F. J. de Heer and J. van Eck, *Atomic Collision Processes: Proc. Third Intern. Conf. on the Physics of Electronic and Atomic Collisions*, London, 1963, p. 635, (M. R. C. McDowell, ed.) North-Holland, Amsterdam, 1964.

78. F. J. de Heer, J. van Eck, and J. Kistemaker, *Proc. Sixth Conf. on the Ionization Phenomena in Gases,* Paris, 1963, Vol. I, p. 73, (P. Huber and E. Crémiue-Alcan, ed.) SERMA, Paris, 1963.

79. R. T. Brackmann, W. L. Fite, and K. E. Hagen, *Rev. Sci. Instr.*, **29,** 125 (1958).

80. G. Bauer, *Measurement of Optical Radiations*, Focal Press, New York, 1965.

81. L. G. Leighton, *Illum. Engng. (U.S.A)*, **57,** 121 (1962).

82. R. D. Larrabee, *J. Opt. Soc. Am.*, **49,** 619 (1959).

83. J. C. de Vos, *Physica*, **20,** 690 (1954).

84. G. A. W. Rutgers and J. C. de Vos, *Physica*, **20,** 715 (1954).

85. R. Stair, R. G. Johnston, and E. W. Halbach, *J. Res. Nat. Bur. of Std.*, **64A,** 291 (1960).

86. J. J. Merril and R. G. Layton, *Appl. Opt.*, **5,** 1818 (1966).

87. C. M. Penchina, *Appl. Opt.*, **6,** 1029 (1967).

88. A. Watanabe and G. C. Tabisz, *Appl. Opt.*, **6,** 1132 (1967).

89. W. G. Griffin and R. W. P. McWhirter, *Proc. Conf. Optical Instruments and Techniques,* London, 1961, p. 14, (K. J. Habell, ed.) Chapman and Hall, 1962.

90. J. van Eck and F. J. de Heer, *Proc. Sixth Conf. on the Ionization Phenomena in Gases*, Paris, 1963, Vol. IV, p. 11, (P. Huber and E. Crémieu-Alcan, eds.) SERMA, Paris, 1963.

91. J. W. McConkey, *J. Opt. Soc. Am.*, **59,** 110 (1969).

92. J. F. M. Aarts and F. J. de Heer, *J. Opt. Soc. Am.*, **58,** 1666 (1968).

93. R. G. Johnston and R. P. Madden, *J. Opt. Soc. Am.*, **55,** 603 (1965).

94. J. A. R. Samson, *J. Opt. Soc. Am.*, **54,** 6 (1964).

95. F. M. Matsunaga, R. S. Jackson, and K. Watanabe, *J. Quant. Spectry. Radiative Transfer*, **5,** 329 (1965).

96. L. R. Canfield, R. G. Johnston, K. Codling, and R. P. Madden, *Appl. Opt.*, **6,** 1886 (1967).

97. K. Watanabe, F. M. Matsunaga, and H. Sakai, *Appl. Opt.*, **6,** 391 (1967).

98. W. L. Fite and R. T. Brackmann, *Phys. Rev.*, **112,** 1151 (1958).

99. G. H. Dunn, R. Geballe, and D. Pretzer, *Phys. Rev.*, **128,** 2200 (1962).

100. B. L. Moiseiwitsch and S. J. Smith, *Rev. Mod. Phys.*, **40,** 238 (1968).

101. J. M. Woolsey and J. W. McConkey, *J. Opt. Soc. Am.*, **58,** 1309 (1968).

102. P. N. Clout and D. W. O. Heddle, *J. Opt. Soc. Am.*, **59,** 715 (1969).

103. T. J. M. Sluyters and J. Kistemaker, *Physica*, **25,** 1389 (1959).

104. S. H. Neff, *Astrophys. J.*, **140,** 348 (1964).

105. J. M. Robinson and H. B. Gilbody, *J. Phys. B*, **2,** 817 (1969).

106. M. Barat, J. Baudon, M. Abignoli, and J-C. Houver, *J. Phys. B.*, **3**, 230 (1970).

107. K. D. Bayes and M. J. Haugh, *Sixth Intern. Conf. on the Physics of Electronic and Atomic Collisions: Abstracts of Papers*, Cambridge, Mass., p. 234, MIT Press, 1969.

108. T. F. Moran and P. C. Cosby, *J. Chem. Phys.*, **51**, 5724 (1969).

109. V. A. Gusev, G. N. Polyakova, V. F. Erko, Ya M. Fogel', and A. V. Zats, *Sixth Intern. Conf. on the Physics of Electronic and Atomic Collisions: Abstracts of Papers*, Cambridge, Mass., p. 809, MIT Press, 1969.

110. G. N. Polyakova, Ya. M. Fogel', V. F. Erko, A. V. Zats, and A. G. Tolstolutskii, *Zh. Eksperim. i Theor. Fiz.*, **54**, 374 (1968); [*Soviet Phys.-JETP*, **27**, 201 (1968)].

111. V. A. Gusev, G. N. Polyakova, and Ya. M. Fogel', *Zh. Eksperim. i Theor. Fiz.*, **55**, 2128 (1968); [*Soviet Phys.-JETP*, **28**, 1126 (1969)].

112. G. N. Polyakova, V. F. Erko, V. A. Gusev, A. V. Zats, and Ya. M. Fogel', *Zh. Eksperim. i Theor. Fiz.*, **56**, 161 (1969); [*Soviet Phys.-JETP*, **29**, 91 (1969)].

113. L. Kurzweg, H. H. Lo, R. T. Brackmann, and W. L. Fite, *Phys. Rev.*, **179**, 55 (1969).

114. H. Andersen, K. Jensen, J. Koch, H. Pedersen, and E. Veje, *Sixth Intern. Conf. on the Physics of Electronic and Atomic Collisions: Abstracts of Papers*, Cambridge, Mass., p. 223, MIT Press, 1969.

115. V. B. Matveyev, S. V. Bobashev, and V. M. Dukelsky, *Sixth Intern. Conf. on the Physics of Electronic and Atomic Collisions: Abstracts of Papers*, Cambridge, Mass., p. 226, MIT Press, 1969.

116. S. V. Bobashev, V. B. Matveyev, and V. A. Ankudinov, *ZhETF Pis'ma Red.*, **9**, 344 (1969); *JETP Letters*, **9**, 201 (1969).

117. D. Pretzer, B. Van Zyl, and R. Geballe, *Atomic Collision Processes: Proc. of the Third Intern. Conf. on the Physics of Electronic and Atomic Collisions*, London, 1963, p. 618, (M. R. C. McDowell, ed.), North-Holland, Amsterdam, 1964.

118. R. N. Il'in, V. A. Oparin, E. S. Solov'ev, N. V. Fedorenko, *ZhETF Pis'ma Red.*, **2**, 310 (1965) [*JETP Letters*, **2**, 197 (1965)].

119. R. N. Il'in, V. A. Oparin, E. S. Solov'ev, and N. V. Fedorenko, *Zh. Tekhn., Fiz.*, **36**, 1241 (1966); [*Soviet Phys.-Tech. Phys.*, **11**, 921 (1967)].

120. R. H. Hughes, H. R. Dawson, B. M. Doughty, D. B. Kay, and C. A. Stigers, *Phys. Rev.*, **146**, 53 (1966).

121. R. H. Hughes, H. R. Dawson, and B. M. Doughty, *Phys. Rev.*, **164**, 166 (1967).

122. E. P. Andreev, V. A. Ankudinov, S. V. Bobashev, and V. B. Matveyev, *Zh. Eksperim. i Theor. Fiz.*, **52**, 357 (1967); [*Soviet Phys.-JETP*, **25**, 232 (1967)].

123. E. P. Andreev, V. A. Ankudinov, S. V. Bobashev, *Fifth Intern. Conf. on the Physics of Electronic and Atomic Collisions: Abstracts of Papers*, Leningrad, p. 307, Publishing House Nauka, Leningrad, 1967.

124. D. Jaecks, F. J. de Heer, and A. Salop, *Physica*, **36**, 606 (1967).

125. M. Dufay, J. P. Buchet, A. Carré, A. Denis, J. Desesquelles, and M. Gaillard, *Fifth Intern. Conf. on the Physics of Electronic and Atomic Collisions: Abstracts of Papers*, Leningrad, 1967, p. 295, Publishing House Nauka, Leningrad, 1967.

126. D. J. Volz and M. E. Rudd, *Sixth Intern. Conf. on the Physics of Electronic and Atomic Collisions: Abstracts of Papers*, Cambridge, Mass., 1969, p. 410, MIT Press, 1969 .

127. E. W. Thomas, *J. Phys. B*, **2**, 625 (1969).

128. P. J. O. Teubner, W. E. Kauppilla, W. L. Fite, and R. J. Girnius, *Sixth Intern. Conf. on the Physics of Electronic and Atomic Collisions: Abstracts of Papers*, Cambridge, Mass., p. 109, MIT Press, 1969.

129. E. W. Thomas, J. L. Edwards, and J. C. Ford, *Sixth Intern. Conf. on the Physics of Electronic and Atomic Collisions: Abstracts of Papers*, Cambridge, Mass., p. 462, MIT Press, 1969.

130. J. E. Bayfield, *Phys. Rev.*, **182**, 115 (1969).

131. D. H. Jaecks, R. M. McKnight, and D. H. Crandall, *Sixth Intern. Conf. on the Physics of Electronic and Atomic Collisions: Abstracts of Papers*, Cambridge, Mass., p. 862, MIT Press, 1969.

132. R. H. McFarland and H. A. Futch, *Sixth Intern. Conf. on the Physics of Electronic and Atomic Collisions: Abstracts of Papers*, Cambridge, Mass., p. 441, MIT Press, 1969.

133. M. Carré and M. Dufay, *Compt. Rend.*, **265**, B259 (1967).

134. R. LeDoucen and J. Guidini, *Sixth Intern. Conf. on the Physics of Electronic and Atomic Collisions: Abstracts of Papers*, Cambridge, Mass., p. 454, MIT Press, 1969.

135. M. C. Poulizac, J. Desesquelles, and M. Dufay, *Ann. Astrophys.*, **30**, 301 (1967).

136. V. A. Oparin, R. N. Il'in, and E. S. Solov'ev, *Zh. Eksperim. i Theor. Fiz.*, **52**, 369 (1967); [*Soviet Phys.-JETP*, **25**, 240 (1967)].

137. B. L. Donnally, T. Clapp, W. Sawyer, and M. Schultz, *Phys. Rev. Letters,* **12**, 502 (1964).

138. A. Cesati, F. Cristofori, L. M. Colli, and P. G. Sona, *Energia Nucl.*, **13**, 649 (1966).

139. I. A. Sellin and L. Granoff, *Phys. Letters*, **25A**, 484 (1967).

140. R. A. Young, private communication.

141. W. E. Kauppila, P. J. O. Teubner, W. L. Fite, and R. J. Girnius, *Sixth Intern. Conf. on the Physics of Electronic and Atomic Collisions: Abstracts of Papers*, Cambridge, Mass., p. 111, MIT Press, 1969.

142. G. Ryding, A. B. Wittkower, and H. B. Gilbody, *Proc. Phys. Soc.*, **89**, 547 (1966).

143. J. E. Bayfield, *Phys. Rev.*, **185**, 105 (1969).

144. L. Madansky and G. E. Owen, *Phys. Rev. Letters*, **2**, 209 (1959).

145. L. Colli, F. Cristofori, G. E. Frigerio, and P. G. Sona, *Phys. Letters*, **3**, 62 (1962).

146. F. Cristofori, private communication.

147. R. H. Hughes, S. Lin, and L. L. Hatfield, *Phys. Rev.*, **130**, 2318 (1963).

148. E. P. Andreev, V. A. Ankudinov, and S. V. Bobashev, *Fifth Intern. Conf. on the Physics of Electronic and Atomic Collisions: Abstracts of Papers*, Leningrad, p. 309, Publishing House Nauka, Leningrad, 1967.

149. B. Van Zyl, D. Jaecks, D. Pretzer, and R. Geballe, *Phys. Rev.*, **158**, 29 (1967).

150. J. L. Edwards and E. W. Thomas, *Phys. Rev.*, **165**, 16 (1968).

151. D. A. Dahlberg, D. K. Anderson, and I. E. Dayton, *Phys. Rev.*, **170**, 127 (1968).

152. J. H. Moore, Jr., and J. P. Doering, *Phys. Rev. Letters*, **23**, 564 (1969).

153. V. Dose and V. Meyer, *Phys. Letters*, **23**, 69 (1966).

154. R. H. Hughes, R. C. Waring, and C. Y. Fan, *Phys. Rev.*, **122**, 525 (1961).

155. J. van Eck, F. J. de Heer, and J. Kistemaker, *Physica*, **28**, 1184 (1962).

156. J. G. Dodd, and R. H. Hughes, *Phys. Rev.*, **135**, A618 (1964).

157. S. V. Bobashev, V. A. Ankudinov, and E. P. Andreev, *Zh. Eksperim. i Theor. Fiz.*, **48**, 833 (1965); [*Soviet Phys.-JETP*, **21**, 554 (1965)].

158. S. V. Bobashev, E. P. Andreev, and V. A. Ankudinov, *Zh. Eksperim. i Theor. Fiz.*, **45**, 1759 (1963); [*Soviet Phys.-JETP*, **18**, 1205 (1964)].

159. V. Dose, *Helv. Phys. Acta*, **39**, 683 (1966).

160. A. Denis, M. Dufay, and M. Gaillard, *Compt. Rend.*, **264**, B440 (1967).

161. E. W. Thomas and G. D. Bent, *Phys. Rev.*, **164**, 143 (1967).

162. J. M. Robinson and H. B. Gilbody, *Proc. Phys. Soc. (London)*, **92**, 589 (1967).

163. H. R. Moustafa Moussa and F. J. de Heer, *Physica*, **36**, 646 (1967).

164. D. Krause and E. A. Soltysik, *Phys. Rev.*, **175**, 142 (1968).

165. J. Van den Bos, G. J. Winter, and F. J. de Heer, *Physica*, **40**, 357 (1968).

166. A. Scharmann and K. H. Schartner, *Z. Physik*, **219**, 55 (1969).

167. A. Scharmann and K. H. Schartner, *Z. Physik,* **228**, 254 (1969).

168. S. V. Bobashev and S. S. Pop, *Opt. i Spektroskopiya*, **18**, 744 (1965); [*Opt. Spectry. (USSR),* **18**, 421 (1965)].

169. D. Pretzer, B. Van Zyl, and R. Geballe, *Phys. Rev. Letters*, **10**, 340 (1963).

170. K. H. Berkner, W. S. Cooper, S. N. Kaplan, and R. V. Pyle, *Phys. Rev.*, **182**, 103 (1969).

171. J. Kingdon, M. F. Payne, and A. C. Riviere, *Sixth Intern. Conf. on the Physics of Electronic and Atomic Collisions: Abstracts of Papers*, Cambridge, Mass., p. 449, MIT Press, 1969.

172. N. P. Carleton and T. R. Lawrence, *Phys. Rev.*, **109**, 1159 (1958).

173. W. F. Sheridan, O. Oldenberg, and N. P. Carleton, *Second Intern. Conf. on the Physics of Electronic and Atomic Collisions: Abstracts of Papers*, Boulder, Colorado, p. 159, W. A. Benjamin, New York, 1961.

174. R. H. Hughes, J. L. Philpot, and C. Y. Fan, *Phys. Rev.*, **123**, 2084 (1961).

175. J. L. Philpot and R. H. Hughes, *Phys. Rev.*, **133**, A107 (1964).

176. J. R. Sheridan and K. C. Clark, *Phys. Rev.*, **140**, A1033 (1965).

177. J. S. Murray, S. J. Young, and J. R. Sheridan, *Phys. Rev. Letters*, **16**, 439 (1966).

178. M. Dufay, J. Desesquelles, M. Druetta, and M. Eidelsberg, *Ann. Geophys.*, **22**, 615 (1966).

179. M. Dufay, J. Desesquelles, M. Carré, G. do Cao, M. Druetta, M. Eidelsberg, and M. C. Poulizac, *Fifth Intern. Conf. on the Physics of Electronic and Atomic Collisions: Abstracts of Papers*, Leningrad, p. 297, Publishing House Nauka, Leningrad, 1967.

180. R. H. Hughes, B. M. Doughty, and A. R. Filippelli, *Phys. Rev.*, **173**, 172 (1968).

181. D. J. Baker, H. A. B. Gardiner, and J. J. Merrill, *J. Chim. Phys.*, **64**, 63 (1967).

182. D. A. Dahlberg, D. K. Anderson, and I. E. Dayton, *Phys. Rev.*, **164**, 20 (1967).

183. E. W. Thomas, G. D. Bent, and J. L. Edwards, *Phys. Rev.*, **165**, 32 (1968).

184. J. H. Moore and J. P. Doering, *Phys. Rev.*, **177**, 218 (1969).

185. R. H. Hughes and D. K. W. Ng, *Phys. Rev.*, **136**, A1222 (1964).

186. J. Baudon, M. Barat, M. Abignoli, and J-C. Houver, *Sixth Intern. Conf. on the Physics of Electronic and Atomic Collisions: Abstracts of Papers*, Cambridge, Mass., p. 314, MIT Press, 1969.

187. A. L. Orbeli, E. P. Andreev, V. A. Ankudinov, and V. M. Dukelskiĭ, *Zh. Eksperim. i Theor, Fiz.*, **57**, 108 (1969); [*Soviet Phys.-JETP*, **30**, 63 (1970)].

188. V. Dose, R. Gunz, and V. Meyer, *Helv. Phys. Acta*, **41**, 269 (1968).

189. V. Dose, R. Gunz, and V. Meyer, *Helv. Phys. Acta*, **41**, 264 (1968).

190. F. J. de Heer and J. Van den Bos, private communication.

191. V. A. Ankudinov, E. P. Andreev, and A. L. Orbeli, *Fifth Intern. Conf. on the Physics of Electronic and Atomic Collisions: Abstracts of Papers*, Leningrad, p. 312, Publishing House Nauka, Leningrad, 1967.

192. E. P. Andreev, V. A. Ankudinov, V. M. Dukelsky, and A. L. Orbeli, *Sixth Intern. Conf. on the Physics of Electronic and Atomic Collisions: Abstracts of Papers*, Cambridge, Mass., p. 800, MIT Press, 1969.

193. K. H. Berkner, J. R. Hiskes, S. N. Kaplan, G. A. Paulikas, and R. V. Pyle, *Atomic Collision Processes: Proc. Third Intern. Conf. on the Physics of Electronic and Atomic Collisions*, London, 1963, p. 726, (M. R. C. McDowell, ed.), North-Holland, Amsterdam, 1964.

194. B. Van Zyl, D. Jaecks, D. Pretzer, and R. Geballe, *Phys. Rev.*, **136**, A1561 (1964).

195. D. Jaecks and E. Tynan, *Fourth International Conference on the Physics of Electronic and Atomic Collisions: Abstracts of Papers*, Québec, 1965, p. 315, Science Bookcrafters, New York, 1965.

196. R. H. Hughes, D. B. Kay, C. A. Stigers, and E. D. Stokes, *Phys. Rev.*, **167**, 26 (1968).

197. R. H. Hughes, *J. Opt. Soc. Am.*, **51**, 696 (1961).

198. L. L. Hatfield and R. H. Hughes, *Phys. Rev.*, **131**, 2556 (1963).

199. W. Hanle and G. A. Voss, *Z. Naturforsch.*, **11a**, 857 (1956).

200. J. Van den Bos, G. J. Winter, and F. J. de Heer, *Physica*, **44**, 143 (1969).

201. E. W. Thomas and G. D. Bent, *J. Opt. Soc. Am.*, **58**, 138 (1968).

202. M. Dufay, M. Druetta, and M. Eidelsberg, *Compt. Rend.*, **261**, 1635 (1965).

203. F. J. de Heer, B. F. J. Luyken, D. Jaecks, L. Wolterbeek Muller, *Physica*, **41**, 588 (1969).

204. M. Lipeles, R. Novick, and N. Tolk, *Phys. Rev. Letters*, **15**, 815 (1965).

205. H. Schlumbohm, *Z. Naturforsch*, **23A**, 970 (1968).

206. R. L. Champion and L. D. Doverspike, *J. Phys. B*, **2**, 1353 (1969).

207. J. Baudon, M. Barat, and M. Abignoli, *J. Phys. B.*, **3**, 207 (1970).

208. M. Lipeles, *Phys. Letters*, **29A**, 297 (1969).

209. A. S. Schlachter, D. H. Loyd, P. J. Bjorkholm, L. W. Anderson, and W. Haeberli, *Phys. Rev.*, **174**, 201 (1968).

210. E. P. Andreev, V. A. Ankudinov, and S. V. Bobashev, *Opt. i Spektroskopiya*, **16**, 187 (1964); [*Opt. Spectry. (USSR)*, **16**, 103 (1964)].

211. V. A. Ankudinov, S. V. Bobashev, and E. P. Andreev, *Zh. Eksperim. i Theor. Fiz.*, **52**, 364 (1967); [*Soviet Phys.-JETP*, **25**, 236 (1967)].

212. F. J. de Heer and J. Van den Bos, *Physica*, **31**, 365 (1965).

213. F. J. de Heer, private communication.

214. C. E. Head and R. H. Hughes, *Phys. Rev.*, **139**, A1392 (1965).

215. S. H. Dworetsky, R. Novick, and N. Tolk, *Sixth Intern. Conf. on the Physics of Electronic and Atomic Collisions: Abstracts of Papers*, Cambridge, Mass., p. 294, MIT Press, 1969.

216. J. Baudon, M. Barat, and M. Abignoli, *J. Phys. B*, **1**, 1083 (1968).

217. E. S. Solov'ev, R. N. Il'in, V. A. Oparin, I. T. Serenkov, and N. V. Fedorenko, *ZhEFT Pis'ma Red.*, **10**, 300 (1969); [*JETP Letters*, **10**, 190 (1969)].

218. N. Tolk and C. W. White, *Sixth Intern. Conf. on the Physics of Electronic and Atomic Collisions, Abstracts of Papers*, Cambridge, Mass., p. 309, MIT Press, 1969.

219. D. Coffey, D. C. Lorents, and F. T. Smith, *Phys. Rev.*, **187**, 201 (1969).

220. V. B. Matveyev, S. V. Bobashev, and V. M. Dukelskiĭ, *Soviet Zh. Eksperim. i Theor. Fiz.*, **55**, 781 (1968); [*Soviet Phys.-JETP*, **28**, 404 (1969)].

221. R. W. Anderson, V. Aquilanti, and D. R. Herschbach, *Chem. Phys. Letters (Netherlands)*, **4**, 5 (1969).

222. J. van Eck, F. J. de Heer, and J. Kistemaker, *Proc. of the Fifth Intern. Conf. on Ionization Phenomena in Gases*, Munich, 1961, Vol. 1, p. 54, (H. Maecker, ed.), North-Holland, Amsterdam, 1962.

223. Von J. Schöttler and J. P. Toennies, *Ber. Bunsengesell. Phys. Chem.*, **72**, 979 (1968).

224. W. Aberth, O. Bernadini, D. Coffey, D. C. Lorents, and R. E. Olson, *Phys. Rev. Letters*, **24**, 345 (1970).

225. W. Aberth and D. C. Lorents, *Phys. Rev.*, **182**, 162 (1969).

226. N. G. Utterback and H. P. Broida, *Phys. Rev. Letters*, **15**, 608 (1965).

227. J. P. Doering, *Phys. Rev.*, **133**, A1537 (1964).

228. H. Schlumbohm, *Z. Naturforsch.*, **23a**, 776 (1968).

229. H. Schlumbohm, *Z. Naturforsch.*, **23a**, 1386 (1968).

230. E. W. Thomas and G. D. Bent, *J. Phys. B*, **1**, 233 (1968).

231. J. E. Bayfield, *Sixth Intern. Conf. on the Physics of Electronic and Atomic Collisions: Abstracts of Papers*, Cambridge, Mass., p. 118, MIT Press, 1969.

232. J. E. Bayfield, *Phys. Rev. Letters*, **20**, 1223 (1968).

233. K. Watanabe, *Advan. Geophys.*, **5**, 153 (1958).

234. T. D. Gaily, *J. Opt. Soc. Am.*, **59**, 536 (1969).

235. H. S. W. Massey, "Excitation of Discrete States," Sec. 19, p. 354 in *Encyclopedia of Physics*, Vol. XXXVI, (S. Flügge, ed.), Springer-Verlag, Berlin, 1956.

236. A. B. Wittkower, G. Ryding, and H. B. Gilbody, *Proc. Phys. Soc. (London)*, **89**, 541 (1966).

237. R. A. Mapleton, *Phys. Rev.*, **126**, 1477 (1962).

238. J. E. Bayfield, *Rev. Sci. Instr.*, **40**, 869 (1969).

239. W. L. Fite, W. E. Kauppila, and W. R. Ott, *Phys. Rev. Letters.*, **20**, 409 (1968).

240. W. R. Ott, W. E. Kauppila, and W. L. Fite, *Phys. Rev. A*, **1**, 1089 (1970).

241. D. R. Bates and G. Griffing, *Proc. Phys. Soc. (London) A*, **66**, 961 (1953).

242. D. R. Bates, A. Dalgarno, *Proc. Phys. Soc. (London) A*, **66**, 972 (1953).

243. A. H. Gabriel and D. W. O. Heddle, *Proc. Roy. Soc. (London) A*, **258**, 124 (1960).

244. J. van Eck and F. J. de Heer, *Atomic Collision Processes: Proc. of the Third Intern. Conf. on the Physics of Electronic and Atomic Collisions*, London, 1963, p. 624, (M. R. C. McDowell, ed.), North-Holland, Amsterdam, 1964.

245. F. J. de Heer, "Experimental Studies of Excitation in Collisions between Atomic and Ionic Systems," p. 327 in *Advances in Atomic and Molecular Physics*, Vol. II, p. 327, (D. R. Bates and I. Estermann, eds.), Academic Press, New York, 1966.

246. J. Van den Bos, Thesis, University of Amsterdam, 1967.

247. W. G. Fastie, *J. Opt. Soc. Am.*, **42**, 641 (1952); **42**, 647 (1952).

248. R. H. Hughes, private communication.

249. A. V. Phelps, *Phys. Rev.*, **110**, 1362 (1958).

250. E. W. Thomas, *Phys. Rev.*, **164**, 151 (1967).

251. J. M. Robinson and H. B. Gilbody, *Fifth Intern. Conf. on the Physics of Electronic and Atomic Collisions: Abstracts of Papers*, Leningrad, p. 291, Publishing House Nauka, Leningrad, 1967.

252. H. B. Gilbody, private communication.

253. A. Denis, Thesis, University of Lyons, 1967.

254. M. G. Belanger, R. L. Gray, D. Krause, and E. A. Soltysik, *Fifth Intern. Conf. on the Physics of Electronic and Atomic Collisions: Abstracts of Papers*, Leningrad, p. 318, Publishing House Nauka, Leningrad, 1967.

255. S. Dworetsky, R. Novick, W. W. Smith, and N. Tolk, *Fifth Intern. Conf. on the Physics of Electronic and Atomic Collisions: Abstracts of Papers*, Leningrad, p. 280, Publishing House Nauka, Leningrad, 1967.

256. A. Scharmann and K-H. Schartner, *Phys. Letters*, **27A**, 43 (1968).

257. A. Scharmann and K-H. Schartner, *Phys. Letters*, **26A**, 51, (1967).

258. A. Scharmann and K-H. Schartner, *Sixth Intern. Conf. on the Physics of Electronic and Atomic Collisions: Abstracts of Papers*, Cambridge, Mass., p. 822, MIT Press, 1969.

259. E. J. Dieterich, *Phys. Rev.*, **103**, 632 (1956).

260. Z. Sternberg and P. Tomaš, *Phys. Rev.*, **124**, 810 (1961).

261. T. Holstein, *Phys. Rev.*, **72**, 1212 (1947); *Phys. Rev.*, **83**, 1159 (1951).

262. E. W. Thomas, unpublished data.

263. R. J. Bell, *Proc. Phys. Soc. (London)*, **78**, 903 (1961).

264. J. van Eck, Thesis, University of Amsterdam, 1964.

265. A. B. Wittkower and H. B. Gilbody, *Proc. Phys. Soc. (London)*, **90**, 353 (1967).

266. G. Carter, J. S. Colligon, *Ion Bombardment of Solids*, p. 66, American Elsevier, New York, 1968.

267. S. K. Allison, *Rev. Mod. Phys.*, **30**, 1137 (1958).

268. F. J. de Heer and J. Van den Bos, *Physica*, **30**, 741 (1964).

269. C. E. Moore, *Atomic Energy Levels*, Vol. I., National Bureau of Standards Circular 467, 1949.

270. C. E. Moore, *Atomic Energy Levels*, Vol. I, p. XI, National Bureau of Standards Circular 467, 1949.

271. A. K. Edwards and M. E. Rudd, *Phys. Rev.*, **170**, 140 (1968).

272. F. J. de Heer, D. Jaecks, A. Salop, L. Wolterbeek Muller, and B. F. J. Luyken, *Fifth Intern. Conf. on the Physics of Electronic and Atomic Collisions: Abstracts of Papers*, Leningrad, p. 283, Publishing House Nauka, Leningrad, 1967.

273. Q. C. Kessel, P. H. Rose, and L. Grodzins, *Phys. Rev. Letters*, **22**, 1031 (1969).

274. G. S. Hurst, T. E. Bortner, and T. D. Strickler, *J. Chem. Phys.*, **49**, 2460(L) (1958); *Phys. Rev.*, **178**, 4 (1969).

275. J. R. Bennett and A. J. L. Collinson, *J. Phys. B*, **2**, 571, (1969).

276. Th. J. M. Sluyters and J. Kistemaker, *Physica*, **25**, 182 (1959).

277. Th. J. M. Sluyters, E. de Haas, and J. Kistemaker, *Rev. Universelle Mines* [9], **15**, 1 (1959).

BIBLIOGRAPHY 409

278. Th. J. M. Sluyters, E. de Haas, and J. Kistemaker, *Proc. of the Fourth Intern. Conf. on Ionization Phenomena in Gases*, Uppsala, 1959, Vol. I, p. 60, North-Holland, Amsterdam, 1960.

279. Th. J. M. Sluyters, *Physica*, **25**, 1376 (1959).

280. E. W. Thomas and H. B. Gilbody, *Atomic Collision Processes: Proc. Third Intern. Conf. on the Physics of Electronic and Atomic Collisions:* London, 1963, p. 644, (M. R. C. McDowell, ed.), North-Holland, Amsterdam, 1964.

281. S. H. Neff and N. P. Carleton, *Atomic Collision Processes: Proc. of the Third Intern. Conf. on the Physics of Electronic and Atomic Collisions*, London, p. 652, (M. R. C. McDowell, ed.), North-Holland, Amsterdam, 1964.

282. M. Lipeles, R. D. Swift, M. S. Longmire, and M. P. Weinreb, *Phys. Rev. Letters*, **24**, 799 (1970).

283. D. T. Stewart, *Proc. Phys. Soc. (London) A*, **69**, 437 (1956).

284. J. W. McConkey and I. D. Latimer, *Proc. Phys. Soc. (London)*, **86**, 463 (1965); J. W. McConkey, J. M. Woolsey, and D. J. Burns, *Planet. Space Sci.*, **15**, 1332 (1967).

285. B. N. Srivastava and I. M. Mirza, *Phys. Rev.*, **168**, 86 (1968); *Phys. Rev.*, **176**, 137 (1968).

286. J. F. M. Aarts, F. J. de Heer, and D. A. Vroom, *Physica*, **40**, 197 (1968).

287. R. F. Holland, Los Alamos Scientific Laboratory Report No. LA-3783 (TID-4500), 1967.

288. R. W. Fink, R. C. Jopson, Hans Mark, and C. D. Swift, *Rev. Mod. Phys.*, **38**, 513 (1966).

289. D. J. Volz, private communication.

290. M. Dufay, private communication.

291. G. A. Govertsen and D. K. Anderson, *J. Opt. Soc. Am.*, **60**, 600 (1970).

292. J. Bakos, J. Szigeti, and L. Varga, *Phys. Letters*, **20**, 503 (1966).

293. G. H. Dieke, "Atomic and Molecular Physics," in *American Institute of Physics Handbook*, 2nd ed., p. 7–1, (D. E. Gray, ed.), McGraw-Hill, New York, 1957.

294. T. L. de Bruin, C. J. Humphreys, and W. F. Meggers, *J. Res. Natl. Bur. Std.*, **11**, 409 (1933).

295. C. J. Humphreys, *J. Res. Natl. Bur. Std.*, **22**, 19 (1939).

296. C. J. Humphreys and W. F. Meggers, *J. Res. Natl. Bur. Std.*, **10**, 139 (1933).

297. H. E. Clark, Thesis, University St. Louis, 1948.

298. *Handbook of Chemistry and Physics*, 46th ed., p. D-96, (R. C. Weast, ed.), Chemical Rubber Co., 1965; *International Critical Tables of Numerical Data, Physics, Chemistry and Technology*, Vol. III, p. 206, (E. W. Washburn, ed.), McGraw-Hill, New York, 1928.

299. E. P. Andreev, private communication.

300. Von. F. Paschen, *Sitzber. Preussichen Akd. Wiss.*, **13**. 536 (1928).

301. E. W. Foster, *Proc. Roy. Soc. London A*, **200**, 429 (1950).

302. R. J. Anderson, E. T. P. Lee, and C. C. Lin, *Phys. Rev.*, **157**, 31 (1967).

303. P. Cahill, R. Schwartz, and A. N. Jette, *Phys. Rev. Letters*, **19**, 283 (1967).

304. V. A. Ankudinov, E. P. Andreev, and S. V. Bobashev, *Fifth Intern. Conf. on the Physics of Electronic and Atomic Collisions: Abstracts of Papers*, Leningrad, p. 304, Publishing House Nauka, Leningrad, 1967.

305. G. Dunn, R. Geballe, and D. Pretzer, *Second Intern. Conf. on the Physics of Electronic and Atomic Collisions: Abstracts of Papers*, Boulder, Colorado, p. 26, W. A. Benjamin, New York, 1961.

306. J. B. Hasted, *Physics of Atomic Collisions*, p. 278, Butterworths, London, 1964.

307. J. L. Edwards and E. W. Thomas, *Fifth Intern. Conf. on the Physics of Electronic and Atomic Collisions: Abstracts of Papers*, Leningrad, p. 288, Publishing House Nauka, Leningrad, 1967.

308. D. A. Dahlberg, private communication.

309. G. N. Polyakova, V. I. Tatus', S. S. Strel'chenko, Ya. M. Fogel', and V. M. Fridman, *Zh. Eksperim. i Theor. Fiz.*, **50**, 1464 (1966); [*Soviet Phys.-JETP*, **23**, 973 (1966)].

310. F. R. Gilmore, *J. Quant. Spectr. Radiative Transfer*, **5**, 369 (1965).

311. L. Wallace, *Astrophys. J. Suppl. Ser.*, **62**, 6, 445 (1962).

312. D. A. Dahlberg, D. K. Anderson, and I. E. Dayton, *Fifth Intern. Conf. on the Physics of Electronic and Atomic Collisions: Abstracts of Papers*, Leningrad, p. 294, Publishing House Nauka, Leningrad, 1967.

313. J. R. Sheridan and K. C. Clark, *Fourth Intern. Conf. on the Physics of Electronic and Atomic Collisions: Abstracts of Papers*, Québec, 1965, p. 276. Science Bookcrafters, New York, 1965.

314. J. H. Moore, Jr. and J. P. Doering, *Phys. Rev.*, **174**, 178 (1968).

315. J. H. Moore, Jr. and J. P. Doering, *Phys. Rev.*, **182**, 176 (1969).

316. L. Kurzweg, H. H. Lo, R. T. Brackmann, and W. L. Fite, *Phys. Rev.*, **185**, 404 (1969).

317. E. W. Thomas, G. D. Bent, and J. L. Edwards, *Fifth Intern. Conf. on the Physics of Electronic and Atomic Collisions: Abstracts of Papers*, Leningrad, p. 286, Publishing House Nauka, Leningrad, 1967.

318. J. L. Philpot and R. H. Hughes, *Phys. Rev.*, **135**, AB3 (1964).

319. A. B. Meinel and C. Y. Fan, *Astrophys. J.*, **115**, 330 (1952).

320. C. Y. Fan and A. B. Meinel, *Astrophys. J.*, **118**, 205 (1953).

321. C. Y. Fan, *Phys. Rev.*, **103**, 1740 (1956).

322. J. R. Sheridan, private communication.

323. W. F. Sheridan, O. Oldenberg, and N. P. Carleton, *Atomic Collision Processes: Proc. Third Intern. Conf. on the Physics of Electronic and Atomic Collisions*, London, 1963, p. 440, (M. R. C. McDowell, ed.), North-Holland, Amsterdam, 1964.

324. N. P. Carleton, *Phys. Rev.*, **107**, 110 (1957).

325. L. M. Branscomb, R. J. Shalek, and T. W. Bonner, *Trans., Am. Geophys. Union*, **35**, 107 (1954).

326. R. W. Nicholls, *J. Res. Natl. Bur. Std.*, **65A**, 451 (1961).

327. R. G. Bennett and F. W. Dalby, *J. Chem. Phys.*, **31**, 434 (1959).

328. R. P. Lowe and H. I. S. Fergusson, *Proc. Phys. Soc. (London)*, **85**, 813 (1965).

329. F. L. Roesler, C. Y. Fan, and J. W. Chamberlain, *J. Atmospheric Terest. Phys.*, **12**, 200 (1958).

330. G. N. Polyakova, V. I. Tatus', and Ya. M. Fogel', *Zh. Eksperim. i Theor. Fiz.*, **52**, 657 (1967); [*Soviet Phys.-JETP*, **25**, 430 (1967)].

331. E. M. Reeves and R. W. Nicholls, *Proc. Phys. Soc. (London)*, **78**, 588 (1961).

332. E. M. Reeves, R. W. Nicholls, and D. A. Bromley, *Proc. Phys. Soc. (London)*, **76**, 217 (1970).

333. J. P. Doering, *Bull. Am. Phys. Soc.*, **13**, 204 (1968).

334. G. N. Polyakova, V. F. Erko, Ya. M. Fogel', A. V. Zats, and B. M. Fizgeev, *Zh. Eksperim. i Theor. Fiz.*, **56**, 1851 (1969); [*Soviet Phys.-JETP*, **30**, 63 (1970)].

335. R. P. Lowe and H. I. S. Ferguson, *Fourth Intern. Conf. on the Physics of Electronic and Atomic Collisions: Abstracts of Papers*, Québec, 1965, p. 285, Science Bookcrafters, New York, 1965.

336. R. W. Nicholls, E. M. Reeves, and D. A. Bromley, *Proc. Phys. Soc.(London)*, **74**, 87 (1959).

337. R. W. Nicholls and E. M. Reeves, *Nature*, **180**, 1188 (1957).

338. R. W. Nicholls and D. Pleiter, *Nature*, **178**, 1456 (1956).

339. L. Herman, H. I. S. Ferguson, and R. W. Nicholls, *Can. J. Phys.*, **39**, 476 (1961).

340. G. N. Polyakova, Ya. M. Fogel', and Chiu Yu-mei, *Soviet Astron.-AJ*, **7**, 267 (1963).

341. M. Dufay, M. Druetta, and M. Eidelsberg, *Compt. Rend.*, **260**, 1123 (1965).

342. W. R. Jarmain, P. A. Fraser, and R. W. Nicholls, *Astrophys. J.*, **122**, 55 (1955).

343. M. Haugh, T. G. Slanger, and K. D. Bayes, *J. Chem. Phys.*, **44**, 837 (1966).

344. M. C. Poulizac, J. Desesquelles, and M. Dufay, *Compt. Rend.*, **263**, B553 (1966).

345. M. Dufay and M. C. Poulizac, *Mem. Soc. Roy. Sci. Liege*, **12**, 427 (1966).

346. M. C. Poulizac, Thesis, University of Lyon, France, 1966.

347. M. Carré, Thesis, University of Lyon, France, 1967.

348. M. Carré, *Physica*, **41**, 63 (1969).

349. M. Carré and M. Dufay, *Compt. Rend.*, **266**, B1367 (1968).

350. T. Gaily and R. Geballe, *Fifth Intern. Conf. on the Physics of Electronic and Atomic Collisions: Abstracts of Papers*, Leningrad, 1967, p. 314, Publishing House Nauka, Leningrad, 1967.

351. E. P. Andreev, V. A. Ankudinov, S. V. Bobashev, V. M. Dukelski, and V. B. Matveyev, *Fifth Intern. Conf. on the Physics of Electronic and Atomic Collisions: Abstracts of Papers*, Leningrad, 1967, p. 302, Publishing House Nauka, Leningrad, 1967.

352. V. Z. Ankudinov, S. V. Bobashev, and E. P. Andreev, *Zh. Eksperim. i Theor. Fiz.*, **48**, 40 (1965); [*Soviet Phys.-JETP*, **21**, 26 (1965)].

353. R. H. Hughes, H. R. Dawson, and B. M. Doughty, *Phys. Rev.*, **164**, 166 (1967).

354. D. Rapp and W. E. Francis, *J. Chem. Phys.*, **37**, 2631 (1962).

355. G. Bassani, L. Colli, F. Cristofori, M. O. Franceschetti, G. E. Frigerio, and P. G. Sona, *Energia Nucl. (Milan)*, **9**, 451 (1962).

356. F. Cristofori, G. E. Frigerio, N. Molko, and P. G. Sona, *Proc. of the Sixth Conf. on the Ionization Phenomena in Gases*, Paris, Vol. I, p. 69, (P. Huber and E. Crémieu-Alcan, eds.), SERMA, Paris, 1963.

357. F. Cristofori, P. Fenici, G. E. Frigerio, N. Molho, and P. G. Sona, *Phys. Letters*, **6**, 171 (1963).

358. D. Blanc and A. Degeilh, *J. Phys. Radium*, **22**, 230 (1961).

359. R. A. Mapleton, *Phys. Rev.*, **122**, 528 (1961).

360. R. Geballe, private communication.

361. F. J. de Heer, L. Wolterbeek Muller, and R. Geballe, *Fourth Intern. Conf. on the Physics of Electronic and Atomic Collisions: Abstracts of Papers*, Québec, 1965, p. 309, Science Bookcrafters, New York, 1965.

362. S. V. Bobashev, B. B. Matveyev, and V. A. Ankudinov, *Sixth Intern. Conf. on the Physics of Electronic and Atomic Collisions: Abstracts of Papers*, Cambridge, Mass., p. 229, MIT Press, 1969.

APPENDIX I

RECENT ADDITIONAL MATERIAL

The following is a list of publications that are dated later than the cutoff date (January 1, 1970) for inclusion in this monograph. The table includes basic information as to the data and coverage of each publication.

The nature of the published data is indicated in the column labeled "Data" as follows: σ_i, level excitation cross section; σ_{ij}, line emission cross section; $\sigma_i(R)$ and $\sigma_{ij}(R)$, relative values of level excitation and line emission cross sections; P_{ij} polarization of emission; $\sigma_i(\theta)$ level-excitation cross sections differential in angle; T_i thresholds for level excitation.

Projectile	Target	Excited Particle	Energy Range (keV)	Data	Reference
Al$^+$	Ar	Ar	5–80	σ_{ij}	Saris[1]
Ar$^+$	Ar	Ar	50–330	σ_{ij}	Cunningham et al.[2]
Ar$^+$	Ar	Ar	10–120	σ_{ij}	Saris et al.[3]
Ar$^+$	C$_2$H$_2$	CH	0.020–1.0	σ_{ij}	Liu et al.[4]
Ar$^+$	CO	CO$^+$	0.020–1.0	σ_{ij}	Liu et al.[4]
Ar$^+$	CO	CO$^+$	2.5	$\sigma_{ij}(R)$	Haugh et al.[5]
Ar$^+$	CS$_2$	CS$_2^+$	2.5	$\sigma_{ij}(R)$	Haugh et al.[5]
Ar$^+$	D$_2$	Ar$^+$	0.011–0.017	$\sigma_i(R)$	Moran et al.[6]
Ar$^+$	D$_2$	D$_2$	0.011–0.017	$\sigma_i(R)$	Moran et al.[6]
Ar$^+$	H$_2$	Ar, Ar$^+$, H	3–40	σ_{ij}	Polyakova et al.[7]
Ar$^+$	HBr	HBr$^+$	2.5	$\sigma_{ij}(R)$	Haugh et al.[5]
Ar$^+$	HCl	HCl$^+$	2.5	$\sigma_{ij}(R)$	Haugh et al.[5]
Ar$^+$	N$_2$	N$_2$	50	σ_i	Schowengerdt et al.[8]
Ar$^+$	N$_2$O	N$_2$O$^+$	2.5	$\sigma_{ij}(R)$	Haugh et al.[5]
Ar	Ar	Unspec.	0.03–0.8	$\sigma_{ij}(R)$	Haugsjaa et al.[9]
Ar	Ar	Ar	0.042–0.25	σ_i	Gerber et al.[10]
Ar	Ar	Ar	0.022–0.032	σ_i	Haugsjaa et al.[11]
Ar	H$_2$	Ar, H	3–40	σ_{ij}	Polyakova et al.[7]
C$^+$	Ar	Ar	5–80	σ_{ij}	Saris[1]
Ca$^+$	Cd	Ca$^+$	0–1.0	$\sigma_{ij}(R)$	Shpenik et al.[12]
Cd$^+$	Zn	Zn	0–1.0	$\sigma_{ij}(R)$	Shpenik et al.[12]
Cd	Zn	Cd	0.05–1.0	$\sigma_{ij}(R)$	Zavilopulo et al.[13]
Cl$^+$	Ar	Ar	5–80	σ_{ij}	Saris[1]
Cs$^+$	Ar	Cs$^+$	2–28	σ_i	Zapesochny et al.[14]

Projectile	Target	Excited Particle	Energy Range (keV)	Data	Reference
Cs^+	He	Cs^+	1.5–12	T_i	Matveyev et al.[15]
Cs^+	He	Cs^+	2–28	σ_i	Zapesochny et al.[14]
Cs^+	Ne	Cs^+	2–28	σ_i	Zapesochny et al.[14]
Cs^+	Ne	Cs^+	1.5–12	T_i, $\sigma_{ij}(R)$	Matveyev et al.[15]
Cu^+	Ar	Ar	5–80	σ_{ij}	Saris[1]
D^+	C_8F_{16}	D	10–40	$\sigma_i(R)$	Kingdon et al.[16]
D^+	H_2	D	10–40	$\sigma_i(R)$	Kingdon et al.[16]
D^+	H_2O	D	10–40	$\sigma_i(R)$	Kingdon et al.[16]
D^+	Mg	D	10–40	$\sigma_i(R,\theta)$	Kingdon et al.[16]
D_2^+	H_2	D	80	$\sigma_i(\theta)$	Kingdon et al.[16]
Fe^+	Ar	Ar	5–80	σ_{ij}	Saris[1]
H^+	Ar	H	10–120	σ_i	Hughes et al.[17]
H^+	Ar	H	30–100	P_{ij}	Hughes et al.[17]
H^+	Ar	H	20–130	σ_i	Hughes et al.[18]
H^+	Ar	H	0.6–24	P_{ij}	Teubner et al.[19]
H^+	Ar	Ar	10–120	σ_{ij}	Saris[3]
H^+	Ar	H	4–26	$\sigma_i(R)$	Fitzwilson et al.[20]
H^+	Ar	Ar	120–300	σ_{ij}	Volz et al.[21]
H^+	Ar	H	75–400	σ_i	Ford et al.[22]
H^+	Ba	H	5–30	σ_i	McFarland et al.[23]
H^+	C_8F_{16}	H	5–30	σ_i	McFarland et al.[23]
H^+	C_2H_2	C_2H_2	0.15–0.5	$\sigma_i(R)$	Moore et al.[24]
H^+	C_2H_4	C_2H_4	0.15–0.5	$\sigma_i(R)$	Moore et al.[24]
H^+	CO	CO	0.15–0.5	$\sigma_i(R)$	Moore et al.[24]
H^+	CO	CO	20–120	σ_i	Park et al.[25]
H^+	Cs	H	0.8	$\sigma_i(R)$	Donnally et al.[26]
H^+	Cs	H	0.5–2.5	σ_i	Spiess et al.[27]
H^+	H	H^f	2–21	P_{ij}	Kauppila et al.[28]
H^+	H	H^e	6–30	$\sigma_i(\theta)$	Bayfield[29]
H^+	H	H^e	7–60	σ_i	Khayrallah et al.[30]
H^+	H_2	H^e	10–120	σ_i	Hughes et al.[17]
H^+	H_2	H^e	30–100	P_{ij}	Hughes et al.[17]
H^+	H_2	H^e	20–130	σ_i	Hughes et al.[18]
H^+	H_2	H^f	2–21	P_{ij}	Kauppila et al.[28]
H^+	H_2	H^e	5–30	σ_i	McFarland et al.[23]
H^+	H_2	H_2	0.010	$\sigma_i(\theta)$	Udseth et al.[31]
H^+	H_2	$H^{e,f,g}$	1.5–30	σ_{ij}	McNeal et al.[32]
H^+	H_2	H^e	75–400	σ_i	Ford et al.[22]
H^+	H_2O	H	5–30	σ_i	McFarland et al.[23]
H^+	He	H	10–120	σ_i	Hughes et al.[17]
H^+	He	H	30–100	P_{ij}	Hughes et al.[17]
H^+	He	He	25–125	σ_i	Park et al.[33]
H^+	He	H	20–130	σ_i	Hughes et al.[18]
H^+	He	H	0.6–24	P_{ij}	Teubner et al.[19]
H^+	He	H	4–26	$\sigma_i(R)$	Fitzwilson et al.[20]
H^+	He	H	75–400	σ_i	Edwards et al.[34]

Projectile	Target	Excited Particle	Energy Range (keV)	Data	Reference
H^+	He	H	6.25	$\sigma_i(\theta)$	Fitzwilson et al.[35]
H^+	He	H	75–400	σ_i	Ford et al.[22]
H^+	He	He	40	$\sigma_{ij}(R)$	Gray et al.[36]
H^+	He	He	200–1000	$\sigma_{ij}(R)$	Hasselkamp et al.[37]
H^+	He	He	20–100	σ_i	Blair et al.[38]
H^+	He	H	4–20	$\sigma_i(R,\theta)$	Jaecks et al.[39]
H^+	K	H	5–30	σ_i	McFarland et al.[23]
H^+	Mg	H	5–30	σ_i	McFarland et al.[23]
H^+	N_2	N_2	0.15–0.5	$\sigma_i(R)$	Moore et al.[24]
H^+	N_2	N_2	20–120	σ_i	Schowengerdt et al.[8]
H^+	N_2	N_2^+	20–100	σ_{ij},σ_i	Gardiner et al.[40]
H^+	N_2	H	10–120	σ_i	Hughes et al.[17]
H^+	N_2	H	30–100	P_{ij}	Hughes et al.[17]
H^+	N_2	H	20–130	σ_i	Hughes et al.[18]
H^+	N_2	H	5–30	σ_i	McFarland et al.[23]
H^+	N_2	N_2	1–150	σ_{ij}	de Heer et al.[41]
H^+	N_2	H	4–26	$\sigma_i(R)$	Fitzwilson et al.[20]
H^+	N_2	H	75–400	σ_i	Edwards et al.[34]
H^+	N_2	N_2^+	5–60	σ_{ij}	Wehrenberg et al.[42]
H^+	N_2	H	75–400	σ_i	Ford et al.[22]
H^+	N_2	H	2–10	σ_i	Suchannek et al.[43]
H^+	N_2	N_2^+	1–25	σ_{ij}	McNeal et al.[44]
H^+	NH_3	H^e	5–30	σ_i	McFarland et al.[23]
H^+	Ne	H	10–120	σ_i	Hughes et al.[17]
H^+	Ne	H	30–100	P_{ij}	Hughes et al.[17]
H^+	Ne	H	20–130	σ_i	Hughes et al.[18]
H^+	Ne	H	0.6–24	P_{ij}	Teubner et al.[19]
H^+	Ne	H	2–20	$\sigma_i(\theta)$	Jaecks et al.[45]
H^+	O_2	H	10–120	σ_i	Hughes et al.[17]
H^+	O_2	H	30–100	P_{ij}	Hughes et al.[17]
H^+	O_2	H	20–130	σ_i	Hughes et al.[18]
H^+	O_2	O_2	20–120	σ_i	Park et al.[46]
H^+	O_2	H	4–26	$\sigma_i(R)$	Fitzwilson et al.[20]
H^+	Tl	H	5–30	σ_i	McFarland et al.[23]
H	H_2	$H^{e,f,g}$	5–20	σ_{ij}	McNeal et al.[32]
H	He	He	20–100	σ_i	Blair et al.[38]
H	N_2	N_2	20–100	σ_{ij}	Gardiner et al.[40]
H	N_2	H	7–35	σ_i	Hughes et al.[47]
H^-	Ar	H	5–40	σ_{ij}	Orbeli et al.[48]
H^-	He	H	5–40	σ_{ij}	Orbeli et al.[48]
H^-	Kr	H	5–40	σ_{ij}	Orbeli et al.[48]
H^-	Ne	H	5–40	σ_{ij}	Orbeli et al.[48]
H^-	Xe	H	5–40	σ_{ij}	Orbeli et al.[48]
H_2^+	Ar	Ar	120–300	σ_{ij}	Volz et al.[21]
H_2^+	C_2H_2	C_2H_2	0.15–0.5	$\sigma_i(R)$	Moore et al.[24]
H_2^+	C_2H_4	C_2H_4	0.15–0.5	$\sigma_i(R)$	Moore et al.[24]

Projectile	Target	Excited Particle	Energy Range (keV)	Data	Reference
H_2^+	CO	CO	0.15–0.5	$\sigma_i(R)$	Moore et al.[24]
H_2^+	He	H	10	$\sigma_i(\theta)$	Jaecks et al.[49]
H_2^+	He	H	12	$\sigma_i(\theta)$	Fitzwilson et al.[35]
H_2^+	He	H	75–400	σ_i	Ford et al.[22]
H_2^+	N_2	N_2	50	σ_i	Schowengerdt et al.[8]
H_2^+	N_2	N_2	0.15–0.5	σ_i	Moore et al.[24]
He^+	Ar	Ar^+	0.01–0.8	σ_{ij}	Lipeless et al.[50]
He^+	Ar	He	0.3–150	σ_i, P_{ij}	Muller et al.[51]
He^+	Ar	He	10–200	σ_i	Gilbody et al.[52]
He^+	Ar	Ar^+	0.001–1.0	σ_{ij}	Lipeless et al.[53]
He^+	Ar	Ar	5–80	σ_{ij}	Saris[1]
He^+	Ar	Ar, He	0.010–1.0	$\sigma_{ij}(R)$	Isler et al.[54]
He^+	H_2	He	10–200	σ_i	Gilbody et al.[52,55]
He^+	H_2	He	14–50	σ_i	Tawara[56]
He^+	H_2	He	0.1–30	$\sigma_{ij}(R)$	Polyakova et al.[57]
He^+	H_2	H, H_2, He	1.1	$\sigma_{ij}(R)$	Hollstein et al.[58]
He^+	H_2	H, He, He^+	3–40	σ_{ij}	Polyakova et al.[7]
He^+	H_2	H	5–30	P_{ij}	Nathan et al.[59]
He^+	He	He^g	0.04–50	$\sigma_{ij}(R)$	Dworetsky et al.[60]
He^+	He	He^g	0.05	T_i	Dworetsky et al.[60]
He^+	He	He^e	0.044–0.23	σ_i	MacVicar-Whelan et al.[61]
He^+	He	He^e	0.3–150	σ_i, P_{ij}	Muller et al.[51]
He^+	He	He^e	10–200	σ_i	Gilbody et al.[52,55]
He^+	He	He^{+e}	20–100	σ_i	Schoonover et al.[62]
He^+	He	He^f	0.2–0.35	$\sigma_i(\theta)$	Rahmat et al.[63,64]
He^+	He	He^f	0.030	$\sigma_i(\theta)$	Olson et al.[65]
He^+	He	He^f	30–60	σ_i	Schoonover et al.[66]
He^+	He	He^g	1.1	$\sigma_{ij}(R)$	Hollstein et al.[58]
He^+	K	K	0.050–1.6	σ_{ij}	Salop et al.[67]
He^+	Kr	He	10–30	σ_i	Miers et al.[68]
He^+	Kr	He	0.3–150	σ_i, P_{ij}	Muller et al.[51]
He^+	Kr	He	10–200	σ_i	Gilbody et al.[52]
He^+	N_2	He	10–30	σ_i	Miers et al.[68]
He^+	N_2	He	10–200	σ_i	Gilbody et al.[52]
He^+	N_2	N_2	1–20	$\sigma_{ij}(R)$	Runstraat et al.[69]
He^+	N_2	He, N^+, N_2^+	1.1	$\sigma_{ij}(R)$	Hollstein et al.[58]
He^+	Ne	He	10–30	σ_i	Miers et al.[68]
He^+	Ne	He	10–200	σ_i	Gilbody et al.[52]
He^+	Ne	He	0.3–150	σ_i, P_{ij}	Muller et al.[51]
He^+	Ne	He, Ne, Ne^+	0.020–1.0	σ_{ij}	Tolk et al.[70]
He^+	Ne	He, Ne	1.1	$\sigma_{ij}(R)$	Hollstein et al.[58]
He^+	Rb	Rb	0.050–1.6	σ_{ij}	Salop et al.[67]
He^+	Xe	He	0.3–150	σ_i, P_{ij}	Muller et al.[51]
He	H_2	H, H_2, He	1.1	$\sigma_{ij}(R)$	Hollstein et al.[58]
He	H_2	H, He, He^+	3–40	σ_{ij}	Polyakova et al.[7]

Projectile	Target	Excited Particle	Energy Range (keV)	Data	Reference
He	He	He^g	1.1	$\sigma_{ij}(R)$	Hollstein et al.[58]
He	N_2	He, N^+, N_2^+	1.1	$\sigma_{ij}(R)$	Hollstein et al.[58]
He	Ne	He, Ne	1.1	$\sigma_{ij}(R)$	Hollstein et al.[58]
K^+	Ar	K^+	0.05–1.5	σ_{ij}	Bobashev et al.[71]
K^+	Ar	Ar	2.1	T_i	Matveyev et al.[72]
K^+	Ar	Ar	1–10	$\sigma_i(\theta)$	Afrosimov et al.[73]
K^+	Ar	K^+	0.020–0.2	σ_{ij}	Bobashev[74]
K^+	Ar	Ar	2	$\sigma_{ij}(R)$	Afrosimov et al.[75]
K^+	Ar	Ar, K^+	1.5–12	T_i	Matveyev et al.[15]
K^+	He	He, K^+	0.5–34	σ_{ij}	Pop et al.[76]
K^+	He	He, K^+	1.5–12	T_i	Matveyev et al.[15]
K^+	Kr	K^+	1.5–12	T_i	Matveyev et al.[15]
K^+	Na	K, Na	0–1.5	σ_{ij}	Aquilanti et al.[77]
K^+	Ne	Ne	1–12	$\sigma_{ij}(R)$, T_i	Matveyev et al.[15,72]
K^+	Ne	Ne, K^+	1–30	σ_{ij}	Pop et al.[78]
K^+	Xe	K^+	1.5–12	T_i	Matveyev et al.[15]
K	CO	K	0.001–0.020	$\sigma_{ij}(R)$	Lacmann et al.[79]
K	CO	K	0.001–0.015	$\sigma_{ij}(R)$	Kempter et al.[80]
K	Cl_2	K	0.001–0.020	$\sigma_{ij}(R)$	Lacmann et al.[79]
K	HCl	K	0.001–0.020	$\sigma_{ij}(R)$	Lacmann et al.[79]
K	N_2	K	0.001–0.020	$\sigma_{ij}(R)$	Lacmann et al.[79]
K	NO	K	0.001–0.015	$\sigma_{ij}(R)$	Kempter et al.[80]
K	NO	K	0.001–0.020	$\sigma_{ij}(R)$	Lacmann et al.[79]
K	NO	K	0.001–0.015	$\sigma_{ij}(R)$	Kempter et al.[80]
K	O_2	K	0.001–0.020	$\sigma_{ij}(R)$	Lacmann et al.[79]
K	O_2	K	0.001–0.015	$\sigma_{ij}(R)$	Kempter et al.[80]
Kr^+	HBr	HBr^+	2.5	$\sigma_{ij}(R)$	Haugh et al.[5]
Li^+	Cs	Li	0.5–20	σ_{ij}	Perel et al.[81]
Li^+	He	He	1.5	$\sigma_i(\theta)$	Lorents et al.[82]
Li^+	K	Li	0.5–20	σ_{ij}	Perel et al.[81]
Li^+	N_2	N_2^+	6	$\sigma_{ij}(R)$	Mickle et al.[83]
Li^+	Na	Li, Na	0–1.5	σ_{ij}	Aquilanti et al.[77]
Li^+	Rb	Li	0.5–20	σ_{ij}	Perel et al.[81]
N^+	Ar	Ar	5–80	σ_{ij}	Saris[1]
N_2^+	CO	CO^+	0.200–1.0	σ_{ij}	Liu et al.[4]
N_2^+	CO_2	CO_2^+	0.020–1.0	σ_{ij}	Liu et al.[4]
N_2^+	C_2H_2	CH	0.020–1.0	σ_{ij}	Liu et al.[4]
N_2^+	K	N_2^+	0.050–1.6	σ_{ij}	Salop et al.[67]
N_2^+	N_2	N_2^{+g}	0.020–1.0	σ_{ij}	Liu et al.[4]
Na^+	He	He	1–13	σ_{ij}	Bobashev et al.[84]
Na^+	N_2	N_2^+	6	$\sigma_{ij}(R)$	Mickle et al.[83]
Na^+	Na	Na^g	0–1.5	σ_{ij}	Aquilanti et al.[77]
Na^+	Ne	Ne	0.2–11	σ_{ij}	Bobashev[85]
Ne^+	Ar	Ar	10–120	σ_{ij}	Saris et al.[3]
Ne^+	H_2	H, Ne, Ne^+	3–40	σ_{ij}	Polyakova et al.[7]
Ne^+	He	He, Ne, Ne^+	0.020–1.0	σ_{ij}	Tolk et al.[70]

Projectile	Target	Excited Particle	Energy Range (keV)	Data	Reference
Ne^+	N_2	N_2^+	1–20	$\sigma_{ij}(R)$	Runstraat et al.[69]
Ne^+	Ne	Ne	10–120	σ_{ij}	Saris et al.[3]
Ne	H_2	H, Ne	3–40	σ_{ij}	Polyakova et al.[7]
O^+	Ar	Ar	5–80	σ_{ij}	Saris[1]
O^+	O_2	O_2	0.008–0.22	$\sigma_i(R)$	Cosby et al.[86]
O_2^+	Ar	O_2^+	0.013	$\sigma_i(R,\theta)$	Cosby et al.[86]
Rb^+	Ar	Rb^+	0.020–0.2	σ_{ij}	Bobashev et al.[74]
Rb^+	He	Rb^+	1.5–12	$\sigma_{ij}(R), T_i$	Matveyev et al.[15]
Ti^+	Ar	Ar	5–80	σ_{ij}	Saris[1]
Zn^+	Cd	Zn	0–1.2	$\sigma_{ij}(R)$	Sphenik et al.[87,88]
Zn^+	Na	Zn	0–0.6	$\sigma_{ij}(R)$	Shpenik et al.[87]
Zn^+	Zn	Zn^g	0–1.0	$\sigma_{ij}(R)$	Zapesochnyi et al.[89]
Zn^+	Zn	Zn^g	0–1.0	$\sigma_{ij}(R)$	Shpenik et al.[12,87]

Note: e = excitation of projectile, f = excitation of target, and g = sum of emissions from target and projectile.

REFERENCES

1. F. W. Saris, *Physica*, **52**, 290 (1971).

2. M. E. Cunningham, R. C. Der, R. J. Fortner, T. M. Kavanagh, J. M. Khan, C. B. Layne, E. J. Zaharis, and J. D. Garcia, *Phys. Rev. Letters*, **24**, 931 (1970).

3. F. W. Saris and D. Onderdelinden, *Physica*, **49**, 441 (1970).

4. C. Liu and H. P. Broida, *Phys. Rev.*, **A2**, 1824 (1970).

5. M. J. Haugh and K. D. Bayles, *Phys. Rev.*, **A2**, 1778 (1970).

6. T. F. Moran and P. C. Cosby, *J. Chem. Phys.*, **51**, 5724 (1969).

7. G. N. Polyakova, V. A. Gusev, V. F. Erko, Ya. M. Fogel, and A. V. Zats, *Zh. Eksperim. i Theor. Fiz.*, **58**, 1186 (1970). [*Soviet Physics JETP*, **31**, 637 (1970)].

8. F. F. Schowengerdt, and J. T. Park, *Phys. Rev.*, **A1**, 848 (1970).

9. P. O. Haugsjaa and R. C. Amme, *Phys. Rev. Letters*, **23**, 633 (1969).

10. G. Gerber, R. Morgenstern, A. Niehaus and M. W. Ruf, Seventh International Conference on the Physics of Electronic and Atomic Collisions: Abstract of Papers, Amsterdam, The Netherlands (1971) p. 610, North Holland Publishing Co., 1971.

11. P. O. Haugsjaa and R. C. Amme, *J. Chem. Phys.*, **52**, 4874 (1970).

12. O. B. Shpenik, I. P. Zapesochny, and A. N. Zavilopulo, Seventh International Conference on the Physics of Electronic and Atomic Collisions: Abstract of Papers, Amsterdam, the Netherlands. (1971), p. 594, North Holland Publishing Co., 1971.

13. A. N. Zavilopulo, I. P. Zapesochny, O. B. Shpenik, and I. P. Kirlik, Seventh International Conference on the Physics of Electronic and Atomic Collisions: Abstract of Papers Amsterdam, The Netherlands. (1971), p. 597, North Holland Publishing Co., 1971.

14. I. P. Zapesochny and S. S. Pop, Seventh International Conference on the Physics of Electronic and Atomic Collisions: Abstract of Papers, Amsterdam, The Netherlands. (1971), p. 591, North Holland Publishing Co., 1971.

15. V. D. Matveyev, S. V. Bobashev, and V. M. Dukelskii, *Zh. Eksperim. i Theor. Fiz.*, **57**, 1534 (1965); [*Soviet Physics JETP*, **30**, 829 (1970).]

16. J. Kingdon, M. F. Payne, and A. C. Riviere, *J. Phys.*, **B3**, 552 (1970).

17. R. H. Hughes, C. A. Stigers, B. M. Doughty, and E. D. Stokes, *Phys. Rev.*, **A1**, 1424 (1970).

18. R. H. Hughes, E. D. Stokes, S. S. Choe, and T. J. King, *Phys. Rev.*, **4**, 1453 (1971).

19. P. J. O. Teubner, W. E. Kauppila, W. L. Fite, and R. J. Girnius, *Phys. Rev.*, **A2**, 1763 (1970).

20. R. L. Fitzwilson and E. W. Thomas, *Phys. Rev.*, **A3**, 1305 (1971).

21. D. J. Volz and M. E. Rudd, *Phys. Rev.*, **A2**, 1395 (1970).

22. J. C. Ford, F. McCoy, and E. W. Thomas, Seventh International Conference on the Physics of Electronic and Atomic Collisions: Abstract of Papers, Amsterdam, The Netherlands. (1971), p. 818, North Holland Publishing Co., 1971.

23. R. H. McFarland and A. H. Futch, *Phys. Rev.*, **A2**, 1795 (1970).

24. J. H. Moore and J. P. Doering, *J. Chem. Phys.*, **52**, 1692 (1970).

25. John T. Park, D. R. Schoonover, and G. W. York, *Phys. Rev.*, **A2**, 2304 (1970).

26. B. L. Donnally and J. E. O'dell, Seventh International Conference on the Physics of Electronic and Atomic Collisions: Abstract of Papers, Amsterdam, The Netherlands. (1971), p. 821, North Holland Publishing Co., 1971.

27. G. Spiess, A. Valance, and P. Pradel, Seventh International Conference on the Physics of Electronic and Atomic Collisions: Abstract of Papers, Amsterdam, The Netherlands. (1971), p. 823, North Holland Publishing Co., 1971.

28. W. E. Kauppila, P. J. O. Teubner, W. L. Fite, and R. J. Girnius, *Phys. Rev.*, **A2**, 1759 (1970).

29. J. E. Bayfield, *Phys. Rev. Letters*, **25**, 1 (1970).

30. G. Khayrallah, R. Karn, P. M. Koch, and J. E. Bayfield, Seventh International Conference on the Physics of Electronic and Atomic Collisions: Abstracts of Papers, Amsterdam, The Netherlands. (1971), p. 813, North Holland Publishing Co., 1971.

31. H. Udseth and C. F. Giese, Seventh International Conference on the Physics of Electronic and Atomic Collisions: Abstract of Papers, Amsterdam, The Netherlands. (1971), p. 255, North Holland Publishing Co., 1971.

32. R. J. McNeal and J. H. Bireley, Seventh International Conference on the Physics of Electronic and Atomic Collisions: Abstracts of Papers, Amsterdam, The Netherlands. (1971), p. 815, North Holland Publishing Co., 1971.

33. J. T. Park and F. D. Schowengerdt, *Phys. Rev.*, **185**, 152 (1969).

34. J. L. Edwards and E. W. Thomas, *Phys. Rev.*, **A2**, 2346 (1970).

35. R. L. Fitzwilson, I. Sauers and E. W. Thomas, Seventh International Conference on the Physics of Electronic and Atomic Collisions: Abstracts of Papers, Amsterdam, The Netherlands. (1971), p. 608, North Holland Publishing Co., 1971.

36. R. L. Gray, H. H. Haselton, D. Krause, Jr., and E. A. Soltysik, Seventh International Conference on the Physics of Electronic and Atomic Collisions: Abstracts of Papers, Amsterdam, The Netherlands. (1971), p. 831, North Holland Publishing Co., 1971.

37. D. Hasselkamp, R. Hippler, A. Scharmann, and K. H. Schartner, Seventh International Conference on the Physics of Electronic and Atomic Collisions: Abstracts of Papers, Amsterdam, The Netherlands, (1971), p. 835, North Holland Publishing Co., 1971.

38. W. F. G. Blair and H. B. Gilbody, Seventh International Conference on the Physics of Electronic and Atomic Collisions: Abstract of Papers, Amsterdam, The Netherlands, (1971), p. 837, North Holland Publishing Co., 1971.

39. D. H. Jaecks, D. H. Crandall, and R. H. McKnight, *Phys. Rev. Letters*, **25**, 491 (1970).

40. H. A. B. Gardiner, W. R. Pendleton, J. J. Merrill, and D. J. Baker, *Phys. Rev.*, **188**, 257 (1969).

41. F. J. de Heer and J. M. F. Aarts, *Physica*, **48**, 620 (1970)

42. P. J. Wehenberg and K. C. Clark, Seventh International Conference of the Physics of Electronic and Atomic Collisions: Abstracts of Papers, Amsterdam, The Netherlands, (1971), p. 387, Holland Publishing Co., 1971.

43. R. Suchannek, J. R. Murray, and J. R. Sheridan, Seventh International Conference on the Physics of Electronic and Atomic Collisions: Abstracts of Papers, Amsterdam, The Netherlands, (1971), p. 820, North Holland Publishing Co., 1971.

44. R. J. McNeal and D. C. Clark, *J. Geophys. Res.*, **74**, 5065 (1969).

45. D. H. Jaecks, W. de Rijk, and P. J. Martin, Seventh International Conference on the Physics of Electronic and Atomic Collisions: Abstracts of Papers, Amsterdam, The Netherlands. (1971), p. 605, North Holland Publishing Co., 1971.

46. J. T. Park, F. D. Schowengerdt, and D. R. Schoonover, *Phys. Rev.*, **A3**, 679 (1971).

47. R. H. Hughes, A. R. Filippeli, and H. M. Petefish, *Phys. Rev.*, **A1**, 21 (1970).

48. A. L. Orbeli, E. P. Andreev, V. A. Ankudinov, and V. M. Dukelskii, *Zh. Eksperim. i Theor. Fiz.*, **58**, 1938 (1970); [*Soviet phys. JETP*, **31**, 1044 (1970)].

49. D. H. Jaecks, W. de Rijk, and P. J. Martin, Seventh International Conference on the Physics of Electronic and Atomic Collisions: Abstracts of Papers, Amsterdam, The Netherlands. (1971), p. 424, North Holland Publishing Co., 1971.

50. M. Lipeless, R. D. Swift, M. S. Longmire, and M. Weinreb, *Phys. Rev. Letters*, **24**, 799 (1970).

51. L. Wolterbeek Muller and F. J. de Heer, *Physica*, **48**, 345 (1970).

52. H. B. Gilbody, K. F. Dunn, R. Browning, and C. J. Latimer, *J. Phys.*, **B4**, 800 (1971)

53. M. Lipeless, *Phys. Rev.*, **A4**, 140 (1971).

54. R. C. Isler and R. D. Natham, *Phys. Rev. Letters* **25**, 3 (1970).

55. H. B. Gilbody, K. F. Dunn, R. Browning, and C. J. Latimer, Seventh International Conference on the Physics of Electronic and Atomic Collisions: Abstracts of Papers, Amsterdam, The Netherlands. (1971), p. 785, North Holland Publishing Co., 1971.

56. H. Tawara, *J. Phys. Soc. Japan*, **31**, 871 (1971).

57. G. N. Polyakova, V. F. Erko, and A. V. Zats, *Opt. i Spektroskopiya*, **29**, 412 (1970; [*Opt Spectry. (USSR)*, **29**, 219 (1970)].

58. M. Hollstein, A. Salop, J. R. Peterson, and D. C. Lorents, *Phys. Letters*, **32A**, 327 (1970).

59. R. D. Nathan and R. C. Isler, *Phys. Rev. Letters*, **26**, 1091 (1971).

60. S. H. Dworetsky and R. Novick, *Phys. Rev. Letters*, **23**, 1484 (1969).

61. P. J. MacVicar-Whelan and W. L. Borst, *Phys. Rev.*, **A1**, 314 (1970).

62. D. R. Schoonover and J. T. Park, *Phys. Rev.*, **A3**, 228 (1971).

63. G. Rahmat, G. Vassiley, J. Beaudon, and M. Barat, *Phys. Rev. Letters*, **26**, 1411 (1971).

64. G. Rahmat, G. Vassilev, J. Baudon, and M. Barat, Seventh International Conference on the Physics of Electronic and Atomic Collisions: Abstracts of Papers, Amsterdam, The Netherlands. (1971), p. 113, North Holland Publishing Co., 1971

65. R. E. Olson and D. C. Lorents, Seventh International Conference on the Physics of Electronic and Atomic Collisions, Abstracts of Papers, Amsterdam, The Netherlands, (1971), p. 115, North Holland Publishing Co., 1971.

66. D. R. Schoonover and J. T. Park, Seventh International Conference on the Physics of Electronic and Atomic Collisions: Abstracts of Papers, Amsterdam, The Netherlands. (1971), p. 839, North Holland Publishing Co., 1971.

67. A Salop, D. C. Lorents, and J. R. Peterson, *J. Chem. Phys.*, **54**, 1187 (1971).

68. R. E. Miers and L. W. Anderson, *Phys. Rev.*, **A1**, 534 (1970).

69. C. A. van de Runstraat, T. R. Govers, P. G. Fournier, and F. J. de Heer, Seventh International Conference on the Physics of Electronic and Atomic Collisions: Abstracts of Papers, Amsterdam, The Netherlands. (1971). p. 382, North Holland Publishing Co., 1971.

70. N. H. Tolk, C. W. White, S. H. Dworetsky, and L. A. Farrow, *Phys. Rev. Letters*, **25**, 1251 (1970).

71. S. V. Bobashev, *Phys. Letters*, **31A**, 204 (1970).

72. V. B. Matveyev, S. V. Bobashev, and V. M. Dukelskii, Sixth International Conference on the Physics of Electronic and Atomic Collisions, Abstracts of Papers, Cambridge, Mass., 1969, p. 226, MIT Press, 1969.

73. V. V. Afrosimov, Yu. S. Gordeev, V. M. Lavrov, and V. Nikulin, Seventh International Conference on the Physics of Electronic and Atomic Collisions: Abstracts of Papers, Amsterdam, The Netherlands, (1971), p. 143, North Holland Publishing Co., 1971.

74. S. Bobashev, Seventh International Conference on the Physics of Electronic and Atomic Collisions: Abstracts of Papers, Amsterdam, The Netherlands, (1971), p. 587, North Holland Publishing Co., 1971

75. V. V. Afrosimov, Yu. S. Gordeev, V. M. Lavrov, and G. N. Ogurtsov, Seventh International Conference on the Physics of Electronic and Atomic Collisions: Abstracts of Papers, Amsterdam, The Netherlands, (1971), p. 825, North Holland Publishing Co., 1971.

76. S . V. Pop, I. Yu. Krivskii, I. P. Zapesochnyi, and M. V. Baletskaya, *Zh. Eksperim. i Theor. Fiz.*, **58**, 810 (1970); [*Soviet Phys. JETP*, **31**, 434 (1970)].

77. V. Aquilanti, G. Liuti, F. Becchio-Cattivi, and G. G. Volpi, Seventh International Conference on the Physics of Electronic and Atomic Collisions: Abstracts of Papers, Amsterdam, The Netherlands, (1971), p. 600, North Holland Publishing Co., 1971.

78. S. V. Pop, I. Y. Krivskii, I. P. Zapesochnyi, and M. V. Baletskaya, *Zh. Eksperimt i Theor. Fiz.*, **59**, 696 (1970); [*Soviet Phys. JETP*, **32**, 380 (1970)].

79. K. Lacmann, D. R. Herschbach, *Chem. Phys. Letters*, **6**, 106 (1970).

80. V. Kempter, W. Meclenbrauck, M. Menzinger, G. Schuller, D. R. Herschbach, and Ch. Schlier, *Chem. Phys. Letters*, **6**, 97 (1970).

81. J. Perel and H. L. Daley, *Phys. Rev.*, **A4**, 162 (1971).

82. D. C. Lorents and G. Conklin, Seventh International Conference on the Physics of Electronic and Atomic Collisions: Abstracts of Papers, Amsterdam, The Netherlands, (1971), p. 130, North Holland Publishing Co., 1971.

83. R. E. Mickle, H. I. S. Ferguson, and R. P. Lowe, Seventh International Conference on the Physics of Electronic and Atomic Collisions: Abstracts of Papers, Amsterdam, The Netherlands, (1971), p. 384, North Holland Publishing Co., 1971.

84. S. V. Bobashev and V. A. Kritshii, *ZhETF Pis'ma Red.*, **12**, 280 (1970); [*JETP Letters*, **12**, 189 (1970)].

85. S. V. Bovashev, *ZhETF Pis'ma Red.*, **11**, 389 (1970); [*JETP Letters*, **11**, 260 (1970)].

86. P. C. Cosby and T. F. Moran, *J. Chem. Phys.*, **52**, 6157 (1970).

87. O. B. Shpenik, A. N. Zavilopulo, and I. P. Zapesochny, *Opt. i Spektroskopiya*, **29**, 1018 (1970); [*Opt. Spectry. (USSR)*, **29**, 541, (1970)].

88. O. B. Shpenik, I. P. Zapesochnyi, and A. N. Zavilopulo, *Zh. Eksperim. i Theor. Fiz.*, **60**, 513 (1971); [*Soviet Phys. JETP*, **33**, 277 (1971)].

89. I. P. Zapesochnyi, A. N. Zavilopulo, and O. B. Shpenik, *Opt. i Spektroskopiya*, **28**, 856 (1970); [*Opt. Spectry. (USSR)*, **28**, 465 (1970)].

APPENDIX II

FORMATION OF FAST EXCITED H AND He ATOMS UNDER MULTIPLE COLLISION CONDITIONS

There are many practical situations where it is important to know the rate at which a projectile excited state is produced when target densities are such that multiple collisions are occurring. A particular range of examples are the formation of beams of excited H and He atoms as a result of the impact of H^+, H_2^+, H_3^+, and He^+ ions on relatively "thick" targets. There has been some investigations of such circumstances. The data has been excluded from the main body of the text because it does not, in general, provide information on microscopic collision cross sections. Because of its potential value in certain practical situations, the sources of such information are listed below in tabular form. No attempt is made to assess the validity of the data.

Projectile	Target	Projectile Excited State	Energy Range keV	Reference
H^+	He	$3 \leq n \leq 6$	10–30	Ankudinov et al.[1]
D^+	He	$8 \leq n \leq 11$	100	Riviere et al.[2]
H^+	Li	$9 < n$	35–42	Futch et al.[3]
H^+	Ne	$3 \leq n \leq 6$	10–30	Ankudinov et al.[1]
H^+	Ne	$9 \leq n \leq 16$	10–120	Il'in et al.[4,5]
D^+	Ne	$8 \leq n \leq 11$	100	Riviere et al.[2]
H^+	Na	$9 \leq n \leq 16$	10–120	Il'in et al.[4,5]
H^+	Mg	$9 \leq n \leq 16$	10–120	Il'in et al.[4]
H^+	Mg	$9 \leq n \leq 16$	15–120	Oparin et al.[6]
H^+	Ar	$10 < n$	30	Sweetman et al.[7]
D^+	Ar	$8 \leq n \leq 11$	100	Riviere et al.[2]
H^+	K	$9 \leq n \leq 16$	20	Il'in et al.[5]
H^+	Ca	$9 \leq n \leq 16$	10–180	Oparin et al.[6]
H^+	Zn	$9 \leq n \leq 16$	10–180	Oparin et al.[6]
D^+	Kr	$8 \leq n \leq 11$	100	Riviere et al.[2]
H^+	Cd	$9 \leq n \leq 16$	10–180	Oparin et al.[6]
D^+	Xe	$8 \leq n \leq 11$	100	Riviere et al.[2]

Projectile	Target	Projectile Excited State	Energy Range keV	Reference
H^+	Cs	$9 \leq n \leq 11$	12.5	Il'in et al.[5]
H^+	Cs	$2s$	2	Cesati et al.[8]
H^+	H_2	$2s$	†	Bassani et al.[9]
H^+	H_2	$2s$	15	Sellin[10]
H^+	H_2	$n = 14$	20–100	Riviere et al.[11]
H^+	H_2	$n = 11$ & 14	5	McFarland et al.[12]
H^+	NH_3	$n = 11$ & 14	5	McFarland et al.[12]
H^+	H_2O	$9 < n$	35–42	Futch et al.[3]
H^+	H_2O	$n = 11$ & 14	5	McFarland et al.[12]
H^+	N_2	$2s$	†	Bassani et al.[9]
H^+	N_2	$n = 11$ & 14	5 & 10	McFarland et al.[12]
H^+	O_2	$2s$	†	Bassani et al.[9]
H^+	CO_2	$2s$	†	Bassani et al.[9]
H^+	CO_2	$9 \leq n \leq 16$	30–120	Il'in et al.[5, 13]
H^+	C_8F_{16}	$n = 11$ & 14	4–25	McFarland et al.[12]
H^+	Toluene	$2s$	†	Bassani et al.[9]
H^+	Ethyl alcohol	$2s$	†	Bassani et al.[9]
H_2^+	He	$3 \leq n \leq 6$	10–30	Ankudinov et al.[1]
H_2^+	Ne	$3 \leq n \leq 6$	10–30	Ankudinov et al.[1]
H_3^+	He	$3 \leq n \leq 6$	10–30	Ankudinov et al.[1]
H_3^+	Ne	$3 \leq n \leq 6$	10–30	Ankudinov et al.[1]
He^+	Cs	$2^1S, 2^3S$	1.5–2.5	Schlachter et al.[14]

† Not quoted.

REFERENCES

1. V. A. Ankudinov, S. V. Bobashev, and E. P. Andreev, *Zh. Tekhn., Fiz.*, **34**, 1645 (1964); [*Soviet Phys.-Tech. Phys.*, **9**, 1272 (1965)].

2. A. C. Riviere and D. R. Sweetman, *Proc. of the Sixth Intern. Conf. on Ionization Phenomena in Gases*, Paris, 1963, Vol. I, p. 105, (P. Huber and E. Crémieu-Alcan, eds.), SERMA, Paris, 1963.

3. A. H. Futch and C. C. Damm, *Nucl. Fusion*, **3**, 124 (1963).

4. R. N. Il'in, V. A. Oparin, E. Solov'ev, and N. V. Fedorenko, *ZhETF Pis'ma Red.*, **2**, 310 (1965); [*JETP Letters*, **2**, 197 (1965)].

5. R. N. Il'in, V. A. Oparin, E. S. Solov'ev, N. V. Fedorenko, *Zh. Tekhn., Fiz.*, **36**, 1241 (1966); [*Soviet Phys.-Tech. Phys.*, **11**, 921 (1967)].

6. V. A. Oparin, R. N. Il'in, and E. S. Solov'ev, *Zh. Eksperim. i Theor. Fiz.*, **52**, 369 (1967); [*Soviet Phys.-JETP*, **25**, 240 (1967)].

7. D. R. Sweetman, *Nuclear Fusion, 1962 Supplement: Proc. of the Conf. on Plasma Physics and Controlled Nuclear Fusion Research*, Salzburg, 1961, Part 1, p. 279, International Atomic Energy Agency, 1962.

8. A. Cesati, F. Cristofori, L. M. Colli, and P. G. Sona, *Energia Nucl. (Milan)*, **13**, 649 (1966).

9. G. Bassani, L. Colli, F. Cristofori, M. O. Franceschetti, G. E. Frigerio, and P. G. Sona, *Energia Nucl. (Milan)*, **9**, 451 (1962).

10. I. A. Sellin, *Phys. Rev.*, **136,** A1245 (1964).

11. A. C. Riviere and D. R. Sweetman, *Atomic Collision Processes: Proc. of the Third Intern. Conf. on the Physics of Electronic and Atomic Collisions*, London, 1963, p. 734, (M. R. C. McDowell, ed.), North-Holland, Amsterdam, 1964.

12. R. H. McFarland, H. A. Futch, *Sixth Intern. Conf. on the Physics of Electronic and Atomic Collisions: Abstracts of Papers*, Cambridge, Mass., p. 441, MIT Press, 1969.

13. R. N. Il'in, B. I. Kikiani, V. A. Oparin, E. S. Solov'ev, and N. V. Fedorenko, *Zh. Eksperim. i Theor. Fiz.*, **47,** 1235 (1964); [*Soviet Phys.-JETP*, **20,** 835 (1965)].

14. A. S. Schlachter, D. H. Loyd, P. J. Bjorkholm, L. W. Anderson, and W. Haeberli, *Phys. Rev.*, **174,** 201 (1968).

AUTHOR INDEX

SUBJECT INDEX

For a detailed index to the reactions on which data is available the reader's attention is drawn to the tabular index contained in Chapter 4 (pages 66 to 85).

Absorption of photons, population of excited states by, 13, 14
  correction for, 117
Acetylene excitation by H^+, 296–298
Additional work published recently, 412–420
Angular distribution of scattered particles, 2
Anisotropy, *see* Polarization
Argon excitation, autoionization electrons, 183, 184
Argon excitation, by Ar^+, 188, 189
  by Cs^+, 189
  data, 190–196
  by H^+, 185
  by H_2^+, 186
  by He^+, 186, 187
  by Ne^+, 188
Argon excitation, effect, of cascade, 187, 188
  of photoabsorption, 186
Argon ions, excitation of, 393–400

Blackbody light source, 46, 50
Branching ratios, tabulated values, 23, 24
Brightness temperature, 47

Calcium, formation of excited atoms by charge transfer, 394–400
Calcium ions, excitation of, 394–400
Carbon monoxide excitation, data, 297
  effect of cascade, 294
  by H^+, 293
  by N_2^+, 295
  rotational structure of emission, 294
Cascade, population of excited states, 13, 14, 15, 17, 18, 19, 21
  in rare gas excitation, 164
Cesium ions, excitation of, 393–400
Charge transfer, definition, 5
Coincidence experiments, 11, 12, 294
Collisional transfer, population of excited states by, 13, 14, 16
Criteria for accurate measurement, of emission, 63–64
  of excited state density, 41
  of polarization fraction, 64
  of projectile beam, 32

of target density, 38
Criteria for valid measurement, of cross sections, 25
  of polarization, 25
Cross section, operational definition, for emission, 8
  for excitation, 7, 8
  general considerations, 6–12
  in coincidence experiments, 11, 12
  in crossed beam experiment, 9–11
Cross section, relationship of emission to excitation, for projectile excited, 16–19
  for static target excited, 13–16
  general considerations, 12–18
Crossed beam experiments, 87, 93

Data sources, catalog of, 65–85
Decay length, 16
Detectors, particle, *see* Projectile beams
Dissociation, definition, 5
Doppler effect, 16, 61

Effective radius in photoabsorption, 118
Elastic collision, definition, 4
Emission, relationship to excitation, 12–18
Emissive power, 47
Emissivity, 47
Energy loss, detection of excitation by, 38, 40
Excitation, definition, 5
Excited state density, direct determination of, by energy loss method, 38, 40
  by field ionization, 39
  by stripping methods, 40, 41, 384–386
  general considerations, 38–41

Faraday cup, 30, 31
Field ionization, 307
Form factor, definition, 10
  in experiments, 91
Furnace, atomic hydrogen, 93
  dissociation of H_2 in, 87
  divergence of beam, 91
  magnetic fields, 98
  measurement, of H density in, 94
  of H_2 density in, 88, 92, 94

433